D1521602

Harald Stumpf

Quantum Processes in Polar Semiconductors and Insulators

Part 2

Harald Stumpf

Quantum Processes in Polar Semiconductors and Insulators

Part 2

Friedr. Vieweg & Sohn Braunschweig / Wiesbaden

CIP-Kurztitelaufnahme der Deutschen Bibliothek

Stumpf, Harald:
Quantum processes in polar semiconductors and insulators / Harald Stumpf. – Braunschweig; Wiesbaden: Vieweg

Pt. 2 (1983)
ISBN 3-528-08526-6

All rights reserved
© Friedr. Vieweg & Sohn Verlagsgesellschaft mbH, Braunschweig 1983

No part of this publication may be reproduced, stored in a retrieval system or transmitted mechanically, by photocopies, recordings or other means, without prior permission of the Copyright holder.

Produced by W. Langelüddecke, Braunschweig
Printed in Germany-West

ISBN 3-528-08526-6

Acknowledgements

The preparation of a book with a comprehensive citation index is not only scientific work but also a matter of organization. For this book, the organization was almost completely done by our secretary, Mrs. *Regine Adler.* She ordered, copied, collected and arranged the great number of original papers. She prepared the references and the author index and corrected the corresponding galley proofs. Furthermore, she typed several versions of the manuscript in an excellent way. Without her help it would have been impossible for me to concentrate on physics and to prepare the book in time. So I want to express to Mrs. *Adler* my deep gratitude for this effectful cooperation.

Concerning the scientific part of the book, Dr. *A. Rieckers* and Dr. *J. Schupfner* read the whole manuscript and gave valuable comments and proposals for improvements. They checked the proofs of the theorems at least partly. Dr. *E. Schöll,* in particular, checked the content of Section 8.6. For the assistance of these younger colleagues I am very grateful.

My English formulation of the text needed a critical supervision. This was done first of all by Mrs. *Irmgard Stumpf,* furthermore by Dr. *M. Brunet* and in a final round by Prof. Dr. *W. Klink* (Iowa). They all contributed to the improvement of the book, not only by discovering faults, but also by trying to improve the general transparency of the text and by pointing out logical jumps in the text. I am very indepted to them for this help.

Finally, I want to express my thanks to all the members of the Vieweg Verlag who made the publication of this book possible through their efforts and cooperation.

H. Stumpf

Tübingen, January 1983

Contents

Part 2

Part 1

intended as a monograph for scientists at universities, research institutes and in the industry doing work in semiconductor and insulator physics and techniques, in linear and nonlinear optics, in laser physics and techniques, in luminescence physics and techniques, in opto-electronics, etc. It gives a review of the literature of the past four decades and offers effective modern methods for a better understanding and for the solution of problems at all levels in great detail. Moreover, due to the clear exposition, the relationship to molecular physics, biochemical reaction theory, physical chemistry, photosynthesis research is so obvious that it might be of considerable interest for scientists in these fields to study the methods developed for semiconductor processes and to transfer them into their own fields.
Owing to the comprehensive literature and the limitations with respect to the size of the book the following topics had to be excluded:
Surface effects and one- and two-dimensional impurities, as for instance dislocations and grain boundaries;
Ternary and quaternary compounds or even amorphous substances and heterogeneous materials, although they play an important role in modern semiconductor technology;
Pure quantum field theoretic techniques, in particular quantum field theoretic Green function methods and algebraic quantum field theory.
All these topics are so comprehensive that they would require a book of their own for a thorough treatment. It should, however, be emphasized that the first two topics are in principle in the range of application of the methods given here. For instance, an earlier book of the author about ionic crystals (Stumpf, 1961) contains a chapter about dislocations which are treated by means of the methods developed further in this book. So the extension to the treatment of technically interesting devices should be possible along the lines outlined here.
Concerning the field theoretic Green function methods it seems that these techniques are more appropriate for the treatment of homogeneous ordered and disordered systems than for the treatment of systems with point defects and one-dimensional defects, etc. Consequently they have primarily been applied to the theory of quasiparticles in homogeneous materials about which excellent monographs already exist.
With respect to the use of this book some further remarks might be of interest:
The subject index does not contain references to the literature as the inclusion of such references would have inflated the index in an unacceptable way. Due to the systematic disposition of the material and the fact that the literature is generally indicated at the end of each section, it is easy to find the literature with respect to a special topic without the help of the subject index.
The transformation to phonon normal coordinates is not performed in a uniform way. This results from the fact that in the literature two versions are also used, and that it was impossible to modify the calculations of various authors to one standard transformation without substantial change of their formulas. In any case the transformation is uniquely defined so that no misunderstanding is possible.
The multiple use of symbols with different meanings in each case could not be avoided. For instance, the letter P in front of a summation sign sometimes means: summation with permutation, but in other cases: summation with Pauli principle for indistinguishable particles. In any case, the meaning becomes obvious from the accompanying text and formulas, so that misunderstandings should be excluded.

Preface

In this book the physics and the corresponding theory of processes and reactions of ideal and non-ideal, i. e., impure polar semiconductors and insulators are discussed and developed. In particular, binary compounds of the type I–VII and the technically interesting II–VI and III–V types are treated. Based on the quantum theoretical microscopic description of crystals as many-particle systems of electrons and nuclei, a complete deduction is given starting at the microscopic level and finally obtaining quantities which can be compared to the experiment, i. e., average equilibrium and non-equilibrium values of quantum statistical reaction kinetic quantities for combined electron, phonon, and photon processes and reactions with and without external fields. At each stage of this deduction the theoretical apparatus is carefully evaluated, in particular the calculation of states, transition probabilities, kinetic equations, etc.
In the past decades an enormous amount of scientific information has been produced. Only deductive theory is able to reduce this material by providing structural insights. It is an aim of this book to give this insight within its field. Concerning the literature on polar semiconductors and insulators, the greatest part of it deals with isolated problems regardless of the need for a deductive theory. In addition, many papers treat the problems at a phenomenological level. Due to the deductive character of the book, it is possible to classify the various theoretical approaches in semiconductor and insulator physics and to show their meaning within the framework of an ab initio unifying theory. More than 6000 papers in this field were collected and taken into account, and most of them are cited in this book. The systematic search for literature was completed at the beginning of 1980. But papers which are of direct importance for the deductions given here were included up to the end of the year 1982.
Another aim of the book is to present theoretical results and methods based on recent research by the author and his coworkers and scientists working on related problems. These results and methods allow a more thorough understanding of the physics and a more effective evaluation in numerical calculations than conventional methods. The major achievements were made, for instance, in the quantum statistical treatment of resonance transitions, in the derivation of rate equations, in the state calculations of impurity center electrons, and in the quantum statistical treatment of processes with inclusion of external fields. But even with respect to conventional 'classical' methods, as for instance in the theory of phonon processes, Raman effect, Jahn-Teller effect, etc., some improvements were made which make these methods and results more transparent.
Thus the book offers a survey of the theory and the corresponding applications in this field and serves as a systematic guide to the understanding of the original literature. Due to its deductive character, it is very well suited for advanced students interested in learning about the major achievements in this field in a strictly systematic way but who cannot work through the comprehensive original literature for themselves. Equally well the book is

Index of Theorems, Lemmas and Definitions

Chapter 7

Chapter 8

6 Symmetry operations

6.1 Symmetries of quantum systems

The effect of symmetries in the case of quantum systems leads to a group theoretical classification of energy spectra and corresponding state spaces, and in addition to selection rules for processes. Hence, the symmetry concept offers an effective method for the treatment of quantum systems. It should be pointed out, however, that the use of group theory is by no means restricted to the case of perfect symmetry, i.e., even for broken symmetries a group theoretical classification is often meaningful. The latter possibility is important, as all real physical systems, e.g., crystals with impurities, have broken symmetries.
For the theoretical investigation of symmetries a mathematical definition of symmetry has to be given. This definition depends on the kind of mathematical description of the system under consideration. We discuss quantum systems which are described by a complete set of simultaneously commuting observables with a corresponding representation space. Quantum systems, in particular crystals, are located in space-time and their symmetries with respect to space-time frames of reference are of interest. Such symmetries are discovered by changing the frame of reference. The change of the frame of reference is described by a coordinate transformation, i.e., by a map of one set of coordinates and coordinate frames to another. As the successive application of several coordinate transformations must again give a coordinate transformation, these coordinate transformations generally constitute a transformation group. For groups the following definition is introduced:

Def. 6.1 : A group is given by a set of elements $\{g_i, i \in I\}$ for which a multiplication may be defined which associates a third element of $\{g_i\}$ with any ordered pair of $\{g_i\}$. This multiplication must satisfy the requirements
α) the set is closed under group multiplication,
β) the associative law holds for group multiplication,
γ) there is a unit element $g_\alpha = e$ with $eg_i = g_i e = g_i$,
δ) there is an inverse element g_i^{-1} for each g_i with $g_i^{-1} g_i = g_i g_i^{-1} = e$.

Among the set of all possible and generally nonlinear coordinate transformations we study linear transformations which are of special interest for symmetry considerations and which lead to linear group representations. These representations are embedded in linear metrical vector spaces which are defined by

Def. 6.2 : A linear metrical affine space is given, if for any two points P, P' a vector $\overrightarrow{PP'} = \mathbf{a}$ exists, and if for the set of all vectors $\mathbf{V}_n$ the operations

α) addition, $\mathbf{a}+\mathbf{b}=\mathbf{c}$
β) scalar multiplication $\mathbf{a}'=\alpha\mathbf{a}$, $\alpha \in \mathbb{R}$ or $\mathbb{C}$
γ) scalarproduct $(\mathbf{a}\cdot\mathbf{b}) = :Q(\mathbf{a},\mathbf{b})$
are defined.

A group representation is specified by

Def. 6.3: A linear representation $\mathfrak{D} := \{D(g)\}$ of a group $G := \{g\}$ in a linear representation space $\mathbf{B}$ by bounded linear transformations in $\mathbf{B}$ exists, if the following holds
α) for $g \in G$ follows $D(g) \in \mathfrak{D}$,
β) from $(g_1 g_2) \in G$ follows $D(g_1)\cdot D(g_2) = D(g_1 g_2)$,
γ) $D(g^{-1}) = D^{-1}(g)$,
δ) $D(e) = 1$.

We apply these definitions first to the coordinate configuration space of a system. In these spaces a vector $\mathbf{a}$ can be represented by a linear combination of base vectors $\{\mathbf{e}_i\}$

$$\mathbf{a} = \sum_{i=1}^{n} a^i \mathbf{e}_i. \tag{6.1.1}$$

Using such a vector space for the coordinate description of a physical system, the various frames of reference can be described by various base vector systems. Denoting one of these systems by $\{\mathbf{e}_i\}$ and another by $\{\bar{\mathbf{e}}_i\}$ according to our assumption, these base systems have to be connected by a linear transformation

$$\bar{\mathbf{e}}_i = \sum_{k=1}^{n} A_i^k \mathbf{e}_k \tag{6.1.2}$$

with $\det|A| \neq 0$. As any vector $\mathbf{a} \in \mathbf{V}_n$ is defined by the directed distance between two space points P, P', this definition refers to absolute quantities, namely to points which are not influenced by a change of the frame of reference used for their mathematical description. Hence, $\mathbf{a}$ is an invariant quantity. As $\mathbf{a}$ can be described in both frames of reference equally well, we have

$$\mathbf{a} = \sum_{i=1}^{n} a^i \mathbf{e}_i = \sum_{i=1}^{n} \bar{a}^j \bar{\mathbf{e}}_j \tag{6.1.3}$$

and from this it follows that

$$a^j = \sum_{i=1}^{n} \bar{a}^i A_i^j. \tag{6.1.4}$$

For a thorough analysis of transformations, the transformations of the basis vectors (6.1.2) as well as the transformations of the vector components (6.1.4) have to be considered. But for the use in quantum mechanics it is sufficient to consider the transformation of the coordinates alone. In this case the vectors $\mathbf{a}$ have to be identified with position vectors $\mathbf{r}$ which are given for crystals by the electronic and nuclear coordinates $(\mathbf{r},\mathbf{R})=(\mathbf{r}_1\ldots\mathbf{r}_N,\mathbf{R}_1\ldots\mathbf{R}_M)$. Then a linear transformation in ordinary space

$$\bar{x}^j=\sum_{i=1}^{3} A_i^j x^i \tag{6.1.5}$$

leads to a simultaneous transformation

$$\bar{x}^j(\alpha)=\sum_{i=1}^{3} A_i^j x^i(\alpha),\quad 1\leqslant\alpha\leqslant N;\quad \bar{X}^j(\beta)=\sum_{i=1}^{3} A_i^j X^i(\beta),\quad 1\leqslant\beta\leqslant M \tag{6.1.6}$$

of the electronic and nuclear coordinates $x^j(\alpha)$, $X^j(\beta)$, while the vectors $\mathbf{r}_1\ldots\mathbf{r}_N$, $\mathbf{R}_1\ldots\mathbf{R}_M$ themselves remain invariant. For brevity we write

$$\mathbf{r}',\mathbf{R}'=g\mathbf{r},g\mathbf{R}\mathrel{\widehat{=}}(6.1.6) \tag{6.1.7}$$

if we perform coordinate transformations. Using this notation the Hamilton operator $H(\mathbf{r},\mathbf{R})$ of a system of electrons and nuclei is an ordinary function of $\mathbf{r}$ and $\mathbf{R}$ and transforms like a scalar if the spin dependence is exluded

$$H'(\mathbf{r}',\mathbf{R}')=H(g^{-1}\mathbf{r}',g^{-1}\mathbf{R}')=H(\mathbf{r},\mathbf{R}) \tag{6.1.8}$$

which expresses the fact that the meaning of $H(\mathbf{r},\mathbf{R})$ is not affected by the choice of different frames of reference. Now a symmetry can be formulated by

Def. 6.4: A group of linear transformations $G=\{g\}$ is a symmetry group of the system, if its Hamilton operator $H(\mathbf{r},\mathbf{R})$ is forminvariant under G, i.e., if $H'(\mathbf{r}',\mathbf{R}')\equiv H(\mathbf{r}',\mathbf{R}')$ is valid.

If such a symmetry group exists this has consequences for the corresponding state space.

Theorem 6.1: If a quantum system has a symmetry group then the corresponding energy eigenspace is a representation space of this group.

Proof: We consider the linear transformations (6.1.7) for a system of electrons and nuclei. The corresponding Hilbert space is given by the eigenstates of

$$H(\mathbf{r},\mathbf{R})\psi_n(\mathbf{r},\mathbf{R})=E_n\psi_n(\mathbf{r},\mathbf{R}) \tag{6.1.9}$$

Performing a transformation, we obtain due to Definition 6.4 from (6.1.9)

$$H(\mathbf{r}',\mathbf{R}')\psi_n(g^{-1}\mathbf{r}',g^{-1}\mathbf{R}')=E_n\psi_n(g^{-1}\mathbf{r}',g^{-1}\mathbf{R}'). \tag{6.1.10}$$

By merely changing the notation from $\mathbf{r}'\mathbf{R}'$ to $\mathbf{r},\mathbf{R}$ if follows that also $\psi_n(g^{-1}\mathbf{r},g^{-1}\mathbf{R})$ has to be a solution of (6.1.9) for the eigenvalue E_n. If E_n is non-degenerate then $\psi_n(g^{-1}\mathbf{r},g^{-1}\mathbf{R})\equiv\psi_n(\mathbf{r},\mathbf{R})$, i.e. ψ_n itself has to be forminvariant. In this case ψ_n transforms as the identity representation. If E_n is degenerate then there are l linear independent eigenfunctions ψ_n^l and any other eigenfunction has to be a linear combination of these base functions. This means

$$\psi_n^l(g^{-1}\mathbf{r},g^{-1}\mathbf{R})\equiv\psi_n^l(\mathbf{r},\mathbf{R})'=\sum_{h=1}^{l}\gamma_{lh}(g)\psi_n^h(\mathbf{r},\mathbf{R}) \tag{6.1.11}$$

for arbitrary g. It can be shown that $\gamma_{lh}(g)$ satisfies all the conditions of Definition 6.3, i.e., ψ_n^l is a basevector for an l-dimensional linear matrix representation of G, Q.E.D.

According to (2.2.1) the energy operator $K = K(\mathbf{r}, \mathbf{R})$ of a polar crystal is given by the kinetic energy operators of the electrons and nuclei and their mutual electromagnetic interactions (2.2.3). These operators are forminvariant under the full Euclidean group of space rotations, space translations and reflections. Considering a one-particle space $\mathbf{V}_3 \equiv \{\mathbf{r}\}$ the corresponding transformations of this group are defined by

$$\mathbf{r}' = \mathscr{A} \cdot \mathbf{r} + \mathbf{c} \tag{6.1.12}$$

where $\mathscr{A}$ effects the reflection or rotation while $\mathbf{c}$ effects the translation. Obviously a representation of the corresponding group element is given by

$$g := (\mathscr{A}, \mathbf{c}) \tag{6.1.13}$$

where the multiplication law follows from (6.1.12) by successive application of several transformations. It is

$$g_1 g_2 = (\mathscr{A}_1, \mathbf{c}_1)(\mathscr{A}_2, \mathbf{c}_2) := (\mathscr{A}_1 \mathscr{A}_2, \mathscr{A}_1 \mathbf{c}_2 + \mathbf{c}_1). \tag{6.1.14}$$

The generalization to many-particle spaces according to (6.1.6) is obvious. This group is not a finite group with a finite number of elements, but a continuous group with a non-denumerable number of elements.

The symmetry groups of crystals, in particular of crystals with point defects, are the crystal point groups which are finite groups, i.e., they are not identical with this full Euclidean group. Nevertheless, the Euclidean group is of special interest for crystal problems as the point groups are subgroups of this group and the representation spaces of this group are the starting point for the construction of representation spaces of the point groups. Hence, we first consider the Euclidean group. But as we are mainly interested in the treatment of defect crystal structures where translational invariance is broken, it is sufficient for a first investigation to consider the subgroup $O(3)$ of rotations and reflections of the full Euclidean group. In Section 6.6, we shall include also translations.

The subgroup $O(3)$ itself is a direct product of the pure rotation group $R(3)$ and the reflection group C, i.e., $O(3) = R(3) \otimes C$. While C is a finite group, the rotation group $R(3)$ is an example of a Lie group, i.e., a group whose group manifold is differentiable. For Lie groups the concept of the infinitesimal generator is of great importance. While finite groups can be classified by their multiplication table, the Lie groups are classified by the multiplication of their infinitesimal generators, i.e., by their Lie algebra. We consider this concept for $R(3)$. For rotations we have

$$\mathbf{r}' = \mathscr{A} \cdot \mathbf{r}. \tag{6.1.15}$$

The rotation group contains three one-parameter subgroups which are defined by rotations around the three space axis. The infinitesimal generators of these subgroups are then given by

$$\mathscr{D}^1 := \begin{pmatrix} 0 & 0 & 0 \\ 0 & 0 & 1 \\ 0 & -1 & 0 \end{pmatrix}; \quad \mathscr{D}^2 := \begin{pmatrix} 0 & 0 & -1 \\ 0 & 0 & 0 \\ 1 & 0 & 0 \end{pmatrix}; \quad \mathscr{D}^3 := \begin{pmatrix} 0 & 1 & 0 \\ -1 & 0 & 0 \\ 0 & 0 & 0 \end{pmatrix} \tag{6.1.16}$$

and a general infinitesimal rotation can be represented by

$$\mathbf{r}' = \left(\mathbb{1} + \sum_{\varrho=1}^{3} \varepsilon_\varrho \mathscr{D}^\varrho \right) \cdot \mathbf{r} \tag{6.1.17}$$

where ε_ϱ, $\varrho = 1, 2, 3$ are the infinitesimal angles around the three space axis. Any finite transformation can be obtained by repeated application of the corresponding infinitesimal rotations. An explicit representation can be obtained by using the Euler-angles α, β, γ. Then any rotation $\mathscr{A}$ is uniquely defined by the three values of α, β, γ and $\mathscr{A}$ can be expressed by infinitesimal generators in the following form

$$\mathscr{A} = \mathscr{A}(\alpha, \beta, \gamma) = e^{\alpha \mathscr{D}^3} e^{\beta \mathscr{D}^2} e^{\gamma \mathscr{D}^3} \tag{6.1.18}$$

where it has to be observed that the $\mathscr{D}^\varrho$, $\varrho = 1, 2, 3$ do not commute. The generalization to multi-particle coordinates is obvious: Any single particle coordinate space is transformed according to (6.1.18) and the infinitesimal generators $\mathscr{D}^\varrho(i)$ of the particle i commute with those of the particle j.
To draw further conclusions it is convenient to represent the effect of linear transformations by linear operators in corresponding function spaces. Excluding spin, the function spaces of interest in quantum theory are constituted by scalar functions. For such function spaces the following theorem holds:

Theorem 6.2: The group $O(3)$ of rotations and reflections has a faithful operator representation $\{\mathfrak{G}(\mathscr{A}, C')\}$ in scalar function spaces.

Proof: For simplicity we consider a one-particle function space $\{f(\mathbf{r})\}$ as the extension to many-particle function spaces $\{f(\mathbf{r}_1 \ldots \mathbf{r}_N)\}$ etc. is obvious. The one-particle state functions satisfy the invariance condition $f'(\mathbf{r}') = f(\mathbf{r})$ from which follows the transformation law

$$f'(\mathbf{r}) = f(\mathscr{A}^{-1} \cdot \mathbf{r}). \tag{6.1.19}$$

For an infinitesimal transformation (6.1.17) we obtain from (6.1.19) by a Taylor expansion

$$f'(\mathbf{r}) = f\left(\mathbf{r} - \sum_{\varrho=1}^{3} \varepsilon_\varrho \mathscr{D}^\varrho \cdot \mathbf{r} \right) = \left(1 + \sum_{i=1}^{3} i\varepsilon_\varrho \mathscr{I}_\varrho \right) f(\mathbf{r}) + \ldots \tag{6.1.20}$$

with

$$\mathscr{I}_l := i(x_k \partial_m - x_m \partial_k), \quad l, k, m \text{ cyclic.} \tag{6.1.21}$$

As in ordinary space also in function space a finite transformation can be expressed by

$$f'(\mathbf{r}) = f(\mathscr{A}^{-1}(\alpha, \beta, \gamma) \cdot \mathbf{r}) = e^{-i\alpha \mathscr{I}_3} e^{-i\beta \mathscr{I}_2} e^{-i\gamma \mathscr{I}_3} f(\mathbf{r}), \tag{6.1.22}$$

since any such transformation is generated by repeated application of transformations of the type (6.1.21). Hence, we have the result that for any rotation there exists a uniquely defined operator $\mathfrak{G}(\mathscr{A})$ in function space which represents $\mathscr{A}$. By successive application of the various group operations it can be shown that $\mathfrak{G}(\mathscr{A})$ is a faithful representation of $R(3)$ in function space. Concerning the reflections in ordinary space, these operations are defined by changing the sign of the various components of $\mathbf{r}$. Considering, for example, the total reflection $\mathbf{r}' = -\mathbf{r}$ we have

$$f'(\mathbf{r}) = f(-\mathbf{r}) = f^+(\mathbf{r}) - f^-(\mathbf{r}) \tag{6.1.23}$$

where f^+ is the even part of f while f^- is the odd part of f. The odd and the even part of f can be separated by using projection operators Π^+, Π^- with

$$\Pi^+ + \Pi^- = \mathbb{1} \tag{6.1.24}$$

and

$$f = (\Pi^+ + \Pi^-) f = f^+ + f^- \tag{6.1.25}$$

Hence

$$f'(\mathbf{r}) = (\Pi^+ - \Pi^-) f(\mathbf{r}) \tag{6.1.26}$$

is valid and we have together with (6.1.24) a representation of the unity operation and the reflection (6.1.23). In a similar way other reflections can also be described by operators in function space. Thus we have a representation of C in function space. Furthermore the direct product of two representations is again a representation, and no transformation leaves all f invariant except the identity transformation. Thus $O(3)$ is faithfully represented, Q.E.D.

Having demonstrated that the group $O(3)$ possesses a faithful operator representation, we assume that this is true also for other groups relevant in solid state physics. Then the following theorem holds:

Theorem 6.3: If the energy operator H of a given system is forminvariant with respect to a symmetry group G, then any element of the operator representation $\{\mathfrak{G}(g)\}$ of G commutes with H, i.e., $[\mathfrak{G}(g), H]_- = 0$.

Proof: We consider the map of an arbitrary Hilbert space vector ψ upon a Hilbert space vector φ defined by

$$\varphi(\mathbf{r}) = H(\mathbf{r})\psi(\mathbf{r}) \tag{6.1.27}$$

where $\mathbf{r}$ denotes symbolically the set of relevant coordinates. Performing a transformation $\mathbf{r}' = g\mathbf{r}$ this map is transformed into

$$\varphi'(\mathbf{r}') = \varphi(g^{-1}\mathbf{r}') = H(g^{-1}\mathbf{r}')\psi(g^{-1}\mathbf{r}') = H'(\mathbf{r}')\psi'(\mathbf{r}') \tag{6.1.28}$$

and with $\varphi(g^{-1}\mathbf{r}') = \mathfrak{G}(g^{-1})\varphi(\mathbf{r}')$ etc. (6.1.28) can be written

$$\varphi(\mathbf{r}') = \mathfrak{G}(g) H'(\mathbf{r}') \mathfrak{G}(g^{-1}) \psi(\mathbf{r}'). \tag{6.1.29}$$

Changing the notation from $\mathbf{r}'$ to $\mathbf{r}$ and comparing (6.1.29) with (6.1.27) it follows that

$$H(\mathbf{r}) = \mathfrak{G}(g) H'(\mathbf{r}) \mathfrak{G}(g^{-1}) \tag{6.1.30}$$

which yields the desired result by observing the forminvariance of H, i.e. $H' \equiv H$, Q.E.D.

6.2 Basic group theoretical theorems

The application of group theory to the solution of problems of quantum theory requires the knowledge of basic theorems of abstract group theory as well as of representation theory. In this section we give without proofs a collection of these relevant theorems. For the proofs we refer to the mathematical group theoretical literature. As the continuous groups such as the $O(3)$ are used only for the construction of appropriate base functions of the finite groups and as the finite groups are the genuinely important groups of solid state physics, we confine ourselves to the presentation of theorems useful for and applicable to finite groups.

Def. 6.5: A finite group G contains a denumerable finite number of elements g, $G := \{g_1 \ldots g_g\}$ where g is called the order of the group G.

Theorem 6.4: For a finite group G the set $\{g_n g_1, \ldots g_n g_g\}$ with arbitrary $g_n \in G$ contains each element of G once and only once, i.e., $\{g_n g_1 \ldots g_n g_g\} = \{g_1 \ldots g_g\}$.

Def. 6.6: If a subset $\mathscr{H} := \{h_1 \ldots h_h\}$ of G exists, such that $\mathscr{H}$ is itself a group, then $\mathscr{H}$ is called a subgroup of G of order h.

Def. 6.7: If $g_k \in G$ is not contained in $\mathscr{H}$, then the right coset of $\mathscr{H}$ is defined by $\mathscr{H} g_k$, and the left coset by $g_k \mathscr{H}$.

Theorem 6.5: The order h of a subgroup must be an integral divisor of g, i.e., $g = hn$ with n integer.

Def. 6.8: If x is an element of G, $x \in G$, then $x^{-1}gx$ is called conjugate to g with respect to x.

Def. 6.9: A subgroup $\mathscr{H}$ is called an invariant subgroup if for all $x \in G$ the relation $x^{-1}\mathscr{H}x = \mathscr{H}$ is valid.

Theorem 6.6: The set $\{\mathscr{H}, g_{\alpha_1}\mathscr{H}, \ldots g_{\alpha_l}\mathscr{H}\}$ of the left cosets of an invariant subgroup $\mathscr{H}$ with respect to G which contains all distinct cosets of $\mathscr{H}$ in G constitutes a group, the factorgroup.

Def. 6.10: The group multiplication of the factorgroup $(g_{\alpha_i}\mathscr{H})(g_{\alpha_j}\mathscr{H})$ is defined by the multiplication of any element of one coset with any element of the other.

Def. 6.11: The set $\{g_1^{-1} g_h g_1, \ldots, g_g^{-1} g_h g_g\}$ is called a class of the group G with respect to $g_h \in G$.

Theorem 6.7: If two classes with respect to g_h and g_l have one element in common then both classes are identical.

Theorem 6.8: The union of all distinct independent classes $K_1 \ldots K_l$ gives G, i.e., $\{K_1 \ldots K_l\} \equiv G$.

Def. 6.12: If for two groups G and G' of the same order there exists a unique map which preserves the multiplication table, then G and G' are called isomorphic.

Def. 6.13: If for two groups G and G', several elements of G can be associated with exactly one element of G' and if this map preserves group multiplication, then G is called homomorphic to G'.

Theorem 6.9: If G is homomorphic to G', then the set $\{g_{\alpha_1} \ldots g_{\alpha_\varrho}\} \subset G$ which is associated to the unity element $g_1' \equiv e'$ of G' is an invariant subgroup of G.

Theorem 6.10: If G is homomorphic to G', then G' is isomorphic to a factorgroup of the group G.

For practical calculations the abstract group operations have to be realized by representations. In the following we give some important theorems concerning these representations.

Def. 6.14: If for a given group $G := \{g_1 \ldots g_g\}$ a set of matrices $\Gamma := \{\gamma(g_1) \ldots \gamma(g_g)\}$ can be found where the group operation of G is homomorphic or isomorphic to the matrix multiplication, then Γ is called a representation of G.

For a fixed group G of order g various dimensions $n_1, n_2 \ldots$ of various representations Γ are possible. Futhermore, due to the existence of the inverse element g_l^{-1} for any g_l the matrices $\gamma(g_l)$ have to be nonsingular.

Def. 6.15: If for two representations Γ and Γ' of the same group G a similarity transformation S exists $\gamma'(g_i) = S^{-1}\gamma(g_i)S$ with the same S for all $i = 1 \ldots g$, then Γ and Γ' are equivalent.

Theorem 6.11: If for two equivalent representations Γ and Γ' one representation Γ is isomorphic to G, then Γ' is also isomorphic to G.

Theorem 6.12: Any representation Γ of a finite group G can be transformed by a similarity transformation S into an equivalent unitary representation Γ' with $\gamma^+ = \gamma^{-1}$.

Def. 6.16: A unitary representation Γ' is reducible if by a similarity transformation S a decomposition $S^{-1}\Gamma' S = \sum_\alpha \oplus \Gamma_\alpha$ is possible, where $\{\Gamma_\alpha,\ 1 \leqslant \alpha \leqslant \varrho\}$ are irreducible, i.e., representations which cannot be decomposed in this way.

Theorem 6.13: If the only matrix which commutes with all matrices $\{\gamma(g_i)\}$ of a representation Γ is the unit matrix multiplied by an arbitrary constant factor unequal zero, then the representation Γ is irreducible.

Theorem 6.14: If two representations Γ_α and Γ_β are irreducible, then the orthogonality relations

$$\sum_{i=1}^{g} \gamma_a(g_i)_{lj}^{\times} \gamma_\beta(g_i)_{kh} = \frac{g}{n_\alpha} \delta_{\alpha\beta} \delta_{lk} \delta_{jh}$$

hold, where n_α is the dimension of Γ_α and l, j, and k, h, are the matrix subscripts.

These orthogonality relations are not valid for reducible representations. They are therefore a criterion for irreducibility.

Theorem 6.15: The trace of a matrix $\mathrm{Tr} A$ is invariant under similarity transformations.

Theorem 6.16: The traces of a given representation Γ for all elements belonging to the same class are equal, i.e.,

$$\mathrm{Tr}\gamma(g_l) = \mathrm{Tr}\gamma(g_h) \qquad \forall g_l, g_h \in K_\alpha.$$

Def. 6.17: The traces of the various classes are called group characters, $\mathrm{Tr}\,\gamma(K_\alpha) =: \chi_\alpha$.

Theorem 6.17: For two irreducible representations Γ_α and Γ_β, the corresponding sets of characters $\{\chi_\alpha(g_i)\}$ and $\{\chi_\beta(g_i)\}$ satisfy the orthogonality relations

$$\sum_{i=1}^{g} \chi_\alpha(g_i)^\times \chi_\beta(g_i) = g\delta_{\alpha\beta}.$$

Theorem 6.18: A necessary and sufficient condition for the equivalence of two irreducible representations Γ_α and Γ_β is the equality of their sets of characters $\{\chi_\alpha(g_i)\} = \{\chi_\beta(g_i)\}$.

Theorem 6.19: The number a_α indicating the repeated occurrence of the irreducible representation Γ_α in the general representation Γ is given by $a_\alpha = \frac{1}{g} \sum_i \chi(g_i) \chi_\alpha(g_i)^\times$. The decomposition is unique up to the arrangement of the irreducible representations; in this sense there exists one and only one decomposition of Γ into irreducible parts.

Theorem 6.20: A necessary and sufficient condition for the irreducibility of Γ is

$$\sum_{i=1}^{g} \chi(g_i)^\times \chi(g_i) = g.$$

Def. 6.18: If the elements of a class K_j are defined by the subset $\{g_1^j \dots g_{\varrho_j}^j\} \subset G$, then class multiplication is defined by

$$K_i K_j := \left(\sum_{z=1}^{\varrho_i} g_z^i \right) \left(\sum_{y=1}^{\varrho_j} g_y^j \right) \equiv \{ g_z^i g_y^j,\ 1 \leqslant z \leqslant \varrho_i,\ 1 \leqslant y \leqslant \varrho_j \}$$

Theorem 6.21 : If a group G contains l different classes, then the class multiplication gives

$$K_i K_j = \sum_{m=1}^{l} c_{ijm} K_m$$

with integer c_{ijm}.

Theorem 6.22 : For the group characters $\chi_\alpha(K_i)$ of an irreducible representation Γ_α, the relations

$$\varrho_i \varrho_j \chi_\alpha(K_i) \chi_\alpha(K_j) = n_\alpha \sum_{m=1}^{l} c_{ijm} \chi_\alpha(K_m)$$

are valid, where $\chi_\alpha(K_i) \equiv \chi_\alpha(g_z^i)$, $g_z^i \in K_i$ and ϱ_i = number of elements of K_i, etc.

Theorem 6.23 : The sum of the squares of the dimensions n_α of the different irreducible representations Γ_α is equal to the order g of the group, i.e.,

$$\sum_{\alpha=1}^{r} n_\alpha^2 = g.$$

Theorem 6.24 : The number of irreducible representations of a finite group is equal to the number of independent classes of this group.

To obtain a starting point for representations, the regular representation can be used:

Def. 6.19 : The regular representation is defined by matrices $A(g_i)$ of dimension g which follow from the group multiplication table according to

$$x g_k = \sum_{l=1}^{g} A_{kl}(x) g_l \quad x \in G.$$

where the entries of $A_{kl}(x)$ are 0 or 1.

Theorem 6.25 : The regular representation is uniquely defined, isomorphic to G, and reducible.

6.3 Basis systems of representations

Concerning the finite groups which are of special interest for crystal physics, these groups are subgroups of the group $O(3)$ of rotations and reflections. Hence these groups possess an operator representation as has been proven in Section 6.1. We shall use this property to derive various theorems about the appropriate construction of symmetry adapted base function systems, i.e., of suitable quantum mechanical representation spaces in crystal physics. For simplicity we will not specialize to electronic and nuclear coordinates $\mathbf{r}$ and $\mathbf{R}$ but will consider only a general symbolic coordinate x which is assumed to transform like a space coordinate with respect to the relevant symmetry groups.

Theorem 6.26 : If the group G corresponds to orthogonal coordinate transformations, then the representations which belong to a complete orthonormal base system of a corresponding forminvariant energy operator are unitary representations.

Proof: If the orthonormal base system is given by $\{\psi_{n\alpha}\}$ where n denumerates the energy levels, while α characterizes the degeneracy, then the most general eigenfunction Ψ_n belonging to E_n is given by $\Psi_n = \sum_\alpha c_\alpha \psi_{n\alpha}$. Due to the orthonormality we have

$$\int \Psi(x)^\times \Psi(x) dx = \sum_{\alpha=1}^{h} c_\alpha^\times c_\alpha \tag{6.3.1}$$

where for brevity we omitted the subscript n. Performing a transformation $x = gy$ we obtain, due to $dx/dy = 1$ for orthogonal transformations,

$$\begin{aligned}
&\int \Psi(x)^\times \Psi(x) dx = \int \Psi(gy)^\times \Psi(gy) dy \\
&= \sum_{\alpha\mu} \int c_\alpha^\times \psi_{n\alpha}(gy)^\times c_\mu \psi_{n\mu}(gy) dy \\
&= \sum_{\substack{\alpha\mu \\ \alpha'\mu'}} c_\alpha^\times \gamma_{\alpha\alpha'}^n(g)^\times c_\mu \gamma_{\mu\mu'}^n(g) \int \psi_{n\alpha'}(y)^\times \psi_{n\mu'}(y) dy
\end{aligned} \tag{6.3.2}$$

and due to the orthonormality of the base system it follows by comparison of (6.3.2) with (6.3.1) that

$$\delta_{\alpha\mu} = \sum_{\alpha'} \gamma_{\alpha\alpha'}^n(g)^\times \gamma_{\mu\alpha'}^n(g) = \sum_{\alpha'} \gamma_{\mu\alpha'}^n(g) \gamma_{\alpha'\alpha}^n(g)^+ \tag{6.3.3}$$

i.e., the representation is unitary, Q.E.D.

The representations gained in this way are not necessarily irreducible. But the following theorem holds:

Theorem 6.27 : The set of eigenfunctions $\{\psi_{n\alpha}\}$ of a Hamilton operator H being forminvariant under an orthogonal group can always be chosen in such a way that it is a base set for irreducible representations of the group G.

Proof: By definition, irreducible representations are obtained by the application of similarity transformations. Thus given a representation Γ', then the corresponding irreducible representation Γ has to be obtained by $\Gamma = S\Gamma' S^{-1}$. According to Theorem 6.12 any representation can be transformed into a unitary representation. Hence Γ and Γ' can be assumed to be unitary without any loss of generality, and due to $\Gamma = S\Gamma' S^{-1}$ the irreducible representation Γ has to be unitarily equivalent to Γ'. Starting with an orthonormal base set $\{\psi_{n\alpha}\}$, the nontrivial representations Γ' are generated in the subsets $\{\psi_{n\alpha}, 1 \leqslant \alpha \leqslant h, n \text{ fixed}\}$. The most general unitary transformation is then given by

$$\varphi_{n\lambda} = \sum_\alpha S_{\lambda\alpha} \psi_{n\alpha}. \tag{6.3.4}$$

Applying this transformation to the representation of a group operation we obtain

$$\varphi'_{n\lambda}=\sum_{\alpha} S_{\lambda\alpha}\psi'_{n\alpha}=\sum_{\alpha\alpha'} S_{\lambda\alpha}\gamma^n_{\alpha\alpha'}\psi_{n\alpha'}$$

$$=\sum_{\alpha\alpha'\lambda'} S_{\lambda\alpha}\gamma^n_{\alpha\alpha'}S^{-1}_{\alpha'\lambda'}\varphi_{n\lambda'} \tag{6.3.5}$$

i.e., if the subset $\{\psi_{n\alpha},\ 1\leqslant\alpha\leqslant h\}$ transforms under the representation Γ, then the set $\{\varphi_{n\lambda},\ 1\leqslant\lambda\leqslant h\}$ transforms under the representation $\Gamma'=S\Gamma S^{-1}$, i.e., both representations are unitarily equivalent. As the base vectors $\{\psi_{n\alpha}, 1\leqslant\alpha\leqslant h\}$ form a complete base of the h-dimensional vector space $\mathbf{V}_h$, by (6.3.4) all possible unitarily equivalent base systems $\{\varphi_{n\lambda}, 1\leqslant\lambda\leqslant h\}$ can be gained. Thus by varying the transformation S all unitarily equivalent representations Γ and Γ' can be obtained. By definition, the irreducible representations are contained in this set, therefore an appropriate orthonormal base system of eigenfunctions of H can be found which are base functions for irreducible representations, Q.E.D.

It is obvious that the classification of the spectrum by division into parts belonging to irreducible representations of the corresponding symmetry group is an effective method for the actual computation as well as for the empirical classification of the observations. Concerning the computation, in particular, the convergence of numerical methods, etc., can be considerably improved by observing from the beginning the symmetry properties of the system states. Even if a symmetry is satisfied only approximately, it is a valuable tool for a zero order classification by considering the symmetry breaking interactions as perturbations. For example, in crystals the impurity centers are influenced by the crystal which leads approximately to the symmetry of the crystal point groups being broken by other irregularities of the crystal and by internal and external forces. In all these cases it is essential to construct symmetry adapted base states. We consider here two methods, namely variational calculations and perturbation calculations and adapt them to symmetry conditions. For the group theoretical treatment of both methods projection operators are required.

Def. 6.20 : If a finite group G of order g with unitary irreducible representations Γ_α is given, then projection operators are defined by

$$R^{\alpha}_{kl}=\sum_{j=1}^{g} \gamma^{\alpha}_{kl}(g_j)^{\times} g_j=\sum_{j=1}^{g} \gamma^{\alpha}_{kl}(g_j)\mathfrak{G}(g_j) \tag{6.3.6}$$

where the sum runs over all group operations.

For these operators the following theorem holds:

Theorem 6.28: The projection operators R^{α}_{kl} transform like the basis functions of an irreducible representation.

Proof: We apply a group operation g_h to R_{kl}^{α}

$$g_h R_{kl}^{\alpha} = \sum_{j=1}^{g} \gamma_{kl}^{\alpha}(g_j)^{\times} g_h g_j. \tag{6.3.7}$$

According to the multiplication table we have $g_h g_j = g_l$, thus giving

$$\begin{aligned} g_h R_{kl}^{\alpha} &= \sum_{l=1}^{g} \gamma_{kl}^{\alpha}(g_h^{-1} g_l)^{\times} g_l \\ &= \sum_{l=1}^{g} \sum_{m} \gamma_{km}^{\alpha}(g_h^{-1})^{\times} \gamma_{ml}^{\alpha}(g_l)^{\times} g_l = \sum_{m} \gamma_{km}^{\alpha}(g_h) R_{ml}^{\alpha} \end{aligned} \tag{6.3.8}$$

which has to be compared with (6.1.11), Q.E.D.

Theorem 6.29 : The projection operators R_{kl}^{α} satisfy the orthogonality relations

$$R_{ij}^{\alpha} R_{kl}^{\beta} = \frac{g}{n_{\beta}} \delta_{\alpha\beta} \delta_{kj} R_{il}^{\beta} \tag{6.3.9}$$

where n_{β} is the matrix dimension of Γ_{β}.

Proof: We have

$$R_{ij}^{\alpha} R_{kl}^{\beta} = \sum_{h} \sum_{m} \gamma_{ij}^{\alpha}(g_h)^{\times} \gamma_{kl}^{\beta}(g_m)^{\times} g_h g_m. \tag{6.3.10}$$

With $g_h g_m = g_n$ it follows that

$$R_{ij}^{\alpha} R_{kl}^{\beta} = \sum_{h} \sum_{n} \gamma_{ij}^{\alpha}(g_h)^{\times} \gamma_{kl}^{\beta}(g_h^{-1} g_n)^{\times} g_n. \tag{6.3.11}$$

Evaluating $\gamma_{kl}^{\beta}(g_h^{-1} g_n)$ and applying theorem 6.14, the formula (6.3.10) follows immediately, Q.E.D.

For variational calculations a set of testfunctions which exhibits the right transformation properties is required, i.e., which transforms as a set of basis vectors of irreducible representations of the corresponding symmetry group. Generally, we may assume that an arbitrary orthonormal set $\{\varphi_h\}$ is given which does not have the right transformation properties. Then the problem is to construct a symmetry adapted set starting from this set. For this task the following theorem is useful.

Theorem 6.30 : For a given orthonormal set $\{\varphi_h\}$, the set $\varphi_{hij}^{\alpha} := R_{ij}^{\alpha} \varphi_h$ transforms as a base set of irreducible representations of the symmetry group G and satisfies the orthonormality relations

$$\langle \varphi_{hij}^{\alpha}, \varphi_{h'kl}^{\beta} \rangle = \frac{g}{n_{\beta}} \delta_{\alpha\beta} \delta_{ki} \langle \varphi_h, R_{jl}^{\beta} \varphi_{h'} \rangle. \tag{6.3.12}$$

A complete orthonormalization conserving the transformation properties is possible.

Proof: Applying R_{ij}^{α} to $\{\varphi_h\}$, we replace g_j in (6.3.6) by its operator representation $\mathfrak{G}(g_j)$. Hence

$$\varphi_{hij}^{\alpha} := \sum_l \gamma_{ij}^{\alpha}(g_l)^{\times}\, \mathfrak{G}(g_l)\, \varphi_h \tag{6.3.13}$$

and using (6.3.8)

$$\begin{aligned} \mathfrak{G}(g_k)\varphi_{hij}^{\alpha} &= \sum_l \gamma_{ij}^{\alpha}(g_l)^{\times}\, \mathfrak{G}(g_k g_l)\varphi_h \\ &= \sum_m \gamma_{im}^{\alpha}(g_k) R_{mj}^{\alpha}\varphi_h \\ &= \sum_m \gamma_{im}^{\alpha}(g_k)\varphi_{hmj}^{\alpha} \end{aligned} \tag{6.3.14}$$

i.e., the columns of the matrix φ_{hij}^{α} $(i,j=1\ldots n_\alpha)$ transform as basis vectors of irreducible representations. Forming the scalar product we have to observe that

$$\varphi_{hij}^{\alpha\,\times} := (R_{ij}^{\alpha}\varphi_h)^{+} = \varphi_h^{\times}\,(R_{ij}^{\alpha})^{+} \tag{6.3.15}$$

is valid. As G is assumed to be an orthogonal group, its representations are unitary. For unitary representations, $A^{+} \equiv A^{-1}$ holds and therefore we have

$$(R_{ij}^{\alpha})^{+} = \sum_{l=1}^{g} \gamma_{ij}^{\alpha}(g_l^{-1})^{\times}\, \mathfrak{G}(g_l^{-1}) = R_{ij}^{\alpha}. \tag{6.3.16}$$

Then the scalar product gives

$$\langle \varphi_{hij}^{\alpha}, \varphi_{h'kl}^{\beta}\rangle = \langle \varphi_h R_{ij}^{\alpha}, R_{kl}^{\beta}\varphi_{h'}\rangle \tag{6.3.17}$$

which leads with (6.3.9) to formula (6.3.12). To achieve a complete orthonormalization we make the ansatz

$$\Phi_{pq}^{\alpha} = \sum_j \sum_h a_{qhj}^{\alpha}\varphi_{hpj}^{\alpha} \tag{6.3.18}$$

where the coefficients a_{qhj}^{α} are to be determined by the conditions

$$\langle \Phi_{pq}^{\alpha}, \Phi_{p'q'}^{\beta}\rangle = \delta_{\alpha\beta}\delta_{pp'}\delta_{qq'}. \tag{6.3.19}$$

It can be easily verified that this procedure leads to a complete orthonormalization. This procedure also preserves the transformation properties since

$$\begin{aligned} \mathfrak{G}(g_l)\Phi_{pq}^{\alpha} &= \sum_{hj} a_{qhj}^{\alpha}\mathfrak{G}(g_l)\varphi_{hpj}^{\alpha} \\ &= \sum_m \gamma_{pm}^{\alpha}(g_l) \sum_{hj} a_{qhj}^{\alpha}\varphi_{hmj}^{\alpha} \\ &= \sum_m \gamma_{pm}^{\alpha}(g_l)\Phi_{mq}^{\alpha} \end{aligned} \tag{6.3.20}$$

holds, Q.E.D.

The successful application of this method requires the explicit knowledge of the various irreducible matrix representations of the symmetry group being considered. Since such representations are well-known for most symmetry groups of physical interest, no further restriction is imposed on the method.
We now turn to the perturbation calculation. We start with an energy operator $H=H_0+V$ where the solutions of H_0 are assumed to be exactly known, and where H is forminvariant under a symmetry group G. The set of eigenfunctions $\{f_j^{n\alpha}\}$ are defined to be eigenstates of H_0 where E_0^n is the corresponding eigenvalue, while α denote the irreducible representation and j denotes its basis vectors. The solutions of the total energy operator are expanded

$$\Psi_\mu=\sum_{n\alpha j} c_{n\alpha j}^\mu f_j^{n\alpha} \tag{6.3.21}$$

and the expansion coefficients $c_{n\alpha j}^\mu$ are to be determined from the equation

$$\sum_{m\beta l} H_{n\alpha j,m\beta l} c_{m\beta l}^\mu = E c_{n\alpha j}^\mu . \tag{6.3.22}$$

Due to the symmetry properties of H and $f_j^{n\alpha}$ this equation can be considerably simplified.

Theorem 6.31 : The general matrix element $H_{n\alpha j,m\beta l}$ is reduced to

$$H_{n\alpha j,m\beta l}=\delta_{\alpha\beta}\delta_{jl}H_{nm}(\alpha) \tag{6.3.23}$$

if H has the symmetry group G and if $H_{n\alpha j,m\beta l}$ is referred to a basis vector system of irreducible representations of G.

Proof: According to the supposition we have

$$\mathfrak{G}(g_h) f_j^{n\alpha}=\sum_{l=1}^{n_\alpha} \gamma_{jl}^\alpha(g_h) f_l^{n\alpha}. \tag{6.3.24}$$

From this it follows that

$$R_{ij}^\alpha f_j^{n\alpha}=\sum_{h=1}^{g} \gamma_{ij}^\alpha(g_h)^\times\, \mathfrak{G}(g_h) f_j^{n\alpha}=\sum_h\sum_k \gamma_{ij}^\alpha(g_h)^\times\, \gamma_{jk}^\alpha(g^h) f_k^{n\alpha} \tag{6.3.25}$$

and due to the orthogonality relations we have

$$R_{ij}^\alpha f_j^{n\alpha}=\frac{g}{n_\alpha} f_j^{n\alpha}. \tag{6.3.26}$$

The matrix element of H can then be written

$$\int f_j^{n\alpha^\times} H f_l^{m\beta}\, dx=\int \frac{n_\alpha n_\beta}{g^2}\,(R_{jj}^\alpha)^+ f_j^{n\alpha^\times} H(R_{ll}^\beta) f_l^{m\beta}\, dx \tag{6.3.27}$$

or equivalently

$$H_{n\alpha j,m\beta l}=\frac{n_\alpha n_\beta}{g^2}\,\langle f_j^{n\alpha}, R_{jj}^\alpha H R_{ll}^\beta f_l^{m\beta}\rangle. \tag{6.3.28}$$

As H commutes with R from (6.3.28) it follows with (6.3.9) that

$$H_{n\alpha j,m\beta l}=\frac{n_\alpha n_\beta}{g^2}\langle f_j^{n\alpha},HR_{jj}^{\alpha}R_{ll}^{\beta}f_l^{m\beta}\rangle$$
$$=\frac{n_\alpha}{g}\delta_{\alpha\beta}\delta_{jl}\langle f_j^{n\alpha},HR_{jj}^{\alpha}f_j^{m\alpha}\rangle. \tag{6.3.29}$$

This matrix element does not depend on the special index j. We have

$$\langle f_j^{n\alpha},HR_{jj}^{\alpha}f_j^{m\alpha}\rangle=\frac{g}{n_\alpha}\langle f_j^{n\alpha},Hf_j^{m\alpha}\rangle$$
$$=\frac{n_\alpha}{g}\langle(R_{jk}^{\alpha})^{+}f_k^{n\alpha},HR_{jk}^{\alpha}f_k^{m\alpha}\rangle=\langle f_k^{n\alpha},HR_{kk}^{\alpha}f_k^{m\alpha}\rangle \tag{6.3.30}$$

where for the last line the evaluation technique used above has been applied. From this follows (6.3.23), Q.E.D.

However, for crystal physics symmetry breaking perturbations V are of more interest than symmetry conserving perturbations. If an impurity atom or ion is placed on a lattice site, the potential V represents the influence of the crystal forces on the impurity particle. While for the free impurity particle the full $O(3)$ group is the symmetry group, the crystal potential V produces a symmetry breaking of the $O(3)$ group, which is reduced to the point group of the crystal. If the electronic structure of the free impurity atom or ion is described by means of the electronic shell model, under the action of V the degenerate angular momentum representations are split into the various crystal point group representations. Then it is of interest which point group representations are contained in the angular momentum representations of the $O(3)$ group. This problem has been treated by Bethe (1929).
First we consider the $O(3)$ angular momentum representations. They are given by

Theorem 6.32 : The set of spherical harmonics

$$Y_m^l(\theta,\varphi)=(2\pi)^{-1/2}P_m^l(\theta)e^{im\varphi} \tag{6.3.31}$$

with

$$P_{-m}^l(\theta)=P_m^l(\theta):=\frac{N_m^l}{2^l l!}\sin^m\theta\,\frac{d^{l+m}\sin^{2l}(\theta)}{d(\cos\theta)^{l+m}},\quad m\geqslant 0$$
$$\{-l\leqslant m\leqslant l\},\quad m\in\mathbb{N} \tag{6.3.32}$$

and

$$N_m^l=\{[(l-m)!/(l+m)!]\,[(2l+1)/2]\}^{1/2} \tag{6.3.33}$$

forms a basis for a $(2l+1)$-dimensional representation $\mathscr{D}^l$ of the rotation group $R(3)$. A representation of the full rotation group $O(3)$ is obtained by defining

$$\mathscr{D}^{l+}(g)=\mathscr{D}^{l+}(\mathscr{I}g);\quad \mathscr{D}^{l-}(g)=-\mathscr{D}^{l-}(\mathscr{I}g) \tag{6.3.34}$$

where $\mathscr{I}$ is an inversion.

For the further analysis the classes and characters of the $R(3)$ representations are required. The classes follow from the theorem:

Theorem 6.33: Any rotation $\mathscr{A}$ from (6.1.15) can be transformed by a rotation $\mathscr{B}$ into

$$\mathscr{B}\mathscr{A}\mathscr{B}^{-1}=\begin{pmatrix}1 & 0 & 0\\ 0 & \cos\varphi' & -\sin\varphi'\\ 0 & \sin\varphi' & \cos\varphi'\end{pmatrix}=\mathscr{L} \tag{6.3.35}$$

i.e., $\mathscr{A}$ and $\mathscr{L}$ belong to the same class.

The set of matrices $\{\mathscr{A}\}$ are a representation of $R(3)$ in the linear vector space $\mathbf{V}_3$. Similar to (6.3.4) and (6.3.5), the transformation $\mathscr{B}$ produces a change of the basis system of $\mathbf{V}_3$. Obviously this basis system of $\mathbf{V}_3$ is changed in such a way that after the change, the axis of rotation of $\mathscr{A}$ coincides with the new z-axis. Hence rotations through the same angle around any axis are equivalent and belong to the same class. As the characters of the elements of one class are equal, we may choose the most convenient representations of any class for the calculation of characters.
Generally we obtain for the group operation $g(\mathscr{A})$ corresponding to a rotation $\mathscr{A}$, the representation

$$g(\mathscr{A})\,Y_m^l(\theta,\varphi)=\sum_{m'} D^l_{mm'}(\mathscr{A})\,Y^l_{m'}(\theta,\varphi) \tag{6.3.36}$$

according to Theorem 6.33. In the special case of a rotation about the z-axis through the angle φ', (6.3.36) can be written

$$g(\mathscr{L})\,Y_m^l(\theta,\varphi)=\sum_{m'} D^l_{mm'}(\mathscr{L})\,Y^l_{m'}(\theta,\varphi)=e^{-im\varphi'}\,Y^l_m(\theta,\varphi). \tag{6.3.37}$$

Hence in this case $D^l_{mm'}(\mathscr{L})$ is a $(2l+1)$-dimensional diagonal matrix

$$D^l_{mm'}(\mathscr{L})=e^{-im\varphi'}\,\delta_{mm'},\qquad -l\leqslant m,m'\leqslant l \tag{6.3.38}$$

and the character of (6.3.38) is

$$\chi^l(\mathscr{L})=\sum_{m=-l}^{l} e^{-im\varphi'}=\frac{\sin\left(l+\dfrac{1}{2}\right)\varphi'}{\sin\,(\varphi'/2)}. \tag{6.3.39}$$

If the full rotational symmetry of a system is reduced to the symmetry of a point group by switching on a corresponding crystal field, the representations D^l which are irreducible with respect to $R(3)$ become reducible with respect to the corresponding point group. As there are 48 different point groups it is impossible to give a complete analysis of the problem. We rather confine ourselves to the treatment of a representative example. As we mainly use NaCl-type lattices for the illustration of explicit calculations of polar crystal processes, we choose the cubic group O_h which is the point symmetry group about a normal lattice site in NaCl-like materials. The classes and their elements are listed below, with the number of elements in the class being the coefficient in the class symbol. The group O_h is of order 48, containing 24 proper rotations, while the doubling is produced by inversion, i.e., $O_h \equiv O \otimes C_i$. The proper rotation classes are given by

E : identity

$3C_4^2$: rotations of π about the three cubic $\langle 100 \rangle$ axis;

$6C_4$: rotations of $\pm\frac{\pi}{2}$ about the three cubic $\langle 100 \rangle$ axis;

$6C_2$: rotations of π about the six twofold $\langle 110 \rangle$ axis;

$8C_3$: rotations of $\pm 2\pi/3$ about the four threefold $\langle 111 \rangle$ axis.

Multiplying by the inversion we obtain the other five classes that contain the remaining 24 elements: I (inversion); $3\,IC_4^2$ (reflection in a [100] plane); $6\,IC_4$; $6\,IC_2$ (reflection in a [110] plane); $8\ IC_3$.
As for the rotation group we distinguish for direct products between Γ^- and Γ^+ representations, where

$$\Gamma^+(\mathscr{A}) = \Gamma^+(\mathscr{I}\mathscr{A}); \quad \Gamma^-(\mathscr{A}) = -\Gamma^-(\mathscr{I}\mathscr{A}) \tag{6.3.40}$$

This leads to the classes and characters

O_h	O	IO
Γ^+	$\{\chi\}$	$\{\chi\}$
Γ^-	$\{\chi\}$	$\{-\chi\}$

(6.3.41)

Thus we need only consider O, the octahedral group in discussing the character table of O_h. The group O has 24 elements and five classes. Therefore, according to Theorem 6.24, there exist five irreducible representations. Applying Theorem 6.23 we have $24 = \sum_{\alpha=1}^{5} n_\alpha^2$ for the corresponding dimensions. The only solution is

$$24 = 3^2 + 3^2 + 2^2 + 1^2 + 1^2.$$

We denote these irreducible representations by Γ_i, $1 \leqslant i \leqslant 5$. The explicit calculation of these representations can be achieved by means of the regular representation and will not be discussed here. We merely give the results. In the notation of Bethe the character table reads

	E	$3\,C_4^2$	$6\,C_4$	$6\,C_2$	$8\,C_3$
Γ_1	1	1	1	1	1
Γ_2	1	1	−1	−1	1
Γ_3	2	2	0	0	−1
Γ_4	3	−1	1	−1	0
Γ_5	3	−1	−1	1	0

Due to the widespread applications many different notations are used as can be seen from Table 6.1.

Table 6.1:
Characters of the full cubic group. References: BSW: Bouckaert, Smoluchowski, and Wigner (1936); Koster: Koster (1957); Bethe: Bethe (1929) (the primes on the Γ_α in this column are those added by Overhauser (1956)); VB: von der Lage and Bethe (1947); Lomont: Lomont (1959); Chemists: Usage exemplified by Eyring, Walter, and Kimball (1944). After Knox and Gold (1964).

				BSW	E	$3C_4^2$	$6C_4$	$16C_2$	$8C_3$	J	$3JC_4^2$	$6JC_4$	$6JC_2$	$8JC_3$
				Koster	E	$3C_2$	$6C_4$	$6C_2'$	$8C_3$	I	$3\sigma_h$	$6S_4$	$6\sigma_d$	$8S_6$
BSW	Chemists	Lomont	VB	Bethe	E	$3C_2$	$6C_3$	$6C_4$	$8C_5$	J	$3JC_2$	$6JC_3$	$6JC_4$	$8JC_5$
Γ_1	A_{1g}	Δ_1	α	Γ_1	1	1	1	1	1	1	1	1	1	1
Γ_2	A_{2g}	Δ_2	β'	Γ_2	1	1	−1	−1	1	1	1	−1	−1	1
Γ_{12}	E_g	Δ_3	γ	Γ_3	2	2	0	0	−1	2	2	0	0	−1
Γ_{15}'	T_{1g}	Δ_5	δ'	Γ_4	3	−1	1	−1	0	3	−1	1	−1	0
Γ_{25}'	T_{2g}	Δ_4	ε	Γ_5	3	−1	−1	1	0	3	−1	−1	1	0
Γ_1'	A_{1u}	...	α'	Γ_1'	1	1	1	1	1	−1	−1	−1	−1	−1
Γ_2'	A_{2u}	...	β	Γ_2'	1	1	−1	−1	1	−1	−1	1	1	−1
Γ_{12}'	E_u	...	γ'	Γ_3'	2	2	0	0	−1	−2	−2	0	0	1
Γ_{15}	T_{1u}	...	δ	Γ_4'	3	−1	1	−1	0	−3	1	−1	1	0
Γ_{25}	T_{2u}	...	ε'	Γ_5'	3	−1	−1	1	0	−3	1	1	−1	0

According to Theorem 6.19 the number of irreducible representations contained in a reducible representation can be calculated by projection of the corresponding characters $a_\alpha = g^{-1} \sum_i \chi(g_i)\chi_\alpha(g_i)^\times$. Substituting those φ'-values corresponding to the discrete rotation operations of $g_i \in O$ in (6.3.39), we obtain

$$\begin{aligned} &\mathscr{D}^0 \to \Gamma_1 \\ &\mathscr{D}^1 \to \Gamma_4 \\ &\mathscr{D}^2 \to \Gamma_3 + \Gamma_5 \\ &\mathscr{D}^3 \to \Gamma_2 + \Gamma_4 + \Gamma_5 \\ &\mathscr{D}^4 \to \Gamma_1 + \Gamma_3 + \Gamma_4 + \Gamma_5 \\ &\mathscr{D}^5 \to \Gamma_3 + 2\Gamma_4 + \Gamma_5 \\ &\mathscr{D}^6 \to \Gamma_1 + \Gamma_2 + \Gamma_3 + \Gamma_4 + 2\Gamma_5 . \end{aligned} \tag{6.3.42}$$

Assuming the Γ_i to be derived explicitly from the regular representation, the projection operator formalism developed in this section can be applied to the Y_m^l set to obtain base functions of the point group representations. The result is

Table 6.2:
Linear combinations of spherical harmonics ($l \leqslant 4$), transforming as basis states in irreducible representations of O_h; after Knox and Gold (1964).

l	Representation	Basis states
0	Γ_1	Y_0^0
1	Γ_4	Y_1^{-1} Y_1^0 $-Y_1^1$
2	Γ_5	$\frac{1}{\sqrt{2}}(Y_2^2 - Y_2^{-2})$ Y_2^1 Y_2^{-1}
2	Γ_3	$\frac{1}{\sqrt{2}}(Y_2^2 + Y_2^{-2})$ Y_2^0
3	Γ_4	$\sqrt{\frac{3}{8}}\,Y_3^1 + \sqrt{\frac{5}{8}}\,Y_3^{-3}$ $\sqrt{\frac{3}{8}}\,Y_3^{-1} + \sqrt{\frac{5}{8}}\,Y_3^3$ Y_3^0
3	Γ_5	$\sqrt{\frac{5}{8}}\,Y_3^1 - \sqrt{\frac{3}{8}}\,Y_3^{-3}$ $\sqrt{\frac{5}{8}}\,Y_3^{-1} - \sqrt{\frac{3}{8}}\,Y_3^3$ $\frac{1}{\sqrt{2}}(Y_3^2 + Y_3^{-2})$
3	Γ_2	$\frac{1}{\sqrt{2}}(Y_3^2 - Y_3^{-2})$
4	Γ_5	$\frac{1}{\sqrt{2}}(Y_4^2 - Y_4^{-2})$ $\sqrt{\frac{1}{8}}\,Y_4^{-1} - \sqrt{\frac{7}{8}}\,Y_4^3$ $\sqrt{\frac{1}{8}}\,Y_4^1 - \sqrt{\frac{7}{8}}\,Y_4^{-3}$
4	Γ_1	$\sqrt{\frac{7}{12}}\,Y_4^0 + \sqrt{\frac{5}{12}}(Y_4^4 + Y_4^{-4})\frac{1}{\sqrt{2}}$
4	Γ_3	$\frac{1}{\sqrt{2}}(Y_4^2 + Y_4^{-2})$ $\sqrt{\frac{5}{12}}\,Y_4^0 - \sqrt{\frac{7}{12}}(Y_4^4 + Y_4^{-4})\frac{1}{\sqrt{2}}$
4	Γ_4	$\frac{1}{\sqrt{2}}(Y_4^4 - Y_4^{-4})$ $\sqrt{\frac{7}{8}}\,Y_4^{-1} + \sqrt{\frac{1}{8}}\,Y_4^3$ $\sqrt{\frac{7}{8}}\,Y_4^1 + \sqrt{\frac{1}{8}}\,Y_4^{-3}$

So far we only considered the states of scalar particles which are represented by the single valued representations of the rotation group or of the point groups. As electrons and nuclei are particles with spin, the use of the single valued representations for the description of such particles is justified only as long as the spin can be neglected. In Sect. 6.4 we will consider examples where the neglection of the spin is not permitted.
Thus we have to look for representations of the rotation group as well as of the point groups which take into account the spin moment of the particles considered. These are the so-called doublevalued representations. Formally such representations can be obtained by investigating the homomorphism between the group of proper rotations in three dimensions and the group SU_2.

Def. 6.21: The group SU_2 is defined as the group of all two-dimensional unitary matrices having determinant unity, with ordinary matrix multiplication as the operation of group multiplication.

The homomorphism is most easily established indirectly by considering the 2×2 traceless Hermitean matrix $m(\mathbf{r})$, whose components are defined in terms of the position vector $\mathbf{r} = (x, y, z)$ by

$$m(\mathbf{r}) := \begin{pmatrix} z & , x - iy \\ x + iy, & -z \end{pmatrix}. \tag{6.3.43}$$

Then, if U is any element of SU_2, $Um(\mathbf{r})U^{-1}$ is also a traceless Hermitean matrix. Thus one may write

$$m(\mathbf{r}') = \begin{pmatrix} z' & , x' - iy' \\ x' + iy', & -z' \end{pmatrix} = Um(\mathbf{r})U^{-1}. \tag{6.3.44}$$

This defines the components of the vector $\mathbf{r}' = (x', y', z')$.
On evaluating the right-hand side of (6.3.44) and equating corresponding matrix elements, one can define a linear transformation from $\mathbf{r}$ to $\mathbf{r}'$ which is given by

$$\mathbf{r}' = \mathcal{A}(U) \cdot \mathbf{r} \tag{6.3.45}$$

with $\mathcal{A}(U)$ being a 3×3 matrix which depends on U. The transformation (6.3.45) represents a rotation, because lengths and angles are preserved. If $\mathbf{r}_1$ and $\mathbf{r}_2$ are any two vectors and $\mathbf{r}_1' = \mathcal{A}(U) \cdot \mathbf{r}_1$ and $\mathbf{r}_2' = \mathcal{A}(U) \cdot \mathbf{r}_2$ is assumed, then

$$\mathbf{r}_1' \cdot \mathbf{r}_2' = \frac{1}{2} \, tr \, [m(\mathbf{r}_1') \cdot m(\mathbf{r}_2')] = \frac{1}{2} \, tr \, [m(\mathbf{r}_1) \cdot m(\mathbf{r}_2)] = \mathbf{r}_1 \cdot \mathbf{r}_2. \tag{6.3.46}$$

The homomorphism is established by showing that if U_1 and U_2 are members of SU_2, then the multiplication property holds also for $\mathcal{A}(U)$. However, the homomorphism is not an isomorphism as $\mathcal{A}(U) = \mathcal{A}(-U)$. It is convenient to invert the relationship between U and $\mathcal{A} = \mathcal{A}(U)$ and to regard U as being dependent on $\mathcal{A}$, i.e., $U = U(\mathcal{A})$ Then $U(\mathcal{A})$ and $-U(\mathcal{A})$ are said to constitute a two-valued representation of $\mathcal{A}$.

An explicit expression for the matrix elements of $\mathscr{A} = \mathscr{A}(U)$ in terms of U can be found by using the three Pauli spin matrices

$$\sigma_1 = \begin{pmatrix} 0 & 1 \\ 1 & 0 \end{pmatrix}; \quad \sigma_2 = \begin{pmatrix} 0 & -i \\ i & 0 \end{pmatrix}; \quad \sigma_3 = \begin{pmatrix} 1 & 0 \\ 0 & -1 \end{pmatrix}. \tag{6.3.47}$$

Then $m(\mathbf{r})$ reads

$$m(\mathbf{r}) = \sigma_1 x + \sigma_2 y + \sigma_3 z \tag{6.3.48}$$

and one obtains

$$A_{jk}(U) = \frac{1}{2}\, tr[\sigma_j U \sigma_k U^{-1}]. \tag{6.3.49}$$

The inversion $U(\mathscr{A})$ is more complicated and will not be given explicitly.
Concerning the states, a spinorial wave function $f_i(\mathbf{r})$ may be defined as a two-component quantity which transforms under a coordinate transformation $\{\mathscr{A}|\mathbf{T}\}$ as

$$\mathfrak{O}(\{\mathscr{A}|\mathbf{T}\}) f_i(\mathbf{r}') = \sum_{j=1}^{2} U_{ij}(\mathscr{A}_p) f_j(\mathbf{r}) \tag{6.3.50}$$

and

$$\bar{\mathfrak{O}}(\{\mathscr{A}|\mathbf{T}\}) f_i(\mathbf{r}') = -\sum_{j=1}^{2} U_{ij}(\mathscr{A}_p) f_j(\mathbf{r}) \tag{6.3.51}$$

with $\mathbf{r}' = \{\mathscr{A}|\mathbf{T}\}\mathbf{r}$ where $\mathscr{A}_p \equiv \mathscr{A}$ for proper rotations and $\mathscr{A}_p \equiv -\mathscr{A}$ for improper rotations.
The transformations laws (6.3.50) and (6.3.51) are the analogue of the transformation law for scalar functions. If a set of ordinary transformations forms a group, then the set of generalized transformations given by (6.3.50) and (6.3.51) also forms a group with twice as many elements. The double group corresponding to a single group G will be denoted by G^d, and the scalar product

$$(f,g) := \sum_{j=1}^{2} \int f_j(\mathbf{r})^{\times} g_j(\mathbf{r})\, d^3r \tag{6.3.52}$$

can be shown to be invariant under G^d, if G corresponds to the real affine group.
As $\mathfrak{O}(\{\mathscr{A}|\mathbf{T}\})$ and $\bar{\mathfrak{O}}(\{\mathscr{A}|\mathbf{T}\})$ differ only in the treatment of the rotational part of the transformation, it is convenient to observe that for every transformation $\{\mathscr{A}|\mathbf{T}\}$, there correspond two generalized transformations $[\mathscr{A}|\mathbf{T}]$ and $[\bar{\mathscr{A}}|\mathbf{T}]$ that are defined to be isomorphic to the operators $\mathfrak{O}(\{\mathscr{A}|\mathbf{T}\})$ and $\bar{\mathfrak{O}}(\{\mathscr{A}|\mathbf{T}\})$, resp. These generalized transformations are distinguished from the ordinary transformations by writing them in square brackets. The product of two generalized transformations is then given by

$$[\mathscr{A}|\mathbf{T}][\mathscr{A}'|\mathbf{T}'] = [\bar{\mathscr{A}}|\mathbf{T}][\bar{\mathscr{A}}'|\mathbf{T}']$$

$$= \begin{cases} [\mathscr{A}\mathscr{A}'|\mathbf{T} + \mathscr{A}\mathbf{T}'] & \text{if} \quad U(\mathscr{A}_p)U(\mathscr{A}'_p) = U(\mathscr{A}_p\mathscr{A}'_p) \\ [\overline{\mathscr{A}\mathscr{A}'}|\mathbf{T} + \mathscr{A}\mathbf{T}'] & \text{if} \quad U(\mathscr{A}_p)U(\mathscr{A}'_p) = -U(\mathscr{A}_p\mathscr{A}'_p) \end{cases} \tag{6.3.53}$$

and by

$$[\bar{\mathscr{A}}|\mathbf{T}][\mathscr{A}'|\mathbf{T}]=[\mathscr{A}|\mathbf{T}][\bar{\mathscr{A}}'|\mathbf{T}'] \tag{6.3.54}$$

$$=\begin{cases}[\mathscr{A}\mathscr{A}'|\mathbf{T}+\mathscr{A}\mathbf{T}'] & \text{if} \quad U(\mathscr{A}_p)U(\mathscr{A}'_p)=-U(\mathscr{A}_p\mathscr{A}'_p)\\ [\overline{\mathscr{A}\mathscr{A}'}|\mathbf{T}+\mathscr{A}\mathbf{T}'] & \text{if} \quad U(\mathscr{A}_p)U(\mathscr{A}'_p)=U(\mathscr{A}_p\mathscr{A}'_p)\end{cases}$$

whichever is appropriate.
In our discussion of symmetry adapted wave functions we have so far not distinguished between one-particle states and many-particle states. In particular the Theorems 6.30 and 6.31 are valid independently of the number of particles involved, provided that we work with point groups, i.e., finite groups. Now the finite point groups are the genuine symmetry groups of crystal physics, but we saw that the rotation group offers a valuable background for the calculation of point group representations. For the rotation group the number of particles decisively influences the construction procedure. Due to the importance of the rotation group we give a brief survey of the corresponding method, which can of course also be applied to point groups.
In many cases the wave functions of many-particle systems are represented by linear combinations of products of one-particle states. To obtain the right symmetry properties of the total state functions, the single one-particle product states must be prepared in a suitable way, i.e., it is sufficient to consider the symmetry properties of these product states. For simplicity we only consider a two-particle product state

$$\psi(\mathbf{r}_1,\mathbf{r}_2):=\varphi_{kl}(\mathbf{r}_1)\varphi_{k'l'}(\mathbf{r}_2) \tag{6.3.55}$$

where $\varphi_{kl}(\mathbf{r}_1)$ and $\varphi_{k'l'}(\mathbf{r}_2)$ are assumed to be base states of representations of the rotation group. Under the action of a group operation $g(\mathbf{A})$, the product state is transformed into

$$\begin{aligned}\psi'(\mathbf{r}'_1,\mathbf{r}'_2)&=\psi_{kl,k'l'}(g(\mathscr{A})\mathbf{r}'_1,g(\mathscr{A})\mathbf{r}'_2)\\ &=\mathfrak{G}_1(\mathscr{A})\varphi_{kl}(\mathbf{r}'_1)\mathfrak{G}_2(\mathscr{A})\varphi_{k'l'}(\mathbf{r}'_2)\\ &=\sum_{\mu}D^k_{l\mu}(\mathscr{A})\varphi_{k\mu}(\mathbf{r}'_1)\sum_{\varrho}D^{k'}_{l'\varrho}(\mathscr{A})\varphi_{k'\varrho}(\mathbf{r}'_2)\\ &=\sum_{\mu\varrho}D^k_{l\mu}(\mathscr{A})D^{k'}_{l'\varrho}(\mathscr{A})\psi_{k\mu,k'\varrho}(\mathbf{r}'_1,\mathbf{r}'_2).\end{aligned} \tag{6.3.56}$$

Therefore, the function (6.3.56) is transformed by $\mathscr{D}^k\otimes\mathscr{D}^{k'}$ which is called a direct product of representations.

Def. 6.22 : If for a given group G two representations $\Gamma_\alpha,\Gamma_\beta$ are defined, then the direct product is defined by

$$\Gamma^{\alpha\beta}_{hljk}(g_i)=\Gamma^{\alpha}_{hj}(g_i)\Gamma^{\beta}_{lk}(g_i) \tag{6.3.57}$$

where $\Gamma^{\alpha\beta}$ acts on the direct product of the representations states.

Without proof we give the following theorem:

Theorem 6.34: If Γ_α and Γ_β are representations of G, then the direct product $\Gamma_{\alpha\beta}$ is also a representation of G.

In general, the direct product representation is not irreducible even if Γ_α and Γ_β are irreducible. As irreducibility means physically the characterization of the corresponding basis states by unique quantum numbers, it is necessary to decompose $\Gamma_{\alpha\beta}$ into irreducible representations and to derive the explicit form of the similarity transformation S which effects this decomposition. For the decomposition we use the Theorem 6.19 about the projection of the characters. The following theorem holds:

Theorem 6.35: The character of the various elements of the product representation $\Gamma_{\alpha\beta}(g_i)$ is given by the product of the characters of the representations $\Gamma_\alpha(g_i)$, $\Gamma_\beta(g_i)$

$$\chi^{\alpha\beta}(g_i)=\chi^\alpha(g_i)\chi^\beta(g_i). \tag{6.3.58}$$

According to Theorem 6.33, the rotation group can be decomposed into various classes, where each class can be represented by an element of the subgroup of rotations about the z-axis. Generalizing the projections of the characters to the one-parameter continuous subgroup of rotations about the z-axis, we obtain according to Theorem 6.19

$$a_j=(2\pi)^{-1}\int \chi^{ll'}(\varphi)\chi_j(\varphi)^\times d\varphi \tag{6.3.59}$$

where $\alpha=l$, $\beta=l'$ and j are the numbers of the irreducible representations $\mathscr{D}$ of $R(3)$, or $O(3)$, resp. The representations of these groups with respect to the subgroup $\mathscr{L}$ are given by (6.3.38) and their characters by (6.3.39). Thus forming $\chi^{ll'}$ and projecting it on $\chi_j^\times$, the number a_j is given by

$$\begin{aligned} a_j&=(2\pi)^{-1}\int \sum_{m=-l}^{l}\sum_{n=-l'}^{l'}\sum_{r=-j}^{j} e^{i\varphi(m+n)}e^{-i\varphi r}\,d\varphi \\ &=\sum_{m=-l}^{l}\sum_{n=-l'}^{l'}\sum_{r=-j}^{j}\delta(m+n-r). \end{aligned} \tag{6.3.60}$$

For a detailed analysis of this expression, we observe that $\mathfrak{D}_j(\mathscr{L})$ is, according to (6.3.38), the direct sum over the representations of the two-dimensional Abelian rotation group. Hence we can look first for the multiple occurrence of these Abelian group representations and then collect the results to obtain the occurrence of $\mathfrak{D}_j(\mathscr{L})$. Doing this it follows that the direct product $\mathscr{D}^l\otimes\mathscr{D}^{l'}$ can be decomposed into

$$\mathscr{D}^l\otimes\mathscr{D}^{l'}\to\mathscr{D}^{l+l'}\otimes\ldots\otimes\mathscr{D}^{l-l'} \tag{6.3.61}$$

where $l>l'$ is assumed, i.e. that $a_j=1$, $l+l'\geqslant j\geqslant l-l'$. The notation of (6.3.61) is only symbolic, as a similarity transformation S is required to bring $\mathscr{D}^l\otimes\mathscr{D}^{l'}$ explicitly into the diagonal form of a sum of irreducible representations. This diagonal form reads in matrix notation

$$M_{j\nu,j'\nu'}=\delta_{jj'}D^j_{\nu\nu'}(\mathscr{A}) \qquad \begin{matrix} l+l'\geqslant j,j'\geqslant l-l' \\ j\geqslant\nu\geqslant -j \end{matrix} \tag{6.3.62}$$

while the original matrix of the direct product is given by

$$(\mathscr{D}^l \otimes \mathscr{D}^{l'})_{\mu\mu'\varrho\varrho'} = D^l_{\mu\mu'}(\mathscr{A}) D^{l'}_{\varrho\varrho'}(\mathscr{A}). \tag{6.3.63}$$

Thus, the similarity transformation $S^{-1}\mathscr{D}^l \otimes \mathscr{D}^{l'} S = M$ reads explicitly

$$\sum_{\substack{\mu\varrho \\ \mu'\varrho'}} S^{-1}_{j\nu\mu\varrho} (\mathscr{D}^l \otimes \mathscr{D}^{l'})_{\mu\mu'\varrho\varrho'} S_{\mu'\varrho' j'\nu'} = M_{j\nu j'\nu'}. \tag{6.3.64}$$

It is possible to calculate S directly. We do not undertake this explicitly, but refer to the literature where the so-called Clebsch-Gordan coefficients are tabulated.
We are only interested in the physical result of this procedure. According to (6.3.5) any similarity transformation of the representation leads to a corresponding transformation of the basis vectors and vice versa, i.e., from $M = S^{-1}(\mathscr{D}^l \otimes \mathscr{D}^{l'})S$ it follows that $\Psi = S\psi$, which gives explicitly

$$\Psi_{j\nu}(\mathbf{r}_1, \mathbf{r}_2) = \sum_{\mu\varrho} S_{\mu\varrho, j\nu} \varphi_{l\mu}(\mathbf{r}_1) \varphi_{l\varrho}(\mathbf{r}_2). \tag{6.3.65}$$

Theorem 6.36: the transformed wave functions (6.3.65) are eigenfunctions of the angular momentum.

Proof: With $\mathfrak{G}(\mathscr{A} \times \mathscr{A}) = \mathfrak{G}(\mathscr{A})\mathfrak{G}(\mathscr{A})$ we obtain

$$\begin{aligned} \Psi'(\mathbf{r}_1', \mathbf{r}_2') &= \mathfrak{G}(\mathscr{A} \times \mathscr{A}) \Psi(\mathbf{r}_1', \mathbf{r}_2') = \mathfrak{G}(\mathscr{A} \times \mathscr{A}) S\psi \\ &S\mathscr{D}^l \otimes \mathscr{D}^{l'} S^{-1} S\psi = M\Psi. \end{aligned} \tag{6.3.66}$$

Specializing now $\mathfrak{G}(\mathscr{A} \times \mathscr{A})$ to the generators of the rotation group and observing the properties of M for these infinitesimal rotations, we immediately derive the statement, Q.E.D.

6.4 Fine structure of impurity levels

In the quantum theory of free atoms, ions or molecules group theoretical methods are particularly helpful for the classification of electronic energy levels and the description of level fine structure under the influence of internal and external perturbations. Thus it is natural to apply such methods also to the calculation of various impurity levels in crystals. For impurities in crystals (in the idealized microblock model) the various point groups may occur as invariance groups, and their representations are required for special calculations. As shown in the preceding chapter, such calculations can be divided into two parts concerned with static and dynamical electron lattice coupling. For dynamical electron-lattice coupling the application of group theoretical methods implies that the phonon modes must be base states of the corresponding representations. While the phonon modes for the space groups are well known, relatively little is known about the phonon modes of the point groups. Only the phonon modes for small molecules which exhibit a point group symmetry have been derived.

Therefore a proper group theoretical treatment of impurities, which includes electron lattice coupling, has so far been restricted to molecular models. In this section we confine ourselves to static coupling calculations by means of such models.
The application of the molecular model means that the interaction of the impurity with the host crystal must be sufficiently well approximated by the interaction of the impurity with nearest neighbours or with very small surroundings which contain only a very small number of neighbouring ions or atoms, resp. This condition is relatively well satisfied for instance for transition metal ion- or rare earth ion-impurities but it is not satisfied, for instance, for F-centers, as the calculations of the preceding chapters show. Nevertheless, as the group theoretical discussion of the levels and their fine structure is very instructive, we shall treat both types of centers by means of the molecular model, disregarding its limited range of applicability.
The molecular model has been applied in many different versions. A rough classification of these approaches with respect to static coupling calculations may be given by the following scheme.

i) The electronic states of the impurity center are explicitly calculated, while the influence of the neighbouring host ions is idealized by a rigid potential being invariant against the corresponding point group operations.

ii) The electronic states of the impurity center as well as of the neighbouring host ions are explicitly calculated by means of molecular orbitals, while the molecular complex is assumed to be rigidly fixed in the ideal lattice positions.

In this section both steps will be discussed for the various physical situations which can be described by these methods.

a) Magnetic ion impurities

Magnetic ions which are used as substitutional impurities have partly unfilled inner shells. This fact is reflected in some magnetic and optical properties when such ions are embedded in a crystal lattice. In particular, the unfilled shells provide a net magnetic moment which may interact with external magnetic fields. The magnetic ions can be divided into five categories, cf. di Bartolo (1968):

i) Transition metal ions of the first series with the electronic configurations (Argon core)^{18} $3d^n 4s^m$, where $m=1,2$ and $n=1,2,\ldots 10$. In their divalent ionic states these elements have configurations of the type (Ar core)^{18} $3d^n$ with the unfilled $3d$ shell.

ii) Transition metal ions of the second series with the electronic configurations (Krypton core)^{36} $4d^n 5s^m$, where $m=1,2$ and $n=1,2,\ldots 10$. In their divalent ionic states these elements have configurations of the type (Kr core)^{36} $4d^n$ with the unfilled $4d$ shell.

iii) Transition metal ions of the third series with the electronic configurations (Palladium core)^{46} $4f^{14}5s^2 5p^6 5d^n 6s^m$ where $m=1,2$ and $n=2,3,\ldots 10$. In their divalent ionic states these elements have configurations with unfilled $5d$ shells.

iv) Rare earth lanthanide ions with the electronic configurations (Palladium core)^{46} $4f^n5s^25p^65d^m6s^2$ where $m=1$ and $n=2,3,\ldots 13$. In their trivalent ionic states these elements have configurations of the type (Pd core)^{46} $4f^n5s^25p^6$ with an unfilled $4f$ shell.

v) Rare earth actinide ions with the electronic configurations (Platinum core)^{78} $5f^n6d^m7s^2$ where $m=1,2$ and $n=0,2,3,\ldots$ In their trivalent ionic states these elements have configurations of the type (Pt core)^{78} $5f^n7s^2$.

The observed spectra of the magnetic ions in crystals in the visible, the ultraviolet and the infrared region mainly result from transitions between levels belonging to the same electronic configuration, the splitting among the different levels being due to the electrostatic and the spin-orbit interaction, spin-spin coupling and the crystal field. Furthermore, by observation of para-magnetic resonance in the radio frequency region, electron-spin interaction with external magnetic fields and with nuclear spins may be detected. This means that for such spectra the fine structure and the hyperfine structure of the levels play an important role. Since these structures are produced by the relativistic corrections to the nonrelativistic adiabatic Hamiltonian (2.2.5), we must use instead of (2.2.5) the improved adiabatic Hamiltonian (2.3.41). We do not attempt to justify the molecular model by ab initio calculations from the microblock model, but rather proceed in accordance with the different approximations of i) and ii).

α) Crystalline field calculations

These calculations are done in the scheme i), where only the electrons of the metal ion are treated explicitly with a rigid potential due to the surroundings. For calculations of this kind the spin-orbit interaction is taken into account. With this approximation in (2.3.41), the following Hamiltonian results

$$H:=\sum_{i=1}^{N}\left[\frac{1}{2m}\mathbf{p}_i^2-e^2zC(\mathbf{r}_i,\mathbf{R}_\alpha^0)+V(\mathbf{r}_i)\right]$$
$$+\frac{e^2}{2}\sum_{i,j}{}'\,C(\mathbf{r}_i,\mathbf{r}_j)-\frac{\mu_0}{2mc}\sum_{i=1}^{N}\mathbf{s}_i\cdot(\mathbf{E}_i\times\mathbf{p}_i). \tag{6.4.1}$$

Here the metal ion core with the charge ze is located at the lattice site $\mathbf{R}=\mathbf{R}_\alpha^0$ and the influence of the host crystal is described by $V(\mathbf{r})$. The last term in (6.4.1) is the spin orbit coupling, where $\mathbf{E}_i$ is the electric field that is felt by the electron i. According to (2.3.25), this electric field is the unquantized part of the external field, if no radiation field is present. This part is given by the electric field of the metal ion core and that produced by $V(\mathbf{r})$. Usually this field is assumed to exhibit approximate radial symmetry, i.e., $\mathbf{E}(\mathbf{r})=E(r)\mathbf{r}/r$. In this case we may write

$$H_{\text{os}}:=-\frac{\mu_0}{2mc}\sum_i\mathbf{s}_i\cdot(\mathbf{E}_i\times\mathbf{p}_i)=\lambda(\mathbf{r})\mathbf{L}\cdot\mathbf{S} \tag{6.4.2}$$

with

$$\mathbf{L}\cdot\mathbf{S}=\sum_i\mathbf{l}_i\cdot\mathbf{s}_i=\sum_i(\mathbf{r}_i\times\mathbf{p}_i)\cdot\mathbf{s}_i. \tag{6.4.3}$$

Although in this model the explicit treatment of electrons is restricted to those of the central ion, for numerical calculations the number of these electrons is still very large as it follows from the configurations which were given above. For comparison with optical absorption and emission measurements only the energy differences are needed. If only these differences have to be calculated, the number of electrons which have to be treated explicitly can further drastically be reduced by the following theorems, which we cite without proofs:

Theorem 6.37 : The only effect of a closed shell on the calculation of the terms arising from certain electrons in configurations outside the closed shell is to contribute an additive constant to the energy.

Theorem 6.38: A shell containing $(N-n)$ electrons has the same term structure as one containing n electrons, where N is the number of electrons in the filled shell. The term separations are the same for a shell lacking n electrons as for a shell with n electrons.

For the proof of these theorems see Slater (1960). As a consequence of these theorems only the electrons of the partly unfilled shells have to be treated explicitly up to the maximum number of half of the number of shell states. The treatment of these electrons runs, in general, along the lines developed for many-electron atoms. A detailed treatment of this "classical" topic of quantum mechanics would exceed the scope of this book. Hence we give only a short survey concerning the symmetry problems and refer for further details to the literature cited in the following.
The crystal field strength resulting from $V(\mathbf{r})$ may be classified by comparing its effects on the ion with that of the various terms in the ionic Hamiltonian. Three cases are customarily distinguished: the weak field, the intermediate field and the strong field case.

In the weak field case the Hamiltonian (6.4.1) is divided into $H=H_I+V$, where H_I is the free ion Hamiltonian, while V is the perturbing crystal potential. The eigenstates of the unperturbed system belong to the complete set of observables H_I, $\mathbf{J}^2$, $\mathbf{J}_z$, P where $\mathbf{J}$ is the total angular momentum and P the parity. Under the influence of V, the $\mathscr{D}_J$ representations are reduced to Γ-representations of the crystal group.

In the intermediate field case, H_I is divided into $H_I=H_o+H_{os}$. The eigenstates of H_o are considered to be base states of a perturbation calculation. They belong to the complete set of observables $H_o, \mathbf{L}^2, \mathbf{S}^2, L_z, S_z, P$, where $\mathbf{L}$ is the total orbital angular momentum, while $\mathbf{S}$ is the total spin operator. Under the influence of V these states are reduced to eigenstates of $H_o+V, S^2, S_z, \Gamma, \Gamma_z$ and afterwards, by switching on the spin orbit coupling, the direct products of $\mathscr{D}_S$ and Γ representations are reduced to Γ-representations alone.

In the strong field case the unperturbed Hamiltonian is given by $H_u=\sum_i [\mathbf{p}_i^2/2m - e^2zC(\mathbf{r}_i\mathbf{R}_\alpha^0)]$ and the system is characterized by the eigenstates of the complete set H_u, $\mathbf{l}_1^2$, $\mathbf{s}_1^2$, l_{1z}, $s_{1z}, \ldots$, $\mathbf{l}_k^2$, $\mathbf{s}_k^2$, l_{kz}, s_{kz}. By switching on the crystal field, the one-electron orbital states are changed from basis states of the full rotation group into basis states of the point group. Afterwards, by switching on the Coulomb interaction, the orbital parts of the point group for the single electrons are changed into basis states of the point group for all electrons, and finally these states together with the spin states are reduced by spin-orbit coupling to basis states of representations of the point group for all electrons.

A simple example of coupling to strong and intermediate crystal fields is afforded by an atom having two electrons in different shells outside of closed shells. Suppose the two electrons have orbital quantum numbers l_1, and l_2. In the strong field case we first make the reductions

$$\mathscr{D}^{l_1}=\sum_k a_{l_1k}\Gamma_k$$
$$\mathscr{D}^{l_2}=\sum_{k'} a_{l_2k'}\Gamma_{k'} \tag{6.4.4}$$

where the a's are coefficients. In this limit the zero-order energy of the system is given as a sum of crystal one-electron eigenvalues $E_k+E_{k'}$. The eigenfunctions are antisymmetric products, which transform according to the product representations $\Gamma_k\otimes\Gamma_{k'}$ of the point group. Turning on the electrostatic repulsion between the electrons, the product representation is reduced as

$$\Gamma_k\otimes\Gamma_{k'}=\sum_i \beta^i_{kk'}\Gamma_i \tag{6.4.5}$$

where the β's are the usual generalized Clebsch-Gordan coefficients.
In an intermediate field we first couple the electrons in the atom in the usual way

$$\mathscr{D}^{l_1}\otimes\mathscr{D}^{l_2}=\sum_{|l_1-l_2|}^{l_1+l_2}\mathscr{D}^{\lambda} \tag{6.4.6}$$

and then reduce D^{λ} according to the point group.

$$\mathscr{D}^{\lambda}=\sum_i a_{\lambda i}\Gamma_i. \tag{6.4.7}$$

The order in which the coupling is performed corresponds to the successive order of importance of the interactions. However, it should be emphasized that the final representations obtained are a consequence of the symmetry alone and are independent of the crystal field strength. In intermediate fields the lowest state in the crystal is connected to the lowest state in the free atom. In strong fields, however, this need not, in general, be the case.
We work out the energy levels of two d-electrons in different shells placed in a cubic crystal field of strong or intermediate strength.

Strong field: According to (6.3.42), the one-electron state with $l=2$ splits into

$$\mathscr{D}^2=\Gamma_3\oplus\Gamma_5. \tag{6.4.8}$$

The possible states of the system are given by $\Gamma_3 \otimes \Gamma_3$, $\Gamma_3 \otimes \Gamma_5$, $\Gamma_5 \otimes \Gamma_3$, and $\Gamma_5 \otimes \Gamma_5$. With the aid of the character table, we obtain eleven terms.

$$\begin{aligned} &\Gamma_3 \otimes \Gamma_3 = \Gamma_1 \oplus \Gamma_2 \oplus \Gamma_3 \\ &\Gamma_3 \otimes \Gamma_5 = \Gamma_5 \otimes \Gamma_3 = \Gamma_4 \oplus \Gamma_5 \\ &\Gamma_5 \otimes \Gamma_5 = \Gamma_1 \oplus \Gamma_3 \oplus \Gamma_4 \oplus \Gamma_5 . \end{aligned} \tag{6.4.9}$$

Intermediate field: We first make the reduction

$$\mathscr{D}^2 \otimes \mathscr{D}^2 = \mathscr{D}^4 \oplus \mathscr{D}^3 \oplus \mathscr{D}^2 \oplus \mathscr{D}^1 \oplus \mathscr{D}^0 \tag{6.4.10}$$

and using (6.3.42) we obtain

$$\begin{aligned} &\mathscr{D}^0 = \Gamma_1 ; \quad \mathscr{D}^3 = \Gamma_2 \oplus \Gamma_4 \oplus \Gamma_5 \\ &\mathscr{D}^1 = \Gamma_4 ; \quad \mathscr{D}^4 = \Gamma_1 \oplus \Gamma_3 \oplus \Gamma_4 \oplus \Gamma_5 \\ &\mathscr{D}^2 = \Gamma_3 \oplus \Gamma_5 . \end{aligned} \tag{6.4.11}$$

These are, as they must be, the same eleven terms given in (6.4.9). Making arbitrary assumptions about the positions of the terms, we illustrate in Figure 6.1 the connection

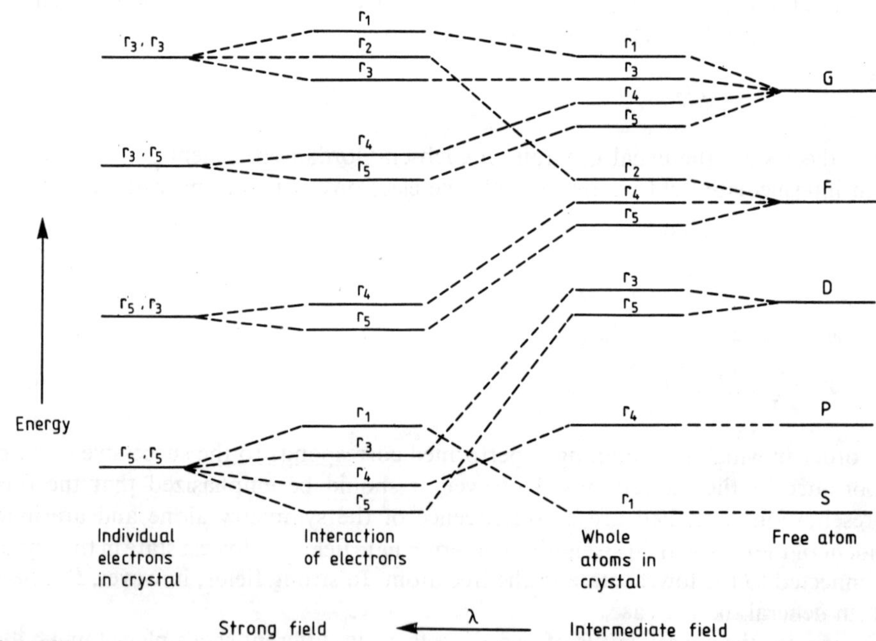

Figure 6.1:
Schematic representations of the terms of an atom with two inequivalent *d*-electrons in a field of symmetry O_h. The crossing points of the dotted lines for various values of λ afford particular examples of accidental degeneracy, not required by symmetry. Two levels of the same symmetry type cannot cross in this scheme. The ground state in the strong field need not be connected to the lowest term of the free atom; after Knox and Gold (1964)

between intermediate and strong fields for two inequivalent d-electrons. The parameter λ describing the crystal field strength increases to the left.
Another example is given by the spin orbit interaction in the crystal field. In a free atom the inclusion of spin gives wave functions that transform according to the representations of the full rotation group as

$$\mathscr{D}^L \otimes \mathscr{D}^S = \sum_{L-S}^{L+S} \mathscr{D}^J. \tag{6.4.12}$$

In a weak crystal field (smaller than spin-orbit interaction), we start with (6.4.12) and decompose the $\mathscr{D}^J$ in the crystalline field via

$$\mathscr{D}^J = \sum_{\mu} a_{J\mu} \Gamma_{\mu}. \tag{6.4.13}$$

On the other hand, in a crystal field of intermediate strength (greater than spin-orbit coupling), the order of steps is first to make the decomposition

$$\mathscr{D}^L = \sum_{\nu} a_{L\nu} \Gamma_{\nu} \tag{6.4.14}$$

and then turn on spin-orbit coupling

$$\Gamma_{\nu} \otimes \mathscr{D}^S = \sum_{\mu} a_{\nu\mu} \Gamma_{\mu}. \tag{6.4.15}$$

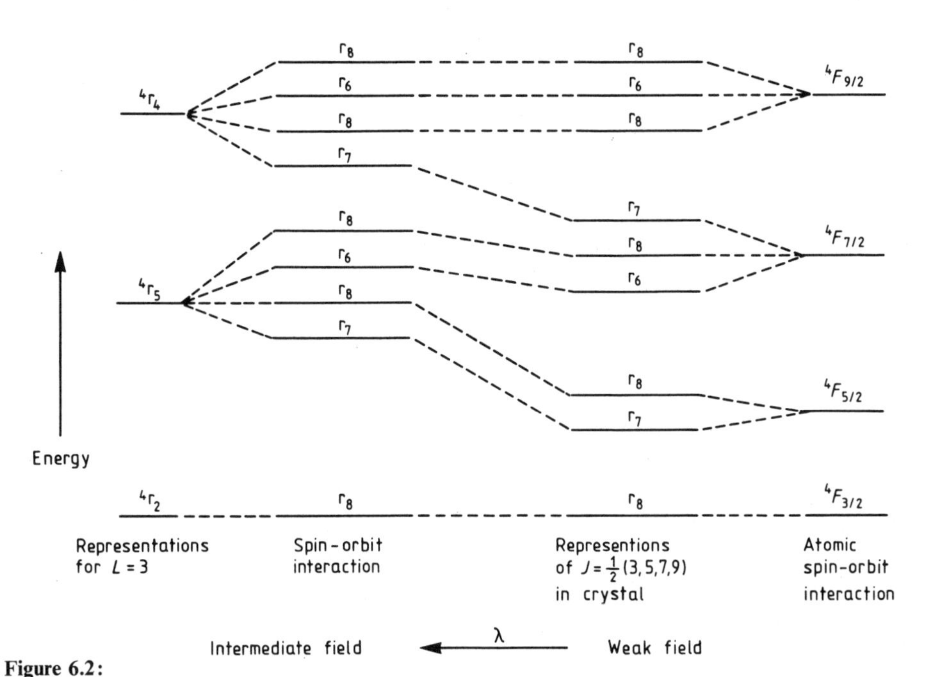

Figure 6.2:
Schematic representation of the splitting of levels arising from a ^4F term due to spin-orbit interaction and a cubic crystal field of weak or intermediate strength; after Knox and Gold (1964)

Here, of course, $\mathscr{D}^S$ denotes a sum of irreducible representations of the point group (or its double group) and the number of components into which (6.4.15) splits is not in general $2S+1$.
As a schematic example, Figure 6.2 illustrates the splitting of a ^4F term due to weak and intermediate strength cubic fields. The derivation of the decomposition is straightforward.

β) Complex ion calculations

Complex ions are transition metal ions which are surrounded by polar molecules or by ions of other elements. They are the topic of coordination compound chemistry and the quantum mechanical calculation of their structures and properties is an essential part of theoretical chemistry. The calculations are done with the model ii) and the essentials of the method can be seen by giving an illustrative example. In the following we will consider a transition metal cation which is surrounded by six anions, i.e., an octahedral complex, which results if a regular cation in an NaCl-lattice is replaced by an impurity cation.
While crystalline field calculations start with the pure ionic configuration of the central ion, the molecular orbital calculations of complex ions start at the opposite end with covalent bonding configurations and the ionic character of the complex then results from the covalent bonding calculation. Hence the first step in the presentation of the calculation procedure must be the explanation of covalent bonding.
Covalent bonding between atoms or ions is effected by the *s*-, *p*- and *d*-orbitals of the partly unfilled outer shells of these particles. It can be illustrated by the bonding of a diatomic molecule. If it is assumed that both atoms of such a molecule possess such partly unfilled outer shells, the combination of the atomic orbitals leads to σ-, π- and δ molecular orbitals. The σ-orbitals are symmetric for rotation about the internuclear axis, the π-orbitals have one-nodal planes containing the internuclear axis, while the δ-orbitals have two-nodal planes containing the internuclear axis. Each type of molecular orbital is realized by several wave functions with lower or with higher energy compared with the energy of the free atoms, i.e., the atomic orbitals can be divided into bonding states (with low energy) and antibonding states (with high energy). If we denote the set of atomic orbitals by $\{s^i, p_x^i, p_y^i, p_z^i, d_{z^2}^i, d_{x^2-y^2}^i, d_{xy}^i, d_{yz}^i, d_{xz}^i, i=1,2\}$ and if we assume the internuclear axis to be the *z*-axis, the σ-orbitals result from combinations of (s^i, s^j) (s^i, p_z^j) (p_z^i, p_z^j) (s^i, d_z^j) $i,j=1,2,\ i\neq j$ atomic orbitals, the π-orbitals result from combinations of (p_x^i, p_x^j) (p_y^i, p_y^j) $i,j=1,2,\ i\neq j$ atomic orbitals, while δ-orbitals result from the combination of the remaining *d*-functions. The sign of the coefficients of the linear combination decides whether a bonding or an antibonding molecular orbital results. For instance, (s^1+s^2) or $(p_z^1-p_z^2)$ are bonding states, while (s^1-s^2) or $(p_z^1+p_z^2)$ are antibonding states. In Figure 6.3 the formation of σ- and π-orbitals is illustrated.
If the bonding of polyatomic configurations is considered, directed bonds must be constructed which tend in the directions of the different internuclear axis. The corresponding atomic orbitals are hybrids and the first step in complex ion calculations is the formation of hybrids.

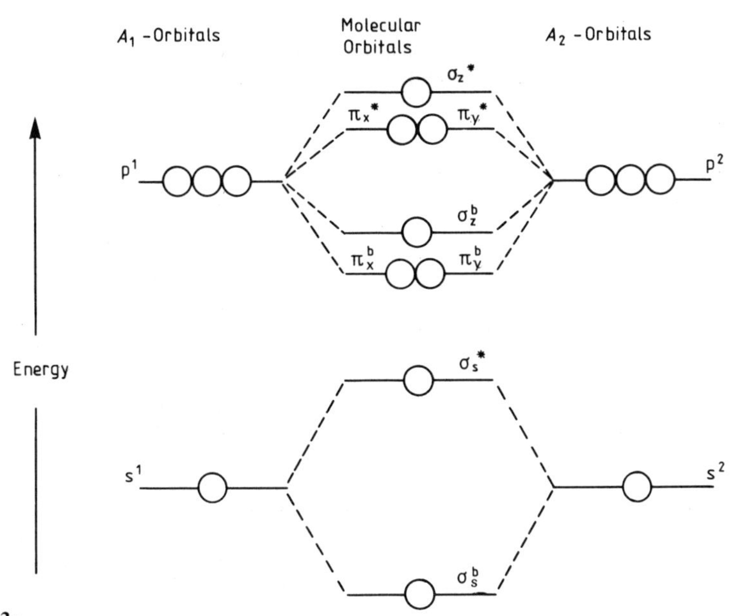

Figure 6.3:
Energy splitting resulting from the formation of atomic orbitals to molecular orbitals. From eight atomic orbitals four bonding molecular orbitals σ_s^b, σ_z^b, π_x^b, π_y^b and four antibonding molecular orbitals σ_s^*, σ_z^*, π_x^*, π_y^* can be constructed. Without regard to normalization it is

$$\sigma_s^b=(s^1+s^2),\ \sigma_z^b=(p_z^1-p_z^2),\ \pi_x^b=(p_x^1+p_x^2),\ \pi_y^b=(p_y^1+p_y^2),$$

while the antibonding orbitals are combinations with the opposite sign; after Dickerson, Gray, and Haight (1974)

The formation of hybrids is closely connected with group theory. For our example of a metal cation surrounded by six anions the relevant symmetry group is the O_h group. Obviously the various internuclear axis play an essential role in the construction procedure. If we consider the bonds of the central cation directed to the six surrounding anions, these directions are given by the set of unit vectors $\{\mathbf{e}_x, -\mathbf{e}_x, \mathbf{e}_y, -\mathbf{e}_y, \mathbf{e}_z, -\mathbf{e}_z\}$. Under the transformation operations of O_h this set of unit vectors behaves like a basis set of vectors for a (reducible) representation of this group. Since the hybrids have to point in the same direction, they must show the same transformation properties as the unit vectors, i.e., hybrid wave functions are characterized by being basis vectors of just this representation. It turns out, cf. di Bartolo (1968), that the corresponding representation Γ_σ can be reduced to the direct sum $\Gamma_\sigma=A_{1g}+E_g+T_{1u}$. The atomic orbitals of the central cation are (s^c, p_x^c, p_y^c, p_z^c, $d_{z^2}^c$, $d_{x^2-y^2}^c$, d_{xy}^c, d_{xz}^c, d_{zy}^c). It can be shown that s transforms under O_h as a basis of A_{1g}; $d_{z^2}^c$, $d_{x^2-y^2}^c$ transforms under O_h as a basis of E_g and p_x^c, p_y^c, p_z^c transforms under O_h as a basis of T_{1u}. The six-dimensional representation for hybrid orbitals of the central ion must therefore be formed by the set $\{s^c, p_x^c, p_y^c, p_z^c, d_{z^2}^c, d_{x^2-y^2}^c\}$. These orbitals lead to the formation of σ-bonds in the corresponding directions and there is no contradiction to

Table 6.3:
The character table and the relevant atomic orbitals for the symmetry group O_h of the central ion in an octahedral complex CA_6 ($C \equiv$cation, $A \equiv$anion). In the last two rows the Γ_σ representation and the Γ_π representation are given which lead to directed atomic hybrid functions for σ-bonding and π-bonding; after di Bartolo (1968).

O_h	E	$8C_3$	$6C_2$	$6C_4$	$3C_2$	I	$6S_4$	$8S_6$	$3\sigma_h$	$6\sigma_d$	Atomic Orbitals
A_{1g}	1	1	1	1	1	1	1	1	1	1	s
A_{2g}	1	1	−1	−1	1	1	−1	1	1	−1	
E_g	2	−1	0	0	2	2	0	−1	2	0	$d_{z^2}, d_{\alpha^2-y^2}$
T_{1g}	3	0	−1	1	−1	3	1	0	−1	−1	
T_{2g}	3	0	1	−1	−1	3	−1	0	−1	1	d_{xy}, d_{yz}, d_{zx}
A_{1u}	1	1	1	1	1	−1	−1	−1	−1	−1	
A_{2u}	1	1	−1	−1	1	−1	1	−1	−1	1	
E_u	2	−1	0	0	2	−2	0	1	−2	0	
T_{1u}	3	0	−1	1	−1	−3	−1	0	1	1	(p_x, p_y, p_z)
T_{2u}	3	0	1	−1	−1	−3	1	0	1	−1	
											Reduced Reps.
Γ_σ	6	0	0	2	2	0	0	0	4	2	$A_{1g}+E_g+T_{1u}$
Γ_π	12	0	0	0	−4	0	0	0	0	0	$T_{1g}+T_{2g}+T_{1u}+T_{2u}$

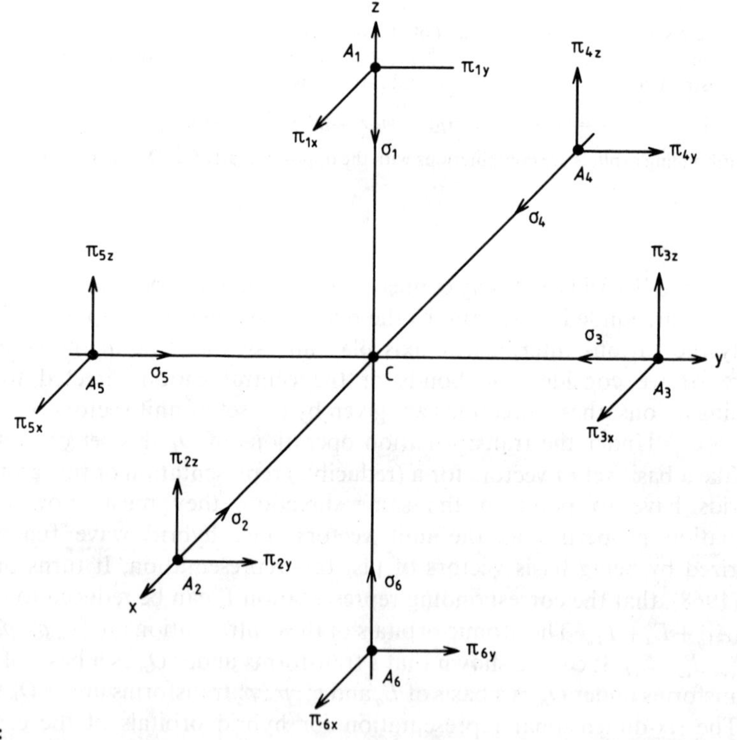

Figure 6.4:
Octahedral complex CA_6. The directed bonds of the surrounding six anions are indicated by arrows, while the directed bonds of the metal ion follow from Table 6.3; after di Bartolo (1968)

the preceding statement that p_x, p_y lead to π-bonds, as here several directions occur, while in the former case only one direction was taken into account. Before we consider the π-bonds of the central cation, we turn to the hybrids of the surrounding six anions. For instance we consider the halogen ions

$$F^- = (\text{Helium core})^2\ 2s^2p^6,\ Cl^- = (\text{Neon core})^{10}\ 3s^2p^6,$$

$$Br^- = (\text{Argon core})^{18}\ 3d^{10}4s^2p^6,\ I^- = (\text{Krypton core})^{36}\ 3d^{10}5s^2p^6$$

to be the surrounding anions. Then it is sufficient to take into account the p-states of these anions for bonding calculations.
Obviously, $p_z^1 \equiv \sigma_1$ leads to a σ-bonding hybrid for the anion A_1, $p_x^2 \equiv \sigma_2$ to a σ-bonding hybrid for A_2, etc. while $p_x^1 = \pi_{1x}$, $p_y^1 = \pi_{1y}$ lead to π-bonding hybrids for the anion A_1, $p_x^2 = \pi_{2x}$, $p_z^2 = \pi_{2z}$ to π-bonding hybrids for the anion A_2 etc. This shows clearly that the representation Γ_π for the π-bonding orbitals must be given by the twelve base states

Table 6.4:
Orbitals for octahedral complexes. The hybrid orbitals of the anions are arranged in appropriate linear combinations in order to provide irreducible representations of the O_h group; after di Bartolo (1968).

Representation	Metal Orbital	Ligand σ	Ligand π
A_{1g}	s	$\frac{1}{\sqrt{6}}(\sigma_1+\sigma_2+\sigma_3+\sigma_4+\sigma_5+\sigma_6)$	
E_g	d_{z^2}	$\frac{1}{\sqrt{3}}(2\sigma_1+2\sigma_6-\sigma_2-\sigma_3-\sigma_4-\sigma_5)$	
	$d_{x^2-y^2}$	$\frac{1}{2}(\sigma_2-\sigma_3+\sigma_4-\sigma_5)$	
T_{1u}	p_x	$\frac{1}{\sqrt{2}}(\sigma_2-\sigma_4)$	$\frac{1}{2}(\pi_{3x}+\pi_{1x}+\pi_{5x}+\pi_{6x})$
	p_y	$\frac{1}{\sqrt{2}}(\sigma_3-\sigma_5)$	$\frac{1}{2}(\pi_{2y}+\pi_{1y}+\pi_{4y}+\pi_{6y})$
	p_z	$\frac{1}{\sqrt{2}}(\sigma_1-\sigma_6)$	$\frac{1}{2}(\pi_{2z}+\pi_{3z}+\pi_{4z}+\pi_{5z})$
T_{2g}	d_{xz}		$\frac{1}{2}(\pi_{2z}+\pi_{1x}-\pi_{4z}-\pi_{6x})$
	d_{yz}		$\frac{1}{2}(\pi_{3z}+\pi_{1y}-\pi_{5z}-\pi_{6y})$
	d_{xy}		$\frac{1}{2}(\pi_{2y}+\pi_{3x}-\pi_{4y}-\pi_{5x})$
T_{1g}			$\frac{1}{2}(\pi_{2z}-\pi_{1x}-\pi_{4z}+\pi_{6x})$
			$\frac{1}{2}(\pi_{3z}-\pi_{1y}-\pi_{5z}+\pi_{6y})$
			$\frac{1}{2}(\pi_{2y}-\pi_{3y}-\pi_{4y}+\pi_{5x})$
T_{2u}			$\frac{1}{2}(\pi_{3x}-\pi_{1x}+\pi_{5x}-\pi_{6x})$
			$\frac{1}{2}(\pi_{2y}-\pi_{1y}+\pi_{4y}-\pi_{6y})$
			$\frac{1}{2}(\pi_{2z}-\pi_{3z}+\pi_{4z}-\pi_{5z})$

$(\pi_{1x}, \pi_{1y}, \pi_{2z}, \pi_{2y}, \ldots)$. The reducible representation Γ_π can be divided into the direct sum $\Gamma_\pi = T_{1g} + T_{2g} + T_{1u} + T_{2u}$. As the central cation is subject to the same symmetry group, its π-bonding representation must be given also by Γ_π, or at least by one of its irreducible parts. As the s^c, p^c, $d^c_{z^2}$, $d^c_{x^2-y^2}$ states have already been used for σ-bonding, only the states d^c_{xy}, d^c_{xz}, d^c_{yz} remain for π-bonding. Their representation is given by $T_{2g} \subset \Gamma_\pi$. After having found the hybrids of the central cation and the surrounding anion, the genuine energy calculation must be carried out.
If a Hamiltonian is invariant under a symmetry group, energy calculations can be simplified, since, according to Theorem 6.31, no matrix elements occur which correspond to two different irreducible representations. This means that the energy calculations of complex ions have to be done with respect to wave functions of the cation and the anion, which belong to irreducible representations of the symmetry group. This task has completely been solved for the cation wave functions which we had already classified into σ- and π-bonding parts which belong to various irreducible representations. For the anions we have so far given a classification into σ- and π-bonding functions, but the corresponding representations are still not irreducible. This task can be solved by applying Theorem 6.30 to the set of σ- and π-orbitals of the anions. As the construction is straightforward we give only the results in Table 6.4. Matrix elements of the Hamiltonian occur only between functions of the same line in the table and lead to bonding and antibonding energy levels if the Hamiltonian of the complex is finally diagonalized. The results of such calculations are given by the Figure 6.5.
The discussion of ion spectra started with a paper of Bethe (1929) who considered ions or atoms at regular lattice positions and idealized the interaction of these particles with their surroundings by a crystal field, i.e., he applied model i) to regular lattice points. Since the publication of this paper, numerous papers have appeared dealing with this topic. We refer to books which summarize the older results and we give some additional references to papers which appeared later. Fick and Joos (1957) gave a group theoretical treatment and a detailed discussion of the calculation techniques in a review article, which was supplemented by a paper of Schulz (1961). In a review article Herzfeld and Meijer (1962) treated the application of group theory to crystal field theory, i.e. to model i). MaClure (1966) published a review article of the theory of electronic states and spectra of ions and imperfections in solids with respect to models i) and ii) with inclusion of electron-phonon coupling. The group theoretical classification and treatment of electronic impurity states in the rigid lattice, i.e., for model i) was reviewed by Bassani (1971). The theory of transition metal ions was summarized by Griffith (1961) (1971). In this book all calculation schemes of model i) are discussed in detail. In connection with electron paramagnetic resonance. Low (1960) and Abragam and Bleaney (1970) gave a detailed presentation of the physical facts and calculation techniques for the model i). In addition the model ii) was treated by Abragam and Bleaney. Di Bartolo (1968) gave a compact review of models i) and ii) in a treatment of optical interactions in solids with respect to applications in laser physics. A review article which contains among other things a discussion about impurity centers in alkali halides was written by Fowler (1968). Short reviews are also presented in books about group theory, cf., for instance, Knox and Gold (1964), Tinkham (1964). Without going

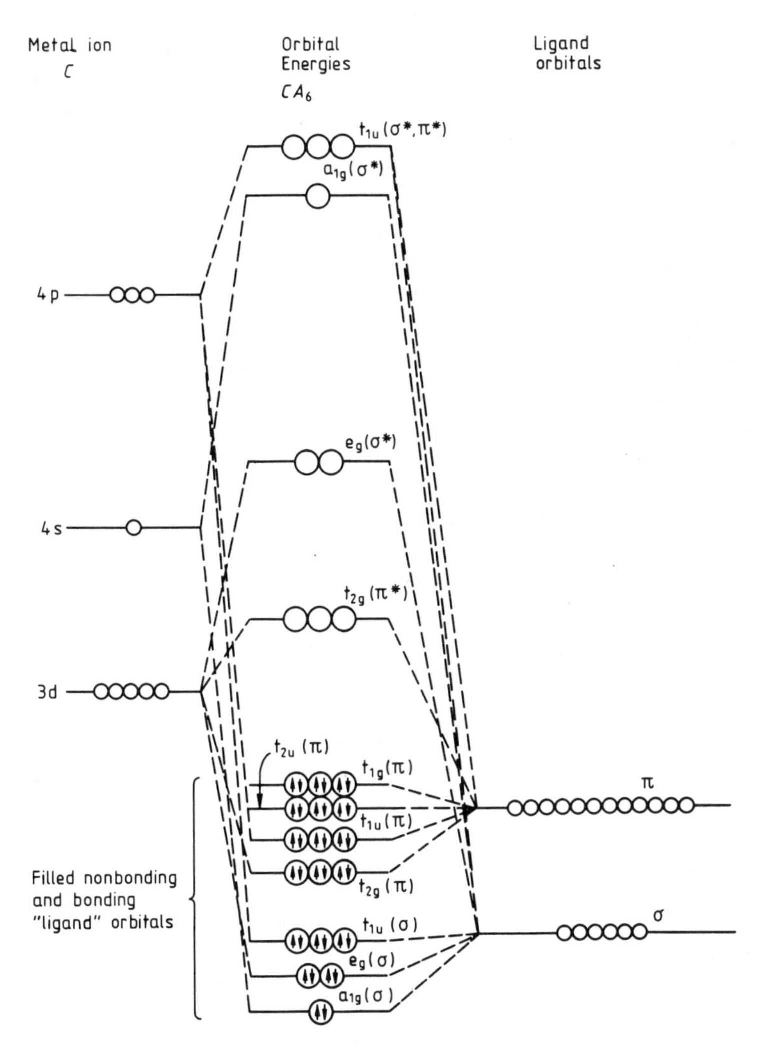

Figure 6.5:
Molecular orbital energy level diagram of a CA_6 octahedral complex with σ and π bonds. If the σ and π bonds of the anions are identified with those of halogen ions, then there are 12 σ-electrons and 24 π-electrons. In addition there are the bonding electrons of the central ion. All these electrons have to fill up the lowest group theoretical energy levels of the CA_6 complex. In the figure the levels of transition metal ions are given, but the electron number is left open; after di Bartolo (1968)

into theoretical details, Stoneham (1974) discussed the principles of calculation procedures and comparison with experiment.

We now cite some papers dealing with magnetic as well as with non-magnetic ions embedded in crystals and we specialize to the case where only two atoms or ions are in the crystal unit cell of the host lattice. Watson (1960) carried out H.F.-calculations for Mn^{2+} ions in cubic fields produced by sets of octahedrally placed point charges. The type and number of atomic energy levels in crystals for all point groups were given in a monograph by Prather (1961). Ballhausen (1962) applied the weak crystalline field model for calculations of a d^2-configuration of first row transition metal ions. Other d^n-configurations which are exhibited by the important transition group ions were treated for cubic symmetry by Liehr (1963) and other authors cited there. Flato (1965) extended the reduction of the rotation group for the cubic field case treated by Bethe to trigonal and tetragonal fields. Druzhinin, Cherepanov and Levin (1966) calculated the energy spectra of ions with $3d^n$ configurations in a cubic crystalline field of ionic crystals. Hagston (1967) discussed the spectra of first transition ion impurities in crystals. Becker, Meek and Dunn (1970) performed perturbation calculations for d^2- and d^8-configurations in trigonal fields. For the calculation of Pr^{3+}- and Tm^{3+}-ion levels in crystalline fields, Erdös and Kang (1972) took into account the core polarization (electronic shielding) which is induced by these fields in the inner electronic shells of these ions. A molecular orbital description of Cu^+ in NaCl were performed by Yip and Fowler (1972), Lowther (1972) examined anisotropic spinorbit coupling of rare earth ions in crystal fields. Bramanti, Mancini and Ranfagni (1971) developed a molecular orbital model for a Tl^+ ion in KCl. Hagston (1972) analyzed the fine structure of the ultraviolet absorption of heavy metal ions in alkali halides employing the strong crystal field formalism. Lowther (1972) performed a molecular orbital calculation for an Er^{3+} ion in copper doped zinc selenide, where Er^{3+} is surrounded by four Cu^+ ions at zinc sites. Relativistic effects in the ground state splitting parameters of Mn^{2+} ions in cubic crystalline fields were derived by Hagston and Lowther (1973). Schlesinger and Nara (1973) studied the Stark effect on the energy levels of f^9-electrons in cubic crystal fields. The d^2 and d^8 trigonal energy levels were investigated by Perumareddi (1973). Ermoshkin and Evarestov (1973) discussed the molecular orbital model of Tl^+ in KCl. Honma (1973) (1974) developed the theory of singlet-singlet and singlet-triplet transitions in the strong crystal field coupling scheme for impurity ions in cubic crystals in particular with application to metal ions in alkali halides. Lulek (1973) (1975) determined the energy level fine structure of impurity paramagnetic ions in the intermediate field model. Lowther and van Wyk (1974) analyzed the effect of the spin-orbit interaction on the zero-field splitting of Mn^{2+} in octahedral symmetry. Basu and Gosh (1973) (1975) investigated configuration interaction effects for Fe^{2+}, Cu^{2+}, Cr^{2+} ions in tetrahedral symmetry. Kikoin and Fleurov (1977) developped a theory for the calculation of transition metal ions in the crystal field of semiconductors which unites the resonance model of deep levels with conventional ligand field theory. Pedrini (1978) treated Cu^+ luminescence centers in LiCl crystals by a molecular orbital method.

b) Electron spin resonance

The discussion of the magnetic ions embedded in a host crystal gives a rough survey of the field splitting of the electronic levels of such ions which is caused by the crystal electric field exerted on the electrons of the impurity ion. By means of this field the degeneracy of the elctronic levels of the free ion is removed and the electrons are forced to occupy states which are basis states of representations of the corresponding

point group. The remaining degeneracy of these states can be removed by applying a static external magnetic field to the system. While the field splitting due to the electric crystal field can be observed optically by irradiation in the visible region, the field splitting due to the external magnetic field can be observed by irradiation in the radio frequency region. The optical excitation in the radio frequency region is followed by a number of radiative and non-radiative secondary transitions. As a result, the reaction kinetics of the system in this energy region is quite analogous to the reaction kinetics in the visible region which was treated in detail in the preceding chapters. A detailed exposition of this reaction kinetics in the radio wave region would require a special monograph. Since in principle the mathematical formalism of Chapter 3 and 4 can be applied equally well to such problems, we do not discuss the reaction kinetics of polar crystals in the radio wave region but refer to the literature. Hence we restrict ourselves to a discussion of the spectra in this region. The physical and theoretical interest in these spectra mainly arise from two reasons: first the structure of impurity centers can be investigated by ESR- (electronic spin resonance) and ENDOR- (electron-nuclear double spin resonance) microwave spectroscopy, secondly the relativistic energy corrections in this region are observable quantities.
For theoretical investigations the latter effects are basic as they depend on the appropriate choice of the Hamiltonians. These Hamiltonians with relativistic corrections were derived in Section 2.3, and we will first adapt them to the problems to be discussed here.
We assume from the beginning that all electromagnetic interactions between the electrons themselves and between electrons and atomic cores are formulated by potentials, which are given in the first relativistic approximation by (2.3.55). In this case the relativistic adiabatic electron Hamiltonian follows from (2.3.41) if $\mathscr{C}'$ is replaced by $\mathscr{C}''$ and the interaction between electrons and atomic cores is added. This gives the expression

$$\begin{aligned}\mathscr{H}(\mathbf{r},\mathbf{R}):=\sum_{i=1}^{k}\Big\{&\frac{1}{2m}\,(\mathbf{p}_i-\frac{e}{c}\,\mathbf{A}_i)^2-\frac{1}{8m^3c^2}\,\mathbf{p}_i^4+e\Phi_i-\mu_e\mathbf{s}_i\cdot\mathbf{B}_i\\&+\frac{\mu_e}{4mc}\,\mathbf{s}_i\cdot(\mathbf{p}_i\times\mathbf{E}_i)-\frac{\mu_e}{2mc}\,\mathbf{s}_i\cdot(\mathbf{E}_i\times\mathbf{p}_i)-\frac{i\mu_e}{4mc}\,\mathbf{p}_i\cdot\mathbf{E}_i\Big\}\\&+\frac{1}{2}\sum_{i,l=1}^{k}{}'\;e^2\mathscr{C}''(\mathbf{r}_i,\mathbf{r}_l)+\sum_{i,j=1}^{k,N}ee_j\mathscr{C}''(\mathbf{r}_i,\mathbf{R}_j)\end{aligned}\tag{6.4.16}$$

which is a spintensor. Its rank depends on the magnitude of the atomic core spin. We first assume a spin value of $s=1/2$ for the atomic cores. Then the rank of (6.4.16) is $2(k+N)$. Later on we will generalize to arbitrary atomic core spins.
Since all electromagnetic interactions between the particles are contained in the relativistic potentials, the fields occurring in (6.4.16) are purely external fields. Furthermore, the transversal parts of the external fields are considered to be perturbations of the system which lead to transitions. Hence they do not appear in (6.4.16) but are shifted into the interaction terms of irreversible statistics. For the remaining longitudinal external fields we assume that only a static homogeneous

magnetic field is present, i.e., $\mathbf{B} = \nabla \times \mathbf{A} \neq 0$ but $\Phi = 0 = \mathbf{E}$. For this case (6.4.16) reads

$$\mathcal{H}(\mathbf{r},\mathbf{R}) = \sum_{i=1}^{k} \left\{ \frac{1}{2m} \left(\mathbf{p}_i - \frac{e}{c}\mathbf{A}_i\right)^2 - \frac{1}{8m^3c^2}\mathbf{p}_i^4 - \mu_e \mathbf{s}_i \cdot \mathbf{B}_i \right\}$$
$$+ \frac{1}{2} \sum_{i,l=1}^{k}{}' e^2 \mathscr{C}''(\mathbf{r}_i,\mathbf{r}_l) + \sum_{i,j=1}^{k,N} ee_j \mathscr{C}''(\mathbf{r}_i,\mathbf{R}_j). \tag{6.4.17}$$

This expression can further be simplified by appropriate approximations. We first consider the electrons. The term $\sum_i (e^2/c^2)\mathbf{A}_i^2$ gives a diamagnetic contribution to the energy. It can be shown, cf. for instance Abragam and Bleaney (1970), that it is much smaller than the paramagnetic contributions. Thus it is neglected. Due to the smallness of the average electron momentum we further neglect all terms of order $(\mathbf{p}/c)^2$. Concerning the atomic cores, all terms containing a core momentum are non-adiabatic terms by definition and must be included in the irreversible interactions. If these approximations are applied to (6.4.17) and (2.3.55) we finally obtain

$$\mathcal{H}(\mathbf{r},\mathbf{R}) \approx \sum_{i=1}^{k} \left[\frac{1}{2m}\mathbf{p}_i^2 - \frac{e}{2mc}(\mathbf{p}_i \cdot \mathbf{A}_i + \mathbf{A}_i \cdot \mathbf{p}_i) - \mu_e \mathbf{s}_i \cdot \mathbf{B}_i \right]$$
$$+ \frac{e^2}{2} \sum_{i,l=1}^{k}{}' \mathscr{C}''(\mathbf{r}_i,\mathbf{r}_l) + \sum_{i,j=1}^{k,N} \Big[ee_j C(\mathbf{r}_i,\mathbf{R}_j)$$
$$+ i\mu_e\mu_j C(\mathbf{r}_i,\mathbf{R}_j)^3 \{[(\mathbf{r}_i - \mathbf{R}_j)\times \nabla_i]\cdot \mathbf{s}_i + 2[(\mathbf{r}_i - \mathbf{R}_j)\times \nabla_i]\cdot \mathbf{t}_j\}$$
$$+ \mu_e\mu_j C(\mathbf{r}_i,\mathbf{R}_j)^5 [(\mathbf{r}_i - \mathbf{R}_j)^2 \mathbf{s}_i \cdot \mathbf{t}_j - 3(\mathbf{r}_i - \mathbf{R}_j)\cdot \mathbf{s}_i (\mathbf{r}_i - \mathbf{R}_j)\cdot \mathbf{t}_j]$$
$$- 4\pi\mu_e\mu_j (1 + \tfrac{2}{3}\mathbf{s}_i \cdot \mathbf{t}_j)\delta(\mathbf{r}_i - \mathbf{R}_j) \Big]. \tag{6.4.18}$$

The next to the last term in (6.4.18) causes a constant energy shift within a group of splitted levels. As optical transitions are observed only between the levels of such groups, those energy contributions drop out and are thus neglected from the beginning. Further simplification can be achieved with the definitions

$$\mathbf{E}(\mathbf{r}) := -\sum_{j=1}^{N} e_j C(\mathbf{r},\mathbf{R}_j)^3 (\mathbf{r} - \mathbf{R}_j) \tag{6.4.19}$$

and

$$\mathbf{l}_i^j := -(\mathbf{r}_i - \mathbf{R}_j) \times i\nabla_i \tag{6.4.20}$$

where $\mathbf{l}_i^j$ is the angular momentum of the electron i with respect to the position $\mathbf{R}_j$ of the atomic core j. Then the electron spin-orbit coupling takes the form

$$\sum_{i,j=1}^{k,N} i\mu_e\mu_j C(\mathbf{r}_i,\mathbf{R}_j)^3 [(\mathbf{r}_i - \mathbf{R}_j)\times \nabla_i]\cdot \mathbf{s}_i = \sum_{i=1}^{k} \frac{i\mu_e\mu_j}{e_j} [\mathbf{E}_i \times \nabla_i]\cdot \mathbf{s}_i \tag{6.4.21}$$

and for the atomic core spin-electron angular momentum coupling we obtain

$$\sum_{ij=1}^{k,N} i\mu_e\mu_j C(\mathbf{r}_i,\mathbf{R}_j)^3 [2(\mathbf{r}_i - \mathbf{R}_j)\times \nabla_i]\cdot \mathbf{t}_j = -\sum_{ij=1}^{k,N} \mu_e\mu_j C(\mathbf{r}_i,\mathbf{R}_j)^3\, 2\mathbf{l}_i^j \cdot \mathbf{t}_j. \tag{6.4.22}$$

If it is assumed that $\mathbf{E}_i$ exhibits approximate radial symmetry, then (6.4.2) can be applied, and with $\mathbf{A}=\mathbf{B}\times\mathbf{r}$ it is easy to calculate the final form of (6.4.18) which is given by

$$\mathscr{H}(\mathbf{r},\mathbf{R})=\sum_{i=1}^{k}\left[\frac{1}{2m}\mathbf{p}_i^2-\mu_e(2\mathbf{l}_i+\mathbf{s}_i)\cdot\mathbf{B}_i+\lambda\mathbf{l}_i\cdot\mathbf{s}_i\right] \tag{6.4.23}$$
$$+\frac{e^2}{2}\sum_{i,l=1}^{k}{}' \mathscr{C}''(\mathbf{r}_i,\mathbf{r}_l)+\sum_{ij=1}^{k,N}\{ee_jC(\mathbf{r}_i,\mathbf{R}_j)-4\pi\mu_e\mu_j\tfrac{2}{3}\mathbf{s}_i\cdot\mathbf{t}_j\delta(\mathbf{r}_i-\mathbf{R}_j)$$
$$+\mu_e\mu_jC(\mathbf{r}_i,\mathbf{R}_j)^5[(\mathbf{r}_i-\mathbf{R}_j)^2(-2\mathbf{l}_i^j+\mathbf{s}_i)\cdot\mathbf{t}_j-3(\mathbf{r}_i-\mathbf{R}_j)\cdot\mathbf{s}_i(\mathbf{r}_i-\mathbf{R}_j)\cdot\mathbf{t}_j]\}.$$

The fine and hyperfine structure of the electronic levels have now to be derived by solving the secular equation corresponding to (6.4.23). Since these structures result from magnetic interactions, it is convenient to assume that the levels and states of the system with electric interactions are known. Then a solution procedure has to be developed based on this set of states. Thus we make the following partition of the Hamiltonian (6.4.23)

$$\mathscr{H}(\mathbf{r},\mathbf{R})=h(\mathbf{r},\mathbf{R})+h_1(\mathbf{r},\mathbf{R})+h_2(\mathbf{r},\mathbf{R}) \tag{6.4.24}$$

with the electronic part

$$h(\mathbf{r},\mathbf{R}):=\sum_{i=1}^{k}\frac{1}{2m}\mathbf{p}_i^2+\frac{e^2}{2}\sum_{i,l=1}^{k}{}' C(\mathbf{r}_i,\mathbf{r}_l)+\sum_{ij=1}^{k,N}ee_jC(\mathbf{r}_i,\mathbf{R}_j) \tag{6.4.25}$$

which is the ordinary nonrelativistic adiabatic Hamiltonian (2.2.5) and the magnetic part

$$h_1(\mathbf{r},\mathbf{R}):=\sum_{i=1}^{k}[-\mu_e(2\mathbf{l}_i+\mathbf{s}_i)\cdot\mathbf{B}_i+\lambda_i\mathbf{l}_i\cdot\mathbf{s}_i]$$
$$+\frac{e^2}{2}\sum_{il=1}^{N}{}' [\mathscr{C}(\mathbf{r}_i,\mathbf{r}_l)-C(\mathbf{r}_i,\mathbf{r}_l)] \tag{6.4.26}$$

which is responsible for the fine structure, while

$$h_2(\mathbf{r},\mathbf{R}):=\sum_{ij=1}^{k,N}\{-4\pi\mu_e\mu_j\tfrac{2}{3}\mathbf{s}_i\cdot\mathbf{t}_j\delta(\mathbf{r}_i-\mathbf{R}_j) \tag{6.4.27}$$
$$+\mu_e\mu_jC(\mathbf{r}_i,\mathbf{R}_j)^5[(\mathbf{r}_i-\mathbf{R}_j)^2(\mathbf{s}_i-2\mathbf{l}_i^j)\cdot\mathbf{t}_j-3(\mathbf{r}_i-\mathbf{R}_j)\cdot\mathbf{s}_i(\mathbf{r}_i-\mathbf{R}_j)\cdot\mathbf{t}_j]\}$$

leads to the hyperfine structure. According to Theorem 6.37, we have to expect that the calculation procedure will strongly depend on the number of electrons outside the closed shells of the impurity ion. Thus the simplest example will be given by one electron outside the closed shells which is realized, for instance, by H, Li, Na, K, Rb, Cs atoms, Be^+, Mg^+, Ca^+, Sr^+, Ba^+ ions with s-states, by B, Al, Ga, In, Tl atoms, and Ge^+, Sn^+, Pb^+ ions with p-states, etc., or by F-center electrons. Obviously, the calculation effort will considerably increase with the number of electrons treated explicitly. In order to show the essentials of the method we will restrict ourselves to the treatment of one electron outside the closed shells which we will apply in particular to the F-center. As the groundstate of the complex-ion is the only stable state, optical measurements of the fine and hyperfine structure are done mainly with this state. Hence we concentrate on the treatment of this state. The method which we apply is due to Pryce (1950) and leads to the derivation of a so-called spin-Hamiltonian, i.e., a Hamiltonian which describes only the magnetic interactions with respect to an initially spin degenerate level.

Theorem 6.39 : Suppose that the unperturbed Hamiltonian $\mathscr{H}_0$ has a complete set of eigenstates $\{|ns\rangle\}$ with eigenvalues $\{E_n\}$ which are degenerate with the order of degeneracy equal to the spin multiplicity. Then from the total Hamiltonian $\mathscr{H} := \mathscr{H}_0 + \mathscr{H}_1$ a spin Hamiltonian

$$\tilde{\mathscr{H}} := E_0 + P_0 \mathscr{H}_1 P_0 - \sum_{n \neq 0} \frac{P_0 \mathscr{H}_1 P_n \mathscr{H}_1 P_0}{(E_n - E_0)} \tag{6.4.28}$$

with

$$P_n := \sum_s |ns\rangle\langle ns| \tag{6.4.29}$$

can be derived which describes the splitting of the unperturbed groundstates $\{|0s\rangle\}$ of $\mathscr{H}_0$ due to the perturbation $\mathscr{H}_1$ in the second order perturbation calculation.

Proof: The eigenvalue equation of the exact eigenstates $\{|\psi\rangle\}$ of $\mathscr{H}$ reads

$$\mathscr{H}|\psi\rangle = E|\psi\rangle. \tag{6.4.30}$$

As the set of eigenstates $\{|ns\rangle\}$ is assumed to be complete, we have $\sum_n P_n = \mathbb{1}$ in the state space of $\mathscr{H}$ or $\mathscr{H}_0$, resp. Hence

$$|\psi\rangle = \sum_n P_n |\psi\rangle =: \sum_n |\psi_n\rangle \tag{6.4.31}$$

holds and the corresponding equation (6.4.30) takes the form

$$(E_0 - E + \mathscr{H}_1)|\psi_0\rangle + \sum_{\substack{n \\ \neq 0}} (E_n - E + \mathscr{H}_1)|\psi_n\rangle = 0. \tag{6.4.32}$$

Operating on (6.4.32) with P_i, $i = 0, 1, \ldots$ we obtain

$$(E_0 - E + P_0 \mathscr{H}_1)|\psi_0\rangle + \sum_{\substack{n \\ \neq 0}} P_0 \mathscr{H}_1 |\psi_n\rangle = 0 \tag{6.4.33}$$

$$(E_n - E)|\psi_n\rangle + P_n \mathscr{H}_1 |\psi_0\rangle + \sum_{\substack{m \\ \neq 0}} P_n \mathscr{H}_1 |\psi_m\rangle = 0 \tag{6.4.34}$$

if the orthonormality of the set $\{|ns\rangle\}$ is observed. Equation (6.4.34) can be solved approximately by

$$|\psi_n\rangle = -\frac{P_n \mathscr{H}_1}{(E_n - E)} |\psi_0\rangle. \tag{6.4.35}$$

Substituting (6.4.35) into (6.4.33) one has, up to second order,

$$\left(E_0+P_0\mathscr{H}_1-\sum_{\substack{n\\ \neq 0}}\frac{P_0\mathscr{H}_1P_n\mathscr{H}_1}{(E_n-E)}-E\right)|\psi_0\rangle=0. \tag{6.4.36}$$

If in the denominator of (6.4.36) E is replaced by E_0, and $P_0^2=P_0$ is observed, then it follows from (6.4.36) that $\tilde{\mathscr{H}}$ as given by (6.4.28) operates entirely within the manifold $\{|0,s\rangle\}$. Hence $\tilde{\mathscr{H}}$ describes the corrections due to $\mathscr{H}_1$ within this manifold. As this manifold is spanned by spin states and as $\mathscr{H}$ is Hermitian, we see that $\tilde{\mathscr{H}}$ is a spin Hamiltonian, Q.E.D.

The formalism can be extended so as to also include higher order corrections, see Pryce (1950). A similar formalism was applied by Löwdin (1951). An extension which allows the inclusion of orbital degeneracy was made by Abragam and Pryce (1951).
We now turn to our special problem. We have to put $\mathscr{H}_0:=h(\mathbf{r},\mathbf{R})$ and $\mathscr{H}_1:=h_1(\mathbf{r},\mathbf{R})+h_2(\mathbf{r},\mathbf{R})$. Then the spin Hamiltonian takes the form

$$\tilde{\mathscr{H}}:=E_0+P_0(h_1+h_2)P_0-\sum_{\substack{n\\ \neq 0}}\frac{P_0(h_1+h_2)P_n(h_1+h_2)P_0}{(E_n-E_0)}. \tag{6.4.37}$$

As h_2 produces hyperfine splitting, we neglect its second order contribution and write approximately

$$\tilde{\mathscr{H}}\approx E_0+P_0h_1P_0-\sum_{\substack{n\\ \neq 0}}\frac{P_0h_1P_nh_1P_0}{(E_n-E_0)}+P_0h_2P_0. \tag{6.4.38}$$

A further simplification can be achieved by applying Theorem 6.37 to a configuration of one electron outside the closed shells. In this case the last term on the right-hand side of (6.4.26), which gives the relativistic corrections to the electron-electron interaction, does not influence the active electron as by Theorem 6.37 only the interactions outside the closed shells have to be taken into account and for one active electron no electron-electron interaction appears. Of course, this Theorem is also only approximately valid and in more sophisticated calculations the core polarization caused by the active electrons is considered, cf. Abragam and Bleaney (1970). Without this core polarization, (6.4.26) is then reduced to

$$\begin{aligned} h_1(\mathbf{r},\mathbf{R}) &= \sum_{i=1}^{k}\left[-\mu_e(2\mathbf{l}_i+\mathbf{s}_i)\cdot\mathbf{B}_i+\lambda\mathbf{l}_i\cdot\mathbf{s}_i\right] \\ &= \mu_e(2\mathbf{L}+\mathbf{S})\cdot\mathbf{B}+\lambda\mathbf{L}\cdot\mathbf{S}. \end{aligned} \tag{6.4.39}$$

For the practical evaluation of the spin Hamiltonian, we assume that all states $|ns\rangle$ can be separated into an orbital part $|n\rangle$ and a spin part $u_n(s)$. Then we have

$$|ns\rangle=|n\rangle u_n(s) \tag{6.4.40}$$

with the completeness relation

$$\sum_s u_n(s)u_n(s)^\times=\mathbb{1}_n \tag{6.4.41}$$

in the spin space of the level n. Without loss of generality we may assume that the orbital parts of the electronic state functions are real. Then the diagonal elements of $\mathbf{L}$ vanish and for the representation of (6.4.38) the following theorem can be derived.

Theorem 6.40 : Suppose that the states $|ns\rangle$ have the form (6.4.40) with (6.4.41) and that the orbital parts $|n\rangle$ are real, then the expectation value $\langle 0|\mathscr{H}|0\rangle$ of (6.4.38) with respect to the orbital part $|0\rangle$ reads

$$\begin{aligned}\langle 0|\mathscr{H}|0\rangle &= E_0+\mu_e\mathbf{S}\cdot\mathbf{B}\\ &-\sum_{\substack{n\\ \neq 0}}\sum_{ij}\frac{\langle 0|L_i|n\rangle\langle n|L_j|0\rangle}{(E_n-E_0)}(\lambda S_i+\mu_e B_i)(\lambda S_j+\mu_e B_j)\\ &+\langle 0|h_2(\mathbf{r},\mathbf{R},\mathbf{s}_i,\mathbf{t}_j)|0\rangle\end{aligned} \tag{6.4.42}$$

where only states $|ns\rangle$ occur with $u_n(s)=u_0(s)$ and where S_i are the Cartesian components of $\mathbf{S}$. Moreover, the operator defined by (6.4.42) acts in the spin space of $\{u_0(s)\}$.

Proof: We first consider the evaluation of the spin parts. The most complicated expression of (6.4.38) with respect to this spin part is

$$\begin{aligned}P_0\mathbf{L}\cdot\mathbf{S}P_n\mathbf{L}\cdot\mathbf{S}P_0 &= \sum_{ij}|0\rangle\langle 0|L_i|n\rangle\langle n|L_j|0\rangle\langle 0|\\ &\times\sum_{s,s',s''}u_0(s)u_0(s)^{\times}S_iu_n(s')u_n(s')^{\times}S_ju_0(s'')u_0(s'')^{\times}.\end{aligned} \tag{6.4.43}$$

Since the spin vector $\mathbf{S}$ is an angular momentum operator, its matrix elements between different angular momentum representations vanish and only matrix elements unequal to zero occur between states belonging to the same j. Hence the transition elements $u_0(s)^{\times}\mathbf{S}u_n(s')$ are unequal to zero only for $u_n(s')\equiv u_0(s')$, i.e., if the spin part of $|ns\rangle$ is equal to the spin part of $|0s\rangle$. Then (6.4.43) takes the form

$$\begin{aligned}P_0\mathbf{L}\cdot\mathbf{S}P_n\mathbf{L}\cdot\mathbf{S}P_0 &= \sum_{ij}|0\rangle\langle 0|L_i|n\rangle\langle n|L_j|0\rangle\langle 0|\\ &\times\sum_{ss's''}u_0(s)u_0(s)^{\times}S_iu_0(s')u_0(s')^{\times}S_ju_0(s'')u_0(s'')^{\times}.\end{aligned} \tag{6.4.44}$$

Using the completeness relation (6.4.41) to write the second factor on the right-hand side of (6.4.44) as $\mathbb{1}_0S_i\mathbb{1}_0S_j\mathbb{1}_0=S_iS_j$ we may rewrite (6.4.43) as

$$P_0\mathbf{L}\cdot\mathbf{S}P_n\mathbf{L}\cdot\mathbf{S}P_0=\sum_{ij}|0\rangle\langle 0|L_i|n\rangle\langle n|L_j|0\rangle\langle 0|S_iS_j. \tag{6.4.45}$$

If the other terms are treated in a similar way and the expectation value of $\mathscr{H}$ is taken, then the result follows, Q.E.D.

To interprete the various terms we define the tensor

$$\Lambda_{ij} := \sum_{\substack{n \\ \neq 0}} \frac{\langle 0|L_i|n\rangle \langle n|L_j|0\rangle}{(E_n - E_0)}. \tag{6.4.46}$$

Then, using the summation convention, we may write for (6.4.42)

$$\langle 0|\tilde{\mathscr{H}}|0\rangle = E_0 + \mu_e(\delta_{ij} - \lambda\Lambda_{ij}) S_i B_j - \lambda^2 \Lambda_{ij} S_i S_j - \frac{\mu_e^2}{4} \Lambda_{ij} B_i B_j + \langle 0|h_2|0\rangle. \tag{6.4.47}$$

The second term in (6.4.47) is the magnetic energy of a spin system in an external field with a g-factor represented by the tensor

$$g_{ij} := (\delta_{ij} - \lambda\Lambda_{ij}). \tag{6.4.48}$$

The third term in (6.4.47) is a second order contribution to the spin-orbit coupling and the fourth term corresponds to a temperature independent paramagnetic susceptibility, Pryce (1950). Usually the third and the fourth term are neglected so that (6.4.47) approximately reads

$$\langle 0|\tilde{\mathscr{H}}|0\rangle \approx E_0 + \mu_e g_{ij} S_i B_j + \langle 0|h_2|0\rangle. \tag{6.4.49}$$

We evaluate the spectrum of (6.4.49) for an F-center electron. In this case no ion core is present and in the molecular model i) only the electron located at the anion vacancy has to be described explicitly, i.e., we have to take $k=1$ in (6.4.39) and (6.4.27). For simplicity we take into account only the six nearest neighbour cations in an octahedral configuration. Furthermore, in a molecular orbital model the groundstate of the F-electron is sufficiently well approximated by s-states centered at the neighbouring cations. Hence in this state the contribution of the orbital angular momentum $\mathbf{l}^j$ of the F-electron with respect to the origin of the cations vanishes and we assume that this holds even if we choose another form of the groundstate function. If we denote this function by $\psi(\mathbf{r})$, then the expectation value of h_2 reads

$$\langle 0|h_2|0\rangle = \sum_{j=1}^{6} \mathbf{s} \cdot \mathscr{A}(j) \cdot \mathbf{t}_j \tag{6.4.50}$$

with

$$\mathscr{A}(j) := a(j)\delta_{il} + b_{il}(j) \tag{6.4.51}$$

and

$$a(j) := \frac{2}{3} g_j \mu_j g_e \mu_e \varrho(\mathbf{R}_j) \tag{6.4.52}$$

$$b_{il}(j) := \frac{1}{4\pi} g_j \mu_j g_e \mu_e \int C(\mathbf{r}, \mathbf{R}_j)^5 \, [(\mathbf{r} - \mathbf{R}_j)^2 \delta_{ij} - 3(\mathbf{r} - \mathbf{R}_j) \otimes (\mathbf{r} - \mathbf{R}_j)] \varrho(\mathbf{r}) d^3 r$$

where $\mu_e := e\hbar/mc$ and $\mu_j := e_j\hbar/M_j c$ are the spin magnetic moments and g_e and g_j are the g-factors of the free electron and ions. One must be careful with the interpretation of the terms $a(j)$ and $b(j)$. In the literature it is frequently maintained that the terms

$a(j)$ correspond to s-states, while $b(j)$ corresponds to non-s-states. This is only true if the atomic core is centered at the origin. In all other configurations s-states as well as p-states contribute to both terms a and b!
With (6.4.50), the spin Hamiltonian (6.4.49) for the F-electron takes the form

$$\langle 0|\mathscr{H}|0\rangle = E_0 + \mu_e \mathbf{S}\cdot\mathscr{G}\cdot\mathbf{B} + \sum_j \mathbf{S}\cdot\mathscr{A}(j)\cdot\mathbf{t}_j. \tag{6.4.53}$$

The spectrum of this operator can be observed by optical transitions. In the general formalism of Chapter 2 optical transitions take place between crystal levels of electrons and atomic cores, i.e., between states $\chi_n(\mathbf{r},\mathbf{R})\varphi_m^n(\mathbf{R})$ with their corresponding energy levels E_m^n. In the simplified model of a spin Hamiltonian the corresponding states are given by the simultaneous spin states of electrons and atomic cores. Hence we have to diagonalize $\langle 0|\mathscr{H}|0\rangle$ in the spin space of both kinds of particles. This diagonalization can be performed by standard algebraic methods which will not be discussed here. To point out the essentials we give only a simple approximation of the eigenvalues of (6.4.53), which, however, reveals the special structure of the spectrum to be expected. Following Slichter (1978) we assume the $\mathscr{G}$-tensor to be already diagonalized, $g_{ik} = g_i\delta_{ik}$ and $\mathbf{B}$ to be a constant magnetic field in $\mathbf{e}_3$-direction. We consider that part of (6.4.53) which is already diagonalized by the spin functions of the unperturbed system and assume that the remaining parts are only a small perturbation. The diagonal part is given by

$$\langle 0|\mathscr{H}|0\rangle_d := E_0 + \mu_e S_3 g_3 B_3 + \sum_{j=1}^{6} S_3 A_{33}(j) t_{3j}. \tag{6.4.54}$$

It commutes with the spin of the electron and its $\mathbf{e}_3$-component as well as with the spin and the $\mathbf{e}_3$ components of the atomic cores or cations, resp. The eigenfunctions of (6.4.54) are

$$\chi\varphi \equiv u_e(s,m) \prod_{j=1}^{6} u_{\mathscr{N}}(s',m_j) = : u(s,m,s',m_1,\ldots,m_6) \tag{6.4.55}$$

where s is the total spin of the electron, m its value in the $\mathbf{e}_3$-direction, s' is the total spin of each cation and m_j the value of the spin projection in the $\mathbf{e}_3$-direction of the cation j. All these numbers are good quantum numbers. For the groundstate of the F-electron it follows that $A_{33}(j) \equiv A_{33}$, $1 \leqslant j \leqslant 6$ and the eigenvalue equation reads

$$[\langle 0|\mathscr{H}|0\rangle_d + E(s,m,s',m_1\ldots m_6)]u(s,m,s',m_1\ldots m_6) = 0 \tag{6.4.56}$$

with

$$E(s,m,s',m_1\ldots m_6) := E_0 + \mu_e g_3 m B_3 + m A_{33}\left(\sum_{j=1}^{6} m_j\right). \tag{6.4.57}$$

From this it follows that the electron can be considered to be coupled to an effective magnetic field

$$B_{\text{eff}} := \left(g_3 B_3 + A_{33}\sum_{j=1}^{6} m_j\right). \tag{6.4.58}$$

If only electronic transitions occur, these transitions can take place from $m=1/2$ to $m=-1/2$ and vice versa. Hence the energy differences observed are

$$\Delta E=E(s,1/2,s',m_1\ldots m_6)-E(s,-1/2,s',m_1\ldots m_6)=\mu_e B_{\text{eff}}. \tag{6.4.59}$$

Obviously, these energy differences depend on the values of the nuclear spin projections $m_1\ldots m_6$. For a single nuclear spin s' the projection can take the values $m=-s', -s'+1,\ldots,s'$. Hence the value of $M=\sum m_j$ ranges from $M=-6s', -6s'+1, \ldots, 6s$. As the nuclear spins are arbitrary, each of these values can be taken by them. Hence, in a statistical ensemble such as F-centers distributed in a crystal, we have to expect that all spin configurations are present. Hence all these lines will occur in the absorption or emission spectrum, giving an equidistant set of lines centered about the mean value $\mu_e g_3 B_3$. The individual values of M have, however, in the statistical ensemble different statistical weights corresponding to the number of possible realizations of a certain value of M by combinations of the individual m_j. These statistical weights are presented in Table 6.5

Table 6.5:
Statistical weights of the states with total nuclear spin M of N equivalent nuclei with spin I; after Seidel and Wolf (1968).

$M=$	0	$\frac{1}{2}$	1	$\frac{3}{2}$	2	$\frac{5}{2}$	3	
$N=1$		1						
2	2		1					$I=\frac{1}{2}$
3		3		1				
4	6		4		1			
5		10		5		1		
6	20		15		6		1	

$M=$	0	$\frac{1}{2}$	1	$\frac{3}{2}$	2	$\frac{5}{2}$	3	$\frac{7}{2}$	4	$\frac{9}{2}$	5	$\frac{11}{2}$	6	$\frac{13}{2}$	7	$\frac{15}{2}$	8	$\frac{17}{2}$	9	
N $=1$		1		1																
2	4		3		2		1													$I=\frac{3}{2}$
3		12		10		6		3		1										
4	44		40		31		20		10		4		1							
5		155		135		101		65		35		15		5		1				
6	580		546		456		336		216		120		56		21		6		1	

and in consequence these lines appear with an approximate Gaussian intensity distribution around the mean value. It is obvious that the inclusion of further nuclei, or ions, resp., leads to an additional splitting of the spectrum which allows one to gain information about the electron density distribution by using the calculation of the $\mathscr{A}(j)$. For details we refer to the literature.

So far we considered the simplest example of the model i). Going over to model ii) we are forced to treat many-electron systems, where the number of electrons depends on the number of nearest neighbour ions of the impurity. In general, such configurations are too complicated to permit explicit numerical calculations. Simpler examples for the

treatment with model ii) are provided by various impurity molecules occurring in ionic crystals. The most frequently investigated center in this respect is the so-called V_k-center. This center is assumed to occur in alkali halides, where one electron is removed from a cation X^-, i.e. $X^- \Rightarrow X + e^-$ and where the neutral X atom and a neighbouring X^- cation form a molecule X_2^-. If, for instance, Cl^- is considered with the electronic configuration $Cl^- = (\text{Neon core})^{10} 3s^2p^6$, in a treatment of Cl_2^- by molecular orbitals 11 electrons in the outer p-shells of both atoms have to be taken into account explicitly. In this case the molecular orbitals have to be formed with respect to two centers and the spin Hamiltonian must be evaluated for the basis of a many-electron system. The evaluation of two center integrals in an external magnetic field which has to be performed in this procedure is facilitated by the gauge invariance of the interactions. Since a detailed review of these calculations was given recently by Slichter (1978) and since these calculations are in principle only a combination of molecular orbital theory with the spin Hamiltonian formalism we do not reproduce them here. We only mention that the most prominent seven lines of the hyperfine structure of this center can be explained by the different spin adjustments of the two atomic cores, which was the clue to the identification of this center.
Electron paramagnetic resonance was discovered by Zavoisky (1946). Since the publication of this paper numerous papers have appeared on this topic. We refer to books and review articles which summarize the older results and supplement them by some references to papers which appeared later. Low (1960) gave a mainly theoretical presentation of this topic which is based on the model i) and contains technical details of the calculation procedures for paramagnetic ions. Slichter (1963) (1978) discussed the principles of magnetic resonance mainly with respect to nuclear magnetic resonance, but the book also contains a chapter about electron paramagnetic resonance. Emphasis is laid on the dynamical processes which are connected with the resonance phenomena, i.e., the theory of rate equations is applied to the radio frequency region. Altschuler and Kosyrev (1964) published a survey of paramagnetic electron resonance. This book contains an extensive list of paramagnetic ions, their crystal field Hamiltonians and properties as well as a description of paramagnetic effects for other impurities, etc. Markham (1966) discussed electron paramagnetic resonance in two chapters of his book about F-centers. Special attention is paid by him to the problem of gauge invariance of the calculation results if the original gauge invariant interaction terms are modified by approximations. Henry and Slichter (1968) described the application of the method of moments to the calculation of absorption and emission intensities with respect to the level fine structure in a review article. Abragam and Bleaney (1970) gave a comprehensive theoretical and phenomenological presentation of electron paramagnetic resonance based on the models i) and ii) and included also spin-phonon interactions. Stoneham (1975) gave a short theoretical survey. Reviews of experiments and phenomenological theory were given by Heuer (1964) for paramagnetic ions, by Seidel and Wolf (1968) for color centers and by Henderson and Garrison (1973) for defects in insulators.
Concerning the papers on this topic, Kahn and Kittel (1953) and Kip, Kittel, Levy and Portis (1953) were the first who theoretically analyzed the hyperfine structure of the F-center by using a molecular orbital wave function of the F-center proposed by Inui and

Uemura (1950). Castner and Känzig (1957) defined the model of a V_k center and derived by means of model ii) a spin Hamiltonian for it. Deigen (1957) used wave functions of the Pekar type to calculate the hyperfine splitting of the F-levels due to interaction with molecular spins. Similar calculations with wave functions orthogonalized to the orbitals of the neighbouring ions were performed by Blumberg and Das (1958). With a molecular orbital model, Shulman (1959) calculated the absorption band of the paramagnetic resonance states of the F-center for KCl. Deigen and Roitsin (1959) investigated the diagonalization of the spin Hamiltonian for an arbitrary external magnetic field. The same problem was treated in the papers by Lord (1957), Wolga and Strandberg (1959), Lewis and Pretzel (1961) and Hughes and Allard (1962). Hughes (1964) gave a critical discussion of these calculations. Van Wieringen (1961) reviewed magnetic resonance in semiconductors. Feuchtwang (1962) calculated an effective spin Hamiltonian for the F-center and took into account the displacements of neighbouring ions which led to a new classification of equivalent ions appearing in the interaction term and therefore a modified spectrum. Suffczynski (1963) calculated in the effective mass approximation the spin-orbit splitting of the first excited state of the F-center in Cesium halides. Vinetskii (1964) published a note concerning the role of singlet and triplet states of F_2 color centers for paramagnetic resonance. Ray (1964) described a method to derive spin Hamiltonians for paramagnetic ions in crystals. Grant and Strandberg (1964) formulated a theory of spin-spin interaction in solids and calculated the magnetic resonance line shape for two interacting spin particles. Rashba and Sheka (1964) investigated the behaviour of magnetic resonances at local centers for closely spaced energy levels based on a one-electron model. Smith (1965) investigated spin-orbit effects in the optical absorption of F-centers based on model i). Kübler and Friauf (1965) calculated with the pseudopotential method for a rigid lattice the isotropic hyperfine interaction constants $a(j)$ defined by (6.4.52). Casselmann and Markham (1965) derived a spin Hamiltonian for a many-center imperfection and applied it to the V_k-center for which they used a one-electron model. They also discussed gauge invariance in connection with a paper of Stone (1963) on this topic. Henry, Schnatterly and Slichter (1965) developed a method for the calculation of the moments of the optical line shape and applied it to the investigation of various effects connected with the level fine structure of colour centers. Smith (1965) discussed the spin-orbit splitting of the first excited state of the F-center in the molecular model i) taking into account the exclusion principle by appropriate orthogonalization of the F-center wave function with respect to the core states. Krupka and Silsbee (1966) investigated the electron-spin resonance of the groundstate of the R-center (three adjacent F-centers in an equilateral triangular arrangement) by using a molecular model of type i) with orthogonalization of the wave functions with respect to external core states. Schumacher and Hollingsworth (1966) discussed energy splitting of rare earth ions in cubic crystalline fields and external magnetic fields. Mattis and Lieb (1966) developed a theory for two-electron paramagnetic impurities in semiconductors based on a two-particle Hamiltonian in the rigid lattice. Sharma, Das and Orbach (1966) studied the zero field splitting of s-state ions by means of a spin Hamiltonian in a rigid point multipole model. Sharma, Das and Orbach (1967) extended their previous calculation to include overlap and covalency with the neighbouring ions. Shuey and Zeller (1967) discussed the fine structure levels of the O_2^--center in alkali halogenides by using a molecular model. Bartram, Swenberg and La (1967) reinvestigated the derivation of F-center g-values by allowing an admixture of new kinds of excited states (charge transfer configurations) in the spin Hamiltonian, which necessitates the employment of many-electron wave functions in a complex ion model of the center. Watanabe (1967) determined the g shift of Cr^+ in NaCl crystals by means of model i). Den Hartog and Arends (1967) calculated the isotropic hyperfine structure constants of F-centers in alkaline earth fluorides by using various wave functions of other authors. With the model i) the linear electric field effect on the electron paramagnetic resonance spectrum of Cu^{2+} ions in tetrahedral crystal fields was investigated by Bates (1968). Similar calculations were performed by Dreybrodt and Silber (1969) for the electric field effect on the groundstate splitting of Mn^{2+} ions in NaCl. Stoneham, Hayes, Smith and Stott (1968) analyzed the hyperfine interactions of F-centers in alkaline earth fluorides. Fong (1968) treated paraelectric resonance transitions of OH^- ions in KCl being the electrical analogon to the paramagnetic transitions. Glinchuk and Deigen (1968) calculated the EPR line shape of impurity centers in non-metallic crystals by means of correlation functions. With a similar method, Smith, Dravnieks and Wertz (1969) treated the EPR line shape of Ni^{2+} in MgO. Chiarotti (1969) gave a survey of the optical properties of color centers in alkali halides. Jette (1969) analyzed hyperfine interactions of V_k-centers in a molecular orbital model. Scherz (1969) derived effective Hamiltonians for transition metal ions in cubic fields with tetragonal and trigonal distortions. In a subsequent paper Scherz (1970) applied this technique to the calculation of g-factors of Cu^{2+}-centers in II–VI-compounds. Hagston (1970) gave critical comments on pseudopotential calculations for F-centers by analyzing the fine and hyperfine interactions resulting from these calculations. Den Hartog (1970) performed a comparison of the results of hyperfine structure calculations with respect to various wave functions used in the literature. Krivoglas and Levenson (1970) discussed the effect of strong spin-electron-spin atomic core interaction on the impurity absorption and emission spectra. Hagston (1970) qualitatively discussed the implications of ligand field theory for substitutional and interstitial H atoms and H^- ions in alkali halides and alkaline earth fluorides with respect to hyperfine interactions. Smith (1970) calculated the spin-orbit splitting of the K band of F-centers in alkali halides with model i).Gupta and Narchal (1970) used a spin Hamiltonian for the calculation of energy eigenvalues, eigenfunctions and transition probabilities of a paramagnetic ion. Deigen, Glinchuk and Korobko (1970) determined the angular and temperature dependence of the EPR line width. Watanabe and Kishishita (1970) applied a molecular orbital model to the description of electron paramagnetic resonance of trivalent ions in II–VI tetrahedral crystals. With the model i), Lowther and Killingbeck (1970) evaluated the g-factor of the groundstate of Yb^{3+} in a cubic field which is modified by a small rhombic distortion. Woodward and Chatterjee (1971) derived higher order electron-muclear interaction terms for the spin Hamiltonian of d^3 ions in cubic symmetry. With a molecular orbital model, Buch and Gelineau (1971) constructed a spin Hamiltonian to calculate the effects of an electric field on the electron paramagnetic resonance spectra of $3d^5$ ions in ZnS. Thuau and Margerie (1971) presented a calculation of the orbital g-factor of the excited state of F-

centers in alkali halides with wave functions orthogonalized to the core orbitals. With the model i) and an additional orthogonalization of the wave functions with respect to core orbitals, Bartram, Harmer and Hayes (1971) investigated the level structure and fine structure of *F*-centers in alkaline earth fluorides. Weber and Lacroix (1971) evaluated with a molecular orbital model the *g*-factor of ions of the iron group in MgO. With the model i) Sharma (1971) studied the zero field splitting of an Mn^{2+} ion in trigonal symmetry. Roitsin (1972) published a review article on electric effects in paramagnetic resonance. With model i), Kambara, Haas, Spedding and Good (1972) calculated the Zeeman effect of rare-earth ions in crystal fields with C_{3h} symmetry. With the model ii), Iida (1972) described electron paramagnetic resonance of group IV ions incorporated into II–VI compounds. With model ii), Biernacki (1972) also studied the fine and hyperfine splitting of levels of a Co^{2+} ion in ZnSe. Smith (1972) investigated the spin orbit interaction for excited *F*-center states in alkali halides with wave functions orthogonalized to the core orbitals. Zwanzger, Muschik and Haug (1972) studied spin effects in the two-particle equation of the exciton. Rumyantsev and Salikhov (1973) published a note on spin exchange between triplet excitons. Leibler and Wilamowski (1973) discussed the *g*-shift of d^5-state ions in II–VI-semiconductors. Matta, Sukheeja and Narchal (1973) gave an exact solution of the spin Hamiltonian for Cr^{3+} ions. With model i) and core orthogonalized wave functions, Smith (1973) investigated the orbital *g*-factor and its influence on the spectral properties for *F*-centers. Roitsin (1973) discussed the same problem for noncentral ions, i.e., ions which are not situated at crystal lattice sites. Chatterjee, Dixon, Lacroix and Weber (1973) calculated the effect of spin-spin contributions to the zero field splitting of Cr^{3+} ions in trigonal symmetry. Matta, Sukheeja and Narchal (1974) derived exact hyperfine states and transition probabilities for the spin Hamiltonian of a Cu^{2+} ion in external magnetic fields. Adrian and Jette (1974) investigated the V_k-center absorption bands by means of valence band wave functions for the halogen molecule anion. Nascimento, Brandi and Ribeiro (1974) discussed spin Hamiltonian and wave functions of the *H*-center. König and Schnakig (1975) considered contributions of excited states to the spin Hamiltonian of Mn^{2+} ions for tetragonal and rhombic symmetries. Ribeiro, Nascimento and Brandi (1975) calculated the spin Hamiltonian of the Na^+ center in LiF with model ii). Baker and Davies (1975) studied with model i) the crystal fields and groundstate wave functions of Yb^{3+} for C_{3v} symmetry by comparison with experimental data. Lulek (1975) determined the splitting of a d^2 orbital singlet configuration in a trigonal crystal field with an external magnetic field. König and Schnakig (1976) calculated the spin Hamiltonian of d^5 ions for tetragonal and rhombic symmetry. Modine, Chen, Major and Wilson (1976) investigated the magnetooptical properties of F^+-centers in alkaline earth oxides with core orthogonalized wave functions. Graf, Maffeo and Brandi (1976) calculated the hyperfine parameters with model ii) for an interstitial hydrogen atom in CaF_2, SrF_2 and BaF_2. Harker (1976) performed calculations of magnetic and optical properties of alkali halide *F*-centers with pseudopotential wave functions for the rigid lattice and compared the values of hyperfine interactions, orbital *g*-factors and spin-orbit coupling constants of all alkali halides with experiment. Grekhov and Roitsin (1976) considered low symmetry effects in paramagnetic resonance with a generalized spin Hamiltonian. Chaney and Lin

(1976) made a molecular orbital approach in order to describe the F-center and included a discussion of hyperfine structure. Soulie and Goodman (1976) investigated with model i) the levels of f^2 ions in a cubic crystal field and calculated their magnetic susceptibility. Bimberg and Dean (1977) treated the magnetic properties of bound excitons in GaP in an external magnetic field based on the model i) and applied to an electron and a hole. Margerie and Martin-Brunetiere (1977) performed a discussion of magnetic circular dichroism of M- and R-centers in KCl based on a phenomenological Hamiltonian. Misra, Faujdar and Kripal (1977) calculated the spin Hamiltonian parameter for the groundstate wave functions of Cu^{2+} ions. Muravev and Yunusov (1977) discussed for the model ii) the change of the hyperfine interaction parameters of impurity ions placed in a crystal environment. Using model ii) Altenberger-Siczek (1977) discussed the electronic level structure of Sn^+ centers in KCl including fine and hyperfine structure. Norgett, Stoneham and Pathak (1977) discussed the electronic structure of the V^- center in MgO. Tasker and Stoneham (1977) studied the molecular model for the V_k center. Dalgaard (1978) gave a derivation of the spin Hamiltonian using Green function techniques. Vanhaelst, Matthys and Boesman (1978) made an analysis of electron-spin transfer in the spin Hamiltonian for V^{2+} in alkali bromides and alkali chlorides. Choh and Yi (1978) discussed the spin Hamiltonian with model i) for Tm^{2+} and Yb^{3+} in fluorite type crystals.

So far we considered the models i) and ii) in the rigid lattice, i.e., we treated the static electron-lattice coupling in its simplest form. Although the results of these calculations are frequently compared with experiment, it is obvious that a complete comparison of theory and experiment requires the inclusion of various kinds of additional electron-lattice interactions, namely static lattice distortions, dynamical electron-lattice coupling, irreversible electron-phonon interactions. Only these interactions allow an adequate treatment of line breadths, phonon absorption and emission, temperature dependence of bands, thermal equilibrium, relaxation processes, etc.

In principle, all corresponding electron-lattice interactions can be obtained by a Taylor expansion of the electron-lattice interaction terms in (6.4.18) about the ideal positions of the atomic cores $\mathbf{R}^0$. The terms arising from this expansion have to be attributed to the various types of interactions, in particular to the reversible and irreversible parts. The criterion that allows such a division is given by the condition that the resulting subsystems have to be stable. However, the magnetic terms are not, in general, required for the stable formation of the subsystems. Hence the corresponding interactions cannot definitely be attributed to the different kinds of interactions and there is a confusion in the literature concerning the application of the various interaction terms.

We first give some references concerning the inclusion of static lattice distortion. Das (1965) included the displacements of the nearest and next nearest neighbor ions into calculations for the crystal field exerted on ions of the iron-group. Wood (1966) considered an electron bound to a V_k center (localized exciton) and took into account the lattice distortion of the nearest neighbour ions for the calculation of optical excitations using a molecular model in the adiabatic coupling scheme. Orbach and Simanek (1967) investigated the zero point phonon contribution to the hyperfine coupling coefficients of d^5 state ions, i.e., they investigated the static effect of the

phonon-electron interaction. Similar calculations were made by Huang (1967) (1968). Wilson, Hatcher, Smoluchowski and Dienes (1969) analyzed the lattice statics of Cu^+ and Ag^+ centers in alkali halides by allowing 26 ions surrounding the defect to move. Song (1969) made calculations of V_k centers where the displacement of the nearest neighbour ions were included. Gazzinelli and Reik (1970) published a note on the small polaron theory of optical absorption and thermal reorientation of V_k centers. Winnacker and Mollenauer (1972) treated spin mixing and magnetic circular dichroism in the absorption band of *F*-centers based on a molecular model with lattice distortion described by molecular symmetry modes. Clack and Williams (1972) showed by a molecular model that Cu^{2+} ions have displaced equilibrium positions with respect to the ideal lattice sites. Iida and Monnier (1976) investigated the static properties of a self-trapped hole in simple cubic CsCl by means of an approximate variational solution of a hole-phonon Hamiltonian in Fock space representation and derived expressions for the corresponding absorption band. Monnier, Song and Stoneham (1977) determined the states of a self-trapped hole in caesium halides using a molecular model and treated the embedding of the molecule in the host lattice by Mott and Littleton's method. Lang (1977) studied the causes of hole localization at two halogen ions in alkali halide crystals based on polaron theory. Bufaical, Maffeo and Brandi (1977) investigated the effect of lattice distortion of the neighbouring ions on the magnetic hyperfine constants of the V_k center. Hizhnyakov and Zazubovich (1978) included lattice relaxation into a molecular model of $Sn^{2+}V_c^-$ centers in alkali halides and calculated absorption and emission energies.
The majority of contributions to this field is concerned with dynamical electron-lattice coupling, where, from the beginning, the static distortion is ignored. This is possible since, in general, the contributions of static lattice distortions to energy differences between neighbouring spin levels are cancelled. The static wave functions of the spin Hamiltonian or of the static Hamiltonian, resp. are then corrected by perturbation theory with respect to dynamical electron-lattice coupling. Afterwards the energies of the spin Hamiltonian are calculated with these functions. Comparison with experiment is then obtained by thermal averaging of the corresponding phonon dependent energies of the spin Hamiltonian. In most cases dynamical coupling leads to the Jahn Teller effect which we will treat in the following section. Here we refer to some papers which are not concerned with the Jahn Teller effect. In most cases only the nonrelativistic electron-lattice coupling arising from the Taylor expansion of the electrostatic forces were taken into account. Van Vleck (1940) used this interaction for the description of spin-lattice relaxation effects, i.e., irreversible processes, while Orbach (1961) and Simanek and Orbach (1966) proposed the use of the phonon perturbed wave functions for the improvement of the calculations of spin energies, i.e., they used the same part in a reversible way. Along the same lines van Heufelen (1967) gave a theory of the groundstate splitting of Mn^{2+} ions by inclusion of relativistic effects. Huang (1967) investigated the temperature dependent hyperfine interactions of Mn^{2+} in alkali halides by using the nonrelativistic orbit-lattice interaction. Similar calculations were made by Shrivastava (1968) for V^{2+} ions in cubic crystalline fields. Menne (1969) improved the thermal averaging of the phonon corrected hyperfine coupling constants for rare earth d^5 ions. Pfister, Dreybrodt and Assmus (1969) studied the temperature

dependent groundstate splitting of Mn^{2+} ions in alkali halides by means of van Henvelen's theory. Shrivastava (1969) (1970) treated the groundstate splitting of Mn^{2+} ions in cubic fields and of Eu^{2+} ions in CaF_2 with the method applied in earlier papers. Huang and Lue (1970) improved the preceding calculations for Mn^{2+} ions of Huang. Menne (1970) also improved his former calculations. Bhattacharyya (1971) considered the adiabatic wave functions of Co^{2+} in an octahedral environment. Calvo, Passeggi and Tovar (1971) analyzed the spin-lattice coefficients of Gd^{3+} in CaF_2 using a point charge model for the crystalline field and the usual orbit-lattice interaction. Schlottmann and Passeggi (1972) considered relativistic contributions to the spin lattice coefficients. Owens (1974) investigated the phonon induced hyperfine interactions for d^5 ions for tetragonal symmetry by using orbit-lattice interaction. The same topic was treated by Koch (1975) for tetrahedral symmetry. Arkhipov and Malkin (1975) considered spin-phonon interaction of V^{3+} ions in CaF_2 crystals. Bates and Szymczak (1976) emphasized the contribution of the anharmonic terms arising from the orbit-lattice coupling to the phonon induced temperature dependence of the spin Hamiltonian parameters for d^5 ions.

Concerning the interactions which are used for irreversible processes we can only refer to the literature. A review article of the theory of spin-lattice relaxation of paramagnetic ions in ionic crystals was given by Stevens (1967). Gill (1975) reviewed the various processes of electron spin-lattice relaxation establishing thermal equilibrium in dielectric crystals containing paramagnetic ions of the transition elements. Vaughan (1975) gave a phenomenological review of electron spin-lattice relaxation. All reviews contain numerous references with respect to this field. Vaughan's review, in particular, contains a survey of review articles with short characterization of their content. In addition, we cite some original papers which deal with these topics and which supplement the review articles:

Spin-lattice relaxation of *F*-center electrons, Blumberg (1960). Spin-lattice relaxation of excitons in an ionic crystal, Paranichev (1960). Spin-lattice interaction of an Mn^{2+} ion in MgO, Kondo (1962). Two-phonon transitions in the spin-lattice relaxation of paramagnetic ions, Aminov (1962). Theory of nuclear spin-lattice relaxation in ionic crystals, Kravchenko (1963). Theory of magnetic resonance saturation in ionic crystals, Provotorov (1963). General methods for the solution of rate equations for multilevel paramagnetic systems, Grant (1964). Multiphonon radiationless transitions between local states of different multiplicity, Perlin, Kovariskii and Tsukerblat (1966). Adiabatic approximation in the theory of spin-lattice interaction between local electron centers in nonmetallic crystals, Vinetskii and Kravchenko (1965). Theory of phonon bottleneck in paramagnetic crystals, Stoneham (1965). Two-phonon process in the spin-lattice relaxation of *F*-centers, Kravchenko and Vinetskii (1965). Theory of paramagnetic relaxation of *F*-centers due to hyperfine and spin-orbit interactions with the lattice, Kravchenko and Vinetskii (1965). Effect of the lattice anharmonicity on electron spin-lattice relaxation, Aleksandrov (1966). Electron spin-lattice relaxation by two-phonon processes, Mills (1966). Theory of many-phonon non-radiative transitions in local paramagnetic centers, Tsukerblat and Perlin (1966). Theory of spin-spin relaxation, Terwiel and Mazur (1966). Spin-lattice relaxation of weakly inhomogeneous conduction-electron distributions in a strong magnetic field,

Kalashnikov (1967). Interference effects in spin-lattice relaxation by two-phonon processes, Stoneham (1967). Optical phonons in electron-spin relaxation, Huang (1967). Theory of paraelectric resonance and relaxation, Sander and Shore (1971). Localized and quasilocalized vibrations in anharmonic Raman spin-lattice relaxation, Zevin and Konovalov (1972). Excitation transfer between localized magnetic impurities in insulators, Kohli and Liu (1974). Kinetic equations for a paramagnetic center in a dielectric medium and the theory of spin-lattice relaxation, Zevin (1975). Stochastic theory of spin-phonon relaxation, Pirc and Krumhansl (1975). Spin relaxation of electrons due to scattering by holes, Bir, Aronov and Pikus (1976). Calculation of the spin-lattice relaxation time of an U^{4+} ion in a cubic crystalline field, Obata and Sasaki (1977). Spin-lattice relaxation of *F*-centers in Na and Cs halides, Carvalho, Terrile and Panepucci (1977).

6.5 Symmetry breaking

In the preceding chapters, the adiabatic coupling scheme was evaluated for nondegenerate electronic states. In this section we extend the formalism to include also degenerate electronic states. We will show that this extension is closely connected with a symmetry breaking of the system under consideration. In the case of degeneracy, a quantum mechanical system of electrons and nuclei is incapable of remaining in a state of given symmetry. This fact has first been discovered for molecules by Jahn and Teller. The corresponding Jahn-Teller effect may also occur in ideal crystals as well as in imperfect crystals. In semiconductor physics we are mainly interested in the reactions of imperfect crystals. Hence we study the symmetry breaking for impurity centers.
To elaborate the essentials we use a very simple model. We assume a low density of defects, so that from the atomistic point of view the average crystal volume containing one defect can be idealized to be an infinite crystal where the boundary effects on the defect are negligible. Then, this microblock containing one defect in its center has a crystal Hamiltonian which is forminvariant under the full rotation group $O(3)$ with the center as its origin. However, we are not interested in the full rotation group, but only in the point groups with respect to this center. As the point groups are subgroups of the full rotation group, it is obvious that an additional assumption has to be made concerning the invariance under a certain point group. This can be done by fixing the nuclei in their equilibrium positions $\{\mathbf{R}_k^0\}$. Then the following definition can be used:

Def. 6.23 : If the set of lattice vectors $\{\mathbf{R}_k^0\}$ is only permuted under the application of a point group, then the array $\{\mathbf{R}_k^0\}$ defines a crystal structure which corresponds to this group.

While the ordinary forminvariance under $O(3)$ only refers to the invariance of relative distances between the particles (and invariance of the kinetic energy), the Definition 6.23 requires the absolute formation of the atomic cores in the crystal. Hence the definition of a crystal structure is much more restrictive than merely defining a forminvariance of the system, and symmetry breaking in the case of a crystal is mainly concerned with the violation of Definition 6.23 by the system under consideration. Concerning impurities this leads to the definition

Def. 6.24 :If in a microblock where the ions are fixed in the ideal crystal positions, an impurity center is incorporated and if this center can be considered to be the origin of a point group which permutes all lattice points, then this point group is called the symmetry group of the microblock with respect to the impurity.

As in reality the microblock immediately reacts to the substitution or incorporation of an impurity center by a rearrangement of electron states and lattice positions, it is evident that the definition of a symmetry group for impurity center states really depends on the states and that symmetry breaking has to be considered as the rule and not as the exception. Nevertheless, this formal symmetry definition is of great theoretical interest. This stems from the fact that any state calculation for an imperfect crystal in the adiabatic coupling scheme starts with the ideal crystal. This can be seen in Chapter 5, where all static calculations refer to ionic displacements $\mathbf{M}_k := (\mathbf{R}_k - \mathbf{R}_k^0)$.

The static calculations of Chapter 5 are based on a variational procedure for electronic states. Such a procedure is very suitable for practical calculations but it is insensitive with respect to symmetry effects. To investigate symmetry effects we supplement the variational procedure by the use of perturbation theory. Insofar as only a formal discussion is intended, we do not reduce the adiabatic electron equation to an effective one-electron or many-electron equation, but work with the original equation (2.2.7). We perform an expansion of all potentials in $h(\mathbf{r},\mathbf{R})$ about the ideal positions $\{\mathbf{R}_k^0\}$ and obtain

$$h(\mathbf{r},\mathbf{R}) = h(\mathbf{r},\mathbf{R}^0) + h_1(\mathbf{r},\mathbf{M}) \qquad (6.5.1)$$

where h_1 contains all parts of $h(\mathbf{r},\mathbf{R})$ which depend on $\{\mathbf{M}_k\}$. Then equation (2.2.7) reads

$$[h(\mathbf{r},\mathbf{R}^0) + h_1(\mathbf{r},\mathbf{M})]|\chi_n\rangle = \tilde{U}_n(\mathbf{M})|\chi_n\rangle. \qquad (6.5.2)$$

If a perturbation calculation is performed, we consider h_1 as a perturbation and start with the equation

$$h(\mathbf{r},\mathbf{R}^0)\chi_n^0(\mathbf{r},\mathbf{R}^0) = \tilde{U}_n^0 \chi_n^0(\mathbf{r},\mathrm{R}^0). \qquad (6.5.3)$$

If an impurity center is placed at a definite lattice side or at an interstitial side, the adiabatic electron operator $h(\mathbf{r},\mathbf{R}^0)$ is forminvariant under the corresponding symmetry group (point group) of the imperfection and the crystal. We do not here give a proof of this statement, as a proof for the more general case of the forminvariance under space groups is given in Section 6.7 and both proofs are completely analogous.

Due to this forminvariance, the set $\{\chi_n^0(\mathbf{r},\mathbf{R}^0)\}$ can be chosen as a basis of irreducible representations of the corresponding point group. If multidimensional representations occur in the set $\{\chi_n^0(\mathbf{r},\mathbf{R}^0)\}$, it is evident that equation (6.5.2) cannot be solved by ordinary perturbation theory and that the perturbation theory for degenerate states has to be applied. Hence, if we perform practical calculations, one has to sharply distinguish between one-dimensional and multi-dimensional representations. Although this distinction seems at first to be only of technical interest, it has further consequences regarding the symmetry properties of the system. The following theorem was proven by Jahn and Teller (1937):

Theorem 6.41: Any system of electrons and nuclei occupying an energy level with electronic degeneracy is unstable against a distortion that removes that degeneracy to the first order.

Jahn and Teller used perturbation theory to prove their theorem. We do not reproduce the proof here. We rather try to give an equivalent formulation of it which emphasizes the connection with symmetry breaking. Since according to Theorem 6.31 the nonadiabatic operator K^t is invariant under $O(3)$ and its subgroups, the symmetry of a system is conserved if the basis states for its description are basis states of the corresponding symmetry groups. Then the transitions effected by K^t cannot destroy the symmetry of the system. Hence, it is sufficient to consider the symmetry properties of the adiabatic coupling scheme.

Theorem 6.42: If a point symmetry group G is supposed to be the symmetry group of an infinite microblock, then the adiabatic coupling states $\chi_n(\mathbf{r},\mathbf{R})\varphi_n^m(\mathbf{R})$ are then and only then basis states of a representation of G, provided that $\chi_n(\mathbf{r},\mathbf{R})$ transforms as a unit representation of G.

Proof: The exact adiabatic states $\chi_n(\mathbf{r},\mathbf{R})$ satisfy the adiabatic equation (2.2.7)

$$h(\mathbf{r},\mathbf{R})|\chi_n\rangle = \tilde{U}_n(\mathbf{R})|\chi_n\rangle \tag{6.5.4}$$

while the phonon parts φ_n^m satisfy equations

$$\left[-\sum_k \frac{\hbar^2}{2M_k}\Delta_k + \tilde{U}_n(\mathbf{R})\right]|\varphi_m^n\rangle = E_m^n|\varphi_m^n\rangle. \tag{6.5.5}$$

If we apply a simultaneous group operation (6.1.7) to $\mathbf{r}$ and $\mathbf{R}$ with $g\in G$, $h(\mathbf{r},\mathbf{R})$ remains forminvariant, i.e.,

$$h(g^{-1}\mathbf{r}',g^{-1}\mathbf{R}') =: h'(\mathbf{r}',\mathbf{R}') \equiv h(\mathbf{r}',\mathbf{R}') \tag{6.5.6}$$

and equation (6.5.4) goes over into

$$\mathrm{h}(\mathbf{r}',\mathbf{R}')\chi_n(g^{-1}\mathbf{r}',g^{-1}\mathbf{R}') = \tilde{U}_n(g^{-1}\mathbf{R}')\chi_n(g^{-1}\mathbf{r}',g^{-1}\mathbf{R}'). \tag{6.5.7}$$

As χ_n has to transform under a unit representation, i.e., $\gamma_{ll'}(g)\equiv 1$, from (6.1.11) it follows that

$$\chi_n(g^{-1}\mathbf{r}',g^{-1}\mathbf{R}') \equiv \chi_n(\mathbf{r}',\mathbf{R}'). \tag{6.5.8}$$

Hence (6.5.7) can be written

$$h(\mathbf{r}',\mathbf{R}')\chi_n(\mathbf{r}',\mathbf{R}') = \tilde{U}_n(g^{-1}\mathbf{R}')\chi_n(\mathbf{r}',\mathbf{R}') \tag{6.5.9}$$

or by a change of variables $\mathbf{r}' \to \mathbf{r}$, $\mathbf{R}' \to \mathbf{R}$, we may further write

$$h(\mathbf{r},\mathbf{R})\chi_n(\mathbf{r},\mathbf{R}) = \tilde{U}_n(g^{-1}\mathbf{R})\chi_n(\mathbf{r},\mathbf{R}). \tag{6.5.10}$$

Using the orthonormality of the χ_n with respect to the variable $\mathbf{r}$, we obtain from (6.5.4) and (6.5.10)

$$\int \chi_n(\mathbf{r},\mathbf{R})^{\times} h(\mathbf{r},\mathbf{R})\chi_n(\mathbf{r},\mathbf{R})\,dr = \tilde{U}_n(\mathbf{R}) = \tilde{U}_n(g^{-1}\mathbf{R}). \tag{6.5.11}$$

Hence $\tilde{U}_n(\mathbf{R})$ is forminvariant with respect to G. In this case the operator of equation (6.5.5) is also forminvariant and the solutions of (6.5.5) are base states of representations of G. Hence, if χ_n is a unit representation of G, the product $\chi_n \varphi_m^n$ is also a base state of a representation of G and symmetry is conserved.
If χ_n is not a unit representation but belongs to a multi-dimensional representation, a derivation along the same lines gives

$$\tilde{U}_n(g^{-1}\mathbf{R}) = \sum_{n'} \gamma_{nn'}(g)\, U_{n'}(\mathbf{R}). \tag{6.5.12}$$

Hence $U_n(\mathbf{R})$ is not forminvariant and consequently the solutions φ_m^n of equation (6.5.5) cannot be base states of irreducible representations of G in this case. But this means that in this case the symmetry of the system is broken, Q.E.D.

Neither the original Jahn-Teller theorem nor this theorem gives any practical hint concerning the consequences for the construction of electronic and nuclear state functions. In order to obtain such information we return to perturbation theory. In perturbation theory the solutions $|\chi_n\rangle$ of (6.5.2) are obtained by a unitary transformation from the solutions $|\chi_n^0\rangle$ of (6.5.3), i.e.,

$$\chi_n = \sum U_{nn'} \chi_{n'}^0 . \tag{6.5.13}$$

The problem is solved if $U_{nn'}$ is known. To simplify the calculation we consider instead of (6.5.1)

$$h(\mathbf{r},\mathbf{R},\lambda) := h(\mathbf{r},\mathbf{R}^0) + \lambda h_1(\mathbf{r},\mathbf{M}) \tag{6.5.14}$$

and put afterwards $\lambda = 1$. Then in place of equation (6.5.2) we may write

$$\sum_{n'} \langle \chi_n^0 | [h(\mathbf{r},\mathbf{R}^0) + \lambda h_1(\mathbf{r},\mathbf{M})] | \chi_{n'}^0 \rangle\, U_{n'l} = \sum_{n'} \tilde{U}_{nn'}\, U_{n'l} \tag{6.5.15}$$

with $\tilde{U} := U_n(\mathbf{M})\delta_{nn'}$. To find an appropriate transformation matrix U, we perform the expansion

$$\begin{aligned} U &= U^0 + \lambda U^1 + \lambda^2 U^2 + \dots \\ \tilde{U} &= \tilde{U}_0 + \lambda \tilde{U}_1 + \lambda^2 \tilde{U}_2 + \dots \end{aligned} \tag{6.5.16}$$

If the eigenvalues of U_0 are degenerate, we divide the matrix $(h_1)_{nn'} := \langle \chi_n^0, h_1 \chi_{n'}^0 \rangle$ into two parts

$$h_1 = h_1^d + h_1^n \tag{6.5.17}$$

where h_1^d contains all elements of h_1 which belong to the degenerate part of the spectrum, while h_1^n contains all other elements. Then the first order equation reads

$$U^0 \tilde{U}_1 + U^1 \tilde{U}_0 = \tilde{U}_0 U^1 + h_1^d U^0 + h_1^n U^0 . \tag{6.5.18}$$

Now we calculate U^0 and $\tilde{U}_1$ from

$$U^0 \tilde{U}_1 = h_1^d U^0 \tag{6.5.19}$$

and U^1 from

$$U^1 \tilde{U}_0 = \tilde{U}_0 U^1 + h_1^n U^0 . \tag{6.5.20}$$

For convenience we replace the single index n by $n \equiv \nu\alpha$, were ν is the principal quantum number, while α characterizes the degeneracy. Then according to our definition of h_1^d, (6.5.19) is equivalent to the systems

$$\sum_{\alpha'} U^0_{\nu\alpha,\nu\alpha'} \tilde{U}_{1\nu\alpha'} = \sum_{\alpha'} h^d_{1\nu\alpha,\nu\alpha'} U^0_{\nu\alpha,\nu\alpha'}, \quad \nu = 1 \ldots \tag{6.5.21}$$

i.e., the set of matrices $\{h^d_{1\nu\alpha,\nu\alpha'}, \nu = 1 \ldots\}$ has to be transformed on principal axis by the matrix set $\{U^0_{\nu\alpha,\nu\alpha'}, \nu = 1 \ldots\}$.

This is only the first step in a systematic treatment of a perturbation calculation of degenerate states. It is, however, the crucial step as it prevents divergencies in the first order calculation. Although many investigations have been made about the Jahn-Teller effect, this first step of a perturbation calculation is the starting point of the calculation in nearly all cases. Hence we describe this step in more detail in order to understand the technique of Jahn-Teller calculations.

If we put

$$U^0_{\nu\alpha,\nu\alpha'} := \{a_\alpha(\nu,1), \ldots, a_\alpha(\nu,r)\} \qquad 1 \leqslant \alpha, \alpha' \leqslant r$$

and observe that $\tilde{U}_1$ is a diagonal matrix, the diagonalization of $h^d_{1\nu\alpha,\nu\alpha'}$ for fixed ν is equivalent to the solution of the equations

$$\tilde{U}_{1\nu\alpha'} a_\alpha(\nu,\alpha') = \sum_\beta h^d_{1\nu\alpha,\nu\beta} a_\beta(\nu,\alpha'), \quad 1 \leqslant \alpha' \leqslant r. \tag{6.5.22}$$

If we add on both sides $\tilde{U}_{0\nu} a_\alpha(\nu,\alpha')$, (6.5.22) goes over into

$$\sum_\beta [\tilde{U}_{0\nu} \delta_{\alpha\beta} + h^d_{1\nu\alpha,\nu\beta}] a_\beta(\nu,\alpha') = E(\nu,\alpha') a_\alpha(\nu,\alpha'). \tag{6.5.23}$$

But this equation can be obtained from (6.5.2) if we simply make the ansatz

$$|\chi_{\nu\alpha'}\rangle = \sum_\beta a_\beta(\nu,\alpha') |\chi^0_{\nu\beta}\rangle \tag{6.5.24}$$

and assume that the states $|\chi^0_{\nu\beta}\rangle$ $\beta = 1 \ldots r$ are orthonormal. Then substitution of (6.5.24) into (6.5.2) and projection on $|\chi^0_{\nu\alpha}\rangle$ gives (6.5.23).

For practical applications we have to divide the crystal electrons into two groups. One group consists of those few electrons that occupy degenerate orbitals and actively participate in the Jahn-Teller effect. We shall call these the active electrons. The

remaining passive electrons form closed shells and the possibility of exciting them really (not virtually) to higher states will be neglected. This supposition leads as in Chapter 5 to the derivation of effective Hamiltonians for the active electrons.
To give a practical demonstration we postpone the use of effective Hamiltonians and work directly with the operator (6.5.1). To evaluate (6.5.23) or (6.5.2), resp., with the ansatz (6.5.24) for special cases, we rewrite the operator (6.5.1). According to (2.2.5) and (2.2.3) we define

$$\begin{aligned} H_e &:= -\sum_i \frac{\hbar^2}{2m} \Delta_i + e^2 \frac{1}{2} \sum_{ij}' C(\mathbf{r}_i, \mathbf{r}_j) + \sum_{ik} e e_k C(\mathbf{r}_i, \mathbf{R}_k^0) \\ &\quad + \frac{1}{2} \sum_{kl}' e_k e_l C(\mathbf{R}_k^0, \mathbf{R}_l^0) \\ V(\mathbf{r}, \mathbf{M}) &:= \sum_{ik} e e_k [C(\mathbf{r}_i, \mathbf{R}_k) - C(\mathbf{r}_i, \mathbf{R}_k^0)] \\ V(\mathbf{M}) &:= \frac{1}{2} \sum_{kl}' e_k e_l [C(\mathbf{R}_k, \mathbf{R}_l) - C(\mathbf{R}_k^0, \mathbf{R}_l^0)] \end{aligned} \tag{6.5.25}$$

and obtain for (6.5.1)

$$h(\mathbf{r}, \mathbf{R}) = H_e + V(\mathbf{M}) + V(\mathbf{r}, \mathbf{M}). \tag{6.5.26}$$

For the simplest information about the Jahn-Teller effect, we reduce the problem theoretically to that of a molecule by assuming that only motions of the impurity or defect and its immediate neighbours are of importance, i.e., they show a deformation $\mathbf{M}_k \neq 0$, while the motions of more distant neighbours can be neglected. For example, we consider the octahedral XY_6 complex which is given by Figure 6.6.

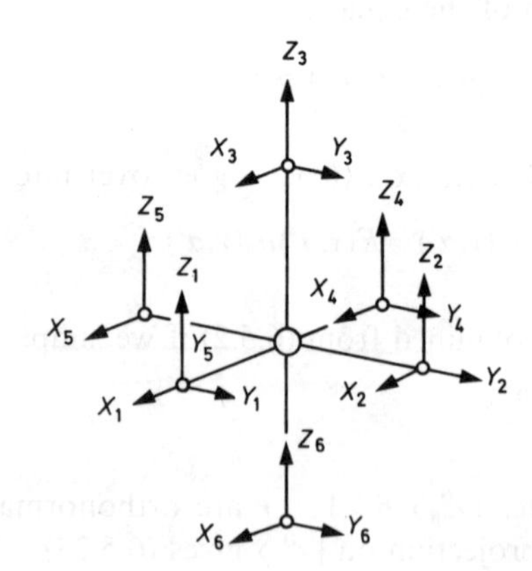

Figure 6.6:
The octahedral XY_6 complex.

This complex shows O_h-symmetry. Its (collective) normal coordinates $\{q_k, k=1\ldots 15\}$ are given in Table 6.6

Table 6.6:
Collective coordinates for the octahedral XY_6 complex. Greek letters are used for the symmetry labels of normal coordinates; after Sturge (1967).

Coordinates	Representations	
q_1	α_{1g}	$(x_1-x_4+y_2-y_5+z_3-z_6)/6^{1/2}$
q_2	$\varepsilon_g(v)$	$(x_1-x_4-y_2+y_5)/2$
q_3	$\varepsilon_g(u)$	$(2z_3-2z_6-x_1+x_4-y_2+y_5)/2(3)^{1/2}$
q_4	$\tau_{2g}(\xi)$	$(z_2-z_5+y_3-y_6)/2$
q_5	$\tau_{2g}(\eta)$	$(x_3-x_6+z_1-z_4)/2$
q_6	$\tau_{2g}(\zeta)$	$(y_1-y_4+x_2-x_5)/2$
q_7	$\tau_{1u}(x)$	$(x_2+x_3+x_5+x_6)/2$
q_8	$\tau_{1u}(y)$	$(y_1+y_3+y_4+y_6)/2$
q_9	$\tau_{1u}(z)$	$(z_1+z_2+z_4+z_5)/2$
q_{10}	$\tau_{1u}(x)$	$(x_1+x_4)/2^{1/2}$
q_{11}	$\tau_{1u}(y)$	$(y_2+y_5)/2^{1/2}$
q_{12}	$\tau_{1u}(z)$	$(z_3+z_6)/2^{1/2}$
q_{13}	$\tau_{2u}(\xi)$	$(x_2+x_5-x_3-x_6)/2$
q_{14}	$\tau_{2u}(\eta)$	$(y_3+y_6-y_1-y_4)/2$
q_{15}	$\tau_{2u}(\zeta)$	$(z_1+z_4-z_2-z_5)/2$

The collective coordinates are connected with the displacements $\mathbf{M}_k$ by a linear transformation and are assumed to have the property that they decouple the potential energy $V(\mathbf{M})$ in the second order term. Hence with these coordinates we obtain

$$\begin{aligned} V(\mathbf{r},\mathbf{M}) &= \sum_{i=1}^{15} v_i(\mathbf{r})q_i + \dots \\ V(\mathbf{M}) &= \sum_{i=1}^{15} \mu_i\omega_i^2 q_i^2 + \dots \end{aligned} \tag{6.5.27}$$

where the dots indicate higher order terms which are taken into account later on. The simplest case where a Jahn-Teller effect has to be expected is given by an octahedrally coordinated impurity ion with a doubly degenerate ground term. Within the harmonic approximation this system can be solved exactly, cf. Sturge (1967). In the harmonic approximation (6.5.26) reads

$$h(\mathbf{r},\mathbf{R})^h := H_e + \sum_{i=1}^{15} v_i(\mathbf{r})q_i + \sum_{i=1}^{15} \mu_i\omega_i^2 q_i^2 . \tag{6.5.28}$$

The two degenerate electronic levels $|\chi_{\nu\beta}^0\rangle$, $\beta=1,2$ are said to belong to an E representation and for the ground state we suppress the label $\nu=0$. For the "true" wave functions (6.5.24) given here by

$$|\chi_\alpha\rangle = \sum_{\beta=1}^{2} a_\beta(\alpha)|\chi_\beta^0\rangle \tag{6.5.29}$$

we obtain the eigenvalue problem

$$\sum_{\beta}\left[E_0\delta_{\alpha\beta}+\sum_{i=1}^{15}\langle\chi_\alpha^0|v_i(\mathbf{r})|\chi_\beta^0\rangle q_i+\sum_{i=1}^{15}\mu_i\omega_i^2 q_i^2\delta_{\alpha\beta}\right]a_\beta(\alpha)=E(\alpha)a_\alpha(\alpha) \tag{6.5.30}$$

with $E_0:=\langle\chi_\beta^0|H_e|\chi_\beta^0\rangle$, $\beta=1,2$. As the normal coordinates are derived for the octahedral complex with O_h symmetry, they are representatives of this point group. This property enables one to apply the Wigner-Eckardt theorem to the reduction of the only nontrivial term in (6.5.30), namely the interaction term between lattice and electrons. We only give the result of this analysis: A nontrivial term splitting only occurs for the q_2 and q_3 mode which leads to a Hamiltonian of the form

$$\left[\begin{pmatrix}1&0\\0&1\end{pmatrix}[E_0+\frac{1}{2}\mu\omega^2(q_2^2+q_3^2)]+A\begin{pmatrix}-q_3&q_2\\q_2&q_3\end{pmatrix}\right]\begin{pmatrix}a_1\\a_2\end{pmatrix}=E\begin{pmatrix}a_1\\a_2\end{pmatrix} \tag{6.5.31}$$

where all contributions of the other normal modes are omitted as they do not influence the term splitting under consideration. A solution of this problem can be found by the introduction of polar coordinates $q_2=\varrho\sin\Theta$, $q_3=\varrho\cos\Theta$. This gives

$$\begin{aligned}|\chi_-\rangle&=|\chi_1^0\rangle\sin\frac{1}{2}\Theta+|\chi_2^0\rangle\cos\frac{1}{2}\Theta\\|\chi_+\rangle&=|\chi_1^0\rangle\cos\frac{1}{2}\Theta+|\chi_2^0\rangle\sin\frac{1}{2}\Theta\end{aligned} \tag{6.5.32}$$

with the corresponding energies

$$U^{\pm}(\varrho)\equiv E(\pm)=E_0\pm A\varrho+\frac{1}{2}\mu\omega^2\varrho^2. \tag{6.5.33}$$

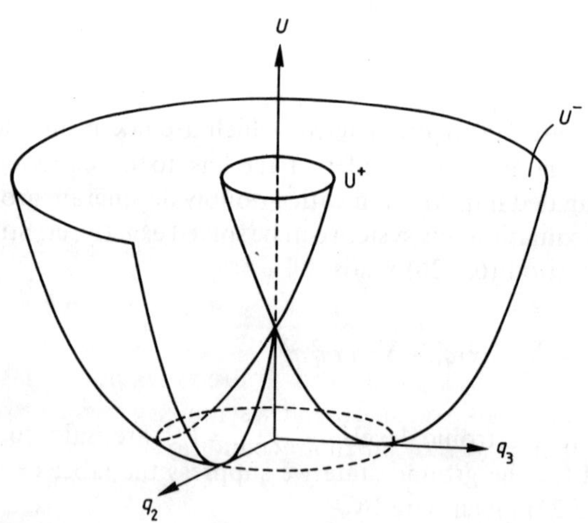

Figure 6.7:
The double valued potential surface $U(q_2, q_3)$ for a doubly degenerate state in the lowest-order theory; after Sturge (1967)

The energy $E(\pm)$ acts as the adiabatic potential for the true eigenmodes of the complex, which is given in Figure 6.7 while the functions $|\chi_-\rangle \equiv \chi_-(\mathbf{r}, q_2, q_3)$ and $|\chi_+\rangle \equiv \chi_+(\mathbf{r}, q_2, q_3)$ are the dynamical adiabatic electronic eigenfunctions of the molecular complex in this approximation. As the interaction has removed the degeneracy, one could continue with the ordinary adiabatic calculation procedure for non-degenerate levels by substitution of the adiabatic lattice potential energy (6.5.33) into the corresponding phonon equation and calculation of the corresponding phonon states. However, in contrast to the phonon potential for primarily non-degenerate electronic levels, the phonon potential (6.5.33) shows some striking peculiarities which prevent a straightforward phonon calculation. These pecularities arise for two reasons, namely instability and symmetry breaking, which can be concluded from the example given in Figure 6.7.

i) Instability occurs because there are no single minimum points of the potential energy, but a complete circle of equivalent minimum points which enables the system to rotate.

ii) Symmetry breaking occurs as the minimum points have coordinates $q_2, q_3 \neq 0$. This means that in such a state the minimum positions $\mathbf{R}_k^n$ do not have those values which are required for the definition of a crystal (or molecule) symmetry group.

The occurrence of i) and ii) for interacting subsystems which are described by representations of a certain symmetry group is called the static Jahn-Teller effect. The notation static has to be emphasized since the "phonon" modes appearing in this calculation are only symmetry displacements of the lattice and not the true phonon coordinates.
Before discussing the phonon mode calculation we consider the situation in insulators and semiconductors. Since in polar insulators and semiconductors we have to expect long-range forces, for impurity state calculations the techniques of Chapter 5 have to be applied also for Jahn-Teller calculations instead of the molecular model. Hence it is desirable to extend our approach to electronic state calculations of Chapter 5 to include also degenerate states. We give a theorem which is concerned with the calculation of one-electron states. But it can be seen from the proof that this theorem can easily be extended to many-electron centers, which we will not explicitly perform here.

Theorem 6.43: If a one-electron impurity center has a symmetry group G according to Def. 6.24, then, for variational calculations of the static equilibrium state, a test function space $\{f\}$ has to be chosen which is a representation space of G. For multi-dimensional representations of G, in general, a symmetry breaking occurs which can be taken into account a priori by auxiliary conditions imposed on the variational calculation of Theorem 5.2.

Proof: We consider a state equation for the electron where the ionic positions are not yet fixed and perform a theoretical perturbation calculation. The energy of this configuration is given by $U_n[\mathbf{M},\psi]$ and the corresponding state equation reads

$$\frac{\delta}{\delta\psi^{\times}} U_n[\mathbf{M},\psi]=E\psi. \tag{6.5.34}$$

If for simplicity the model of Theorem 5.2 is used, $U_n[\mathbf{M},\psi]$ is given by (5.2.56) and (6.5.34) yields

$$\begin{aligned}
&\left[-\frac{\hbar^2}{2m}\Delta+\sum_{\{l\mu\}} ea_{l\mu}C(\mathbf{r},l\mu)\right]\psi(\mathbf{r})\\
&\quad-\sum_{\{l\mu\}}\sum_{\alpha_{l\mu}}\int e^2 C(\mathbf{r},\mathbf{r}')\psi(\mathbf{r}')\varrho_{l\mu\alpha_{l\mu}}(\mathbf{r},\mathbf{r}')d^3r'\\
&\quad-\frac{\alpha^e e}{1+\alpha^e}\int\hat{\varrho}(\mathbf{r}')F(\mathbf{r}',\mathbf{r})\psi(\mathbf{r})d^3r'\\
&\quad+\frac{e}{1+\alpha^e}\sum_{\{l\mu\}}[a_{l\mu}\mathbf{M}_{l\mu}\cdot\nabla_{l\mu}C(\mathbf{r},l\mu)]\psi(\mathbf{r})\\
&\quad=:(H_0+V)\psi(\mathbf{r})=E\psi(\mathbf{r}).
\end{aligned} \tag{6.5.35}$$

According to definition (5.2.28) the expansion for the ionic coordinates $\{\mathbf{R}_{l\mu}\}$ has been performed at the ideal lattice sites $\{\mathbf{R}^0_{l\mu}\}$. Hence, if $\mathbf{M}_{l\mu}=0$ is assumed, equation (6.5.35) has to be forminvariant with respect to the point group G. The self-energy term $\hat{\varrho}F\psi$ seems to contradict this statement. But this term results directly from the Hartree-Fock interaction and by standard procedures it can be prepared to conserve symmetry, cf. Ripka (1968). For simplification, we consider $\hat{\varrho}$ to be a given density which is the first step in a Hartree-Fock iteration procedure. Defining now

$$V(\mathbf{r}):=\frac{e}{1+\alpha^e}\sum_{\{l\mu\}}[a_{l\mu}\mathbf{M}_{l\mu}\cdot\nabla_{l\mu}C(\mathbf{r},l\mu)]+\text{exch. terms} \tag{6.5.36}$$

as the perturbing potential, a perturbation calculation can be started, where the solutions $\{\psi^0_m\}$ of H_0 are base states of a representation space of G. If multi-dimensional representations of G occur, then H_0 is degenerate with respect to these representations and the perturbation calculation for degenerate states has to be performed. Denoting the degenerate state families by $\{\psi^0_{m\kappa}\}$ where κ is the index of degeneracy, due to the energy degeneracy, all systems $\{\psi^{0\prime}_{m\kappa}\}$ obtained by a unitary transformation $\psi^{0\prime}_{m\lambda}=\sum_{\kappa'} a_{\lambda\kappa'}\psi^0_{m\kappa'}$ are equivalent. To secure a meaningful perturbation calculation from all possible linear combinations of the initial states $\{\psi^0_{m\kappa}\}$, those combinations have to be used for which

$$\sum_{\kappa'}\langle\psi^0_{m\kappa}|V|\psi^0_{m\kappa'}\rangle a_{\kappa'}(\lambda)=v_\lambda a_\kappa(\lambda) \tag{6.5.37}$$

is satisfied, i.e., for which the non-diagonal parts of V vanish. Here we have put $a_{\lambda\kappa'}\equiv a_{\kappa'}(\lambda)$. It is obvious that in (6.5.37) as many coefficients $a_\kappa(\lambda)$ exist as there are conditions which have to be satisfied. We now consider the special case where $V(\mathbf{r})$ is given by (6.5.36). According to the definition of the adiabatic coupling, the displacements **M** are arbitrary. In this case the method fails because many more conditions have to be satisfied than there are coefficients available. This can be seen by writing V in the form $V=\sum_{\{l\mu\}} \mathbf{V}_{l\mu}\cdot\mathbf{M}_{l\mu}$. As $\mathbf{M}_{l\mu}$ is arbitrary, this leads to the conditions

$$\sum_{\kappa'} \langle\psi^0_{m\kappa}|\mathbf{V}_{l\mu}|\psi^0_{m\kappa'}\rangle=\mathbf{v}_{\lambda l\mu}a_\kappa(\lambda) \qquad \forall\ l\mu \tag{6.5.38}$$

which should be satisfied simultaneously. As this is not possible, it follows that a consistent perturbation calculation starting from a given symmetry with nontrivial representations is, in general, impossible. Nevertheless, it can be applied consistently for definite fixed ion positions. We choose these positions to be $\mathbf{M}_{l\mu}=\mathbf{M}^n_{l\mu}$, i.e., the equilibrium positions of the static configuration. Then only the condition

$$\sum_{\kappa'}\sum_{\{l\mu\}} \langle\psi^0_{m\kappa}|\mathbf{V}_{l\mu}|\psi^0_{m\kappa'}\rangle\cdot\mathbf{M}^n_{l\mu}a_{k'}(\lambda)=a_\kappa(\lambda)v_\lambda(n) \tag{6.5.39}$$

has to be satisfied. But in contrast to ordinary potentials (6.5.39) is not a linear set of equations. From (5.2.59), (5.2.68) and (5.2.71) it follows that

$$\begin{aligned}\mathbf{M}^n_{l\mu}= &-\frac{\hat{\chi}^g\tau}{1+\alpha^e}\,V_{l\mu}\int\hat{\varrho}(\mathbf{r})a_{l\mu}C(\mathbf{r},l\mu)d^3r\\ &-\tau a_{l\mu}\sum_{k\varrho}\sum_{n=0}^{\infty}(\hat{\chi}^g)^{n+1}(\hat{\mathcal{V}}^n)_{l\mu k\varrho}\cdot V_{k\varrho}\hat{C}(l^i\mu^i,k\varrho)^n+\dots\end{aligned} \tag{6.5.40}$$

where, according to (5.2.36), $\hat{\varrho}(\mathbf{r})$ contains the term $\varrho(\mathbf{r})=|\psi(\mathbf{r})|^2$, i.e., the true electronic wave function at the position $\mathbf{M}=\mathbf{M}^n$. In a first approximation we put $\psi(\mathbf{r})\equiv\psi^{0\prime}_{m\lambda}=\sum_{\kappa'}a_{\kappa'}(\lambda)\psi^0_{m\kappa'}$. Then $\mathbf{M}^n$ becomes a function of the $a_\kappa(\lambda)$, i.e.,

$$\mathbf{M}^n\equiv\mathbf{M}^n[a_\kappa(\lambda)] \tag{6.5.41}$$

and equations (6.5.39) go over into

$$\sum_{\kappa'}\sum_{\{l\mu\}} \langle\psi^0_{m\kappa}|\mathbf{V}_{l\mu}|\psi^0_{m\kappa'}\rangle\cdot\mathbf{M}^n[a_\kappa(\lambda)]a_{\kappa'}(\lambda)=v_\lambda(n)a_\kappa(\lambda). \tag{6.5.42}$$

Using the full perturbation series for $\psi(\mathbf{r})$, the representation of $\mathbf{M}^n$ in terms of $a_\kappa(\lambda)$ can be improved and gives a consistent set of equations in any order for the calculation of the $a_\kappa(\lambda)$, Q.E.D.

This theorem has an obvious consequence in regard to symmetry breaking: Since $\mathbf{M}^n\neq 0$ means $\mathbf{R}^n\neq\mathbf{R}^0$, in general, a symmetry breaking will occur as the symmetry point group with respect to $\mathbf{R}^n$ will differ from that with respect to $\mathbf{R}^0$. The instability effect, however, is more difficult to describe in this representation as we work in a multidimensional space of high dimension. Nevertheless, as our model contains the molecular model as a special case, we expect that the set of equilibrium points $\mathbf{M}^n$ is not a discrete set, but is rather a continuous manifold $\{\mathbf{M}^n\}$ of points of equal energy which leads to the instability of the equilibrium state of the static adiabatic electron-lattice

configuration. Since such instabilities have not yet been investigated for the high dimensional configuration space, we assume that symmetry coordinates are introduced so that the further discussion can be performed in the low dimensional space of these coordinates as, for instance, in our example (6.5.33).
Concerning the phonon mode calculation, it follows from our example that the adiabatic energy $U_n(\mathbf{R})$ cannot be represented by a quadratic form of the normal coordinates as in the case of non-degenerate electron levels. This fact gives rise to the dynamical Jahn-Teller effect. Two ways are open to treat the dynamical Jahn-Teller effect:

α) The energy expression for degenerate electronic levels is divided into a quadratic form of the normal coordinates and an interaction term which contains the non-quadratic coordinate mixtures. While the former term is used for the phonon state calculation, the latter term is assumed to be responsible for irreversible transitions between the phonon states. If such transitions occur between levels which are originally degenerate, then they are occasionally called hopping processes. Due to the resonance character of these processes, additional corrections to the static energy levels have to be expected. Hence the dynamical Jahn-Teller effect in the irreversible picture results in hopping processes and energy splitting. In addition to the genuine Jahn-Teller terms, of course also the other non-radiative interaction terms contribute to these processes.
β) The adiabatic energy expression for degenerate electronic levels is not divided into different terms, but is used as a whole. Then one gets mixtures of the original symmetry coordinates which are called vibrons and which can be considered to be a new kind of quasiparticle. Instead of irreversible hopping motion such vibrons perform a wave packet like motion through the instability region.

In the literature the Jahn-Teller effect is frequently discussed by the simultaneous diagonalization of the entire crystal Hamiltonian (in the Jahn Teller approximation) with respect to electronic as well as to lattice states. In contrast to the state calculation techniques in the adiabatic coupling scheme, in this case no division between static and dynamical electron-lattice coupling is made and the non-adiabatic interaction terms as well as the phonon-phonon interaction terms resulting from degeneracy are incorporated into the pure state calculation, while in the adiabatic coupling scheme these interactions mix the adiabatic states in an irreversible way. The decision whether the adiabatic coupling scheme or the vibronic quasiparticle states have to be applied, rests finally on comparison with experiment. In the literature it is argued, cf. for instance Ham (1965), that, if the coupling between the electrons and the lattice displacements is sufficiently strong relative to the zero point energy of the associated vibrational modes, then the complex undergoes a static distortion to a new configuration of minimal energy, i.e., the adiabatic coupling scheme is applicable. If, however, the coupling is less strong, or if the zero point vibrational energy is comparable with the energy barrier separating equivalent configurations, no static distortions occur and the electron-lattice complex has to be treated in the quasiparticle picture. As we have consistently used the adiabatic coupling scheme, we prefer to apply this scheme also to Jahn-Teller calculations, i.e., to perform the calculations according to α). In this case care has to be taken with respect to the choice of the unperturbed adiabatic energy according to the peculiarities of this potential with respect to instability. Fortunately, the inclusion of anharmonic terms of the phonon potential at least partly removes this difficulty, as these terms reduce the continuous manifold of energetically equivalent equilibrium positions to a discrete set, Sturge (1967), for which a harmonic zero order Hamiltonian can be defined. This is illustrated by Figure 6.8.

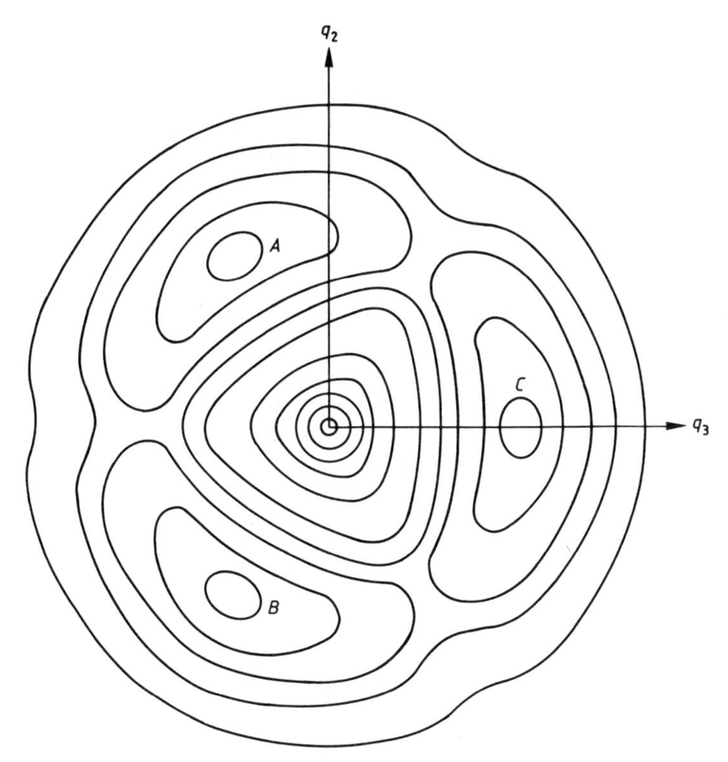

Figure 6.8:
Schematic contour map of $U(q_2, q_3)$ when anharmonic terms are included. The rotational symmetry is removed, the three ellipses A, B, C are potential minima; after Sturge (1967)

Concerning the vibron states, Renner (1934) proposed that instead of the ordinary adiabatic state functions (2.2.4) the states

$$\Psi_n(\mathbf{r},\mathbf{R}) = \sum_{\alpha} \chi_{n\alpha}(\mathbf{r},\mathbf{R})\,\varphi_m^n(\mathbf{R},\alpha) \tag{6.5.43}$$

should be used for degenerate electronic levels. Here α is the index of degeneracy of the electronic functions. But this formulation suffers from the fact, that for arbitrary $\mathbf{R} \neq \mathbf{R}^0$, the degeneracy of the electronic wave functions cannot properly be defined.

Hence in order to give this formulation a proper meaning, one has to perform a static and a dynamical adiabatic calculation, which leads to the techniques just discussed. In this case the static Jahn-Teller effect produces a level splitting of the electronic states which removes the electronic degeneracy and thus (under the assumption that the degeneracy is totally removed) the formulation (2.2.4) can be used again in connection with a properly defined zero order harmonic Hamiltonian. Hence (6.5.43) is neither applicable to adiabatic coupling calculations nor to vibron state calculations. For vibronic state calculations, (6.5.43) was therefore abandoned in favour of the formulation

$$\Psi_n(\mathbf{r},\mathbf{R})=\sum_{\alpha}\chi_{n\alpha}(\mathbf{r})\,\varphi_m^n(\mathbf{R},\alpha) \tag{6.5.44}$$

where now the set $\{\chi_{n\alpha}(\mathbf{r})\}$ is referred to the ideal lattice positions for which electronic degeneracy is well defined. Then from the vibron-state equation $K\Psi_n=E\Psi_n$, we obtain a set of equations

$$\sum_{\alpha}\langle\chi_{n\beta}|K|\chi_{n\alpha}\rangle\,\varphi_m^n(\mathbf{R},\alpha)=E\varphi_m^n(\mathbf{R},\alpha) \tag{6.5.45}$$

from which the corresponding phonon states can be derived.

Apart from molecular models, in practically all papers in this field, the authors did not perform ab initio calculations but used idealized effective model Hamiltonians which were derived by means of group theoretical selection rules and adaptive parameters. In particular in the last decade, Hamiltonians in the Fock-space representation were introduced which were stripped down to the essentials of the coupling of a degenerate electron system (with operators a_i^+, a_i) to phonons (with operators b_j^+, b_j) and which are for instance of the form

$$\begin{aligned} H=&\,\varepsilon\sum_{i=1}^{3}a_i^+a_i+\omega\sum_{j=1}^{3}b_j^+b_j\\ &+\kappa\sum_{i=1}^{3}(a_i^+a_{i+1}+a_{i+1}^+a_i)(b_{i+2}^++b_{i+2}) \end{aligned} \tag{6.5.46}$$

for triply degenerate Jahn-Teller levels, etc. Such systems were extensively studied by Wagner and coworkers, but it would exceed the scope of this book to give a detailed account of these field theoretic techniques. Hence we refer to the literature which we cite in the following.

Review articles about this topic were published by Longuet-Higgins (1961), Sturge (1967), Gebhardt (1969) and by Cianchi, Mancini and Moretti (1975); a book about the Jahn-Teller effect was published by Englman (1972). Concerning original papers, a clear separation between molecular physics and solid state physics is not possible as in solid state physics the molecular model is frequently used. Hence we cite papers from both fields. Moreover, a large number of very special calculation procedures has been applied which cannot be explained in detail in our review. Furthermore, no unique notation of the group representations is used. We cite the symbols as used by the authors and refer for comparison to Table 6.1. The following topics were treated: Static

J.T.- (Jahn-Teller) effect for the octahedral complex with secular equation resulting from first order perturbation theory, Van Vleck (1939). Application of the octahedral complex to the Copper fluosilicate spectrum, Abragam and Pryce (1950). Survey of the static problem for linear molecules and octahedral complexes, Öpik and Pryce (1957). Calculation of vibronic states with the formulation (6.5.44) for molecules with six identical nuclei and twofold electronic degeneracy, Moffitt and Liehr (1957), and the same procedure for octahedral complexes, Moffitt and Thorson (1957). Discussion of the dynamical J.T.-effect for a symmetrical non-linear molecule, Longuet-Higgins, Öpik, Pryce and Sack (1958). Vibronically induced absorption of forbidden infrared transitions for nonlinear molecules, Thorson (1958). Dynamical J.T.-effect in hydrocarbon radicals with formulation (6.5.43), Hobey and McLachlan (1960). Interaction of the vibrational and electronic motions in some simple conjugated hydrocarbons by means of a semiempirical formulation, Liehr (1961). Dynamical J.T.-effect in aromatic ions with formulation (6.5.43), McConnell and McLachlan (1961). A review of the consequences of the J.T.-theorem, Ham (1962). Adiabatic calculation of J.T.-states for $Cu^{2+}Y_6$ complexes and the influence on the paramagnetic resonance of Cu^{2+}, Avvakumov (1960). Studies of the J.T.-effect for molecules with functions of the kind (6.5.43), Child and Longuet-Higgins (1962), and Child (1963). Theory of the dynamical J.T.-effect with vibronic states of the kind (6.5.44) in the molecular model, Slonczewski (1963). EPR-spectrum of Cu^{2+} ion complexes with vibronic states of the kind (6.5.44), Bersuker (1963). Topological aspects of the conformational stability problem for symmetric molecules with degenerate electronic states, Liehr (1963). Vibronic interactions in molecules with a fourfold symmetry axis, Hougen (1964). Adiabatic calculation of the dynamical J.T.-effect in octahedrally coordinated d^9 ions, O'Brien (1964). Low temperature polarization of the radiation from isotropic impurity centers and the J.T.-effect by phenomenological analysis, Kristofel (1965). Vibronic effects in paramagnetic resonance spectra in the molecular model, Ham (1965). J.T.-effect in the $^4T_{2g}$ excited state of V^{2+} in MgO by phenomenological analysis, Sturge (1965). Adiabatic calculation for bound impurity electrons coupled to localized and nonlocalized modes and corresponding vibronic spectra for Sm^{2+}, Eu^{2+} and Yb^{2+} in alkali halides, Bron and Wagner (1965), Wagner and Bron (1965), Bron (1965). Vibrational coupling of nearly degenerate electronic states in the adiabatic coupling scheme for Eu^{2+} ions in NaCl, KCl and RbCl, Bron and Wagner (1966). Dynamical J.T.-effect in alkali halide phosphors containing heavy metal ions in the adiabatic coupling scheme and the molecular model, Toyozawa and Inoue (1966). J.T.-distortions and inversion splitting for transition metal complexes and local centers in crystals with degenerate electronic terms in the adiabatic coupling scheme and the molecular model, Bersuker and Vekhter (1966). J.T.-effects in the electron spin resonance spectra of ions with spin 1 by means of vibron states and a Fock-Hamiltonian of the kind (6.5.46), Stevens and Persico (1966). J.T.-coupling of $3d^6$ ions in a cubic crystal in the adiabatic coupling scheme and the molecular model, O'Brien (1965). Vibrational levels of the ^2D-E state of Ce^{3+} in CaF_2 with functions of the kind (6.5.44) in the molecular model, Struck and Herzfeld (1966). Shape of F-aggregate bands in KCl and KBr with phenomenological analysis, Warren (1967). J.T.-effect in the groundstate of the V^{2+} ion in CaF_2 in the adiabatic coupling scheme and the molecular model, Aminov and Malkin (1967). J.T.-effect on impurity centers in semiconductors by means of a continuum description of the lattice, Bir (1967). Optical bands in paramagnetic crystals with degenerate impurity terms and wave functions of the kind (6.5.43), Tsukerblat (1967). Theory of J.T.-induced optical transition in the adiabatic coupling scheme, Kamimura and Yamaguchi (1968). Dynamic J.T.-effect in octahedrally coordinated d^1 impurity systems, Macfarlane, Wong and Sturge (1968). Effect of linear J.T.-coupling on paramagnetic resonance in a ^2E state in the adiabatic coupling scheme and a molecular model, Ham (1968). J.T.-splitting of electronic spin levels with a coupled spin-phonon Hamiltonian in Fock-representation and field theoretic calculational technique of vibronic response for Ni^{2+} in MgO, Böttger (1968). Polarization of luminescence and its dependence on the frequency of exciting light for J.T.-states of impurity ions in the adiabatic coupling scheme, Vekhter and Tsukerblat (1968). Dynamic J.T.-effect in trigonally distorted cubic systems with states of the kind (6.5.44) in the molecular model, Stephens (1969). Vibronic state calculations of J.T.-effects in the far-infrared, EPR and Mössbauer spectra of Fe^{2+} in MgO, Ham, Schwarz and O'Brien (1969). J.T.-effect of octahedrally coordinated $3d^4$ ions with vibronic state calculation and elastic continuum corrections to the molecular model, Fletcher and Stevens (1969). Derivation of an effective Hamiltonian for the J.T.-coupling of T_{1g} and T_{2g} ions, Stevens (1969). Dynamic J.T.-effect in an orbital triplet state coupled to both E_g and T_{2g} vibrations in the molecular model with vibronic state calculation, O'Brien (1969). Dynamic J.T.-effect in the orbital ^5E state of Fe^{2+} in CdTe in the molecular model with vibronic state calculation, Vallin (1970). Influence of the J.T.-effect on the U-center local mode absorption with Fock-space Hamiltonian of the type (6.5.46) and evaluation by Green function techniques, Boese and Wagner (1970). A method of moments for the optical absorption bands of J.T.-vibronic states with wave functions of the kind (6.5.44) as well as with Fock-space representation, Wagner (1970). Interaction between two distant $T_{1g}(T_{2g})$ J.T.-ions at a phenomenological level, Novak and Stevens (1970). The J.T.-effect for an F^+-center in CaO at a phenomenological level, Hughes (1970). Vibronic coupling in semiconductors treated phenomenologically, Morgan (1970). Manifestations of the dynamic J.T.-effect in trigonally distorted defect centers in a molecular model, Hagston (1970). Level shifting effect without removal of electronic degeneracy, Shrivastava (1970). Lattice distortion and polarisation near color centers with inclusion of J.T.-effects by means of the Kanzaki method, Stoneham and Bartram (1970). Optical band shape of an orbital triplet vibronically coupled with a phonon triplet, Englman, Caner and Toaff (1970). Electronic structure and dynamic J.T.-effect of R'-centers ($\equiv$three adjacent F-centers forming an equilateral triangle in a (111) plane plus an extra electron) in alkali halides based on a molecular model, Inoue, Sati and Wang (1970). Calculation of the optical absorption between J.T.-levels with vibronic states of the kind (6.5.44) and subsequent evaluation in the interaction representation, Wagner (1971). Lattice-ion interactions of Fe^{2+} ions in a tetrahedral environment by vibronic state calculation in a molecular model, Bhattacharyya (1971). J.T.-effect in an electronic p-state coupled to E_g and T_{2g} vibrations, O'Brien (1971). J.T.-interactions and $3d^9$ ions at tetrahedral sites in a molecular model with effective spin Hamiltonian, Bates and Chandler (1971) (1973). J.T.-effect of an orbital triplet coupled to both E_g

and T_{2g} vibrations in a molecular model, Romestain and d'Aubigné (1971). Infrared absorption and luminescence spectra of Fe^{2+} in cubic ZnS and the role of the J.T.-coupling treated phenomenologically, Ham and Slack (1971). Dynamical J.T.-effects for a V^{3+} ion in MgO by means of a molecular model, Ray (1972). J.T.-effect with inclusion of anharmonic lattice potentials in a molecular model with vibronic state calculation, Dixon and Smith (1972). Adiabatic potential surfaces for the J.T.-effect of Tl in KJ in a molecular model, Ranfagni (1972). Dynamic J.T.-effect with coupling to many lattice modes of different frequencies in Fock-space representation, O'Brien (1972). Calculation of the vibronic groundstate for a J.T.-system with many different lattice modes in Fock-space representation, Fletcher (1972). J.T.-effect due to electron-photon coupling instead of electron-phonon coupling, Tsukerblat, Rosenfeld and Vekhter (1972). Dynamical J.T.-interaction in optical and paramagnetic resonance spectra with application to a Cu impurity in II–VI semiconductors in a molecular model, Yamaguchi and Kamimura (1972). Optical absorption bands of J.T.-systems with orbital doublets in a semiclassical approximation, Vekhter, Perlin, Polinger, Rosenfeld and Tsukerblat (1972). Dynamical J.T.-effect in a relaxed excited state of the F-center with a molecular model, Iida, Kurata and Muramatsu (1972). Solution of a molecular vibronic model for the relaxed excited state of the F-center, Ham (1972) (1973), Ham and Grevsmühl (1973). Exchange interactions between J.T.-ions, Passeggi and Stevens (1973) (1974). Reduction factors for the J.T.-effect in solids, Halperin and Englman (1973). Dynamical J.T.-effect for an electronic E state coupled to a phonon continuum in Fock-space representation and with Green function techniques, Gauthier and Walker (1973). Influence of the J.T.-effect upon EPR spectra of an ^2E_g state in a molecular model, Vekhter (1973). Dynamical resonance phenomena in non-adiabatic vibronic systems with Fock-space Hamiltonians of the kind (6.5.46), Wagner (1973), Sigmund and Wagner (1973). The linear J.T.-effect for an orbital triplet with wave functions of the kind (6.5.44), Bersuker and Polinger (1973). Calculation of vibronic selfenergies of coupled impurity centers by means of canonical transformations applied to the Hamiltonian in Fock-space representation, Kristoffel, Sigmund and Wagner (1973). Effect of vibronic interactions on the energy spectrum of an impurity pair of J.T.-ions, Tsukerblat and Vekhter (1973). A simple proof of the J.T.-theorem, Raghavacharyulu (1973). Multiphonon non-radiative relaxation in impurity-phonon systems caused by the static J.T.-effect, Perlin, Tsukerblat and Perepelitsa (1973). Derivation of effective Hamiltonians for impurity ions coupled to a large number of lattice modes with different frequencies, Englman and Halperin (1973). Influence of isotropic mass changes in the neighbourhood of J.T.-ions on their spectrum, Halperin and Englman (1973). Effects of electron-phonon interaction on luminescence spectra of rare earth and transition metal impurity ions in crystals with inclusion of J.T.-levels, Perlin (1973). Theory of the J.T.-effect on the optical spectra of degenerate excitons based on an exciton-phonon Hamiltonian in Fock space representation with quasimolecular localized phonon modes at each lattice site, Sakoda and Toyozawa (1973). Application of a molecular model to stress-induced polarization of F-center emission, Muramatsu (1973). An extensive analysis of the application of the theory of Lie groups to the description of the J.T.-effect, Judd (1974). Theory of the optical band shape for $A_{1g} \rightarrow T_{1u}$ transitions, Nasu and Kojima

(1974), Nasu (1975). Dynamical J.T.- and reorientation effects in the EPR spectrum of O^- in CaF_2 in a molecular model, Bill and Silsbee (1974). Optical response function of a dynamical J.T.-system, Mulazzi and Terzi (1974). Dynamic J.T.-effect of Cu^{2+} in cubic ZnS, Sauer, Scherz and Maier (1974). J.T.-effect on excited orbital triplet states of $3d^n$ ions in eightfold cubic coordination in a molecular model, Ulrici (1974). Zero phonon lines of singlet-doublet transitions in systems with a static J.T.-distortion with wave functions of the type (6.5.44), Polinger, Rosenfeld, Vekhter and Tsukerblat (1974). Adiabatic potentials for an octahedral complex with $^3T_1$ and $^1T_1$-electronic states coupled to E_g vibrational modes, Ranfagni and Viliani (1974), and for Tl impurities in KJ, Ranfagni, Pazzi, Fabeni, Viliani and Fontana (1974). Vibronic levels for triply degenerate electron states and vibrational states with Fock-space Hamiltonian of the type (6.5.46), Schultz and Silbey (1974). Derivation of effective Hamiltonians for magnetic ions with orbital triplets in trigonal symmetry, Abou-Ghantous, Bates and Stevens (1974), and in cubic crystals, Abou-Ghantous, Bates, Chandler and Stevens (1974), Bates, Chandler, and Stevens (1974). Variational treatment of the multimode coupling of a J.T.-impurity ion with wave functions of the type (6.5.44), Halperin and Englman (1974). Dynamical J.T.-effects in $^4T_{1g}$ and $^4T_{2g}$ orbital triplets of a Co^{2+} ion in an MgO crystal, Ray and Regnard (1974). Quadratic J.T.-effect of an orbital triplet state coupled to a T_{1u} vibrational mode, Lacroix (1974). Adiabatic potential for an orbital T term coupled to all active e and t_2 displacements and related physical effects, Bersuker and Polinger (1974). Local dielectric modes arising from the dynamical J.T.-effect of impurity centers, Vekhter, Polinger, Rosenfeld and Tsukerblat (1974). Coherent states and the J.T.-effect, Judd and Vogel (1975). Phonon scattering at local vibronic states by means of Green function techniques, Halperin and Englman (1975). J.T.-effect in ESR-spectra of Cr^{2+} in II–VI semiconductors, Bhattacharyya (1975). Comparison of the molecular cluster model with the phonon model for J.T.-active impurities in crystals, Ray, Ray and Sangster (1975). First order calculation of spin Hamiltonian parameters for MX_4^{2-} J.T.-complexes, Moreno and Barriuso (1975). Localized modes coupled to J.T.-defects, Halperin (1975). Magnetic susceptibility and the J.T.-effect for a T_g ions coupled to E_g displacements, Chandler (1975). Dynamic, intermediate, and static J.T.-effect in the EPR spectra of ^2E orbital states by means of a molecular model applied to Ag^{2+} in CaO and MgO and to Cu^{2+} in CaO, Reynolds and Boatner (1975), and for ^2E orbital states in cubic symmetry, Setser, Barksdale and Estle (1975). Symmetric J.T.-distortions around O_2^- in alkali halides, Lowther (1975). Zero-phonon line of J.T.-systems for weak electron-phonon coupling, Tsukerblat, Rozenfeld, Polinger and Vekhter (1976). Moments and absorption line of an antiresonant electron-phonon system in the strong coupling limit with a Hamiltonian of type (6.5.46), Sigmund (1976). Lie groups and the J.T.-effect for color centers, Judd (1976). Calculation of moments and of the zero phonon line for $T-t$ J.T.-systems with a Hamiltonian of type (6.5.46), Sigmund (1976). Quadratic J.T.-effect in octahedral clusters, Bacci (1976). Vibronic model for Au^- in KCl, Lemoyne, Duran, Billardon and Dang (1976), Lemoyne, Duran and Badoz (1977). Relaxed excited states determined by the J.T.-effect in Tl^+ centers in alkali halides, Fukuda, Matsushima and Masunaga (1976). Static J.T.-effect for an E state of an ion in a tetragonal crystal field, Bir (1976).

Structure of the J.T.-induced B absorption band of Tl^+-type centers in alkali halides, Matsushima and Fukuda (1976). Role of the J.T.-effect in the excited states of Tl centers in KBr, Giorgianni, Mondio, Saitta and Vermiglio (1976). J.T.-effect in $^4T_1(1)$ and $^4T_2(1)$ states of tetrahedrally coordinated Mn^{2+}, Koidl (1976). Dichroism effects in the absorption spectra of J.T.-centers, Kushkulei, Perlin, Tsukerblat and Engelgardt (1976). Microscopic determination of the J.T.-interaction energy of a Co^{2+} in ZnO, Vasilev, Malkin, Natadze and Ryskin (1976). Dynamic J.T.-effect for an orbital doublet term coupled to phonon modes with non-vanishing dispersion, Rozenfeld and Polinger (1976). An operator method for optical lineshape calculations in non-adiabatic electron-phonon systems with Hamiltonians of type (6.5.46), Sigmund and Wagner (1976). Fractional parentage coefficients for R_5 and their application to $T\otimes(\varepsilon+\tau_2)$ J.T.-systems such as F and F^+ centers, O'Brien (1976). Relationship among J.T.-reduction factors for a Γ_8 state in cubic symmetry, Ham, Leung and Kleiner (1976). Molecular vibronic model of the relaxed excited state of the F-center in alkali halides, Kayanuma and Toyozawa (1976), Kayanuma (1976). Approximation of resonance structure of J.T.- and Fano-systems, Sigmund (1976). Calculations of the low-lying energy levels of $E-e$ J.T.-systems, Rueff and Wagner (1977). Application of canonical transformations to an antiresonant electron-phonon system, Sigmund (1977). Coupling of a T_g ion to E_g modes with dispersion and calculation of levels and distortions, Steggles (1977), Bates and Steggles (1977). Resonance scattering of phonons at trigonal J.T.-centers with a Hamiltonian of type (6.5.46), Rueff, Sigmund and Wagner (1977). Calculation of the internal dynamics of a J.T.-system of $E-e$ type by time-dependent occupation numbers, Sigmund (1977). Optical response studies of $E-e$ J.T.-systems, Rueff and Sigmund (1977). Dynamical treatment of quenching factors for the linear $E-e$ J.T.-problem, Sigmund, Wagner, Birkhold (1977). Quadratic J.T.-effect and tunneling splitting in octahedral clusters applied to V^{2+} in MgO, Bacci (1977). Tunneling splitting in a $T\otimes\varepsilon_g$ J.T.-problem with spin-orbit interaction, Ranfagni and Viliani (1977). Multimode J.T.-coupling for an orbital triplet in cubic crystals, Biernacki (1977). Orbital ordering in crystals with orbitally degenerate centers due to direct Coulomb interaction, Ostrovskii and Kharkyanen (1977). J.T.-effect in a Γ_8 quartet with equal coupling to ε and τ_2 vibrations, Pooler and O'Brien (1977). Systems with nearly dynamic J.T.-effect, Natadze and Ryskin (1977). Vibronic problem of the higher excited states of the F-center in alkali halides by means of a polaron-phonon Hamiltonian in Fock-space representation, Kayanuma and Kondo (1977) (1978). Vibronic theory of magnetic properties of the relaxed excited states of the F-center, Imanaka, Iida and Ohkura (1978). Depression of vibronic levels and transition from the dynamic to static J.T.-effect in the $^4T_1$ multiplet applied to Co^{2+} in ZnSe, Uba and Baranowski (1978). J.T.-effect in EPR spectra with a multistate theory for ^2E orbital states in cubic symmetry, Setser and Estle (1978). J.T.-effect in the fluorescent level of Mn^{2+} in ZnSe and ZnS, Parrot, Naud, Porte, Fournier, Boccara and Rivoal (1978). Canonical transformation approach to the linear J.T.-effect in $E-e$ systems, Barentzen and Polansky (1978). J.T.-effect on the optical absorption spectra of MnO, Takaoka, Suzuki and Motizuki (1978). J.T.-coupling of Cr^{2+} in ZnS, ZnSe and ZnTe, Biernacki (1978). Review of recent developments with respect to the J.T.-effect in ionic crystals, Duran (1978).

The Jahn-Teller effect does not only occur for the coupling of phonons to electrons located at impurity center ions, but also for the electron-phonon coupling in the regular lattice. The symmetry breaking and the instability connected with the latter cause a change in the symmetry properties of the entire crystal and thus lead to phase transitions of the material. We cite some recent papers about this topic, which is called the cooperative J.T.-effect. A review was given by Gehring and Gehring (1975). Concerning original papers the following problems were treated: Structural phase transitions induced by a pseudo J.T.-effect, Halperin (1973). Orbital magnetism and the structural phase transitions in crystals with J.T.-ions, Vekhter and Kaplan (1973) (1974). Dynamics of J.T. phase transitions, Pytte (1973). Statistical mechanics of the collective J.T. phase transition, Feder and Pytte (1973). Cooperative J.T.-effect for several active ions in the unit cell, Gehring (1974). Anharmonic effects in a J.T. phase transition, Cowan and Zuckermann (1975). Theory of sidebands in a cooperative J.T.-material, Gehring and Laugsch (1976). Cooperative J.T.-systems with locally triply degenerate electron states, Brühl and Wagner (1976). Renormalization group approach to cooperative J.T.-systems, Brühl and Sigmund (1977). Cooperative J.T.-effect and related problems, Elliott (1977).

6.6 Space group representations

The ideal crystal is defined to be a strictly periodic structure of atoms or ions in space. To apply this definition, the meaning of the periodic structure has to be explained. This can be done on a purely classical level. We assume the atoms or ions to be at rest at the ideal position $\{\mathbf{R}_k = \mathbf{R}_k^0\}$. Then the structure of the crystal is defined by the set of lattice vectors $\{\mathbf{R}_k^0\}$ and the periodicity can be defined by the application of space group elements.

Def. 6.25 : A space group is a discrete subgroup of the real affine group. Its translation elements $\{0|\mathbf{T}\}$ are defined by $\mathbf{T} = \sum_{i=1}^{3} n_i \mathbf{a}_i$ (n_i integers) and form an invariant subgroup T of the space group.

Def. 6.26 : If the set of lattice vectors $\{\mathbf{R}_k^0\}$ is only permuted under the application of a space group, then the array $\{\mathbf{R}_k^0\}$ defines an ideal crystal structure.

By this definition the geometrical investigation of ideal crystal structures is reduced to that of the investigation of space groups. As the translation group is a noncompact group, any point in the crystal can be transformed to infinity. Hence, the ideal crystal has to be infinite. Quite apart from imperfections, such crystals never occur in reality. Hence, the ideal crystal is an idealization. However, as will be demonstrated in the following chapters, the ideal crystal plays an important role in the theoretical description even of imperfect crystals.
In order to prepare the way for application later on, we will discuss some properties of space groups in the following. Instead of giving proofs, we refer to the corresponding literature, cf. Streitwolf (1967), Cornwell (1969).

Def. 6.27: The elements of the invariant subgroup T of a space group G are called primitive translations. The set of elements of T is called the Bravais lattice of G.

The Bravais lattice is characterized by the set of formal lattice points $\{\mathbf{R} = \sum_{i=1}^{3} n_i \mathbf{a}_i,$ n_i integers$\}$ and must not be confused with the real lattice points $\{\mathbf{R}_k^0\}$. It is a mathematical tool.
By the Bravais lattice the space can be divided into congruent regions, which are called elementary cells. Two points $\mathbf{r}$ and $\mathbf{r}'$ are called equivalent if they are connected by $\mathbf{r}' = \mathbf{r} + \mathbf{R}, \mathbf{R} \in T$. Any elementary cell contains only inequivalent points. However, the construction of elementary cells is not unique. The same lattice can be produced by infinitely many different elementary cells.

Def. 6.28: The symmetrical elementary cell is called a Wigner-Seitz cell. It consists of all points which are nearer to a definite lattice point than to all other points.

Concerning the other elements of a space group G the following theorem holds.

Theorem 6.44: The rotational elements of a space group G form itself a group G_0, which leaves the corresponding Bravais lattice invariant, if they are combined with appropriate translations from T.

These simple properties of a space group have far-reaching consequences. It follows that only 14 different Bravais lattices exist and that the corresponding point groups G_0 must contain only rotations through multiples of $\frac{1}{3}\pi$ or $\frac{1}{2}\pi$ and inversions. Due to these restrictions only 32 crystallographic point groups exist. Not any point group is compatible with any Bravais lattice. Hence, only special combinations occur. Space groups having the same point group G_0 are said to belong to the same crystal class. There exist 230 different space groups. An important division is the distinction between symmorphic and non-symmorphic space groups.

Def. 6.29: A space group G is called symmorphic if its corresponding point group G_0 is a subgroup of G.

If G does not exhibit this property, it is called non-symmorphic. Hence, in general, G_0 is a group but not a subgroup of G. There are 73 symmorphic space groups. Symmorphic groups can equivalently be characterized by the property that G_0 leaves the Bravais lattice invariant without the application of additional translations from T.
For applications we are interested in representations of these groups. In this section, representations will be discussed in electron state space. As we construct the many-electron states by appropriate combinations of one-electron orbitals, the symmetry properties of the many-electron states depend on those of the one-electron orbitals. Hence, it is sufficient to consider representations of the space groups for one-particle states. It is convenient (but not necessary) to assume that the one-particle states are solutions of a corresponding one-particle Schrödinger equation

$$H(\mathbf{r})\varphi(\mathbf{r}) = E\varphi(\mathbf{r}) \tag{6.6.1}$$

which is invariant under a definite space group G, i.e. $[G,H]_- = 0$. It is further convenient to assume that for each energy eigenfunction $\varphi(\mathbf{r})$, cyclic boundary conditions hold, i.e., that

$$\varphi(\mathbf{r}) = \varphi(N_1\mathbf{a}_1\mathbf{r}) = \varphi(N_2\mathbf{a}_2\mathbf{r}) = \varphi(N_3\mathbf{a}_3\mathbf{r}) \tag{6.6.2}$$

is valid, where N_1, N_2, N_3 are very large positive integers, and $\mathbf{a}_1, \mathbf{a}_2, \mathbf{a}_3$ are the basic lattice vectors of T. This assumption implies that the infinite crystal is assumed to be a periodic array of parallelepipeds and that the physical situation is identical in corresponding points of different blocks. This is a physically reasonable assumption as the situation in the interior of a crystal must be almost independent on what happens at the boundaries and hence on the boundary conditions. The eigenfunctions of (6.6.1) then have to be normalized over the parallelepiped. To give an appropriate description of the eigenfunctions of (6.6.1) we introduce the reciprocal lattice vectors $\mathbf{b}_1, \mathbf{b}_2, \mathbf{b}_3$ by

$$\mathbf{a}_i \cdot \mathbf{b}_j = 2\pi\delta_{ij}, \qquad 1 \leqslant i,j \leqslant 3 \tag{6.6.3}$$

so that explicitly

$$\mathbf{b}_i = 2\pi\mathbf{a}_l \times \mathbf{a}_j(\det |\mathbf{a}_1\mathbf{a}_2\mathbf{a}_3|)^{-1}, \quad i,l,j = \text{cyclic perm. of } 1,2,3 \tag{6.6.4}$$

and define the wave vector

$$\mathbf{k} := k_1\mathbf{b}_1 + k_2\mathbf{b}_2 + k_3\mathbf{b}_3 \tag{6.6.5}$$

where $k_j = p_j/N_j$ with p_j an integer. Then the following theorem, derived by Bloch (1928) holds

Theorem 6.45: The solutions of (6.6.1) are given by

$$\varphi(\mathbf{k},\mathbf{r}) := \exp(i\mathbf{k} \cdot \mathbf{r})u(\mathbf{k},\mathbf{r}) \tag{6.6.6}$$

where $u(\mathbf{k},\mathbf{r})$ is an invariant function under T, and the energy values depend on $\mathbf{k}$

$$H(\mathbf{r})\varphi(\mathbf{k},\mathbf{r}) = E(\mathbf{k})\varphi(\mathbf{k},\mathbf{r}) \tag{6.6.7}$$

while $\varphi(\mathbf{k},\mathbf{r})$ spans a one-dimensional representation of T.

To avoid superfluous states, the domain of $\mathbf{k}$ has to be restricted. In the reciprocal space we define in analogy to the Bravais lattice a reciprocal lattice by

$$\mathbf{K}_m := m_1\mathbf{b}_1 + m_2\mathbf{b}_2 + m_3\mathbf{b}_3 \tag{6.6.8}$$

where $m_i, i = 1,2,3$ are integers. Then

$$\exp(i\mathbf{K}_m \cdot \mathbf{T}_n) = 1 \tag{6.6.9}$$

and for $\mathbf{k}' = \mathbf{k} + \mathbf{K}_m$ we obtain, by applying an element $\mathbf{T}_n$ of T to $\varphi(\mathbf{k},\mathbf{r})$

$$\mathfrak{T}_n\varphi(\mathbf{k},\mathbf{r}) = \exp(-i\mathbf{k} \cdot \mathbf{T}_n)\varphi(\mathbf{k},\mathbf{r}) \tag{6.6.10}$$

$$\mathfrak{T}_n\varphi(\mathbf{k}',\mathbf{r}) = \exp(-i\mathbf{k}' \cdot \mathbf{T}_n)\varphi(\mathbf{k}',\mathbf{r}) = \exp(-i\mathbf{k} \cdot \mathbf{T}_n)\varphi(\mathbf{k}',\mathbf{r}).$$

Hence, $\varphi(\mathbf{k},\mathbf{r})$ and $\varphi(\mathbf{k}',\mathbf{r})$ belong to the same (irreducible) representation of T and for an investigation of these representations it is sufficient to consider only the domain of $\mathbf{k}$ mod $\mathbf{K}_m$. This leads to the definition of the k-space analogy to the Wigner-Seitz cell

Def. 6.30 : The symmetrical elementary cell in k-space is called Brillouin zone. It consists of all points $\mathbf{k}$ which are nearer to $\mathbf{k}=0$ than to any other lattice point $\mathbf{K}_m$.

To any $\mathbf{k}$-vector in the Brillouin zone corresponds an infinite set of energy eigenvalues $E_n(\mathbf{k})$, which leads to the so-called energy bands if $\mathbf{k}$ is varied. We now extend these considerations to include the complete space group representations.

a) Symmorphic space groups

Def. 6.31 : The point group $G_0(\mathbf{k})$ is that subgroup of the point group G_0 of the corresponding space group G that consists of all rotations $\{\mathscr{A}|0\}$ of G_0 which rotate $\mathbf{k}$ into itself or into $\mathbf{k}+\mathbf{K}_m$ with arbitrary $\mathbf{K}_m$.

Def. 6.32 : If $G_0(\mathbf{k})$ is the trivial group consisting only of the identity transformation $\{\mathbb{1}|0\}$, then $\mathbf{k}$ is called a general point of the Brillouin zone. If $\mathbf{k}$ has a value such that $G_0(\mathbf{k})$ is larger than those corresponding to all neighbouring points of the Brillouin zone, then $\mathbf{k}$ is called a symmetry point. If all the points on a line or in a plane have the same nontrivial $G_0(\mathbf{k})$, then this line or plane is called a symmetry axis or plane.

If the components of $\mathbf{k}$ in the reciprocal lattice are denoted by $\mathbf{k}=(k_1,k_2,k_3)$, the symmetry points, axis and planes can generally be classified and are given in the Table 6.7a, b, c.
If $\{\mathscr{A}|0\}$ is a member of G_0, but not a member of $G_0(\mathbf{k})$, then $\mathscr{A}\mathbf{k}$ is not equal or equivalent to $\mathbf{k}$. Let $\mathbf{k}_2, \mathbf{k}_3, \ldots$ be a set of distinct and nonequivalent vectors which we obtain by application of G_0 to $\mathbf{k}$ and which we denote by $\mathbb{M}(\mathbf{k})$. Then we introduce

Def. 6.33: The set $\mathbb{M}(\mathbf{k})$ (or any equivalent set) is called a star of $\mathbf{k}$.

The link between the irreducible representations of the space group G and those of T and $G_0(\mathbf{k})$ is then provided by the following theorem, cf. Cornwell (1969):

Theorem 6.46: Suppose that $\mathbf{k}$ is an allowed $\mathbf{k}$-vector and $\varphi_s^p(\mathbf{k},r)$ is a Bloch function with wave vector $\mathbf{k}$, which transforms as the s-th row of the irreducible unitary representation Γ_p of $G_0(\mathbf{k})$ and this representation being of dimension l. Suppose that the star of $\mathbf{k}$ consists of the $M(\mathbf{k})$ vectors $\mathbf{k}_1(=\mathbf{k}), \mathbf{k}_2, \ldots$ with which the $M(\mathbf{k})$ rotations $\mathscr{A}_j$ are associated. Then the set of $l \cdot M(\mathbf{k})$ functions

$$\mathfrak{G}(\{\mathscr{A}_i|0\})\,\varphi_s^p(\mathbf{k},\mathbf{r}) \qquad \begin{array}{l} i=1,2\ldots M(\mathbf{k}) \\ s=1,2\ldots l \end{array} \tag{6.6.11}$$

forms a basis for an $l \cdot M(\mathbf{k})$-dimensional unitary irreducible representation of the

Table 6.7a:
Symmetry points. The abstract point group is given in brackets. After Streitwolf (1967).

Point	$\mathbf{k}$	$G_0(\mathbf{k})$	Elements α of $G_0(\mathbf{k})$		
			$\alpha \in T_d$	$\alpha \notin T_d$	
Γ	0	$O_h(O \times C_2)$	ε	i	
			$6\sigma_{4i}^{\pm 1}$	$6\delta_{4i}^{\pm 1}$	$i=x,y,z$
			$6\varrho_i$	$6\delta_{2i}$	$i=xy,yz,zx$
					$x\bar{y},y\bar{z},z\bar{x}$
			$8\delta_{3i}^{\pm 1}$	$8\sigma_{6i}^{\pm 1}$	$i=xyz,x\bar{y}\bar{z},\bar{x}y\bar{z},\bar{x}\bar{y}z$
			$3\delta_{2i}$	$3\varrho_i$	$i=x,y,z$
X	$\frac{\pi}{a}(0,1,0)$	$D_{4h}(D_4 \times C_2)$	ε	i	
			$2\sigma_{4y}^{\pm 1}$	$2\delta_{4y}^{\pm 1}$	
			δ_{2y}	ϱ_y	
			$\varrho_{zx},\varrho_{z\bar{x}}$	$\delta_{2zx},\delta_{2z\bar{x}}$	
			δ_{2z},δ_{2x}	ϱ_z,ϱ_x	
L	$\frac{\pi}{2a}(1,1,1)$	$D_{3v}(D_6)$	ε	i	
			$2\delta_{3xyz}^{\pm 1}$	$2\sigma_{6xyz}^{\pm 1}$	
			$3\varrho_i$	$3\delta_{2i}$	$i=x\bar{y},y\bar{z},z\bar{x}$
W	$\frac{\pi}{2a}(1,2,0)$	$D_{2d}(D_4)$	ε	δ_{2yz}	
			σ_{4x}	ϱ_y	
			δ_{2x}	δ_{2yz}	
			σ_{4x}^{-1}	ϱ_z	

corresponding symmorphic space group G. Moreover, all the irreducible representations of G may be obtained in this way, by working through all the irreducible representations of $G_0(\mathbf{k})$ for all the allowed k-vectors that are in different stars.

b) Non-symmorphic space groups

Def. 6.34: The group $G(\mathbf{k})$ is that subgroup of the space group G that consists of all transformations $\{\mathscr{A}|\mathbf{T}\}$ having the property $\mathscr{A}\mathbf{k}=\mathbf{k}+\mathbf{K}_m$ with arbitrary $\mathbf{K}_m$.

$G(\mathbf{k})$ contains translations as well as rotations. In particular, $G(\mathbf{k})$ contains the group T of primitive translations as a subgroup.

Def. 6.35: If $G(\mathbf{k})$ is merely equal to T, then $\mathbf{k}$ is said to be a general point of the Brillouin zone. If $\mathbf{k}$ is such that $G(\mathbf{k})$ is a larger group than those corresponding to all neighbouring points, then $\mathbf{k}$ is called a symmetry point of the Brillouin zone. If all points on a line or plane have the same $G(\mathbf{k})$ which is larger than T, then this line or plane is called a symmetry axis or plane.

Table 6.7b:
Symmetry lines. The abstract point group is given in brackets. After Streitwolf (1967).

Lines	$\mathbf{k}$	$G_0(\mathbf{k})$	Elements α of $G_0(\mathbf{k})$	
			$\alpha \in T_d$	$\alpha \notin T_d$
Δ	$\frac{\pi}{a}(0,\lambda,0)$	$C_{4v}(D_4)$	ε	$2\delta_{4y}^{\pm 1}$
			δ_{2y}	ϱ_z, ϱ_x
			$\varrho_{zx}, \varrho_{z\bar{x}}$	
Λ	$\frac{\pi}{2a}(\lambda,\lambda,\lambda)$	$C_{3v}(D_3)$	ε	
			$2\delta_{3xyz}^{\pm 1}$	
			$3\varrho_i \quad i = x\bar{y}, y\bar{z}, z\bar{x}$	
Σ, K	$\frac{3\pi}{4a}(\lambda,\lambda,0)$	$C_{2v}(D_2)$	ε	δ_{2xy}
			$\varrho_{x\bar{y}}$	ϱ_z
Z	$\frac{\pi}{2a}(\lambda,2,0)$	$C_{2v}(D_2)$	ε	ϱ_y
			δ_{2x}	ϱ_z
S	$\frac{\pi}{a}\left(\frac{1}{4}\lambda,1,\frac{1}{4}\right)$	$C_{2v}(D_2)$	ε	δ_{2zx}
			$\varrho_{z\bar{x}}$	ϱ_y
Q	$\frac{\pi}{2a}(1,2-\lambda,\lambda)$	$C_2(C_2)$	ε	δ_{2yz}

Table 6.7c:
Symmetry planes. The abstract point group is given in brackets. After Streitwolf (1967).

Planes	$\mathbf{k}$	$G_0(\mathbf{k})$	Elements α of $G_0(\mathbf{k})$
		$\alpha \in T_d$	$\alpha \notin T_d$
$\Gamma\Lambda\Delta$	$C_8(C_2)$	ε	
		$\varrho_{z\bar{x}}$	
$\Gamma\Sigma\Lambda, KL$	$C_8(C_2)$	ε	
		$\varrho_{x\bar{y}}$	
$\Gamma\Delta\Sigma, KW$	$C_8(C_2)$	ε	ϱ_z
WUX, WU	$C_8(C_2)$	ε	ϱ_y

Def. 6.36: Let $\{\mathscr{A}_1|\mathbf{T}_1\}\equiv\{\mathbb{1}|0\}$, $\{\mathscr{A}_2|\mathbf{T}_2\},\ldots$ be a set of coset representatives for the decomposition of the space group G into left cosets with respect to $G(\mathbf{k})$. Then the set of wave vectors $\mathscr{A}_1\mathbf{k}=\mathbf{k}_1$, $\mathscr{A}_2\mathbf{k}=\mathbf{k}_2,\ldots$ is called the star of $\mathbf{k}$.

It follows from the general theorems of Section 6.2 that the number of coset representatives $M(\mathbf{k})$ is given by $g=g(\mathbf{k})M(\mathbf{k})$ where g and $g(\mathbf{k})$ are the orders of G and $G(\mathbf{k})$, resp. This relation makes sense, as due to the cyclic boundary conditions it can be shown that the full space group G is of finite order.

Theorem 6.47: Suppose that $\mathbf{k}$ is an allowed k-vector and $\varphi_s^p(\mathbf{k},\mathbf{r})$ is a Bloch function transforming according to the s-th row of the irreducible unitary representation Γ_p of $G(\mathbf{k})$ of dimension l and that this representation satisfies the equation

$$\Gamma_p(\{\mathbb{1}|\mathbf{T}_n\})=\exp(-i\mathbf{k}\cdot\mathbf{T}_n)\Gamma_p(\{\mathbb{1}|0\}) \tag{6.6.12}$$

for every primitive translation $\{\mathbb{1}|\mathbf{T}_n\}$. Then the set of $l\cdot M(\mathbf{k})$ functions

$$\mathfrak{G}(\{\mathscr{A}_i|\mathbf{T}_i\})\varphi_s^p(\mathbf{k},\mathbf{r}) \qquad \begin{array}{l} i=1\ldots M(\mathbf{k}) \\ s=1\ldots l \end{array} \tag{6.6.13}$$

forms a basis for an $l\cdot M(\mathbf{k})$-dimensional irreducible unitary representation of the space group G. Moreover, all the irreducible representations of G may be obtained in this way, by working through all the irreducible representations of $G(\mathbf{k})$ which satisfy (6.6.12) for all the allowed values of $\mathbf{k}$ that are in different stars, cf. Cornwell (1969).

The theorem reduces the problem of finding irreducible representations of G to that of finding irreducible representations of $G(\mathbf{k})$. However, $G(\mathbf{k})$ is itself a very large group as it contains the group T of primitive translations and has a complicated structure when $\mathbf{k}$ is a non-general point. Hence, a further reduction is necessary in order to obtain more manageable results.

Def. 6.37: The group $T(\mathbf{k})$ consists of all primitive translations $\{\mathbb{1}|\mathbf{T}_n\}$ that satisfy the equation $\exp(-i\mathbf{k}\cdot\mathbf{T}_n)=1$.

Obviously $T(\mathbf{k})$ is a subgroup of the group T of all primitive translations and therefore it is also a subgroup of $G(\mathbf{k})$. It is, moreover, an invariant subgroup of $G(\mathbf{k})$.

Theorem 6.48: Let Γ' be an irreducible representation of the factorgroup $G(\mathbf{k})/T(\mathbf{k})$ that satisfies

$$\Gamma'(\{\mathbb{1}|\mathbf{T}_n\}T(\mathbf{k}))=\exp(-i\mathbf{k}\cdot\mathbf{T}_n)\Gamma'(\{\mathbb{1}|0\}T(\mathbf{k})) \tag{6.6.14}$$

for every coset $\{\mathbb{1}|\mathbf{T}_n\}T(\mathbf{k})$ formed from primitive translations. Then the set of matrices $\Gamma(\{\mathscr{A}|\mathbf{T}\})$ defined by

$$\Gamma(\{\mathscr{A}|\mathbf{T}\})=\Gamma'(\{\mathscr{A}|\mathbf{T}\}T(\mathbf{k})) \tag{6.6.15}$$

for every $\{\mathscr{A}|\mathbf{T}\}$ of $G(\mathbf{k})$ forms an irreducible representation of $G(\mathbf{k})$ which satisfies equation (6.6.12). Moreover, all such irreducible representations of $G(\mathbf{k})$ can be constructed in this way.

The factor group $G(\mathbf{k})/T(\mathbf{k})$ is not, in general, isomorphic to a point group, so that its irreducible representations are not so easily found. The case $\mathbf{k}=0$ is exceptional, for $G(0)/T(0)$ is isomorphic to the point group G_0. A comprehensive list of the tabulations of characters for the groups $G(\mathbf{k})/T(\mathbf{k})$ belonging to the non-symmorphic space groups can be found in Cornwell (1969).

c) Double groups

So far the group theoretical investigation has been developed for scalar wave functions. We now extend the formalism to spinorial wave functions. As in the scalar case, we consider a one-particle equation and its solutions. The simplest spinorial equation follows from (2.3.41) if one only takes into account the contributions of the scalar potential Φ. Then after some algebra the equation

$$H(\mathbf{r})\begin{pmatrix}\varphi_1(\mathbf{r})\\ \varphi_2(\mathbf{r})\end{pmatrix}=E\begin{pmatrix}\varphi_1(\mathbf{r})\\ \varphi_2(\mathbf{r})\end{pmatrix} \tag{6.6.16}$$

follows with

$$H(\mathbf{r})=H_0(\mathbf{r})+H_1(\mathbf{r}) \tag{6.6.17}$$

and

$$H_0(\mathbf{r}):=-\frac{\hbar^2}{2m}\Delta+e\Phi(\mathbf{r});\quad H_1(\mathbf{r})=\frac{\hbar^2 e}{4im^2c^2}[\nabla\Phi\times\nabla]\cdot\sigma. \tag{6.6.18}$$

As the energy operator is a spin tensor of second rank in spin space, its components $H(\mathbf{r})\equiv H(\mathbf{r})_{ij}$ transform like the direct product of two spin vectors (6.3.50), i.e., the transformation law reads

$$\mathfrak{G}(\{\mathscr{A}|\mathbf{T}\})\,H(\mathbf{r}')_{ij}\,\mathfrak{G}(\{\mathscr{A}|\mathbf{T}\})^{-1}=\sum_{hl=1}^{2}U(\mathscr{A}^{-1})_{ih}H(\mathbf{r})_{hl}U_{lj}(\mathscr{A}) \tag{6.6.19}$$

with $\mathbf{r}'=\{\mathscr{A}|\mathbf{T}\}\mathbf{r}$. The forminvariance under the group operations is guaranteed if

$$\mathfrak{G}(\{\mathscr{A}|\mathbf{T}\})H(\mathbf{r})=H(\mathbf{r})\mathfrak{G}(\{\mathscr{A}|\mathbf{T}\}). \tag{6.6.20}$$

This follows from (6.6.19) if

$$H(\mathbf{r}')_{ij}=\sum_{hl}U(\mathscr{A}^{-1})_{ih}H(\mathbf{r})_{hl}U(\mathscr{A})_{lj} \tag{6.6.21}$$

is satisfied. But the latter equation follows just by definition from the transformation properties of the σ-matrices. Similar formulas hold for the other branch of the double group $\bar{O}(\{\mathscr{A}|\mathbf{T}\})$.

If the transformations $\{\mathscr{A}|\mathbf{T}\}$ are elements of a space group, then for a forminvariant $H(\mathbf{r})$ the equation (6.6.16) has the solutions

$$\varphi_i(\mathbf{k},\mathbf{r})=\exp\,(i\mathbf{k}\cdot\mathbf{r})u_i(\mathbf{k},\mathbf{r}) \tag{6.6.22}$$

where the functions $u_i(\mathbf{r},\mathbf{k})$ have the periodicity of the lattice. Concerning the representations the following theorem holds, cf. Cornwell (1969):

Theorem 6.49: Suppose that $\mathbf{k}$ is an allowed $\mathbf{k}$-vector, and $\varphi^p_{js}(\mathbf{r},\mathbf{k})$ are spinorial Bloch functions which transform as the s-th row of the extra unitary irreducible representation Γ_p of dimension l of the double point group $G_0^D(\mathbf{k})$ of $G_0(\mathbf{k})$ corresponding to a symmorphic space group G. Suppose further that the star of $\mathbf{k}$ consists of the $M(\mathbf{k})$ wave vectors $\mathbf{k}_1(=\mathbf{k})\mathbf{k}_2 \ldots$ which are associated with the $M(\mathbf{k})$ rotations $\mathscr{A}_i$. Then the set of $l \cdot M(\mathbf{k})$ spinors with components

$$\mathfrak{G}([\mathscr{A}_i|0])\varphi^p_{js}(\mathbf{r},\mathbf{k}) \quad \begin{array}{l} j=1,2 \\ 1 \leqq i \leqq M(\mathbf{k}) \\ 1 \leqq s \leqq l \end{array} \tag{6.6.23}$$

forms a basis for an $l \cdot M(\mathbf{k})$-dimensional extra unitary irreducible representation of the double space group G^D. Moreover, all the extra irreducible representations of G^D may be obtained in this way, by working through all the extra irreducible representations of $G_0^D(\mathbf{k})$ for the allowed $\mathbf{k}$-vectors that are in different stars.

It is important to note that, in general, the single group is not isomorphic to a subgroup of the double group. This is a consequence of the fact that the product of two generalized transformations of G^D does not necessarily correspond to the product of the corresponding ordinary transformations. Double point groups were first investigated by Bethe (1929) and then, more thoroughly, by Opechowski (1940). For non-symmorphic double groups the following theorem holds, cf. Cornwell (1969):

Theorem 6.50: Suppose that $\mathbf{k}$ is an allowed $\mathbf{k}$-vector and $\varphi^p_{js}(\mathbf{r},\mathbf{k})$ are spinorial Bloch functions which transform as the s-th row of the extra irreducible representation Γ_p of dimension l of the double group $G^D(\mathbf{k})$. Suppose further, that this representation satisfies the equation

$$\Gamma_p([\mathbb{1}|\mathbf{T}_n]) = \exp(-i\mathbf{k}\cdot\mathbf{T}_n)\Gamma_p([\mathbb{1}|0]) \tag{6.6.24}$$

for every generalized primitive translation $[\mathbb{1}|\mathbf{T}_n]$. Then the set of $l \cdot M(\mathbf{k})$ spinors with components

$$\mathfrak{G}([\mathscr{A}_i|\mathbf{T}_i])\varphi^p_{js}(\mathbf{r},\mathbf{k}) \quad \begin{array}{l} j=1,2 \\ 1 \leqq i \leqq M(\mathbf{k}) \\ 1 \leqq s \leqq l \end{array} \tag{6.6.25}$$

forms a basis for an $l \cdot M(\mathbf{k})$-dimensional extra unitary irreducible representation of the double space group G^D. Moreover, all the extra irreducible representations of G^D may be obtained in this way by working through all the extra irreducible representations of $G^D(\mathbf{k})$ that satisfy (6.6.24) for all the allowed values of $\mathbf{k}$ that are in different stars.

Further the theorem holds, cf. Cornwell (1969):

Theorem 6.51: Let Γ' be an extra irreducible representation of the factor group $G^D(\mathbf{k})/T(\mathbf{k})$ that satisfies

$$\Gamma'([\mathbb{1}|\mathbf{T}_n]T(\mathbf{k})) = \exp(-i\mathbf{k}\cdot\mathbf{T}_n)\Gamma'([\mathbb{1}|0]T(\mathbf{k})) \tag{6.6.26}$$

for every $\mathbf{T}_n$. Then the set of matrices $\Gamma([\mathscr{A}|\mathbf{T}])$ defined by

$$\Gamma([\mathscr{A}|\mathbf{T}])=\Gamma'([\mathscr{A}|\mathbf{T}]\,T(\mathbf{k})) \tag{6.6.27}$$

for every $[\mathscr{A}|\mathbf{T}]$ of $G^D(\mathbf{k})$ forms an extra irreducible representation of $G^D(\mathbf{k})$ which satisfies (6.6.24). Moreover, all such extra irreducible representations of $G^D(\mathbf{k})$ can be constructed in this way.

6.7 Conduction band states

We consider conduction band states of ideal crystals as these states are required for the further evaluation of the reaction kinetics in imperfect crystals. Conduction band states are excited states of the crystal. However, it depends on the physical situation what kinds of excited states appear. In extrinsic semiconductors the electrons which contribute to extrinsic conduction and hence are in excited crystal states come from impurities, whereas in intrinsic semiconductors the excited electrons come from the lattice itself. In the first case the extrinsic electrons act as polarons, in the second case the intrinsic electrons act like ionized excitons. Hence, the formalism of quasi-particle states which was discussed in Section 5.5 should be applied to the calculation of conduction band states. In the development of the theory of conduction band states, however, one attempted first to describe such states by one-particle and not by quasiparticle equations. In this formalism it is scarcely possible to take into account the differences between polaronic and excitonic states. Only in recent calculations were attempts made to understand the conduction band states as many-particle states. Special calculations which lead to numerical results have not exceeded a perturbation treatment of the many-particle effects. In this case the zeroth order calculation is done by a Hartree-Fock treatment and the configuration interaction is taken into account by perturbation theory of second order. Before we go into details of such calculations, we consider the symmetry properties. The discussion is given for the ideal crystal where no phonons are present.

Theorem 6.52: If in the adiabatic electron operator $h(\mathbf{r},\mathbf{R})$ of an ideal crystal the lattice coordinates are supposed to be in their ideal positions $\{\mathbf{R}_k=\mathbf{R}_k^0\}$, then $h(\mathbf{r},\mathbf{R}^0)$ is forminvariant under the corresponding space group G of the crystal.

Proof: We apply a space group transformation $\{\mathscr{A}|\mathbf{T}\}$ to $h(\mathbf{r},\mathbf{R}^0)$. Since the $\mathbf{R}^0$ coordinates have fixed values, only the electron coordinates are transformed. Hence we have $h'(\mathbf{r}'\mathbf{R}^0)=h(\{\mathscr{A}|\mathbf{T}\}\mathbf{r},\mathbf{R}^0)$, where $h(\mathbf{r},\mathbf{R}^0)$ is given by (5.1.1) if we put $\mathbf{R}_k=\mathbf{R}_k^0$. Clearly, the pure electronic terms of (5.1.1) are invariant against the space group. So only those terms have to be discussed which contain nuclear coordinates. As the space group of the crystal applied to $\{\mathbf{R}_k^0\}$ only produces a permutation $\{\{\mathscr{A}|\mathbf{T}\}\mathbf{R}_k^0=\mathbf{R}_l^0\}$ of all lattice points, we may write

$$\sum_l ee_l C(\mathbf{r},\mathbf{R}_l^0)=\sum_k ee_k C(\mathbf{r},\{\mathscr{A}|\mathbf{T}\}\mathbf{R}_k^0) \tag{6.7.1}$$

and from this it follows that

$$\sum_l ee_l C(\{\mathscr{A}|\mathbf{T}\}\mathbf{r},\mathbf{R}_l^0) = \sum_k ee_k C(\mathbf{r},\mathbf{R}_k^0) \tag{6.7.2}$$

i.e., the electron-nucleus interactions are forminvariant. As the nucleus-nucleus interaction is not transformed at all, all terms contained in $h(\mathbf{r},\mathbf{R}^0)$ are forminvariant, Q.E.D.

Due to this theorem, the eigensolutions $|\chi\rangle$ can be chosen as basis functions of the representations of the corresponding space group G. If a configuration interaction is performed, any determinant of the expansion should have the same symmetry properties as the complete state. This can be achieved by choosing the one-electron states to be Bloch states. Then any determinant has to be built up by a special set of Bloch functions being a subset of $\{\varphi_s^p(\mathbf{r}|\mathbf{k},n), n=1,2\ldots\}$ where n denotes the various energy bands. For simplicity we do not introduce the spin states explicitly and suppress the representation indices p and s. It is assumed that the Bloch states are orthonormal, i.e.

$$\int \varphi(\mathbf{r}|\mathbf{k},n)^\times \varphi(\mathbf{r}|\mathbf{k}',n')d^3r = \delta_{\mathbf{k}\mathbf{k}'}\delta_{nn'}. \tag{6.7.3}$$

Then if we consider a single determinant of N Bloch states

$$|\chi(\mathbf{k}_1,n_1\ldots\mathbf{k}_N,n_N)\rangle := N^{-1/2} \sum_{\lambda_1\ldots\lambda_N} (-1)^p \varphi(\mathbf{r}_{\lambda_1}|\mathbf{k}_1,n_1)\ldots\varphi(\mathbf{r}_{\lambda_N}|\mathbf{k}_N,n_N) \tag{6.7.4}$$

we obtain by applying a primitive translation $\mathbf{T}_n$

$$\mathfrak{T}_n|\chi(\mathbf{k}_1,n_1\ldots\mathbf{k}_N,n_N)\rangle = \exp\left(-i\mathbf{T}_n\cdot\sum_{i=1}^N \mathbf{k}_i\right)|\chi(\mathbf{k}_1,n_1\ldots\mathbf{k}_N,n_N)\rangle \tag{6.7.5}$$

due to the transformation property (6.6.10). A configuration interaction is then performed by a superposition of states of the type (6.7.4) leading to a wave function

$$|\chi\rangle = \sum_{\mathbf{k}_1 n_1\ldots\mathbf{k}_N n_N} c(\mathbf{k}_1,n_1\ldots\mathbf{k}_N,n_N)|\chi(\mathbf{k}_1,n_1\ldots\mathbf{k}_N,n_N)\rangle \tag{6.7.6}$$

where the summation runs over all possible configurations of $\mathbf{k}_1,n_1\ldots\mathbf{k}_N,n_N$ which belong to the same representation of T. Such configuration interaction states were treated for excitons by the derivation of effective Hamiltonians which contain effective mass approximations, etc. But these procedures are not applicable to the calculation of conduction band states, since for these states far more detailed information about the electronic binding is necessary than that which is compatible with the effective mass approximation. Hence band calculations should be done by using quasiparticle state methods beyond the effective mass approximation.
Obviously the realization of such a program depends on the binding type of the solid. For instance, in weakly polar substances with considerable partially covalent bonding the binding states are quite complicated, whereas in strongly polar substances the binding states with closed shells are comparatively simple. Hence the greatest progress were made for the latter substances while the former substances are still treated at the one-particle level.

We will describe below some of the recent many-particle approaches to band calculations. Afterwards we will also discuss one-particle solutions since the weakly polar substances are an important class of semiconductors.
In a series of papers Kunz and collaborators developed methods for successfully working with states of the type (6.7.6). In order to reduce the calculation effort, according to Kunz, it is useful to choose only one configuration $|\chi(\mathbf{k}_1^0, n_1^0 \ldots \mathbf{k}_N^0, n_N^0)\rangle$ from this expansion as a zero order representation of the state to be calculated, then to minimize the expectation value of $h(\mathbf{r}, \mathbf{R}^0)$ with respect to this state and afterwards to take into account the remaining configurations by a perturbation calculation. In this procedure the Bloch orbitals $\{\varphi(\mathbf{r}|\mathbf{k}, n)\}$ are not given a priori, but they are themselves objects of the calculation. In the following we write for brevity $\mathbf{k}_1^0, n_1^0, \ldots, \mathbf{k}_N^0, n_N^0 \equiv \mathbf{k}_1, n_1 \ldots \mathbf{k}_N, n_N$. Then, according to Section 5.1, the Hartree-Fock equations for the one-particle states of (6.7.4) read

$$F\varphi(\mathbf{r}|\mathbf{k},j) = \varepsilon_j \varphi(\mathbf{r}\,|\,\mathbf{k},j) \tag{6.7.7}$$

with

$$F := -\frac{\hbar^2}{2m}\Delta + \sum_l 2ee_l C(\mathbf{r}, \mathbf{R}_l^0) + 2e^2 \int [\varrho(\mathbf{r}',\mathbf{r}') - \varrho(\mathbf{r},\mathbf{r}')] C(\mathbf{r},\mathbf{r}') d^3r' \tag{6.7.8}$$

and

$$\varrho(\mathbf{r},\mathbf{r}') := \sum_{\substack{h \\ \neq j}} \varphi(\mathbf{r}|\mathbf{k},h)^\times \varphi(\mathbf{r}'|\mathbf{k},h). \tag{6.7.9}$$

We now consider the situation of a polaron conduction band in a polar crystal. In this case all valence bands are completely filled up and only one electron is in the conduction band. The equation for this electron then reads

$$F(\varrho_0)\varphi(\mathbf{r}|\mathbf{k},c) = \varepsilon_c(\mathbf{k})\varphi(\mathbf{r}|\mathbf{k},c) \tag{6.7.10}$$

where $\varphi(\mathbf{r}|\mathbf{k},c)$ is the Bloch function of the conduction band while ϱ_0 is defined by (6.7.9) with a sum over the completely filled valence bands. As conduction band states are to be calculated, only equation (6.7.10) is of real interest while the simultaneous equations for the valence band states are of minor interest. Nevertheless, some information about the valence band states is required as (6.7.10) contains ϱ_0. Therefore, the very first step of the calculation is that of ϱ_0.
As the band functions depend on $\mathbf{k}$ in a very complicated way, a direct calculation of (6.7.9) for $\varrho = \varrho_0$ is not practicable. To remove this difficulty we represent the Bloch functions by Wannier functions leading to a representation

$$\varphi(\mathbf{r}|\mathbf{k},n) = N^{-1/2} \sum_{l\kappa} e^{i\mathbf{k}\cdot\mathbf{R}_{l\kappa}^0} w_n(\mathbf{r} - \mathbf{R}_{l\kappa}^0) \tag{6.7.11}$$

where $\mathbf{R}_{l\kappa}^0$ runs over all lattice points. Then the orthonormality of the Bloch functions (6.7.3) leads immediately to the orthonormality of the Wannier functions,

$$\int w_n(\mathbf{r} - \mathbf{R}_{l\kappa}^0)^\times w_{n'}(\mathbf{r} - \mathbf{R}_{l\kappa'}^0) d^3r = \delta_{l\kappa, l\kappa'} \delta_{nn'} \tag{6.7.12}$$

and in terms of Wannier functions the density $\varrho = \varrho_0$ can be written

$$\varrho_0(\mathbf{r},\mathbf{r}') = \sum_{v,\, l\kappa} w_v(\mathbf{r} - \mathbf{R}_{l\kappa}^0)^\times w_v(\mathbf{r}' - \mathbf{R}_{l\kappa}^0) \tag{6.7.13}$$

where the index v runs over the various valence bands that occur in the crystal. As the Bloch functions (6.7.11) are not yet determined the same is true for the Wannier functions. The only restriction on these functions is that they be orthonormalized. It is reasonable to identify the Wannier functions with atomic orbitals, provided that we take into account that the Wannier functions are orthonormal, while the atomic orbitals are not, as can be seen from (5.1.7). Denoting the set of original atomic orbitals by $\{\varphi(\mathbf{r}|l,\kappa,\alpha_{l\kappa})\}$, according to (5.1.8) and (6.1.10) orthonormalized orbitals can be obtained by putting

$$\psi(\mathbf{r}\,|\,l,\kappa,\alpha_{l\kappa}) = \sum_{k\varrho\beta_{k\varrho}} (\delta_{l\kappa\alpha_{l\kappa},\,k\varrho\beta_{k\varrho}} + S_{l\kappa\alpha_{l\kappa},\,k\varrho\beta_{k\varrho}})^{-1/2}\,\varphi(\mathbf{r}|k,\varrho,\beta_{k\varrho}). \tag{6.7.14}$$

If we now make the identifications

$$w_v(\mathbf{r}-\mathbf{R}^0_{l\kappa}) \to \psi(\mathbf{r}|l,\kappa,\alpha_{l\kappa}) \tag{6.7.15}$$

where $\mathbf{R}^0_{l\kappa}$ has to be that lattice vector which leads to the lattice point with number $l\kappa$, and $v \equiv \alpha_{l\kappa}$, then the density ϱ_0 can be written in terms of atomic orbitals

$$\varrho_0(\mathbf{r},\mathbf{r}') = \sum_{\substack{l\kappa\alpha_{l\kappa}\\ k\varrho\beta_{k\varrho}}} \varphi(\mathbf{r}|l,\kappa,\alpha_{l\kappa})\,(\delta_{l\kappa\alpha_{l\kappa},\,k\varrho\beta_{k\varrho}} + S_{l\kappa\alpha_{l\kappa},\,k\varrho\beta_{k\varrho}})^{-1}\,\varphi(\mathbf{r}'|k,\varrho,\beta_{k\varrho}). \tag{6.7.16}$$

This procedure is still exactly valid and a Hartree-Fock calculation using the set $\varphi(\mathbf{r}|l,\kappa,\alpha_{l\kappa})$, $\varphi(\mathbf{r}|\mathbf{k},c)$ could be performed. However, Kunz proposed to proceed in two steps:

i) In the first step atomic orbitals for the ideal crystal are calculated:
ii) in the second step these atomic orbitals are used to define the effective potential for the conduction band equation (6.7.10) and this equation is solved.

To perform the first step Kunz used an equation for localized atomic orbitals in a complex system which was derived by Adams (1961) (1962), Gilbert (1964) (1970), Anderson (1968). Kunz (1969) (1971) adapted this equation to the calculation of crystal atomic orbitals for crystals with closed shell constituents and performed special calculations. It would exceed the scope of this book to give an account of this theory. Hence we turn to the second step (ii).

After having calculated the potential ϱ_0 by means of (6.7.16) with self-consistent local orbitals, Kunz made the ansatz

$$\begin{aligned}\varphi(\mathbf{r}|\mathbf{k},c) &= \sum_{nlm} a^c_{nlm}(\mathbf{k})\,N^{-1/2}\sum_{\mu}\varphi_{nlm}(\mathbf{r}-\mathbf{R}^0_\mu)\exp(i\mathbf{k}\cdot\mathbf{R}^0_\mu)\\ &\quad+\sum_{nlm} b^c_{nlm}(\mathbf{k})\,N^{-1/2}\sum_{\mu}\xi_{nlm}(\mathbf{r}-\mathbf{R}^0_\mu)\exp(i\mathbf{k}\cdot\mathbf{R}^0_\mu)\end{aligned} \tag{6.7.17}$$

for the solution of (6.7.10).

Here $\mu=(l,\kappa)$; $\alpha_{l\kappa}=(n,l,m)$, where n is the principal quantum number and l,m are the azimuthal quantum numbers of the localized orbital centered at $\mathbf{R}^0_{l\kappa}$. In the second part of the sum (6.7.17) Slater orbitals are used which are defined by

$$\xi_{nlm}(\mathbf{r}-\mathbf{R}^0_\mu) = |\mathbf{r}-\mathbf{R}^0_\mu|^{n-1}\,Y^m_l(\mathbf{r}-\mathbf{R}^0_\mu)\exp(-\alpha_{nl}|\mathbf{r}-\mathbf{R}^0_\mu|). \tag{6.7.18}$$

The normalization is made with respect to a periodicity region which is characterized by N. Substitution of (6.7.17) into (6.7.10) leads with the usual procedures to a variational problem for the determination of the coefficients $\{a_{nlm}^c(\mathbf{k})\}$ and $\{b_{nlm}^c(\mathbf{k})\}$. For the solution of this problem two-center, three-center and four-center integrals between atomic orbitals and Slater orbitals have to be calculated. Several hundred expansion functions were used for actual calculations. Since the ideal crystal orbitals are known from selfconsistency calculations, by formation of Bloch waves with atomic orbitals the valence band states and their energies can also be calculated. The disadvantage of this procedure is the fact that only the energies are evaluated numerically, while no simple analytic expressions for the wave functions can be derived. For NaCl, Lipari and Kunz (1971) obtained the results given in Figure 6.9.

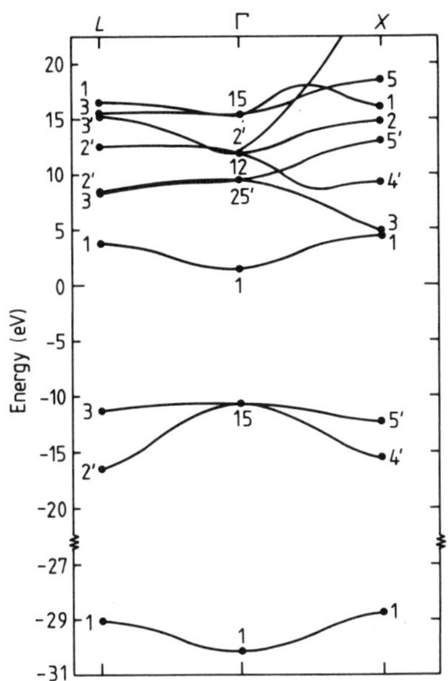

Figure 6.9:
Hartree-Fock energy bands for NaCl. Only the points Γ, X, L, and the midpoint of Δ were actually computed. The numbers in the figure refer to the irreducible representations corresponding to particular k values. After Lipari and Kunz (1971).

We summarize the main results as follows:
The computed band gap is 12.08 eV, the width for the 3s-like valence bands is 5.88 eV. The experimental band gap is 8.97 eV, i.e., the computed Hartree-Fock energy band gap is considerably larger than the experimental one. Also the valence band widths are very large. Another example is the calculation of Hartree-Fock energy bands for LiCl by Kunz (1970) given in Figure 6.10.
In order to correct such discrepancies, Kunz et al. improved their calculations by additionally performing a configuration interaction calculation in which the electronic polaron model (EPM) of Toyozawa (cf. Section 5.5) was used. Further improvements

were achieved by a short-range force correction of Toyozawa's model (IEPM). The results can be seen in Figure 6.11.
Subsequently Pantelides, Mickisch and Kunz (1974) gave a unified treatment of correlation effects in configuration interaction calculations which did not rest on any

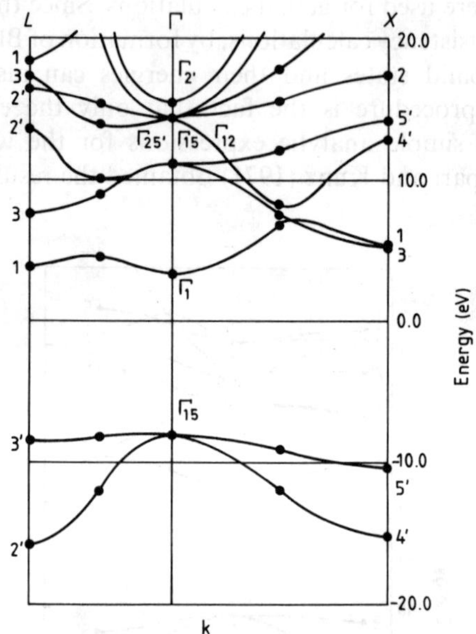

Figure 6.10:
Self-consistent Hartree-Fock energy bands for LiCl. Spin-orbit effects are negelected and the BSW notation is used. After Kunz (1970).

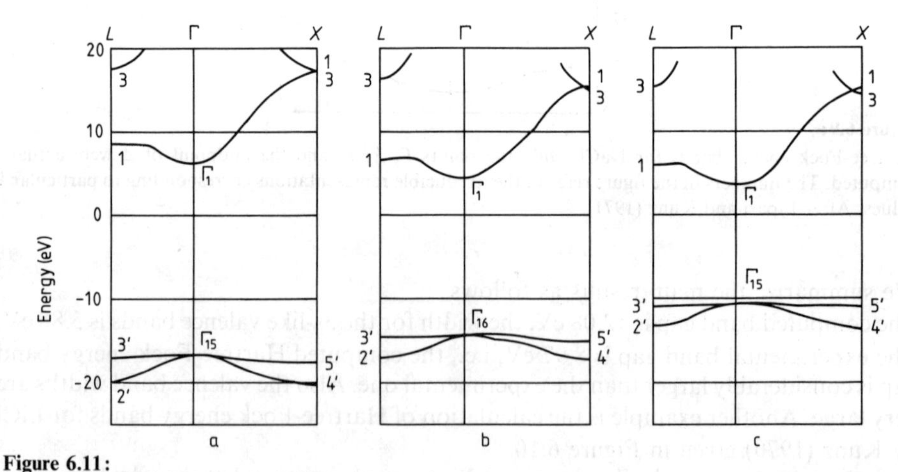

Figure 6.11:
Relevant part of the LiF band structure. In part (a) the Hartree-Fock bands are given. In part (b) the bands corrected by the EPM, in part (c) the bands computed using IEPM are given. After Mickisch, Kunz, and Collins (1974).

special model and allowed a systematic classification of the correlations which are of importance for band calculations. In addition, a clear distinction between the various types of excitations (polaronic, excitonic etc.) was made. This is a necessary condition for a successful discussion of the observed spectra.
Another H.F.-approach which was put forward recently by Drost and Fry (1972) was the selfconsistent calculation of linear combinations of atomic orbitals (LCAO-method). This method is the solid state equivalent of H.F.-selfconsistent calculations for atoms. It starts with an initial test potential for the crystal and one-particle wave functions being linear combinations of atomic orbitals. By the H.F.-equations the corresponding expansion coefficients are calculated and the resulting wave functions are used for an improved derivation of the crystal potential. This iteration is continued until selfconsistency is achieved. The method was applied by Drost and Fry (1972) to LiF, and by Brener and Fry (1972) to NaF.
Rahman, Raschid and Chowdhury (1975) (1976) also applied linear combinations of atomic orbitals to the solution of H.F.-equations for KCl. However, they avoided the laborious selfconsistency procedure by using some results of Howland (1958) and Löwdin (1956). They expressed the one-electron orbitals $\{\varphi(\mathbf{r}|\mathbf{k},j)\}$ by a linear combination of Bloch orbitals in the tight binding approximation.

$$\varphi(\mathbf{r}|\mathbf{k},j)=\sum_{m} c(\mathbf{k},j|m)\,b(\mathbf{r}\,|\,\mathbf{k},m) \tag{6.7.19}$$

with

$$b(\mathbf{r}\,|\,\mathbf{k},m)=N^{-1/2}\sum_{l\kappa} e^{i\mathbf{k}\cdot\mathbf{R}^{0}_{l\kappa}}\,\varphi(\mathbf{r}\,|\,l,\kappa,m) \tag{6.7.20}$$

where $\varphi(\mathbf{r}|l,\kappa,m)$ is an atomic orbital located at $\mathbf{R}^{0}_{l\kappa}$ with quantum numbers m. Then the density (6.7.9) of the H.F.-equations depends on the wave functions $\varphi(\mathbf{r}|\mathbf{k},j)$ and thus also on the variational parameters $c(\mathbf{k},j|m)$. However, Löwdin succeeded in showing that even in this general case the density can be rewritten as (6.7.16), i.e., that the variational parameter drops out. This allows one to solve the problem in one calculation step without further iteration.
We now turn to the one-particle equations. Such equations, first studied by Slater (1934) and Slater and Krutter (1935), are thought to result from H.F.-equations by averaging procedures. For a general set $\{\psi_i\}$ of wave functions the H.F.-equations (6.7.7) can be written

$$\begin{aligned} H_1\psi_i(\mathbf{r})&+\sum_{k=1}^{n} e^2\int \psi_k(\mathbf{r}')^{\times}\psi_k(\mathbf{r}')C(\mathbf{r}',\mathbf{r})d^3r'\,\psi_i(\mathbf{r})\\ &-\sum_{k=1}^{n} e^2\int \psi_k(\mathbf{r}')^{\times}\psi_i(\mathbf{r}')C(\mathbf{r}',\mathbf{r})d^3r'\,\psi_k(\mathbf{r})=\varepsilon_i\psi_i(\mathbf{r}). \end{aligned} \tag{6.7.21}$$

This set of equations can be cast into the form

$$[H_1+V(\mathbf{r})+V_i^{\mathrm{ex}}]\,\psi_i(\mathbf{r})=\varepsilon_i\psi_i(\mathbf{r}) \tag{6.7.22}$$

with

$$V(\mathbf{r}):=\sum_{k=1}^{n} e^2\int \varrho_k(\mathbf{r}')\,C(\mathbf{r}',\mathbf{r})\,d^3r' \tag{6.7.23}$$

and

$$V_i^{\text{ex}} := -\sum_{k=1}^{n} e^2 \frac{\int \psi_k(\mathbf{r}')^\times \psi_k(\mathbf{r}) C(\mathbf{r},\mathbf{r}') \psi_i(\mathbf{r}') \psi_i(\mathbf{r})^\times d^3r'}{\psi_i(\mathbf{r})^\times \psi_i(\mathbf{r})}. \tag{6.7.24}$$

Obviously it is the exchange potential that leads to differences between the one-electron equations. In order to obtain an exchange potential which is the same for all electron states, one can form the weighted mean of the exchange charge density by multiplying equation (6.7.24) by the probability that an electron at **r** is in the state ψ_i, namely

$$p_i(\mathbf{r}) = \psi_i(\mathbf{r})^\times \psi_i(\mathbf{r}) \left[\sum_j \psi_j(\mathbf{r})^\times \psi_j(\mathbf{r}) \right]^{-1}. \tag{6.7.25}$$

When this is done one obtains the average exchange potential

$$\bar{V}^{\text{ex}} := -\sum_{j,k} \int \psi_k(\mathbf{r}')^\times \psi_k(\mathbf{r}) e^2 C(\mathbf{r}',\mathbf{r}) \psi_j(\mathbf{r})^\times \psi_j(\mathbf{r}') d^3r' \left[\sum_j \psi_j(\mathbf{r})^\times \psi_j(\mathbf{r}) \right]^{-1} \tag{6.7.26}$$

which is the same for all electron states. Then the $\{\psi_i\}$ are assumed to satisfy the equation

$$[H_1 + V(\mathbf{r}) + \bar{V}^{\text{ex}}] \psi_i(\mathbf{r}) = \varepsilon_i \psi_i(\mathbf{r}). \tag{6.7.27}$$

The further evaluation of (6.7.27) depends on the approximation which is used for the exchange potential (6.7.26). Numerous methods for the approximation of $\bar{V}^{\text{ex}}$ and the solution of (6.7.27) have been developed. We refer to the literature which will be cited in the following.

One-particle equations were mainly used for band calculations of weakly polar semiconductors, most of which crystallize in the zincblende structure. This structure consists of two face-centered sublattices whose points are occupied by different types of atoms, displaced with respect to each other by the vector $d(1/4, 1/4, 1/4)$, where d is the edge of the cube. The corresponding space group is T_d^2 with which the face-centered cubic translation group and the tetrahedral point group T_d are associated. This space group is symmorphic and hence, concerning the band calculation, we have to deal only with the tetrahedral point group T_d according to Theorem 6.46, 6.49 and to Definition 6.31. This point group has 24 elements, the corresponding double group 48 elements. According to Definition 6.31, all wave vector groups $G_0(\mathbf{k})$ are subgroups of the point group T_d. These groups have been investigated by Dresselhaus (1955) and Parmenter (1955). The character tables of the single and the double groups for the Γ-point ($\equiv \mathbf{k} = 0$) are:

Table 6.8:
Character table for the representations of the single group of Γ. After Parmenter (1955).

24	Γ	Γ_1	Γ_2	Γ_{12}	Γ_{15}	Γ_{16}
1	E	1	1	2	3	3
3	C_4^2	1	1	2	−1	−1
8	C_2	1	1	−1	0	0
6	JC_4	1	−1	0	−1	1
6	JC_2	1	−1	0	1	−1

Table 6.9:
Character table of the double group of Γ; $k=(000)$. After Dresselhaus (1955).

Γ	E	$\bar{E}$	$6C_4^2$	$8C_2$	$8\bar{C}_2$	$6I\times C_4$	$6I\times\bar{C}_4$	$12I\times C_2$
Γ_1	1	1	1	1	1	1	1	1
Γ_2	1	1	1	1	1	-1	-1	-1
Γ_3	2	2	2	-1	-1	0	0	0
$\Gamma_4(x,y,z)$	3	3	-1	0	0	-1	-1	1
Γ_5	3	3	-1	0	0	1	1	-1
Γ_6	2	-2	0	1	-1	$\sqrt{2}$	$-\sqrt{2}$	0
Γ_7	2	-2	0	1	-1	$-\sqrt{2}$	$\sqrt{2}$	0
Γ_8	4	-4	0	-1	1	0	0	0

Γ_i :	Γ_1	Γ_2	Γ_3	Γ_4	Γ_5
$\Gamma_i\times D_j$:	Γ_6	Γ_7	Γ_8	$\Gamma_7+\Gamma_8$	$\Gamma_6+\Gamma_8$

Selection rules

Γ_i :	Γ_1	Γ_2	Γ_3	Γ_4	Γ_5	Γ_6	Γ_7	Γ_8
$\Gamma_i\times\Gamma_4$:	Γ_4	Γ_5	$\Gamma_4+\Gamma_5$	$\Gamma_1+\Gamma_3+\Gamma_4+\Gamma_5$	$\Gamma_2+\Gamma_3+\Gamma_4+\Gamma_5$	$\Gamma_7+\Gamma_8$	$\Gamma_6+\Gamma_8$	$\Gamma_6+\Gamma_7+2\Gamma_8$

The most successful calculations of the band structures of these substances were made by means of one-particle equations with pseudopotentials and plane wave representations. The pseudopotential concept was discussed in Section 5.3. The peculiarities of its application to band structure calculations in plane wave representations were discussed in detail by Sandrock (1970); the group theoretical technique with respect to

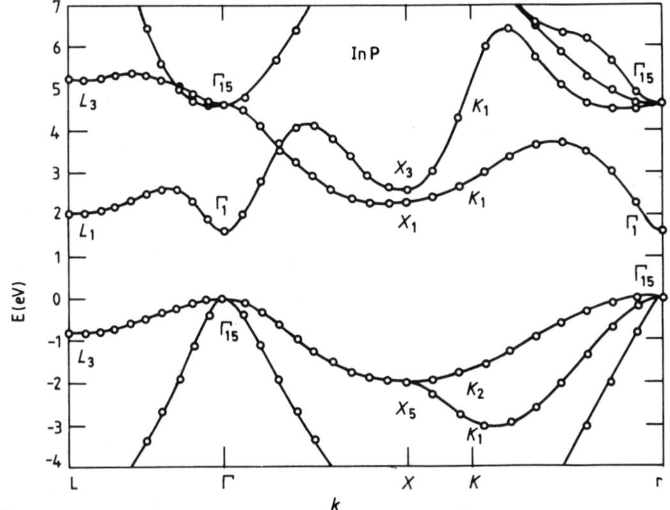

Fig. 6.12.a

Figure 6.12.a-i:
Band structures for polar semiconductors. The symmetry assignments are according to Herring and to Parmenter. The bands are computed along the symmetry directions Λ, Δ, Σ, and the line between X and $U(K)$. After Cohen and Bergstresser (1966).

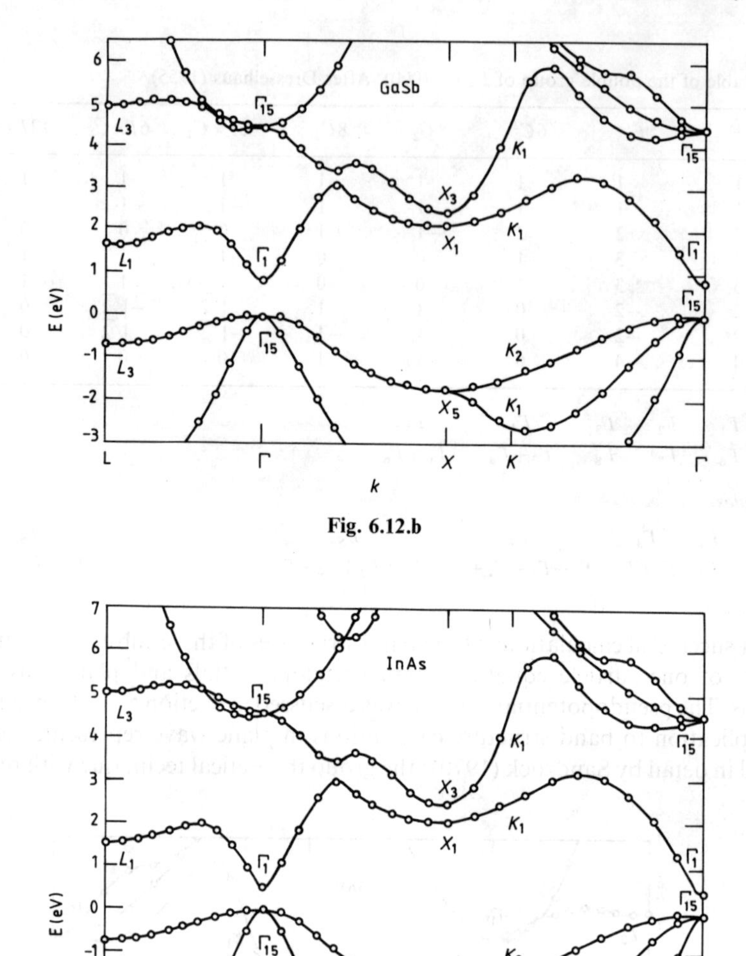

Fig. 6.12.b

Fig. 6.12.c

such representations by Streitwolf (1967). In the Figures 6.12a–6.12i we give some of the results obtained by Cohen and Bergstresser (1966).
The majority of such calculations are purely numerical. On the other hand, as will be seen in Chapter 8 for reaction kinetic calculations, analytic expressions for the band structure and the corresponding wave functions are needed. Based on the work of Dresselhaus (1955) and Parmenter (1955), Kane (1957) developed a calculation

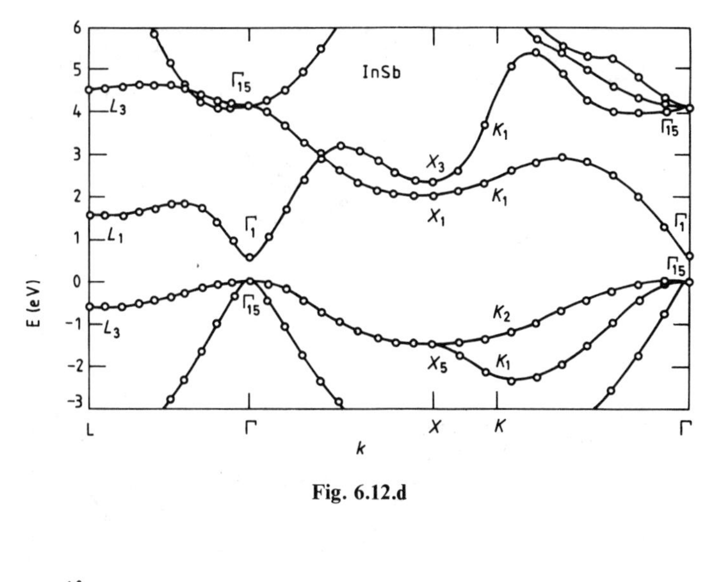

Fig. 6.12.d

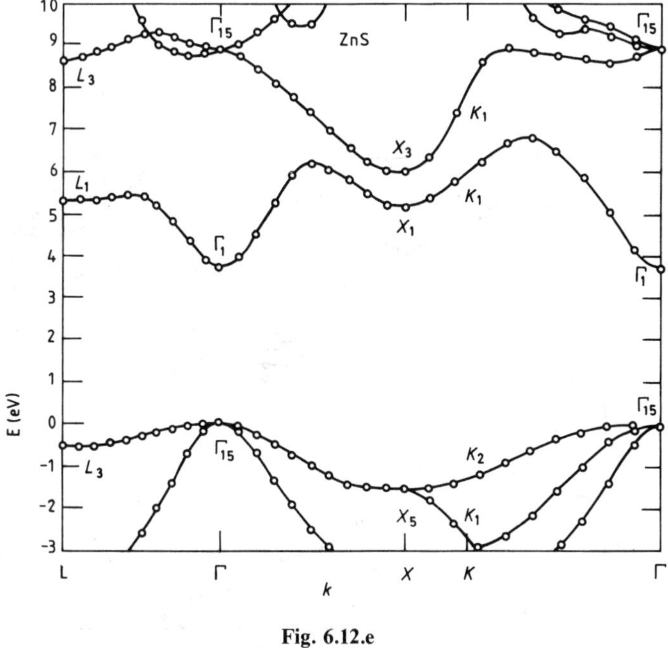

Fig. 6.12.e

scheme for the bands of InSb which allowed a derivation of approximate analytic expressions for all quantities of interest. These calculations were improved by Gorczyca and Miasek (1975) and Gorczyca (1977). In order to see the essentials of this method it is sufficient to discuss Kane's original approach. It starts with a one-particle

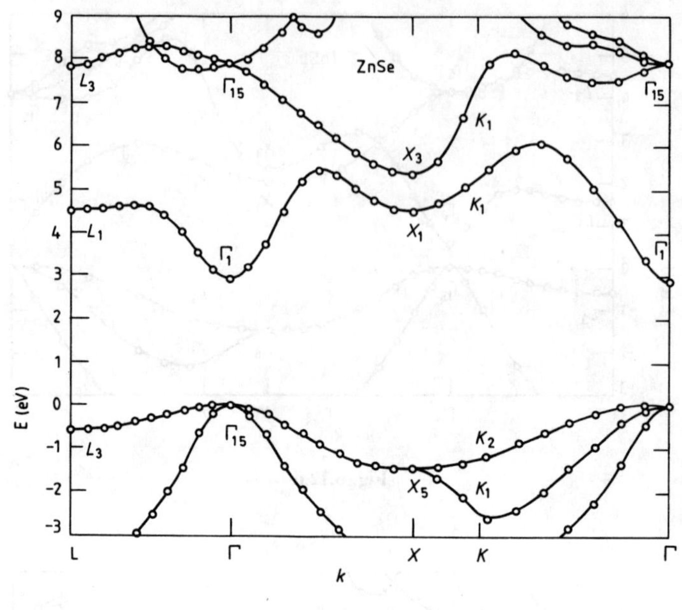

Fig. 6.12.f

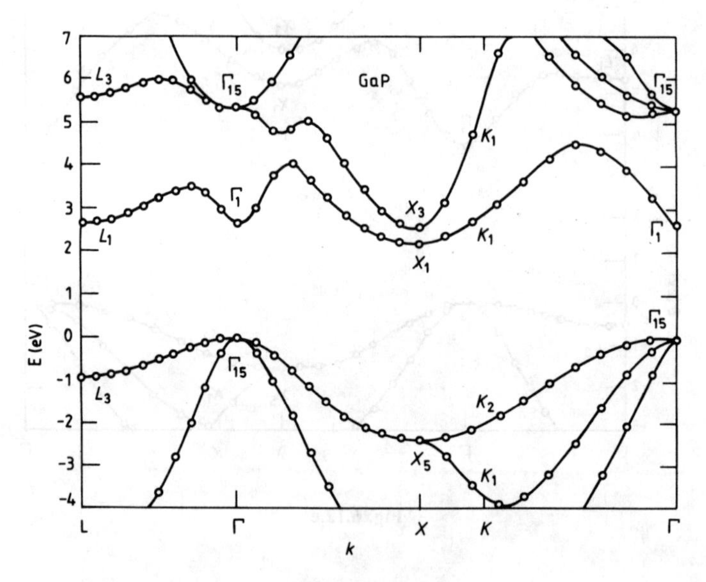

Fig. 6.12.g

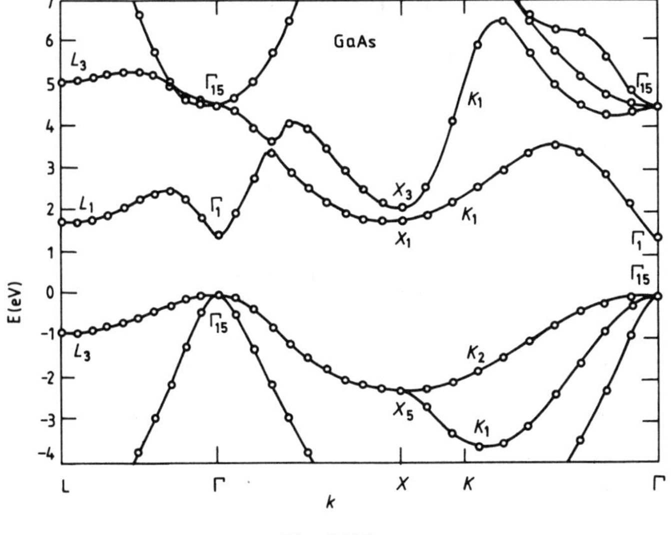

Fig. 6.12.h

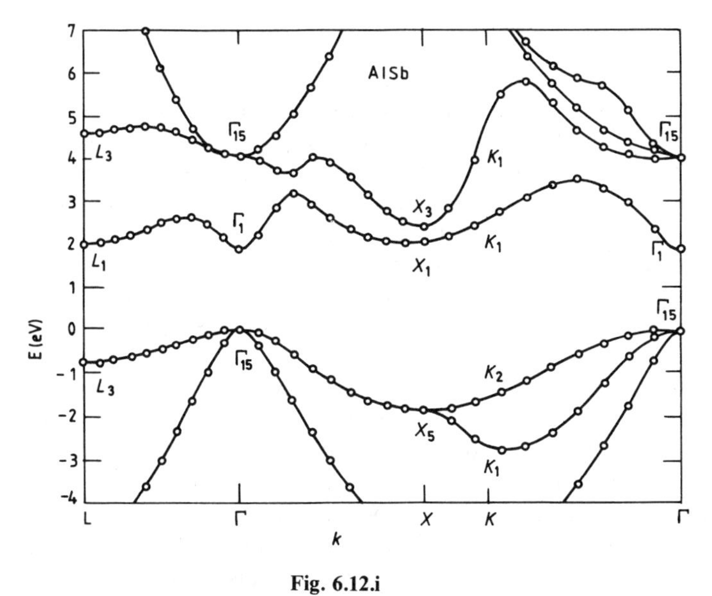

Fig. 6.12.i

Hamiltonian

$$H := \frac{1}{2m}\,\mathbf{p}^2 + V(\mathbf{r}) + \frac{\hbar}{4m^2c^2}\,\sigma\cdot[\nabla V(\mathbf{r})\times\mathbf{p}] \tag{6.7.28}$$

with spin-orbit coupling which can be derived as a special case of (2.3.41). The eigenfunctions of (6.7.28) are of the Bloch form

$$\psi(\mathbf{r} \mid \mathbf{k}) = \exp(i\mathbf{k} \cdot \mathbf{r})\, u(\mathbf{r} \mid \mathbf{k}) \tag{6.7.29}$$

with

$$H\psi(\mathbf{r}|\mathbf{k}) = E(\mathbf{k})\psi(\mathbf{r}|\mathbf{k}). \tag{6.7.30}$$

By eliminating the plane wave part of (6.7.29) from (6.7.30), we see that $u(\mathbf{r}|\mathbf{k})$ satisfies the equation

$$\left\{\frac{1}{2m}\mathbf{p}^2 + V(\mathbf{r}) + \frac{\hbar}{m}\mathbf{k}\cdot\mathbf{p} + \frac{\hbar}{4m^2c^2}\boldsymbol{\sigma}\cdot[\nabla V(\mathbf{r})\times\mathbf{p}] + \frac{\hbar^2}{4m^2c^2}\mathbf{k}\cdot[\boldsymbol{\sigma}\times\nabla V(\mathbf{r})]\right\} u(\mathbf{r}|\mathbf{k}) = E'(\mathbf{k})u(\mathbf{r}|\mathbf{k}) \tag{6.7.31}$$

where $E'(\mathbf{k})$ is defined by

$$E'(\mathbf{k}) := E(\mathbf{k}) - \frac{\hbar^2}{2m}\mathbf{k}^2. \tag{6.7.32}$$

We now assume that the solutions u_i of the equation

$$\left[\frac{1}{2m}\mathbf{p}^2 + V(\mathbf{r})\right] u_i = E_i u_i \tag{6.7.33}$$

are known and we use the complete set as a basis for an expansion of the solutions of (6.7.31). Group theory gives the symmetry properties of the functions u_i. Eq. (6.7.33) is equal to (6.7.31) at the Γ-point $\mathbf{k}\equiv 0$ without spin-orbit coupling. Representations of the Γ-point are given in Tables 6.8, 6.9 for the single and the double point group. From Fig. 6.12d it follows that the valence band has at the Γ-point a Γ_{15} representation, while the lowest conduction band has a Γ_1-representation for the single point group. For a successfull band calculation of valence bands and the lowest conduction band, at least both these representations have to be taken into account. A table of basis functions of these representations for the single group and for the corresponding representations of the double group is given in Table 6.10.

In this table the arrows mean the spin eigenfunctions, s is a function which transforms as an atomic s-function under the point group, while x, y, z transform as atomic p-functions under this group. In applying equation (6.7.33), it has to be observed that, although these functions have to transform in this way, they must be simultaneously periodic functions with the lattice periodicity. It is not necessary to use the basis functions of Table 6.10 directly for the diagonalization of (6.7.31). Rather any linear combination of them can be used equally well. In his calculations Kane starts with $i|s\downarrow\rangle$, $i|s\uparrow\rangle$, $|z\downarrow\rangle$, $|z\uparrow\rangle$, $(2)^{-1/2}|x-iy\uparrow\rangle$, $(2)^{-1/2}|x-iy\downarrow\rangle$, $(2)^{-1/2}|x+iy\uparrow\rangle$, and $(2)^{-1/2}|x+iy\downarrow\rangle$. For a first investigation the last term on the right-hand side of (6.7.31) is omitted. Then the Hamiltonian of (6.7.31) for $\mathbf{k}=(0,0,\mathbf{k})$ takes, with respect to this system, the form

$$\langle a|H|b\rangle = \begin{pmatrix} \langle a|H'|b\rangle & 0 \\ 0 & \langle a|H'|b\rangle \end{pmatrix} \tag{6.7.34}$$

Table 6.10:
Functions for single-group and double-group representations with notation according to Parmenter and Dresselhaus. After Gorczyca (1977).

Functions for single-group representations		Functions for double-group representations	
Γ_1	S	Γ_6	$S\vert\uparrow\rangle$
			$S\vert\downarrow\rangle$
Γ_{15}	X	Γ_8	$\frac{1}{\sqrt{2}}(X+iY)\vert\uparrow\rangle$
	Y		$\frac{1}{\sqrt{6}}(X+iY)\vert\downarrow\rangle-\sqrt{\frac{2}{3}}Z\vert\uparrow\rangle$
	Z		$-\frac{1}{\sqrt{6}}(X-iY)\vert\uparrow\rangle-\sqrt{\frac{2}{3}}Z\vert\downarrow\rangle$
			$\frac{1}{\sqrt{2}}(X-iY)\vert\downarrow\rangle$
		Γ_7	$\frac{1}{\sqrt{3}}(X+iY)\vert\downarrow\rangle+Z\vert\uparrow\rangle$
			$\frac{1}{\sqrt{3}}(X-iY)\vert\uparrow\rangle-Z\vert\downarrow\rangle$

with

$$\langle a|H'|b\rangle := \begin{pmatrix} E_s & 0 & kP & 0 \\ 0 & E_p-\Delta/3 & 2^{1/2}\Delta/3 & 0 \\ kP & 2^{1/2}\Delta/3 & E_p & 0 \\ 0 & 0 & 0 & E_p+\Delta/3 \end{pmatrix} \tag{6.7.35}$$

and

$$P := -i\frac{\hbar}{m}\langle s|p_z|s\rangle$$

$$\Delta := \frac{3\hbar i}{4m^2c^2}\langle x|\frac{\partial V}{\partial x}p_y-\frac{\partial V}{\partial y}p_x|y\rangle. \tag{6.7.36}$$

Here E_s and E_p refer to the eigenvalues of the Hamiltonian of equation (6.7.33), where E_p belongs to a threefold degenerate level as Γ_{15} is three dimensional. If the **k**-vector is not in the z-direction, the Hamiltonian is more complicated but it can be transformed to the form of equations (6.7.34) (6.7.35) by a rotation of the basis functions. The roots of the secular equation det $|H'-E'|=0$ are given by the equations $E'=0$ and

$$E'(E'-E_G)(E'+\Delta)-k^2P^2(E'+2\Delta/3)=0 \tag{6.7.37}$$

with $E_G = E_s$ the band gap energy at $\mathbf{k}=0$. For small values of k^2 the solutions of these equations give parabolic bands

$$E_c = E_G + \frac{\hbar^2 k^2}{2m} + \frac{P^2 k^2}{3}\left(\frac{2}{E_G} + \frac{1}{E_G + \Delta}\right) \tag{6.7.38}$$

$$E_{v1} = \frac{\hbar^2 k^2}{2m}$$

$$E_{v2} = \frac{\hbar^2 k^2}{2m} - \frac{2 P^2 k^2}{3 E_G} \tag{6.7.39}$$

$$E_{v3} = -\Delta + \frac{\hbar^2 k^2}{2m} - \frac{P^2 k^2}{3(E_G + \Delta)}$$

where (6.7.38) is the conduction band, while (6.7.39) are the three valence bands which split up for $\mathbf{k} \neq 0$.

Another approximation can be made if the spin-orbit coupling is dominant, i.e., if $\Delta \gg kP, E_G$. The solutions then become

$$E_c = \frac{\hbar^2 k^2}{2m} + \frac{1}{2}\left[E_G + (E_G^2 + 8 P^2 k^2/3)^{1/2}\right] \tag{6.7.40}$$

$$E_{v1} = \frac{\hbar^2 k^2}{2m}$$

$$E_{v2} = \frac{\hbar^2 k^2}{2m} + \frac{1}{2}\left[E_G - (E_G^2 + 8 P^2 k^2/3)^{1/2}\right] \tag{6.7.41}$$

$$E_{v3} = -\Delta + \frac{\hbar^2 k^2}{2m} - \frac{P^2 k^2}{3(E_G + \Delta)}.$$

These equations simply exhibit the non-parabolic nature of the bands. For further improvements of these results by perturbation theory, etc., we refer to both the original paper and the work of Gorczyca.

We now give a review of the literature in this field. As energy band calculations belong to the fundamentals of solid state physics, their methods are widely applied and we will not give references to original papers concerning the general development of these methods. With respect to this general development and the numerous approaches to the treatment of the one-electron equation with periodic potential we refer rather to some books, for instance to those of Callaway (1964), Ziman (1960), Tinkham (1964), Streitwolf (1967), Alder, Fernbach and Rotenberg (1968), Madelung (1972), Lax (1974), and to some reviews articles, for instance to those of Herman (1958), Pincherle (1960), Streitwolf (1962), Heine (1970), Sandrock (1970), Dimmock (1971).

With respect to original papers we cite only those papers which are directly concerned with or which are closely related to the energy band calculation of binary compounds of polar insulators and polar semiconductors. The following topics were treated: Electronic band structure calculations of NaCl by means of the Wigner and Seitz cellular method with OEH (= one-electron Hamiltonian), Shockley (1936); similar calculations for LiF and LiH with an attempt to establish a selfconsistent solution of the corresponding H.F.-equations, Ewing and Seitz (1936). Energy calculation of the lowest conduction band level in NaCl and determination of the band width with OEH, Tibbs (1939). Band structure calculation of PbS with the cellular method and with OEH, Bell, Hum, Pincherle, Sciama and Woodward (1953). Symmetry effects of spin-orbit coupling on the band structure in Zincblende materials with OEH, Dresselhaus (1955). Symmetry properties of the energy bands in Zincblende materials with OEH, Parmenter (1955). Tight binding calculation of the width of the halogen band in NaCl with OEH, Casella (1956). Band structure of InSb with OEH, Kane (1957). Valence band structure calculation of crystals with NaCl type lattice with H.F.-equations and LCAO-method, Grimley (1957) (1958). Calculation of the crystal potential of ZnS for OEH, Birman (1958). Tight binding calculation of the band structure of KCl with OEH, Howland (1958). Valence band calculation of crystals with NaCl structure based on MEH (many-electron Hamiltonian) with linear combinations of local hole wave functions, Kucher (1957) (1958). Hole bands in NaCl type crystals based on H.F.-states and tight binding approximation, Kucher (1958). Symmetry of energy bands in crystals of wurtzite type, Rashba (1959). Simplified tight binding method for Zincblende, Wurtzite and mixed crystal structure with OEH, Birman (1959). Derivation of effective one-electron equations in insulators and semiconductors based on MEH in the rigid lattice, Klein (1959). Toroidal energy surfaces in crystals with Wurtzite symmetry with OEH, Casella (1960). Group theoretical discussion of the band structure of wurtzite type crystals and corresponding optical absorption, Balkanski and Cloizeaux (1960). Investigation of the structure of hole bands of some alkali halide crystals based on H.F.-states, Kucher and Tolpygo (1961). Solution of H.F.-equations in terms of localized orbitals, Adams (1961). Theoretical estimates of band structures in CdS and CdSe with OEH, Hopfield (1961). Mathematical fundamentals of the one-electron equation in infinite lattices, Tamaschke (1961). Spin-orbit splitting of valence bands in CdS and CdSe with OEH, Pedrotti and Reynolds (1962). Calculation of d-bands in cubic lattices of binary ionic compounds with OEH, Callaway (1959) (1960) (1961). Orbital theory of electronic structure based on H.F.-equations, Adams (1962). A review of band theory, valence band and tight-binding calculations for non-metallic crystals, Löwdin (1962). Theory of valence band splittings of $\mathbf{k}=0$ in Zincblende and Wurtzite structures with OEH, Adler (1962). A review of energy bands in semiconductors, Long (1962). Computation of the electronic band structure of CsAu based on OEH and the cellular method, Wood and Reitz (1962). Band calculation for an excess electron in an NaCl crystal by means of a linear combination of H.F.-states, Evseev and Tolpygo (1963). Electronic band structure of group IV elements and III–V compounds with the orthogonalized plane wave method and OEH, Bassani and Yoshimine (1963). GaAs band structure calculations with OEH, Marcus (1964). Electron correlations in narrow energy bands with MEH in Fock representation, Hubbard (1963) (1964). Spin-orbit splitting of hole bands in alkali halide crystals, Tolpygo and Sheka (1964). Calculation of energy bands in PbTe by means of the augmented-plane wave method with OEH, Conklin, Johnson and Pratt (1965). Discussion of the band structure of CdSb, Frei and Velicky (1965). The electronic band structure of a model of a sodium chloride type crystal with OEH, Hassan (1965). Valence bands and conduction bands calculations by means of the augmented plane wave with OEH for AgCl and AgBr, Scop (1965). Tight binding calculation of band structures of AgCl and AgBr with OEH, Bassani, Knox and Fowler (1965). Effects of spin-orbit coupling on the valence bands of ionic crystals with Zincblende and Wurtzite structures with OEH, Shindo, Morita and Kamimura (1965). Band structures and pseudopotential form factors for polar semiconductors of the diamond and Zincblende structure with OEH, Cohen and Bergstresser (1966). Tight binding band calculation for LiCl and NaJ with OEH, Kunz and Van Sciver (1966), Kunz (1966). Conduction band structure calculation for KCl with the orthogonalized plane wave method and OEH, Oyama and Miyakawa (1966). Relationship between the augmented plane wave method and Korringa-Kohn-Rostoker methods of band theory with OEH, Johnson (1966). Energy bands of PbTe, PbSe, PbS with the pseudopotential method and with OEH, Lin and Kleinman (1966). Band structure of cubic and hexagonal ZnS with OEH and augmented plane wave method, Rössler and Lietz (1966). Band structure of hexagonal ZnS with OEH and pseudopotential method, Bergstresser and Cohen (1966). Electron correlations in narrow energy bands and the connexion with many-body perturbation theory, Hubbard (1966). Relativistic theory for the energy band calculation with OEH and application to KJ, Onodera and Okazaki (1966), Onodera, Okazaki and Inui (1966). Relativistic formulation of the Green functions method for OEH band calculations, Takada (1966). Energy band structure of CdS using self-consistent OPW and pseudopotential methods, Collins, Euwema and de Witt (1966). Band structure of GaAs, GaP, InP and AlSb based on the $\mathbf{k} \cdot \mathbf{p}$ method, Pollak, Higginbotham and Cardona (1966). Calculation of energy bands of Cu_2O with APW-method and OEH, Dahl and Switendick (1966). Structure of energy bands for CuCl with a combined LCAO- and OPW-method and with OEH, Song (1967). Computation of the points Γ and X for the valence and conduction bands of NaCl with OPW-method and with OEH, Kunz (1967). Tight binding method with three-center corrections applied to the valence band of NaCl, Kunz (1967). Calculation of the spin-orbit parameters for the valence bands of the face-centered alkali chlorides, alkali bromides and alkali iodides with OEH, Kunz (1967). Electronic structure and optical properties of hexagonal CdSe, CdS and ZnS with pseudopotential method and with OEH, Bergstresser and Cohen (1967). Spin-polarized electronic energy band structure in EuS with APW-method and with OEH, Cho (1967). Convergence study of a selfconsistent orthogonalized plane wave band calculation for hexagonal CdS with OEH, Euwema, Collins, Shankland and de Witt (1967). Band structure of cubic ZnS with Korringa-Kohn-Rostoker method with OEH, Eckelt, Madelung and Treusch (1967). Valence band structure of Wurtzite type crystals with OEH and $\mathbf{k} \cdot \mathbf{p}$ approximation, Jahne and Gutsche (1967). Energy band structure of cubic ZnS, ZnSe, ZnTe and CdTe with OEH and the Korringa-Kohn-Rostoker method, Eckelt (1967). Approximate calculation of electronic band structure with OEH, Hubbard (1967). Summary of the various methods of band structure calculations, in particular with respect to pseudopotentials, Gautier (1967). Electronic energy bands for RbCl and the face-centered cubic alkali bromides with OEH and OPW-method, Kunz (1968). APW-calculation of the energy bands of NaCl with OEH, Clark and Kliewer (1968). Application of the OPW-method to energy band calculation of LiCl, NaCl and KCl with OEH, Kunz (1968). Investigation of energy band structures and electronic properties of PbS and PbSe with OEH and APW-method, Rabii (1968). Note on the band structure of InSb and CdTe including spin-orbit effects, Bloom and Bergstresser (1968). Relativistic band structure of GeTe, SnTe, PbTe, PbSe and PbS based on OPW method with OEH, Herman, Kortum, Ortenburger and van Dyke (1968). Microscopic theory of band states and polaron states in ionic crystals based on linear combinations of H.F.-states and inclusion of electron-phonon coupling, Evseev and Tolpygo (1969). Modification of the APW method for the calculation of energy bands, Kleinman and Shurtleff (1969). Relativistic band structure and electronic properties of SnTe, GeTe and PbTe based on OEH, Tung and Cohen (1969). Energy band structure and electronic properties of SnTe with APW method and OEH, Rabii (1969). Effective potentials for exchange interactions and correlations in band structure calculations, Ferreira (1969). Energy band structure of CdO with OEH and APW-method, Maschke and Rössler (1968). Mixed basis method for energy band calculations with OEH, Kunz (1968). Energy bands of hexagonal II–VI semiconductors with relativistic corrections and with OEH, Rössler (1969). Equations for localized orbitals in polyatomic systems derived from MEH, Kunz (1969). Calculation of the valence band of NaF by the tight binding method with OEH, Gout, Frandon and Sadaca (1969). A $\mathbf{k} \cdot \mathbf{p}$-calculation of effective masses in Zincblende semiconductors with OEH, Bowers and Mahan (1969). Combined plane wave tight binding method for energy band calculations with application to LiJ and NaJ with OEH, Kunz (1969), and with application to CsF, Fowler and Kunz (1969). Orthogonalized plane wave and empirically refined orthogonalized plane wave energy band models for cubic ZnS, CdS and CdSe with OEH and selfconsistent calculated potentials, Stukel, Euwema, Collins, Herman and Kortum (1969). Energy bands of CsJ with OEH by means of relativistic Green function method, Rössler (1969). Review of bonds and bands in semiconductors, Phillips (1970). Energy bands and optical properties of LiCl with H.F.-state calculation, Kunz (1970). Local pseudopotential model for GaSb with OEH, Cahn and Cohen (1970). Calculation of energy bands with OEH and the APW-method, Page and Hygh (1970). Selfconsistent local orbitals for Lithium Halide crystals with the Adam-Gilbert method, Kunz (1970). Energy band structure calculation of BeS, BeSe and BeTe with OEH and selfconsistent determination of potentials by APW-method, Stukel (1970), and for GaAs, Collins, Stukel and Euwema (1970). Convergence of the orthogonalized plane wave method with OEH for the tetrahedral semiconductors ZnSe, CdTe, ZnO, ZnS, BAs, BeTe, ZnTe, Euwema and Stukel (1970). A review of the evaluation of electronic energy band structures of GaAs and GaP, Gray (1970). Electronic band structure calculation of PbS, PbSe and PbTe with OEH and relativistic Green function method, Overhof and Rössler (1970). Relativistic electronic-band structure in crystals with symmorphic space group with OEH, Buzano and Rasetti (1970). An optimized LCAO-method for crystal electrons, Halpern (1970). Electron energy bands of RbJ and KJ with OEH and OPW-technique, Kunz (1970). Note on the band structure of GaAs from a self-consistent pseudopotential approach, Brust (1970). g-Factor and effective mass of electrons in the conduction and valence bands of CuCl, Khan (1970). Calculation of the electronic band structure of ScN (semiconductor with partially ionic character) based on APW method and OEH, Weinberger, Schwarz and Neckel (1971). Calculation of the temperature dependence of the energy gaps in PbTe and SnTe based on empirical pseudopotential method, Tsang and Cohen (1971). Spin-orbit splitting of the valence and conduction bands of III–V and II–VI semiconductors with OEH and relativistic OPW method, Wepfer, Collins and Euwema (1971). Energy band calculations of NaBr and NaCl by H.F.-equations and nonrelativistic mixed basis method, Kunz and Lipari (1971), Lipari and Kunz (1971). Local orbital equations for excited states derived from H.F.-procedure, Kunz (1971). Band structure calculation of TlCl by the Heine-Abarankov model potential method with OEH, Inoue and Okazaki (1971). Energy bands of KCl, KBr and KJ with OEH and relativistic Green function method, Overhof (1971). Energy band structure of LiF in the tight binding approximation with OEH and selfconsistent potential, Chaney, Lafon and Lin (1971). H.F.-determination of the energy bands and the optical properties of LiBr and KCl with selfconsistent mixed basis method, Kunz and Lipari (1971), Lipari and Kunz (1971). Pseudopotential calculations of electronic charge densities in GaAs, ZnSe, InSb, CdTe, Walter and Cohen (1971). Energy bands of TlCl and TlBr with OEH and KKR-method, Overhof and Treusch (1971). Band ordering in PbTe by use of pseudopotential method with OEH, Tsang and Cohen (1971). Band structure of III–V semiconductors with Zincblende structure from a selfconsistent pseudopotential approach, Brust (1971). Valence band structure of AgF with OEH and LCAO-method, Birtcher, Deutsch, Wendelken and Kunz (1972). A selfconsistent procedure for the LCAO-method with application to LiF, Drost and Fry (1972). Convergence of reciprocal-lattice expansions and selfconsistent energy bands of NaF with OEH and Slater exchange potentials, Brener and Fry (1972). Electron band structure changes associated with optic phonons in NaCl with OEH and APW-method, Melvin and Smith (1972). OPW-calculation of the conduction band of CuCl with OEH, Khan (1972). Dependence of Silver halide energy bands on the lattice constant with the mixed basis method of Kunz, Fowler (1972). Pseudopotential approach to the band structure of alkali halides with OEH, Bassani and Giuliano (1972). H.F.-band structure of alkali fluorides and chlorides with APW-method, Perrot (1972). A $\mathbf{k} \cdot \mathbf{p}$ approximation calculation of the valence band and the conduction band of PbSe with OEH, Bangert and Kästner (1972). Band structure of GaN with OEH, Bourne and Jacobs (1972). Band structure calculation of TiO based on the tight binding method with OEH, Honig, Wahnsiedler and Dimmock (1972). Determination of the Hartree potential of a perfect solid or semiconductor, resp., and formulation of corresponding screened ionic charge potentials, Morgan (1972), Morgan and Shahtahmasebi (1973). Energy bands of LiF with H.F.-procedure and local orbital equations, Mickish and Kunz (1973). A simplified H.F.-method for calculating groundstate properties and energy band structures, Kunz (1973). Energy bands of CuCl with the mixed basis method of Kunz and with OEH, Calabrese and Fowler (1973). Selfconsistent energy bands of LiF with the method of Drost and Frey and improvements thereof, Brener (1973). LCAO-calculations of the band structure of TlBr with selfconsistent Slater exchange potentials, Overton and Hernandez (1973). Pseudopotential band structure of ZnO with OEH, Bloom and Ortenburger (1973).

Empirical pseudopotential band structure of CdSb with OEH, Yamada (1973), and of GaSe, Schlüter (1973). Pseudopotential approach to energy band structure of RbCl and CsCl, Donato, Giulano and Ruggeri (1973). Calculation of band structures, optical constants and electronic charge densities for InAs and InSb with empirical pseudopotential method and with OEH, de Alvarez, Walter, Boyd and Cohen (1973). The electronic band structure of ZnSe in two phases with pseudopotential method and with OEH, Brust (1973). Band structure of CuBr with OEH and combined T.B.- and OPW-method, Khan (1973). Calculation of the exchange contribution to the band gap in semiconductors, Inkson (1973). Note on the temperature dependence of the band gap in semiconductors, Baumann (1974). Derivation of pseudopotentials for an extra electron in an ionic crystal, Abarenkov (1974). Application of Harrison's pseudopotential method to semiconductors, Kotova and Vörös (1974). Correlation effects in energy band theory based on the H.F.-approach, Pantelides, Mickish and Kunz (1974). Optical properties of LiF and band calculation with selfconsistent H.F.-procedure and subsequent configuration interaction calculation, Mickish, Kunz and Collins (1974). Electronic structure of MgO with selfconsistent H.F.-procedure and subsequent perturbation calculation, Pantelides, Mickish and Kunz (1974). Conduction bands of LiF with pseudopotential method and OEH, Jouanin, Albert and Gout (1974). H.F.-calculations of the energy bands of LiF, Euwema, Wepfer, Surratt and Wilhite (1974). Pseudopotential calculation of the band structure of GaAs, Chelikowsky and Cohen (1974). Electronic energy band structure of EuO with OEH, and APW-method, Tewari (1974). Electronic structure of LiH and NaH with selfconsistent H.F.-procedure and subsequent perturbation calculation, Kunz and Mickish (1975). Energy band structure calculation of KCl with H.F.-equations and LCAO-method, Rahman, Rashid and Chowdhury (1975). A critical survey of theoretical calculations and of experimental results of the electronic band structure of alkali halides, Poole, Liesegang, Leckey and Jenkin (1975). Electronic band structure of Lithium, Sodium and Potassium Fluorides with combined tight binding and pseudopotential method and with OEH, Jouanin, Albert and Gout (1975). Screened-exchange and Coulomb-hole correlation corrections to H.F.-energy bands for LiF with field theoretic Green function technique, Brener (1975). Electronic structure of PbSe and PbTe with OEH and pseudopotential method, Martinez, Schlüter and Cohen (1975). Energy band structure of ZnS, ZnSe and ZnTe with OEH and tight binding approximation, Chatterjee and Sinha (1975). Structural stability and trends in band structures of covalent-ionic compounds, Altshuler, Vekilov and Umarov (1975). Molecular cluster approach to binding and energy structure of magnesium and calcium oxide crystals, Ermoshkin, Kotomin and Evarestov (1975). Energy band structure of CaO by the tight binding method, Seth and Chaney (1975). Structure of the valence bands of zincblende type semiconductors in the tight binding approximation, Pantelides and Harrison (1975). Electronic charges densities in PbSe and PbTe from pseudopotential calculations, Schlüter, Martinez and Cohen (1975). Band structure and electronic properties of InP with OEH and a combination of the empirical pseudopotential method and $\mathbf{k} \cdot \mathbf{p}$-method, Neumann, Hess and Topol (1975). Pressure and temperature dependence of electronic energy levels in PbSe and PbTe, Schlüter, Martinez and Cohen (1975). Wave functions for Γ_1 and Γ_{15} states in

InSb using pseudopotential method, Gorczyca and Miasek (1975). Note on calculations of hydrostatic pressure coefficients of energy gaps based on empirical pseudopotential method, Bazhenov, Mutal and Soloshenko (1976). Pseudopotential constructed by a complete set of orthogonalized plane waves, Gurskij and Gurskij (1976). A limitation on the pseudopotential method for band calculations, Janak (1976). Effect of electron-electron interaction on the band gap of extrinsic semiconductors, Inkson (1976). Mixed approach of LCAO- and OPW-methods to the band structure calculation of alkali halides with reduced H.F.-equations, Rahman, Rashid and Chowdhury (1976). Tight binding approach to the electronic band structure of silver halides, Smith (1976). Pseudopotential calculation of the electronic valence charge density of NaCl and MgO, Nagel, Maschke and Baldereschi (1976), and for homopolar and heteropolar semiconductors, Chelikowsky and Cohen (1976). Selfconsistent local orbitals for silver halide crystals derived from H.F.-equations, Mickish and Kunz (1976). Empirical pseudopotential calculation of the energy bands, densities of states and charge densities of AgCl with OEH, Wang, Schlüter and Cohen (1976). Relativistic electronic structure of the NaCl polymorph of CdS with OEH and APW-method, Liu and Rabii (1976). Orbital non-orthogonality effects in band structures with MEH, Tejeda and Shevchik (1976). Non-local pseudopotential calculations for the electronic structure of GaP, GaAs, GaSb, InP, InSb, ZnSe and CdTe with OEH, Chelikowsky and Cohen (1976). Band structure and optical properties of MgF_2 with OEH and combined tight binding and pseudopotential method, Jouanin, Albert and Gout (1976). Effect of a non-local potential on the hole spectrum in narrow gap semiconductors, Gelmont (1976). Semiempirical LCAO calculation of band structures of InSb with zincblende and rocksalt structure with OEH, Shimizu and Ishii (1977). Spin-dependent wave functions and band parameters at the Γ-point of InSb with OEH and empirical pseudopotential method, Gorczyca (1977). A review of band structures and optical properties of tetrahedrally coordinated Cu- and Ag-halides, Goldmann (1977). Ground- and excited state properties of LiF in the local-density formalism with OEH, Zunger and Freeman (1977). Electronic band structure of MgO and CaO with OEH and combined tight binding and pseudopotential method, Daude, Jouanin and Gout (1977). Electronic energy bands of BeS, BeSe and BeTe with OEH and combined APW- and LCAO-method, Sarkar and Chatterjee (1977). Band structures of GaSe and InSe with OEH and semiempirical tight binding method, McCanny and Murray (1977), Williams, McCanny, Murray, Ley and Kemeny (1977). Electronic band structure of ZnO with OEH and selfconsistent pseudopotentials, Chelikowsky (1977). Conduction bands in CuCl and CuBr, Khan (1977). Electronic energy bands of CaF_2 and CdF_2 based on the OPW method and OEH, Albert, Jouanin and Gout (1977). Effect of deformation on the conduction band of III–V semiconductors, Howlett and Zukotynski (1977). Band structure of semiconductors under high stress, Tsay and Bendow (1977). Band structure calculation of ZnSb with OEH and empirical pseudopotential method, Yamada (1978). Valence band structures of III–V compounds with the bond orbital model, Chen and Sher (1978). Local-density selfconsistent energy band structure of cubic CdS with OEH, Zunger and Freeman (1978).

7. Phonon states and processes

7.1 Ideal lattice energy and symmetry

To a great extent electronic processes in polar crystals are accompanied by phonon excitations. In addition, pure phonon processes are important for electric and thermal conduction, light absorption and emission, etc. Both the corresponding transition probabilities and the corresponding energy conservation depend very sensitively on the various phonon configurations involved. Hence a detailed investigation of phonon states is indispensible.
We will give the most important results of phonon calculations which are relevant for crystal processes. As these processes are mainly concerned with impurity reactions, phonons of imperfect crystals are of special interest. The calculations of phonon states for imperfect crystals are always reduced to problems where the phonons of the ideal crystal are used as appropriate base systems in the calculation method. Hence it is necessary first to consider the phonon states and spectra of ideal crystals.
We begin the discussion of lattice dynamics of ideal crystals with the assumption of an infinitely extended crystal, since the perfect lattice periodicity greatly simplifies the formulation of the theory. We thus consider a crystal composed of an infinite number of unit cells, each of which is a parallelepiped bounded by three noncoplanar vectors $\mathbf{a}_1, \mathbf{a}_2, \mathbf{a}_3$. We denote the equilibrium position vector of the l-th unit cell relative to an origin located at some atom by

$$\mathbf{R}_l^0 := l_1\mathbf{a}_1 + l_2\mathbf{a}_2 + l_3\mathbf{a}_3 \tag{7.1.1}$$

where l_1, l_2, l_3 are any three integers. If there is only one atom per unit cell, we can take Eq. (7.1.1) to define the atomic positions. Such crystals are called primitive or Bravais crystals. If there are r atoms in the unit cell, $r > 1$ the locations of the r atoms within the unit cell in their ideal equilibrium positions are given by the vectors $\mathbf{x}(\kappa)$, $1 \leqslant \kappa \leqslant r$, and the position of the κ-th atom in the l-th unit is given by

$$\mathbf{R}_{l\kappa}^0 := \mathbf{R}_l^0 + \mathbf{x}(\kappa) \tag{7.1.2}$$

Crystals containing more than one atom per unit cell are referred to as nonprimitive crystals.
In order to perform any kind of calculation for phonons, the adiabatic energy expectation value $U_n(\mathbf{R})$ has to be known explicitly since by (2.2.10) and (2.2.11) this energy for arbitrary $\mathbf{R}$ is the potential energy of the lattice. The calculation of such energy expectation values for both perfect and imperfect crystals was discussed in Chapters 5 and 6. In this chapter we assume these expectation values to be known and draw conclusions concerning the phonons. Denoting in particular the adiabatic energy

of the ideal crystal by $U_0(\mathbf{R})$, the harmonic part $U_0^h(\mathbf{R})$ is given according to (2.2.13) (2.2.14) (2.2.17) by

$$U_0^h(\mathbf{R}) := U_0 + \sum_{l\kappa} \Phi^0(l\kappa) \cdot \mathbf{u}_{l\kappa}^0 + \frac{1}{2} \sum_{\substack{l\kappa \\ l'\kappa'}} \Phi^0(l\kappa, l'\kappa') \cdot\cdot\, \mathbf{u}_{l\kappa}^0 \mathbf{u}_{l'\kappa'}^0 \tag{7.1.3}$$

where $\mathbf{u}_{l\kappa}^0$ are the atomic (or ionic) displacements

$$\mathbf{u}_{l\kappa}^0 := \mathbf{R}_{l\kappa} - \mathbf{R}_{l\kappa}^0 = \mathbf{u}(l\kappa) \tag{7.1.4}$$

while the expansion coefficients are defined according to (2.2.12) (2.2.15) by

$$\Phi^0(l\kappa) := \left(\frac{\partial U_0^h}{\partial \mathbf{u}(l\kappa)} \right)_{\mathbf{R}=\mathbf{R}^0} \equiv 0 \tag{7.1.5}$$

$$\Phi^0(l\kappa, l'\kappa') := \left(\frac{\partial U_0^h}{\partial \mathbf{u}(l\kappa) \partial \mathbf{u}(l'\kappa')} \right)_{\mathbf{R}=\mathbf{R}^0} \tag{7.1.6}$$

From its definition we see that (7.1.6) satisfies the symmetry condition

$$\Phi^0(l\kappa, l'\kappa')_{\alpha\beta} = \Phi^0(l'\kappa', l\kappa)_{\beta\alpha} \tag{7.1.7}$$

where the subscripts denote the tensor components α, β of the second rank tensor (7.1.6).

There are, however, additional relations among the atomic force constants which can be divided into two categories. To the first category the general relations belong which are imposed by the invariance of the potential energy against the Euclidean group. To the second category those relations belong which are imposed by the special structure and symmetry of a particular crystal. We will discuss these two kinds of relations successively.

Theorem 7.1 : If $U_0(\mathbf{R})$ is assumed to be invariant against the Euclidean group, then in particular from the invariance under translations and rotations the relations

$$\sum_{l'\kappa'} \Phi^0(l\kappa, l'\kappa') = 0 \tag{7.1.8}$$

$$\sum_{l'\kappa'} [\Phi^0(l\kappa, l'\kappa')_{\alpha\beta} (\mathbf{R}_{l'\kappa'}^0)_\gamma - \Phi^0(l\kappa, l'\kappa')_{\alpha\gamma} (\mathbf{R}_{l'\kappa'}^0)_\beta] = 0 \tag{7.1.9}$$

follow.

Proof: For a translation each displacement vector $\mathbf{u}_{l\kappa}^0$ is set equal to an arbitrary constant vector $\mathbf{v}$ which is independent of l and κ. According to our assumption such a translation does not change the value of the potential energy U_0. However, formally $U_0(\mathbf{R}^0 + \mathbf{v})$ becomes

$$U_0(\mathbf{R}^0 + \mathbf{v}) = U_0(\mathbf{R}^0) + \frac{1}{2} \sum_{\substack{l\kappa \\ l'\kappa'}} \Phi^0(l\kappa, l'\kappa') \cdot\cdot\, \mathbf{v}\mathbf{v} + \ldots \tag{7.1.10}$$

As $\mathbf{v}$ is arbitrary, from $U_0(\mathbf{R}^0+\mathbf{v})=U_0(\mathbf{R}^0)$ it follows that the coefficients of the expansion (7.1.10) have to vanish in any power of $\mathbf{v}$. In particular, for the second order terms we obtain

$$\sum_{\substack{l\kappa\\ l'\kappa'}} \Phi^0(l\kappa, l'\kappa')_{\alpha\beta}=0. \tag{7.1.11}$$

As, in addition, the forces exerted on a particular atom have to be invariant under translations from

$$\left(\frac{\partial U_0}{\partial \mathbf{u}(l\kappa)}\right)_{\mathbf{R}=\mathbf{R}^0}=\left(\frac{\partial U_0}{\partial \mathbf{u}(l\kappa)}\right)_{\mathbf{R}=\mathbf{R}^0+\mathbf{v}} \tag{7.1.12}$$

(7.1.8) follows directly by substitution of (7.1.10) into (7.1.12).
As rotations can be characterized by their infinitesimal generators, it suffices to consider an infinitesimal rotation

$$u(l\kappa)_\alpha=\sum_\beta \omega_{\alpha\beta}(\mathbf{R}^0_{l\kappa})_\beta \tag{7.1.13}$$

with $\omega_{\alpha\beta}=-\omega_{\beta\alpha}$. Due to the invariance of U_0 we must have

$$U_0(\mathbf{R}^0)=U_0(\mathbf{R}^0+\mathbf{u}) \tag{7.1.14}$$

where the displacements are given by (7.1.13). Using again the Taylor expansion in any power of $\omega_{\alpha\beta}$ the corresponding coefficients have to vanish separately, leading to (7.1.9), Q.E.D.

In the proof of this theorem we made use only of the invariance of the crystal energy against the transformations of the restricted Euclidean group. In the next statement we consider the consequences which result from symmetry identities of symmetry transformations. Since the symmetry operations are operations which take the crystal into itself, it is clear that we must deal with an infinitely extended crystal.
While the general invariance transformations are valid for finite crystals and give results which depend upon the special atomic positions, cf. (7.1.9), it is a peculiarity of the symmetry transformations that they give results which depend only upon the force constants and not upon the special atomic positions. The most general symmetry transformations of infinite crystals are classified by the set of 230 crystallographic space groups. According to Section 6.6, any symmetry operation can be represented by $\{\mathcal{S}|\mathbf{v}(s)+\mathbf{x}(m)\}$ where $\mathcal{S}$ is a three-dimensional real orthogonal matrix representation of one of the proper or improper rotations of the point group of the space group, $\mathbf{v}(s)$ is a vector which is smaller than any primitive translation vector of the crystal and $\mathbf{x}(m)$ is a translation vector of the crystal. Non-zero values of the vector $\mathbf{v}(s)$ are associated with the symmetry elements called glide planes and screw axes. Space groups for which $\mathbf{v}(s)$ is identically zero for every rotation $\mathcal{S}$ of the point group of the space group are called symmorphic. All other space groups are called non-symmorphic. Applied to the position vector of the equilibrium position of the κ-th atom in the l-th unit cell, this operation has to produce an equivalent equilibrium position of another atom as the

crystal has to be taken into itself, i.e., we have

$$\{\mathscr{S}|\mathbf{v}(s)+\mathbf{x}(m)\}\,\mathbf{R}^0_{l\kappa}\equiv\mathscr{S}\mathbf{R}^0_{l\kappa}+\mathbf{v}(s)+\mathbf{x}(m)=\mathbf{R}^0_{h\varrho}. \tag{7.1.15}$$

It is this peculiarity of symmetry transformations which gives rise to the definition of symmetry identities.

Considering an atomic position $\mathbf{R}_{l\kappa}$ which is not the ideal lattice position we obtain

$$\{\mathscr{S}|\mathbf{v}(s)+\mathbf{x}(m)\}\,\mathbf{R}_{l\kappa}=\mathbf{R}'_{l\kappa} \tag{7.1.16}$$

i.e., the symmetry transformation of a general position gives only its transformed value, while applied to an ideal equilibrium point it leads to a permutation.

The definitions of invariance and forminvariance refer to the general transformations of type (7.1.16). Invariance refers to the numerical equivalence between transformed and untransformed quantities, forminvariance refers to the functional equivalence. But if transformations lead to permutations a new definition of invariance is required. This is given by the following definition:

Def. 7.1 : If a function $F(\mathbf{R}^0)$ is transformed by all symmetry permutations of a given space group into itself, it is called a symmetry identity with respect to this space group.

By means of this definition the following theorem can be derived:

Theorem 7.2 : If the adiabatic energy $U_0(\mathbf{R}^0)$ is a symmetry identity with respect to a given space group, then the following relations hold for the force constants

$$\begin{aligned}\Phi^0(h\varrho)_\alpha&=\sum_\mu S_{\alpha\mu}\Phi^0(l\kappa)_\mu\\ \Phi^0(h\varrho,h'\varrho')_{\alpha\beta}&=\sum_{\mu\nu} S_{\alpha\mu}S_{\beta\nu}\Phi^0(l\kappa,l'\kappa')_{\mu\nu}\end{aligned} \tag{7.1.17}$$

and

$$\begin{aligned}&\Phi^0(l\kappa)=\Phi^0(0\kappa)\\ &\Phi^0(l\kappa,l'\kappa')=\Phi^0(l-l',\kappa,0,\kappa')=\Phi^0(0,\kappa,l'-l,\kappa')\end{aligned} \tag{7.1.18}$$

where $h\varrho$, $l\kappa$ and $h'\varrho'$, $l'\kappa'$ are corresponding points of the symmetry operations.

Proof: The instantaneous position of the atom $l\kappa$ is denoted by $\mathbf{R}_{l\kappa}=\mathbf{R}^0_{l\kappa}+\mathbf{u}^0_{l\kappa}$. The potential energy $U_0(\mathbf{R})$ is a function depending on all instantaneous positions of the atoms, $U_0(\mathbf{R})\equiv U(\ldots\mathbf{R}_{l_1\kappa_1},\ldots\mathbf{R}_{l_2\kappa_2}\ldots)$. The invariance of U_0 against the space groups means

$$U_0(\mathbf{R})=U_0(\{\mathscr{S}|\mathbf{v}(s)+\mathbf{x}(m)\}\mathbf{R}). \tag{7.1.19}$$

To evaluate (7.1.19) we first observe that $\mathbf{u}^0_{l\kappa}$ transforms only with the homogeneous part of $\{\mathscr{S}|\mathbf{v}(s)+\mathbf{x}(m)\}$ as it is a difference of two position vectors $\mathbf{R}_{l\kappa}$ and $\mathbf{R}^0_{l\kappa}$. Further, the application of $\{\mathscr{S}|\mathbf{v}(s)+\mathbf{x}(m)\}$ to a position vector $\mathbf{R}^0_{l\kappa}$ of an atom produces then and only then another position vector $\mathbf{R}^0_{h\varrho}$ of a crystal atom if the set of position vectors $\{\mathbf{R}^0_{l\kappa}\}$ is the set of ideal lattice positions. This condition is not trivial as these position vectors cannot be assumed to be arbitrary but have to be calculated by means of the minimum conditions (7.1.5) of the crystal. Assuming first that instead of (7.1.15) the general relation (7.1.16) holds, then the instantaneous positions transform like

$$\{\mathscr{S}|\mathbf{v}(s)+\mathbf{x}(m)\}\,\mathbf{R}_{l\kappa}=\mathbf{R}^{0\prime}_{l\kappa}+\mathbf{u}^{0\prime}_{l\kappa}=\mathbf{R}^{0\prime}_{l\kappa}+\mathscr{S}\,\mathbf{u}^{0}_{l\kappa}. \tag{7.1.20}$$

If now $U_0(\mathbf{R}^0)$ is assumed to be not only an invariant but a symmetry identity, this means that the space group transformations have to satisfy, instead of (7.1.16), the special formula (7.1.15). Then the ideal lattice points undergo only a permutation to equivalent positions and the effect of $\{\mathscr{S}|\mathbf{v}(s)+\mathbf{x}(m)\}$ on $U_0(\mathbf{R}^0)$ is only a rearrangement between equivalent points which leaves $U_0(\mathbf{R}^0)$ unchanged, i.e., $U_0(\mathbf{R}^0)\equiv\{\mathscr{S}|\mathbf{v}(s)+\mathbf{x}(m)\}\;U_0(\mathbf{R}^0)$. In this case (7.1.20) goes over into

$$\{\mathscr{S}|\mathbf{v}(s)+\mathbf{x}(m)\}\,\mathbf{R}_{l\kappa}=\mathbf{R}^{0}_{h\varrho}+\mathscr{S}\,\mathbf{u}^{0}_{l\kappa} \tag{7.1.21}$$

and by substitution of (7.1.21) into (7.1.19) we obtain

$$\begin{aligned}&U_0(\ldots\mathbf{R}^0_{l_1\kappa_1}+\mathbf{u}^0_{l_1\kappa_1},\ldots,\mathbf{R}^0_{l_2\kappa_2}+\mathbf{u}^0_{l_2\kappa_2},\ldots)\\&=U_0(\ldots\mathbf{R}^0_{h_1\varrho_1}+\mathscr{S}\,\mathbf{u}^0_{l_1\kappa_1},\ldots,\mathbf{R}^0_{h_2\varrho_2}+\mathscr{S}\,\mathbf{u}^0_{l_2\kappa_2},\ldots).\end{aligned} \tag{7.1.22}$$

If we now expand both sides of (7.1.19) in powers of $\{\mathbf{u}^0_{l\kappa}\}$ or $\{\mathscr{S}\mathbf{u}^0_{l\kappa}\}$, resp., and observe that the displacements $\mathbf{u}^0_{l\kappa}$ are arbitrary, the only way in which the resulting equation can be satisfied is the equality of the corresponding coefficients. By observing the orthogonality of $\mathscr{S}$ this gives the equations (7.1.17). From these equations it follows immediately that if $\{\mathbf{R}^0_{l\kappa}\}$ are equilibrium points of the untransformed crystal, they have to be equilibrium points also of the transformed crystal.
Important special cases of Eq. (7.1.17) are obtained when we specialize the general space group operation $\{\mathscr{S}|\mathbf{v}(s)+\mathbf{x}(m)\}$ to the operation $\{\mathbb{1}|\mathbf{x}(m)\}$ which is the operation of displacing the crystal through the lattice translation vector $\mathbf{x}(m)$. The effect of this operation on $\mathbf{R}^0_{l\kappa}$ is

$$\{\mathbb{1}|\mathbf{x}(m)\}\,\mathbf{R}^0_{l\kappa}=\mathbf{R}^0_{l\kappa}+\mathbf{x}(m)=\mathbf{R}^0_{l+m,\kappa}. \tag{7.1.23}$$

If we substitute this result into Eq. (7.1.17) we obtain

$$\begin{aligned}&\Phi^0(l+m,\,\kappa)=\Phi^0(l,\,\kappa)\\&\Phi^0(l+m,\,\kappa,\,l'+m,\,\kappa')=\Phi^0(l\kappa,\,l'\kappa').\end{aligned} \tag{7.1.24}$$

By setting $m=-l$ or $m=-l'$, resp., we obtain Eq. (7.1.18), Q.E.D.

The formulas (7.1.17) (7.1.18) can be extended so as to hold also for the higher order expansion coefficients of $U_0(\mathbf{R})$, however, we will not discuss this explicitly here. We rather consider some consequences of the general formalism for the special case where the interatomic potentials are central force potentials. In this case the atoms are assumed to interact pairwise through a potential function which is a function only of the magnitude of their separation. If we denote the potential function for the interaction of an atom of type κ with an atom of type κ' by $\varphi_{\kappa\kappa'}(\mathbf{r})$, with the distance of separation $\mathbf{r}=\mathbf{R}_{l\kappa}-\mathbf{R}_{l'\kappa'}$, then the potential energy of the crystal can be written as

$$U_0(\mathbf{R})=\frac{1}{2}\sum_{\substack{l\kappa\\l'\kappa'}}{}' \varphi_{\kappa\kappa'}(\mathbf{R}_{l\kappa}-\mathbf{R}_{l'\kappa'}). \tag{7.1.25}$$

In this expression, the prime on the sum indicates that terms in it with $(l\kappa)=(l'\kappa')$ are to be omitted, while the factor $\frac{1}{2}$ corrects for the fact that all interactions are counted twice in the sum. For this representation of $U_0(\mathbf{R})$ the following theorem holds:

Theorem 7.3 : If $U_0(\mathbf{R})$ can be represented by central potentials, then the second order force constants are given by

$$\begin{aligned}\Phi^0(l\kappa,l'\kappa')_{\alpha\beta} &:= -\varphi^2_{\alpha\beta}(l\kappa,l'\kappa'),\quad l\kappa\neq l'\kappa'\\ \Phi^0(l\kappa,l\kappa)_{\alpha\beta} &:= \sum_{l'\kappa'}{}' \varphi^2_{\alpha\beta}(l\kappa,l'\kappa')\end{aligned} \tag{7.1.26}$$

where the quantities $\varphi^2_{\alpha\beta}(l\kappa,l'\kappa')$ are defined by (7.1.31).

Proof: When each atom in the crystal undergoes a displacement $\mathbf{u}(l\kappa)$ from its rest position, we can express $r(l\kappa,\, l'\kappa'):=|\mathbf{R}_{l\kappa}-\mathbf{R}_{l'\kappa'}|$ as

$$r(l\kappa,l'\kappa')=[\mathbf{r}^0(l\kappa,l'\kappa')^2+2\mathbf{r}^0(l\kappa,l'\kappa')\cdot\mathbf{u}^0(l\kappa,l'\kappa')+\mathbf{u}^0(l\kappa,\,l'\kappa')^2]^{1/2} \tag{7.1.27}$$

where we have used the abbreviations

$$\begin{aligned}\mathbf{r}^0(l\kappa,l'\kappa') &:= \mathbf{R}^0_{l\kappa}-\mathbf{R}^0_{l'\kappa'}\\ \mathbf{u}^0(l\kappa,l'\kappa') &:= \mathbf{u}^0_{l\kappa}-\mathbf{u}^0_{l'\kappa'}.\end{aligned} \tag{7.1.28}$$

We can now expand the potential energy in powers of the components $u^0_\alpha(l\kappa,l'\kappa')$ as

$$\begin{aligned}U_0(\mathbf{R}) &= \frac{1}{2}\sum_{\substack{l\kappa\\l'\kappa'}}{}' \varphi_{\kappa\kappa'}(\mathbf{R}^0_{l\kappa}-\mathbf{R}^0_{l'\kappa'})\\ &+\frac{1}{2}\sum_{\substack{l\kappa\\l'\kappa'\\\alpha}}{}' \varphi^1_\alpha(l\kappa,\,l'\kappa')\,u^0_\alpha(l\kappa,\,l'\kappa')\\ &+\frac{1}{2}\sum_{\substack{l\kappa\\l'\kappa'\\\alpha\beta}}{}' \varphi^2_{\alpha\beta}(l\kappa,\,l'\kappa')\,u^0_\alpha(l\kappa,\,l'\kappa')\,u^0_\beta(l\kappa,\,l'\kappa').\end{aligned} \tag{7.1.29}$$

The expansion coefficients are given by

$$\begin{aligned}\varphi^1_\alpha(l\kappa,l'\kappa') &:= \frac{\partial}{\partial x_\alpha}\varphi_{\kappa\kappa'}(r)_{/r=r^0(l\kappa,l'\kappa')}\\ &= \frac{x_\alpha}{r}\varphi'_{\kappa\kappa'}(r)_{/r=r^0(l\kappa,l'\kappa')}\end{aligned} \tag{7.1.30}$$

$$\varphi^2_{\alpha\beta}(l\kappa,l'\kappa') := \frac{\partial^2}{\partial x_\alpha \partial x_\beta} \varphi_{\kappa\kappa'}(r)_{/r=r^0(l\kappa,l'\kappa')}$$
$$= \left\{ \frac{x_\alpha x_\beta}{r^2} \left[\varphi''_{\kappa\kappa'}(r) - \frac{1}{r} \varphi'_{\kappa\kappa'}(r) \right] + \frac{\delta_{\alpha\beta}}{r} \varphi'_{\kappa\kappa'}(r) \right\}_{/r=r^0(l\kappa,l'\kappa')} \tag{7.1.31}$$

where the primes denote differentiation with respect to the argument. We now relate the potential derivatives $\varphi^1_\alpha(l\kappa,l'\kappa')$ and $\varphi^2_{\alpha\beta}(l\kappa,l'\kappa')$ to the atomic force constants $\Phi^0(l\kappa)$ and $\Phi^0(l\kappa,l'\kappa')$, resp.

The first order term in the displacements in Eq. (7.1.29) can be written in the form

$$U_1 := \frac{1}{2} \sum_{\substack{l\kappa \\ l'\kappa' \\ \alpha}}' [\varphi^1_\alpha(l\kappa,l'\kappa')(u^0_{l\kappa})_\alpha - \varphi^1_\alpha(l\kappa,l'\kappa')(u^0_{l'\kappa'})_\alpha]$$
$$= \frac{1}{2} \sum_{\substack{l\kappa \\ l'\kappa' \\ \alpha}}' [\varphi^1_\alpha(l\kappa,l'\kappa') - \varphi^1_\alpha(l'\kappa',l\kappa)]\,(u^0_{l\kappa})_\alpha \tag{7.1.32}$$
$$= \sum_{\substack{l\kappa \\ l'\kappa' \\ \alpha}}' \varphi^1_\alpha(l\kappa,l'\kappa')\,(u^0_{l\kappa})_\alpha .$$

In going from the first to the second line of this equation we relabelled the summation variables $(l\kappa)$ and $(l'\kappa')$ as $(l'\kappa')$ and $(l\kappa)$, resp. In going from the second to the third line we have used the fact that

$$\varphi^1_\alpha(l\kappa,l'\kappa') = -\varphi^1_\alpha(l'\kappa',l\kappa) \tag{7.1.33}$$

which follows directly from Eq. (7.1.30). From the results given by Eq. (7.1.32) we directly obtain the result that for central forces

$$\Phi^0(l\kappa)_\alpha = \sum_{l'\kappa'}' \varphi^1_\alpha(l\kappa,l'\kappa'). \tag{7.1.34}$$

The second order contribution to the potential energy in Eq. (7.1.29) can be rewritten as

$$U_2 = \frac{1}{4} \sum_{\substack{l\kappa\alpha \\ l'\kappa'\beta}}' \varphi^2_{\alpha\beta}(l\kappa,l'\kappa')\,[(u^0_{l\kappa})_\alpha (u^0_{l\kappa})_\beta - (u^0_{l\kappa})_\alpha (u^0_{l'\kappa'})_\beta - (u^0_{l'\kappa'})_\alpha (u^0_{l\kappa})_\beta - (u^0_{l'\kappa'})_\alpha (u^0_{l'\kappa'})_\beta]$$
$$= \frac{1}{2} \sum_{\substack{l\kappa\alpha \\ l'\kappa'\beta}}' \varphi^2_{\alpha\beta}(l\kappa,l'\kappa')\,[(u^0_{l\kappa})_\alpha (u^0_{l\kappa})_\beta - (u^0_{l\kappa})_\alpha (u^0_{l'\kappa'})_\beta]. \tag{7.1.35}$$

This result yields the conclusion that for central forces the formulas (7.1.26) have to be valid, Q.E.D.

7.2. Ideal lattice phonon modes

In the preceding section we discussed some properties of the potential energy of the lattice. The corresponding harmonic phonon quantum states are obtained by the solution of Equation (2.2.18). If we restrict ourselves to the ideal lattice configuration, then Equation (2.2.18) reads

$$\left[-\sum_{l\kappa} \frac{\hbar^2}{2M_\kappa} \Delta_{l\kappa} + \frac{1}{2} \sum_{\substack{l\kappa \\ l'\kappa'}} \Phi^0(l\kappa, l'\kappa') \cdot\cdot\, \mathbf{u}^0_{l\kappa} \mathbf{u}^0_{l'\kappa'}\right] |\varphi^0_m\rangle^h = (E^0_m - U^0_0)\, |\varphi^0_m\rangle^h. \tag{7.2.1}$$

Transforming to the new coordinates

$$\bar{\mathbf{u}}(l\kappa) = M_\kappa^{1/2} \mathbf{u}^0_{l\kappa}. \tag{7.2.2}$$

Eq. (7.2.1) goes over into

$$\left[-\sum_{l\kappa} \frac{\hbar^2}{2} \bar{\Delta}_{l\kappa} + \frac{1}{2} \sum_{\substack{l\kappa \\ l'\kappa'}} M_\kappa^{-1/2} M_{\kappa'}^{-1/2} \Phi^0(l\kappa, l'\kappa') \cdot\cdot\, \bar{\mathbf{u}}(l\kappa) \bar{\mathbf{u}}(l'\kappa')\right] |\varphi^0_m\rangle^h$$

$$= (E^0_m - U^0_0)\, |\varphi^0_m\rangle^h. \tag{7.2.3}$$

The solution of (7.2.3) can be found by transforming (7.2.3) to a normal-coordinate representation. The corresponding transformation matrix is obtained by solving the eigenvalue equation

$$\omega_\varrho^2 \mathbf{a}(l\kappa|\varrho) = \sum_{l'\kappa'} \Phi^0(l\kappa, l'\kappa') \cdot M_\kappa^{-1/2} M_{\kappa'}^{-1/2} \mathbf{a}(l'\kappa'|\varrho) \tag{7.2.4}$$

which corresponds to Eq. (2.2.24). Obviously, the matrix Φ^0 in (7.2.4) is of the same dimension as the coordinate space $\{\bar{\mathbf{u}}(l\kappa)\}$ of the atoms. This enormous dimension can drastically be reduced by making use of the translational invariance for ideal lattices. Making the ansatz

$$\mathbf{a}(l\kappa|\varrho) = \mathbf{e}(\kappa|\varrho) \exp\,(i\mathbf{k} \cdot \mathbf{R}^0_l) \tag{7.2.5}$$

we obtain by substitution of (7.2.5) into (7.2.4) the equation

$$\omega_\varrho^2 \mathbf{e}(\kappa|\varrho) = \sum_{\kappa'} \mathscr{D}^0(\kappa, \kappa'|\mathbf{k}) \cdot \mathbf{e}(\kappa'|\varrho) \tag{7.2.6}$$

with the so-called dynamical matrix $\mathscr{D}^0 \equiv D^0_{\alpha\beta}$ and

$$D^0(\kappa, \kappa'|\mathbf{k})_{\alpha\beta} := (M_\kappa M_{\kappa'})^{-1/2} \sum_{l'} \Phi^0(l\kappa, l'\kappa')_{\alpha\beta} \exp\,[-i\mathbf{k} \cdot (\mathbf{R}^0_l - \mathbf{R}^0_{l'})]. \tag{7.2.7}$$

Then the following Lemma holds:

Lemma 7.1: The matrix $\mathscr{D}^0(\kappa, \kappa'|\mathbf{k})$ does not depend on the index l in Eq. (7.2.7).

Proof: According to (7.2.7) $\mathscr{D}^0(\kappa\kappa'|\mathbf{k})$ depends formally on l. To show its independence of l we first observe (7.1.1). Then we have

$$\mathbf{R}_l^0 - \mathbf{R}_{l'}^0 = \sum_i (l_i - l_i')\mathbf{a}_i \tag{7.2.8}$$

and by using (7.1.18), the Definition (7.2.7) goes over into

$$\mathscr{D}^0(\kappa,\kappa'|\mathbf{k}) = (M_\kappa M_{\kappa'})^{-1/2} \sum_{l_1' l_2' l_3'} \Phi^0(l-l',\kappa,0,\kappa') \exp\left[-i\sum_j k_j(l_j - l_j')\right]. \tag{7.2.9}$$

Due to the translational invariance, the sums over l' and $l-l'$ are equivalent. Thus from (7.2.9) it follows that

$$\mathscr{D}^0(\kappa,\kappa'|\mathbf{k}) = (M_\kappa M_{\kappa'})^{-1/2} \sum_{l'} \Phi^0(l'\kappa,0,\kappa') \exp(-i\mathbf{k}\cdot\mathbf{l}') \tag{7.2.10}$$

which shows explicitly the independence of l, Q.E.D.

We remark that, since l is the index of $\mathbf{R}_l$, Lemma 7.1 can be interpreted as stating that the dynamical matrix is independent of the special choice of the origin.
By (7.2.5) we have reduced the problem of solving the infinite set of equations (7.2.4) to the problem of solving a set of $3r$ linear homogeneous equations in $3r$ unknowns. The condition for the set of equations (7.2.6) to have a nontrivial solution is that the determinant of the coefficients vanishes

$$\det\left|D^0(\kappa,\kappa'|\mathbf{k})_{\alpha\beta} - \omega^2\delta_{\alpha\beta}\delta_{\kappa\kappa'}\right| = 0. \tag{7.2.11}$$

Equation (7.2.11) is an equation of $3r$-th degree in ω^2 and the $3r$ solutions for each value of $\mathbf{k}$ will be denoted by $\omega^2(\varrho) \equiv \omega^2(j,\mathbf{k})$, where $j = 1\ldots r$. The properties of the solutions of (7.2.6) are influenced by the properties of the dynamical matrix (7.2.7). By direct calculation it can be verified that $\mathscr{D}^0$ is a Hermitian matrix

$$D^0(\kappa',\kappa|\mathbf{k})_{\beta\alpha} = D^0(\kappa,\kappa'|\mathbf{k})_{\alpha\beta}^{\times} \tag{7.2.12}$$

where it has to be observed that the matrix properties are defined with respect to the indices α, κ, and β, κ'. The Hermiticity of $\mathscr{D}^0$ immediately implies that the $\{\omega^2(j,\mathbf{k})\}$ are real, so that $\omega(j,\mathbf{k})$ is either real or purely imaginary. Since a purely imaginary value of $\omega(j,\mathbf{k})$ would imply a classical motion of the lattice which erupts exponentially either into the past or into the future, the microscopic condition for the stability of the lattice is that each $\omega^2(j,\mathbf{k})$ be positive. The condition that this be satisfied is that the principal minors of the matrix $D^0(\kappa,\kappa'|\mathbf{k})_{\alpha\beta}$ be positive. This condition imposes additional restrictions on the force constants $\Phi^0(l\kappa,l'\kappa')$ and in all that follows we assume that these additional restrictions are satisfied. The $3r$ functions $\omega^2(j,\mathbf{k})$ for each value of $\mathbf{k}$ can be regarded as the branches of a multivalued function $\omega^2(j,\mathbf{k})$. The relation expressed by the equation

$$\omega = \omega(j,\mathbf{k}); \quad 1 \leqslant j \leqslant 3r \tag{7.2.13}$$

is known as the dispersion relation. In general it is not possible to obtain a closed expression for the function $\omega(j,\mathbf{k})$.

For each of the $3r$ values of $\omega^2(j,\mathbf{k})$ corresponding to a given value of $\mathbf{k}$ there exists a vector $\mathbf{e}(\kappa|j,\mathbf{k})$ whose components are the solutions of equations (7.2.6) which we can now write as

$$\omega^2(j,\mathbf{k})\mathbf{e}(\kappa|j,\mathbf{k})=\sum_{\kappa'} \mathscr{D}^0(\kappa,\kappa'|\mathbf{k}) \cdot \mathbf{e}(\kappa'|j,\mathbf{k}). \tag{7.2.14}$$

Equation (7.2.14) defines $\mathbf{e}(\kappa|j,\mathbf{k})$ to within a constant factor, and we can choose this factor in such a way that $\mathbf{e}(\kappa|j,\mathbf{k})$ satisfies the orthonormality and closure conditions

$$\begin{aligned}&\sum_{\kappa} \mathbf{e}(\kappa|j,\mathbf{k})^{\times} \cdot \mathbf{e}(\kappa|j',\mathbf{k})=\delta_{jj'}\\ &\sum_{j} e_\beta(\kappa'|j,\mathbf{k})^{\times} e_\alpha(\kappa|j,\mathbf{k})=\delta_{\alpha\beta}\delta_{\kappa\kappa'}.\end{aligned} \tag{7.2.15}$$

We see further from Eq. (7.2.10) that

$$D^0(\kappa,\kappa'|-\mathbf{k})_{\alpha\beta}=D^0(\kappa,\kappa'|\mathbf{k})^{\times}_{\alpha\beta}. \tag{7.2.16}$$

If we replace $\mathbf{k}$ by $-\mathbf{k}$ in Eq. (7.2.14), take the complex conjugate of the resulting equation, and make use of Eq. (7.2.16), we obtain

$$\sum_{\kappa'} \mathscr{D}^0(\kappa,\kappa'|\mathbf{k}) \cdot \mathbf{e}(\kappa'|j,-\mathbf{k})^{\times}=\omega^2(j,-\mathbf{k})\mathbf{e}(\kappa|j,-\mathbf{k})^{\times} \tag{7.2.17}$$

where we have used the fact that because $\mathscr{D}^0(\mathbf{k})$ is a Hermitian matrix, $\omega^2(j,\mathbf{k})$ is real, i.e., $\{\omega^2(j,-\mathbf{k})\}$ and $\{\omega^2(j,\mathbf{k})\}$ are eigenvalues of the same matrix $\mathscr{D}^0(\mathbf{k})$. If $\mathbf{k}$ is not a point in the first Brillouin zone at which $\mathscr{D}^0(\mathbf{k})$ has degenerate eigenvalues, we, therefore, see that

$$\omega^2(j,-\mathbf{k})=\omega^2(j,\mathbf{k}). \tag{7.2.18}$$

Since the vector $\mathbf{e}(j,-\mathbf{k})^{\times}$ satisfies the same equation as the eigenvector $\mathbf{e}(j,\mathbf{k})$, then as long as $\mathbf{k}$ is not a point of degeneracy, the two vectors can differ at most by an arbitrary factor of modulus unity

$$\mathbf{e}(j,-\mathbf{k})^{\times}=e^{i\alpha}\mathbf{e}(j,\mathbf{k}). \tag{7.2.19}$$

We follow Born and Huang and choose the phase factor to equal unity

$$\mathbf{e}(\kappa|j,-\mathbf{k})=\mathbf{e}(\kappa|j,\mathbf{k}). \tag{7.2.20}$$

Equation (7.2.20) was derived on the assumption that $\mathbf{k}$ is not a point of degeneracy. When $\mathbf{k}$ is a point of degeneracy, the most we can infer from Eq. (7.2.17) is that $\mathbf{e}(\kappa|j,\mathbf{k})^{\times}$ is an arbitrary linear combination of the eigenvectors $\{\mathbf{e}(\kappa|j',\mathbf{k})\}$ for which $\omega^2(j',\mathbf{k})=\omega^2(j,\mathbf{k})$.

For the solution of certain lattice dynamical problems, in particular for discussing long wavelength acoustic and optical vibration modes, it is convenient to introduce a modified dynamical matrix $C(\kappa,\kappa'|\mathbf{k})_{\alpha\beta}$ which differs from $D^0(\kappa,\kappa'|\mathbf{k})_{\alpha\beta}$ by a phase factor of modulus unity,

$$\begin{aligned}\mathscr{C}(\kappa,\kappa'|\mathbf{k}) &:=\exp\,[-i\mathbf{k}\cdot\mathbf{x}(\kappa)]\mathscr{D}^0(\kappa,\kappa'|\mathbf{k})\exp\,[i\mathbf{k}\cdot\mathbf{x}(\kappa')]\\ &=(M_\kappa M_{\kappa'})^{-1/2}\sum_{l'} \Phi^0(l\kappa,l'\kappa')\exp\,[-i\mathbf{k}\cdot(\mathbf{R}^0_{l\kappa}-\mathbf{R}^0_{l'\kappa'})].\end{aligned} \tag{7.2.21}$$

The corresponding eigenvectors, which we denote by $\{\mathbf{w}(\kappa|j,\mathbf{k})\}$, differ from the $\{\mathbf{e}(\kappa|j,\mathbf{k})\}$ by a compensating factor $\exp[-i\mathbf{k}\cdot\mathbf{x}(\kappa)]$

$$\mathbf{w}(\kappa|j,\mathbf{k})=\exp\,[-i\mathbf{k}\cdot\mathbf{x}(\kappa)]\mathbf{e}(\kappa|j,\mathbf{k}). \tag{7.2.22}$$

The equation determining the $\{\mathbf{w}(\kappa|j,\mathbf{k})\}$ becomes

$$\omega^2(j,\mathbf{k})\mathbf{w}(\kappa|j,\mathbf{k})=\sum_{\kappa'}\mathscr{C}(\kappa,\kappa'|\mathbf{k})\cdot\mathbf{w}(\kappa'|j,\mathbf{k}) \tag{7.2.23}$$

and the eigenvalues $\{\omega^2(j,\mathbf{k})\}$ of the matrices $\mathscr{C}(\mathbf{k})$ and $\mathscr{D}^0(\mathbf{k})$ are the same. The eigenvectors $\{\mathbf{w}(\kappa|j,\mathbf{k})\}$ satisfy the orthonormality and completeness conditions

$$\begin{aligned}&\sum_{\kappa}\mathbf{w}(\kappa|j,\mathbf{k})^{\times}\cdot\mathbf{w}(\kappa|j,\mathbf{k})=\delta_{jj'}\\ &\sum_{j}w_\alpha(\kappa|j,\mathbf{k})^{\times}\,w_\beta(\kappa'|j,\mathbf{k})=\delta_{\alpha\beta}\delta_{\kappa\kappa'}.\end{aligned} \tag{7.2.24}$$

Turning now to the solutions of Eq. (7.2.14) the following Lemma holds:

Lemma 7.2: From the $3r$ solutions for each value of $\mathbf{k}$ three eigenvalues $\omega^2(j,\mathbf{k})$ tend with $\mathbf{k}$ to zero, and the corresponding modes describe acoustic waves of the crystal

Proof: Setting $\mathbf{k}$ equal to zero in Eqs. (7.2.10) and (7.2.14), we have that

$$\sum_{l'\kappa'}M_\kappa^{-1}\Phi^0(l\kappa,l'\kappa')\cdot M_{\kappa'}^{-1/2}\mathbf{e}(\kappa'|j,0)=\omega^2(j,0)M_\kappa^{-1/2}\mathbf{e}(\kappa|j,0). \tag{7.2.25}$$

If, for each β, $M_{\kappa'}^{-1/2}e_\beta(\kappa'|j,0)$ is independent of κ', then due to Eq. (7.1.8) we find that the left-hand side of the equation vanishes, implying the vanishing of $\omega^2(j,0)$. This argument could fail only if all three components $\{e_\alpha(\kappa|j,0)\}$ vanished for each value of κ; however, we exclude the trivial solution $\mathbf{e}(\kappa|j,\mathbf{k})\equiv 0$. Thus we have three solutions, one for each value of α, which vanish with vanishing $\mathbf{k}$. Such modes are called acoustic modes, since by Eqs. (7.2.5) and (7.2.2) the condition which characterizes them, namely

$$M_\kappa^{-1/2}\mathbf{e}(\kappa|j,0)=M_{\kappa'}^{-1/2}\mathbf{e}(\kappa'|j,0)=u^0_{l\kappa}(j,0)=u^0_{l\kappa'}(j,0) \tag{7.2.26}$$

means that all r particles in each unit cell move in parallel and with equal amplitudes, and this is characteristic of the displacement in an elastic continuum upon which a sound wave is impressed, Q.E.D.

The remaining $3r-3$ modes whose frequencies do not vanish at $\mathbf{k}=0$ are called optical modes. This appelation stems from the following circumstance. For ionic crystals of the NaCl type ($r=2$), Eq. (7.2.15) for $\mathbf{k}=0$ can be written in vector form as

$$\mathbf{e}(+|j,0)\cdot\mathbf{e}(+|j',0)+\mathbf{e}(-|j,0)\cdot\mathbf{e}(-|j',0)=0 \tag{7.2.27}$$

where the plus (+) and minus (−) signs refer to the alkali and halide ions, while j refers to any one of the acoustic branches, and j' to any one of the optical branches, resp. This result together with Eq. (7.2.26) implies that

$$\mathbf{e}(+|j,0)\cdot[\mathbf{e}(+|j',0)+(M_-/M_+)^{1/2}\mathbf{e}(-|j',0)]=0 \tag{7.2.28}$$

The possibility that $\mathbf{e}(+|j,0)=0$ has been ruled out by the discussion following Eq. (7.2.25), so that either

$$\mathbf{e}(+|j,0) \perp [\mathbf{e}(+|j',0)+(M_-/M_+)^{1/2}\mathbf{e}(-|j',0)] \tag{7.2.29}$$

or

$$M_+^{1/2}\mathbf{e}(+|j',0) = -M_-^{1/2}\mathbf{e}(-|j',0). \tag{7.2.30}$$

The first alternative can be dismissed since it would require the right-hand vector to be simultaneously perpendicular to all three of the (noncoplanar) polarization vectors for the acoustic modes for each value of j'. In terms of the displacement vectors $\mathbf{u}(l\kappa)$, Eq. (7.2.30) can be written as

$$M_+\mathbf{u}^0_{l1}(j',0)+M_-\mathbf{u}^0_{l2}(j',0)=0. \tag{7.2.31}$$

This result means that the two ions in each unit cell vibrate, while the center of mass of the cell remains fixed. Since the two ions have charges of opposite signs, a net fluctuating dipole moment for the crystal is induced by these modes of vibration.

A further reduction of the eigenvalue problem (7.2.6) can be achieved by applying the symmetry operations of the space group of the crystal under consideration. Without proof we give some theorems concerning this topic. cf. Maradudin, Montroll, Weiss and Ipatova (1971).

Theorem 7.4 : If $\{\mathscr{S}\,|\,\mathbf{v}(s)+\mathbf{x}(m)\}\equiv\{g_i\}$ is the symmetry space group of a crystal, then the transformation matrices

$$G(\kappa,\kappa'|\mathbf{k},g_i)_{\alpha\beta} := S_{\alpha\beta}\delta(g_i\kappa,\kappa')\exp\{i\mathbf{k}\cdot[g_i\mathbf{x}(\kappa)-\mathbf{x}(\kappa')]\} \tag{7.2.32}$$

where $g_i\kappa=\varrho$ is the image of κ with respect to g_i, are unitary matrices, but they do not form a representation of this space group.

Theorem 7.5 : If a symmetry operation g_i is applied to the potential energy $U_0(\mathbf{R})$ of a crystal, then the dynamical matrix is transformed simultaneously by

$$D^0(\varrho,\varrho'|\mathscr{S}\cdot\mathbf{k})_{\alpha\beta} = \sum_{\substack{\kappa\mu\\ \kappa'\nu}} G_{\alpha\mu}(\varrho,\kappa|\mathbf{k},g_i)\,D_{\mu\nu}(\kappa,\kappa'|\mathbf{k})\,G^{-1}_{\nu\beta}(\kappa',\varrho'|\mathbf{k},g_i) \tag{7.2.33}$$

and the secular equation (7.2.11) remains unchanged.

Theorem 7.6 : If $\mathbf{k}$ is a wave vector for which none of the eigenvalues of $\mathscr{D}^0(\mathbf{k})$ is degenerate, then the relation

$$\omega^2(j,\mathscr{S}\cdot\mathbf{k})=\omega^2(j,\mathbf{k}) \tag{7.2.34}$$

holds for any symmetry operation g_i of the crystal, i.e., $\mathscr{D}^0(\mathscr{S}\cdot\mathbf{k})$ and $\mathscr{D}^0(\mathbf{k})$ have the same eigenvalues.

The question whether $\mathbf{k}$ leads to degenerate eigenvalues or not can be answered by considering symmetry operations which leave not only the crystal, but also $\mathbf{k}$ within the first Brillouin zone invariant.

Def. 7.2: The symmetry space group $G(\mathbf{k})$ associated with a definite wavevector $\mathbf{k}$ is defined to be the set of operations $\{\mathscr{R}|\mathbf{v}(r)+\mathbf{x}(m)\}\equiv\{r_i\}$ which take the crystal into itself and whose purely rotational elements $\{\mathscr{R}\}$ have the property

$$\mathscr{R}\mathbf{k}=\mathbf{k}-\tau(\mathbf{k},\mathscr{R}) \tag{7.2.35}$$

where $\tau(\mathbf{k},\mathscr{R})$ is a translation vector of the reciprocal lattice.

Theorem 7.7: If the transformation matrices (7.2.32) are defined for $G(\mathbf{k})$, then they are unitary and transform the dynamical matrix $\mathscr{D}^0(\mathbf{k})$ into itself

$$\mathscr{G}(\mathbf{k},r_i)\mathscr{D}^0(\mathbf{k})\mathscr{G}^{-1}(\mathbf{k},r_i)=\mathscr{D}^0(\mathbf{k}) \tag{7.2.36}$$

but they do not form a representation of $G(\mathbf{k})$.

Theorem 7.8: The matrices

$$\mathscr{T}(\mathbf{k},r_i):=\exp\{i\mathbf{k}\cdot[\mathbf{v}(r)+\mathbf{x}(m)]\}\mathscr{G}(\mathbf{k},r_i) \tag{7.2.37}$$

are unitary and form a multiplier representation of $G(\mathbf{k})$. Under the action of $G(\mathbf{k})$ the dynamical matrix is transformed equivalently according to

$$\mathscr{T}(\mathbf{k},r_i)\mathscr{D}^0(\mathbf{k})\mathscr{T}^{-1}(\mathbf{k},r_i)=\mathscr{D}^0(\mathbf{k}). \tag{7.2.38}$$

Also reflections can be included leading to antiunitary matrices $\mathscr{T}(\mathbf{k},r_i)$ and the same transformation property.

Theorem 7.9: The degenerate eigenvalues $\omega(j,\mathbf{k})$ occur if the associated symmetry group $G(\mathbf{k})$ is nontrivial. The corresponding eigenvectors transform according to the various irreducible representations of $G(\mathbf{k})$.

The time reversal invariance also leads to degeneracy of eigenvalues. But it would be beyond the scope of the book to give more details. We rather discuss the consequences of the wave vector representations (7.2.5) of the normal coordinate transformation matrix (2.2.23). Observing that the index ϱ is specified according to (7.2.13) by $\varrho\equiv j,\mathbf{k}$, the principal axis transformation (2.2.23) is then generated by the following expansion of $\mathbf{u}^0_{l\kappa}$ in terms of plane waves

$$\mathbf{u}^0_{l\kappa}=(NM_\kappa)^{-1/2}\sum_{j,\mathbf{k}}\mathbf{e}(\kappa|j,\mathbf{k})\exp(i\mathbf{k}\cdot\mathbf{R}^0_l)Q(j,\mathbf{k}). \tag{7.2.39}$$

As this transformation expresses real coordinates $\mathbf{u}^0_{l\kappa}$ by complex coordinates $Q(j,\mathbf{k})$ some care has to be taken in applying the transformation (7.2.39) to equation (7.2.1). To do this it is convenient to return to the corresponding classical expressions which can be obtained by the substitution

$$T=-\sum_{l\kappa}\frac{\hbar^2}{2M_\kappa}\Delta_{l\kappa}\to\sum_{l\kappa}\frac{1}{2}M_\kappa(\dot{\mathbf{u}}^0_{l\kappa})^2 \tag{7.2.40}$$

and

$$\dot{\mathbf{u}}^0_{l\kappa}=(NM_\kappa)^{-1/2}\sum_{j,\mathbf{k}}\mathbf{e}(\kappa|j,\mathbf{k})\exp(i\mathbf{k}\cdot\mathbf{R}^0_l)\dot{Q}(j,\mathbf{k}). \tag{7.2.41}$$

After having performed the transformation we return to the quantized theory. With (7.2.41) the expression for the kinetic energy (7.2.40) becomes

$$T=(2N)^{-1}\sum_{l\kappa}\sum_{\substack{\mathbf{k}j\\ \mathbf{k}'j'}}\mathbf{e}(\kappa|j,\mathbf{k})\,\mathbf{e}(\kappa|j',\mathbf{k}')\dot{Q}(j,\mathbf{k})\dot{Q}(j',\mathbf{k}')$$
$$\times\exp\left[i(\mathbf{k}+\mathbf{k}')\cdot\mathbf{R}_l^0\right]. \tag{7.2.42}$$

For the further evaluation we use the formula

$$\sum_l e^{i\mathbf{k}\cdot\mathbf{R}_l^0}=N\Delta(\mathbf{k})=\begin{cases}N, & \mathbf{k}=2\pi\mathbf{g}\\ 0, & \mathbf{k}\neq 2\pi\mathbf{g}\end{cases} \tag{7.2.43}$$

which we do not prove here. Observing that the summations over $\mathbf{k}$ and $\mathbf{k}'$ in (7.2.42) are restricted to the first Brillouin zone and using (7.2.43), the kinetic energy reads

$$T=\frac{1}{2}\sum_{j\mathbf{k}}\dot{Q}(j,-\mathbf{k})\dot{Q}(j,\mathbf{k})=\frac{1}{2}\sum_{j,\mathbf{k}}\dot{Q}(j,\mathbf{k})^\times\dot{Q}(j,\mathbf{k}) \tag{7.2.44}$$

where the second term follows from the condition

$$Q(j,-\mathbf{k})=Q(j,\mathbf{k})^\times \tag{7.2.45}$$

which has to be imposed on $Q(j,\mathbf{k})$ in order to obtain real $\mathbf{u}(l\kappa)$. The potential energy is transformed in a similar way.

$$U_0^h=\frac{1}{2}\sum_{\substack{l\kappa\\ l'\kappa'}}\Phi^0(l\kappa,l'\kappa')N^{-1}(M_\kappa M_{\kappa'})^{-1/2}\cdot\cdot\sum_{\substack{j\mathbf{k}\\ j'\mathbf{k}'}}\mathbf{e}(\kappa|j,\mathbf{k})\mathbf{e}(\kappa'|j',\mathbf{k}')$$
$$\times Q(j,\mathbf{k})Q(j',\mathbf{k}')\exp\left[i\mathbf{k}\cdot\mathbf{R}_l^0+i\mathbf{k}'\cdot\mathbf{R}_{l'}^0\right]. \tag{7.2.46}$$

If we multiply the right-hand side of (7.2.46) by $1=\exp\left[i\mathbf{k}'\cdot(\mathbf{R}_l^0-\mathbf{R}_l^0)\right]$ and recall the definition (7.2.7), then with (7.2.43) we obtain for (7.2.46)

$$U_0^h=\frac{1}{2}\sum_{\kappa\kappa'}\sum_{\mathbf{k}jj'}\mathbf{e}(\kappa'|j',\mathbf{k})\cdot\mathscr{D}^0(\kappa,\kappa'|\mathbf{k})\cdot\mathbf{e}(\kappa|j,-\mathbf{k})Q(j,-\mathbf{k})Q(j',\mathbf{k}). \tag{7.2.47}$$

With equation (7.2.6) this gives together with (7.2.45)

$$U_0^h=\frac{1}{2}\sum_{j,\mathbf{k}}\omega^2(j,\mathbf{k})Q(j,\mathbf{k})^\times Q(j,\mathbf{k}). \tag{7.2.48}$$

Therefore, the lattice Hamiltonian in the harmonic approximation reads

$$H_{\text{latt}}^0=\frac{1}{2}\sum_{j,\mathbf{k}}\left[\dot{Q}(j,\mathbf{k})^\times\dot{Q}(j,\mathbf{k})+\omega^2(j,\mathbf{k})Q(j,\mathbf{k})^\times Q(j,\mathbf{k})\right]. \tag{7.2.49}$$

According to the Lagrangian formalism the conjugate momentum to $Q(j,\mathbf{k})^\times$ is

$$P(j,\mathbf{k})=\partial L/\partial\dot{Q}(j,\mathbf{k})^\times=\dot{Q}(j,\mathbf{k}). \tag{7.2.50}$$

Hence the Hamiltonian (7.2.49) can be written

$$H^0_{\text{latt}} = \frac{1}{2} \sum_{j\mathbf{k}} [P(j,\mathbf{k})^\times P(j,\mathbf{k}) + \omega^2(j,\mathbf{k}) Q(j,\mathbf{k})^\times Q(j,\mathbf{k})]. \tag{7.2.51}$$

The quantized version of (7.2.51) can be obtained by inversion of (7.2.39). This leads to

$$Q(j,\mathbf{k}) = N^{-1/2} \sum_{l\kappa} \mathbf{e}(\kappa|j,\mathbf{k})^\times \cdot M_\kappa^{1/2} \mathbf{u}^0_{l\kappa} \exp(i\mathbf{k}\cdot\mathbf{R}^0_l) \tag{7.2.52}$$

for the $Q(j,\mathbf{k})$ variables and to

$$P(j,\mathbf{k}) = N^{-1/2} \sum_{l\kappa} \mathbf{e}(\kappa|j,\mathbf{k})^\times \cdot M_\kappa^{-1/2} \mathbf{p}^0_{l\kappa} \exp(i\mathbf{k}\cdot\mathbf{R}^0_l) \tag{7.2.53}$$

for the corresponding conjugate variables with $\mathbf{p}(l\kappa) = M_\kappa \dot{\mathbf{u}}(l\kappa)$. The quantization for the Cartesian variables reads

$$[(u^0_{l\kappa})_\alpha (p^0_{l'\kappa'})_\beta]_- = i\hbar \delta_{ll'} \delta_{\kappa\kappa'} \delta_{\alpha\beta} \tag{7.2.54}$$

while all other commutators have to vanish. From (7.2.54) and (7.2.43) it follows that the quantized variables (7.2.52) and (7.2.53) obey the commutation relations

$$[Q(j,\mathbf{k}), P(j',\mathbf{k}')^\times]_- = i\hbar \Delta(\mathbf{k}-\mathbf{k}')\delta_{jj'} \tag{7.2.55}$$

and

$$[Q(j,\mathbf{k})^\times, P(j',\mathbf{k}')]_- = i\hbar \Delta(\mathbf{k}-\mathbf{k}')\delta_{jj'} \tag{7.2.56}$$

while all other commutators vanish.
The quantized Hamiltonian following from these coordinates is not very convenient for the further calculations. The standard form of decoupled harmonic oscillators can be achieved by a transformation to real normal coordinates. We apply the transformation

$$Q(j,\mathbf{k}) = \frac{1}{2}[q(j,-\mathbf{k}) + q(j,\mathbf{k})] + \frac{i}{2\omega(j,\mathbf{k})}[\dot{q}(j,\mathbf{k}) - \dot{q}(j,-\mathbf{k})] \tag{7.2.57}$$

$$P(j,\mathbf{k}) = \dot{Q}(j,\mathbf{k}) = \frac{i}{2}\omega(j,\mathbf{k})[q(j,-\mathbf{k}) - q(j,\mathbf{k})] + \frac{1}{2}[\dot{q}(j,-\mathbf{k}) + \dot{q}(j,\mathbf{k})] \tag{7.2.58}$$

and obtain by substitution into (7.2.49)

$$H^0_{\text{latt}} = \frac{1}{2} \sum_{j\mathbf{k}} [\dot{q}(j,\mathbf{k})^2 + \omega^2(j,\mathbf{k}) q(j,\mathbf{k})^2] \tag{7.2.59}$$

which shows that this transformation is canonical as $q(j,\mathbf{k})$ and $\dot{q}(j,\mathbf{k})$ are canonical conjugates. The transition to quantum variables with

$$\dot{q}(j,\mathbf{k}) = p(j,\mathbf{k}) \to -i\hbar\partial/\partial q(j,\mathbf{k}) \tag{7.2.60}$$

then leads to a Schrödinger equation

$$\frac{1}{2}\sum_{j,\mathbf{k}} \left[-\hbar^2 \frac{\partial^2}{\partial q(j,\mathbf{k})^2} + \omega^2(j,\mathbf{k}) q(j,\mathbf{k})^2\right] |\varphi^0_m\rangle^h = (E^0_m - U^0_0)|\varphi^0_m\rangle^h \tag{7.2.61}$$

which is of the type (2.2.25).

Another very useful representation is given by creation and destruction operators which were already used in the Sections 2.6, 3.4 and 5.5. We introduce this representation here by defining the transformation for the classical operators

$$b(j,\mathbf{k})^\times = [\omega(j,\mathbf{k})/2\hbar]^{1/2}\left[-\frac{i}{\omega(j,\mathbf{k})}\dot{q}(j,\mathbf{k})+q(j,\mathbf{k})\right] \tag{7.2.62}$$

$$b(j,\mathbf{k}) = [\omega(j,\mathbf{k})/2\hbar]^{1/2}\left[\frac{i}{\omega(j,\mathbf{k})}\dot{q}(j,\mathbf{k})+q(j,\mathbf{k})\right]. \tag{7.2.63}$$

With this transformation the Hamiltonian (7.2.59) is changed into

$$H^0_{\text{latt}} = \sum_{j,\mathbf{k}} \hbar\omega(j,\mathbf{k})b(j,\mathbf{k})^\times b(j,\mathbf{k}) \tag{7.2.64}$$

and quantization can be performed by the substitution of $b(j,\mathbf{k})^\times$ through the Hermitian conjugate $b^+(j,\mathbf{k})$ with the commutation relation

$$[b(j,\mathbf{k}),\ b^+(j',\mathbf{k}')]_- = \delta_{jj'}\delta_{\mathbf{kk}'} \tag{7.2.65}$$

while all other commutators vanish. In terms of these operators the transformation (7.2.39) can be written

$$\mathbf{u}^0_{l\kappa} = \sum_{j\mathbf{k}} \eta(\kappa|j,\mathbf{k})\mathbf{e}(\kappa|j,\mathbf{k})N^{-1/2}\exp(i\mathbf{k}\cdot\mathbf{R}^0_l)\,[b(j,\mathbf{k})-b^+(j,-\mathbf{k})] \tag{7.2.66}$$

with

$$\eta(\kappa|\mathrm{j},\mathbf{k}) := [\hbar/2M_\kappa\omega(j,\mathbf{k})]^{1/2} \tag{7.2.67}$$

and the eigenvalue equation (7.2.61) reads

$$\sum_{j\mathbf{k}} \hbar\omega(j,\mathbf{k})b^+(j,\mathbf{k})\,b(j,\mathbf{k})|\varphi^0_m\rangle^h = (E^0_m - U^0_0)|\varphi^0_m\rangle^h \tag{7.2.68}$$

where of course the eigenvalues are the same as those of equations (2.2.26). The corresponding eigenfunctions are of a very simple form. They read

$$|\varphi^0_m\rangle^h = \left\{\prod_{\alpha=1}^{N} m_\alpha^{-1/2}\,[b^+(j_\alpha,\mathbf{k}_\alpha)]^{m_\alpha}\right\}|\varphi^0_0\rangle^h \tag{7.2.69}$$

where $|\varphi^0_0\rangle^h$ is the groundstate and the product runs over all possible **k**-vectors and branches which are enumerated by a single index α. If the crystal has periodic boundary conditions and a finite volume, then the set $\{j_\alpha,\ \mathbf{k}_\alpha,\ 1\leqslant\alpha\leqslant N\}$ has only a finite number of elements. The quantities $m_\alpha \equiv m(j_\alpha,\ \mathbf{k}_\alpha)$ are the occupation numbers of the modes.

7.3 Phonon modes of polar crystals

To derive phonon modes of ideal crystals, the potential energy $U_0(\mathbf{R})$ for the groundstate of the crystal has to be calculated. As in $U_0(\mathbf{R})$ the lattice coordinates $\mathbf{R}:=\{\mathbf{R}_{l\mu}\}$ may be chosen arbitrarily, the notation "ideal" refers to the condition that the evaluation of the equilibrium condition $\partial U_0/\partial\mathbf{R}_{l\mu}=0\forall l\mu$ leads to the equilibrium positions $\{\mathbf{R}_{l\mu}=\mathbf{R}^0_{l\mu}\}$ of the ideal crystal. For polar crystals the formalism of Chapter 5 enables us to perform such calculations. The corresponding formulas are given by (5.1.32) (5.1.33) which read

$$U_0(\mathbf{R})=\underset{\{\mathbf{m}\}}{\mathrm{Min}}\ U_0(\mathbf{R},\mathbf{m}) \tag{7.3.1}$$

with

$$U_0(\mathbf{R},\mathbf{m}):=\sum_{l\mu} H^0_{l\mu}+\frac{1}{2}\sum_{\substack{l\mu\\k\varrho}}{}' [a_{l\mu}a_{k\varrho}C(\mathbf{R}_{l\mu},\mathbf{R}_{k\varrho})+bC(\mathbf{R}_{l\mu},\mathbf{R}_{k\varrho})^n]+\frac{1}{2}\sum_{l\mu}\varepsilon_{l\mu}\mathbf{m}^2_{l\mu}$$
$$+\frac{1}{2}\sum_{\substack{l\mu\\k\varrho}}{}' [a_{k\varrho}\mathbf{m}_{l\mu}\cdot\nabla_{l\mu}+a_{l\mu}\mathbf{m}_{k\varrho}\cdot\nabla_{k\varrho}+(\mathbf{m}_{l\mu}\cdot\nabla_{l\mu})(\mathbf{m}_{k\varrho}\cdot\nabla_{k\varrho})]C(\mathbf{R}_{l\mu},\mathbf{R}_{k\varrho}) \tag{7.3.2}$$

where the definitions are given in Section 5.1.
Potentials of this kind were used for phonon calculations in two steps. In the first step the rigid ion model was considered, i.e., an approximation of (7.3.1) and (7.3.2) where the electronic dipoles were assumed to be zero. In this case the lattice energy reads

$$U_0(\mathbf{R})=\sum_{l\mu} H^0_{l\mu}+\frac{1}{2}\sum_{\substack{l\mu\\k\varrho}}{}' [a_{l\mu}a_{k\varrho}C(\mathbf{R}_{l\mu},\mathbf{R}_{k\varrho})+bC(\mathbf{R}_{l\mu},\mathbf{R}_{k\varrho})^n] \tag{7.3.3}$$

i.e., it contains central forces which are composed of a Coulomb part and a repulsive part. While the repulsive part leads to short-range interactions, the Coulomb part exerts long-range forces on the lattice. Due to this long-range character of the Coulomb forces, special care has to be taken in evaluating the force constants as the Coulomb series are only conditionally convergent.
We investigate the equations of motion of a crystal lattice (7.2.23) for the case where $U_0(\mathbf{R})$ is given by (7.3.3). According to the division into Coulomb forces and repulsive forces we write the force constants $\Phi^0(l\kappa,l'\kappa')$ as the sum of two terms, one representing the contribution from the Coulomb forces $\Phi^{\mathrm{C}}(l\kappa,l'\kappa')$, and the other representing the contribution from the repulsive forces $\Phi^{\mathrm{R}}(l\kappa,l'\kappa')$. The elements of the dynamical matrix $\mathscr{C}(\kappa,\kappa'|\mathbf{k})$ separate correspondingly into a Coulomb and a non-Coulomb part

$$\mathscr{C}(\kappa,\kappa'|\mathbf{k})=\mathscr{T}(\kappa,\kappa'|\mathbf{k})+\mathscr{R}(\kappa,\kappa'|\mathbf{k}) \tag{7.3.4}$$

with

$$\mathscr{T}(\kappa,\kappa'|\mathbf{k}):=(M_\kappa M_{\kappa'})^{-1/2}\sum_{l'}\Phi^{\mathrm{C}}(l\kappa,l'\kappa')\exp[-i\mathbf{k}\cdot(\mathbf{R}^0_{l\kappa}-\mathbf{R}^0_{l'\kappa'})] \tag{7.3.5}$$

and

$$\mathscr{R}(\kappa,\kappa'|\mathbf{k}):=(M_\kappa M_{\kappa'})^{-1/2}\sum_{l'}\Phi^{\mathrm{R}}(l\kappa,l'\kappa')\exp[-i\mathbf{k}\cdot(\mathbf{R}^0_{l\kappa}-\mathbf{R}^0_{l'\kappa'})]. \tag{7.3.6}$$

According to equations (7.1.31) (7.1.26), the contribution to the force constants from the Coulomb potential between ions is given by

$$\Phi^{\mathrm{C}}(l\kappa,l'\kappa')_{\alpha\beta}=-[\partial_\alpha\partial_\beta e_\kappa e_{\kappa'}\,C(\mathbf{r})]_{/\mathbf{r}=\mathbf{R}^0_{l\kappa}-\mathbf{R}^0_{l'\kappa'}},\quad l\kappa\neq l'\kappa' \tag{7.3.7}$$

$$\Phi^{\mathrm{C}}(l\kappa,l\kappa)_{\alpha\beta}=\sum_{\substack{l'\kappa'\\ \neq l\kappa}}[\partial_\alpha\partial_\beta e_\kappa e_{\kappa'}\,C(\mathbf{r})]_{/\mathbf{r}=\mathbf{R}^0_{l\kappa}-\mathbf{R}^0_{l'\kappa'}}. \tag{7.3.8}$$

Hence for $\kappa \neq \kappa'$ the matrix elements (7.3.5) can be written

$$T(\kappa,\kappa'|\mathbf{k})_{\alpha\beta} = -e_\kappa e_{\kappa'}(M_\kappa M_{\kappa'})^{-1/2} \sum_{l'} [\partial_\alpha \partial_\beta C(\mathbf{r})] \exp(-i\mathbf{k}\cdot\mathbf{r})_{/\mathbf{r}=\mathbf{R}^0_{l\kappa}-\mathbf{R}^0_{l'\kappa'}}. \tag{7.3.9}$$

For $\kappa = \kappa'$ we obtain from (7.3.5) due to (7.3.8) the formula

$$\begin{aligned} T(\kappa,\kappa|\mathbf{k})_{\alpha\beta} = {} & M_\kappa^{-1} \sum_{\substack{l'\kappa' \\ \neq l\kappa}} [e_\kappa e_{\kappa'} \partial_\alpha \partial_\beta C(\mathbf{r})]_{/\mathbf{r}=\mathbf{R}^0_{l\kappa}-\mathbf{R}^0_{l'\kappa'}} \\ & - M_\kappa^{-1} \sum_{\substack{l' \\ l'\neq l}} [e_\kappa e_{\kappa'} \partial_\alpha \partial_\beta C(\mathbf{r})] \exp(-i\mathbf{k}\cdot\mathbf{r})_{/\mathbf{r}=\mathbf{R}^0_{l\kappa}-\mathbf{R}^0_{l'\kappa}}. \end{aligned} \tag{7.3.10}$$

The first calculations for ionic crystals fully taking into account the difficulties with the long-range Coulomb forces were made by Kellermann (1940). His calculation technique is based on a summation technique of Ewald (1917), (1921) and is still considered to be basic for the treatment of Coulomb forces. Hence we describe it in detail. We use a slightly modified notation, etc., which was given by Maradudin, Montroll, Weiss, Ipatova (1971).

Theorem 7.10: The elements of the Coulomb part $\mathcal{T}$ of the dynamical matrix $\mathscr{C}$ can be represented by

$$\begin{aligned} T(\kappa,\kappa'|\mathbf{k})_{\alpha\beta} = {} & -e_\kappa e_{\kappa'}(M_\kappa M_{\kappa'})^{-1/2}\left[Q(\kappa,\kappa'|\mathbf{k})_{\alpha\beta} + \frac{4\pi}{v}\frac{k_\alpha k_\beta}{\mathbf{k}^2}\right] \\ & + \delta_{\kappa\kappa'} \sum_{\kappa''} e_\kappa e_{\kappa''} M_\kappa^{-1} Q(\kappa,\kappa''|0)_{\alpha\beta} \end{aligned} \tag{7.3.11}$$

with

$$\begin{aligned} Q(\kappa,\kappa'|\mathbf{k})_{\alpha\beta} = {} & -\frac{4\pi}{v}\frac{k_\alpha k_\beta}{\mathbf{k}^2}\left[\exp\left(-\frac{\mathbf{k}^2}{4P}\right)-1\right] \\ & -\frac{4\pi}{v}\sum_{\substack{\mathbf{g} \\ \neq 0}} \frac{(\mathbf{g}+\mathbf{k})_\alpha(\mathbf{g}+\mathbf{k})_\beta}{(\mathbf{g}+\mathbf{k})^2} \exp\left[-\frac{(\mathbf{g}+\mathbf{k})^2}{4P}\right] \exp(i\mathbf{g}\cdot\mathbf{r}_{\kappa\kappa'}) \\ & + P^{3/2}\sum_{l'} \partial_\alpha \partial_\beta H(P^{1/2}|\mathbf{R}^0_{l\kappa}-\mathbf{R}^0_{l'\kappa'}|) \exp[-i\mathbf{k}\cdot(\mathbf{R}^0_{l\kappa}-\mathbf{R}^0_{l'\kappa'})] \end{aligned} \tag{7.3.12}$$

and $H(x) = x^{-1}\,\mathrm{erf}\,c(x)$, where both summations in (7.3.12) are rapidly converging series.

Proof: Ewald has developed a summation procedure for the Coulomb potentials of periodic arrays of charges which change the slowly convergent expressions into rapidly converging ones. In order to apply this method to the present problem we have to change (7.3.10) into a periodic expression with respect to the point charge distribution, whereas (7.3.9) already has the appropriate periodic distribution. The former can be done if we remove the restriction $l'\kappa' \neq l\kappa$ in the first sum and the restriction $l' \neq l$ in the second sum of the right-hand side of equation (7.3.10). The two additional terms are singular but they are equal and cancel each other. Hence $\mathscr{T}(\kappa,\kappa|\mathbf{k})$ remains unchanged but is now represented by the difference of two expressions referring to periodic charge arrays. It is now possible to combine (7.3.9) and (7.3.10) into the single expression

$$\begin{aligned} T(\kappa,\kappa'|\mathbf{k})_{\alpha\beta} = &-(M_\kappa M_{\kappa'})^{-1/2} \sum_{l'} [e_\kappa e_{\kappa'} \partial_\alpha \partial_\beta C(\mathbf{r})] \exp(-i\mathbf{k}\cdot\mathbf{r})_{/\mathbf{r}=\mathbf{R}^0_{l\kappa}-\mathbf{R}^0_{l'\kappa'}} \\ &+\delta_{\kappa\kappa'} M_\kappa^{-1} \sum_{l'\kappa''} [e_\kappa e_{\kappa''} \partial_\alpha \partial_\beta C(\mathbf{r})]_{/\mathbf{r}=\mathbf{R}^0_{l\kappa}-\mathbf{R}^0_{l'\kappa''}}. \end{aligned} \tag{7.3.13}$$

If we define

$$\Phi^{\mathrm{C}}(\kappa,\kappa'|\mathbf{k})_{\alpha\beta} := -e_\kappa e_{\kappa'} \sum_{l'} [\partial_\alpha \partial_\beta C(\mathbf{r})] \exp(-i\mathbf{k}\cdot\mathbf{r})_{/\mathbf{r}=\mathbf{R}^0_{l\kappa}-\mathbf{R}^0_{l'\kappa'}} \tag{7.3.14}$$

then equation (7.3.13) can be written

$$\mathscr{T}(\kappa,\kappa'|\mathbf{k}) = (M_\kappa M_{\kappa'})^{-1/2} \Phi^{\mathrm{C}}(\kappa,\kappa'|\mathbf{k}) - \delta_{\kappa\kappa'} M_\kappa^{-1} \sum_{\kappa''} \Phi^{\mathrm{C}}(\kappa,\kappa''|0). \tag{7.3.15}$$

To evaluate (7.3.15) we perform a Fourier transformation of (7.3.14). If we denote the Fourier transform of $e_\kappa e_{\kappa'} C(\mathbf{r}) = \varphi^{\mathrm{C}}_{\kappa\kappa'}(\mathbf{r})$ by $\tilde{\varphi}^{\mathrm{C}}_{\kappa\kappa'}(\mathbf{k})$ and if we apply the formula

$$\sum_l \exp(i\mathbf{k}\cdot\mathbf{R}^0_l) = \frac{(2\pi)}{v} \sum_{\mathbf{g}} \delta(\mathbf{g}-\mathbf{k}) \tag{7.3.16}$$

where v is the unit cell volume and $\{\mathbf{g}\}$ the set of vectors of the reciprocal lattice, we obtain from (7.3.14) the expression

$$\Phi^{\mathrm{C}}(\kappa,\kappa'|\mathbf{k}) = \frac{1}{v} \sum_{\mathbf{g}} (\mathbf{g}+\mathbf{k})_\alpha (\mathbf{g}+\mathbf{k})_\beta \tilde{\varphi}^{\mathrm{C}}_{\kappa\kappa'}(\mathbf{g}+\mathbf{k}) \exp(i\mathbf{g}\cdot\mathbf{r}_{\kappa\kappa'}) \tag{7.3.17}$$

with $\mathbf{r}_{\kappa\kappa'} := \mathbf{x}(\kappa) - \mathbf{x}(\kappa')$. To obtain rapid convergence we decompose $\varphi^{\mathrm{C}}_{\kappa\kappa'}(\mathbf{r})$ into

$$\varphi^{\mathrm{C}}_{\kappa\kappa'}(\mathbf{r}) = \varphi^1_{\kappa\kappa'}(\mathbf{r}) + [\varphi^{\mathrm{C}}_{\kappa\kappa'}(\mathbf{r}) - \varphi^1_{\kappa\kappa'}(\mathbf{r})] =: \varphi^1_{\kappa\kappa'}(\mathbf{r}) + \varphi^2_{\kappa\kappa'}(\mathbf{r}) \tag{7.3.18}$$

where $\varphi^1_{\kappa\kappa'}(\mathbf{r})$ is assumed to be generated by a Gaussian charge distribution $\varrho(\mathbf{r}) = e_{\kappa'}(P/\pi)^{3/2} \exp(-Pr^2)$. The Fourier transform of $\varphi^1_{\kappa\kappa'}(\mathbf{r})$ reads

$$\tilde{\varphi}^1_{\kappa\kappa'}(\mathbf{k}) = (4\pi e_\kappa e_{\kappa'}/\mathbf{k}^2) \exp(-\mathbf{k}^2/4P) \tag{7.3.19}$$

while for $\varphi^2_{\kappa\kappa'}(\mathbf{r})$ the expression

$$\varphi^2_{\kappa\kappa'}(\mathbf{r}) = e_\kappa e_{\kappa'} 2(\pi)^{-1/2} \int_{(Pr)^{1/2}}^{\infty} \exp(-s^2)\, ds \tag{7.3.20}$$

in ordinary space results. Substitution of (7.3.18) into (7.3.14) leads correspondingly to

$$\Phi^{\mathrm{C}}(\kappa,\kappa'|\mathbf{k}) = \Phi^{\mathrm{C}}_1(\kappa,\kappa'|\mathbf{k}) + \Phi^{\mathrm{C}}_2(\kappa,\kappa'|\mathbf{k}). \tag{7.3.21}$$

While Φ_1^C is evaluated by means of (7.3.17), the original formula (7.3.14) is used for Φ_2^C. This gives

$$\Phi_1^C(\kappa,\kappa'|\mathbf{k})_{\alpha\beta} = \frac{4\pi}{v} \sum_{\mathbf{g}} e_\kappa e_{\kappa'} \frac{(\mathbf{g}+\mathbf{k})_\alpha (\mathbf{g}+\mathbf{k})_\beta}{(\mathbf{g}+\mathbf{k})^2} \times \exp\left[-\frac{(\mathbf{g}+\mathbf{k})^2}{4P}\right] \exp(i\mathbf{g}\cdot\mathbf{r}_{\kappa\kappa'}) \tag{7.3.22}$$

and

$$\Phi_2^C(\kappa,\kappa'|\mathbf{k})_{\alpha\beta} = -e_\kappa e_{\kappa'} P^{3/2} \sum_{l'} H_{\alpha\beta}(P^{1/2}|\mathbf{R}_{l\kappa}^0 - \mathbf{R}_{l'\kappa'}^0|) \times \exp[-i\mathbf{k}\cdot(\mathbf{R}_{l\kappa}^0 - \mathbf{R}_{l'\kappa'}^0)] \tag{7.3.23}$$

with

$$H_{\alpha\beta}(x) = \frac{x_\alpha x_\beta}{x^2}\left[\frac{3}{x^3}\operatorname{erf} c(x) + 2\pi^{-1/2}\left(\frac{3}{x^2}+2\right)\exp(-x^2)\right] - \delta_{\alpha\beta}\left[\frac{1}{x^3}\operatorname{erf} c(x) + 2\pi^{-1/2}\exp(-x^2)\right] \tag{7.3.24}$$

and

$$\operatorname{erf} c(x) := 2\pi^{-1/2}\int_x^\infty \exp(-s^2)\,ds. \tag{7.3.25}$$

If we separate the summation in (7.3.22) into the singular term with $\mathbf{g}=0$ and the rest, then (7.3.15) with (7.3.21) (7.3.22) and (7.2.23) gives (7.3.11), Q.E.D.

Concerning the dicussion of the convergence of the various expressions introduced by the application of the Ewald procedure, we refer to the literature. Another point which has to be discussed are the singularities. Because the function $\partial_\alpha\partial_\beta H(x)$ is singular for $x=0$, we note from the definition of $\mathcal{Q}(\kappa,\kappa'|\mathbf{k})$ given by (7.3.12) that it is singular when $\kappa=\kappa'$, the divergence coming from the term $l'=l$ in the last sum. This is due to the inclusion of the energy of interaction of an ion with itself in obtaining the appropriate expression for $\mathcal{T}(\kappa,\kappa'|\mathbf{k})$. But this singularity is cancelled between the first and the third term in (7.3.11). Of course $\mathcal{Q}(\kappa,\kappa'|\mathbf{k})$ can be redefined in order to avoid such divergencies, but also for this procedure we refer to the literature, cf. Maradudin, Montroll, Weiss and Ipatova (1971).
For $\mathbf{k}=0$ the second term in (7.3.11) gives rise to a divergence, so that the lattice dynamical derivation of the equations of motion breaks down. The reason for this breakdown is that the problem of describing the motion of the lattice in polar crystals is not a purely electrostatic problem as is assumed by accepting the expression for $U_0(\mathbf{R})$. Because of the ion charges, there will be in general an electromagnetic field in the crystal. The proper way, therefore, of determining the dynamical coupling coefficients must be to start from Maxwell's equations and to find a proper solution of the problem. This leads to polaritons and the problem will be treated in the next section in complete generality.

The technique of deriving the coupling constants for polar crystals has been improved, cf. Maradudin et al. (1971), although its essentials remain the same. We prefer to give an elementary derivation in order to refer to examples of numerical calculations which were done by Kellermann (1940) and Karo (1959) in this way.
To perform numerical calculations the repulsive forces have also to be known. They will be derived after having specified the lattice type. The NaCl-lattices were investigated by Kellermann (1940).
In case of NaCl there are two different particles in the cell ($s=2$); but, as far as symmetry is concerned, it may be regarded as a simple lattice, i.e., the Na sites are entirely equivalent to the Cl sites. Therefore, to any lattice vector $\mathbf{r}^l_{\kappa\kappa'}$ there corresponds another one of the same magnitude and pointing in the opposite direction. The same is true for the reciprocal lattice. For this reason the coupling coefficients are all real

$$\begin{aligned} t(\kappa,\kappa'|\mathbf{k}) &:= -(M_\kappa M_{\kappa'})^{1/2}\mathscr{T}(\kappa,\kappa'|\mathbf{k}) \\ &= \sum_l \Phi^{\mathrm{C}}(l\kappa,0\kappa')\cos(2\pi\mathbf{k}\cdot\mathbf{r}^l_{\kappa\kappa'}), \quad \mathbf{r}^l_{\kappa\kappa'} := \mathbf{R}^0_{l\kappa}-\mathbf{R}^0_{0\kappa'} \end{aligned} \tag{7.3.26}$$

and since $\mathbf{r}_{\kappa\kappa'} = -\mathbf{r}_{\kappa'\kappa}$ we have

$$t(\kappa,\kappa'|\mathbf{k})_{\alpha\beta} = t(\kappa',\kappa|\mathbf{k})_{\alpha\beta}. \tag{7.3.27}$$

From (7.3.14) (7.3.15) it can be seen that the coefficients are entirely symmetrical in α and β so that

$$t(\kappa,\kappa'|\mathbf{k})_{\alpha\beta} = t(\kappa',\kappa|\mathbf{k})_{\beta\alpha}. \tag{7.3.28}$$

Furthermore, the NaCl lattice is entirely symmetrical in Na and Cl, therefore

$$t(1,1|\mathbf{k})_{\alpha\beta} = t(2,2|\mathbf{k})_{\alpha\beta}. \tag{7.3.29}$$

The cell vectors of the NaCl lattice are

$$\begin{aligned} &\mathbf{a}_1 = r_0(0,1,1); \quad \mathbf{r}_{21} = -\mathbf{r}_{12} = r_0(1,1,1) \\ &\mathbf{a}_2 = r_0(1,0,1); \quad v = 2r_0^3 \\ &\mathbf{a}_3 = r_0(1,1,0) \end{aligned} \tag{7.3.30}$$

where r_0 is the distance between the nearest neighbours i.e., the lattice constant and v the volume of the cell. A lattice vector is therefore of the form

$$\begin{aligned} \mathbf{r}^l_{22} = \mathbf{r}^l_{11} &= r_0(l_2+l_3, l_3+l_1, l_1+l_2) = r_0(l_x, l_y, l_z) \\ \mathbf{r}^l_{21} &= r_0(l_2+l_3+1, l_3+l_1+1, l_1+l_2+1) = r_0(m_x, m_y, m_z). \end{aligned} \tag{7.3.31}$$

Here l_1, l_2, l_3 cover all integers. Therefore, l_x, l_y, l_z cover all sets of integers for which $l_x+l_y+l_z = \sum_x l_x$ is even and m_x, m_y, m_z cover all sets of integers for which $\sum_x m_x$ is odd.

The reciprocal vectors are given by

$$\mathbf{b}_1=\frac{1}{2r_0}(-1,\,1,\,1)$$
$$\mathbf{b}_2=\frac{1}{2r_0}(1,\,-1,1) \qquad (7.3.32)$$
$$\mathbf{b}_3=\frac{1}{2r_0}(1,1,\,-1)$$

so that a vector in the reciprocal lattice reads

$$\mathbf{b}_h=\frac{1}{2r_0}(h_2+h_3-h_1,h_3+h_1-h_2,h_1+h_2-h_3)$$
$$=\frac{1}{2r_0}(h_x,h_y,h_z) \qquad (7.3.33)$$

where h_x, h_y, h_z cover all sets of integers which are either all odd or all even, i.e., the reciprocal lattice is body-centered. The sums (7.3.11) representing the coupling coefficients may now be written for the wave vector $\mathbf{k}$:

$$\mathbf{k}=k_1\mathbf{b}_1+k_2\mathbf{b}_2+k_3\mathbf{b}_3=\frac{1}{2r_0}(q_x,q_y,q_z)$$
$$q_x=k_2+k_3-k_1\,;\; q_y=k_3+k_1-k_2\,;\; q_z=k_1+k_2-k_3\,. \qquad (7.3.34)$$

From the condition of the cyclic lattice $\mathbf{u}(l+L,\kappa)=\mathbf{u}(l\kappa)$ it follows that k_1,k_2,k_3 range from 0 to 1. In order to make full use of the symmetry properties of the coefficients it is more convenient to consider the region of allowed wave vectors in the q_x,q_y,q_z space. Since the reciprocal lattice is of the body-centered type, this region is of the form of an octahedron with its vertices cut off. The boundaries are given by the equations

$$q_x\pm q_y\pm q_z=\pm\tfrac{3}{2}\,;\; q_x=\pm1\,;\; q_y=\pm1\,;\; q_z=\pm1\,. \qquad (7.3.35)$$

In view of the properties just considered one may restrict the calculation to positive values of q_x,q_y,q_z. Furthermore, the calculation can be restricted to sets of numbers such that $q_x\geqslant q_y\geqslant q_z$. Thus only those values of $\mathbf{q}$ such that

$$0\leqslant q_z\leqslant q_y\leqslant q_x\leqslant 1$$
$$q_x+q_y+q_z\leqslant\tfrac{3}{2} \qquad (7.3.36)$$

need be considered.

Apart from the Coulomb forces there are other forces present in an ionic lattice, mainly the repulsive forces and also the Van der Waals forces. Let us collect all those other forces in a potential $\varphi^{\mathrm{R}}(\mathbf{r})$. These forces decrease very rapidly with distance, with the exception of the Van der Waals forces. But the latter forces form only a very small percentage of the potential $\varphi^{R}(\mathbf{r})$, so that with respect to this potential it is sufficient to consider only the interaction between nearest neighbours. In the NaCl lattice each ion is surrounded by six nearest neighbours so that the energy per cell is given by

$$\frac{1}{2}u_0 = -\alpha\frac{e^2}{r_0} + 6\varphi^R(r_0) \tag{7.3.37}$$

where α is the Madelung constant and r_0 the distance between nearest neighbours. The first two derivatives of $\varphi^R(\mathbf{r})$ can be obtained from the condition of equilibrium and the compressibility; putting for abbreviation

$$\begin{aligned} \frac{e^2}{2r_0^2}B &= 2\left[\frac{d\varphi^R(\mathbf{r})}{dr}\right]_{/\mathbf{r}=\mathbf{r}_0} \\ \frac{e^2}{2r_0^3}A &= 2\left[\frac{d^2\varphi^R(\mathbf{r})}{dr^2}\right]_{/\mathbf{r}=\mathbf{r}_0} \end{aligned} \tag{7.3.38}$$

the condition of equilibrium is

$$\frac{du_0}{dr_0} = 0 = 2\alpha\frac{e^2}{r_0^2} + 3B\frac{e^2}{r_0^2}$$

so that

$$B = -\frac{2\alpha}{3} = -1{,}165 \tag{7.3.39}$$

using Madelung's constant $\alpha = 1{,}7476$. The compressibility is given by

$$\kappa^{-1} = \frac{1}{18v}r_0^2\frac{d^2u_0}{dr_0^2} = \frac{1}{18r_0}\left[-2\alpha\frac{e^2}{r_0^3} + \frac{3}{2}A\frac{e^2}{r_0^3}\right]$$

so that

$$A = \frac{12r_0^4}{\kappa e^2} + \frac{4}{3}\alpha. \tag{7.3.40}$$

The constant A depends, of course, on the particular crystal considered. For NaCl the constants are $\kappa = 4.16 \times 10^{-12}$ cm^2/dyne, $r_0 = 2.814 \times 10^{-8}$ cm. With $e = 4.8 \times 10^{-10}$ e.s.u.,

$$A = 10.18 \quad \text{(for NaCl)}. \tag{7.3.41}$$

We consider now those parts of the coupling coefficients which are due to the repulsive forces. They are given by

$$\begin{aligned} \imath(\kappa,\kappa'|\mathbf{k}) &:= -(M_\kappa M_{\kappa'})^{1/2}\mathscr{R}(\kappa,\kappa'|\mathbf{k}) \\ &= \sum_l \Phi^R(l\kappa, 0\kappa')\exp(i\mathbf{k}\cdot\mathbf{r}^l_{\kappa\kappa'}). \end{aligned} \tag{7.3.42}$$

In the case $\kappa' \neq \kappa$, the sum extends over the six nearest neighbours given by the vectors

$$r_0(\pm 1, 0, 0)$$
$$r_0(0, \pm 1, 0) \qquad (7.3.43)$$
$$r_0(0, 0, \pm 1).$$

The differentiation in (7.3.42) yields with the definition (7.3.38)

$$r(1,2|\mathbf{k})_{\alpha\beta} = \frac{1}{2}\frac{e^2}{v}\sum_l \left[A\,\frac{(\mathbf{r}_{12}^l)_\alpha\,(\mathbf{r}_{12}^l)_\beta}{(\mathbf{r}_{12}^l)^2} + B\left(\delta_{\alpha\beta} - \frac{(\mathbf{r}_{12}^l)_\alpha(\mathbf{r}_{12}^l)_\beta}{(\mathbf{r}_{12}^l)^2}\right)\right] e^{i\mathbf{k}\cdot\mathbf{r}_{12}^l}. \qquad (7.3.44)$$

The sum over the vectors (7.3.43) gives

$$r(1,2|\mathbf{k})_{\alpha\beta} = \frac{v}{e^2}\,\delta_{\alpha\beta}\,[A\cos 2\pi k_x r_0 + B(\cos 2\pi k_y r_0 + \cos 2\pi k_z r_0)]. \qquad (7.3.45)$$

In the case $\kappa' = \kappa$ only the term with $l = 0$ in (7.3.42) remains, while all other terms representing interactions between distant ions are omitted:

$$r(1,1|\mathbf{k}) = \Phi^{\mathrm{R}}(0\kappa, 0\kappa). \qquad (7.3.46)$$

This expression is independent of $\mathbf{k}$ and can therefore be calculated from

$$r(1,1|0) + r(1,2|0) = 0$$

so that

$$r(1,1|\mathbf{k})_{\alpha\beta} = -\delta_{\alpha\beta}\,\frac{e^2}{v}\,(A + 2B). \qquad (7.3.47)$$

If for $\mathbf{k}$ the dimensionless vector $\mathbf{q}$ given by (7.3.34) is introduced, (7.3.45) and (7.3.47) may be written as

$$r(1,1|\mathbf{k})_{\alpha\beta} = -\delta_{\alpha\beta}\,\frac{e^2}{v}\,(A + 2B)$$
$$r(1,2|\mathbf{k})_{\alpha\beta} = \delta_{\alpha\beta}\,\frac{e^2}{v}\,[A\cos\pi q_x + B(\cos\pi q_y + \cos\pi q_z)]. \qquad (7.3.48)$$

Of course the calculation of A and B is equivalent to the calculation of the power coefficient η and the constant b in the repulsive potential of (7.3.3).
By this method the frequency distributions of the various branches can also be calculated by simply counting the number of calculated frequencies in each interval. The calculations for NaCl done by Kellermann were extended by Karo (1959) for Lithium- and Sodium-halides. Some of his results are given in Table 7.1:

Table 7.1:
Sodium halide frequencies as a function of position in the reciprocal lattice. la = long. ac.; lo = long. opt.; a = transv. ac.; to = transv. opt.; after Karo (1959).

q_x, q_y, q_z	NaCl					
	la	lo	ta	ta	to	to
	(10^{13} sec^{-1})					
000	0.00	5.99	0.00	0.00	2.78	2.78
200	0.99	5.83	0.54	0.54	2.81	2.81
400	1.86	5.41	1.03	1.03	2.87	2.87
600	2.54	4.86	1.42	1.42	2.94	2.94
800	2.93	4.37	1.68	1.68	3.00	3.00
1000	3.05	4.19	1.77	1.77	3.02	3.02
111	0.83	5.91	0.50	0.50	2.76	2.76
222	1.66	5.67	0.98	0.98	2.69	2.69
333	2.46	5.30	1.39	1.39	2.58	2.58
444	3.20	4.83	1.73	1.73	2.44	2.44
555	3.62	4.49	1.89	1.89	2.34	2.34
220	1.33	5.75	0.78	0.83	2.68	2.83
440	2.36	5.06	1.61	1.69	2.43	2.96
660	2.85	4.02	1.92	2.57	2.48	3.25
333	2.46	5.30	1.39	1.39	2.58	2.58
442	2.68	5.04	1.60	1.64	2.43	2.77
551	2.80	4.58	1.87	2.08	2.36	3.02
660	2.85	4.02	1.92	2.57	2.48	3.25
771	2.98	3.76	1.85	2.69	2.62	3.29

Subsequent calculations by Karo and Hardy (1966) with the rigid ion model (and other models) were improved by a group theoretical classification of the wave vectors and by an investigation of critical points which are closely related to symmetries. Critical points in the Brillouin zone of phonon wave vectors had first been discussed by Van Hove (1953) and Phillips (1956). They occur in the frequency distribution of the phonons. If we define $g(\omega)d\omega$ as the number of phonon modes which have a frequency between ω and $\omega+d\omega$, then in three dimensions we have

$$g(\omega)d\omega = \frac{Nv}{(2\pi)^3}\int d^3q \tag{7.3.49}$$

where the integral is extended over that region of the Brillouin zone for which the frequency $\omega(q)$ lies between ω and $\omega+d\omega$, the factor $Nv/(2\pi)^3$ being the density of points representing the wave vectors $\mathbf{q}$. Let dS be an element of the area on the surface for which the frequency is equal to ω. The element of the volume in reciprocal space between constant frequency surfaces ω and $\omega+d\omega$ is

$$d^3q = dS d\omega |\nabla_{\mathbf{q}}\omega(\mathbf{q})|^{-1}; \tag{7.3.50}$$

thus allowing r different branches of the dispersion curve we have

$$g(\omega)=\frac{Nv}{(2\pi)^3}\sum_{j=1}^{3r}\int_{\omega=\omega_j(\mathbf{q})} dS|\nabla_{\mathbf{q}}\omega(j\mathbf{q})|^{-1}. \tag{7.3.51}$$

The critical points are those for which $\nabla_{\mathbf{q}}\omega(j\mathbf{q})=0$. Obviously this gives rise to a singularity of the integrand and thus to singularities in the frequency distribution, which is characteristic of the phonon spectrum of a substance. For further discussion of such points we refer to the literature and give only an example of improved calculations in the rigid ion model for NaCl which were made by Karo and Hardy (1966) cf. Figures 7.1 and 7.2:

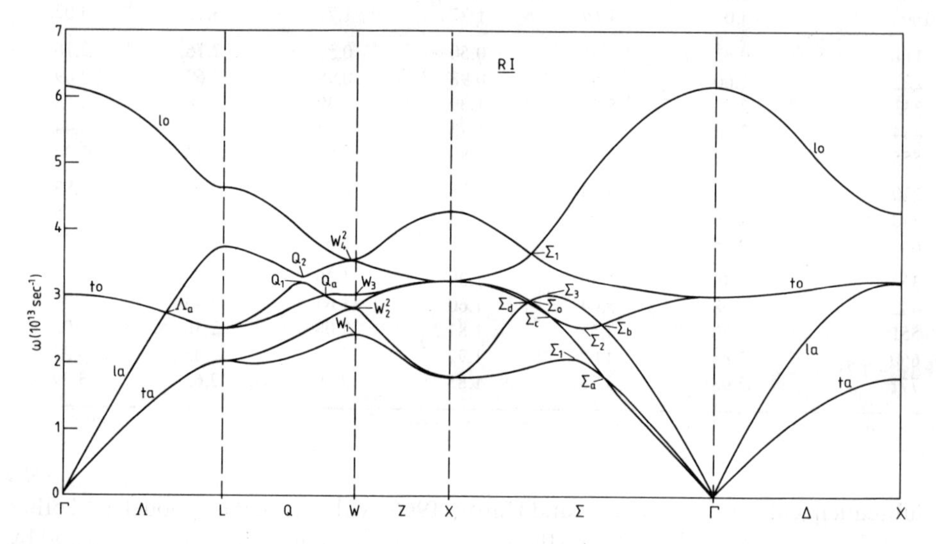

Figure 7.1:
Single-phonon dispersion curves showing the group-theoretic identification of the various directions and symmetry points in the rigid ion model = RI. The phonon branches are classified as optic and acoustic (o and a) and transverse and longitudinal (t and l) along Λ and Δ, and at L, Γ, and X. Then the critical points (c.p.) are labeled by the branch and direction, or point; e.g. la (Δ) or la (X). For other c.p. a similar q vector identification is used with numerical subscripts labeling the extrema in order of increasing frequency, and superscripts denoting the degeneracy; e.g. W_1^2 is the lowest frequency c.p. at W, and it is twofold degenerate. In a similar manner alphabetic subscripts are used to identify crossover singularities; after Karo and Hardy (1966).

After having discussed the rigid ion model based on the expression (7.3.3) for the potential energy $U_0(\mathbf{R})$ of the ideal crystal, we proceed to the treatment of the complete energy expression (7.3.2) including the electronic dipole moments $\{\mathbf{m}_{l\mu}\}$. As in the preceding sections we expand $U_0(\mathbf{R},\mathbf{m})$ in powers of $\{\mathbf{u}_{l\mu}^0\}$ and consider only the harmonic part $U_0^h(\mathbf{R},\mathbf{m})$ of the full energy. This leads to the following theorem:

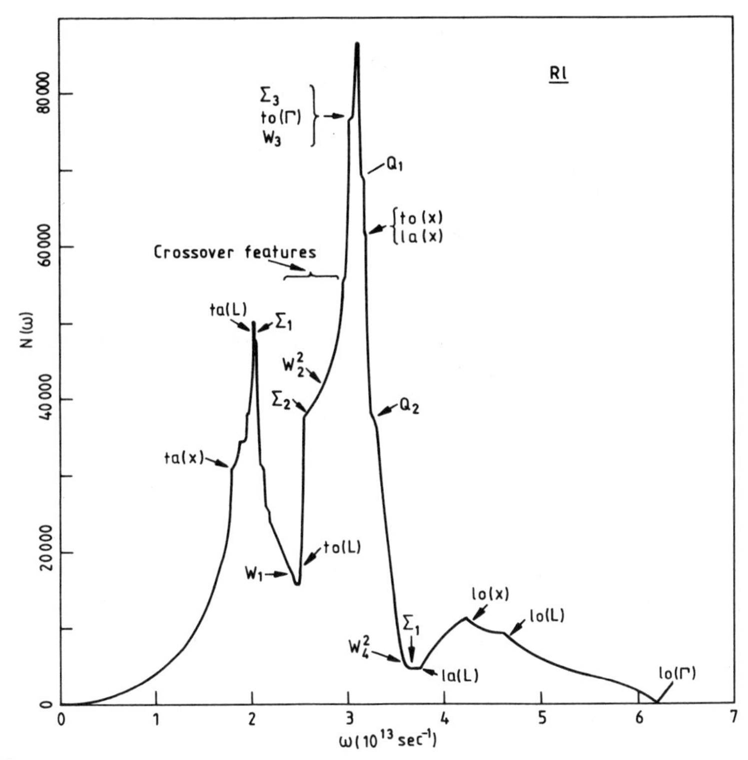

Figure 7.2:
Single-phonon frequency distributions for the rigid ion model = RI showing the various critical points inferred from the dispersion curves of Fig. 7.1 using the notation of that figure; after Karo and Hardy (1966).

Theorem 7.11: The phonon modes of an ideal crystal whose potential energy contains electronic polarizations follow from the solution of the secular equation

$$M_\kappa \omega^2(j,\mathbf{k})\mathbf{q}(\kappa|j,\mathbf{k}) = \frac{1}{2}\sum_{\kappa'} c(\kappa,\kappa'|\mathbf{k}) \cdot \mathbf{q}(\kappa'|j,\mathbf{k}) + \frac{1}{2}\sum_{\lambda,\varrho,\varrho'} c(\kappa,\lambda|\mathbf{k})a_\lambda^{-1} \cdot [1-\mathscr{L}(\lambda,\varrho|\mathbf{k})]^{-1} \cdot \mathscr{U}(\varrho,\varrho'|\mathbf{k}) \cdot \mathbf{q}(\varrho'|j,\mathbf{k}) \tag{7.3.52}$$

where the symbols are explained by (7.3.64).

Proof: We expand (7.3.3) in powers of the dynamical displacements (7.1.4) and obtain

$$U_0^h(\mathbf{u}) = U_0 + \frac{1}{2}\sum_{\substack{l\kappa \\ l'\kappa'}} \Phi^{0r}(l\kappa, l'\kappa') \cdot\cdot\, \mathbf{u}_{l\kappa}^0 \mathbf{u}_{l'\kappa'}^0 \tag{7.3.53}$$

where U_0 is defined by (5.2.33) and the superscript r means rigid. By definition the repulsive term in (7.3.3) is assumed to have parameters b and η such that the equilibrium condition for the rigid ion model at the ideal lattice sites $\Phi^{0r}(l\kappa)=0$ is satisfied. Hence in (7.3.53) no linear term occurs. Expanding also the last term on the right-hand side of (7.3.2) in powers of **u**, we obtain by observing (5.2.29)

$$U_0^h(\mathbf{u},\mathbf{m})=U_0+\frac{1}{2}\sum_{\substack{l\kappa\\ l'\kappa'}}\Phi^{0r}(l\kappa,l'\kappa')\cdot\cdot\,\mathbf{u}_{l\kappa}^0\mathbf{u}_{l'\kappa'}^0+\frac{1}{2}\sum_{l\mu}\varepsilon_\mu\mathbf{m}_{l\mu}^2$$
$$+\frac{1}{2}\sum_{\substack{l\mu\\ k\varrho}}{}'\,[(\mathbf{m}_{l\mu}\cdot\nabla_{l\mu})(\mathbf{m}_{k\varrho}\cdot\nabla_{k\varrho})+2a_{k\varrho}(\mathbf{m}_{l\mu}\cdot\nabla_{l\mu})(\mathbf{u}_{k\varrho}^0\cdot\nabla_{k\varrho})]C(l\mu,k\varrho). \tag{7.3.54}$$

The equilibrium condition of the lattice energy including electronic polarization leads to (7.3.1) which gives for the harmonic energy part (7.3.54) the equations

$$\mathbf{m}_{r\kappa}=-\varepsilon_\kappa^{-1}\sum_{k\varrho}{}'\,[\varphi^{C}(r\kappa,k\varrho)\cdot\mathbf{m}_{k\varrho}+a_{k\varrho}\varphi^{C}(r\kappa,k\varrho)\cdot\mathbf{u}_{k\varrho}^0] \tag{7.3.55}$$

with

$$\varphi^{C}(r\kappa,k\varrho):=e_\kappa^{-1}e_\varrho^{-1}\Phi^{C}(r\kappa,k\varrho). \tag{7.3.56}$$

In Section 5.2 these equations were solved directly for imperfect lattices by applying an appropriate Green function. For ideal lattices we use the translational invariance instead of Green functions. Since $\mathbf{m}_{k\varrho}$ as well as $\mathbf{u}_{k\varrho}^0$ are dynamical variables we can apply the ansatz

$$\mathbf{u}_{l\kappa}^0=\sum_{\mathbf{k}}\mathbf{q}(\kappa|\mathbf{k})\exp(i\mathbf{k}\cdot\mathbf{R}_l^0)$$
$$\mathbf{m}_{r\kappa}=\sum_{\mathbf{k}}\mathbf{m}(\kappa|\mathbf{k})\exp(i\mathbf{k}\cdot\mathbf{R}_r^0) \tag{7.3.57}$$

where **k** is summed over the first Brillouin zone and $\mathbf{q}(\kappa|\mathbf{k})$ as well as $\mathbf{m}(\kappa|\mathbf{k})$ are vector coordinates. Substitution of (7.3.57) into (7.3.55) gives then

$$\sum_{\mathbf{k}}\mathbf{m}(\kappa|\mathbf{k})\exp(i\mathbf{k}\cdot\mathbf{R}_r^0)=$$
$$-\sum_{k\varrho}{}'\sum_{\mathbf{k}}[\varepsilon_\kappa^{-1}\varphi^{C}(r\kappa,k\varrho)\mathbf{m}(\varrho|\mathbf{k})+a_\varrho\varepsilon_\kappa^{-1}\varphi^{C}(r\kappa,k\varrho)\mathbf{q}(\varrho|\mathbf{k})]\exp(i\mathbf{k}\cdot\mathbf{R}_k^0). \tag{7.3.58}$$

We multiply Eq. (7.3.58) by $\exp(-i\mathbf{k}'\cdot\mathbf{R}_r^0)$ and sum over r. With

$$\sum_l\exp(i\mathbf{k}\cdot\mathbf{R}_l^0)=\frac{(2\pi)^3}{v}\sum_{\mathbf{g}}\delta(\mathbf{g}-\mathbf{k}) \tag{7.3.59}$$

where v is the unit cell volume and **g** a vector of the reciprocal lattice, we obtain for the left-hand side of (7.3.58)

$$\sum_{\mathbf{k}}\sum_r\mathbf{m}(\kappa|\mathbf{k})\exp[-i(\mathbf{k}'-\mathbf{k})\cdot\mathbf{R}_r^0]$$
$$=\frac{(2\pi)^3}{v}\sum_{\mathbf{g},\mathbf{k}}\delta(\mathbf{g}-\mathbf{k}'+\mathbf{k})\mathbf{m}(\kappa|\mathbf{k})=\mathbf{m}(\kappa|\mathbf{k}') \tag{7.3.60}$$

by observing the domain of summation over $\mathbf{k}$. Then Eq. (7.3.58) reads

$$\mathbf{m}(\kappa|\mathbf{k}')\,\frac{(2\pi)^3}{v}= \sum_r\sum_{k\varrho}\sum_{\mathbf{k}} \varepsilon_\kappa^{-1}\varphi^C(r\kappa,k\varrho)\cdot[\mathbf{m}(\varrho|\mathbf{k})+a_\varrho\mathbf{q}(\varrho|\mathbf{k})]\exp[-i\mathbf{k}'\cdot\mathbf{R}_r^0+i\mathbf{k}\cdot\mathbf{R}_k^0]. \tag{7.3.61}$$

To evaluate the right-hand side of (7.3.61) we introduce the new summation coordinates

$$\mathbf{R}_h^0:=\tfrac{1}{2}(\mathbf{R}_k^0+\mathbf{R}_r^0)$$
$$\mathbf{R}_l^0:=\tfrac{1}{2}(\mathbf{R}_k^0-\mathbf{R}_r^0) \tag{7.3.62}$$

and observe (7.1.18) as well as (7.3.59). Then we obtain from (7.3.61) the following equation

$$\mathbf{m}(\kappa|\mathbf{k}')=\sum_\varrho[\mathscr{L}(\kappa,\varrho|\mathbf{k}')\cdot\mathbf{m}(\varrho|\mathbf{k}')+\mathscr{U}(\kappa,\varrho|\mathbf{k}')\cdot\mathbf{q}(\varrho|\mathbf{k}')] \tag{7.3.63}$$

with

$$\mathscr{L}(\kappa,\varrho|\mathbf{k}):=\sum_l{}'\,\varepsilon_\kappa^{-1}\varphi^C(0\kappa,2l,\varrho)\exp(2i\mathbf{k}\cdot\mathbf{R}_l^0)$$
$$\mathscr{U}(\kappa,\varrho|\mathbf{k}):=\sum_l{}'\,\varepsilon_\kappa^{-1}\varphi^C(0\kappa,2l,\varrho)\exp(2i\mathbf{k}\cdot\mathbf{R}_l^0). \tag{7.3.64}$$

It has to be emphasized that, in contrast to the lattice sums for the ions (7.3.8), for $\varrho=\kappa$ no divergence occurs, as the interactions of the electronic dipoles with their surroundings do not lead to central forces. This simplifies the treatment of equation (7.3.63) and its formal solution reads

$$\mathbf{m}(\kappa|\mathbf{k})=\sum_{\varrho\varrho'}[\delta_{\kappa\varrho}\mathbb{1}-\mathscr{L}(\kappa,\varrho|\mathbf{k})]^{-1}\cdot\mathscr{U}(\varrho,\varrho'|\mathbf{k})\cdot\mathbf{q}(\varrho'|\mathbf{k}). \tag{7.3.65}$$

This expression has to be substituted into (7.3.54). To do this we observe that the equation

$$\frac{1}{2}\sum_{r\kappa}\varepsilon_\kappa\mathbf{m}_{r\kappa}^2=-\frac{1}{2}\sum_{\substack{k\varrho\\ r\kappa}}{}'\,[\mathbf{m}_{r\kappa}\cdot\varphi^C(r\kappa,k\varrho)\cdot\mathbf{m}_{k\varrho}+\mathbf{m}_{r\kappa}\cdot a_{k\varrho}\varphi^C(r\kappa,k\varrho)\cdot\mathbf{u}_{k\varrho}^0] \tag{7.3.66}$$

follows from (7.3.55). Hence we obtain for the equilibrium value of (7.3.54)

$$\operatorname*{Min}_{\{\mathbf{m}\}}U_0^h(u^0,\mathbf{m})=U_0+\frac{1}{2}\sum_{\substack{l\kappa\\ l'\kappa'}}\Phi^0(l\kappa,l'\kappa')\cdot\cdot\,\mathbf{u}_{l\kappa}^0\mathbf{u}_{l'\kappa'}^0$$
$$+\frac{1}{2}\sum_{\substack{l\mu\\ k\varrho}}a_{k\varrho}\varphi^C(r\kappa,k\varrho)\cdot\cdot\,\mathbf{m}_{r\kappa}\mathbf{u}_{k\varrho}^0. \tag{7.3.67}$$

Substitution of (7.3.57) and (7.3.65) into (7.3.67) and observing the definition (7.2.7) gives with $c(\kappa,\kappa'|\mathbf{k})=(M_\kappa M_{\kappa'})^{1/2}\mathscr{D}^0(\kappa,\kappa'|\mathbf{k})$ the expression

$$U_0^h(\mathbf{q})=\underset{\{\mathbf{m}\}}{\mathrm{Min}}\, U_0^h(\mathbf{u}^0,\mathbf{m})$$

$$=U_0+\frac{1}{2}\sum_{\mathbf{k}}\sum_{\kappa\kappa'} c(\kappa,\kappa'|\mathbf{k})\cdot\cdot\,\mathbf{q}(\kappa|\mathbf{k})\mathbf{q}(\kappa'|\mathbf{k}) \tag{7.3.68}$$

$$+\frac{1}{2}\sum_{\mathbf{k}}\sum_{\substack{\kappa\lambda\\ \varrho\varrho'}} c(\kappa,\lambda|\mathbf{k})a_\lambda^{-1}\cdot[\delta_{\kappa\varrho}\mathbb{1}-\mathscr{L}(\kappa,\varrho|\mathbf{k})]^{-1}\cdot\mathscr{U}(\varrho,\varrho'|\mathbf{k})\cdot\cdot\,\mathbf{q}(\lambda|\mathbf{k})\mathbf{q}(\varrho'|\mathbf{k}).$$

If we now diagonalize the adiabatic energy operator

$$H_{\mathrm{latt}}^0=\sum_{\mathbf{k}\kappa}\frac{1}{2M_\kappa}\,\mathbf{p}(\kappa|\mathbf{k})^2+U_0^h(\mathbf{q}) \tag{7.3.69}$$

we have to solve the secular equation (7.3.52), Q.E.D.

For simplicity in the derivation of the modified secular equation (7.3.52) given above only the effect of the long-range Coulomb forces but not that of the short-range forces were taken into account. Therefore, the formulas obtained here are slightly different from those used by other authors. But this difference is not essential as long as no direct calculation of the coefficients is undertaken. So far these formulas were only used in order to have an appropriate scheme for the introduction of additional parameters into the secular equation so that a more satisfactory agreement between theory and experiment may be obtained.
In order to relate our secular equations (7.3.52) to those of other authors we reformulate them somewhat. In doing so it is convenient to express the diagonalization conditions (7.2.4) of the harmonic part U_0^h of U_0 by the equivalent equations

$$\omega_\varrho^2 M_\kappa \mathbf{u}_{l\kappa}^0=\sum_{l'\kappa'}\Phi^0(l\kappa,l'\kappa')\cdot\mathbf{u}_{l'\kappa'}^0=\frac{\partial U_0^h(\mathbf{R})}{\partial \mathbf{u}_{l\kappa}^0}. \tag{7.3.70}$$

If these conditions are combined with (7.3.1) we obtain

$$\omega_\varrho^2 M_\kappa \mathbf{u}_{l\kappa}^0=\frac{\partial U_0^h(\mathbf{R},\mathbf{m})}{\partial \mathbf{u}_{l\kappa}^0};\quad \frac{\partial U_0^h(\mathbf{R},\mathbf{m})}{\partial \mathbf{m}_{l\kappa}}=0\quad \forall l\kappa \tag{7.3.71}$$

as diagonalization conditions which have to be fulfilled simultaneously. Solving these equations one observes that any system of equations resulting from the substitution of equilibrium values in (7.3.71) is equivalent to this system.
We first consider the adiabatic equilibrium conditions (7.3.63) for the electronic dipoles which we rewrite in the form

$$\sum_\varrho[\delta_{\kappa\varrho}\varepsilon_\kappa+\mathscr{C}(\kappa,\varrho|\mathbf{k}]\cdot\mathbf{m}(\varrho|\mathbf{k})+\sum_\varrho\mathscr{C}(\kappa,\varrho|\mathbf{k})a_\varrho\cdot\mathbf{q}(\varrho|\mathbf{k})=0 \tag{7.3.72}$$

with the Coulomb interaction matrix

$$\mathscr{C}(\kappa,\varrho|\mathbf{k}):=\sum_l{}'\,\varphi^{\mathrm{C}}(0\kappa,2l,\varrho)\,\exp(2i\mathbf{k}\cdot\mathbf{R}_l^0). \tag{7.3.73}$$

We now introduce effective charges y_ϱ for the electronic dipoles

$$\mathbf{m}(\varrho|\mathbf{k}) = y_\varrho \mathbf{d}(\varrho|\mathbf{k}) \tag{7.3.74}$$

and with this definition and multiplication of (7.3.72) with y_κ we obtain

$$\sum_\varrho [\delta_{\kappa\varrho}\varepsilon_\kappa + \tilde{\mathscr{C}}(\kappa,\varrho|\mathbf{k})] y_\kappa y_\varrho \cdot \mathbf{d}(\varrho|\mathbf{k}) + \sum_\varrho \tilde{\mathscr{C}}(\kappa,\varrho|\mathbf{k}) y_\kappa a_\varrho \cdot \mathbf{q}(\varrho|\mathbf{k}) = 0. \tag{7.3.75}$$

If the equation (7.3.66) following from the adiabatic equilibrium condition is substituted into (7.3.54), then (7.3.67) results and for eigenmodes $\mathbf{u}^0_{l\kappa}(j,\mathbf{k})$ the first set of equations (7.3.71) reads

$$M_\kappa \omega^2(j,\mathbf{k}) \mathbf{u}^0_{l\kappa}(j,\mathbf{k}) = \sum_{l'\kappa'} \Phi^0(l\kappa, l'\kappa') \cdot \mathbf{u}^0_{l'\kappa'}(j,\mathbf{k}) + \frac{1}{2} \sum_{l'\kappa'} a_\kappa \varphi^{\mathrm{C}}(l'\kappa', l\kappa) \cdot \mathbf{m}_{l'\kappa'}(j,\mathbf{k}). \tag{7.3.76}$$

With (7.3.57) this set of equations can be transformed into

$$M_\kappa \omega^2(j,\mathbf{k}) \mathbf{q}(\kappa|j,\mathbf{k}) = \sum_{\kappa'} \mathscr{C}(\kappa,\kappa'|\mathbf{k}) \cdot \mathbf{q}(\kappa'|j,\mathbf{k}) + \frac{1}{2} \sum_{\kappa'} a_\kappa \tilde{\mathscr{C}}(\kappa,\kappa'|\mathbf{k}) y_\kappa \cdot \mathbf{d}(\kappa'|j,\mathbf{k}). \tag{7.3.77}$$

Equations (7.3.75) and (7.3.77) can be written as matrix equations if diagonal charge tensors are introduced

$$\mathfrak{Z} \Rightarrow \delta_{\kappa\kappa'} a_{\kappa'} \delta_{ij}; \quad \mathfrak{Y} \Rightarrow \delta_{\kappa\kappa'} y_{\kappa'} \delta_{ij} \tag{7.3.78}$$

where the Kronecker symbol refers to the cartesian components of the vectors $\mathbf{q}(\kappa|j,\mathbf{k})$ and $\mathbf{d}(\kappa|j,\mathbf{k})$, and if the notation

$$\mathfrak{S} \Rightarrow \delta_{\kappa\kappa'} \varepsilon_{\kappa'} \delta_{ij}; \quad \mathfrak{M} \Rightarrow \delta_{\kappa\kappa'} M_{\kappa'} \delta_{ij} \tag{7.3.79}$$

together with the decomposition (7.3.4) is used. Then equation (7.3.75) reads

$$(\mathfrak{S} + \mathfrak{Y} \cdot \tilde{\mathfrak{T}} \cdot \mathfrak{Y}) \cdot \mathfrak{d} + \mathfrak{Y} \cdot \tilde{\mathfrak{T}} \cdot \mathfrak{Z} \cdot \mathfrak{q} = 0 \tag{7.3.80}$$

and equation (7.3.77) takes the form

$$\mathfrak{M} \cdot \omega^2 \mathfrak{q} = (\mathfrak{R} + \mathfrak{Z} \cdot \tilde{\mathfrak{T}} \cdot \mathfrak{Z}) \cdot \mathfrak{q} + \tfrac{1}{2} \mathfrak{Z} \cdot \tilde{\mathfrak{T}} \cdot \mathfrak{Y} \cdot \mathfrak{d}. \tag{7.3.81}$$

Elimination of $\mathfrak{d}$ from (7.3.81) (7.3.80) gives

$$\begin{aligned} \mathfrak{M} \cdot \omega^2 \mathfrak{q} &= (\mathfrak{R} + \mathfrak{Z} \cdot \tilde{\mathfrak{T}} \cdot \mathfrak{Z}) \cdot \mathfrak{q} \\ &\quad - \tfrac{1}{2} \mathfrak{Z} \cdot \tilde{\mathfrak{T}} \cdot \mathfrak{Y} \cdot (\mathfrak{S} + \mathfrak{Y} \cdot \tilde{\mathfrak{T}} \cdot \mathfrak{Y})^{-1} \cdot \mathfrak{Y} \cdot \tilde{\mathfrak{T}} \cdot \mathfrak{Z} \cdot \mathfrak{q} \end{aligned} \tag{7.3.82}$$

which is equivalent to (7.3.52). Equations (7.3.82) are identical with the so-called shell model equations if short-range forces are neglected. If such forces are included they have to be added to the Coulomb term in a way corresponding to (7.3.4). This gives the set of equations

$$\begin{aligned} \mathfrak{M} \cdot \omega^2 \mathfrak{q} &= (\mathfrak{R} + \mathfrak{Z} \cdot \tilde{\mathfrak{T}} \cdot \mathfrak{Z}) \cdot \mathfrak{q} \\ &\quad - \tfrac{1}{2} (\mathfrak{R}' + \mathfrak{Z} \cdot \tilde{\mathfrak{T}} \cdot \mathfrak{Y}) \cdot (\mathfrak{S} + \mathfrak{Y} \cdot \tilde{\mathfrak{T}} \cdot \mathfrak{Y})^{-1} \cdot (\mathfrak{R}'^{+} + \mathfrak{Y} \cdot \tilde{\mathfrak{T}} \cdot \mathfrak{Z}) \cdot \mathfrak{q} \end{aligned} \tag{7.3.83}$$

which corresponds to the complete shell model equations. As the calculations of Chapter 5 show, it is possible to derive such equations from an H.F.-procedure, but numerical calculations with this model have mainly been done by fitting parameters to experimental values of neutron scattering data, etc. For instance, dispersion curves were calculated by Woods, Brockhouse, Cowley and Cochran (1963) for NaI and KBr with the following simplifying assumptions:

i) Only the negative ion is polarizable.
ii) The short-range interaction is between the positive ion and the shell of the negative ion, not the core.

Other assumptions were as in Kellermann's work, i.e., short-range forces extend only to nearest neighbours, and the combined potential is central. The theory then involves three parameters which were chosen to agree with the elastic constant c_{11}, the static dielectric constant ε_0 and the high-frequency dielectric constant ε_∞. The comparison with experiment is shown by the full lines in Figs. 7.3 and 7.4. The general shape of the experimental curves is well reproduced, although discrepancies remain.
Further refinements have been aimed at this semiphenomenological level. The most popular approach has been initiated by Schröder (1966) and Nüsslein and Schröder

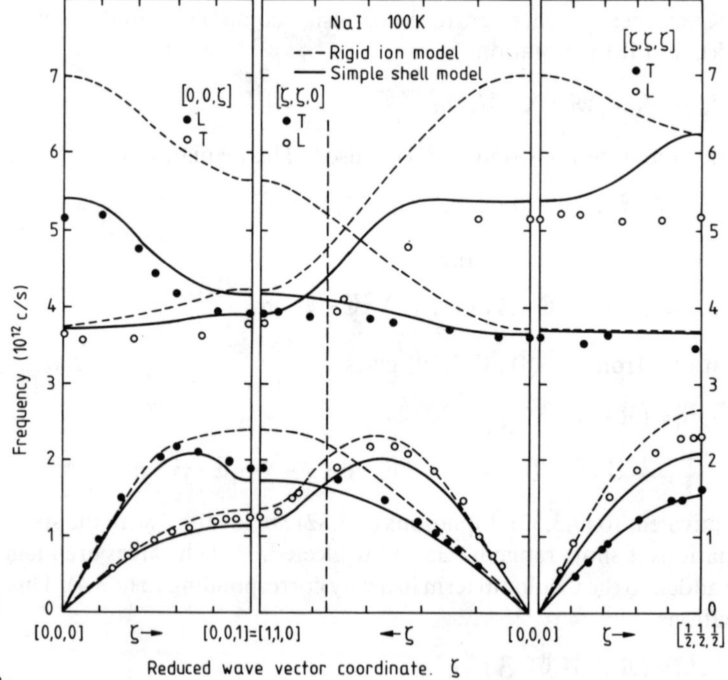

Figure 7.3:
Dispersion curves for NaI at 100 K; after Cochran and Cowley (1967).

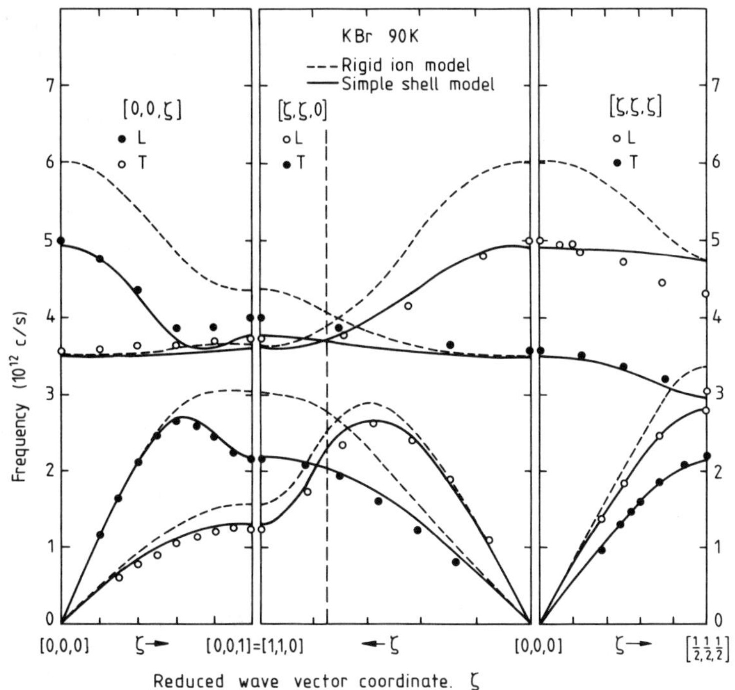

Figure 7.4:
Dispersion curves for KBr at 90 K; after Cochran and Cowley (1967).

(1967) with the so-called breathing shell model. Its equations read

$$\begin{aligned}\mathfrak{M}\cdot\omega^2\mathfrak{q} &= (\mathfrak{R}+\mathfrak{Z}\cdot\tilde{\mathfrak{T}}\cdot\mathfrak{Z})\cdot\mathfrak{q}+\mathfrak{Q}\cdot\mathfrak{v}+(\mathfrak{R}+\mathfrak{Z}\cdot\tilde{\mathfrak{T}}\cdot\mathfrak{Y})\cdot\mathfrak{d}\\ 0 &= \mathfrak{Q}^+\cdot(\mathfrak{q}+\mathfrak{d})+\mathfrak{H}\cdot\mathfrak{v}\\ 0 &= (\mathfrak{R}+\mathfrak{Y}\cdot\tilde{\mathfrak{T}}\cdot\mathfrak{Z})\cdot\mathfrak{q}+\mathfrak{Q}\cdot\mathfrak{v}+(\mathfrak{S}+\mathfrak{R}+\mathfrak{Y}\cdot\tilde{\mathfrak{T}}\cdot\mathfrak{Y})\cdot\mathfrak{d}\end{aligned} \tag{7.3.84}$$

where $\mathfrak{q}$ and $\mathfrak{d}$ are the core displacement vector and the displacement vector of the shells relative to their own cores as in the usual shell model, while $\mathfrak{v}$ is the vector of the isotropic deformations of the shells with only two components. The parameters of this model are also obtained by fitting to the experimental data.
In principle, the quality of the calculation of phonon spectra and states depends upon the use of appropriate adiabatic or more general energy expressions for the lattice considered. In Section 5.1 an extensive discussion has been given of the various approaches to derive such energy expressions. Before going on to refer to original papers we give a short review of the development of these procedures which overlaps partly with the discussion of Section 5.1 but now emphasizes the phonon aspect. The first model that was introduced to describe the lattice statics and dynamics was the rigid ion model. Literature about its introduction and early calculations of lattice energies and phonon modes can be found in the book of Born and Huang (1954).

Concerning phonon mode calculations the difficulties mainly arose from the summation of the long range Coulomb forces. The first correct phonon mode calculation for polar crystals was performed by Kellermann (1940). The next step was the extension to polarizable lattice ions, i.e., to the point dipole model. Lyddane and Herzfeld (1938) included ionic polarization in the equations of motion of the ions in order to study the optical branches of the phonon modes of polar crystals. In this model the ionic polarization is caused by the electric field produced by the distorted lattice. Szigeti (1949) (1950) suggested another mechanism of ionic polarization. Instead of induced ionic polarization by the microscopic electric field he proposed the short range repulsive forces between close neighbors to be associated with deformations of the ionic charge clouds. A formal expression of the lattice energy with all kinds of monopole and dipole interactions was given by Tolpygo (1950). Huang (1951) reformulated Szigeti's work to provide a general description of the long wave optical vibrations of a diatomic ionic crystal. The connection between the polarization of ions and the repulsive force between them was studied by Yamashita and Kurosawa (1955), Dick and Overhauser (1958) and Hanlon and Lawson (1959). These latter authors have independently suggested a "shell model" for an ion having a closed electron configuration. In particular, Dick and Overhauser postulated that the ions should be regarded as "cores" to which a massless "shell", representing the outer valence electrons, is bound by isotropic springs. The shell model was further developed, generalized and applied by Woods, Cochran and Brockhouse (1960). Schröder (1966). Nüsslein and Schröder (1967) introduced the breathing shell model, which also admits isotropic deformations of the shells. Basu and Sengupta (1968), Verma and Singh (1969), and Singh and Verma (1969) (1970) incorporated into the shell model the many-body forces originally derived by Löwdin (1948) and Lundqvist (1952) (1955) (1957) and arrived independently at a breathing shell model. An alternative approach is the deformation dipole model of Hardy (1959) (1962) and Karo and Hardy (1960) (1963) which allows for the existence of two possible types of polarization corresponding to the two different polarization mechanisms mentioned above, and which is a generalization of the work of Szigeti and Huang. With the exception of Löwdin's and Lundqvist's work the derivation of phonon models discussed so far is at least partially based on phenomenological considerations, with the consequence that the models contain a lot of adaptable parameters. In contrast to these models, in the last decade a number of authors derived energy expressions by ab initio calculations with the many-electron problem of the crystal. These approaches were discussed in Section 5.1 in detail and references concerning a unified approach to lattice statics and dynamics are also given there. We further refer to the review article of Bilz, Gliss and Hanke (1974) which relates some of these ab initio calculations to phonon mode calculations and to the various phonon models used.

With respect to reviews we refer to the books of Born and Huang (1954), Maradudin, Montroll, Weiss, and Ipatova (1971) and the review articles of Blackman (1955), Leibfried (1955), Cochran (1963), Cochran and Cowley (1967), Cochran (1971), Maradudin (1974), Birman (1974), Hardy (1974), Bilz, Gliss and Hanke (1974). The connection between vibration spectra and the symmetry of crystals was treated by Poulet and Mathieu (1976). A conference report on lattice dynamics, edited by

Balkanski (1978) contains numerous contributions to the microscopic theory of phonons, anharmonic phonon-phonon interactions, phase transitions, phonon-electron interactions and lattice dynamics of imperfect crystals. In the book of Bilz and Kress (1979) a collection of calculated phonon disperion curves is given.

With respect to original papers we give a review of the literature starting with Kellermann's paper and concentrate on phonon calculations. The following topics were treated: Theory and derivation of phonon modes and frequencies of NaCl with the R.I. (≡ rigid ion)-model, Kellermann (1940). Distribution of lattice vibrations of the KCl crystal with the R.I.-model, Iona (1941). Long wave modes of the optical branches with a phenomenological model of lattice polarization, Szigeti (1949) (1950) and Huang (1951). Frequencies and anharmonicities of the normal modes in alkali halides with the R.I.-model, Krishnan and Roy (1951) (1952). Long wave limit of frequencies of optical modes and general derivation of effective forces by means of Löwdin's method of binding calculation with inclusion of three-body forces, Lundqvist (1954) (1957). Frequency and lattice mode calculation for KCl by the method of Iona with the R.I.-model, Tenerz (1956). Frequencies and amplitudes of normal modes for KCl by means of Tolpygo's P.D. (≡ polarization dipole)-model, Kucher (1957). Calculation of lattice modes of crystals with zincblende structure in the Born model, Merten (1958). Sum rules for lattice vibrations in ionic crystals, Brout (1959). Lattice dynamics of alkali halides by means of the R.S. (≡ rigid shell)-model, Cochran (1959). Lattice vibrations in lithium and sodium halides with the R.I.-model, Karo (1959). Derivation of effective ionic charges based on the introduction of a shell model for the coupled electronic and ionic motion, Hanlon and Lawson (1959). Equilibrium conditions for the force constants of lattice potentials, Leibfried and Ludwig (1960). Lattice vibrations for potassium and rubidium halides and cesium fluoride with the R.I.-model, Karo (1960). Lattice dynamics of alkali halide crystals with the R.S.-model, Woods, Cochran and Brockhouse (1960). Derivation of the equations of Tolpygo's P.D.-model with inclusion of electronic dispersion, Mashkevich (1961). Discussion of effective ionic charges in relation to lattice, vibrations, Hardy (1961). Calculation of optical, elastic and piezoelectric properties of ionic and valence crystals with ZnS type structure based on Tolpygo's P.D.-model, Tolpygo (1961). Limiting optical frequencies in ionic crystals with the R.I.-model, Rosenstock (1961), Maradudin and Weiss (1961), and Barron (1961). Lattice modes of crystals with Wurtzite structure with the Born-model, Merten (1961). Lattice dynamics of alkali halide crystals in relation to specific heat data with the D.D. (≡ deformation dipole)-model, Hardy (1962). Review of lattice vibrations in solids, Krumhansl (1962). Normal vibrations of alkali halide crystals with ions differing considerably in size by the P.D.-model, Demidenko and Tolpygo (1962). Lattice dynamics of ionic and covalent crystals and connection between the R.S.-model and the P.D.-model, Cowley (1962). Dispersion of phonons in AgCl with the R.S.-model, Joshi and Gupta (1962). Lattice dynamics of alkali halide crystals with the R.S.-model and comparison with the P.D.-model and D.D.-model, Cowley, Cochran, Brockhouse and Woods (1963). Lattice dynamics of anharmonic crystals and derivation of macroscopic thermodynamic, dielectric, electric and mechanical properties by means of thermodynamic Green functions for phonons, Cowley (1963). Long wave modes of the Cu_2O lattice in the Born model with group theoretical analysis, Huang (1963). Lattice dynamics and specific heat data for rocksalt-structure alkali halides with the D.D.-model, Karo and Hardy (1963) (1966). Lattice vibrations of zincblende structure crystals with the R.S.-model, Kaplan and Sullivan (1963). Discussion of trends in the characteristic phonon frequencies of I–VII, II–VI, III–V and IV–IV compounds, Mitra and Marshall (1964). Lattice vibrations of wurtzite structure crystals based on the shell model, Sullivan (1964). Long wave lattice dynamics of the fluorite structure with the R.S.-model, Axe (1965). Lattice dynamics of MgO based on Lundqvist's model of three-body interactions, Verma and Dayal (1967). Calculations of dispersion curves and specific heat for LiF and NaCl with the B.S. (≡ breathing shell)-model, Nüsslein and Schröder (1967). Lattice dynamics of CdS in the Born model with phenomenological short range interactions, Nusimovici and Birman (1967). Lattice dynamics and thermal properties of LiH and LiD crystals based on the D.D.-model, Benedek (1967). Lattice dynamics of the exchange charge model of alkali halides, Marston and Dick (1967). Symmetry properties of the normal vibrations of crystals, Maradudin and Vosko (1968). Lattice modes and thermodynamic functions of ionic crystals with partially covalent bonding, Schäfer and Ludwig (1968). Lattice dynamics and specific heat data of CsCl, CsBr and CsJ in the R.S.-, P.D.-, and D.D.-model and comparison of the results, Karo and Hardy (1968). Lattice dynamics and second-order Raman spectrum of CsF in the D.D.-model, Hardy and Karo (1968). Lattice dynamics of NaF, NaCl, NaBr, and NaJ with a shell model in which the shells are allowed to deform, Melvin, Pirie and Smith (1968). Dispersion curves of NaCl based on the shell model, Lagu and Dayal (1968). Lattice dynamics of III–V compounds based on the R.I.-model, Banerjee and Varshni (1969). Lattice dynamics of cubic ZnS in the R.I.-model, Vetelino, Mitra, Brafman and Damen (1969). Discussion of the applicability of a rigid ion model to ZnS crystals, Vetelino and Mitra (1969). Quantum theory of lattice dynamics based on the adiabatic coupling scheme and on H.F.-states of the electrons, Johnson (1969). R.S.-model calculation of microscopic Grüneisen parameters for rocksalt-type materials, Barsch and Achar (1969). Calculation of transversal optical mode frequencies for $\mathbf{q} = 0$ with the D.S. (≡ deformable shell)-model based on many-body interactions, Roy, Basu and Sengupta (1969). Analysis of the forces in the D.S.-model, Sarkar and Sengupta (1969). Effect of the three-body forces on the shell model of alkali halides and application to the calculation of phonon dispersion curves in KBr and KJ, Singh and Verma (1969). Contribution of three-body overlap forces to the dynamical matrix of alkali halides, Verma and Singh (1969). Note on the vibrations of ionic lattices, Lidiard (1969). Dynamics of the Wurtzite lattice with extensive group theoretical analysis and phenomenological discussion of forces, Nusimovici (1969). Analysis of long wave vibrations in ionic crystals by means of phenomenological equations for polarization and distortion, Bryksin and Firsov (1969). Derivation of the shell model of lattice dynamics based on a selfconsistent Born-Oppenheimer perturbation expansion in powers of the lattice displacement, Sinha (1969). Lattice dynamics of NaF with the R.I.- and the D.D.-model, Karo and Hardy (1969). Lattice dynamics of MgO in the R.S.- and B.S.-model, Sangster, Peckham and Saunderson (1970). Lattice dynamics of ionic crystals based on a quantum mechanical treatment of the ionic deformation by means of an H.F.-approximation and a tight binding approximation and derivation of

B.S.-model equations, Kühner (1970). Effect of three-body interactions on the shell model of alkali halides and application to KF, KCl and sodium halides, Singh and Verma (1970). Lattice dynamics of NaCl, KCl, RbCl and RbF in the extended shell model, Raunio and Rolandson (1970). Lattice dynamics, Grüneisen parameters and coefficients of thermal expansion of CsBr and CsJ in the R.I.-model with an appropriate effective charge, Vetelino, Mitra and Namjoshi (1970). Lattice dynamics of CdS with electronic and ionic polarization, homopolar forces and Coulomb forces, Nusimovici, Balkanski and Birman (1970). Lattice dynamics of ZnO and BeO in the shell model, Hewat (1970). Lattice dynamics of RbCl with a modified shell model, Lal and Verma (1970). Covalent bonding in rocksalt structure crystals and application to the lattice dynamics of MgO with the phenomenological model of dielectric screening, Gillis (1971). Formulation of lattice dynamics in terms of rigidly moving Wannier functions, Ferreira (1971). Microscopic theory of lattice dynamics for ionic crystals in a selfconsistent H.F.-approximation of the adiabatic lattice energy, Gliss, Zeyher and Bilz (1971). Contribution of long-range three-center potentials to the phonon dispersion of alkali halides, Zeyher (1971). Simple shell model calculation of lattice dynamics and thermal expansion of alkali halides, Namjoshi, Mitra and Vetelino (1971). R.S.-model with effective ionic charges and calculation of phonon dispersion curves in NaCl and CsCl, Namjoshi, Mitra and Vetelino (1971). Lattice dynamics of LiH, LiF and LiCl in the T.B.F.S. ($\equiv$ three-body force shell)-model, Verma and Singh (1971). Lattice dynamics of Cu_2O and Ag_2O in the R.I.-model, Carabatos and Prevot (1971), Prevot and Carabatos (1971). Phonon dispersion curves and densities of states of CsCl, CsBr and CsI in the B.S.-model, Mahler and Engelhardt (1971). Incorporation of charge transfer in the shell model of lattice dynamics of II–VI and III–V compounds, Feldkamp (1972). Covalency and deformability of Ag^+-ions in an extended shell model of lattice dynamics of AgCl and AgBr, Fischer, Bilz, Haberkorn and Weber (1972). Derivation of shell models for the lattice dynamics of ionic crystals based on adiabatic electronic state functions beyond the H.F.-approximation in Fockspace representation, Wagner and Kühner (1972), Kühner and Wagner (1972). Lattice dynamics of rubidium halides and of CsCl, CsBr and CsI in the T.B.F.S.-model, Lal and Verma (1972). Dispersion curves of LiF crystals in the D.S.-model, Basu and Sengupta (1972). Long-wave length normal modes of crystals in the R.I.-model, Davies (1972). Lattice dynamics of AgBr in the D.D.-model, Hawranek and Lowndes (1972). Generalization of the shell model of lattice dynamics, Singh (1972). Lattice dynamics of MgO in the T.B.F.S.-model, Singh and Upadhyaya (1972). Shell-model lattice dynamics of CsCl, CsBr and CsI, Carabatos and Prevot (1972). Static equilibrium conditions for rigid-ion crystals and constraints on short range forces, Boyer and Hardy (1973). Extension of the B.S.-model by inclusion of axial deformations and application to KBr, Bluthardt, Schneider and Wagner (1973). Dispersion curves of AgCl in the T.B.F.S.-model with inclusion of van der Waals interactions, Upadhyaya and Mahesh (1973). Lattice dynamics of MgO in the T.B.F.S.-model, Verma and Agarwal (1973). Lattice dynamics of NaCl, NaBr, KI, KCl and KBr in the D.S.-model, Basu and Sengupta (1973). Lattice dynamics of CsCl, CsBr and CsI in the modified R.I.-model, Vetelino, Namjoshi and Mitra (1973). Lattice statics and dynamics of the NaF crystal in the D.S.-model, i.e., a unified approach to the calculation of crystal properties, Ghosh, Basu and Sengupta (1974). Derivation of lattice dynamical equations by means of energy calculation with molecular orbitals, Singh (1974). Lattice dynamics of MgF_2 and MnF_2 in a R.S.-model, Almairac and Benoit (1974), Cran and Sangster (1974). Phonon dispersion curves and mode Grüneisen parameters of RbI in the B.S.-model, Kress (1974). Dispersion relations of mode Grüneisen parameters in rubidium halides by means of an extended anharmonic shell model, Jex (1974). Lattice dynamics of II–VI compounds in the R.I.-model, Altshuler, Vekilov, Kadyshevich and Rusakov (1975). Lattice dynamics of MnO in the D.D.-model, Sharma (1974). Lattice dynamics of CuBr and CuI by a modified R.I.-model, Pandey and Dayal (1974). Lattice dynamics of CsCl, CsBr and CsI in the D.D.-model, Agrawal and Hardy (1974). Lattice dynamics of CuCl by a modified R.I.-model, Pandey (1974). Derivation of a generalized D.D.-model of lattice dynamics from microscopic theory, Jaswal (1975). Lattice dynamics of zincblende structure compounds in the D.D.-model, Kunc, Balkanski and Nusimovici (1975). Microscopic theory of lattice dynamics of non-conducting crystals, Mitskevich (1974) (1975). Microscopic theory of lattice dynamics of ionic crystals in terms of non-orthogonal highly localized wave functions, Zeyher (1975). Calculation of the relative stability of the structure of alkali halide crystals in the B.S.-model, Ghosh, Sarkar and Basu (1975). Lattice dynamics of NaF and RbF by the R.S.-model, Sneh and Dayal (1975). Lattice dynamics of covalent ionic compounds with pseudopotential method and the dielectric model, Altshuler, Vekilov and Izotov (1975). Three-body interactions in the B.S.-model, Sarkar and Chakrabarti (1975). Long-wavelength normal mode vibrations of infinite ionic crystal lattices in the R.I.-model, Davies and Mainville (1975), Davies (1976). Derivation and application of the local mode formulation in the theory of ideal lattice vibrations, Tindemans and Kroese (1975). Group theoretic investigation of the phonon spectrum of MnO in the magnetically ordered state, Gasanov (1975). Dielectric approximation in crystal lattice dynamics, Zaretskii, Kucher and Tolpygo (1975). Lattice dynamics of CsF in the extended T.B.F.S.-model, Singh and Agarwal (1975). Theoretical framework of an extended T.B.F.S.-model for crystals with rocksalt structure, Singh and Gupta (1975). Lattice dynamics of CaO and SrO in the T.B.F.S.-model, Upadhyaya and Singh (1975). Born-Oppenheimer potential and the polarizable model of lattice dynamics, Sarkar and Sengupta (1975). Derivation of a microscopic theory of lattice dynamics in harmonic and adiabatic approximation, Czachor and Holas (1976). A unified study of the lattice statics and dynamics of the caesium halides in the D.S.-model, Ghosh and Basu (1976). Consistent shell model parameters for the rocksalt structure alkali halides, Sangster (1976). Spectrum of phonons in NaCl with allowance for the quadrupole deformation of the ions, Bolonin and Tolpygo (1976), Bolonin (1977). Calculation of lattice statics and dynamics of the CsBr crystal in the D.S.-model, Ghosh and Basu (1976). Effects of polarization on equilibrium and dynamic properties of ionic system, Jacucci, McDonald and Rahman (1976). Lattice statics and dynamics of alkali halides by an H.F.-representation of the ion state and subsequent perturbation calculation, Basu and Sengupta (1976). Extended T.B.F.S.-model dynamics of sodium halide crystals, Singh and Chandra (1976). Group theoretical analysis of long-wavelength vibrations of polar crystals, Maradudin (1976). Lattice dynamics of CsF in the T.B.F.S.-model, Agarwal (1976). An extended T.B.F.S.-model for the lattice dynamics of ionic crystals, Singh and Gupta (1976). Adiabatic bond charge model for the phonons in III–V semiconductor compounds, Rustagi and Weber (1976). Analysis of the phonon spectra of polar semiconductors by the method of a model pseudopotential, Altshuler, Vekilov and Izotov (1976). Lattice dynamics of AgBr in the T.B.F.S.-model, Singh and Singh (1976). On the quantum mechanical foundation of the classical shell models in lattice dynamics, Niedermann and Wagner (1976). Lattice dynamics of BaO in the T.B.F.S.-model, Pandey and Upadhyaya (1976). Lattice dynamics of $FeCl_2$ in the displaced shell model, Pasternak (1976). Lattice dynamics of AgBr with a modified shell model, Dorner, von der Osten and Bührer (1976). Zero and first sound in NaF and RbI by the B.S.-model, Loidl, Jex, Daubert and Müllner (1976). Lattice dynamics of SrO in T.B.F.S.-model, Agarwal (1977). Lattice dynamics of the superionic conductor AgI, Alben and Burns (1977). Lattice dynamics of FeO in R.I.-models and shell models, Kugel, Carabatos, Hennion, Prevot, Revcolevschi and Tocchetti (1977). Lattice dynamics of LiD in the deformable ion model, Jaswal and Dilly (1977). Lattice dynamics of rubidium halide crystals in the T.B.F.S.-model, Goel and Dayal (1977). Phenomenological discussion of anisotropic phonon dispersion in GaS, Powell, Jandl, Brebner and Levy (1977). Deformable ion model and lattice dynamics of II–VI compounds, Jaswal (1978). Microscopic theory of the crystal lattice dynamics, Altshuler, Vekilov and Kacherets (1978). Lattice dynamics of Wurtzite type crystals by a polarizable point ion model, Miura, Murata and Shiro (1977) (1978). Crystal independent shell parameters and fitted Born-Mayer potentials, Sangster, Schröder and Atwood (1978), Sangster and Atwood (1978). Lundqvist three-body interaction and lattice dynamics of ionic crystals, Sarkar and Sengupta (1978).

7.4 Lattice-polaritons

As ions are charged particles, their collective motions are accompanied by electromagnetic fields according to the laws of electrodynamics. In the preceding sections such electromagnetic fields have been ignored and only the electrostatic interactions have been taken into account. In this section we will take the electromagnetic field into consideration. Then the mechanical motion of the ions is expected to be closely connected with the motion of the electromagnetic field, i.e., the collective motions of both systems must be treated simultaneously. Such a procedure leads to combined field-lattice states and the corresponding excitations lead to quasi-particle states which are called polaritons. Polaritons can be described by means of a phonon-photon Hamiltonian with general phonon-photon coupling terms in a Fock space representation. However, this representation does not contain an analysis of the forces acting on ions which is important for the understanding of polariton states. It is only a formal description of the possibility of quasiparticle formation which should primarily be based on an explanation of the microscopic mechanism. Such an explanation must be sought in the framework of a microscopic crystal theory, which will at least partially be done in the following in order to get some insight into this mechanism.

We will treat these states at a semi-classical level by extending the adiabatic coupling to include "external" fields and by using the harmonic approximation for the interaction between crystal and radiation field and for the internal interaction of the crystal itself. The energy operator of the crystal and its physical environment is given by (2.1.1). As we are not interested in damping processes and thermal exchange, we omit W and H^W. Then the Hamiltonian of the reduced system reads

$$H^r := K + S + H^S \tag{7.4.1}$$

where S and H^S contain only the transverse parts of the radiation field, while the Coulomb part of the electromagnetic interaction is contained in K. For the treatment of polaritons it is convenient to rewrite the Hamiltonian (7.4.1) by expressing all electromagnetic interactions by the electromagnetic fields. In this case H^r can by equivalently written

$$\begin{aligned} H^r = &\sum_i \left\{ (2m)^{-1} \left[\mathbf{p}_i - \frac{e}{c} \mathbf{A}(\mathbf{r}_i) \right]^2 + e\varphi(\mathbf{r}_i) \right\} \\ &+ \sum_k \left\{ (2M_k)^{-1} \left[\mathbf{P}_k - \frac{e_k}{c} \mathbf{A}(\mathbf{R}_k) \right]^2 + e_k \varphi(\mathbf{R}_k) \right\} \\ &+ \frac{1}{8\pi} \int [\mathbf{E}^2(\mathbf{r}) + \mathbf{B}^2(\mathbf{r})] d^3 r \end{aligned} \tag{7.4.2}$$

where the vector potential $\mathbf{A}(\mathbf{r})$ and the scalar potential $\varphi(\mathbf{r})$ are connected with the field strengths by

$$\mathbf{E} = -\nabla\varphi - \frac{1}{c}\frac{\partial}{\partial t}\mathbf{A}; \quad \mathbf{B} = \nabla \times \mathbf{A}. \tag{7.4.3}$$

The proof of this equivalence is given, for instance, by Heitler (1949). We now represent the scalar and vector potentials by series of orthonormal eigenfunctions

$$\begin{aligned} \mathbf{A} &= \sum_\lambda q_\lambda(t) \mathbf{A}_\lambda(\mathbf{r}) + \sum_\sigma q_\sigma(t) \mathbf{A}_\sigma(\mathbf{r}) \\ \varphi &= \sum_\sigma a_\sigma(t) \varphi_\sigma(\mathbf{r}) \end{aligned} \tag{7.4.4}$$

where $\{\mathbf{A}_\lambda\}$ are transverse waves, $\{\mathbf{A}_\sigma\}$ are longitudinal waves and $\{\varphi_\sigma\}$ are scalar waves which are solutions of the homogeneous equations

$$\Box(\mathbf{A}_\lambda, \mathbf{A}_\sigma, \varphi_\sigma) = 0. \tag{7.4.5}$$

Then the Hamiltonian (7.4.2) goes over into

$$H^r = \sum_i \frac{1}{2m} \mathbf{p}_i^2 + \sum_k \frac{1}{2M_k} \mathbf{P}_k^2 + \tilde{H}^s(\mathbf{r}, \mathbf{R}, z) + \tilde{S} \tag{7.4.6}$$

with

$$\tilde{S}=\frac{1}{2}\sum_{\lambda}(p_{\lambda}^2+v_{\lambda}^2q_{\lambda}^2)+\frac{1}{2}\sum_{\sigma}(p_{\sigma}^2+v_{\sigma}^2q_{\sigma}^2)+\frac{1}{2}\sum_{\sigma}(b_{\sigma}^2+v_{\sigma}^2a_{\sigma}^2) \tag{7.4.7}$$

and $z\equiv\{p_\lambda,q_\lambda,p_\sigma,q_\sigma,b_\sigma,a_\sigma\}$, where $\tilde{H}^S$ is the interaction energy between fields and particles resulting directly from (7.4.2).
To include the electromagnetic field into the calculation, we assume that by the ordinary adiabatic equation (2.2.7) an equilibrium state $|\chi_n\rangle$ with energy $U_n(\mathbf{R})$ at the equilibrium positions $\{\mathbf{R}_k=\mathbf{R}_k^n\}$ has been found. To this state corresponds a certain amount of electrostatic field energy with certain values of the classical field variables $z=z^n$. We expand $\tilde{H}^S(\mathbf{r},\mathbf{R},z)$ into a power series about $\mathbf{R}^n$ and z^n and write formally

$$\tilde{H}^S(\mathbf{r},\mathbf{R},z)=\sum_{\varrho\nu}\left[\frac{\partial^\varrho}{\partial\mathbf{R}^\varrho}\frac{\partial^\nu}{\partial z^\nu}\tilde{H}^S\right]_{\mathbf{R}=\mathbf{R}^n,z=z^n}\cdot(\mathbf{R}-\mathbf{R}^n)^\varrho(z-z^n)^\nu \tag{7.4.8}$$

Now we define the modified adiabatic energy operator by

$$h(\mathbf{r},\mathbf{R},z):=\sum_i\frac{1}{2m}\mathbf{p}_i^2+\sum_{\varrho\nu}\left[\frac{\partial^\varrho}{\partial\mathbf{R}^\varrho}\frac{\partial^\nu}{\partial z^\nu}\tilde{H}^S\right]\cdot(\mathbf{R}-\mathbf{R}^n)^\varrho(z-z^n)^\nu. \tag{7.4.9}$$

In our semiclassical treatment we consider the quantities z^n, $(z-z^n)$ to be classical parameters and define the electronic state functions to be solutions of the adiabatic equation

$$h(\mathbf{r},\mathbf{R},z)\chi_n(\mathbf{r},\mathbf{R}-\mathbf{R}^n,z-z^n)=U_n(\mathbf{R}-\mathbf{R}^n,z-z^n)\chi_n(\mathbf{r},\mathbf{R}-\mathbf{R}^n,z-z^n). \tag{7.4.10}$$

In the preceding sections the adiabatic potential was used to define an equation for the ionic motion which is given by (2.2.11). As we expect strong correlations between field and lattice, we replace this equation by

$$\left[\sum_k\frac{1}{2M_k}\mathbf{P}_k^2+U_n(\mathbf{R}-\mathbf{R}^n,z-z^n)+\tilde{S}\right]|\varphi_n^{m\nu}\rangle=E_n^{m\nu}|\varphi_n^{m\nu}\rangle. \tag{7.4.11}$$

In this case the wave function $|\varphi_n^{m\nu}\rangle$ contains not only the lattice coordinates but also the field coordinates, i.e., we have

$$|\varphi_n^{m\nu}\rangle\equiv\varphi_n^{m\nu}(\mathbf{R}-\mathbf{R}^n,z-z^n) \tag{7.4.12}$$

and the quantum numbers m,ν describe the whole set of excitations of transverse fields as well as of lattice waves. Hence, instead of (2.1.2), a state $|a\rangle$ of the complete system is now given by

$$|a\rangle=|nm\nu\rangle\otimes|\varrho\rangle. \tag{7.4.13}$$

A solution of the extended field-lattice equation (7.4.11) can be found if we restrict ourselves to the harmonic approximation. As usual we expand $U_n(\mathbf{R}-\mathbf{R}^n,z-z^n)$ about its equilibrium positions. These are given by $\mathbf{R}=\mathbf{R}^n$ for the crystal and by $z=z^n$ for the fields. Therefore the expansion of U_n reads with $\mathbf{u}^n := \mathbf{R} - \mathbf{R}^n$; $v^n := z - z^n$

$$U_n(\mathbf{u}^n, v^n) = U_n(\mathbf{R}^n, z^n) + \sum_r \left(\frac{\partial}{\partial z_r} U_n\right)_n v_r^n + \sum_k \left(\frac{\partial}{\partial \mathbf{R}_k} U_n\right)_n \cdot \mathbf{u}_k^n$$
$$+ \frac{1}{2} \sum_{rs} \left(\frac{\partial^2}{\partial z_r \partial z_s} U_n\right)_n v_r^n v_s^n + \sum_{ks} \left(\frac{\partial^2}{\partial \mathbf{R}_k \partial z_s} U_n\right)_n \cdot \mathbf{u}_k^n v_s^n \tag{7.4.14}$$
$$+ \frac{1}{2} \sum_{kj} \left(\frac{\partial^2}{\partial \mathbf{R}_k \partial \mathbf{R}_j} U_n\right)_n \cdot \cdot\, \mathbf{u}_k^n \mathbf{u}_j^n + \dots$$

For simplicity we exclude the possibility that the lattice is driven by external fields. Then only those electromagnetic fields appear in $U_n(\mathbf{u}^n, v^n)$ which are connected with the ionic motions. This means that if no motion of ions exists, then also no electromagnetic fields are present, i.e., we have the simultaneous equilibrium conditions

$$\left(\frac{\partial}{\partial z_r} U_n\right)_n = 0; \quad \left(\frac{\partial}{\partial \mathbf{R}_k} U_n\right)_n = 0; \quad \begin{matrix} 1 \leqslant r < \infty \\ 1 \leqslant k < \infty \end{matrix}. \tag{7.4.15}$$

It is, of course, possible to include also external fields, but since the subsequent calculations run along similar lines as those for intrinsic fields we do not treat this extension explicitly. Then with (7.4.15) the harmonic adiabatic energy is given by

$$U_n^h(\mathbf{u}^n, v^n) = U_n(\mathbf{R}^n, z^n) + \frac{1}{2} \sum_{rs} \left(\frac{\partial^2}{\partial z_r \partial z_s} U_n\right)_n v_r^n v_s^n$$
$$+ \sum_{ks} \left(\frac{\partial^2}{\partial \mathbf{R}_k \, \partial z_s} U_n\right)_n \cdot \mathbf{u}_k^n v_s^n + \frac{1}{2} \sum_{kj} \left(\frac{\partial^2}{\partial \mathbf{R}_k \partial \mathbf{R}_j} U_n\right)_n \cdot \cdot\, \mathbf{u}_k^n \mathbf{u}_j^n \tag{7.4.16}$$

and if we substitute this expression into equation (7.4.11), then in the harmonic approximation we have

$$\left[\sum_k \frac{1}{2 M_k} \mathbf{P}_k^2 + U_n^h(\mathbf{u}^n, v^n) + \tilde{S}\right] |\varphi_n^{mv}\rangle^h = E_n^{mv} |\varphi_n^{mv}\rangle^h. \tag{7.4.17}$$

For equations with harmonic potentials the following Lemma holds:

Lemma 7.3: The solutions of state equations with harmonic potentials can be obtained by solving the corresponding classical equations of motion for all periodic processes.

Proof: We consider a test Hamilton operator

$$H := \frac{1}{2} \sum_k p_k^2 + \frac{1}{2} \sum_{ki} A_{ki} x_k x_i \tag{7.4.18}$$

with $A_{ki} = A_{ik}$ real. Then the classical equations of motion are

$$\frac{\partial H}{\partial x_k} = -\dot{p}_k; \quad \frac{\partial H}{\partial p_k} = \dot{x}_k \tag{7.4.19}$$

from which the equations

$$-\ddot{x}_k = \sum_i A_{ki} x_i \tag{7.4.20}$$

follow. For periodic motions $x_k = x_k(\varrho) \exp(i\omega_\varrho t)$ equations (7.4.20) go over into

$$\omega_\varrho^2 x_k(\varrho) = \sum_i A_{ki} x_i(\varrho) \tag{7.4.21}$$

If det $|A| \neq 0$, then a complete set $\{x_i(\varrho)\}$ of orthonormal solutions for the various eigenfrequencies $\{\omega_\varrho\}$ can be obtained. We now introduce normal coordinates by

$$x_k = \sum_\varrho x_k(\varrho) \zeta_\varrho \tag{7.4.22}$$

and observe $p_k = i\hbar \partial/\partial x_k$. Then H can be expressed in the normal coordinates by

$$H := -\frac{1}{2} \sum_\varrho \hbar^2 \frac{\partial^2}{\partial \zeta_\varrho^2} + \frac{1}{2} \sum_\varrho \omega_\varrho^2 \zeta_\varrho^2 \tag{7.4.23}$$

i.e., by a set of uncoupled harmonic oscillators for which the state functions follow analogously to (2.2.25) (2.2.26), Q.E.D.

This rather simple theorem can be applied to the polariton operator of (7.4.17) with (7.4.16). As, according to (7.4.4), the field variables $\{z\}$ depend linearly on the field strengths, the equations of motion can be equivalently formulated also for these quantities. Hence the polariton problem can be solved by solving the corresponding Maxwell-Lorentz-equations for linear coupling between matter and field. After having found the eigensolutions of these equations for periodic motions, the field quantities of the normalized eigensolutions can be projected on the field variables $\{z\}$ which leads, together with the corresponding expansion of the set $\{\mathbf{u}_k^n\}$, to a complete diagonalization of the polariton Hamiltonian of (7.4.17).
This plausible program, however, has not fully been realized so far. Rather further approximations have been used to obtain physically meaningful solutions. We give a version which stresses the connection with Lemma 7.3 and for which it is assumed that the adiabatic coupling constants between field and lattice are given by those of classical ions in fields.
According to Lemma 7.3 we consider the classical equations of the electromagnetic field

$$\begin{aligned} \nabla \times \mathbf{B} &= \frac{1}{c}\frac{\partial}{\partial t}\mathbf{E} + \frac{4\pi}{c} j; \quad \nabla \cdot \mathbf{B} = 0 \\ \nabla \times \mathbf{E} &= -\frac{1}{c}\frac{\partial}{\partial t}\mathbf{B}; \quad \nabla \cdot \mathbf{E} = 4\pi\varrho \end{aligned} \tag{7.4.24}$$

and the classical equations of the lattice

$$M_\mu \ddot{\mathbf{u}}_{m\mu} = -\sum_{n\nu} \tilde{\Phi}^0(m\mu,n\nu)\cdot \mathbf{u}_{n\nu} + e_\mu \tilde{\mathbf{E}}(\mathbf{R}^0_{m\mu}+\mathbf{u}_{m\mu}) + \frac{e_\mu}{c}[\dot{\mathbf{u}}_{m\mu}\times \tilde{\mathbf{B}}(\mathbf{R}^0_{m\mu}+\mathbf{u}_{m\mu})] + \tilde{\Phi}^0(m\mu) \tag{7.4.25}$$

where $\tilde{\mathbf{E}}$ and $\tilde{\mathbf{B}}$ are the effective fields which arise from **E** and **B** by subtracting the eigenfield of the ion under consideration, while $\tilde{\Phi}^0(m\mu,n\nu)$ are the purely "mechanical" coupling constants, i.e., in this version all electromagnetic interactions are included in the fields.

It can be expected that it will be very difficult to detect the essential physical effects at this level. Hence we try to simplify this model before going on to explicit calculations. This simplification can be justified from the physical point of view. As any ionic motion is accompanied by electric dipoles we expect that for organized collective motions of the lattice, i.e., lattice vibrations, not only local electric and magnetic fields arise, but also collective, i.e., macroscopic electromagnetic fields. It is this effect that we are interested in. To separate the macroscopic fields from the "uninteresting" microscopic fields it has to be observed that macroscopic fields survive an averaging procedure, while microscopic fields are cancelled by such a procedure. In performing such an averaging one observes that only the Maxwell equations (7.4.24) can be meaningfully averaged, while it makes no sense to average the mechanical equations (7.4.25). To prepare for such an averaging we first consider the original systems (7.4.24) (7.4.25).

In the point ion model the charge and current density are given by

$$\begin{aligned}\varrho(\mathbf{r},t) &= \sum_{m\mu} e_\mu \delta(\mathbf{r}-\mathbf{R}^0_{m\mu}-\mathbf{u}_{m\mu}) \\ &= \sum_{m\mu} e_\mu \delta(\mathbf{r}-\mathbf{R}^0_{m\mu}) - \sum_{m\mu} e_\mu \mathbf{u}_{m\mu}\cdot \nabla_{m\mu}\delta(\mathbf{r}-\mathbf{R}^0_{m\mu}) + \dots\end{aligned} \tag{7.4.26}$$

$$\begin{aligned}j(\mathbf{r},t) &= \sum_{m\mu} e_\mu \dot{\mathbf{u}}_{m\mu}\delta(\mathbf{r}-\mathbf{R}^0_{m\mu}+\mathbf{u}_{m\mu}) \\ &= \sum_{m\mu} e_\mu \dot{\mathbf{u}}_{m\mu}\delta(\mathbf{r}-\mathbf{R}^0_{m\mu}) + \dots\end{aligned} \tag{7.4.27}$$

if for small displacements $\mathbf{u}_{m\mu}$ only the linear terms in $\mathbf{u}_{m\mu}$ are taken into account. As the magnetic field depends linearly on the current and since according to (7.4.27) the current depends in lowest order approximation linearly on $\dot{\mathbf{u}}_{m\mu}$, the lowest order term of the magnetic field-lattice interaction in (7.4.25) is proportional to $\dot{\mathbf{u}}^2_{m\mu}$. Hence in the harmonic coupling approximation the magnetic interaction of lattice and fields drops out, and the dynamical equations (7.4.25) are reduced to

$$M_\mu \ddot{\mathbf{u}}_{m\mu} = -\sum_{n\nu} \tilde{\Phi}^0(m\mu,n\nu)\cdot \mathbf{u}_{n\nu} + e_\mu \tilde{\mathbf{E}}(\mathbf{R}^0_{m\mu}+\mathbf{u}_{m\mu}) + \tilde{\Phi}^0(m\mu) \tag{7.4.28}$$

It is convenient to represent the fields by potentials which are introduced by (7.4.3) but for which we do not use expansions (7.4.4) as we perform explicit calculations. Then the field equations (7.4.24) can be replaced by the set of equations

$$\Box\varphi(\mathbf{r},t) = -4\pi\varrho(\mathbf{r},t) \tag{7.4.29}$$

$$\Box\mathbf{A}(\mathbf{r},t) = -\frac{4\pi}{c}\,j(\mathbf{r},t) \tag{7.4.30}$$

and the Lorentz condition

$$\nabla\cdot\mathbf{A} + \frac{1}{c}\frac{\partial}{\partial t}\,\varphi = 0. \tag{7.4.31}$$

As for the charge and current density (7.4.26) (7.4.27) we have the continuity equation

$$\nabla\cdot j + \frac{\partial}{\partial t}\,\varrho = 0; \tag{7.4.32}$$

the Lorentz condition (7.4.31) is automatically satisfied and can be omitted for the further evaluation. By the use of Green functions the equations (7.4.29) and (7.4.30) can directly be integrated and give

$$\begin{aligned} \varphi &= -4\pi G\varrho \\ \mathbf{A} &= -\frac{4\pi}{c}\,G\mathbf{j}. \end{aligned} \tag{7.4.33}$$

For periodic motions with a small frequency ω, the ratio ω/c is a small quantity and we shall neglect all terms with powers of ω/c in the field lattice interaction. This approximation is equivalent to the assumption $\lim c\to\infty$. If we denote the selfpotentials of the ions by the labels $m\mu$, we obtain for the effective field acting on the $m\mu$-th ion

$$\begin{aligned} \tilde{\mathbf{E}}(\mathbf{r}) &= \lim_{c\to\infty}\left[-\nabla(\varphi-\varphi_{m\mu}) - \frac{1}{c}\frac{\partial}{\partial t}(\mathbf{A}-\mathbf{A}_{m\mu})\right] \\ &= -\nabla(\varphi^{\mathrm{C}}-\varphi^{\mathrm{C}}_{m\mu}) \end{aligned} \tag{7.4.34}$$

where φ^{C} is the Coulomb potential

$$\begin{aligned} \varphi^{\mathrm{C}}(\mathbf{r}) &= \sum_{n\nu}\int e_\nu\delta(\mathbf{r}'-\mathbf{R}^0_{n\nu}-\mathbf{u}_{n\nu})\,C(\mathbf{r}',\mathbf{r})\,d^3r \\ \varphi^{\mathrm{C}}_{m\mu}(\mathbf{r}) &= \int e_\mu\delta(\mathbf{r}'-\mathbf{R}^0_{m\mu}-\mathbf{u}_{m\mu})\,C(\mathbf{r}',\mathbf{r})\,d^3r. \end{aligned} \tag{7.4.35}$$

We now turn to the averaging procedure. If we average, e.g., over the volume of the unit cell $\Delta V = v$ and a suitable small time intervall Δt, the averaged field quantities are defined by

$$\bar{f}(\mathbf{r},t) := \frac{1}{\Delta V\Delta t}\int_{\Delta V}\int_{\Delta t} f(\mathbf{r}+\xi, t+\tau)\,d^3\xi\,d\tau. \tag{7.4.36}$$

Then for these quantities phenomenological field equations can be derived. The following theorem holds:

Theorem 7.12: Suppose that the elementary cells of the crystal are electrically neutral. Then one has the phenomenological equations

$$\nabla\times\bar{\mathbf{E}}+\frac{1}{c}\frac{\partial}{\partial t}\bar{\mathbf{B}}=0;\quad \nabla\cdot\bar{\mathbf{D}}=0$$
$$\nabla\times\bar{\mathbf{H}}-\frac{1}{c}\frac{\partial}{\partial t}\bar{\mathbf{D}}=0;\quad \nabla\cdot\bar{\mathbf{B}}=0 \tag{7.4.37}$$

with

$$\bar{\mathbf{D}}+\bar{\mathbf{E}}+4\pi\bar{\mathbf{P}};\quad \bar{\mathbf{H}}=\mathbf{B}-4\pi\bar{\mathbf{M}} \tag{7.4.38}$$

and

$$\bar{\mathbf{P}}(\mathbf{r},t)=\frac{1}{\varDelta V\varDelta t}\int\int\sum_{l\mu}\mathbf{m}^{e}_{l\mu}(t+\tau)\delta(\mathbf{r}+\boldsymbol{\xi}-\mathbf{R}^{0}_{l\mu})d^{3}\xi d\tau$$
$$\bar{\mathbf{M}}(\mathbf{r},t)=\frac{1}{\varDelta V\varDelta t}\int\int\sum_{l\mu}\mathbf{m}^{m}_{l\mu}(t+\tau)\delta(\mathbf{r}+\boldsymbol{\xi}-\mathbf{R}^{0}_{l\mu})d^{3}\xi d\tau \tag{7.4.39}$$

and

$$\mathbf{m}^{e}_{l\mu}(t):=\int\mathbf{r}'\varrho_{l\mu}(\mathbf{r}'+\mathbf{R}^{0}_{l\mu})d^{3}r'$$
$$\mathbf{m}^{m}_{l\mu}(t):=\frac{1}{2c}\int\mathbf{r}'\times j_{l\mu}(\mathbf{r}'+\mathbf{R}^{0}_{l\mu})d^{3}r' \tag{7.4.40}$$

where $\varrho_{l\mu}(\mathbf{r})$ is given by $e_{\mu}\delta(\mathbf{r}-\mathbf{R}^{0}_{l\mu}-\mathbf{u}_{l\mu})$ etc. These equations result from averaging the Maxwell equations (7.4.24) (7.4.26).

The proof is not trivial. As its length exceeds the scope of this book we refer to the literature, cf. Stumpf and Schuler (1973). According to our assumptions we neglect magnetic effects of matter. Hence we may put $\bar{\mathbf{M}}\equiv 0$, and equations (7.4.37) go over into the system

$$\nabla\times\bar{\mathbf{E}}+\frac{1}{c}\frac{\partial}{\partial t}\bar{\mathbf{B}}=0;\quad \nabla\cdot(\bar{\mathbf{E}}+4\pi\bar{\mathbf{P}})=0 \tag{7.4.41}$$

$$\nabla\times\bar{\mathbf{B}}-\frac{1}{c}\frac{\partial}{\partial t}(\bar{\mathbf{E}}+4\pi\bar{\mathbf{P}})=0;\quad \nabla\cdot\bar{\mathbf{B}}=0. \tag{7.4.42}$$

Clearly we must be able to detect the dependence of (7.4.28) on the averaged field. To achieve this we decompose $\tilde{\mathbf{E}}$ of (7.4.28) formally into

$$\tilde{\mathbf{E}}(\mathbf{R}^{0}_{m\mu}+\mathbf{u}_{m\mu})=\tilde{\mathbf{E}}_{1}(\mathbf{R}^{0}_{m\mu}+\mathbf{u}_{m\mu})+\bar{\mathbf{E}}(\mathbf{R}^{0}_{m\mu}+\mathbf{u}_{m\mu}) \tag{7.4.43}$$

where $\tilde{\mathbf{E}}_{1}$ is the local field, while $\bar{\mathbf{E}}$ is the average field. An evaluation of (7.4.43) can be only performed if we give an explicit expression for $\bar{\mathbf{E}}$. We use the Coulomb approximation (7.4.34). Then the following theorem holds, Born and Huang (1966), Maradudin, Montroll, Weiss and Ipatova (1971):

Theorem 7.13: For periodic motions

$$\mathbf{u}_{m\mu}(t) = M_\mu^{-1/2} \exp\,[i(\mathbf{k}\cdot\mathbf{R}^0_{m\mu} - \omega t)]\mathbf{w}(\mu|j,\mathbf{k}) \tag{7.4.44}$$

$$\bar{\mathbf{E}}(\mathbf{r},t) = \bar{\mathbf{E}}(\mathbf{k}) \exp\,[i(\mathbf{k}\cdot\mathbf{r} - \omega t)]; \quad \bar{\mathbf{B}}(\mathbf{r},t) = \bar{\mathbf{B}}(\mathbf{k}) \exp\,[i(\mathbf{k}\cdot\mathbf{r} - \omega t)] \tag{7.4.45}$$

with $|\mathbf{k}| \ll |\mathbf{g}|$ and for $c \to \infty$ in the field equations, the dynamical equation of the lattice becomes

$$\begin{aligned}\omega^2 \mathbf{w}(\kappa|j,\mathbf{k}) = &\sum_{\kappa'} \mathscr{R}(\kappa,\kappa'|\mathbf{k})\cdot\mathbf{w}(\kappa'|j,\mathbf{k}) \\ &+ \sum_{\kappa''} \frac{e_\kappa e_{\kappa''}}{M_\kappa} \mathscr{Q}(\kappa,\kappa''|0)\cdot\mathbf{w}(\kappa|j,\mathbf{k}) \\ &- \sum_{\kappa'} \frac{e_\kappa e_{\kappa'}}{(M_\kappa M_{\kappa'})^{1/2}} \mathscr{Q}(\kappa,\kappa'|\mathbf{k})\cdot\mathbf{w}(\kappa'|j,\mathbf{k}) - \frac{e_\kappa}{M_\kappa^{1/2}} \bar{\mathbf{E}}(\mathbf{k})\end{aligned} \tag{7.4.46}$$

where the quantities $\mathscr{R}$ and $\mathscr{Q}$ are defined by (7.3.6) (7.3.12) and $\bar{\mathbf{E}}$ is given by (7.4.60).

Proof: We expand $\tilde{\mathbf{E}}$ of equations (7.4.28) in powers of $\mathbf{u}_{m\mu}$. For $c \to \infty$ we use (7.4.34). The lowest order term in (7.4.28) is given by $e_\mu \tilde{\mathbf{E}}(\mathbf{R}^0_{m\mu})$. Due to the equilibrium condition of the lattice we must have

$$\tilde{\Phi}^0(m\mu) + e_\mu \tilde{\mathbf{E}}(\mathbf{R}^0_{m\mu}) = 0. \tag{7.4.47}$$

Hence in the harmonic coupling approximation equations (7.4.28) read

$$\begin{aligned}M_\mu \ddot{\mathbf{u}}_{m\mu} &= -\sum_{n\nu} [\tilde{\Phi}^0(m\mu,n\nu) + e_\mu \nabla \tilde{\mathbf{E}}(\mathbf{R}^0_{m\mu})]\cdot\mathbf{u}_{n\nu} \\ &\equiv -\sum_{n\nu} \Phi^0(m\mu,n\nu)\cdot\mathbf{u}_{n\nu}.\end{aligned} \tag{7.4.48}$$

As $\tilde{\mathbf{E}}$ is derived from a potential, the coupling constants Φ^0 of (7.4.48) can be derived by means of Theorem 7.3 if the two-particle interaction potential φ is composed of a repulsive and a Coulomb potential. Substitution of (7.4.44) into (7.4.48) leads to equation (7.2.23) and the elements of the dynamical matrix $\mathscr{C}(\kappa,\kappa'|\mathbf{k})$ separate correspondingly into a non-Coulomb and a Coulomb part given by (7.3.4). Then the equations of motion can be written

$$\omega^2 \mathbf{w}(\kappa|j,\mathbf{k}) = \sum_{\kappa'} [\mathscr{T}(\kappa,\kappa'|\mathbf{k}) + \mathscr{R}(\kappa,\kappa'|\mathbf{k})]\cdot\mathbf{w}(\kappa'|j,\mathbf{k}). \tag{7.4.49}$$

If we now substitute (7.3.11) into (7.4.49) we obtain

$$\begin{aligned}\omega^2 \mathbf{w}(\kappa|j,\mathbf{k}) = &\sum_{\kappa'} \mathscr{R}(\kappa,\kappa'|\mathbf{k})\cdot\mathbf{w}(\kappa'|j,\mathbf{k}) \\ &+ \sum_{\kappa''} \frac{e_\kappa e_{\kappa''}}{M_\kappa} \mathscr{Q}(\kappa,\kappa''|0)\cdot\mathbf{w}(\kappa|j,\mathbf{k}) \\ &- \sum_{\kappa'} \frac{e_\kappa e_{\kappa'}}{(M_\kappa M_{\kappa'})^{1/2}} \mathscr{Q}(\kappa,\kappa'|\mathbf{k})\cdot\mathbf{w}(\kappa'|j,\mathbf{k}) \\ &+ \sum_{\kappa'} \frac{e_\kappa e_{\kappa'}}{(M_\kappa M_{\kappa'})^{1/2}} \frac{4\pi}{v} \frac{\mathbf{k}\otimes\mathbf{k}}{\mathbf{k}^2}\cdot\mathbf{w}(\kappa'|j,\mathbf{k}).\end{aligned} \tag{7.4.50}$$

The last term on the right-hand side of equation (7.4.50) is connected with the average field $\bar{\mathbf{E}}$, as may be shown by substituting (7.4.44) into (7.4.39). With $\bar{\mathbf{P}}(\mathbf{r},t)=\bar{\mathbf{P}}(\mathbf{r})\times \exp(i\omega t)$ from (7.4.39) and (7.4.40) observing the definition of $\varrho_{l\mu}$, we get the expression

$$\bar{\mathbf{P}}(\mathbf{r})=\frac{1}{v}\sum_{l\mu}\int e_\mu \mathbf{w}(\mu|j,\mathbf{k})M_\mu^{-1/2}\exp(i\mathbf{k}\cdot\mathbf{R}_{l\mu}^0)\delta(\mathbf{r}+\xi-\mathbf{R}_{l\mu}^0)d^3\xi$$
$$=\frac{1}{v}\exp(i\mathbf{k}\cdot\mathbf{r})\sum_{l\mu}\int \mathbf{p}(\mu|j,\mathbf{k})\exp(i\mathbf{k}\cdot\xi)\delta(\mathbf{r}+\xi-\mathbf{R}_{l\mu}^0)d^3\xi. \tag{7.4.51}$$

The periodic function in (7.4.51) can be represented by

$$\sum_l \delta(\mathbf{y}-\mathbf{R}_{l\mu}^0)=\sum_{\mathbf{g}} a(\mathbf{g})\exp(i\mathbf{g}\cdot\mathbf{y}). \tag{7.4.52}$$

Its coefficients $a(\mathbf{g})$ follow by inversion and read

$$a(\mathbf{g})=\frac{1}{v}\exp[-i\mathbf{g}\cdot\mathbf{x}(\mu)]. \tag{7.4.53}$$

If we substitute this expansion into (7.4.51), then for small $\mathbf{k}$ only the term with $\mathbf{g}=0$ gives an essential contribution to (7.4.51). This leads to

$$\bar{\mathbf{P}}(\mathbf{r})=\exp(i\mathbf{k}\cdot\mathbf{r})\frac{1}{v}\sum_\mu e_\mu \mathbf{w}(\mu|j,\mathbf{k})M_\mu^{-1/2}$$
$$=:\exp(i\mathbf{k}\cdot\mathbf{r})\sum_\mu \mathbf{p}(\mu|j,\mathbf{k}). \tag{7.4.54}$$

For $c\to\infty$ the field equations (7.4.41) for the electric field read

$$\nabla\times\bar{\mathbf{E}}=0;\quad \nabla\cdot(\bar{\mathbf{E}}+4\pi\bar{\mathbf{P}})=0. \tag{7.4.55}$$

In these equations the factor $\exp(-i\omega t)$ can be eliminated so that they hold also for the time independent fieldstrengths $\bar{\mathbf{E}}(\mathbf{r})$, $\bar{\mathbf{P}}(\mathbf{r})$. If the polarization $\bar{\mathbf{P}}(\mathbf{r})$ is decomposed into a part $\bar{\mathbf{P}}_\parallel(\mathbf{r})$ being parallel to $\mathbf{k}$ and a part $\bar{\mathbf{P}}_\perp(\mathbf{r})$ being perpendicular to $\mathbf{k}$

$$\bar{\mathbf{P}}(\mathbf{r})=\bar{\mathbf{P}}_\parallel(\mathbf{r})+\bar{\mathbf{P}}_\perp(\mathbf{r}) \tag{7.4.56}$$

then for these parts the equations

$$\nabla\times\bar{\mathbf{P}}_\parallel(\mathbf{r})=0;\qquad \nabla\cdot\bar{\mathbf{P}}_\perp(\mathbf{r})=0 \tag{7.4.57}$$

hold. Therefore, equations (7.4.55) can be rewritten to give

$$\nabla\cdot(\bar{\mathbf{E}}+4\pi\bar{\mathbf{P}}_\parallel)=0;\quad \nabla\times(\bar{\mathbf{E}}+4\pi\bar{\mathbf{P}}_\parallel)=0. \tag{7.4.58}$$

From these equations it follows that

$$\bar{\mathbf{E}}(\mathbf{r})=-4\pi\bar{\mathbf{P}}_\parallel(\mathbf{r})=-\exp(i\mathbf{k}\cdot\mathbf{r})4\pi\frac{\mathbf{k}\otimes\mathbf{k}}{k^2}\cdot\sum_\mu \mathbf{p}(\mu) \tag{7.4.59}$$

or, with (7.4.45) and (7.4.54)

$$\bar{\mathbf{E}}(\mathbf{k}) = -4\pi \frac{\mathbf{k}\otimes\mathbf{k}}{\mathbf{k}^2} \cdot \sum_{\mu} e_\mu v^{-1} M_\mu^{-1/2} \mathbf{w}(\mu | j, \mathbf{k}). \tag{7.4.60}$$

Substitution in (7.4.50) gives (7.4.46), Q.E.D.

The result of Theorem 7.13 is an illustration of the connection between the average field and the electric forces at the microscopic level in the crystal i.e., the result of the application of the Ewald-technique given by Theorem 7.10 can be interpreted as the separation of the effects of these two fields. A more realistic description of the forces has to take into account

i) the ion deformation and exchange forces,
ii) the electronic polarizability of the ions,
iii) the retardation effects of the electric interaction and fields.

A complete deduction from first principles including all these effects has so far not yet been given. Hence if we try to understand the formation and reactions of polaritons, we have to use semiphenomenological models. According to Lemma 7.3 we restrict our attention to the harmonic approximation in the atomic displacements and the macroscopic field. If we suppose, in accordance with Theorem 7.13 that the field variables of (7.4.10) are macroscopic, then the adiabatic energy of crystal and field can be written equivalently as

$$U_n^h(\mathbf{R}, \bar{\mathbf{E}}) = U_n(\mathbf{R}^0) \tag{7.4.61}$$

$$+ \sum_{\substack{l\kappa \\ l'\kappa'}} [\tfrac{1}{2} \Phi(l\kappa, l'\kappa') \cdot\cdot\, \mathbf{u}_{l\kappa} \mathbf{u}_{l'\kappa'} - \mathscr{M}(l\kappa, l'\kappa') \cdot\cdot\, \bar{\mathbf{E}}_{l\kappa} \mathbf{u}_{l'\kappa'} - \tfrac{1}{2} \mathscr{P}(l\kappa, l'\kappa') \cdot\cdot\, \bar{\mathbf{E}}_{l\kappa} \bar{\mathbf{E}}_{l'\kappa'}]$$

where $\bar{\mathbf{E}}_{l\kappa}$ is the macroscopic electric field at $\mathbf{r} = \mathbf{R}_{l\kappa}^0$. According to Theorem 7.13, in this expansion the force constants $\{\Phi(l\kappa, l'\kappa')\}$ describe the effects of all short-range forces, including the ones associated with the local electric fields at the microscopic level, but contain no contribution associated with the macroscopic field. The coefficients $\{\mathscr{M}(l\kappa, l'\kappa')\}$ are the transverse effective charge tensors, whereas the tensors $\{\mathscr{P}(l\kappa, l'\kappa')\}$ are the electronic polarizability tensors. Concerning their physical interpretation we refer to the literature e.g. Maradudin, Montroll, Weiss and Ipatova (1971).

We postpone the derivation of the complete Hamiltonian of lattice and averaged fields and consider first the equations of motion for these quantities. These are given for the lattice by

$$M_\kappa \ddot{\mathbf{u}}_{l\kappa} = -\frac{\partial U_n^h(\mathbf{u}, \bar{\mathbf{E}})}{\partial \mathbf{u}_{l\kappa}}$$

$$= -\sum_{l'\kappa'} \Phi(l\kappa, l'\kappa') \cdot \mathbf{u}_{l'\kappa'} + \sum_{l'\kappa'} \mathscr{M}(l'\kappa', l\kappa) \cdot \bar{\mathbf{E}}_{l'\kappa'} \tag{7.4.62}$$

with the following definition of the dipole moments

$$\mathbf{p}_{l\kappa} := -\frac{\partial U_n^h(\mathbf{u}, \bar{\mathbf{E}})}{\partial \bar{\mathbf{E}}_{l\kappa}} = \sum_{l'\kappa'} \mathscr{M}(l\kappa, l'\kappa') \cdot \mathbf{u}_{l'\kappa'} + \sum_{l'\kappa'} \mathscr{P}(l\kappa, l'\kappa') \cdot \bar{\mathbf{E}}_{l'\kappa'}. \tag{7.4.63}$$

The field equations for the averaged fields are given by (7.4.41). As the magnetic field does not take part in the lattice-field interaction it is convenient to eliminate it from the beginning. With (7.4.38) we obtain from (7.4.41)

$$\nabla \times \nabla \times \bar{\mathbf{E}} + \frac{1}{c^2}\frac{\partial^2}{\partial t^2}\bar{\mathbf{E}} = -\frac{4\pi}{c^2}\frac{\partial^2}{\partial t^2}\bar{\mathbf{P}}. \tag{7.4.64}$$

With

$$\bar{\mathbf{E}}(\mathbf{r},t) = \bar{\mathbf{E}}(\mathbf{k}) \exp(i\mathbf{k}\cdot\mathbf{r} - \omega t); \quad \bar{\mathbf{P}}(\mathbf{r},t) = \bar{\mathbf{P}}(\mathbf{k}) \exp(i\mathbf{k}\cdot\mathbf{r} - \omega t) \tag{7.4.65}$$

and

$$\begin{aligned} \mathbf{u}_{l\kappa} &= M_\kappa^{-1/2}\, \mathrm{w}(\kappa|\mathbf{k}) \exp(i\mathbf{k}\cdot\mathbf{R}_{l\kappa}^0 - i\omega t); \\ \mathbf{p}_{l\kappa} &= \mathbf{p}(\kappa|\mathbf{k}) \exp(i\mathbf{k}\cdot\mathbf{R}_{l\kappa}^0 - i\omega t) \end{aligned} \tag{7.4.66}$$

equations (7.4.62) (7.4.63) go over into

$$\omega^2 \mathbf{w}(\kappa|\mathbf{k}) = \sum_{\kappa'} \mathscr{C}(\kappa,\kappa'|\mathbf{k}) \cdot \mathbf{w}(\kappa'|\mathbf{k}) - M_\kappa^{-1/2} \sum_{\kappa'} \mathscr{M}(\kappa',\kappa|\mathbf{k})^\times \cdot \bar{\mathbf{E}}(\mathbf{k}) \tag{7.4.67}$$

$$\mathbf{p}(\kappa|\mathbf{k}) = \sum_{\kappa'} \mathscr{M}(\kappa,\kappa'|\mathbf{k}) M_{\kappa'}^{-1/2} \cdot \mathbf{w}(\kappa'|\mathbf{k}) + \sum_{\kappa'} \mathscr{P}(\kappa,\kappa'|\mathbf{k}) \cdot \bar{\mathbf{E}}(\mathbf{k}) \tag{7.4.68}$$

where

$$\begin{aligned} \mathscr{C}(\kappa,\kappa'|\mathbf{k}) &:= (M_\kappa M_{\kappa'})^{1/2} \sum_{l'} \Phi(l\kappa, l'\kappa') \exp[-i\mathbf{k}\cdot(\mathbf{R}_{l\kappa}^0 - \mathbf{R}_{l'\kappa'}^0)] \\ \mathscr{M}(\kappa,\kappa'|\mathbf{k}) &:= \sum_{l'} \mathscr{M}(l\kappa, l'\kappa') \exp[-i\mathbf{k}\cdot(\mathbf{R}_{l\kappa}^0 - \mathbf{R}_{l'\kappa'}^0)] \\ \mathscr{P}(\kappa,\kappa'|\mathbf{k}) &:= \sum_{l'} \mathscr{P}(l\kappa, l'\kappa') \exp[-i\mathbf{k}\cdot(\mathbf{R}_{l\kappa}^0 - \mathbf{R}_{l'\kappa'}^0)]. \end{aligned} \tag{7.4.69}$$

The matrices $\mathscr{C}(\kappa,\kappa'|\mathbf{k})$ and $\mathscr{P}(\kappa,\kappa'|\mathbf{k})$ are Hermitian.
The amplitude of the macroscopic polarization in the crystal is given by eq. (7.4.54) which in the present case takes the form

$$\bar{\mathbf{P}}(\mathbf{k}) = v^{-1} \sum_\kappa \mathbf{p}(\kappa|\mathbf{k}) = v^{-1} \sum_{\kappa'} z(\kappa'|\mathbf{k}) M_\kappa^{-1/2} \cdot \mathbf{w}(\kappa'|\mathbf{k}) + \chi^\infty(\mathbf{k}) \cdot \bar{\mathbf{E}}(\mathbf{k}) \tag{7.4.70}$$

where we have introduced the **k**-dependent effective charge tensor for the κ-th kind of ion

$$z(\kappa|\mathbf{k}) := \sum_{\kappa'} \mathscr{M}(\kappa',\kappa|\mathbf{k}) \tag{7.4.71}$$

and the **k**-dependent optical frequency susceptibility

$$\chi^{\infty}(\mathbf{k}) = v^{-1} \sum_{\kappa\kappa'} \mathscr{P}(\kappa, \kappa' | \mathbf{k}). \tag{7.4.72}$$

With (7.4.65) the field equations (7.4.64) can be written

$$\sum_{\beta} \left[\delta_{\alpha\beta} - \hat{k}_{\alpha} \hat{k}_{\beta} - \frac{\omega^2}{c^2 k^2} \delta_{\alpha\beta} \right] \bar{E}_{\alpha}(\mathbf{k}) = \frac{4\pi\omega^2}{c^2 k^2} \bar{P}_{\alpha}(\mathbf{k}) \tag{7.4.73}$$

where $\hat{\mathbf{k}}$ is a unit vector in the direction of **k** and their inversion reads

$$\bar{\mathbf{E}}(\mathbf{k}) = 4\pi \left[-\hat{\mathbf{k}} \otimes \hat{\mathbf{k}} + \frac{1}{n^2 - 1} (1 - \hat{\mathbf{k}} \otimes \hat{\mathbf{k}}) \right] \cdot \bar{\mathbf{P}}(\mathbf{k}) \tag{7.4.74}$$

with $n = ck/\omega$. To evaluate these equations further we have to use symmetry properties of the system cf. Maradudin, Montroll, Weiss and Ipatova (1971).
The coefficients $\Phi(l\kappa, l'\kappa')$ and $\mathscr{P}(l\kappa, l'\kappa')$ are symmetric in the indices $(l\kappa\alpha)$ and $(l'\kappa'\beta)$

$$\Phi(l\kappa, l'\kappa')_{\alpha\beta} = \Phi(l'\kappa', l\kappa)_{\beta\alpha}; \quad P(l\kappa, l'\kappa')_{\alpha\beta} = P(l'\kappa', l\kappa)_{\beta\alpha}. \tag{7.4.75}$$

The invariance of the force on an atom and of the dipole moment of the crystal against a rigid body displacement of the crystal as a whole (infinitesimal translational invariance) yields the conditions

$$\sum_{l\kappa} \Phi(l\kappa, l'\kappa') = \sum_{l'\kappa'} \Phi(l\kappa, l'\kappa') = 0; \quad \sum_{l'\kappa'} \mathscr{M}(l\kappa, l'\kappa') = 0. \tag{7.4.76}$$

The transformation properties of the force on an ion and of the crystal dipole moment under a rigid body rotation of the crystal as a whole (infinitesimal rotational invariance) yields the conditions

$$\begin{aligned} &\sum_{l'\kappa'} \Phi(l\kappa, l'\kappa')_{\alpha\beta} (\mathbf{R}^0_{l'\kappa'})_{\gamma} && \text{symmetric in } \beta \text{ and } \gamma \\ &\sum_{l\kappa} (\mathbf{R}^0_{l\kappa})_{\gamma} \Phi(l\kappa, l'\kappa')_{\alpha\beta} && \text{symmetric in } \alpha \text{ and } \gamma \\ &\sum_{l'\kappa'} M(l\kappa, l'\kappa')_{\alpha\beta} (\mathbf{R}^0_{l'\kappa'})_{\gamma} && \text{symmetric in } \beta \text{ and } \gamma. \end{aligned} \tag{7.4.77}$$

Finally, the transformation laws for these coefficients when the crystal is subjected to an operation $\{\mathscr{S} | \mathbf{v}(s) + \mathbf{x}(m)\}$ from its space group are

$$\begin{aligned} \Phi(h\varrho, h'\varrho')_{\alpha\beta} &= \sum_{\mu\nu} S_{\alpha\mu} S_{\beta\nu} \Phi(l\kappa, l'\kappa')_{\mu\nu} \\ M(h\varrho, h'\varrho')_{\alpha\beta} &= \sum_{\mu\nu} S_{\alpha\mu} S_{\beta\nu} M(l\kappa, l'\kappa')_{\mu\nu} \\ P(h\varrho, h'\varrho')_{\alpha\beta} &= \sum_{\mu\nu} S_{\alpha\mu} S_{\beta\nu} P(l\kappa, l'\kappa')_{\mu\nu}. \end{aligned} \tag{7.4.78}$$

It follows from these transformation laws that each of the coefficients $\Phi(l\kappa, l'\kappa')$, $\mathscr{M}(l\kappa, l'\kappa')$, $\mathscr{P}(l\kappa, l'\kappa')$ depends only on the difference of the cell indices l and l'.

To obtain a qualitative understanding of the polariton states we consider the long-wave length limit $\mathbf{k}\to 0$. In this case, equations (7.4.67) (7.4.70) read with $\lim_{\mathbf{k}\to 0} \mathbf{w}(\kappa|\mathbf{k}) \equiv \mathbf{w}(\kappa)$

$$\omega^2 \mathbf{w}(\kappa) = \sum_{\kappa'} (M_\kappa M_{\kappa'})^{-1/2} f(\kappa,\kappa') \cdot \mathbf{w}(\kappa') - \bar{\mathbf{E}} \cdot f(\kappa) M_\kappa^{-1/2} \tag{7.4.79}$$

and

$$\bar{\mathbf{P}} = v^{-1} \sum_{\kappa} M_\kappa^{-1/2} f(\kappa) \cdot \mathbf{w}(\kappa) + \chi^\infty \cdot \bar{\mathbf{E}} \tag{7.4.80}$$

with

$$f(\kappa,\kappa') := \sum_{l'} \Phi(l\kappa, l'\kappa')$$

$$f(\kappa) := \sum_{l'\kappa'} \mathscr{M}(0\kappa, l'\kappa') \tag{7.4.81}$$

$$\chi^\infty := \sum_{\substack{l'\kappa' \\ \kappa}} \mathscr{P}(0\kappa, l'\kappa')$$

if $\bar{\mathbf{E}}_{l\kappa} \approx \bar{\mathbf{E}}_{l\kappa'}$ is assumed in the long-wave length limit. For these quantities from (7.4.76) (7.4.77) (7.4.78) the symmetry properties

$$f(\kappa,\kappa')_{\alpha\beta} = f(\kappa',\kappa)_{\beta\alpha}; \quad \chi^\infty_{\mu\nu} = \chi^\infty_{\nu\mu}$$

$$\sum_{\kappa'} f(\kappa,\kappa')_{\alpha\beta} = \sum_{\kappa} f(\kappa,\kappa')_{\alpha\beta} = 0 \tag{7.4.82}$$

$$\sum_{\kappa} f(\kappa)_{\mu\alpha} = 0$$

can be derived. They also possess the following transformation properties

$$f(\lambda,\lambda')_{\alpha\beta} = \sum_{\mu\nu} S_{\alpha\mu} S_{\beta\nu} f_{\mu\nu}(\kappa,\kappa')$$

$$f(\lambda)_{\mu\alpha} = \sum_{\nu\beta} S_{\mu\nu} S_{\alpha\beta} f(\kappa)_{\nu\beta} \tag{7.4.83}$$

$$\chi^\infty_{\mu\nu} = \sum_{\alpha\beta} S_{\mu\alpha} S_{\nu\beta} \chi^\infty_{\alpha\beta}.$$

If we now combine equation (7.4.74) with (7.4.80), we obtain the relation between the macroscopic field and the displacement field

$$\mathbf{E} = -\frac{4\pi}{v} \sum_{\kappa'} \mathscr{T} \cdot \left[\hat{\mathbf{k}} \otimes \hat{\mathbf{k}} - \frac{1}{n^2 - 1} (1 - \hat{\mathbf{k}} \otimes \hat{\mathbf{k}}) \right] \cdot M_{\kappa'}^{-1/2} f(\kappa') \cdot \mathbf{w}(\kappa') \tag{7.4.84}$$

with

$$\mathcal{T}^{-1} := 1 + 4\pi\hat{\mathbf{k}}\otimes\hat{\mathbf{k}} \cdot \chi^{\infty} - \frac{4\pi}{n^2-1}(1-\hat{\mathbf{k}}\otimes\hat{\mathbf{k}}) \cdot \chi^{\infty}. \tag{7.4.85}$$

Substituting equation (7.4.84) into (7.4.79) we obtain

$$\omega^2 \mathbf{w}(\kappa) = \sum_{\kappa'} (M_\kappa M_{\kappa'})^{-1/2} f(\kappa,\kappa') \cdot \mathbf{w}(\kappa') \tag{7.4.86}$$
$$+ \frac{4\pi}{v} \sum_{\kappa'} M_\kappa^{-1/2} f(\kappa) \cdot \mathcal{T} \cdot \left[\hat{\mathbf{k}}\otimes\hat{\mathbf{k}} - \frac{(1-\hat{\mathbf{k}}\otimes\hat{\mathbf{k}})}{(n^2-1)}\right] \cdot M_{\kappa'}^{-1/2} f(\kappa') \cdot \mathbf{w}(\kappa').$$

The normal modes for $\mathbf{k}\to 0$ of the coupled photon-phonon system are the solutions of this equation.
If χ^{∞} is diagonal and the crystal is optically isotropic, then $\mathcal{T}$ can be obtained by inversion of (7.4.85) and reads

$$\mathcal{T} = \frac{1}{\varepsilon_\infty} \hat{\mathbf{k}}\otimes\hat{\mathbf{k}} + \frac{n^2-1}{n^2-\varepsilon_\infty}(1-\hat{\mathbf{k}}\otimes\hat{\mathbf{k}}) \tag{7.4.87}$$

with

$$\varepsilon_\infty := 1 + 4\pi\chi^{\infty}. \tag{7.4.88}$$

These conditions are satisfied for cubic crystals.
Furthermore, if we assume that the crystal contains two inequivalent ions in the unit cell (labeled by + and −), then the cubic symmetry of these crystals yields

$$f(\kappa)_{\mu\alpha} = \delta_{\mu\alpha} e(\kappa) \tag{7.4.89}$$

and from translational invariance (7.4.82) it follows

$$e(+) = e_T^* = -e(-) \tag{7.4.90}$$

where e_T^* is the transverse effective charge. From (7.4.83) it follows further

$$f(\kappa,\kappa')_{\alpha\beta} = \delta_{\alpha\beta} f(\kappa,\kappa') \tag{7.4.91}$$

which leads together with (7.4.82) to

$$f(+,+) = -f(+,-) = -f(-,+) = f(-,-) =: \mu\omega_T^2 \tag{7.4.92}$$

where μ is the reduced mass

$$\mu = \frac{M_+ M_-}{M_+ + M_-}. \tag{7.4.93}$$

If these formulas are used, the equations (7.4.86) are reduced to

$$\omega^2 \mathbf{w}(\kappa) = \mu\omega_T^2 \sum_{\kappa'} \frac{\operatorname{sgn}\kappa \operatorname{sgn}\kappa'}{(M_\kappa M_{\kappa'})^{1/2}} \mathbf{w}(\kappa) \tag{7.4.94}$$
$$+ \frac{4\pi(e_T^*)^2}{v} \sum_{\kappa'} \frac{\operatorname{sgn}\kappa \operatorname{sgn}\kappa'}{(M_\kappa M_{\kappa'})^{1/2}} \left[\frac{\hat{\mathbf{k}}\otimes\hat{\mathbf{k}}}{\varepsilon_\infty} - \frac{(1-\hat{\mathbf{k}}\otimes\hat{\mathbf{k}})}{n^2-\varepsilon_\infty}\right] \cdot \mathbf{w}(\kappa').$$

For diatomic crystals a set of 6 equations results from (7.4.94). They possess acoustic and optical vibrations as eigenmodes. The acoustic modes are characterized by $\lim_{\mathbf{k}\to 0} \omega(\mathbf{k})=0$. They can easily be obtained from (7.4.88) if we put $\mathbf{k}=0$ and $\hat{\mathbf{k}}=0$. Then for $\omega(\mathbf{k})=0$ the determinant of (7.4.94) vanishes and the corresponding solutions are

$$\mathbf{w}(\kappa)=(M_\kappa/M_c)^{1/2}\mathbf{e}(j) \tag{7.4.95}$$

where $M_c=M_1+M_2$ is the mass in the unit cell and $\{\mathbf{e}(j), j=1, 2, 3\}$ are any three mutually perpendicular unit vectors. This result corresponds to Theorem 7.5 where the acoustic modes for diatomic crystals have been derived without the participation of the averaged fields. The optical solutions are given in that former case by (7.2.30). We take over this result for the present case and write

$$\mathbf{w}(\kappa)=\operatorname{sgn}\kappa(\mu/M_\kappa)^{1/2}\mathbf{e}. \tag{7.4.96}$$

Substitution of (7.4.96) into (7.4.94) leads to

$$\omega^2\mathbf{e}=\omega_T^2\mathbf{e}+\frac{4\pi(e_T^*)^2}{\mu v}\left(\frac{\hat{\mathbf{k}}\otimes\hat{\mathbf{k}}}{\varepsilon_\infty}-\frac{1-\hat{\mathbf{k}}\otimes\hat{\mathbf{k}}}{n^2-\varepsilon_\infty}\right)\cdot\mathbf{e}. \tag{7.4.97}$$

The solutions of this equation are pure longitudinal and pure transverse oscillations. For the longitudinal case we obtain with $\mathbf{e}\parallel\hat{\mathbf{k}}$

$$\omega^2\mathbf{e}=\omega_T^2\mathbf{e}+\frac{4\pi(e_T^*)^2}{\varepsilon_\infty\mu v}\mathbf{e} \tag{7.4.98}$$

which yields for the corresponding frequency with $\omega_P^2=4\pi(e_T^*)^2/\varepsilon_\infty\mu v$

$$\omega^2=\omega_T^2+\omega_P^2=\omega_L^2. \tag{7.4.99}$$

By comparison with calculations where the retardation has not been taken into account it can be seen that its inclusion does not affect this branch. For the transverse case we find with $\mathbf{e}\perp\mathbf{k}$

$$\omega^2\mathbf{e}=\omega_T^2\mathbf{e}-\omega_P^2\varepsilon_\infty\left(\frac{c^2k^2}{\omega^2}-\varepsilon_\infty\right)^{-1}\mathbf{e}. \tag{7.4.100}$$

The solutions are

$$\omega_\pm^2=\frac{1}{2}\left\{\omega_L^2+\frac{c^2k^2}{\varepsilon_\infty}\pm\left[\left(\omega_L^2+\frac{c^2k^2}{\varepsilon_\infty}\right)^2-\frac{4\omega_T^2c^2k^2}{\varepsilon_\infty}\right]^{1/2}\right\}. \tag{7.4.101}$$

The two branches $\omega_\pm$ of this dispersion relation are the two branches of the polariton dispersion relation.

The discussion for small $\mathbf{k}$ provides only a very simple example of the calculation of polariton states. Of course such calculations can also be done for finite values of $\mathbf{k}$, but then they are rather complicated; cf. e.g. Maradudin, Montroll, Weiss, Ipatova (1971). We show such a dispersion relation in the following Figure 7.5.

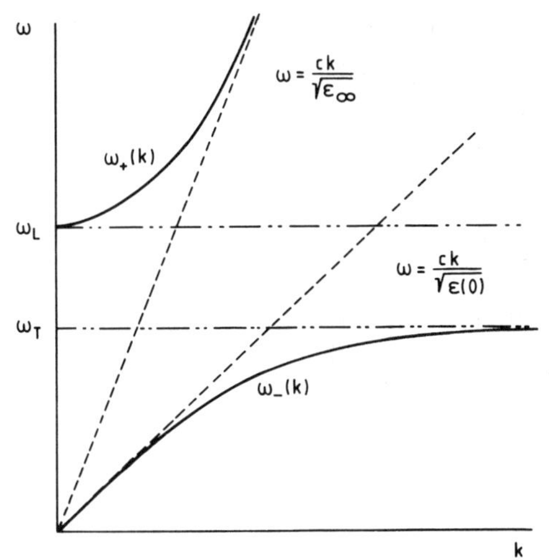

Figure 7.5:
A schematic representation of the polariton dispersion curves with two branches ω_+ and ω_- for a cubic ionic crystal with two inequivalent ions in a primitive unit cell; after Maradudin (1974).

We now turn to the energy calculation. If the interaction between lattice particles and the genuine electromagnetic field is considered according to (7.4.2), then the coupling term contains only the vector potentials. From this it follows that the Lagrangian conjugate momentum variables of the field variables (7.4.4) are not affected by the interaction term. If, however, the interaction energy between lattice and field contains the electromagnetic field itself as in the case of the averaged field with the energy expression (7.4.61), then the conjugate momentum variables depend on this interaction. Hence the derivation of an energy expression for the system with non-averaged and with averaged fields by means of the Lagrangian formalism runs along different lines, and in general this derivation is much more complicated for averaged fields than for non-averaged fields. Therefore, instead of applying the Lagrangian formalism we describe a phenomenological approach for averaged fields by means of the Poynting theorem which is due to Huang (1951). This approach is restricted to small **k** vectors. But due to its simplicity the essential physical effect of the energy distribution between lattice and electromagnetic field can be seen clearly.

Theorem 7.14: In the limit $\mathbf{k}\to 0$ the energy density of the lattice polariton is given by

$$U_P := \frac{1}{2v}\sum_{\kappa} M_\kappa \dot{\mathbf{u}}_\kappa^2 + \frac{1}{2v}\sum_{\kappa\kappa'} f(\kappa,\kappa') \cdot\cdot\, \mathbf{u}_\kappa \mathbf{u}_{\kappa'} + \frac{1}{8\pi}\,\bar{\mathbf{E}}\cdot\varepsilon_\infty\cdot\bar{\mathbf{E}} + \frac{1}{8\pi}\,\mathbf{B}^2 \tag{7.4.102}$$

where ε_∞ is defined by (7.4.88).

Proof: From the Maxwell equations (7.4.41) and (7.4.42) the following equation for the Poynting vector $\mathbf{S}$ can be derived

$$\begin{aligned} \nabla \cdot \mathbf{S} :&= \frac{c}{4\pi} \nabla \cdot (\bar{\mathbf{E}} \times \bar{\mathbf{B}}) \\ &= -\left[\frac{1}{4\pi}(\bar{\mathbf{E}} \cdot \dot{\bar{\mathbf{E}}} + \bar{\mathbf{B}} \cdot \dot{\bar{\mathbf{B}}}) + \bar{\mathbf{E}} \cdot \dot{\bar{\mathbf{P}}}\right]. \end{aligned} \tag{7.4.103}$$

As, according to (7.4.79), for long waves the various unit cells of the crystal are only coupled by the electric field, we see that the energy flow in the crystal can only be caused by the electromagnetic field. Hence the rate of energy change of the total energy in a certain volume V must be

$$\frac{d}{dt} \int_V U_P d^3r = \int_{O(V)} \mathbf{S} \cdot d\mathbf{s} = \int_V \nabla \cdot \mathbf{S} d^3r. \tag{7.4.104}$$

Combining equation (7.4.104) with (7.4.103) we obtain

$$\frac{d}{dt} U_P = \frac{1}{4\pi}(\bar{\mathbf{E}} \cdot \dot{\bar{\mathbf{E}}} + \bar{\mathbf{B}} \cdot \dot{\bar{\mathbf{B}}}) + \bar{\mathbf{E}} \cdot \dot{\bar{\mathbf{P}}}. \tag{7.4.105}$$

This relation is satisfied by the expression

$$\begin{aligned} U_P :=& \frac{1}{2v} \sum_\kappa M_\kappa \dot{\mathbf{u}}_\kappa^2 + \frac{1}{2v} \sum_{\kappa\kappa'} f(\kappa\kappa') \cdot \cdot \, \mathbf{u}_\kappa \mathbf{u}_{\kappa'} \\ & - \frac{1}{v} \sum_\kappa f(\kappa) \cdot \cdot \, \bar{\mathbf{E}} \mathbf{u}_\kappa - \frac{1}{2} \bar{\mathbf{E}} \cdot \chi^\infty \cdot \bar{\mathbf{E}} \\ & + \bar{\mathbf{E}} \cdot \bar{\mathbf{P}} + \frac{1}{8\pi}(\bar{\mathbf{E}}^2 + \bar{\mathbf{B}}^2). \end{aligned} \tag{7.4.106}$$

For, if we differentiate U_P with respect to t and if we observe (7.4.79) and (7.4.80) with $\mathbf{u}(\kappa) = M_\kappa^{-1/2} \mathbf{w}(\kappa) \exp(-i\omega t)$ for $\mathbf{k} \to 0$, then from (7.4.106) the equation (7.4.105) results. If we further substitute (7.4.80) into (7.4.106), then formula (7.4.102) results, Q.E.D.

The application of this energy formula leads to the following Figure 7.6.
A review of lattice polariton theory is given in the book of Maradudin, Montroll, Weiss and Ipatova (1971). Review articles of polariton theory were published by Merten (1972) and by Wallis (1977). Light scattering by lattice polaritons is treated in the book of Claus, Merten and Brandmüller (1975). Concerning the original literature, Huang (1951) first considered the combined collective motion of lattice modes and electromagnetic modes at a classical level and the book of Born and Huang (1954) contains a survey of this approach. The notation polariton was introduced later by Hopfield (1958). The combined collective motion of material polarization waves and electromagnetic waves is not restricted to lattice polarization waves. Fano (1956)

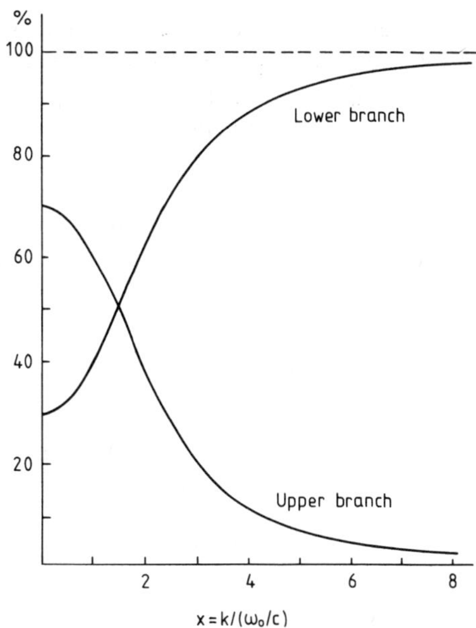

Figure 7.6:
Percentage of the mechanical energy in the transverse modes; after Born and Huang (1954).

discussed such generalized excitation states at the quantum level and described the matter polarization by a set of oscillators. Hopfield (1958) treated electronic polaritons at the quantum level based on a Fock space representation for photons and excitons, Merten (1960) applied Huang's model to the calculation of long-wave optical lattice modes for diatomic polar crystals with trigonal, tetragonal and hexagonal structure. Kurosawa (1961) studied polarization waves in solids at a phenomenological level. Loudon (1963) (1964) calculated the frequencies and branches of uniaxial crystals in the longwave region. Merten (1967) (1968) investigated the infrared-dispersion of uniaxial and biaxial crystals based on Huang's model. Lamprecht and Merten (1969) studied the crossing of polariton branches in uniaxial crystals. Opie (1968) (1969) developed a theory of lattice polaritons with inclusion of anharmonic lattice interactions based on a Fock-space Hamiltonian. The same problem was treated by Ohtaka and Fujiwara (1969). Benson and Mills (1970) formulated a theory of light scattering from polaritons including lattice damping and based on field theoretic Green function techniques. Inomata and Horie (1970) published a note on the use of the Master equation for polariton systems. Borstel and Merten (1971) (1972) calculated eigenfrequencies and eigenvectors of polaritons with application to $LiNbO_3$ based on Huang's model. Merten and Borstel (1972) treated damped polaritons at the same level. Giallorenzi (1972) developed a quantum theory of light scattering by damped polaritons. Emelyanov and Klimontovich (1972) treated parametric scattering of light by polaritons. Barker and Loudon (1972) gave a review of the application of response functions in the theory of Raman scattering by vibrational and polariton modes in dielectric crystals. Merten (1973) reported on sum rules of damping factors of polaritons. Posledovich, Winter, Borstel and Claus (1973) discussed properties of extraordinary polaritons in $LiNbO_3$. Rath, Borstel, Lamprecht and Merten (1975) derived the density of states of polaritons in cubic and uniaxial crystals. Maddox and Mills (1975) investigated polariton scattering by point defects in spatially dispersive media; Boardman, Parker and Allos (1977) reported on optical generation of longitudinal polariton modes in ionic crystals. For further references with respect to polariton scattering we refer to Section 7.6.d.

7.5 Phonon modes of imperfect crystals

The general adiabatic lattice equation is given by (2.2.18). The label n of the coupling constants $\{\Phi_{jk}(n)\}$ of the potential energy in this equation indicates that phonon states have to be calculated for various crystal states denoted by n. In our context such crystal states of interest are defined by crystals with various imperfections in their electronic ground states as well as in excited electronic states. Hence we have to generalize the calculations of the preceding sections which have been done for ideal crystals so as to also include the case of imperfect crystals. According to Section 2.2 to solve this problem it is sufficient to treat the classical equations (2.2.24), which read in our special notation

$$\sum_{l\kappa} (M_{l\kappa} M_{l'\kappa'})^{-1/2} \Phi(l\kappa, l'\kappa'|n) \cdot \mathbf{a}(l'\kappa'|j) - \omega^2(j|n)\mathbf{a}(l\kappa|j) = 0. \tag{7.5.1}$$

As we have confined ourselves to the discussion of zero dimensional imperfections (point imperfections including molecular impurity configurations), the index n characterizes the state and the local distribution of such imperfections. For a low density of these imperfections it is reasonable to study first the phonon spectrum of an isolated impurity in a microblock and to include afterwards the effects of a finite distribution density. By direct calculation it can be shown afterwards that this assumption is justified as long as the impurity distribution density is not too large. Hence we start with this simplified model and assume for simplicity that the microblock is of infinite volume. If an infinite crystal contains only one imperfection, it may be assumed that this crystal still has a symmetry group G which transforms the lattice equilibrium positions $\{\mathbf{R}_k^n\}$ into themselves. In this case the adiabatic energy $U_n(\mathbf{R}^n)$ at the equilibrium positions must also be transformed into itself for all group operations of this symmetry group, i.e., according to Definition 7.1 it must be a symmetry identity with respect to G. If $U_n(\mathbf{R}^n)$ is a symmetry identity, then Theorem 7.2 holds for the transformation properties of the coupling constants $\{\Phi_{jk}(n)\}$. Further conclusions can be drawn from this theorem. They are contained in the following theorem which is due to Maradudin, Montroll, Weiss, Ipatova (1971)

Theorem 7.15: If the adiabatic energy $U_n(\mathbf{R}^n)$ is supposed to be a symmetry identity with respect to a point group G, then any eigenvector $\mathbf{a}(l\kappa|j)$ of (7.5.1) is a basis vector of a corresponding irreducible unitary representation of G, and the complete set of eigenvectors $\{\mathbf{a}(l\kappa|j), 1 \leqslant j \leqslant 3N\}$ can be decomposed into a union of complete sets of base vectors with respect to these representations.

Proof: As the eigenvectors of (7.5.1) can be orthonormalized and form a complete set, equation (7.5.1) can be considered to be a linear equation in a Hilbert space. Hence the theorem can be proven if we prove the forminvariance of (7.5.1) with respect to G and apply Theorem 6.1. We define

$$\mathscr{D}(l\kappa,l'\kappa'|n) := (M_{l\kappa}M_{l'\kappa'})^{-1/2}\Phi(l\kappa,l'\kappa'|n). \tag{7.5.2}$$

Because a point group operation S sends the crystal into itself, the atoms at the sites $(l\kappa)$ and $(h\varrho)$ must be the same, from which it follows that $M_{h\varrho}=M_{l\kappa}$. Hence equations (7.1.17) can be taken over for (7.5.2) to give

$$D(h\varrho,h'\varrho')_{\alpha\beta}=\sum_{\mu\nu} S_{\alpha\mu}S_{\beta\nu}D(l\kappa,l'\kappa')_{\mu\nu}. \tag{7.5.3}$$

We rewrite (7.5.3) as

$$D(h\varrho,h'\varrho')_{\alpha\beta}=\sum_{l\kappa\mu}\sum_{l'\kappa'\nu} S(h\varrho,l\kappa)_{\alpha\mu}D(l\kappa,l'\kappa')_{\mu\nu}S(h'\varrho',l'\kappa')_{\beta\nu} \tag{7.5.4}$$

by defining

$$S(l\kappa,l'\kappa')_{\alpha\beta} := S_{\alpha\beta}\delta(\mathscr{S}^{-1}\mathbf{R}^0_{l\kappa},\mathbf{R}^0_{l'\kappa'}). \tag{7.5.5}$$

The matrix (7.5.5) is real and orthogonal. These properties follow directly from the corresponding properties of $S_{\alpha\beta}$. In particular we obtain

$$\sum_{l''\kappa''\gamma} S(l\kappa,l''\kappa'')_{\alpha\gamma}S(l'\kappa',l''\kappa'')_{\beta\gamma}=\delta_{\alpha\beta}\delta(\mathbf{R}^0_{l\kappa},\mathbf{R}^0_{l'\kappa'}) \tag{7.5.6}$$

and

$$\sum_{l''\kappa''\gamma} S(l''\kappa'',l\kappa)_{\gamma\alpha}S(l''\kappa'',l\kappa)_{\gamma\beta}=\delta_{\alpha\beta}\delta(\mathbf{R}^0_{l\kappa},\mathbf{R}^0_{l'\kappa'}). \tag{7.5.7}$$

Hence we have

$$S^{-1}(l''\kappa'',l'\kappa')_{\gamma\beta}\equiv S(l'\kappa',l''\kappa'')_{\beta\gamma} \tag{7.5.8}$$

and equation (7.5.4) may be rewritten as

$$D(h\varrho,h'\varrho')_{\alpha\beta}=\sum_{l\kappa\mu}\sum_{l'\kappa'\nu} S(h\varrho,l\kappa)_{\alpha\mu}D(l\kappa,l'\kappa')_{\mu\nu}S^{-1}(l'\kappa',h'\varrho')_{\nu\beta}. \tag{7.5.9}$$

But this means $[\mathscr{D},\mathscr{S}]_-=0$, i.e., forminvariance of (7.5.2). Application of Theorem 6.1 implies that the $\{\mathbf{a}(l\kappa|j)\}$ must be base vectors of representations of G. The unitarity of these representations follows from the orthogonality of the $\{\mathbf{a}(l\kappa|j)\}$ and the orthogonality of (7.5.5), Q.E.D.

As the solutions of (7.5.1) belong to representations of the corresponding point groups and not to those of space groups, we have to expect phonon modes which are quite different from those of the ideal lattice. The following types of modes may occur:

i) Bound state modes, which are localized at and around the imperfection;
ii) Scattering state modes, which correspond to ideal lattice phonon modes which are scattered by the modified potential due to the imperfection.

This division follows if (7.5.2) is decomposed into

$$\mathscr{D}(l\kappa,l'\kappa'|n)=\mathscr{D}^0(l\kappa,l'\kappa')+\Delta(l\kappa,l'\kappa'|n) \tag{7.5.10}$$

where $\mathscr{D}^0(l\kappa,l'\kappa')$ is given by (7.5.2) for the ideal lattice configuration. Substitution of (7.5.10) into (7.5.1) then leads to

$$\sum_{l'\kappa'}[\mathscr{D}^0(l\kappa,l'\kappa')+\Delta(l\kappa,l'\kappa'|n)-\omega^2(j,n)\delta(l\kappa,l'\kappa')]\cdot\mathbf{a}(l\kappa|j,n)=0. \tag{7.5.11}$$

From this system the two types of states i) and ii) can be derived if the Greenfunction

$$\mathscr{G}^0(l\kappa,l'\kappa'|\omega):=[\mathscr{D}^0(l\kappa,l'\kappa')-\omega^2\delta(l\kappa,l'\kappa')]^{-1}\equiv\mathfrak{G}^0 \tag{7.5.12}$$

is applied. To obtain an explicit expression for $\mathfrak{G}^0$, we first make the formal ansatz

$$\mathfrak{G}^0=N^{-1}\sum_{j,\mathbf{k}}\frac{\mathbf{w}(\kappa|j,\mathbf{k})\otimes\mathbf{w}(\kappa'|j,\mathbf{k})^\times\exp[i\mathbf{k}\cdot(\mathbf{R}^0_{l\kappa}-\mathbf{R}^0_{l'\kappa'})]}{\omega^2-\omega^2(j,\mathbf{k})} \tag{7.5.13}$$

where $\mathbf{w}(\kappa|j,\mathbf{k})$ are the solutions of equations (7.2.23) for periodic lattice volumes of N particles. From (7.5.13) together with the formulae of Section 7.2 it can easily be verified that

$$\mathfrak{G}^0(\mathfrak{D}^0-\omega^2)=\mathbb{1}. \tag{7.5.14}$$

If the volume V is assumed to become infinite, then the summation in (7.5.13) is changed into an integration over the first Brillouin zone in **k**-space. From this it follows that two cases have to be distinguished for the evaluation of $\mathscr{G}^0$

i) the frequency ω is outside of the bands $\{\omega(\mathrm{j},\mathbf{k})\}$
ii) the frequency ω is inside of a band $\{\omega(\mathrm{j},\mathbf{k})\}$

We note that in i), the integrand of (7.5.13) remains finite, whereas in ii) the integrand of (7.5.13) exhibits a singularity. Both cases have to be treated separately. This situation corresponds to the quantum theory of bound and scattering states and the evaluation of the Greenfunction can be performed analogously leading to the bound and to the scattered mode solutions i) and ii). The corresponding equations read

$$\mathbf{a}(l\kappa|j,n)=-\sum_{\substack{l'\kappa'\\l''\kappa''}}\mathscr{G}^0(l\kappa,l''\kappa'')\Delta(l''\kappa'',l'\kappa'|n)\mathbf{a}(l'\kappa'|j,n) \tag{7.5.15}$$

for bound states, and

$$\mathbf{a}(l\kappa|j,n)^{(\pm)}=\mathbf{a}^0(l\kappa|j,\mathbf{k}) -\sum_{\substack{l'\kappa'\\l''\kappa''}}\mathscr{G}^0(l\kappa,l''\kappa'')^{(\pm)}\Delta(l''\kappa'',l'\kappa'|n)\mathbf{a}(l'\kappa'|j,n)^{(\pm)} \tag{7.5.16}$$

for scattering states, where $(\pm)$ indicates the two different types of integration around the singularity (advanced and retarded Greenfunctions) and $\mathbf{a}^0(l\kappa|j,\mathbf{k})$ are the phonon mode solutions of the ideal lattice.
For the explicit solution of equations (7.5.15) and (7.5.16) the Green functions (7.5.13) have to be evaluated. According to Maradudin, Montroll, Weiss, Ipatova (1971) the following expressions can be derived:

Theorem 7.16: If the microblock is supposed to be infinitely large and if ω is supposed to have a value in a gap of the ideal phonon spectrum, then the bound state Green function (7.5.13) can be approximately represented by

$$\mathscr{G}^0(l\kappa,l'\kappa'|\omega)=\sum_j \frac{v}{2\pi}(M_\kappa M_{\kappa'})^{-1/2}\mathbf{w}(\kappa|j,\mathbf{k}_{0j})\otimes\mathbf{w}(\kappa'|j,\mathbf{k}_{0j})^\times$$
$$\times|\mathscr{A}(j)|^{-1/2}\exp\,[i\mathbf{k}_{0j}\cdot(\mathbf{R}^0_{l\kappa}-\mathbf{R}^0_{l'\kappa'})] \tag{7.5.17}$$
$$\times\exp\{-[2|\omega^2-\omega_0^2(j)|]^{1/2}R(l\kappa,l'\kappa')\}R(l\kappa,l'\kappa')^{-1}$$

with

$$R^2(l\kappa,l'\kappa'):=(\mathbf{R}^0_{l\kappa}-\mathbf{R}^0_{l'\kappa'})\cdot\mathscr{A}(j)\cdot(\mathbf{R}^0_{l\kappa}-\mathbf{R}^0_{l'\kappa'}). \tag{7.5.18}$$

The definitions of the other quantities are given in the proof.

Proof: We assume that the crystal under consideration has $j=n$ bands which are separated by gaps, and that the numbering of the bands corresponds to increasing frequencies. We further assume that ω has a value in the gap between the bands $j=i$ and $j=i+1$. If we denote the wave vector corresponding to the maximum frequency $\omega_0(j)$ in the bands $1\leqslant j\leqslant i$ and to the minimum frequency $\omega_0(j)$ in the bands $i+1\leqslant j\leqslant n$ by $\mathbf{k}_{0j}$, then we have the expansion about this point

$$\omega^2(j,\mathbf{k})=\omega_0^2(j)\pm\tfrac{1}{2}(\mathbf{k}-\mathbf{k}_{0j})\cdot\mathscr{A}(j)\cdot(\mathbf{k}-\mathbf{k}_{0j}) \tag{7.5.19}$$

where the $+$ sign has to be used for $i+1\leqslant j\leqslant n$, while the $-$ sign has to be used for $1\leqslant j\leqslant i$. The linear terms in this expansion vanish because we are expanding about a maximum or a minimum. For simplicity the case of several maxima or minima, resp. is excluded. Since the expansion is made about a maximum or a minimum, the matrix $\mathscr{A}(j)$ has to be positive definite. In the Green function

$$\mathfrak{G}^0=\sum_j\int\frac{\mathbf{w}(\kappa|j,\mathbf{k})\otimes\mathbf{w}(\kappa'|j,\mathbf{k})^\times}{\omega^2-\omega^2(j,\mathbf{k})}\exp\,[i\mathbf{k}\cdot(\mathbf{R}^0_{l\kappa}-\mathbf{R}^0_{l'\kappa'})]\,d^3k \tag{7.5.20}$$

we make an expansion about $\mathbf{k}_{0j}$ and obtain with $\mathbf{k}'=(\mathbf{k}-\mathbf{k}_{0j})$ and the integral representation

$$\frac{1}{x}=\int_0^\infty e^{-xt}\,dt \tag{7.5.21}$$

the approximate expression

$$\mathfrak{G}^0 \approx \sum_j \frac{v}{(2\pi)^3} (\pm)(M_\kappa M_{\kappa'})^{-1/2} \mathbf{w}(\kappa|j,\mathbf{k}_{0j}) \otimes \mathbf{w}(\kappa'|j,\mathbf{k}_{0j})^\times$$
$$\times \exp [i\mathbf{k}_{0j} \cdot (\mathbf{R}^0_{l\kappa} - \mathbf{R}^0_{l'\kappa'})] \tag{7.5.22}$$
$$\times \int d^3k' \int_0^\infty dt \exp \{i\mathbf{k}' \cdot (\mathbf{R}^0_{l\kappa} - \mathbf{R}^0_{l'\kappa'}) - t[|\omega^2 - \omega_0^2(j)| + \tfrac{1}{2}\mathbf{k}' \cdot \mathscr{A}(j) \cdot \mathbf{k}']\}$$

where the positive sign holds for $1 \leqslant j \leqslant i$, while the negative sign holds for $i+1 \leqslant j \leqslant n$ in the sum (7.5.22). If in the integral (7.5.22) the integration is extended from the Brillouin zone to infinity, which produces a negligible error, the integrals can exactly be solved giving (7.5.17), Q.E.D.

Similar considerations can be performed for the case that ω lies within a band. We give only the results, but not the proof, cf. Maradudin, Montroll, Weiss and Ipatova (1971).

Theorem 7.17: If the microblock is supposed to be infinitely large and if ω is supposed to have a value inside of the i-th frequency band of the ideal phonon spectrum, then the scattering state Greenfunction (7.5.13) can be approximately represented by

$$\mathfrak{G}^{0(\pm)} \approx \frac{v}{2\pi} (M_\kappa M_{\kappa'})^{-1/2} \sum_\nu \mathbf{w}(\kappa|i,\mathbf{k}_\nu) \otimes \mathbf{w}(\kappa'|i,\mathbf{k}_\nu)^\times$$
$$\times \frac{\exp [i\mathbf{k}_\nu \cdot (\mathbf{R}^0_{l\kappa} - \mathbf{R}^0_{l'\kappa'})]}{|\mathbf{R}^0_{l\kappa} - \mathbf{R}^0_{l'\kappa'}|} \exp [-i\pi(\varphi_{\nu x} + \varphi_{\nu y}) + \tfrac{1}{2}](\varphi_{\nu x}\varphi_{\nu y})^{-1/2} \tag{7.5.23}$$
$$+ \mathscr{G}^0_i(l\kappa, l'\kappa'|\omega)$$

where $\mathscr{G}^0_i(l\kappa, l'\kappa'|\omega)$ is defined by omitting the term $j=i$ in the sum of (7.5.17). Further $\varphi_{\nu x} := \partial^2\omega(i,\mathbf{k}_\nu)/\partial^2 k'$ and $\mathbf{k}_\nu$ are the points at the frequency surface $\omega^2 = \omega^2(i,\mathbf{k})$ at which the phonon group velocity is parallel to and points in the same direction as $(\mathbf{R}^0_{l\kappa} - \mathbf{R}^0_{l'\kappa'})$.

We now turn to the solution of the equations (7.5.15) (7.5.16). We first consider (7.5.16). The methods of solving scattering equations for ordinary scattering processes and for resonance scattering have been completely developed in quantum theory, cf. Goldberger, Watson (1964). They can be taken over for the solution of equations (7.5.16). The only difference which has to be observed is between the structure of the Greenfunctions (7.5.12) and those of potential scattering in quantum theory. As a thorough deduction of these methods is rather lengthy and does not lead to essential improvement on the results of potential scattering theory we refer to the literature.

Concerning bound states, Lifshitz (1943) (1944) was the first who applied the Greenfunction (7.5.12) to equation (7.5.11). He used the fact that if the perturbation $\Delta(l\kappa, l'\kappa'|n)$ in (7.5.10) is of short-range character, a direct algebraic solution of the bound state equation (7.5.15) can be achieved. This can be seen by representing $\mathscr{G}^0$ and

Δ in the form

$$\mathfrak{G}^0 = \begin{pmatrix} \mathfrak{G}^0_{11} \mathfrak{G}^0_{12} \\ \mathfrak{G}^0_{21} \mathfrak{G}^0_{22} \end{pmatrix}; \quad \Delta = \begin{pmatrix} \Delta_{11} & 0 \\ 0 & 0 \end{pmatrix} \tag{7.5.24}$$

where the index 1 refers to that subspace for which $\Delta \neq 0$, while the index 2 refers to the complementary space for which Δ vanishes. Obviously the solution of (7.5.15) is then reduced to the subspace 1

$$\mathbf{a}_1 = \mathfrak{G}^0_{11} \cdot \Delta_{11} \cdot \mathbf{a}_1 \tag{7.5.25}$$

while the remaining part $\mathbf{a}_2$ follows from

$$\mathbf{a}_2 = \mathfrak{G}^0_{21} \cdot \Delta_{11} \cdot \mathbf{a}_1 \tag{7.5.26}$$

i.e., it is a quantity which depends on the subspace solution $\mathbf{a}_1$. This method works effectively only for very small dimensions of the subspace 1. For instance, if only the mass of a substitutional atom is different from that of a regular lattice particle, or if only the nearest neighbour force constants around an impurity are changed. In polar crystals, however, this situation occurs very seldom as in most cases the imperfections are accompanied by electric monopoles or multipoles whose forces are of long-range character. Hence a more systematic approach to the bound state problem is needed. A method which does not depend on the dimension of the subspace 1 and which has already been successfully applied to quantum theory is the Fredholm method. The results of this method can be formulated in two theorems (without proof):

Theorem 7.18: If the kernel K and the function f of the integral equation

$$\varphi(x) = \lambda \int K(x,y)\varphi(y)dy + f(x) \tag{7.5.27}$$

are elements of $\mathfrak{L}^2$, then the Green function $G := (1-\lambda K)^{-1}$ of (7.5.27) can be represented by

$$G(x,y,\lambda) := (1-\lambda K)^{-1} = \frac{D(x,y,\lambda)}{d(\lambda)} \tag{7.5.28}$$

where $D(x,y,\lambda)$ and $d(\lambda)$ are entire analytic functions of λ. For $d(\lambda) \neq 0$ the inhomogeneous equation has a unique solution

$$\varphi(x) = f(x) + \lambda \int G(x,y,\lambda) f(y) dy \tag{7.5.29}$$

while for $d(\lambda) = 0$, the homogeneous equation with $f \equiv 0$ is solved by

$$\varphi(x) = D(x,y,\lambda_0)_{/y=y_0} \tag{7.5.30}$$

with y_0 arbitrary.

Theorem 7.19: For the representation (7.5.28) the Green function G of the Fredholm equation admits the series expansion

$$d(\lambda) = \sum_{n=0}^{\infty} (-1)^n a_n \lambda^n \tag{7.5.31}$$

and

$$D(x,y,\lambda)=\sum_{n=0}^{\infty}(-1)^n A_n(x,y)\lambda^n \tag{7.5.32}$$

with

$$a_n:=\frac{1}{n!}\int\dots\int\sum_{\lambda_1\dots\lambda_n}K(z_1,z_{\lambda_1})\dots K(z_n,z_{\lambda_n})dz_1\dots dz_n \tag{7.5.33}$$

and

$$A_n(x,y)=K(x,y)a_n-\int K(x,z)A_{n-1}(z,y)dz \tag{7.5.34}$$

where (7.5.31) and (7.5.32) converge uniformly for $0\leqslant\lambda<\infty$ if $K(x,y)$ is supposed to be an element of $\mathfrak{L}^2$.

Obviously the Fredholm theory admits the calculation of eigenvalues of the coupling constant λ, whereas for the physical problems considered here, the coupling constant is a fixed parameter while the eigenvalues of ω or E, resp. are of interest. This can be taken into account by observing that for our problems $d(\lambda)$ depends also on ω, i.e., $d(\lambda)\equiv d(\lambda,\omega)$, and that the condition $d(\lambda)=0$ for eigensolutions of the homogeneous equation reads then $d(\lambda,\omega)=0$ which can be satisfied for $\lambda=\text{const}$ by various ω values $\{\omega=\omega_\nu,\ \nu=1,\ 2,\ \dots\}$. This change of the role of eigenvalues between λ and ω influences of course also the analytic properties of the function $d(\lambda,\omega)$ which now depends on two variables. Theorems about the analytic behaviour of d in the (λ,ω) complex space have been derived for potential scattering in quantum theory by Jost and Pais (1951) and by Khuri (1957). Similar theorems for equations (7.5.15) (7.5.16) are at present not yet available. But if only eigenvalues and eigenstates are of interest, it is sufficient to show that $K\in\mathfrak{L}^2$. As no general proofs are available, we discuss a special case explicitly, namely that which exhibits the most extreme character of long-range forces and which is given by monopole perturbations. As several excess charges do not change the qualitative character of the perturbation, we consider for simplicity only one excess charge of an imperfection being located at the origin. The simplest model of such an excess charge is a vacancy. In this case the following changes have to be taken into account:

i) The mass at the position of the vacancy has to vanish,
ii) the coupling constants between the vacancy and all lattice ions vanish,
iii) the equilibrium positions of the lattice ions are changed.

As the deviations of the new equilibrium positions $\{\mathbf{R}^n_{l\kappa}\}$ from the ideal positions are small, we neglect this effect and put $\{\mathbf{R}^n_{l\kappa}\approx\mathbf{R}^0_{l\kappa}\}$. This means that the remaining coupling constants are those of the ideal crystal. Furthermore, in order to demonstrate the essentials we work with a point ion model and neglect electronic polarization effects. Then we obtain for i)

$$M'_\kappa=\begin{cases}M_\kappa & (l\kappa)\neq(0,0)\\ 0 & (l\kappa)=(0,0)\end{cases} \tag{7.5.35}$$

and for ii) according to the neglection of iii)

$$\Phi'(l\kappa,l'\kappa')=\begin{cases}\Phi^0(l\kappa,l'\kappa'); & (l\kappa)\neq(0,0)\neq(l'\kappa')\\ 0; & (l\kappa)=(0,0) \text{ and/or } (l'\kappa')=(0,0).\end{cases} \tag{7.5.36}$$

With these assumptions the eigenvalue equations for the crystal with vacancy read

$$\omega^2 M_\kappa \mathbf{u}_{l\kappa}=\sum_{\substack{l'\kappa'\\ \neq 0,0}} \Phi^0(l\kappa,l'\kappa')\cdot\mathbf{u}_{l'\kappa'};\quad (l\kappa)\neq(0,0). \tag{7.5.37}$$

These equations are defined for a space, whose dimension is reduced compared with that of the ideal crystal. According to Stumpf (1961) we introduce fictitious displacements $\mathbf{u}(0,0)$ in order to restore the dimensions of the displacement space of the ideal crystal and thus to be able to apply the Green function $\mathfrak{G}^0$ to these equations. Then equations (7.5.37) can be rewritten to give for $(l\kappa)\neq(0,0)$

$$\omega^2 M_\kappa \mathbf{u}_{l\kappa}-\sum_{l'\kappa'}\Phi^0(l\kappa,l'\kappa')\cdot\mathbf{u}_{l'\kappa'} = -\Phi^0(l\kappa,0,0)\cdot\mathbf{u}_{0,0}+\Phi^0(0,0,l\kappa)\cdot\mathbf{u}_{l\kappa} \tag{7.5.38}$$

where the last term of the right-hand side of (7.5.38) comes from the completion of the diagonal term of the sum on the right-hand side of (7.5.37). For $(l\kappa)=(0,0)$ we obtain the identity

$$\omega^2 M_0 \mathbf{u}_{0,0}-\sum_{l'\kappa'}\Phi^0(0,0,l'\kappa')\cdot\mathbf{u}_{l'\kappa'} =\omega^2 M_0 \mathbf{u}_{0,0}-\sum_{l'\kappa'}\Phi^0(0,0,l'\kappa')\cdot\mathbf{u}_{l'\kappa'}. \tag{7.5.39}$$

All summations which occur in these equations now run over the indices of the ideal lattice and for any solution with $\mathbf{u}(0,0)\equiv 0$ this system is equivalent to the system (7.5.37). Equations (7.5.38) (7.5.39) can be combined to give

$$\omega^2 M_\kappa \mathbf{u}_{l\kappa}-\sum_{l'\kappa'}\Phi^0(l\kappa,l'\kappa')\cdot\mathbf{u}_{l'\kappa'}=-\sum_{l'\kappa'}\Delta(l\kappa,l'\kappa')\cdot\mathbf{u}_{l'\kappa'} \tag{7.5.40}$$

with

$$\Delta(l\kappa,l'\kappa'):=[\Phi^0(l\kappa,l'\kappa')\delta_{l'0}\delta_{\kappa'0}-\Phi^0(0,0,l'\kappa')\delta_{ll'}\delta_{\kappa\kappa'}]\,(1-\delta_{l0}\delta_{\kappa0}) + [\Phi(0,0,l'\kappa')-\omega^2 M_0\delta_{l0}\delta_{\kappa0}\delta_{l'0}\delta_{\kappa'0}]\delta_{l0}\delta_{\kappa0}. \tag{7.5.41}$$

With the transformation

$$\mathbf{u}_{l\kappa}=M_\kappa^{-1/2}\mathbf{a}(l\kappa) \tag{7.5.42}$$

and application of the Green function $\mathfrak{G}^0$ to equations (7.5.40), we obtain after some rearrangements

$$\mathbf{a}(l\kappa)=\mathbf{a}(l\kappa)\delta_{l0}\delta_{\kappa0} -\sum_{\substack{l'\kappa'\\ \neq 0,0}}\left[\mathscr{G}^0(l\kappa,0,0)-\left(\frac{M_0}{M_{\kappa'}}\right)^{1/2}\mathscr{G}^0(l\kappa,l'\kappa')\right]\cdot\mathscr{D}^0(0,0,l'\kappa')\cdot\mathbf{a}(l'\kappa') \tag{7.5.43}$$

where $\mathscr{D}^0(0,0,l'\kappa')$ is defined by (7.5.2) for $n=0$. This system reads explicitly for $(l\kappa)\neq(0,0)$

$$\mathbf{a}(l\kappa)=-\sum_{\substack{l'\kappa'\\ \neq 0,0}}\left[\mathscr{G}^0(l\kappa,0,0)-\left(\frac{M_0}{M_{\kappa'}}\right)^{1/2}\mathscr{G}^0(l\kappa,l'\kappa')\right]\cdot\mathscr{D}^0(0,0,l'\kappa')\cdot\mathbf{a}(l'\kappa') \tag{7.5.44}$$

and for $(l\kappa)=(0,0)$

$$0=\sum_{\substack{l'\kappa'\\ \neq 0,0}}\left[\mathscr{G}^0(0,0,0,0)-\left(\frac{M_0}{M_\kappa}\right)^{1/2}\mathscr{G}^0(0,0,l'\kappa')\right]\cdot\mathscr{D}^0(0,0,l'\kappa')\cdot\mathbf{a}(l'\kappa'). \tag{7.5.45}$$

As the equations (7.5.43) are equivalent to equations (7.5.40), the subsidiary condition (7.5.45) is satisfied if $\mathbf{a}(0,0)\equiv 0$. In the following we shall satisfy this condition and may therefore omit (7.5.45) in the further calculations.
For equations (7.5.44) we now give an estimate for K by the following lemma:

Lemma 7.4: The kernel K which is defined by

$$\mathscr{K}(l\kappa,l'\kappa'):=\left[\mathscr{G}^0(l\kappa,0,0)-\left(\frac{M_0}{M_{\kappa'}}\right)^{1/2}\mathscr{G}^0(l\kappa,l'\kappa')\right]\cdot\mathscr{D}^0(0,0,l'\kappa') \tag{7.5.46}$$

is an element of $\mathfrak{L}^2$ for ω having values in a gap.

Proof: We give only a rough estimate. From (7.5.13) it follows that for ω in a gap, due to the finite Brillouin zone $|\mathscr{G}^0(l\kappa,l'\kappa')|\leq g<\infty$. Hence we obtain for the norm of $\mathscr{K}$ the following inequality

$$\sum_{\substack{l\kappa\\ l'\kappa'}}|\mathscr{K}(l\kappa,l'\kappa')|^2\leq\sum_{l\kappa}|\mathscr{G}^0(l\kappa,0,0)|^2\sum_{l'\kappa'}|\mathscr{D}^0(0,0,l'\kappa')|^2\left[1+2\left(\frac{M_0}{M_{\kappa'}}\right)^{1/2}g\right]$$

$$+\sum_{\substack{l\kappa\\ l'\kappa'}}|\mathscr{G}^0(l\kappa,l'\kappa')|^2\,|\mathscr{D}^0(0,0,l'\kappa'|^2\,\frac{M_0}{M_{\kappa'}}. \tag{7.5.47}$$

From the definition (7.1.26) we obtain for a rough estimate

$$|\mathscr{D}^0(0,0,l'\kappa')|^2\approx|\mathbf{R}^0_{l'\kappa'}|^{-6}\,e_0^2e_{\kappa'}^2M_0^{-1}M_{\kappa'}^{-1}. \tag{7.5.48}$$

Together with the estimate (7.5.17) it can be concluded that the sums of the first term of the right-hand side of (7.5.47) converge and give finite values as $(l\kappa)=(0,0)$ is excluded by definition. The second term of the right-hand side of (7.5.47) can be represented approximately by an integral

$$\int a(x-y)b(y)dxdy\equiv\tilde{a}(0)\tilde{b}(0) \tag{7.5.49}$$

with $a(x):=|\mathscr{G}^0(x)|^2$ and $b(y):=|\mathscr{D}^0(0,y)|^2$ and

$$\begin{aligned}|\mathscr{G}^0(l\kappa,l'\kappa')|^2&=|\mathscr{G}^0(x-y)|^2{}_{/x=\mathbf{R}^0_{l\kappa},y=\mathbf{R}^0_{l'\kappa'}}\\ |\mathscr{D}^0(0,0,l'\kappa')|^2&=|\mathscr{D}^0(0,y)|^2{}_{/y=\mathbf{R}^0_{l'\kappa'}}\end{aligned} \tag{7.5.50}$$

where $\tilde{a}(0)$ and $\tilde{b}(0)$ are the Fourier transforms at the origin. Due to (7.5.48) and (7.5.17) it can be shown that $\tilde{a}(0)$ and $\tilde{b}(0)$ have finite values, Q.E.D.

Hence the Fredholm theory can be applied to equation (7.5.44). In principle, the Fredholm theory should give the proper eigenfunctions even if no group theoretical reduction is performed. It is, however, known that for low order approximations the results are essentially improved if such a reduction is performed before the calculation is done.
The symmetry group of a vacancy at a regular lattice site of an NaCl-lattice is the cubic point group O_h, which was discussed in Section 6.3. This point group has a simple structure. Nevertheless, it is quite a problem to find the representations of this group which can be used for an expansion of eigensolutions of equations (7.5.44) as these representations have to be given for an infinite-dimensional space. So far, in the literature, only representations for spaces with a very low number of dimensions have been discussed. To derive infinite dimensional representations of the point groups, we make use of the representations of the full rotation group $O(3)$ about the origin. We suppose that $\{\varphi_j^l(\Theta,\varphi)\}$ is a complete set of vectorial spherical harmonics, where l is the number of the representation and j the index of the basis vectors. Then from this set, for any element $G_i \in O(3)$ we obtain a corresponding set of representation matrices $\{D_{jj'}^l(G_i)\}$ which is supposed to be irreducible. If we now consider a subgroup $\mathfrak{G} \subset O(3)$, then $\mathfrak{G}$ is defined by a subset of elements $\{G_{\lambda_i}\, i=1\ldots n\} \subset O(3)$. In general, the representation matrices $\{D_{jj}^l(G_{\lambda_i}), i=1\ldots n\}$ are then no longer irreducible. As the representations of $O(3)$ and of $\mathfrak{G}$ can be assumed to be unitary, a unitary similarity transformation U exists, which transforms the set $\{D_{jj'}^l(G_{\lambda_i})\}$ into the direct sum of irreducible parts. This transformations can also be applied to the basis vectors and gives a new system of basis vectors

$$\chi_{i\alpha}^{l\varrho} = \sum_j U(i,\alpha,\varrho|l,j)\,\varphi_j^l \tag{7.5.51}$$

where i is the number of the representation, α the basis index, while ϱ counts the multiplicity with which the i-th irreducible representation of $\mathfrak{G}$ occurs in the l-th irreducible representation of $O(3)$. Due to the unitarity of U, the orthonormality relations

$$\langle \varphi_j^l, \varphi_{j'}^{l'} \rangle = \delta_{ll'}\delta_{jj'} \tag{7.5.52}$$

go over into

$$\langle \chi_{i\alpha}^{l\varrho}, \chi_{i'\alpha'}^{l'\varrho'} \rangle = \delta_{ll'}\delta_{\varrho\varrho'}\delta_{ii'}\delta_{\alpha\alpha'} \tag{7.5.53}$$

and any vectorial function $\mathbf{z}(\mathbf{r}) \in \mathfrak{L}^2$ can be expanded into a series with respect to the set $\{\varphi_j^l\}$ or equivalently to the set $\{\chi_{i\alpha}^{l\varrho}\}$. This leads to

$$\mathbf{z}(\mathbf{r}) = \sum_{lj} b_j^l(r)\,\varphi_j^l(\Theta,\varphi) = \sum_{\substack{l\varrho\\ i\alpha}} a_{i\alpha}^{l\varrho}(r)\,\chi_{i\alpha}^{l\varrho}(\Theta,\varphi). \tag{7.5.54}$$

If such a function is supposed to be a basis state j of the l-th irreducible representation of the full rotation group, then $\mathbf{z}(\mathbf{r})$ must have the form

$$\mathbf{z}(\mathbf{r}|l,j)=b_j^l(r)\varphi_j^l(\Theta,\varphi) \tag{7.5.55}$$

whereas if $\mathbf{z}(\mathbf{r})$ is supposed to be a basis state α of the i-th irreducible representation of the point group $\mathfrak{G}$, it must have the form

$$\mathbf{z}(\mathbf{r}\mid i,\alpha)=\sum_{l\varrho} a^{l\varrho}(r)\chi_{i\alpha}^{l\varrho}(\Theta,\varphi). \tag{7.5.56}$$

Of course, such expansions can be taken over also for discrete-valued functions. If we represent $\mathbf{R}_{l\kappa}^0$ in polar coordinates $\mathbf{R}_{l\kappa}^0\equiv(r_{l\kappa},\Theta_{l\kappa},\varphi_{l\kappa})$ then we obtain an analogous expansion

$$\mathbf{z}(l\kappa|i,\alpha)=\sum_{l\varrho} a^{l\varrho}(r_{l\kappa})\chi_{i\alpha}^{l\varrho}(\Theta_{l\kappa},\varphi_{l\kappa}) \tag{7.5.57}$$

for the representation states of the point group $\mathfrak{G}$.
In contrast to the expansion of continuous functions, the expansions (7.5.57) have a disadvantage: If we want to calculate the expansion coefficients $a^{l\varrho}(r)$ or $a^{l\varrho}(r_{h\kappa})$, resp. we obtain from (7.5.56) by means of (7.5.53)

$$a^{l'\varrho'}(r)=\int \mathbf{z}(\mathbf{r}\mid i,\alpha)\chi_{i\alpha}^{l'\varrho'}(\Theta,\varphi)r^2\sin\Theta\,d\Theta\,d\varphi \tag{7.5.58}$$

whereas in the discrete case the sum runs only over those points $\mathbf{R}_{h'\kappa'}^0$ for which $|\mathbf{R}_{h'\kappa'}^0| =|\mathbf{R}_{h\kappa}^0|$. The number of points for the spherical surfaces centered at the origin have been calculated by Scholz (1967) for the lowest orders

Surface	1	2	3	4	5	6	7	8	9	10	11	12
Ion	100	110	111	200	210	211	220	221	300	310	311	222
Number	6	12	8	6	24	24	12	24	6	24	24	8

They show no striking regularity which would allow the extension of the orthonormality to the discrete case. As the value of any expansion depends on an easy prescription of its projection properties, the expansion (7.5.57) seems to be of little use. The difficulty just mentioned can, however, easily be circumvented, if the use of polar coordinates is avoided. To do this we define a complete orthonormal system $\{\psi_k(r)\}$ for the radial coordinate which satisfies the orthonormality conditions

$$\int \psi_k(r)r^2\psi_{k'}(r)\,dr=\delta_{kk'}. \tag{7.5.59}$$

Then for the set $\{\psi_k(r)\chi_{i\alpha}^{l\varrho}(\Theta,\varphi)\}$ the orthonormality relations

$$\int \psi_k(r)\chi_{i\alpha}^{l\varrho}(\Theta,\varphi)\psi_{k'}(r)\chi_{i'\alpha'}^{l'\varrho'}(\Theta,\varphi)\,d^3r=\delta_{kk'}\delta_{ll'}\delta_{\varrho\varrho'}\delta_{ii'}\delta_{\alpha\alpha'} \tag{7.5.60}$$

hold. If we now expand (7.5.57) in a series of the form

$$\mathbf{z}(l\kappa|i\alpha)=\sum_{l\varrho k} a_k^{l\varrho}\psi_k(r_{l\kappa})\chi_{i\alpha}^{l\varrho}(\Theta_{l\kappa},\varphi_{l\kappa}) \tag{7.5.61}$$

we obtain by projection

$$\sum_{h\kappa} \psi_{k'}(r_{h\kappa}) \chi_{i\alpha}^{l'\varrho'}(\Theta_{h\kappa}, \varphi_{h\kappa}) \mathbf{z}(h\kappa|i\alpha)$$
$$= \sum_{l\varrho k} a_k^{l\varrho} \sum_{h\kappa} \psi_{k'}(r_{h\kappa}) \chi_{i\alpha}^{l'\varrho'}(\Theta_{h\kappa}, \varphi_{h\kappa}) \psi_k(r_{h\kappa}) \chi_{i\alpha}^{l\varrho}(\Theta_{h\kappa}, \varphi_{h\kappa}). \tag{7.5.62}$$

In contrast to the summation over the points on a spherical surface the lattice sum on the right-hand side of equation (7.5.62) can now be considered to be an approximate expression for the Riemann integral (7.5.60). Hence we are allowed to assume approximate orthonormality so that from (7.5.62) the projection

$$a_{k'}^{l'\varrho'} = \sum_{h\kappa} \psi_{k'}(r_{h\kappa}) \chi_{i\alpha}^{l'\varrho'}(\Theta_{h\kappa}, \varphi_{h\kappa}) \mathbf{z}(h\kappa|i\alpha) \tag{7.5.63}$$

results. As the linear algebra of the vectors $\{\mathbf{z}(l\kappa|i\alpha)\} \subset \mathfrak{L}^2$ and $\{a_{k'}^{l'\varrho'}\} \subset \mathfrak{L}^2$ is completely equivalent, the Fredholm method can be applied equally well to the corresponding equations of the set $\{a_{k'}^{l'\varrho'}\}$.
The calculation can be simplified if in the lowest approximation only the lowest term $l\varrho \equiv l^0\varrho^0$ in the expansion (7.5.57) is taken into account. Then the use of the additional radial orthonormal system $\{\psi_k(r)\}$ is not necessary. Calculations of this kind were done by Heyna (1974) for the lattice eigenmodes of a vacancy. We give only a short review of these calculations. The basis system of the $O(3)$ representation is given by the set of spherical harmonics $\{Y_{Il1}^M(\Omega)\}$, cf. Blatt and Weisskopf (1966). Here I is the value of the total angular momentum $\mathbf{J}^2$, M the value of $\mathbf{J}_z$, l the value of the orbital angular momentum $\mathbf{L}^2$ and 1 the spin value. The orthonormality relations read

$$\langle Y_{Il1}^M, Y_{I'l'1}^{M'} \rangle = \delta_{II'} \delta_{MM'} \delta_{ll'} \tag{7.5.64}$$

The number of the irreducible representation is given by the positive integer I. For $I \neq 0$, any of these representations occurs with multiplicity 3, namely with $l = I-1$, I, $I+1$. Hence the representations are characterized by (I, l) whereas the number of the corresponding basis vectors M with $-I \leqslant M \leqslant I$ is integer. As the parity of the spin part is the same for all representations, the parity, i.e., the quantum number with respect to space inversion, is determined by the orbital part alone. Its value is $(-1)^l$. Since the potential energy of the crystal is invariant under space inversion, the parity is a good quantum number which has to be observed for expansions. The general expansion of a solution of $\mathbf{a}(l\kappa)$ of equation (7.5.15) reads

$$\mathbf{a}(h\kappa) = \sum_{I=0}^{\infty} \sum_{M=-I}^{I} \sum_{l=I-1}^{I+1} a(I, M, l, h\kappa) Y_{Il1}^M(h\kappa). \tag{7.5.65}$$

The reduction to irreducible representations of the O_h group can be achieved by means of the decomposition of Bethe (1929). The set $\{Y_{Il1}^M\}$ is a basis for the representations $\{D_{MM'}^I(g)\}$ of the $O(3)$ group, i.e., these representations are the same as those of the scalar spherical harmonics. Hence the reduction formula of Bethe which was already given in Section 6.3 can be taken over for the present case. In particular, we obtain $D^0 \equiv \Gamma_1$, $D^1 \equiv \Gamma_4$, while all other representations D^I are reducible with respect to O_h.

As the spin parity is -1, the total parity of the state $I=0$, $M=0$, $l=1$ is 1, while for $I=1$, $M=-1,0,1$, $l=0,2$ the total parity is -1. The corresponding O_h representations are then denoted by Γ_1^+ and Γ_4^-. If we neglect all higher order contributions to Γ_1^+ and Γ_4^-, the expansion (7.5.65) can be written

$$\mathbf{a}(h\kappa|\Gamma_1^+)=a(0,0,1,r)\,Y_{011}^0(\Omega)_{/r=r_{h\kappa}} \tag{7.5.66}$$

$$\mathbf{a}(h\kappa|\Gamma_4^-,M)=a(1,M,0,r)\,Y_{101}^M(\Omega)_{/r=r_{h\kappa}}+a(1,M,2,r)\,Y_{121}^M(\Omega)_{/r=r_{h\kappa}} \tag{7.5.67}$$

It was demonstrated by Heyna that the orthonormality for these expansion functions holds, even if the summation is performed only over the discrete points of the various spherical surfaces. Hence (7.5.66) can be substituted into (7.5.15) and the resulting integral equations can be calculated directly. They read for the Γ_1^+ state (7.5.66)

$$a(0,0,1,r)=\sum_{r'}\frac{n_{r'}}{4\pi}K(0,0,1,r,0,0,1,r')a(0,0,1,r') \tag{7.5.68}$$

and for the Γ_4^- state (7.5.67)

$$a(1,M,l,r)=\sum_{r'}\frac{n_{r'}}{4\pi}\sum_{l'=0,2}K(1,M,l,r,1,M,l',r')a(1,M,l',r'),\quad l=0,2 \tag{7.5.69}$$

Here the summation over r' runs over all the radii of the various spherical surfaces on which ions are located in their ideal positions, while $n_{r'}$ is the number of these ions in any shell. For the lowest shells these numbers are given above, and it follows that $n_{r'}\neq cr'^2$, i.e., in the discrete lattice the occupation numbers are not proportional to the surface area of a sphere. The kernels are defined by the projections

$$K(0,0,1,r,0,0,1,r')=\sum_{\left\{ {l\kappa \atop l'\kappa'} \right\}} Y_{011}^0(\Omega)\cdot\mathscr{K}(l\kappa,l'\kappa')\cdot Y_{011}^0(\Omega') \tag{7.5.70}$$

and

$$K(1,M,l,r,1,M,l',r')=\sum_{\left\{ {l\kappa \atop l'\kappa'} \right\}} Y_{1l1}^M(\Omega)\cdot\mathscr{K}(l\kappa,l'\kappa')\cdot Y_{1l'1}^M(\Omega'),\ l,l'=1,2 \tag{7.5.71}$$

where the brackets $\{\ \}$ indicate that the summation has to be performed over all ions of a spherical surface with $|\mathbf{R}_{l\kappa}^0|=r=\text{const.}$ or $|\mathbf{R}_{l'\kappa'}^0|=r'=\text{const.}$, resp.
The explicit evaluation of these expressions is achieved by expanding $\mathscr{G}^0$ as well as $\mathscr{D}^0$ of (7.5.46) into series of spherical harmonics. For the calculation of the coupling matrices contained in $\mathscr{D}^0$ the potential (2.6.1) was used. An explicit expression for $\mathscr{G}^0$ was derived by using the results of Theorem 2.7 for the acoustic phonon modes, while the optical phonon modes were assumed to be degenerate. The Fredholm theory can be applied to the equations (7.5.68) and (7.5.69). According to (7.5.31) the lowest approximation for eigenvalue equation of (7.5.68) reads

$$1-\sum_{r'}\frac{n_{r'}}{4\pi}K(0,0,1,r',0,0,1,r')=0 \tag{7.5.72}$$

whereas the corresponding eigenvalue equation of (7.5.69) reads

$$1-\sum_{r'}\sum_{l'=0,2}\frac{n_{r'}}{4\pi}K(1,M,l',r',1,M,l',r')=0. \tag{7.5.73}$$

In this approximation, Heyna found no solution with ω in the gap for equation (7.5.72), whereas for equation (7.5.73) he obtained the gap solution

$$\omega=1{,}62\cdot 10^{13}\ \mathrm{s}^{-1}. \tag{7.5.74}$$

If the influence of the widely spread *F*-center electron states on the lattice dynamics is neglected in a rough approximation, these results can be compared with measurements of the *F*-center modes. In this case the nonexistence of a Γ_1^+ gap solution is in agreement with the investigation of the Raman spectrum of *F*-centers in KI by Benedek and Mulazzi (1969) who found only Γ_1^+ resonance modes within the bands, but no local mode in the gap. On the other hand, a local mode which is harmonically stable can be optically excited and gives very sharp lines. Bäuerle and Fritz (1968), and Bäuerle and Hübner (1970) found a Γ_4^- mode with $\omega=1{,}556\cdot 10^{13}\,\mathrm{s}^{-1}$ for the *F*-center in KI. This result is also in good agreement with the theoretical value.
The occurrence of harmonic phonon scattering or resonance states, resp. leads to damping effects related with these states. Obviously such effects are not contained in our line breadth functions as the set of interactions which produce these line breadths contains only anharmonic forces etc., but not the harmonic phonon forces. Hence the harmonic phonon resonance states require an extension of the formalism of irreversible statistics for the calculation of observables used so far. The straightforward extension of this formalism runs in the following way:
The harmonic interactions resulting from the terms $\{\Delta(l\kappa,l'\kappa'|n)\}$ in (7.5.10) can be included in the set of interactions leading to irreversible motion of the system. Then, according to Chapter 1 a contribution to the general line breadth function results which is due to these phonon resonances. If in addition bound local phonon modes occur, the corresponding quantum states have to be added to the phonon band states and the non-diagonal terms of the Hamiltonian corresponding to (7.5.10) with respect to this system have to be taken as the harmonic irreversible interactions.
This systematic procedure has not yet been applied and in the literature less systematic approaches have been made to treat the damping effects of phonon resonances. An example is the calculation of the infrared absorption coefficient by means of Green functions, which we will briefly discuss here, using Theorem 7.22 of the following section.
The initial formula which is of interest for our discussion is given by the quantum number absorption rate of infrared photons (7.6.116). If the quantum numbers of the incident radiation $b(\lambda,\mathbf{k})$ are much larger than the mean phonon occupation numbers $\bar{m}_l$ and if the transverse optical phonon modes $\zeta(\lambda,\mathbf{k})$ are substituted into (7.6.116), then the rate formula

$$\Delta B(\lambda,\mathbf{k})=\sum_l b(\lambda,\mathbf{k})\left(\frac{2\pi c\hbar}{kV}\right)\varrho(\lambda,\mathbf{k}|z)_{/z=\omega(\lambda\mathbf{k})\hbar} \tag{7.5.75}$$

results with the projected density

$$\varrho(\lambda,\mathbf{k}|z)=\sum_l |\sum_j \zeta_j(\lambda,\mathbf{k})\cdot \mathbf{a}_j(l)|^2\delta_\gamma(\hbar\omega_l-z). \tag{7.5.76}$$

This formula means that the absorption rate of infrared radiation is proportional to the phonon density at that frequency projected on the optically active phonon modes of the ideal crystal. Hence we obtain a superposition of Lorentzian absorption lines weighted with corresponding projection factors. The damping factor γ contained in (7.5.76) arises from anharmonic interactions etc., but does not take into account the harmonic contributions. This can be achieved if we rewrite (7.5.75) according to Theorem 7.23 in the form (7.6.103)

$$\varDelta B(\lambda,\mathbf{k})=b(\lambda,\mathbf{k})\,\frac{4\pi\omega(\lambda,\mathbf{k})}{V}\sum_{jj'}\zeta_j(\lambda,\mathbf{k})\cdot \operatorname{Im}\mathscr{G}_{jj'}(z)\cdot\zeta_{j'}(\lambda,\mathbf{k})^{\times} \tag{7.5.77}$$

with z given by $z=\omega(\lambda,\mathbf{k})^2+i\gamma/\hbar$, where $\mathscr{G}_{jj'}(z)$ is the Greenfunction matrix in ordinary configuration space of the lattice. As is shown in Theorem 7.23 this expression is equivalent to (7.5.75) if the Green function $\mathscr{G}(z)$ is of the form

$$\mathscr{G}_{jj'}(z)=\sum_l \mathbf{a}_j(l)\otimes\mathbf{a}_{j'}(l)^{\times}(\omega_l^2-z)^{-1}. \tag{7.5.78}$$

Hitherto formula (7.5.78) has been verified only in (7.5.13) for $\mathscr{G}_{jj'}(z)=\mathscr{G}^0_{jj'}(z)$, i.e., for an ideal undamped lattice. Hence it must be shown that (7.5.78) holds for a lattice with damped vibrations too. In order to do this the techniques of Theorem 1.17 can be applied to the construction of classical lattice Green functions. In this case all perturbations, i.e., harmonic and anharmonic ones etc., can be taken into account. For brevity we confine ourselves to the harmonic perturbations. Then in the proof of Theorem 1.17 the Hamiltonian H_0 has to be replaced by $\mathscr{D}^0$ and λV by $\varDelta$ of equation (7.5.11). If the projectors Π_l are identified with projectors in the space of eigensolutions of $\mathscr{D}^0$ and if E_l is replaced by ω_l^2 then from (1.6.22) it follows the equation

$$(\Pi_l\mathscr{G}^-(z)\Pi_l)=\left[z-\omega_l^2-d_l(z)-\frac{i}{2}\gamma(z)-i\varepsilon\right]^{-1} \tag{7.5.79}$$

with infinitesimal ε and with γ_l and d_l following from (1.6.20) (1.6.21) where now the calculation has to be made with $\mathscr{D}^0$ and $\varDelta$! According to (2.2.24) we may denote the eigenvectors of $\mathscr{D}^0$ by $\mathbf{a}_j(l)$. Hence in the diagonal approximation (7.5.79) yields

$$\mathscr{G}_{jj'}(z)=\sum_{jj'}\mathbf{a}_j(l)\otimes\mathbf{a}_{j'}(l)^{\times}\left(\omega_l^2+d_l+\frac{i}{2}\gamma_l-z\right)^{-1} \tag{7.5.80}$$

if we approximately put $d_l(z)\approx d_l$, $\gamma_l(z)\approx\gamma_l$. Therefore we obtain the infrared absorption (7.5.77) with harmonic as well as with anharmonic phonon resonance damping and line shift by taking z to be

$$z=\omega(\lambda,\mathbf{k})^2+d_l+\frac{i}{2}\gamma_l+\frac{i}{2}\gamma\hbar^{-1} \tag{7.5.81}$$

If additionally the $\mathbf{a}_j(l)$ are assumed to be undamped states of a perturbed lattice and if $\varDelta$ is only that part of the harmonic interaction which causes damping we have

$$\Delta B(\lambda,\mathbf{k})=\sum_l b(\lambda,\mathbf{k})\left(\frac{2\pi c\hbar}{kV}\right)\hat{\varrho}(\lambda,\mathbf{k}|z)_{/z=\omega(\lambda\mathbf{k})h} \tag{7.5.82}$$

with

$$\hat{\varrho}(\lambda,\mathbf{k}\,|\,z)=\sum_l|\sum_j \zeta_j(\lambda,\mathbf{k})\cdot\mathbf{a}_j(l)|^2\delta_{(\gamma+\gamma_l)}(\hbar\omega_l+d_l-z). \tag{7.5.83}$$

In this way we obtain a clear description of the influence of phonon resonances on infrared absorption.

In the literature, however, other methods for the evaluation of resonance lattice Green functions are preferred which are particularly useful if Δ extends only over a small lattice region around the impurity. From the Neumann series it follows that $\mathscr{G}=(\mathscr{D}^0+\Delta-z)^{-1}$ can be written in the form

$$\mathscr{G}=\mathscr{G}_0-\mathscr{G}_0\mathscr{T}\mathscr{G}_0 \tag{7.5.84}$$

with

$$\mathscr{T}:=\Delta-\Delta\mathscr{G}_0\Delta+\Delta\mathscr{G}_0\Delta\mathscr{G}_0\Delta-\ldots\equiv\Delta(1+\mathscr{G}_0\Delta)^{-1} \tag{7.5.85}$$

and $\mathscr{G}_0=(\mathscr{D}^0-z)^{-1}$. If (7.5.84) is substituted into (7.5.77), we obtain

$$\Delta B(\lambda,\mathbf{k})=b(\lambda,\mathbf{k})\frac{4\pi\omega(\lambda,\mathbf{k})}{V}\left\{\sum_{jj'}\zeta_j(\lambda,\mathbf{k})\cdot\operatorname{Im}\mathscr{G}^0_{jj'}(z)\cdot\zeta_{j'}(\lambda,\mathbf{k})^\times\right.$$
$$\left.-\sum_{jj'}\zeta_j(\lambda,\mathbf{k})\cdot\operatorname{Im}(\mathscr{G}_0\mathscr{T}\mathscr{G}_0)_{jj'}\cdot\zeta_{j'}(\lambda,\mathbf{k})^\times\right\}. \tag{7.5.86}$$

Hence in this form the infrared absorption is separated into a part representing the absorption of the ideal lattice and a part resulting from absorption of resonance states. The latter part can easily be calculated if Δ is of small dimension. Then according to (7.5.85), $\mathscr{T}$ is of the same dimension, i.e., corresponds to a few degrees of freedom in configuration space and thus the second part of (7.5.86) can easily be evaluated. This advantage is, of course, lost if long-range forces have to be taken into account. Then our systematic approach seems to be more appropriate for performing effective calculations.

We now give some references with respect to local phonon modes which are connected to numerous experimental effects in semiconductors, metals, etc. Review articles which take these connections into account were given by Maradudin (1962) (1966) and Klein (1968). Rather than giving a complete list of references we concentrate on mode calculations for three-dimensional lattices of semiconductors and insulators, omitting the numerous attempts of mode calculations for one- or two-dimensional lattice, etc. Some further references are given in Section 7.6. Most of the calculations cited here, are done with L.G.F.T. (≡lattice Green functions techniques). The quality of such calculations depends upon the particular model of the impurity and also on the representation of the Green functions of the ideal lattice. We refer to the original papers for the various approaches to the calculation of such representations. The following topics were treated in literature: Introduction of classical lattice Green functions for the algebraic calculation of lattice modes connected with strongly localized impurities, Lifshitz (1943) (1944). A note on the treatment of local perturbations by L.G.F.T., Lax (1954). A perturbation theoretical calculation of the frequencies and states of a three-dimensional lattice with an incorporated impurity system, Fues and Stumpf (1954), Stumpf (1961). Calculation of local modes by L.G.F.T., Montroll and Potts (1955). A review of the treatment of non-ideal crystal lattice vibrations by L.G.F.T. and the influence of local modes on thermodynamic quantities, etc., Lifshitz (1956). Criteria for the occurrence of localized vibrations in a simple cubic lattice with one atom replaced by an impurity atom, Litzman (1958). Thermodynamic properties due to defect vibrations in solids with L.G.F.T., Mahanty, Maradudin and Weiss (1960). Formal method of calculation of the energy spectra of lattices with defects for vibrational states as well as for electronic states by L.G.F.T. and integral equation methods, Takeno (1961). Connection between the calculation of localized vibrations and random walk problems, Toda, Kotera and Kogure (1962). A review of lattice vibrations in solids with inclusion of local modes, Krumhansl (1962). Frequencies of localized modes in one- and three-dimensional lattices by direct algebraic solution and by L.G.F.T., Takeno, Kashiwamura and Teramoto (1962). A solution procedure for the lattice equations considered as systems of difference equations for two- and three-dimensional lattices with defects, Asahi (1962). Dynamics of cubic crystals with a local change of mass by L.G.F.T., Nardelli and Tettamanzi (1962). Energy spectrum of lattices containing isotopic impurities of a single kind by L.G.F.T., Takeno (1962). Frequencies and modes of the *F*-center by direct calculation of a molecular model and their influence on the *F*-center absorption bands, McCombie, Matthew and Murray (1962). Lattice vibrations in crystals with molecular impurity centers by L.G.F.T., Wagner (1963). Localized modes of point defects in simple cubic lattices by L.G.F.T., Lengeler and Ludwig (1963). Vibrations of an atom of different mass in a cubic crystal with L.G.F.T., Dawber and Elliott (1963). Lattice dynamics of NaCl-type crystals with foreign ions by L.G.F.T., Kristofel and Zavt (1963). Phonon scattering by lattice defects with L.G.F.T., Klein (1963). Lattice modes of impurity atoms in NaCl and KCl with the R.I.- and the D.D.-model and L.G.F.T., Jaswal and Montgomery (1964). Localized mode calculation in an anharmonic crystal, Visscher (1964). Localized modes in cubic crystals with point defects by a variational procedure and group theoretical derivation of localized eigenmodes, Dettmann and Ludwig (1964). Resonance scattering of phonons by molecular impurity centers with L.G.F.T. and Lippmann-Schwinger equation combined with group theoretical analysis, Wagner (1964). Frequency calculation for a single mass defect by the formula of L.G.F.T., Jaswal and Montgomery (1964). Frequency calculation of local modes for a vacancy pair in alkali halides based on a molecular model, Kristofel and Tyurkson (1964). Review of the calculation of effects due to point defects on the vibrations of crystal lattices, Maradudin (1965). A high frequency resonant mode of a two-impurity system in the lattice by L.G.F.T., Takeno (1965). Local mode frequencies due to light substitutional impurities in alkali halide crystals with application to the *U*-center by direct solution of the secular equation for the molecular model of the impurity and its nearest neighbors, Fieschi, Nardelli and Terzi (1965). Localized lattices modes due to a substitutional mass defect in NaI by L.G.F.T. with ideal lattice modes derived from the D.D.-model, Jaswal (1965). Localized phonon modes of the *F*-center groundstate and first excited states by group theoretical analysis of the eigenmodes in the molecular model (with application to absorption band calculation), Lemos and Markham (1965). Local modes for *U*-

centers in NaCl and KCl in the molecular model, Jaswal (1965). Pseudolocalized vibrational frequencies of rare earth ions in alkali halides by L.G.F.T., Wagner and Bron (1965). Frequencies of localized vibrations in a diatomic simple cubic lattice with a point imperfection by L.G.F.T., Mitani and Takeno (1965). Lattice vibration effects due to impurities in NaI by L.G.F.T., Gunther (1965). Localized vibrations of point defects in body centered cubic lattices with L.G.F.T. and comparison with the results of the corresponding molecular modes, Kunc (1965). Vibrations of large imperfections in a crystal by direct evaluation of the corresponding secular problem, Litzman and Rozsa (1965). Calculation of frequencies of localized modes by L.G.F.T., Balcar (1966). Localized modes of *U*-centers of CsCl in a molecular model, Krishnamurty (1966). Localized modes due to impurities in KBr with L.G.F.T., Krishnamurthy and Haridasan (1966). Localized vibrations of impure NaCl-type ionic crystals with L.G.F.T., Zavt (1966), Zavt and Kristofel (1967). Dynamics of non-ideal crystals and a quasimolecular model of impurity center vibrations, Kristofel and Zavt (1966). A survey of the dynamics of a crystal lattice with defects, Lifshitz and Kosevich (1966). Evaluation of observables by Green functions of a crystal lattice with defects, Litzman (1966). Approximate calculation of lattice Green functions, Kunc (1966). Vibrations of crystals containing substitutional defects by L.G.F.T. in the R.S.-model, Page and Strauch (1967). Force constant changes and local modes in a molecular model of the *U*-center in KCl, KBr and KI, Wood and Gilbert (1967). A new method for the calculation of Green functions of a disturbed lattice dynamics, Wagner (1967). Evaluation of the Fourier coefficients in the expansion of the Green function and the frequency spectrum, Tewary (1967). Vibrational properties of imperfect crystals with large defect concentration by L.G.F.T. combined with phonon scattering calculation and averaging over the defect distributions, Taylor (1967). Discussion of vibrational instabilities of impurity centers and non-symmetric distortions of NaCl-type lattices, Kristofel and Zavt (1967). Discussion of rotational motions of diatomic impurities in alkali halides, Lawless (1967). Note on local mode calculations of *U*-centers in mixed alkali halides, Jaswal and Striefler (1968). Asymptotic description of localized lattice modes and low frequency resonances by division of the lattice into an impurity system and its surroundings and evaluation of the coupled secular problems, Krumhansl and Matthew (1968). *T*-matrix approach to lattice dynamics of imperfect alkali halides according to formula (7.5.84) etc., Benedek and Nardelli (1968). Strict bounds for out-of-band frequencies due to impurities and defects in solids, Dean (1968). Theory of localized vibrations of interstitial atoms in bcc lattices by means of field theoretic double-time Green functions of the lattice, Blaesser, Peretti and Toth (1968). Localized modes due to impurities in III–V semiconductors by L.G.F.T., Govindarajan and Haridasan (1969). Vibrations of point defects in Wurtzite lattices by L.G.F.T., Nusimovici (1969). The vibrational self-entropy of a point defect by a pseudomolecular model, Mahanty (1969). Lattice dynamics of an isotropic defect by L.G.F.T. and response formalism, Eisenriegler (1969). Local modes of an *U*-center in KBr in the molecular model, Striefler and Jaswal (1969). Calculation of the electric field produced by local modes due to substitutional impurities in ionic crystals, Page (1969). Discussion of collective oscillations of a finite concentration of impurities, Maradudin and Oitmaa (1969). Calculation of local modes of substitutional impurities in cubic crystals based on a molecular shell model, Krishnamurthy and Haridasan (1969). Evaluation of the lattice dynamics of crystals with extended point defects based on L.G.F.T. and Fredholm theory and application to a linear chain with a substitutional defect, Oitmaa and Maradudin (1969). Effect of changes of atomic force constants on localized vibrational modes by a perturbation theoretical approach, Croitoru and Grecu (1970). Vibrations of a mass defect in CdS by L.G.F.T., Nusimovici, Balkanski and Birman (1970). *F*-center gap mode in alkali halides by a molecular model, Singh and Mitra (1970). Local oscillations in an ionic crystal in the presence of a long-range defect by a continuum approximation, Bryksin and Firsov (1970). Investigation of the lattice dynamics of Sm^{2+} in KBr by means of vibronic spectra derived from formula (7.5.77), Kühner and Wagner (1970). Derivation of phenomenological differential equations for the description of localized oscillations of defects with long-range forces in an ionic crystal from a microscopic theory, Bryksin and Firsov (1970). Local mode frequencies for an interstitial hydrogen atom in alkaline earth fluorides by L.G.F.T., Hartmann, Gilbert, Kaiser and Wahl (1970). Localized mode frequencies for substitutional impurities in zincblende type crystals of II–VI and III–V compounds by L.G.F.T., Gaur, Vetelino and Mitra (1971). Localized modes of alkali halides with molecular impurities by L.G.F.T. and a variational approach, Jüngst and Kuri (1971). Elastic relaxation around impurities in KI and its influence on local gap modes by L.G.F.T., de Jong (1971). Calculation of effective forces constants and Green functions in the lattice dynamics of CsI containing impurity atoms, Agrawal and Ram (1971). Impurity vibrations of copper defect complexes in GaS by L.G.F.T. and group theoretical analysis, Grimm, Maradudin, Ipatova and Subashiev (1972). Lattice dynamics of a crystal containing a point defect with a long-range interaction by a special evaluation procedure of the Green function, Litzmann, Bartusek and Zavadil (1972). Phenomenological analysis of resonant phonon modes of Ag^+ in potassium halides, Jain and Prabhakaran (1972). Defect induced lattice vibrations in zincblende type crystals of II–VI and III–V compounds by a molecular model calculation, Singh and Mitra (1972). Impurity modes and effective force constants in ionic crystals, Ram and Agrawal (1972) (1973). Phenomenological analysis of resonant phonon modes of Ag^+ and Au^+ in alkali halides, Jain and Prabhakaran (1972). Calculation of hydrogen vibrations in alkaline earth fluorides by L.G.F.T., Hayes and Wiltshire (1973). Resonant modes due to impurity pairs in NaCl and KCl by L.G.F.T., Haridasan, Gupta and Ludwig (1973), Gupta and Haridasan (1974), Gupta (1974). Long-range forces in the dynamics of crystals with defects, Litzman and Rozsa (1973). Vibrations of ionic crystals containing substitutional atoms in the R.S.-model, Lacina (1974). Discussion of an anharmonic oscillator model for local modes of Li^+ ions in KCl, Pandey, Shukla and Pandey (1974). Molecular model calculations of pair modes in alkali halides, Ward and Clayman (1974). Discussion of the lattice dynamics of crystals with molecular impurities based on L.G.F.T., Sahoo and Venkataraman (1975). Lattice vibrations induced by a polarizable impurity atom in polar crystals by generalized L.G.F.T., Mahanty and Paranjape (1974) (1975) (1976). Theory of local phonon modes resulting from electron phonon interaction at impurity centers, Rashba and Zimin (1974). Local mode frequencies of isoelectronic impurities in zincblende type crystals of II–VI and

III–V compounds by L.G.F.T., Talwar and Agrawal (1974) (1975). Effects of static distortions on the properties of impurity resonances in the asymptotic limit of low frequencies in a molecular model, Page and Helliwell (1975). Calculation of the frequency shifts of the optical modes of a crystal due to polarizable impurity atoms by Mahanty and Paranjape's method, Johri and Paranjape (1976). Spectral distribution of resonant localized impurity vibrations at finite temperatures by a field theoretic method, Dykman and Ivanov (1976). Description of resonant and localized defect vibrations by means of the Krumhansl-Matthew method, Dederichs and Zeller (1976). Localized modes due to *U*-center pairs in alkali halides by L.G.F.T., Kalyani and Haridasan (1976). An exponential transformation of perturbed harmonic lattice Hamiltonians for the successive calculation of perturbed lattice modes, Brühl, Sigmund and Wagner (1977). *U*-center modes in KBr containing additive cation impurities by L.G.F.T., Gupta and Singh (1977) (1978). Local modes of *U*-center pairs in RbCl by L.G.F.T., Gupta, Mathur and Singh (1977). Lattice modes and frequencies of substitutional impurities in InP, GaP and ZnS by L.G.F.T. with a rigid ion model, Vandevyver and Plumelle (1977). Gap modes due to impurity modes in KI with L.G.F.T., Kalyani and Haridasan (1977). Localized modes due to *U*-centers in alkali halides based on the Ram-Agrawal effective force constant method, Ram (1977). Dynamical properties of NO_2^- impurity molecules in alkali halides by Wagner's molecular L.G.F.T., Zavt (1977). The influence of the static lattice distortion on the localized modes of atomic impurities in alkali halides in the molecular model, Frey and Jüngst (1977). Local force variations due to substitutional impurities in nine compounds with zincblende structure with L.G.F.T. for a rigid ion model, Vandevyver and Plumelle (1978). Localized vibrations due to substitutional impurities in calcium fluoride by L.G.F.T., Krishnamurthy and Haridasan (1978). The effect of uncertainty in the eigendata of the pure host lattice on Green function calculations, de Geus and Gijzeman (1978).
A review article of impurity induced vibrations which treats also the effect of a definite concentration of impurities on the vibrational spectrum was published by Taylor (1975).

7.6 Phonon processes

Phonon processes occur if by the action of external forces a crystal is removed from thermal equilibrium. In this case in addition to the external forces also internal forces result since any deviation from thermal equilibrium is connected with internal anharmonic interactions of the phonons. These anharmonic interactions entail energy dissipation, i.e., damping, so that under the action of external driving forces and internal forces of friction new dynamical equilibrium states arise. In Section 2.5 and 2.6 the damping effect of anharmonic forces on local phonon modes was already studied. In this section we investigate the combined effects of external and internal forces in particular for the phonon band states. From the great variety of possible experimental arrangements we consider three important cases, namely thermal conduction and optical dispersion and absorption.

a) Thermal conduction

The external forces for thermal conduction are given by a temperature gradient which creates a heat flow in the crystal. Heat energy can be transmitted through a crystal via the motion of quasiparticles such as phonons, photons, free electrons and holes, excitons etc. The electronic components of heat conduction are usually the largest components in a metal, but almost all of the thermal current in a non-metal is carried by the phonons, except at the highest temperatures when photons may become important. The electron population in a semiconducting material is usually too small to produce a very large electronic thermal conductivity, though such conduction can be measured in some semiconductors at fairly high temperatures when there are enough thermally excited electrons. In this section we restrict ourselves to the dominating phonon contribution, reserving the treatment of the electronic contribution to thermal conduction in semiconductors for the next chapter. The internal forces which are connected with the phonon flow in a crystal are mainly the anharmonic forces which lead to phonon-phonon scattering. In addition to these phonon scattering processes, phonons may also experience scattering by point defects, line defects, grain boundaries in a polycrystal, outer surfaces of a monocrystal, random distributions of different isotopes, etc.. In accordance with the introduction, line defects and boundaries are excluded from the discussion as they exceed the scope of this book. Hence only the effects of the genuine anharmonic forces and of point defects on the phonon distribution are to be studied. Since according to experimental and theoretical evidence the boundaries influence only the low temperature limit of the heat conduction, for monocrystals with a fairly small number of dislocations it is to be expected that anharmonic interactions and point defect scattering lead to a realistic model valid for wide temperature ranges.
We first treat the anharmonic forces. In the quantized version of the theory these forces arise from the anharmonic potential (2.2.16) of the lattice energy. We define

$$\Phi^n(l_1\kappa_1 \dots l_r\kappa_r) := \left(\frac{\partial^r U_n}{\partial \mathbf{R}_{l_1\kappa_1} \dots \partial \mathbf{R}_{l_r\kappa_r}}\right)_{/\mathbf{R}=\mathbf{R}^n} \tag{7.6.1}$$

and restrict ourselves to the discussion of the third order terms of (2.2.16) which are given by

$$U_n^3(\mathbf{u}) := \frac{1}{3} \sum_{l\kappa, l'\kappa', l''\kappa''} \Phi^n(l\kappa, l'\kappa', l''\kappa'') \cdots \mathbf{u}_{l\kappa} \mathbf{u}_{l'\kappa'} \mathbf{u}_{l''\kappa''}. \tag{7.6.2}$$

In the first step of the investigation of phonon transport properties we consider an ideal crystal, i.e., we work with $U_0^3(\mathbf{u})$. Since the total potential energy $U_0(\mathbf{R})$ is invariant with respect to transformations of the corresponding space group and since $U_0(\mathbf{R}^0)$ is a symmetry identity for this group, equation (7.1.22) holds. By expansion in powers of the displacements $\mathbf{u}_{l\kappa}$ symmetry properties for the coupling constants (7.6.1) can be derived to any order.
In Theorem 7.2 the symmetry properties of the first and second order coupling constants were given. Along the same lines the symmetry properties of the third order

terms follow. We give only the results for translational invariance

$$\Phi^0(l\kappa, l'\kappa', l''\kappa'') = \Phi^0(l+\mathbf{g}, \kappa, l'+\mathbf{g}, \kappa', l''+\mathbf{g}, \kappa'') \tag{7.6.3}$$

which is of general significance for the further discussion. By substituting expansion (7.2.66) into (7.6.2) we obtain

$$U_0^3 = \frac{1}{3} \sum_{\substack{\mathbf{k},\mathbf{k}',\mathbf{k}'' \\ j,j',j''}} \Phi(j,\mathbf{k},j',\mathbf{k}',j'',\mathbf{k}'') \times [b(j,\mathbf{k}) - b^+(j,-\mathbf{k})]\,[b(j',\mathbf{k}') - b^+(j',-\mathbf{k}')]\,[b(j'',\mathbf{k}'') - b^+(j'',-\mathbf{k}'')] \tag{7.6.4}$$

with

$$\begin{aligned} \Phi(j,\mathbf{k},j',\mathbf{k}',j'',\mathbf{k}'') = & \sum_{\kappa,\kappa',\kappa''} \eta(\kappa|j,\mathbf{k})\eta(\kappa'|j',\mathbf{k}')\eta(\kappa''|j'',\mathbf{k}'') N^{-3/2} \\ & \times \mathbf{e}(\kappa|j,\mathbf{k})\mathbf{e}(\kappa'|j',\mathbf{k}')\mathbf{e}(\kappa''|j'',\mathbf{k}'') \\ & \times \sum_{l,l',l''} \Phi^0(l\kappa,l'\kappa',l''\kappa'') \exp[i(\mathbf{k}\cdot\mathbf{R}_l^0 + \mathbf{k}'\cdot\mathbf{R}_{l'}^0 + \mathbf{k}''\cdot\mathbf{R}_{l''}^0)]. \end{aligned} \tag{7.6.5}$$

By means of (7.6.3) and (7.2.43) the second sum in the product on the right-hand side of (7.6.5) can be rewritten as

$$\begin{aligned} & \sum_{l,l',l''} \Phi^0(l\kappa,l'\kappa',l''\kappa'') \exp[i(\mathbf{k}\cdot\mathbf{R}_l^0 + \mathbf{k}'\cdot\mathbf{R}_{l'}^0 + \mathbf{k}''\cdot\mathbf{R}_{l''}^0)] \\ & = N\Delta(\mathbf{k}+\mathbf{k}'+\mathbf{k}'') \sum_{h,h'} \Phi^0(0\kappa,h'\kappa',h''\kappa'') \exp[i(\mathbf{k}'\cdot\mathbf{R}_{h'}^0 + \mathbf{k}''\cdot\mathbf{R}_{h''}^0)]. \end{aligned} \tag{7.6.6}$$

From this it follows that for a phonon interaction of the third order either the quasimomentum $\mathbf{K}$ has to be conserved $\mathbf{K} := \mathbf{k} + \mathbf{k}' + \mathbf{k}'' = 0$ (normal process) or $\mathbf{K}$ has to be a reciprocal lattice vector $\mathbf{K} = \mathbf{g}$ (Umklapp-process). It will be shown later that for phonon transport processes at low temperatures only the Umklapp-processes contribute to the thermal resistance, while the normal processes do not influence this resistance.

As we do with all other processes, we study the transport processes of phonons in the framework of irreversible quantum statistical mechanics, the basic equations of which were given in Chapters 1 and 2. In Theorem 2.2 we distinguished between active and passive phonon modes. This distinction was made in order to separate those phonon modes which are mainly coupled to electronic states from the other modes whose coupling to electronic states can be neglected in good approximation. Since in this chapter pure phonon processes are considered we need not make use of this distinction. Hence all phonon modes can be treated on the same footing and for the ideal crystal the only internal interaction that appears is U_0^a, i.e., in our approximation U_0^3. In Section 2.4, the irreversible phonon motion was separated from the electronic processes leading to the Pauli-Master equation (2.4.14). If according to Theorem 2.2 the phonon-heat bath interaction is identified with the anharmonic interaction in the crystal itself and if the distinction between active and passive phonons is omitted, then the transition probabilities $W_{mm'}^W$ of (2.4.14) arise from U_0^3 alone. Hence for the investigation of phonon transport properties we have to investigate equation (2.4.14) with the U_0^3-interaction. If only the decay of a definite local mode has to be calculated, one can try to solve equations (2.4.14) directly as in Section 2.5. If, however, all phonon modes are involved in transport and decay processes, another solution technique has to be applied. In accordance with Chapter 3 and 4, this technique consists in the derivation of kinetic equations for the mean quantum numbers, which, in the case of band states leads to the Boltzmann equation as has been demonstrated for conduction band electrons in Theorem 3.6. Formula (3.6.9) and formula (3.6.14) show that the Boltzmann distribution function has to be identified with the mean electron occupation number. In analogy to this definition, for a phonon state occupation probability

$$P_m(t) \equiv P(m_1 \dots m_N, t) \tag{7.6.7}$$

the mean phonon occupation number

$$\bar{m}_\alpha(t) := \sum_{m_1 \dots m_N = 0}^{\infty} m_\alpha P(m_1 \dots m_N, t) \tag{7.6.8}$$

can be defined and kinetic equations can be derived, Stumpf (1957) (1961). Consequently one would expect that kinetic equations for these quantities are called phonon Boltzmann equations. This, however, it not the case. In the literature a type of kinetic equation is used which, in contrast to the general Boltzmann equation, is only valid for stationary phonon processes and which is nevertheless called a phonon Boltzmann equation. The discrepancy between these two types of kinetic equations is obscured by the fact that the "phonon Boltzmann" equation is introduced only by heuristic considerations. Peierls (1929) was the first who worked with such equations and used them to explain thermal conduction, cf. also Leibfried (1957), Ziman (1960). As all work done on this topic is concerned with these equations (provided that no Green function technique is used) we do not derive equations for the mean quantum numbers (7.6.8) but give a more rigorous deduction of the common approach which clearly shows its limits of applicability.
The external driving forces are defined by a macroscopic temperature gradient. To incorporate such rough macroscopic forces in the reaction kinetics the ensemble definition has to be extended.

i) In the first step in which only processes at single impurities are to be considered, the ensemble can be thought as being constituted by a set of uncorrelated microblocks, the center of each containing one impurity (μ-space of Boltzmann).
ii) In the second step in which processes between various imperfections compete and in which charge and energy transport in microscopic regions is admitted, the ensemble can be thought as being constituted by a set of mosaicblocks each containing already a large number of imperfections.
iii) In the third and final step correlations between mosaicblocks which lead to the possibility of charge and energy exchange over macroscopic distances are admitted. In this case the macroscopic crystal is only one member of the ensemble and it has to be expected that only when the conditions for the thermodynamic limit are sufficiently well approximated by the real situation, can meaningful predictions for one ensemble member be made (Γ-space of Boltzmann).

For this ensemble hierarchy the existence of imperfections is only needed in order to obtain localizable processes. If charge and energy transport over macroscopic distances are considered, imperfections, whether present or not, play no role for the definition of ensembles. Hence this concept can also be used for the ideal crystal where the mosaicblocks have to be identified with the volume V of the large periodicity range. The set of phonon occupation numbers then has to be labelled by the phonon mode indices $j,\mathbf{k}$ and by the index of the mosaicblock α, and is given therefore by $\{m(j,\mathbf{k},\alpha)\}$. The corresponding occupation probabilities then read

$$P_m(t) \equiv P(\ldots m(j_1,\mathbf{k}_1,\alpha)\ldots m(j_N,\mathbf{k}_N,\alpha)\ldots t). \tag{7.6.9}$$

To distinguish the kinetic equations for phonons which are used in the literature from the general Boltzmann equation, we call them stationary Boltzmann equations. Then by means of the ensemble definition iii) and the Pauli Master equation for the occupation probabilities (7.6.9), the following theorem can be derived

Theorem 7.20: Suppose that an ideal crystal is considered to be a combination of mosaicblocks and the phonons are allowed to pass through the entire crystal. If then the internal mosaicblock interaction is defined by U_0^3 and if the mean quantum numbers $\bar{m}(j,\mathbf{k},\mathbf{r})$ are considered in a coarse sense to be functions of the macroscopic position vector $\mathbf{r}$, for these functions $\bar{m}(j,\mathbf{k},\mathbf{r})$ the stationary Boltzmann equations

$$-\mathbf{v}(j,\mathbf{k})\cdot\nabla m(j,\mathbf{k},\mathbf{r}) = G[\bar{m},j,\mathbf{k}] :=$$

$$\sum_{\substack{\mathbf{k}'\mathbf{k}''\\ j'j''}} \{m(j,\mathbf{k},\mathbf{r})m(j',\mathbf{k}',\mathbf{r})[1+m(j'',\mathbf{k}'',\mathbf{r})]$$
$$-[1+m(j,\mathbf{k},\mathbf{r})][1+m(j',\mathbf{k}',\mathbf{r})]m(j'',\mathbf{k}'',\mathbf{r})\}\,A^{j''\mathbf{k}''}_{j\mathbf{k},\,j'\mathbf{k}'} \tag{7.6.10}$$

$$\sum_{\substack{\mathbf{k}'\mathbf{k}''\\ j'j''}} \{m(j,\mathbf{k},\mathbf{r})[1+m(j',\mathbf{k}',\mathbf{r})][1+m(j'',\mathbf{k}'',\mathbf{r})]$$
$$-[1+m(j,\mathbf{k},\mathbf{r})]m(j',\mathbf{k}',\mathbf{r})\,m(j'',\mathbf{k}'',\mathbf{r})\}\,A^{j'\mathbf{k}',\,j''\mathbf{k}''}_{j\mathbf{k}}$$

hold, where $\mathbf{v}(j,\mathbf{k})$ is the velocity of the phonons of the mode $j,\mathbf{k}$.

Proof: According to the supposition, the entire crystal is involved in phonon propagation. Hence for irreversible quantum statistics the entire crystal has to be considered as one member of a hypothetic ensemble whose occupation probabilities are given by (7.6.9). These quantities have to satisfy the Pauli Master equation (2.4.14). The general transition probability is given by (2.4.4), where in our case the anharmonic interaction U_n^a is the sum over the internal interactions $U_0^3(\alpha)$ of the various mosaicblocks, i.e.,

$$U_n^a := \sum_\alpha U_0^3(\alpha) =: U_0^3 \tag{7.6.11}$$

while the interaction with the heat reservoir U_n^w has to be replaced by the transition operators for the phonon transitions from the mosaicblock α to a neighbouring mosaicblock α'. Hence we write

$$U^t := \sum_{\alpha\alpha'} \sum_{j,\mathbf{k}} U(j,\mathbf{k},\alpha,\alpha') \tag{7.6.12}$$

where $U(j,\mathbf{k},\alpha,\alpha')$ is the transition operator for the phonons of the mode $j,\mathbf{k}$ between the mosaicblocks α and α' and we replace $\overset{w}{U_n}$ by U^t. Then the definition (2.4.4) takes the form

$$W^w_{mm'} := \frac{2\pi}{\hbar} \,|\langle m|U^3_0 + U^t|m'\rangle|^2 \delta_\gamma(E^0_m - E^0_{m'}) \tag{7.6.13}$$

with $m := \{m(j,\mathbf{k},\alpha)\}$. In this approximation the transitions with respect to U^3_0 do not interfere with those with respect to U^t. Hence (7.6.13) can be decomposed into

$$W^w_{mm'} = W^3_{mm'} + W^t_{mm'} \tag{7.6.14}$$

and equation (2.4.14) goes over into

$$\dot{P}_m = \sum_{m'} W^3_{mm'}(P_{m'} - P_m) + \sum_{m'} W^t_{mm'}(P_{m'} - P_m). \tag{7.6.15}$$

We treat the two parts of equation (7.6.15) separately.

i) We first consider the contribution of U^3_0 to the interaction in the mosaicblock α. It is given by (7.6.4) if we add the label α. Making use of the symmetry of the coupling constants (7.6.5) with respect to permutations of the variables $(j,\mathbf{k})$ $(j',\mathbf{k}')$ $(j'',\mathbf{k}'')$, by a suitable change of the variables expression (7.6.4) can be rewritten as

$$\begin{aligned} U^3_0(\alpha) = \sum_{\substack{\mathbf{k}\mathbf{k}'\mathbf{k}''\\ jj'j''}} \{&\Phi(j,\mathbf{k},j',\mathbf{k}',j'',\mathbf{k}''|\alpha) b(j,\mathbf{k},\alpha) b(j',\mathbf{k}',\alpha) b^+(j'',\mathbf{k}'',\alpha)\\ &+ \Phi(j,\mathbf{k},j',-\mathbf{k}',j'',-\mathbf{k}''|\alpha) b(j,\mathbf{k},\alpha) b^+(j',\mathbf{k}',\alpha) b^+(j'',\mathbf{k}'',\alpha)\}\\ &+\frac{1}{3} \sum_{\substack{\mathbf{k}\mathbf{k}'\mathbf{k}''\\ jj'j''}} \Phi(j,\mathbf{k},j',\mathbf{k}',j'',\mathbf{k}'')\ [b(j,\mathbf{k},\alpha) b(j',\mathbf{k}',\alpha') b(j'',\mathbf{k}'',\alpha)\\ &+ b^+(j,\mathbf{k},\alpha) b^+(j',\mathbf{k}',\alpha) b^+(j'',\mathbf{k}'',\alpha)] \end{aligned} \tag{7.6.16}$$

If the states $|m\rangle$ or $|m'\rangle$, resp. are represented by

$$|m\rangle = \left\{\prod_{j\mathbf{k}\alpha} [m(j,\mathbf{k},\alpha)!]^{-1/2} b^+(j,\mathbf{k},\alpha)^{m(j,\mathbf{k},\alpha)}\right\}|0\rangle \tag{7.6.17}$$

and if the states $|m'\rangle$are characterized with respect to $|m\rangle$ by the definition $m'(j,\mathbf{k},\alpha):= m(j,\mathbf{k},\alpha) + \Delta m(j,\mathbf{k},\alpha)$ for the quantum numbers m', then by taking into account (7.6.11) and by applying (7.6.16) to $|m'\rangle$, we obtain

$$\begin{aligned} &\sum_{m'} W^3_{mm'}(P_{m'} - P_m)\\ &= \sum_{\alpha} \sum_{\substack{\mathbf{k}\mathbf{k}'\mathbf{k}''\\ jj'j''}} \{A_1(j,\mathbf{k},j',\mathbf{k}',j'',\mathbf{k}''|\alpha)[m(j,\mathbf{k},\alpha)+1]\,[m(j',\mathbf{k}',\alpha)+1] m(j'',\mathbf{k}'',\alpha)\} \end{aligned}$$

$$\times [P(\ldots m(j,\mathbf{k},\alpha)+1,\ m(j',\mathbf{k}',\alpha)+1,\ m(j'',\mathbf{k}'',\alpha)-1,\ldots)-P(m)] \tag{7.6.18}$$

$$+\sum_{\alpha}\sum_{\substack{\mathbf{k}\mathbf{k}'\mathbf{k}''\\ jj'j''}} \{A_2(j,\mathbf{k},j',\mathbf{k}',j'',\mathbf{k}''|\alpha)\,[m(j,\mathbf{k},\alpha)+1]\,m(j',\mathbf{k}',\alpha)\,m(j'',\mathbf{k}'',\alpha)\}$$

$$\times [P(\ldots m(j,\mathbf{k},\alpha)+1,\ m(j',\mathbf{k}',\alpha)-1,\ m(j'',\mathbf{k}'',\alpha)-1,\ldots)-P(m)]$$

where the definitions

$$A_1=|\Phi(j,\mathbf{k},j',\mathbf{k}',j'',-\mathbf{k}''|\alpha)|^2\,\delta_\gamma[-\omega(j,\mathbf{k})-\omega(j',\mathbf{k}')+\omega(j'',\mathbf{k}'')] \tag{7.6.19}$$

and

$$A_2=|\Phi(j,\mathbf{k},j',-\mathbf{k}',j'',-\mathbf{k}''|\alpha)|^2\,\delta_\gamma[-\omega(j,\mathbf{k})+\omega(j',\mathbf{k})+\omega(j'',\mathbf{k}'')] \tag{7.6.20}$$

are used. Those terms of (7.6.16) which do not appear in (7.6.18) vanish due to energy conservation.

ii) The operator $U(j,\mathbf{k},\alpha,\alpha')$ for the transition of a phonon from the mosaicblock α to the mosaicblock α' can in general be expressed by

$$U(j,\mathbf{k},\alpha,\alpha')=u(j,\mathbf{k},\alpha,\alpha')\,b(j,\mathbf{k},\alpha)b^+(j,\mathbf{k},\alpha') \tag{7.6.21}$$

where $u(j,\mathbf{k},\alpha,\alpha')$ is an appropriate classical quantity, i.e., by this operator one phonon is destroyed in the mosaicblock α and one phonon is generated in the mosaicblock α'. Then, if we use the state representation (7.6.17) again, the second part on the right-hand side of equation (7.6.15) becomes

$$\begin{aligned}&\sum_{m'} W^t_{mm'}(P_{m'}-P_m)\\ &=\sum_{\alpha\alpha'}\sum_{jk} w(j,\mathbf{k},\alpha,\alpha')\,[m(j,\mathbf{k},\alpha)+1]\,m(j,\mathbf{k},\alpha')\\ &\times [P(\ldots m(j,\mathbf{k},\alpha)+1,\ldots,m(j,\mathbf{k},\alpha')-1,\ldots)-P(m)]\end{aligned} \tag{7.6.22}$$

where $w(j,\mathbf{k},\alpha,\alpha')$ is defined by

$$w(j,\mathbf{k},\alpha,\alpha'):=|u(j,\mathbf{k},\alpha,\alpha')|^2\,\delta_\gamma(0) \tag{7.6.23}$$

and will be evaluated later.

iii) We now sum equations (7.6.15) over m. Then due to probability conservation, $\sum_m \dot{P}_m$ vanish and (7.6.15) yields

$$-\sum_{m'm} W^t_{mm'}(P_{m'}-P_m)=\sum_{m'm} W^3_{mm'}(P_{m'}-P_m). \tag{7.6.24}$$

According to (7.6.18) we split $W^3_{mm'}$ into two parts $W^3_{mm'}=W^3_{mm'}(1)+W^3_{mm'}(2)$ and consider the term $\sum_{mm'} W^3_{mm'}P_{m'}$ in more detail. By an appropriate change of the summation variables this term becomes

$$\sum_{mm'} W^3_{mm'}(1)P_{m'} = \sum_m \sum_\alpha \sum_{\substack{\mathbf{kk'k''}\\ jj'j''}} A_1[m(j,\mathbf{k},\alpha)+1]\,[m(j',\mathbf{k}',\alpha)+1]\,m(j'',\mathbf{k}'',\alpha)$$
$$\times P(\ldots m(j,\mathbf{k},\alpha)+1, \ldots m(j',\mathbf{k}',\alpha)+1, \ldots m(j'',\mathbf{k}'',\alpha)-1, \ldots) \tag{7.6.25}$$
$$= \sum_m \sum_\alpha \sum_{\substack{\mathbf{kk'k''}\\ jj'j''}} A_1 m(j,\mathbf{k},\alpha)m(j',\mathbf{k}',\alpha)\,[m(j'',\mathbf{k}'',\alpha)+1]P(m)$$

if we neglect the changes at the lower limit of the summations. If we proceed in the same way with the term $W^3_{mm'}(2)$, then the right-hand side of (7.6.24) reads

$$\sum_{mm'} W^3_{mm'}(P_{m'}-P_m)$$
$$= \sum_m \sum_\alpha \sum_{\substack{\mathbf{kk'k''}\\ jj'j''}} \{m(j,\mathbf{k},\alpha)\,m(j',\mathbf{k}',\alpha)[m(j'',\mathbf{k}'',\alpha)+1]$$
$$-[m(j,\mathbf{k},\alpha)+1]\,[m(j',\mathbf{k}',\alpha)+1]\,m(j'',\mathbf{k}'',\alpha)\}\,A_1\,P(m) \tag{7.6.26}$$
$$+\sum_m \sum_\alpha \sum_{\substack{\mathbf{kk'k''}\\ jj'j''}} \{m(j,\mathbf{k},\alpha)[m(j',\mathbf{k}',\alpha)+1]\,[m(j'',\mathbf{k}'',\alpha)+1]$$
$$-[m(j,\mathbf{k},\alpha)+1]m(j',\mathbf{k}',\alpha)m(j'',\mathbf{k}'',\alpha)\}\,A_2\,P(m).$$

With the same procedure we obtain for the left-hand side of (7.6.24)

$$\sum_{mm'} W^t_{mm'}(P_{m'}-P_m)$$
$$= \sum_m \sum_{\alpha\alpha'} \sum_{j\mathbf{k}} \{m(j,\mathbf{k},\alpha)\,[m(j,\mathbf{k},\alpha')+1]-[m(j,\mathbf{k},\alpha)+1]\,m(j,\mathbf{k},\alpha')\}\,wP(m) \tag{7.6.27}$$
$$= \sum_m \sum_{\alpha\alpha'} \sum_{j\mathbf{k}} \{m(j,\mathbf{k},\alpha)-m(j,\mathbf{k},\alpha')\}\,wP(m).$$

As both expressions (7.6.26) and (7.6.27) contain a summation over α and $j,\mathbf{k}$, equation (7.6.24) is satisfied if the equations

$$\sum_m \sum_{\substack{\mathbf{k'k''}\\ j'j''}} [\{\quad\}A_1+\{\quad\}A_2]P(m)$$
$$= -\sum_{m\alpha'} [m(j,\mathbf{k},\alpha)-m(j,\mathbf{k},\alpha')]\,w(j,\mathbf{k},\alpha,\alpha')\,P(m) \tag{7.6.28}$$

hold separately for any α and $j,\mathbf{k}$. The probability for a phonon transition from the mosaicblock α to the mosaicblock α' follows from a phenomenological consideration. If d is the distance between both elements, w must be proportional d^{-1}. If further $\mathbf{e}(\alpha,\alpha')$ is the direction from the element α to the element α', then a transition takes place only if $\mathbf{v}(j,\mathbf{k}) \parallel \mathbf{e}(\alpha,\alpha')$. Hence we obtain $w \approx \mathbf{v}(j,\mathbf{k}) \cdot \mathbf{e}(\alpha,\alpha')d^{-1}$ with the subsidiary condition $\mathbf{v}(j,\mathbf{k}) \parallel \mathbf{e}(\alpha,\alpha')$. This condition reduces the summation over α' on the right-hand side of

(7.6.28) to a single term. Therefore, this expression takes the form

$$\sum_m \sum_{\alpha'} [m(j,\mathbf{k},\alpha) - m(j,\mathbf{k},\alpha')]\, w(j,\mathbf{k},\alpha,\alpha')\, P(m)$$
$$\approx \frac{\mathbf{e}(\alpha,\alpha')}{d} \cdot \mathbf{v}(j,\mathbf{k})\,[\bar{m}(j,\mathbf{k},\alpha) - \bar{m}(j,\mathbf{k},\alpha')] \tag{7.6.29}$$
$$\approx \mathbf{v}(j,\mathbf{k}) \cdot \nabla \bar{m}(j,\mathbf{k},\mathbf{r})$$

where $\mathbf{r}$ is a mean value of the position of the mosaicblock α. As the dimensions of a mosaicblock are still very small (edge length of a mosaicblock $\approx 10^{-5}$ cm) in comparison with macrocrystals, for a coarse-grained phonon density, $\mathbf{r}$ can be considered to be a continuous variable. If, further, on the left-hand side of equation (7.6.28) the average of a product is replaced by the product of the averaged factors, from (7.6.29) and (7.6.28) equation (7.6.10) results, Q.E.D.

As the central equation of this proof is given by (7.6.24) where the explicit time derivative of $P(m)$ has been removed, it is clear that the resulting equation (7.6.10) can only be applied to stationary nonequilibrium states, but not to time dependent ones. This restriction has to be observed when looking for solutions of equation (7.6.10). The phonon rate equation is an integro differential equation with partial derivatives and a rather complicated kernel.
In contrast to the electron case where the nonlinearities are essential for the derivation of physically meaningful results, the restriction of stationarity for the phonon equation allows the possibility of stationary small deviations from equilibrium for which the nonlinearities are not important. In the literature only cases which permit a linearization have been treated in detail. It is obvious that by such assumptions the discussion of effects of a nonlinear phonon hydrodynamics is excluded, but this has already been done by chosing the stationary phonon equations (7.6.10) for the description of phonon processes. Hence these equations can only be used for stationary thermal conduction not far from thermal equilibrium. Before continuing we define thermal conductivity.
Thermal conductivity depends on energy flow, in which the energy is transported by the phonon waves. Usually the energy flow of waves is defined by the average velocity (group velocity) of a wave packet and not by the velocity of a single wave. If all waves have the same velocity the wave velocity and the group velocity coincide. If, however, dispersion occurs as is the case for phonons, the group velocity is different from the wave velocity and is given for phonons by

$$\mathbf{v}(j,\mathbf{k}) = \nabla_{\mathbf{k}}\, \omega(j,\mathbf{k}) \tag{7.6.30}$$

where $\omega(j,\mathbf{k})$ is considered to be a continuous function of $\mathbf{k}$. If $\varepsilon(j,\mathbf{k},\alpha)$ is the corresponding energy of the wave j, $\mathbf{k}$, then the density of the energy flow is defined by

$$\mathbf{q}(j,\mathbf{k},\mathbf{r}) = V^{-1}\varepsilon(j,\mathbf{k},\alpha)\mathbf{v}(j,\mathbf{k}) \tag{7.6.31}$$

where V is the volume of the mosaicblock. The total energy flow density then becomes

$$\mathbf{q}(\mathbf{r}) = V^{-1} \sum_{j\mathbf{k}} \varepsilon(j,\mathbf{k},\alpha)\mathbf{v}(j,\mathbf{k}) \tag{7.6.32}$$

and the corresponding mean value reads

$$\bar{\mathbf{q}}(\mathbf{r}) = V^{-1} \sum_{j\mathbf{k}} \bar{\varepsilon}(j,\mathbf{k},\alpha)\,\mathbf{v}(j,\mathbf{k}). \tag{7.6.33}$$

For thermal equilibrium $\bar{\varepsilon}(j,\mathbf{k},\alpha) = \varepsilon[\omega(j,\mathbf{k}), T(\alpha)]$, the energy flow (7.6.33) vanishes, as $\bar{\varepsilon}(j,\mathbf{k},\alpha)$ is symmetric in $\mathbf{k}$, $\bar{\varepsilon}(j,\mathbf{k},\alpha) = \bar{\varepsilon}(j,-\mathbf{k},\alpha)$ while $\mathbf{v}(j,\mathbf{k})$ is antisymmetric in $\mathbf{k}$, $\mathbf{v}(j,\mathbf{k}) = -\mathbf{v}(j,-\mathbf{k})$ and the summation in (7.6.33) has to be extended over the whole Brillouin zone.
For stationary nonequilibrium states we write $\bar{\varepsilon}(j,\mathbf{k},\alpha)$ in the form

$$\bar{\varepsilon}(j,\mathbf{k},\alpha) = \hbar\omega(j,\mathbf{k})\,[\bar{m}(j,\mathbf{k},\alpha) + \tfrac{1}{2}] \tag{7.6.34}$$

and obtain for (7.6.33)

$$\bar{\mathbf{q}}(\mathbf{r}) = V^{-1} \sum_{j\mathbf{k}} \hbar\omega(j,\mathbf{k})\bar{m}(j,\mathbf{k},\alpha)\mathbf{v}(j,\mathbf{k}) \tag{7.6.35}$$

where the zero point energy drops out. By referring $\bar{m}(j,\mathbf{k},\alpha)$ to the thermal equilibrium $\bar{m}_0(j,\mathbf{k},\alpha)$, we may write

$$\bar{m}(j,\mathbf{k},\alpha) = \bar{m}_0(j,\mathbf{k},\alpha) + \eta(j,\mathbf{k},\alpha) \tag{7.6.36}$$

with

$$\bar{m}_0(j,\mathbf{k},\alpha) := \{\exp\,[\hbar\omega(j,\mathbf{k})/k_B T(\alpha)] - 1\}^{-1}. \tag{7.6.37}$$

where it has to be noted that the quantities $\eta(j,\mathbf{k},\alpha)$ are not to be confused with the quantities $\eta(\kappa|j,\mathbf{k})$ defined by (7.2.67). Then by substitution of (7.6.36) in (7.3.35) this expression takes the form

$$\bar{\mathbf{q}}(\mathbf{r}) = V^{-1} \sum_{j\mathbf{k}} \hbar\omega(j,\mathbf{k})\eta(j,\mathbf{k},\alpha)\mathbf{v}(j,\mathbf{k}). \tag{7.6.38}$$

We now formally calculate the quantities $\{\eta(\mathrm{j},\mathbf{k},\alpha)\}$ for the case of an external temperature gradient acting on the crystal and relate the result to the thermal conductivity. For the left-hand side of (7.6.10) we assume that $\bar{m}(j,\mathbf{k},\alpha)$ depends only on T, so that $\bar{m}(j,\mathbf{k},\alpha) \equiv \bar{m}(j,\mathbf{k},T(\alpha))$ and

$$-\mathbf{v}(j,\mathbf{k}) \cdot \nabla \bar{m}(j,\mathbf{k},T) \equiv -\frac{d\bar{m}(j,\mathbf{k},T)}{dT}\,\mathbf{v}(j,\mathbf{k}) \cdot \nabla T \tag{7.6.39}$$

hold and in this way the external force is incorporated into the dynamic equation. For the right-hand side of equation (7.6.10) we assume that the $\eta(j,\mathbf{k},\alpha)$ of (7.6.36) are so small, that only the linear terms of an expansion in $\{\eta(\mathrm{j},\mathbf{k},\alpha)\}$ have to be taken into account, i.e.,

$$G(\bar{m}_0 + \eta) \approx \sum_{j'\mathbf{k}'} G(j,\mathbf{k},j',\mathbf{k}')\eta(j',\mathbf{k}',\alpha) + G(\bar{m}_0). \tag{7.6.40}$$

Since in thermal equilibrium the temperature gradient and the gradient of $\bar{m}(j,\mathbf{k},\alpha)$ vanish, equation (7.6.10) is reduced to $0 = G(m_0)$. Hence in (7.6.40) the zero order term drops out, and equation (7.6.10) becomes with (7.6.39) and (7.6.40)

$$-\frac{d\bar{m}(j,\mathbf{k},T)}{dT}\mathbf{v}(j,\mathbf{k})\cdot\nabla T=\sum_{j'\mathbf{k}'}G(j,\mathbf{k},j',\mathbf{k}')\eta(j',\mathbf{k}',\mathbf{r}). \tag{7.6.41}$$

The quantity $\eta(j,\mathbf{k},\mathbf{r})$ is expected to be proportional to the external force (linear response). If we substitute (7.6.36) into the left-hand side of (7.6.41) then the term with $\eta(j,\mathbf{k},\mathbf{r})$ is of second order in the external force and hence must be neglected. Thus (7.6.41) goes over into

$$-\frac{d\bar{m}_0(j,\mathbf{k},T)}{dT}\mathbf{v}(j,\mathbf{k})\cdot\nabla T=\sum_{j'\mathbf{k}'}G(j,\mathbf{k},j',\mathbf{k}')\eta(j',\mathbf{k}',\mathbf{r}). \tag{7.6.42}$$

From (7.6.37) it follows by direct calculation that

$$\frac{d\bar{m}_0(j,\mathbf{k},T)}{dT}=\frac{\hbar\omega(j,\mathbf{k})}{4k_BT^2}\left[\sinh\frac{\hbar\omega(j,\mathbf{k})}{2k_BT}\right]^{-2} \tag{7.6.43}$$

and it is convenient to rewrite equation (7.6.42) in the form

$$-\frac{\hbar\omega(j,\mathbf{k})}{T^2}\mathbf{v}(j,\mathbf{k})\cdot\nabla T=\sum_{j'\mathbf{k}'}P(j,\mathbf{k},j',\mathbf{k}')\eta(j',\mathbf{k}',\mathbf{r}) \tag{7.6.44}$$

with

$$P(j,\mathbf{k},j',\mathbf{k}'):=4k_B\left[\sinh\frac{\hbar\omega(j,\mathbf{k})}{2k_BT}\right]^2G(j,\mathbf{k},j',\mathbf{k}') \tag{7.6.45}$$

A formal solution of equation (7.6.44) can be obtained by multiplication with P^{-1} which leads to

$$\eta(j,\mathbf{k},\mathbf{r})=-\sum_{j'\mathbf{k}'}P^{-1}(j,\mathbf{k},j',\mathbf{k}')\frac{\hbar\omega(j',\mathbf{k}')}{T^2}\mathbf{v}(j',\mathbf{k}')\cdot\nabla T. \tag{7.6.46}$$

If (7.6.46) is substituted in (7.6.38), the relation

$$\bar{\mathbf{q}}(\mathbf{r})=-\boldsymbol{\Lambda}\cdot\nabla T(\mathbf{r}) \tag{7.6.47}$$

results, where the thermal conductivity tensor $\boldsymbol{\Lambda}$ is defined by

$$\boldsymbol{\Lambda}:=\hbar^2(VT^2)^{-1}\sum_{\substack{\mathbf{k}\mathbf{k}'\\jj'}}\omega(j,\mathbf{k})\mathbf{v}(j,\mathbf{k})P^{-1}(j,\mathbf{k},j',\mathbf{k}')\omega(j',\mathbf{k}')\mathbf{v}(j',\mathbf{k}'). \tag{7.6.48}$$

Obviously the difficulty is to calculate P^{-1}. For the development of appropriate solution procedures of the first order integral equation (7.6.44), the properties of its kernel P have to be investigated. We give a lemma which was proven by Leibfried (1957)

Lemma 7.5: The kernel $P(j,\mathbf{k},j',\mathbf{k}')$ is symmetric and positive semidefinite, i.e.,

$$\sum_{\substack{\mathbf{k}\mathbf{k}'\\jj'}}P(j,\mathbf{k},j',\mathbf{k}')\eta(j,\mathbf{k})\eta(j',\mathbf{k}')\geqslant 0 \tag{7.6.49}$$

is valid for any real $\eta(j,\mathbf{k})$.

Proof: The collision term $G[\bar{m}]$ of (7.6.10) can be equivalently written

$$G[\bar{m}] = \sum_{\substack{\mathbf{k}'\mathbf{k}'' \\ j'j''}} \{\bar{m}(j,\mathbf{k})\bar{m}(j',\mathbf{k}')[1+\bar{m}(j'',-\mathbf{k}'')] - [1+\bar{m}(j,\mathbf{k})][1+\bar{m}(j',\mathbf{k}')]\bar{m}(j'',-\mathbf{k}'')\} \times |\Phi(j,\mathbf{k},j',\mathbf{k}',j'',\mathbf{k}'')|^2 \delta_\gamma[-\omega(j,\mathbf{k})-\omega(j',\mathbf{k}')+\omega(j'',\mathbf{k}'')] + \sum_{\substack{\mathbf{k}'\mathbf{k}'' \\ j'j''}} \{\bar{m}(j,\mathbf{k})[1+\bar{m}(j',-\mathbf{k}')][1+\bar{m}(j'',-\mathbf{k}'')] - [1+\bar{m}(j,\mathbf{k})]\bar{m}(j',-\mathbf{k}')\bar{m}(j'',-\mathbf{k}'')\} \times |\Phi(j,\mathbf{k},j',\mathbf{k}',j'',\mathbf{k}'')|^2 \delta_\gamma[-\omega(j,\mathbf{k})+\omega(j',\mathbf{k}')+\omega(j'',\mathbf{k}'')]. \tag{7.6.50}$$

To simplify this expression we introduce additional auxiliary variables $\{\bar{m}(j,\mathbf{k})\}$ which are defined for negative values of the branch- and polarisation-number j. These variables are not independent, but are related to the original set by the definitions

$$\bar{m}(-j,-\mathbf{k}) := -\bar{m}(j,\mathbf{k})-1; \quad j>0. \tag{7.6.51}$$

In addition, frequencies and coupling constants for these new variables are defined by

$$\omega(-j,\mathbf{k}) := -\omega(j,\mathbf{k}); \quad j>0 \tag{7.6.52}$$

and

$$D(j,\mathbf{k},j',\mathbf{k}',j'',\mathbf{k}'') := |\Phi(j,\mathbf{k},j',\mathbf{k}',j'',\mathbf{k}'')|^2; \quad j,j',j''>0 \tag{7.6.53}$$

$$D(-j,\mathbf{k},j',\mathbf{k}',j'',\mathbf{k}'') := -|\Phi(j,\mathbf{k},j',\mathbf{k}',j'',\mathbf{k}'')|^2; \quad j,j',j''>0 \tag{7.6.54}$$

with corresponding expressions for $-j, -j', -j, -j', -j''$. With these definitions (7.6.50) takes the form

$$G[m,j,\mathbf{k}] = \frac{2\pi}{\hbar^2} \sum_{\substack{\mathbf{k}'\mathbf{k}'' \\ j'j''\gtrless 0}} D(j,\mathbf{k},j',\mathbf{k}',j'',\mathbf{k}'')\,\delta_\gamma[\omega(j,\mathbf{k})+\omega(j',\mathbf{k}')+\omega(j'',\mathbf{k}'')] \times \{[\bar{m}(j,\mathbf{k})+1][\bar{m}(j'.\mathbf{k}')+1][\bar{m}(j'',\mathbf{k}'')+1] - \bar{m}(j,\mathbf{k})\bar{m}(j',\mathbf{k}')\bar{m}(j'',\mathbf{k}'')\}. \tag{7.6.55}$$

In (7.6.55) the summation is extended over all combinations of positive and negative j,j',j'' values. Such a summation is possible as all terms of (7.6.55) which are not contained in (7.6.50) drop out due to energy conservation. For negative j we will define

$$\bar{m}_0(-j,-\mathbf{k}) := -\bar{m}_0(j,\mathbf{k})-1; \quad j>0 \tag{7.6.56}$$

and with (7.6.51) and (7.6.36) we have

$$\eta(-j,-\mathbf{k}) := -\eta(j,\mathbf{k}); \quad j>0. \tag{7.6.57}$$

Hence (7.6.36) can be applied also for negative j. Substitution of (7.6.36) in (7.6.56) gives the first order term

$$G[m,j,\mathbf{k}] \approx \frac{\pi}{\hbar^2} \sum_{\substack{\mathbf{k}'\mathbf{k}'' \\ j'j'' \gtrless 0}} D(j,\mathbf{k},j',\mathbf{k}',j'',\mathbf{k}'')\, \delta_\gamma[\omega(j,\mathbf{k})+\omega(j',\mathbf{k}')+\omega(j'',\mathbf{k}'')]$$

$$\times \{\eta(j,\mathbf{k})[\bar{m}_0(j',\mathbf{k}')+\bar{m}_0(j'',\mathbf{k}'')+1] \tag{7.6.58}$$

$$+\eta(j',\mathbf{k}')[\bar{m}_0(j,\mathbf{k})+\bar{m}_0(j'',\mathbf{k}'')+1]$$

$$+\eta(j'',\mathbf{k}'')[\bar{m}_0(j,\mathbf{k})+\bar{m}_0(j',\mathbf{k}')+1]\}$$

whereas the zero order term vanishes, and the higher order terms are neglected. By direct calculation it follows, with $x(j,\mathbf{k}) := \hbar\omega(j,\mathbf{k})/2k_B T$, that

$$[\bar{m}_0(j',\mathbf{k}')+\bar{m}_0(j'',\mathbf{k}'')+1] = \frac{1}{2}\,\frac{\sinh[x(j',\mathbf{k}')+x(j'',\mathbf{k}'')]}{\sinh[x(j',\mathbf{k}')]\sinh[x(j'',\mathbf{k}'')]} \tag{7.6.59}$$

and analogous expressions for the other factors result. If we use energy conservation $-\omega=\omega'+\omega''$ and substitute (7.6.59) into (7.6.58), then we obtain

$$G[m,j,\mathbf{k}] \approx$$

$$-\frac{\pi}{\hbar^2} \sum_{\substack{\mathbf{k}'\mathbf{k}'' \\ j'j'' \gtrless 0}} \frac{D(j,\mathbf{k},j',\mathbf{k}',j'',\mathbf{k}'')\,\delta_\gamma[\omega(j,\mathbf{k})+\omega(j',\mathbf{k}')+\omega(j'',\mathbf{k}'')]}{\sinh[x(j,\mathbf{k})]\sinh[x(j',\mathbf{k}')]\sinh[x(j'',\mathbf{k}'')]} \tag{7.6.60}$$

$$\times \{\eta(j,\mathbf{k})\sinh^2[x(j,\mathbf{k})]+\eta(j',\mathbf{k}')\sinh^2[x(j',\mathbf{k}')]+\eta(j'',\mathbf{k}'')\sinh^2[x(j'',\mathbf{k}'')]\}.$$

If we now apply (7.6.40) and (7.6.45), then we have

$$\sum_{\substack{\mathbf{k}\mathbf{k}' \\ jj'>0}} P(j,\mathbf{k},j',\mathbf{k}')\,\eta(j,\mathbf{k})\,\eta(j',\mathbf{k}') =$$

$$-\sum_{\mathbf{k},j>0} 4k_B \sinh^2[x(j,\mathbf{k})]\,\eta(j,\mathbf{k})\,G[m,j,\mathbf{k}]. \tag{7.6.61}$$

It is possible to change the right-hand side of (7.6.61) into a summation over both positive and negative j. First we write (7.6.55) in the form

$$G[m,j,\mathbf{k}] = \sum_{\substack{\mathbf{k}'\mathbf{k}'' \\ j'j'' \gtrless 0}} D(j,\mathbf{k},j',\mathbf{k}',j'',\mathbf{k}'')\,\{j,\mathbf{k},j',\mathbf{k}',j'',\mathbf{k}''\} \tag{7.6.62}$$

where the bracket expression $\{j,\mathbf{k},j',\mathbf{k}',j'',\mathbf{k}''\}$ is defined by (7.6.58). We now replace in (7.6.62) j by $-j$. As the summations are not changed if we simultaneously replace also j' by $-j'$ and j'' by $-j''$ this yields

$$G[m,-j,\mathbf{k}] = \sum_{\substack{\mathbf{k}'\mathbf{k}'' \\ j'j'' \gtrless 0}} D(-j,\mathbf{k},-j',\mathbf{k}',-j'',\mathbf{k}'')\,\{-j,\mathbf{k},-j',\mathbf{k}',-j'',\mathbf{k}''\}. \tag{7.6.63}$$

With the aid of (7.6.54) and (7.6.52) we may rewrite the above as

$$G[m,-j,\mathbf{k}] = -\sum_{\substack{\mathbf{k}'\mathbf{k}''\\ j'j''\gtrless 0}} D(j,\mathbf{k},j',\mathbf{k}',j'',\mathbf{k}'')\,\{j,-\mathbf{k},j',-\mathbf{k}',j'',-\mathbf{k}''\}. \tag{7.6.64}$$

For brevity we write $c(j,\mathbf{k}) := 4k_B \sinh^2[x(j,\mathbf{k})]$ and consider

$$\begin{aligned}
&\sum_{\mathbf{k},j>0} c(j,\mathbf{k})\,\eta(j,\mathbf{k})\,G[m,j,\mathbf{k}]\\
&= \sum_{\mathbf{k},j>0} c(-j,\mathbf{k})\,\eta(-j,\mathbf{k})\,G[m,-j,\mathbf{k}] \qquad (7.6.65)\\
&= \sum_{\mathbf{k},j>0} c(j,\mathbf{k}) \sum_{\substack{\mathbf{k}'\mathbf{k}''\\ j'j''}} \eta(j,-\mathbf{k})\,D(j,\mathbf{k},j',\mathbf{k}',j'',\mathbf{k}'')\,\{j,-\mathbf{k},j',-\mathbf{k}',j'',-\mathbf{k}''\}.
\end{aligned}$$

If we change the variables by $\{\mathbf{k},\mathbf{k}',\mathbf{k}''\}\to\{-\mathbf{k},-\mathbf{k}',-\mathbf{k}''\}$ and use the relation $D(j,-\mathbf{k},j',-\mathbf{k}',j'',-\mathbf{k}'') = D(j,\mathbf{k},j',\mathbf{k}',j'',\mathbf{k}'')$ as well as $c(j,\mathbf{k}) = c(-\mathbf{k},j)$, then from (7.6.65) it follows with the aid of (7.6.62) that

$$\sum_{\mathbf{k},j<0} c(j,\mathbf{k})\,\eta(j,\mathbf{k})\,G[m,j,\mathbf{k}] = \sum_{\mathbf{k},j>0} c(j,\mathbf{k})\,\eta(j,\mathbf{k})\,G[m,j,\mathbf{k}]. \tag{7.6.66}$$

Thus (7.6.61) may be written in the equivalent form

$$\begin{aligned}
&\sum_{\substack{\mathbf{k}\mathbf{k}'\\ jj'>0}} P(j,\mathbf{k},j',\mathbf{k}')\,\eta(j,\mathbf{k})\,\eta(j',\mathbf{k}')\\
&= -\sum_{\mathbf{k},j\gtrless 0} \frac{1}{2}\,c(j,\mathbf{k})\,\eta(j,\mathbf{k})\,G[m,j,\mathbf{k}].
\end{aligned} \tag{7.6.67}$$

Substituting (7.6.60) into (7.6.67) we see that (7.6.67) is completely symmetric under the summation over all variables. By corresponding changes of the variables it follows that (7.6.67) can be equivalently expressed as

$$\begin{aligned}
&\sum_{\substack{\mathbf{k}\mathbf{k}'\\ jj'>0}} P(j,\mathbf{k},j',\mathbf{k}')\,\eta(j,\mathbf{k})\,\eta(j',\mathbf{k}') =\\
&\frac{2\pi k_B}{3\hbar^2} \sum_{\substack{\mathbf{k},\mathbf{k}',\mathbf{k}''\\ jj'j''\gtrless 0}} \frac{|\Phi(j,\mathbf{k},j',\mathbf{k}',j'',\mathbf{k}'')|^2\,\delta_\gamma[\omega(j,\mathbf{k})+\omega(j',\mathbf{k}')+\omega(j'',\mathbf{k}'')]}{|\sinh x \sinh x' \sinh x''|}\\
&\times\{\eta(j,\mathbf{k})\sinh^2 x + \eta(j',\mathbf{k}')\sinh^2 x' + \eta(j'',\mathbf{k}'')\sinh^2 x''\}^2.
\end{aligned} \tag{7.6.68}$$

If we set $\eta(j,\mathbf{k}) = \omega(j,\mathbf{k})/\sinh^2 x$, the right-hand side of (7.6.68) vanishes (for $\gamma\to 0$) and this form is therefore semidefinite. This is due to energy conservation.
Concerning the symmetry of $P(j,\mathbf{k},j',\mathbf{k}')$, we observe that due to the completely symmetric expression on the right-hand side of (7.6.68), these terms can be rewritten with the factors $\eta^2(j,\mathbf{k})$ and $\eta(j,\mathbf{k})\,\eta(j',\mathbf{k}')$. A further inspection then reveals the symmetry of P, Q.E.D.

It should be noted immediately that this theorem has only a structural meaning. Concerning explicit numerical calculations it is not possible to use the symmetric expression (7.6.68) directly, since in this formula the auxiliary quantities for negative j are involved. Hence, while (7.6.68) is suitable for demonstrating the positive semidefiniteness of P, for any practical calculation one must return to the original formula (7.6.50).
The positive semidefiniteness of P allows the application of two approaches for the solution of equation (7.6.44) which are particularly useful for rough estimates, namely

i) the variational method,
ii) the relaxation time method.

The variational method was developed by Kohler (1948) (1949), Sondheimer (1950), Leibfried and Schlömann (1954), Leibfried (1957), Ziman (1960). The relaxation time method was used by Leibfried (1957), Klemens (1956), and other authors. We first discuss the variational method. We use instead of (7.6.44) the condensed form

$$\sum_j P_{ij}\eta_j = -x_i. \tag{7.6.69}$$

For this equation the following theorem holds

Theorem 7.21: If P is symmetric and positive semidefinite, and if the set $\{\eta_i\}$ is supposed to be a solution of (7.6.69), then for all test-function sets $\{\psi_i\}$ which satisfy the condition $\sum_{ij} P_{ij}\psi_i\psi_j = -\sum_i x_i\psi_i$, the functional

$$L[z] := -\frac{1}{2}\sum_{ij} P_{ij}z_i z_j - \sum_i x_i z_i \tag{7.6.70}$$

possesses for $z=\eta$ a maximum value which is strictly greater than $L[\psi]$, provided the zero solutions are excluded.

Proof: We first show that $L[\eta]$ has an extremum. The conditions are

$$-\frac{\partial L[\eta]}{\partial \eta_i} = \sum_j P_{ij}\eta_j + x_i = 0. \tag{7.6.71}$$

These conditions are satisfied according to the supposition. We now show that $L[\eta]$ has a maximum. According to the assumptions we can conclude

$$\begin{aligned} 0 &< \langle(\eta-\psi), P(\eta-\psi)\rangle \\ &= \langle\eta, P\eta\rangle + \langle\psi, P\psi\rangle - 2\langle\psi, P\eta\rangle \\ &= \langle\eta, P\eta\rangle + \langle\psi, P\psi\rangle + 2\langle\psi, x\rangle \\ &= \langle\eta, P\eta\rangle - \langle\psi, P\psi\rangle \end{aligned} \tag{7.6.72}$$

provided that zero solutions are excluded. According to the assumptions we may further write

$$L[\eta]=\tfrac{1}{2}\langle\eta, P\eta\rangle; \quad L[\psi]=\tfrac{1}{2}\langle\psi, P\psi\rangle. \tag{7.6.73}$$

Then from the last line of (7.6.72) we have

$$L[\eta]>L[\psi]>0 \tag{7.6.74}$$

Q.E.D.

While in principle this method allows an exact solution of the problem, the relaxation time method is only a rough approximation. In this method the nondiagonal parts of P are suppressed, so that equations (7.6.69) can be immediately solved by the expression

$$\eta_i=-\frac{x_i}{P_{ii}}=:-\tau_i x_i \tag{7.6.75}$$

where τ_i are the so-called relaxation times. This name comes from the time-dependent theory where, for instance, in the Pauli Master equation the diagonal term approximation leads to an exponential decay. Also this approach depends on the positive semidefiniteness of P since with this approximation the thermal conductivity tensor becomes

$$\mathbb{A}=\hbar^2(VT^2)^{-1}\sum_{j\mathbf{k}}\omega^2(j,\mathbf{k})\,\tau(j,\mathbf{k})\,\mathbf{v}(j,\mathbf{k})\otimes\mathbf{v}(j,\mathbf{k}). \tag{7.6.76}$$

If the zero solutions which lead to infinite conductivity are excluded, then $\tau(j,\mathbf{k})$ is always >0 and this is the condition for damping processes.
For the explicit numerical evaluation of the set $\{\tau(j,\mathbf{k})\}$ we must return to formula (7.6.50). Due to "momentum"- and energy conservation the sums or integrals, resp. in **k**-space are difficult to evaluate, even if the coupling constants are derived explicitly from a point ion model. In explicit calculations so far only rough estimates have been used with adjusted parameters, etc. Hence we consider qualitatively only two special cases, namely the high and the low temperature limit.
We note that the right-hand side of (7.6.44) reads with (7.6.45)

$$\sum_{j'\mathbf{k}'}P(j,\mathbf{k},j',\mathbf{k}')\eta(j',\mathbf{k}')=\sum_{j'\mathbf{k}'}4k[\sinh x(j.\mathbf{k})]^2G(j,\mathbf{k},j',\mathbf{k}')\eta(j',\mathbf{k}'). \tag{7.6.77}$$

According to Haug (1964) we now substitute

$$\eta(j,\mathbf{k})=\bar{m}_0(j,\mathbf{k})\,[\bar{m}_0(j,\mathbf{k})+1]\,\sigma(j,\mathbf{k}) \tag{7.6.78}$$

and evaluate the bracket factors of (7.6.50). Using the abbreviations $\bar{m}(j,\mathbf{k})=m$, $\bar{m}(j',\mathbf{k}')=m'$, $\bar{m}(j'',\mathbf{k}'')=m''$, etc., this yields

$$\begin{aligned} &mm'(1+m'')-(1+m)(1+m')m'' \\ &\quad=m_0m_0'(1+m_0'')[1+\sigma(m_0+1)][1+\sigma'(m_0'+1)][1+\sigma''m_0''] \\ &\quad\quad-(1+m_0)(1+m_0')m_0''[1+\sigma m_0][1+\sigma' m_0'][1+\sigma''(m_0''+1)]. \end{aligned} \tag{7.6.79}$$

By direct calculation it can be verified that for thermal equilibrium

$$\{m_0 m_0'(1+m_0'')-(1+m_0)(1+m_0')m_0''\}\delta_\gamma(-\omega-\omega'+\omega'')=0 \tag{7.6.80}$$

holds. Hence in (7.6.79) both factors are equal and we obtain up to a linear approximation

$$\begin{aligned} mm'(1+m'')-(1+m)(1+m')m'' \\ &= m_0 m_0'(m_0''+1)[\sigma+\sigma'-\sigma'') \\ &= \frac{m_0'(m_0''+1)}{(m_0+1)}\eta+\frac{m_0(m_0''+1)}{(m_0'+1)}\eta'-\frac{m_0 m_0'}{m_0''}\eta''. \end{aligned} \tag{7.6.81}$$

By the same procedure the expansion

$$\begin{aligned} m(1+m')(1+m'')-(1+m)m'm'' \\ = \frac{(1+m_0')(1+m_0'')}{(m_0+1)}\eta-\frac{m_0(m_0''+1)}{m_0'}\eta'-\frac{m_0(m_0'+1)}{m_0''}\eta \end{aligned} \tag{7.6.82}$$

follows. With $(2\sinh x)^{-2}=m_0(m_0+1)$, the diagonal part of P therefore takes the form

$$\begin{aligned} P(j,\mathbf{k},j,\mathbf{k}) = \sum_{\substack{\mathbf{k}'\mathbf{k}'' \\ j'j''}} |\Phi(j,\mathbf{k},j',\mathbf{k}',j'',\mathbf{k}'')|^2\,\delta_\gamma(-\omega-\omega'+\omega'')\,\frac{m_0'(m_0''+1)}{m_0(m_0+1)(m_0+1)} \\ + \sum_{\substack{\mathbf{k}'\mathbf{k}'' \\ j'j''}} |\Phi(j,\mathbf{k},j',\mathbf{k}',j'',\mathbf{k}'')|^2\,\delta_\gamma(-\omega+\omega'+\omega'')\,\frac{(m_0'+1)(m_0''+1)}{m_0(m_0+1)(m_0+1)}. \end{aligned} \tag{7.6.83}$$

For large T we expand the exponential in (7.6.37) in a power series, and obtain

$$\lim_{T\to\infty} m_0 = \lim_{T\to\infty} \frac{k_B T}{\hbar\omega}\left[1+\sum_\nu \frac{1}{(\nu+1)!}\left(\frac{\hbar\omega}{k_B T}\right)^\nu\right]^{-1} \approx \frac{k_B T}{\hbar\omega}. \tag{7.6.84}$$

Hence from (7.6.83) it follows that

$$\lim_{T\to\infty} P(j,\mathbf{k},j,\mathbf{k}) = \lim_{T\to\infty} \tau(j,\mathbf{k})^{-1} \sim (k_B T)^{-1} \tag{7.6.85}$$

and therefore in the high temperature limit

$$\lim_{T\to\infty} \Lambda \sim T^{-1} \tag{7.6.86}$$

holds. For very low T (7.6.37) leads to

$$\lim_{T\to 0} m_0 = \exp\,[-\hbar\omega/k_B T]. \tag{7.6.87}$$

With this expression we obtain for small T

$$\lim_{T\to\infty} G(j,\mathbf{k},j,\mathbf{k}) = \sum_{\substack{\mathbf{k}'\mathbf{k}'' \\ j'j''}} |\Phi(j,\mathbf{k},j',\mathbf{k}',j'',\mathbf{k}'')|^2\,\delta_\gamma(-\omega+\omega'+\omega'') \tag{7.6.88}$$

and therefore, if we replace $m_0(m_0+1)$ by its original definition, then

$$\lim_{T\to 0} \mathbb{\Lambda} \sim \sum_{\mathbf{k}j} \exp\,[\hbar\omega(j,\mathbf{k})/k_B T]\,\{\exp\,[\hbar\omega(j,\mathbf{k})/k_B T]-1\}^2 \times \frac{\hbar\omega^2\,(j,\mathbf{k})}{VT^2} G^{-1}. \tag{7.6.89}$$

If we suppress the factor G^{-1}, this expression coincides with the expression for the specific heat. The low temperature behavior of the specific heat is $\sim T^3$ and thus

$$\lim_{T\to\infty} \mathbb{\Lambda} \sim T^3. \tag{7.6.90}$$

Since the **k**-summation plays an essential role, this estimate is very rough. However, a more profound analysis shows that the low temperature behaviour must be essentially influenced by scattering of phonons at the boundaries, cf. Casimir (1938).

In the literature special attention is paid to the distinction between normal processes and Umklapp-processes. Their role can be illustrated if we use the idealizations concerning the frequency distribution of Theorem 2.8. The same assumptions are often made for calculations of thermal conductivity, cf. Klemens (1958), Carruthers (1961). In this procedure we have $\omega(j,\mathbf{k})=\omega_0$ for the optical branches and $\omega(j,\mathbf{k})=v_j|\mathbf{k}|$ for the acoustic branches where additionally $v_j\equiv v$ is assumed. Then from (7.6.30) it follows that the optical branches do not contribute to the thermal conductivity in this approximation. Since for the acoustic branches we obtain $\nabla\omega(j,\mathbf{k})=v^2\mathbf{k}/\omega(j,\mathbf{k})$, the heat current (7.6.35) reads

$$\bar{\mathbf{q}}(\mathbf{r})=V^{-1}\sum_{\mathbf{k}j}\hbar v^2\bar{m}(j,\mathbf{k})\mathbf{k}. \tag{7.6.91}$$

An elementary anharmonic phonon interaction leads, for instance, from $m(j,\mathbf{k})$, $m(j',\mathbf{k}')$, $m(j'',\mathbf{k}'')$ to $m(j,\mathbf{k})+1$, $m(j',\mathbf{k}')+1$, $m(j'',\mathbf{k}'')-1$. If we consider the instantaneous heat current $\mathbf{q}(\mathbf{r})$, then, after the collision, the difference between the initial heat current $\mathbf{q}_i(\mathbf{r})$ and the final heat current $\mathbf{q}_f(\mathbf{r})$ is

$$\mathbf{q}_f(\mathbf{r})-\mathbf{q}_i(\mathbf{r})=\Delta\mathbf{q}(\mathbf{r})=V^{-1}\hbar v^2\,[\mathbf{k}+\mathbf{k}'-\mathbf{k}'']. \tag{7.6.92}$$

If now the process is a normal process, then $\mathbf{k}+\mathbf{k}'+\mathbf{k}''=0$, while for an Umklapp-process this value is $\mathbf{g}$. Hence an elementary normal collision process does not contribute to the instantaneous conductivity, and as this holds for any normal process, these processes do not contribute to the average conductivity. It should, however, be emphasized that this is only true for the idealized assumptions about the spectrum. These assumptions are only justified for low temperatures where only those long-wave length phonons are in interaction which show the assumed dispersion law. For higher temperature ranges no general statements can be made, as all results depend on the spectrum. If an imperfect crystal is considered, new scattering effects arise. There are two approaches:

i) The imperfections are considered to be only static perturbations of the ideal crystal structure, where the ideal phonons are scattered;
ii) the phonon modes of the imperfect crystal are considered to be the basic quantities, and anharmonic interactions between these modes are taken into account.

The first approach was initiated by Klemens (1951) (1955) and the second approach by Wagner (1963). In the former approach perturbation energies occur which are quadratic in the ion displacements, while in the latter approach the lowest order perturbation is still connected with third order powers of the displacements. Of course, the second approach is the more systematic one from our point of view. It is, however, more difficult for numerical evaluation since the spectrum and the eigenmodes of the imperfect crystal are required for the calculation. In any case, there appear processes which are different from those appearing in the ideal crystal. In the lowest order approximation of the Pauli Master equation these processes are then again additive, so that we obtain a stationary Boltzmann equation with several collision term operators $\{P_\alpha\}$. Thus in its generalized form this equation reads

$$\left(\sum_\alpha P_\alpha\right)\eta = x. \tag{7.6.93}$$

According to the definition of the relaxation time $\tau_\alpha := P_\alpha^{-1}$, the total relaxation time τ for equation (7.6.93) is given by

$$\tau^{-1} = \sum_\alpha \tau_\alpha^{-1}. \tag{7.6.94}$$

Callaway (1959) showed that with the assumptions of Theorem 2.8 about the phonon spectrum, which were mentioned above, a relatively simple calculation of the thermal conductivity can be performed, if the relaxation times are given in terms of frequencies and temperature. Such calculations with appropriate simplifications have been made for normal- and Umklapp-processes, cf. Klemens (1958), Peierls (1955), with the result

$$\tau_n^{-1} \sim N_s \omega^s T^{5-s} \tag{7.6.95}$$

and

$$\tau_u^{-1} \sim U_s \exp(-\Theta/\alpha T)\omega^s T^{5-s} \tag{7.6.96}$$

where Θ is the Debye temperature of the acoustic branches and s depends on the crystal symmetry and phonon polarization. A typical value of s is $s=2$. These processes determine the conductivity of the ideal crystal for mean and high temperatures. In the non-ideal crystal we have boundary scattering, Casimir (1938), which leads to the relaxation time

$$\tau_b^{-1} \approx v/L \tag{7.6.97}$$

and which is important for low temperatures. Further there occurs the isotopic scattering due to mass defects with

$$\tau_i^{-1} \sim I\omega^4 \tag{7.6.98}$$

which was first mentioned by Pomeranchuk (1942) and calculated by Klemens (1955); further there is scattering at imperfections with

$$\tau_P^{-1} \sim P\omega^4 \tag{7.6.99}$$

as calculated by Klemens (1955) (1951) and scattering at dislocations

$$\tau_d^{-1} \sim D\omega \tag{7.6.100}$$

as treated by Klemens (1955), Bross (1962). Using these assumptions Callaway calculated the thermal conductivity of Ge with and without impurities and obtained good results.

Benin (1970) improved the variational method for the Boltzmann equation in order to find appropriate bounds for the transport coefficients and applied this to polar crystals. He calculated the matrix elements of the perturbation directly with adjustable parameters and applied this technique to the case of ideal (highly purified) LiF and NaF crystals, Benin (1971). The phonon distribution trial function was chosen to be a displaced Planck distribution plus another term reducing the deviation from thermal equilibrium for high frequency phonons. An isotropic Debye approximation for the phonon spectra of LiF and NaF, i.e., the model discussed above, gives a good fit to the conductivity data, with only two semiadjustable parameters (Grüneisen-constant and a zone-edge longitudinal phonon frequency) for the anharmonic contributions. His results are given in the Figures 7.7a and 7.7b

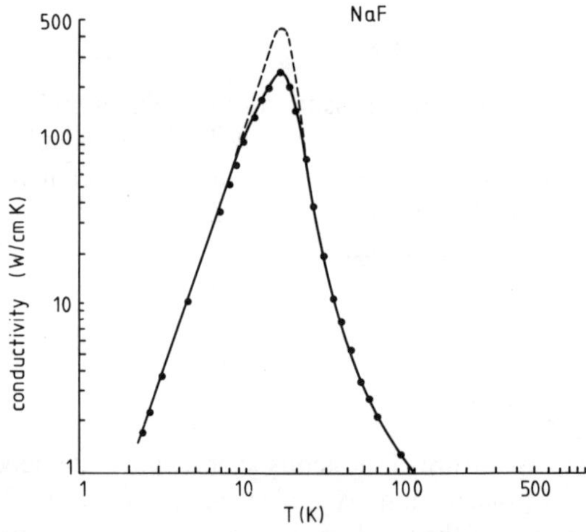

Figure 7.7.a:
Thermal conductivity of NaF. Circles: experimental values after Jackson and Walker (1971). Lines: theoretical values after Benin (1971).

For all calculations mentioned so far the influence of the resonance function δ_γ was idealized to a strict energy conservation, i.e., $\delta_\gamma \sim \delta$. Pohl (1962) discovered that for impurity scattering a relaxation time is given by the formula

$$\tau^{-1} \sim \frac{A\omega^2}{(\omega_0^2 - \omega^2)^2 + (\Lambda/\pi)^2 \omega_0^2 \omega^2}. \tag{7.6.101}$$

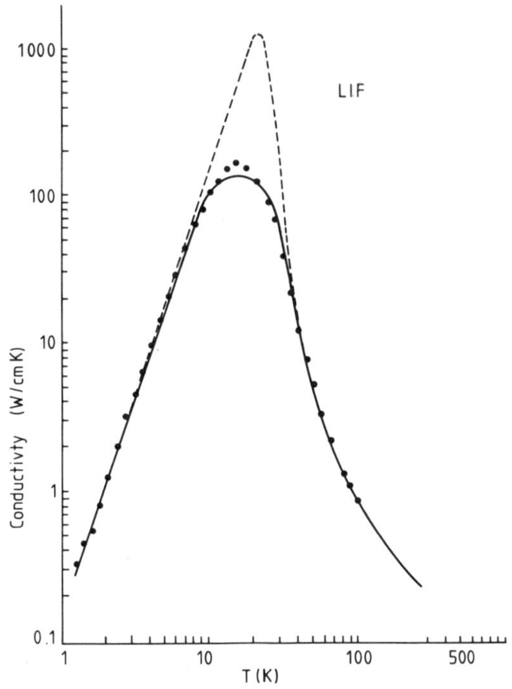

Figure 7.7.b:
Thermal conductivity of LiF. Circles: experimental values after Thacher (1967). Lines: theoretical values after Benin (1971).

He measured the thermal conductivities as functions of impurity concentration, see Fig. 7.8, and found a striking variation for the various doping concentrations. It is obvious that (7.6.101) is a consequence of the resonance function δ_γ. Wagner (1963) developed the modification for resonance transitions which includes the modification for resonance transitions between local and non-local modes.
Concerning the literature, the early attempts to derive thermal conductivity were already cited in the text. Additional information and references are given in the review articles of Klemens (1956) (1958) and Carruthers (1961). Since the appearence of these review articles, the following topics with respect to semiconductors and insulators have been treated: Inclusion of four-phonon processes in the calculation of the relaxation time, Carruthers (1962). Modification of three-phonon processes by isotopic scattering, Carruthers (1962). Influence of localized modes on thermal conductivity, Wagner (1963). Phonon scattering by lattice defects with L.G.F.T. and its connection with thermal conductivity via relaxation times, Klein (1963) (1966). Derivation of the energy flux operator for a three-dimensional lattice, Hardy (1963). Resonance scattering of phonons by molecular impurity centers, Wagner (1964). Derivation of the Boltzmann equation in a phonon system by means of field theoretic Green functions, Horie and Krumhansl (1964). Calculation of the thermal resistivity due to phonon scattering by vacancies and vacancy pairs based on perturbation theory and the

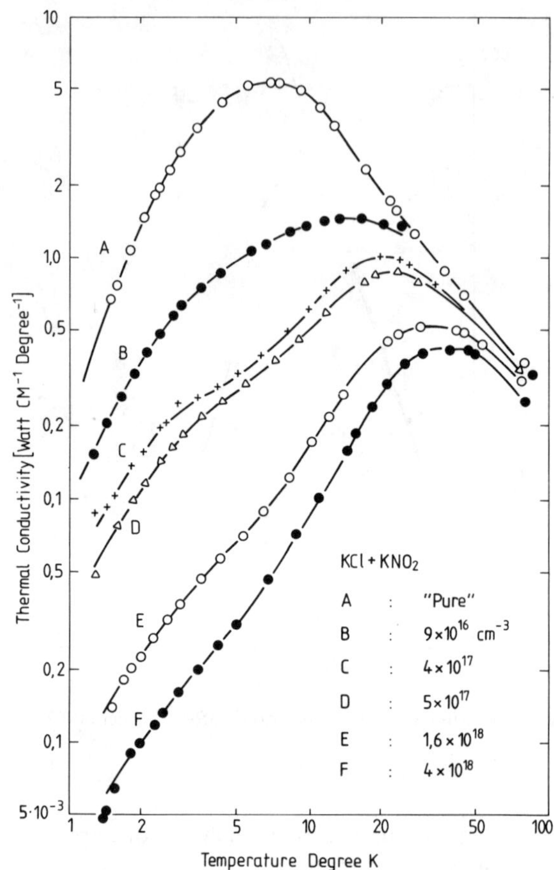

Figure 7.8:
Thermal conductivity of KCl doped with different concentrations of KNO_2 as an example of the influence of inelastic phonon scattering at localized modes, with measurements by Pohl (1962); after Wagner (1963).

variational solution of the Peierls-Boltzmann equation, Gallina and Omini (1964). Scattering of phonons by point defects, McCombie and Slater (1964). Three-phonon processes in an electron-phonon system which predominate over the three-phonon processes due to anharmonic interactions, Mikoshiba (1965). Calculation of relaxation times of phonon scattering due to the interaction with polarization of the electron-hole vacuum, Uritsky and Novikov (1965). Third-order anharmonic interactions of infrared phonons in alkali halides and calculation of relaxation times, Vredevoe (1965). Lowest order contribution to the lattice thermal conductivity with transport equations equivalent to the Boltzmann equation, describing both anharmonic forces and lattice imperfections, Hardy (1965). Scattering of long wave phonons by point imperfections in crystals, Krumhansl and Matthew (1965). Concise formulation of the formulas for lattice thermal conductivity by means of response functions, Hardy (1963) (1966). Variational estimates of the scattering of phonons from static lattice imperfections, Morgan (1966). Review of the theories of thermal conductivity, Tavernier (1967). Treatment of heat transfer in crystals of finite dimensions with relaxation due to phonon-impurity scattering, Kazakov and Nagaev (1967). Derivation of a frequency- and wavelength dependent Peierls-Boltzmann equation by means of field theoretic Green functions techniques for an interacting phonon system and analysis of propagation of temperature waves (second sound), Sham (1967). Scattering of phonons by U-centers in alkali halides and calculation of the corresponding relaxation times, Radosevich and Walker (1968). Derivation of a generalized Peierls-Boltzmann equation by means of linear response function theory and field theoretic evaluation of the corresponding Green functions, Ranninger (1967) (1968). Derivation of transport equations for phonon systems with anharmonic interactions based on field theoretic Green functions techniques, Klein and Wehner (1968) and Niklasson and Sjölander (1968). Umklapp processes during spontaneous decay of phonons, Novikov (1969). Treatment of impurity induced phonon scattering resonances in thermal conductivity by calculation of the relaxation rates with L.G.F.T., Klein (1969). Resonant scattering of phonons in CN-doped alkali halides and calculation of relaxation times, Kumar, Srivastava and Verma (1969). Variational solutions of the Peierls-Boltzmann equation applied to the thermal conductivity in Ge, Hamilton and Parrott (1969). Study of the influence of the Umklapp processes on the hydrodynamic motions of phonons in crystals by means of the Peierls-Boltzmann equation, Thellung and Weiss (1969). Selection rules for anharmonic interactions of nearest neighbors in crystals with zincblende or diamond structure, Nedoluha (1970). Discussion of a possible Bose condensation of optical modes in polar semiconductors based on an analysis of phonon transport equations, Paranjape and Krishnamurthy (1970). Improved variational principles for transport coefficients, Benin (1970). Theory of transport properties of anharmonic crystals based on the evaluation of a generalized Boltzmann equation by Niklasson and Sjölander's method, Niklasson (1970). Comments on Klein's theory of the thermal conductivity of doped alkali halides, Singh and Verma (1971). Temperature dependence of the bounds on thermal resistance due to U-processes in the variational treatment of the Peierls-Boltzmann equation, Srivastava and Verma (1971). Thermal conductivity of LiF and NaF and the Ziman limit by means of the variational method, Benin (1971). Three-phonon scattering strengths and the Ziman limit of resistivity in Ge, Srivastava, Singh and Verma (1972). Thermal conductivity derived from linear response functions theory by means of field theoretic Green functions, Conan (1972). Role of three-phonon normal processes in the phonon thermal conductivity of an insulator, Dubey and Verma (1972). Theory of lattice thermal conductivity based on the derivation of relaxation times, Hamilton (1973). Variational calculation of the three-phonon Umklapp resistivity, Singh and Verma (1973). Theory of thermal conductivity in anharmonic crystals with response function technique and field theoretic Green functions, Wilson and Kim (1973). Description of transport phenomena in dielectric crystals by evaluation of the statistical operator, Sergeev and Pokrovsky (1973). Effect of point imperfections on lattice thermal conductivity of an insulator at high temperatures by an analysis of corresponding relaxation times, Dubey (1974). Theory of lattice thermal conductivity of anharmonic crystals with response function technique and field theoretic Green functions, Altukhov (1974). A modification of the generalized Callaway thermal conductivity equation to allow for phonon dispersion, Dubey (1974). Resonant scattering of phonons by iron impurities in ZnS and calculation of the corresponding relaxation times, Srivastava and Verma (1974). Derivation of a collision operator for phonon Umklapp processes based on the linearized Peierls-Boltzmann equation, Simons (1975). Thermal conductivity for phonon scattering by substitutional defects in crystals using field theoretic double time thermal Green functions and the Kubo formula, Sharma and Bahadur (1975). Theory of phonon relaxation due to multilevel impurities by thermal Green functions, Joshi (1975), Joshi and Singh (1976). Resonance scattering of phonons by interstitial impurity atoms using field theoretical double time Green function technique, Ohashi and Ohashi (1976). Derivation and calculation of a sequence of lower bound results for lattice thermal conductivity by the variational method, Srivastava (1976). Influence of the electron-phonon scattering on the scattering of phonons by phonon in heavily doped semiconductors calculated by relaxation times, Korzhevykh (1976). Note on photon thermal conductivity of polar semiconductors at high temperatures, Dzhaksimov (1976). Note on three-phonon scattering strengths for semiconductors and alkali halides, Singh (1976). Analytical expression of phonon-phonon scattering strength and corresponding relaxation times and its application to Ge and KCl in the low and high temperature regions, Singh and Verma (1977). Derivation of a non-linear Peierls-Boltzmann equation for phonons and investigation of transport phase transitions in phonon systems and of structure and stability of corresponding solutions, Kaiser and Wagner (1977), Kaiser (1977). Phonon scattering by paramagnetic ions using field theoretical Green function technique, Care and Tucker (1976) (1977). Thermal resistivity of dielectric crystals due to four-phonon processes and optical modes by calculation of relaxation times, Ecsedy and Klemens (1977). Contribution to the lattice thermal conductivity due to the correction term in the Callaway integral, Dubey (1977). Role of point-defect scattering in the lattice thermal conductivity of an insulator at low temperatures, Dubey (1977). Resonance scattering of phonons by paramagnetic ions using field theoretic Green function technique, Altukhov and Zavt (1977). Three-phonon relaxation rate and phonon conductivity, Dubey and Misho (1977), applied to InSb and GaAs, Al-Edani and Dubey (1978). Lattice thermal resistivity due to the presence of electrons, Kassim and Dubey (1978). Boundary scattering and phonon conductivity with application to GaAs, Al-Edani and Dubey (1978). Relaxation times for resonant scattering of phonons by bound holes and application to the calculation of thermal conductivity for Mn-doped GaAs, Singh (1978). Derivation of generalized Peierls-Boltzmann equations for phonons in interaction with local systems with excitations by using a generalized Langevin equation, Michel and Wagner (1978).

b) Anharmonic crystals

The calculations of phonon decay in Section 2.6 as well as of thermal phonon conductivity in Section 7.6 a) show that the lowest order transition probabilities of the third order anharmonic terms do not suffice for a proper explanation of these phenomena. Rather, we had, at least partially, to use resonance transition probabilities. Such probabilities are beyond a perturbation theoretical treatment and produce energy corrections which, in addition to the decay rates connected with the imaginary parts, contain also level shifts as can be seen from Theorem 1.18. In a field theoretic treatment such corrections formally correspond to an infinite series of perturbation theoretical diagramms, i.e., to a summation of perturbation terms of all orders. Corrections of this kind are observable: for instance, the decay rates lead to finite linebreadths of the electromagnetic phonon absorption and emission lines, while the level shifts lead to energy corrections of the absorbed and emitted frequency spectrum. These optical effects are only one example of the striking influence of higher order anharmonic interactions on the physical behaviour of crystals. Phase transitions, thermal expansion, elastic and dielectric properties, thermodynamic functions, etc., also depend strongly on such higher order anharmonic interactions. Thus in the last two decades the theory of anharmonic phonon-phonon interactions was further developed in order to include more systematically the effects of the higher order contributions of the anharmonic terms to the transition probabilities etc. The development of this theory was mainly performed by field theoretical methods, in particular by field theoretical Green functions. Since we do not treat such techniques in this book we give only some references in this field.

In addition to thermal conductivity, early approaches to the incorporation of anharmonic terms in crystal theory were concerned with the calculation of thermodynamic state functions based on their representations in statistical mechanics, and with the effect of the anharmonic potential on dispersion in the infrared region. These methods are described in the book by Born and Huang (1954) where also references concerning preceding work are given. The review article of Leibfried (1957) of the mechanical and thermal properties of crystals contains, besides thermal conductivity theory, also the calculation of thermodynamic state functions with inclusion of anharmonic terms. Fletcher (1959) (1961) investigated the thermal expansion of solids, in particular of NaCl, and noticed that anharmonic terms are indispensible for a proper description of this effect. He took into account such terms by a perturbation calculation. In a review article, Leibfried and Ludwig (1961) gave an extensive discussion of the derivation of thermodynamic state functions from anharmonic lattice theory and applied perturbation theory to include the anharmonic terms. Wallis and Maradudin (1962) improved infrared dispersion theory by applying the Kubo formula for the representation of the dielectric susceptibility and by direct integration of the time-dependent Schrödinger equation with anharmonic phonon interactions in order to obtain damping solutions. Maradudin (1962) studied the connection between thermal expansion and phonon frequency shifts for anharmonic crystals. Hanamura and Inui (1963) investigated the line broadening of infrared absorption lines due to localized modes of lattice vibrations of U-centers by applying the Kubo formula and performing a damping calculation with anharmonic terms. Wallace (1963) derived the anharmonic free energy of crystals at high temperatures for crystal Hamiltonians containing anharmonicities up to fourth order terms by applying perturbation theory. The same topic was treated along similar lines by Maradudin, Flim and Coldwell-Horsfall (1961) and Flinn and Maradudin (1963). Cowley (1963) gave an extensive treatment of the thermal, electric, and mechanical properties of anharmonic crystals. He used the potential energy of the crystal in the adiabatic coupling scheme and applied field theoretical thermodynamic Green functions, calculated by an appropriate perturbation theory, for the evaluation of the interesting physical quantities. Similarly, Cowley and Cowley (1965) (1966) investigated the effects of anharmonic interactions in alkali halides. Wallace (1965) applied the results of his previous work to calculate the thermal expansion and other anharmonic properties of crystals. Pathak (1965) used thermodynamic Green functions to formulate a theory of anharmonic crystals. Ludwig (1967) published a review of developments in lattice theory which also contained the discussion of anharmonic effetcs. Choquard (1967) treated the theory of anharmonic crystals in a book and applied this theory to the discussion of thermal properties and the problem of dynamical stability. Götze and Michel (1967) developed a Green functions approach to derive in lowest order of the anharmonic interaction transport equations for a pure Bravais lattice. Barron and Batana (1968) discussed the thermal expansion of alkali halides at low temperatures. Srivastava, Sharma and Madan (1968) calculated anharmonic terms of alkali halides and made an estimate of their influence on lattice modes. Werthamer (1969) gave a review of the theory of quantum crystals with large amplitude and highly anharmonic motion, which is based on the use of field theoretic Green functions technique. Klein and Wehner (1969) derived transport equations for anharmonic lattices by means of Green functions technique and obtained equations which are analogous to the Peierls-Boltzmann equation. Krivoglaz and Pinkevich (1969) studied the influence of anharmonicity on the spectral distribution of quasilocal modes. Niklasson (1970) developed a theory of transport properties of anharmonic crystals based on field theoretic Green functions techniques. Ida (1970) discussed the anharmonic contributions to the heat capacity of solids up to the critical temperature of lattice instability. Hardy and Karo (1970) extended the deformation dipole model for alkali halides to include anharmonic effects and calculated the coefficient of thermal expansion for KBr. Werthamer (1970) derived a selfconsistent phonon theory of anharmonic lattice dynamics. Ruvalds and Zawadowski (1970) studied two-phonon resonances due to anharmonic phonon-phonon interaction based on an analysis of field theoretic Green functions. Benedek (1971) calculated the selfenergy shift of a local phonon mode of Li^+ in KCl due to anharmonic interactions and the influence of the lattice potential allowing for tunneling. Laplaze, Vergnoux and Benoit (1971) published a review of the effects of anharmonicities in non-conducting crystals with a comprehensive list of references. Siklos (1971) treated the theory of anharmonic crystals in pseudoharmonic approximation. Cowley (1971) calculated the anharmonic contributions to the thermodynamic properties of NaCl using the breathing shell model. Glyde and Klein (1971) gave a review of anharmonic effects in the lattice dynamics of insulators based on the application of field theoretic Green functions techniques. Paul and Takeno (1972) developed a theory of the anharmonic vibrational properties of impurities in alkali halide crystals by performing a damping calculation. Siklos and Aksienov (1972) investigated the thermodynamics of strongly anharmonic crystals by means of field theoretic Green functions. Aggarwal and Pathak (1973) evaluated the free energy of an anharmonic crystal using perturbation theory for the inclusion of anharmonicities.

Pandey and Dayal (1973) studied the thermal expansion of RbCl by the shell model. Hardy and Karo (1973) also studied various anharmonic models and their influence on thermal expansion of KBr. Barron and Klein (1974) published a review article on perturbation theory performed with field theoretic Green functions calculus for anharmonic crystals and on its application to mechanical, thermal and electric properties. Horner (1974) treated in a review article the case of strongly anharmonic crystals with hard core interaction by means of field theoretic Green function technique, while Götze and Michel (1974) gave a review of selfconsistent phonon calculations in anharmonic crystals. These reviews contain numerous references in this field. Shukla and Wilk (1974) calculated the Helmholtz free energy of an anharmonic crystal by perturbation theory in higher orders. Rastogi, Hawranek, and Lowndes (1974) calculated the anharmonic selfenergies and relaxation rates of phonons for the lithium halides. Altukhov and Zavt (1974) studied the interference of the anharmonic and impurity scattering of phonons in a theory of lattice thermal conductivity of anharmonic crystals. Fischer (1974) investigated the lattice dynamics and anharmonic effects in AgCl. Tripathi and Pathak (1974) derived the selfenergy of phonons in an anharmonic crystal by field theoretic Green functions technique valid for all temperatures. Knauss and Wilson (1974) extended the Wilson-Kim theory of thermal conductivity of anharmonic crystals to include the nondiagonal Peierls contribution. Cowley, Jacucci, Klein and McDonald (1976) analyzed anharmonic effects in the phonon spectra of NaCl in terms of perturbation theory.

c) Infrared absorption

Infared absorption arises by conversion from photons into phonons and vice versa. Owing to energy conservation, for such processes the photons to be absorbed or emitted are in the infrared region. To calculate the absorption rate, various formulae have been derived by various methods. Since we are interested in deducing all results from the basic formalism of irreversible rate equations, we derive the expressions for infrared absorption directly from the basic formula (3.7.27) which describes the absorption or emission rate of radiation by a crystal. In order to obtain simple formulae, some simplifying assumptions have to be made which are physically reasonable. It would be of interest to study the absorption rate without these assumptions but no results are available at present.
We first assume that in the infrared region no direct electronic excitations are induced by the photon field. According to our discussion in section 3.7 this means that no Stokes shifts occur and therefore no differences arise between absorption and emission energies. Hence the full balance (3.7.27) has to be used. We write it with $\Delta B_j(\mathbf{k}) := B_j^-(\mathbf{k}) - B_j^+(\mathbf{k})$ in the form

$$\Delta B(j,\mathbf{k}) = W^-(j,\mathbf{k},\Delta E,s)\, b(j,\mathbf{k}) - W^+(j,\mathbf{k},\Delta E,s)\ [b(j,\mathbf{k})+1] \tag{7.6.102}$$

where $B_j^-(\mathbf{k})$ is the compensating emitter and $B_j^+(\mathbf{k})$ the compensating absorber. If $\Delta B_j(\mathbf{k}) > 0$ holds the compensating emission is larger than the compensating absorption. This means that absorption takes place. Conversely, if $\Delta B_j(\mathbf{k}) < 0$ the physical system emits radiation.

Theorem 7.22: Suppose that
i) no direct electron excitations by photon processes occur,
ii) the differences of the phonon modes and frequencies in dependence on n can be neglected,
iii) the photon field mean value of the occupation number is far over the thermal equilibrium value of the phonon field mean values (incident radiation).
Then the infrared absorption rate reads

$$\Delta B(\lambda,\mathbf{k})=b(\lambda,\mathbf{k})4\pi\,\frac{\omega(\lambda,\mathbf{k})}{V}$$
$$\times\,\mathrm{Im}\sum_{jj'}\zeta_j(\lambda,\mathbf{k})\cdot\mathscr{G}_{jj'}(z)\cdot\zeta_j(\lambda,\mathbf{k})^{\times}_{/z=\omega(\lambda,\mathbf{k})^2+i\gamma/\hbar} \tag{7.6.103}$$

where $\mathscr{G}_{jj'}(z)$ is the classical lattice Green function and $\zeta_j(\lambda,\mathbf{k})$ are the transverse optical modes of the ideal lattice with $\{j=k\varrho=\text{lattice point numbers}\}$.

Proof: If no direct electronic excitations occur, (3.7.2) reads

$$W^{\mp}=\frac{2\pi}{\hbar}\sum_{mm'}\sum_{n}|\langle nm|\mathfrak{P}^{\mp}(\lambda,\mathbf{k})|nm'\rangle|^2 f_{m'}p_n\delta_\gamma[E^n_m-E^n_{m'}\pm\hbar\omega(\lambda,\mathbf{k})] \tag{7.6.104}$$

with $\mathfrak{P}^{\mp}(\lambda,\mathbf{k})$ given by (3.7.6). We now assume that the one-electron states can be attributed to the various ions and introduce the denumeration $\{i\equiv j,\alpha_j\}$. Then $\mathfrak{P}^{\mp}(\lambda,\mathbf{k})$ can be equivalently written

$$\mathfrak{P}^{\mp}(\lambda,\mathbf{k})=-i\frac{\hbar}{c}\sum_j\{M_j^{-1}\nabla_j e_j+\sum_{\alpha_j}m^{-1}\nabla_{j\alpha_j}e\,\exp\,[\mp i\mathbf{k}\cdot(\mathbf{r}_{j\alpha_j}-\mathbf{R}_j)]\}$$
$$\times\exp\,(\mp i\mathbf{k}\cdot\mathbf{R}_j)\mathbf{e}(\lambda,\mathbf{k})\left(\frac{2\pi c\hbar}{|\mathbf{k}|V}\right)^{1/2}. \tag{7.6.105}$$

The phonon frequencies at which substantial absorption takes place are of order of magnitude $10^{13}\,\mathrm{s}^{-1}$. Due to energy conservation, the photon frequencies must be of the same magnitude. From $|\mathbf{k}|=\omega(\lambda,\mathbf{k})c^{-1}$ it follows that the photon wave vector is of order $|\mathbf{k}|\approx 3\cdot 10^2\,\mathrm{cm}^{-1}$. Now the differences $(\mathbf{r}_{j\alpha_j}-\mathbf{R}_j)$ and $(\mathbf{R}_j-\mathbf{R}_j^0)$ are of order of magnitude 10^{-8} cm. Hence to a good approximation we may assume $\mathbf{k}\cdot(\mathbf{r}_{j\alpha_j}-\mathbf{R}_j)\approx 0$ and $\mathbf{k}\cdot(\mathbf{R}_j-\mathbf{R}_j^0)\approx 0$. Then (7.6.105) reads

$$\mathfrak{P}^{\mp}(\lambda,\mathbf{k})\approx -i\frac{\hbar}{c}\sum_j[M_j^{-1}\nabla_j e_j+\sum_{\alpha_j}m^{-1}\nabla_{j\alpha_j}e]$$
$$\times\exp\,(\mp i\mathbf{k}\cdot\mathbf{R}_j^0)\mathbf{e}(\lambda,\mathbf{k})\left(\frac{2\pi c\hbar}{kV}\right)^{1/2}. \tag{7.6.106}$$

According to (3.2.7), the Heisenberg representation of the momentum operators has to be used for the further evaluation of (7.6.103) with (7.6.106). Since the adiabatic states $|nm\rangle$ do not satisfy a Schrödinger equation no Heisenberg representation can be used for them. It is rather necessary to consider the $|n\rangle$ states and the $|m\rangle$ states separately as, due to the adiabatic coupling, these states separately satisfy Schrödinger equations and hence a corresponding Heisenberg representation exists only for them. From (3.2.7) it then follows that

$$\langle n|\mathbf{p}_i|n\rangle = \langle n|m\dot{\mathbf{r}}|n\rangle = (E^n - E^n)\langle n|m\mathbf{r}|n\rangle\hbar^{-1} = 0. \tag{7.6.107}$$

Further, if we consider only bound electron states (which includes also conduction band states in the crystal!), these states can be represented by real wave functions. For such functions $|n\rangle \equiv \chi(\mathbf{r},\mathbf{R})$ we have

$$\langle n|\nabla_j|n\rangle|m\rangle = \langle n|n\rangle\nabla_j|m\rangle = \nabla_j|m\rangle \tag{7.6.108}$$

since $2\langle n|\nabla_j|n\rangle = \nabla_j\langle n|n\rangle = 0$ owing to $\langle n|n\rangle = 1$. Hence (7.6.104) can be reduced to

$$W^{\mp} = \frac{2\pi}{\hbar}\sum_{mm'}|\langle m|\sum_j \frac{e_j}{c} M_j^{-1}\mathbf{P}_j\cdot\mathbf{e}(\lambda,\mathbf{k})\exp(\mp i\mathbf{k}\cdot\mathbf{R}_j^0)|m'\rangle|^2 f_{m'}\left(\frac{2\pi c\hbar}{kV}\right)$$
$$\times\sum_n \langle n|n\rangle p_n\delta_\gamma[E_m^n - E_{m'}^n \pm \hbar\omega(\lambda,\mathbf{k})]. \tag{7.6.109}$$

The energies E_m^n, and $E_{m'}^n$, are given by (3.4.5). We now assume that the phonon parts of these energies ε_m^n and $\varepsilon_{m'}^n$ do not strongly depend on n. If we make an approximation and neglect this dependence, then the summation over n can be performed. Then with (3.2.7) applied to the ion momentum and with (2.2.14) we obtain from (7.6.109)

$$W^{\mp} = \frac{2\pi}{\hbar}\sum_{mm'}\omega_{mm'}^2|\langle m|\sum_j \frac{e_j}{c}\mathbf{u}_j\cdot\mathbf{e}(\lambda,\mathbf{k})\exp(\mp i\mathbf{k}\cdot\mathbf{R}_j^0)|m'\rangle|^2 f_{m'}$$
$$\times\left(\frac{2\pi c\hbar}{kV}\right)\delta_\gamma[\varepsilon_m - \varepsilon_{m'} \pm \hbar\omega(\lambda,\mathbf{k})] \tag{7.6.110}$$

if we observe that $\langle m|\mathbf{R}_j^0|m'\rangle = 0$ for $m \neq m'$. With the normal coordinate transformation (2.2.19) (2.2.23), (7.6.110) becomes

$$W^{\mp} = \frac{2\pi}{\hbar}M\sum_{mm'}f_{m'}\omega_{mm'}^2|\langle m|\sum_{jl}\frac{e_j}{c}M_j^{-1/2}q_l\mathbf{a}_j(l)\cdot\mathbf{e}(\lambda,\mathbf{k})\exp(\mp i\mathbf{k}\cdot\mathbf{R}_j^0)|m'\rangle|^2$$
$$\times\left(\frac{2\pi c\hbar}{kV}\right)\delta_\gamma[\varepsilon_m - \varepsilon_{m'} \pm \hbar\omega(\lambda,\mathbf{k})]. \tag{7.6.111}$$

With $m_r = m_r'$, $r = 1,2,\ldots, r \neq h$ for the matrix elements of q_l the following relations hold

$$\langle\ldots m_h'+1\ldots|q_l|\ldots m_h'\ldots\rangle = (m_h'+1)^{1/2}\delta_{lh}(\hbar/2\omega_l M)^{1/2} \tag{7.6.112}$$

$$\langle\ldots m_h'-1\ldots|q_l|\ldots m_h'\ldots\rangle = (m_h')^{1/2}\delta_{lh}(\hbar/2\omega_l M)^{1/2}. \tag{7.6.113}$$

Equation (7.6.112) describes a process where in the final state one additional phonon is generated as compared with the initial state, while (7.6.113) describes a process where in the final state one phonon is annihilated as compared with the initial state. In connection with energy conservation equation (7.6.112) leads to photon absorption, while (7.6.113) leads to photon emission. Due to energy conservation photon absorption is possible only in W^+ while photon emission is possible only in W^-. We therefore obtain from (7.6.111)

$$W^+ = \sum_l (\bar{m}_l' + 1)\omega_l \sum_{jj'} \exp(-i\mathbf{k}\cdot\mathbf{R}_j^0) \frac{e_j}{c} M_j^{-1/2} \mathbf{e}(\lambda,\mathbf{k}) \cdot \mathbf{a}_j(l) \tag{7.6.114}$$
$$\times \mathbf{a}_{j'}(l)^\times \cdot \mathbf{e}(\lambda,\mathbf{k}) \frac{e_{j'}}{c} M_{j'}^{-1/2} \exp(i\mathbf{k}\cdot\mathbf{R}_{j'}^0) \left(\frac{2\pi c\hbar}{kV}\right) \delta_\gamma[\hbar\omega_l - \hbar\omega(\lambda,\mathbf{k})]\pi$$

and

$$W^- = \sum_l (\bar{m}_l')\omega_l \sum_{jj'} \exp(i\mathbf{k}\cdot\mathbf{R}_j^0) \frac{e_j}{c} M_j^{-1/2} \mathbf{e}(\lambda,\mathbf{k}) \cdot \mathbf{a}_j(l) \tag{7.6.115}$$
$$\times \mathbf{a}_{j'}(l)^\times \cdot \mathbf{e}(\lambda,\mathbf{k}) \frac{e_{j'}}{c} M_{j'}^{-1/2} \exp(-i\mathbf{k}\cdot\mathbf{R}_{j'}^0) \left(\frac{2\pi c\hbar}{kV}\right) \delta_\gamma[-\hbar\omega_l + \hbar\omega(\lambda,\mathbf{k})]\pi.$$

As, according to the original formula (7.6.111), the absolute values of the sums over j and j' have to be taken and as $\delta_\gamma(x) = \delta_\gamma(-x)$, the transition rates W^+ and W^- are different only in the average occupation numbers $\bar{m}_l$ or $\bar{m}_l + 1$, resp. Then by substitution of (7.6.114) and (7.6.115) in (7.6.102) the expression

$$\Delta B(\lambda,\mathbf{k}) = \sum_l [b(\lambda,\mathbf{k}) - \bar{m}_l'] \left(\frac{2\pi c\hbar}{kV}\right) \delta_\gamma[\hbar\omega_l - \hbar\omega(\lambda,\mathbf{k})]\omega_l\pi$$
$$\times \sum_{jj'} \exp(-i\mathbf{k}\cdot\mathbf{R}_j^0) \frac{e_j}{c} M_j^{-1/2} \mathbf{e}(\lambda,\mathbf{k}) \cdot \mathbf{a}_j(l) \tag{7.6.116}$$
$$\times \mathbf{a}_{j'}(l)^\times \cdot \mathbf{e}(\lambda,\mathbf{k}) \frac{e_{j'}}{c} M_{j'}^{-1/2} \exp(i\mathbf{k}\cdot\mathbf{R}_{j'}^0)$$

results.

According to our supposition we have $b(\lambda,\mathbf{k}) \gg \bar{m}_l'$. Hence the $\bar{m}_l'$-term in (7.6.116) can be neglected. Owing to (approximate) energy conservation by δ_γ in (7.6.116) the factor ω_l can be replaced by $\omega(\lambda,\mathbf{k})$. If we further assume that the relation

$$\delta(\hbar\omega_l - \hbar\omega) = \hbar^{-1}\delta(\omega_l - \omega) = 2\hbar\omega\delta(\omega_l^2 - \omega^2), \quad \omega_l, \ \omega > 0 \tag{7.6.117}$$

also approximately holds for δ_γ if γ is small, and if we use

$$\delta_\gamma(\omega_l^2 - \omega^2) = \frac{1}{\pi} Im(\omega_l^2 - \omega^2 - i\gamma)^{-1} \tag{7.6.118}$$

then with the Green function (7.5.80) for real $\mathbf{a}_j(l)$ and $d_l = 0$ formula (7.6.116) reads

$$\Delta B(\lambda,\mathbf{k}) = b(\lambda,\mathbf{k}) \frac{2\omega^2(\lambda,\mathbf{k})}{\hbar} \sum_{jj'} \exp(-i\mathbf{k}\cdot\mathbf{R}_j^0) \frac{e_j}{c} M_j^{-1/2} \tag{7.6.119}$$
$$\times \mathbf{e}(\lambda,\mathbf{k}) \cdot \mathrm{Im}\, \mathscr{G}_{jj'}(z) \cdot \mathbf{e}(\lambda,\mathbf{k}) \frac{e_{j'}}{c} M_{j'}^{-1/2} \exp(-i\mathbf{k}\cdot\mathbf{R}_{j'}^0) \left(\frac{2\pi c\hbar}{kV}\right)$$

with $z = \omega(\lambda,\mathbf{k})^2 + i\gamma/\hbar$. In the diatomic case, for $\mathbf{k}\to 0$ the transverse optical phonon modes of the ideal lattice $\zeta_j(\lambda,\mathbf{k}) := \mathbf{e}_j(\lambda,\mathbf{k}) \exp(-i\mathbf{k}\cdot\mathbf{R}_j^0)$ satisfy the equation (7.2.31) and from this it follows that

$$\lim_{\mathbf{k}\to 0} \mathbf{e}_j(\lambda,\mathbf{k}) = \mathbf{e}(\lambda,\mathbf{k}) e_j M_j^{-1/2}. \tag{7.6.120}$$

If these formulae are applied to (7.6.119), then (7.6.103) results. Q.E.D.

Formula (7.6.103) was first derived by Burstein (1963) by phenomenological arguments. The derivation given here has the advantage that the various assumptions which are required to derive it can clearly be seen. The formula (7.6.103) can be applied to both ideal lattices and imperfect lattices. For ideal lattices it leads to an absorption line for the transverse optical mode with $\mathbf{k}\to 0$ broadened by damping. For an imperfect lattice many eigenmodes may have nonvanishing projections on $\mathbf{e}(\lambda,\mathbf{k})$ and the absorbed photon energy may vary over the range of all those frequencies which belong to the eigenmodes. Special techniques have been developed to calculate such projections. Before going into details of such an evaluation, we give yet another version of the absorption rate formula, which follows from the response theory of Kubo (1957). We do not use the original approach of Kubo, but rather generalize the approach of Ipatova, Subashiev and Maradudin (1969) which was made for $\gamma\to 0$ and which is now given for finite γ.

Theorem 7.23: With the suppositions of Theorem 7.22 for macroscopic incident fields, the infrared absorption rate reads

$$\Delta B(\lambda,\mathbf{k})=\pi(2\hbar^2)^{-1}\sum_{m'} f_{m'}\ \mathrm{Im}\ i\int_{-\infty}^{\infty}\Theta(t)\exp\left[-i\omega(\lambda,\mathbf{k})t-\gamma\hbar^{-1}t\right]$$
$$\times\left[\langle m'|\mathbf{M}(\mathbf{k},t)\otimes\mathbf{M}(-\mathbf{k},0)|m'\rangle\right]_{-}\cdot\cdot\,\mathbf{E}(\lambda,\mathbf{k})\otimes\mathbf{E}(\lambda,\mathbf{k})\,dt \tag{7.6.121}$$

where $\mathbf{M}(\mathbf{k},0)$ is the Fourier transform of the crystal ionic dipole moment.

Proof: The absorption rate is given by (7.6.102). The first steps of the proof run along the same lines as in the foregoing theorem, in particular we obtain formula (7.6.110). The energy of a homogeneous macroscopic field of frequency $\omega(\lambda,\mathbf{k})$ and field strength $\mathbf{E}(\lambda,\mathbf{k})$ in a volume V is given by $u_{\mathrm{em}}(\omega)=|\mathbf{E}(\lambda,\mathbf{k})|^2 V/8\pi$. The corresponding quantum energy is $u_{\mathrm{em}}(\omega)=\hbar b(\lambda,\mathbf{k})\omega(\lambda,\mathbf{k})$ where we neglected the zero point contribution. Hence we have $(b(\lambda,\mathbf{k}))^{1/2}=(V/8\pi\hbar\omega(\lambda,\mathbf{k}))^{1/2}E(\lambda,\mathbf{k})$ with $E(\lambda,\mathbf{k})=|\mathbf{E}(\lambda,\mathbf{k})|$. As in (7.6.110), only one-photon and one-phonon absorption and emission processes are allowed, energy conservation leads approximately to $\omega_{mm'}^2=\omega^2(\lambda,\mathbf{k})$. Substituting these expressions in (7.6.110) and using (7.6.102), for macroscopic fields with $b(\lambda,\mathbf{k})\approx b(\lambda,\mathbf{k})+1$ we obtain the formula

$$\Delta B(\lambda,\mathbf{k})=\frac{\pi}{2\hbar}\sum_{mm'} f_{m'}\{|\langle m|\mathbf{M}(\mathbf{k})\cdot\mathbf{E}(\lambda,\mathbf{k})|m'\rangle|^2\delta_\gamma[\varepsilon_m-\varepsilon_{m'}+\hbar\omega(\lambda,\mathbf{k})]$$
$$-|\langle m|\mathbf{M}(-\mathbf{k})\cdot\mathbf{E}(\lambda,\mathbf{k})|m'\rangle|^2\delta_\gamma[\varepsilon_m-\varepsilon_{m'}-\hbar\omega(\lambda,\mathbf{k})]\} \tag{7.6.122}$$

with the Fourier transform of the dipole moment

$$\mathbf{M}(\mathbf{k})=\sum_j e_j\mathbf{u}_j\exp(-i\mathbf{k}\cdot\mathbf{R}_j^0)\equiv\mathbf{M}(\mathbf{k},0). \tag{7.6.123}$$

If now equation (7.6.117) is applied to (7.6.122) and if we use

$$(\varepsilon+\omega-i\gamma)^{-1}=i\int_{-\infty}^{\infty}\Theta(t)\exp\left[-i(\varepsilon+\omega)-\gamma t\right]dt \tag{7.6.124}$$

then from (7.6.122) we see that

$$\Delta B(\lambda,\mathbf{k})=\frac{\pi}{2\hbar^2}\sum_{mm'} f_{m'}\ \mathrm{Im}\ i\int_{-\infty}^{\infty}\Big\{|\langle m|\mathbf{M}(\mathbf{k})\cdot\mathbf{E}(\lambda,\mathbf{k})|m'\rangle|^2\exp\left[-i\frac{1}{\hbar}(\varepsilon_m-\varepsilon_{m'})\right]$$
$$-|\langle m|\mathbf{M}(-\mathbf{k})\cdot\mathbf{E}(\lambda,\mathbf{k})|m'\rangle|^2\exp\left[i\frac{1}{\hbar}(\varepsilon_m-\varepsilon_{m'})\right]\Big\}\ \Theta(t)\exp\left[-i\omega(\lambda,\mathbf{k})t-\gamma\hbar^{-1}t\right]dt \tag{7.6.125}$$

Since $\mathbf{M}(\mathbf{k})$ is equivalent to the Heisenberg operator $\mathbf{M}(\mathbf{k},0)$ at $t=0$ we have

$$\langle m|\mathbf{M}(\mathbf{k})|m'\rangle\ \exp\left[i(\varepsilon_m-\varepsilon_{m'})t\hbar^{-1}\right]=\langle m|\mathbf{M}(\mathbf{k},t)|m'\rangle \tag{7.6.126}$$

etc. Using this formula we obtain the formula (7.6.121) from (7.6.125), Q.E.D.

According to (3.7.27) the absorbed energy is

$$W=\frac{V}{8\pi}\ \omega(\lambda,\mathbf{k})\sum_{\mu\nu}\varepsilon^{(2)}_{\mu\nu}(\omega)E_\mu(\lambda,\mathbf{k})E_\nu(\lambda,\mathbf{k})=\Delta B(\lambda,\mathbf{k})\hbar\omega(\lambda,\mathbf{k}). \tag{7.6.127}$$

By comparison with (7.6.121) we see that the imaginary part of $\varepsilon(\omega)$ is given by

$$\varepsilon^{(2)}=\frac{4\pi}{V\hbar}\sum_{m'} f_{m'}\ \mathrm{Im}\ i\int_{-\infty}^{\infty}\Theta(t)\exp\left[-i\omega t-\frac{\gamma}{\hbar}t\right]\left[\langle m'|\mathbf{M}(\mathbf{k},t)\otimes\mathbf{M}(-\mathbf{k},0)|m'\rangle\right]_{-}dt \tag{7.6.128}$$

As $\mathrm{Im}\,\varepsilon(\omega)=\mathrm{Im}\,\chi(\omega)$ holds, formula (7.6.128) can be used to calculate the imaginary part of the dielectric susceptibility. This formula was used by Knohl (1972) to calculate the values of $\mathrm{Im}\,\chi(\omega)$ for ideal crystals by a method of Wehner (1966). The states $|m\rangle$ used for this calculation were chosen to be breathing shell model states including third order anharmonic Coulomb interactions which lead to damping. It would exceed the scope of this book to give a detailed account of these shell model calculations. We give only a result for ideal LiF-crystals in the Figures 7.9a and 7.9b.

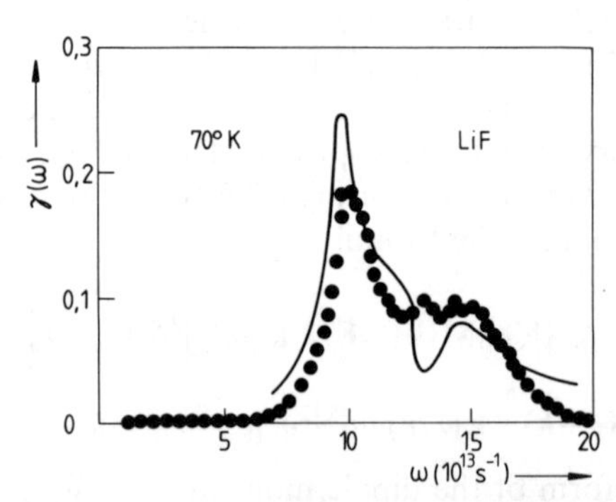

Figure 7.9.a:
Comparison of the calculated damping factor $\gamma(\omega)$ with experimental measurements taken from Wehner (1966). The damping factor is proportional to $\mathrm{Im}\ \chi(\omega)$; (●●●●) theory (——) experiment, after Knohl (1972).

While formula (7.6.121) was used for ideal crystals, the formula (7.6.103) seems to be appropriate for imperfect crystals. This stems from the fact that the lattice Greenfunction $\mathscr{G}(z)$ allows an expansion in powers of the perturbation $\Delta(l\kappa,l'\kappa'|n)$

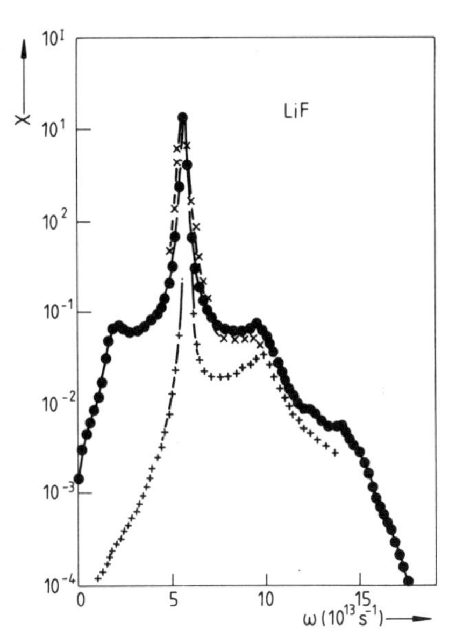

Figure 7.9.b:
Imaginary part of the calculated dielectric susceptibility of LiF for 300 K (●●●) and for 7,5 K (+ + +) in comparison with experimental values for 300 K, given by Fröhlich (1962) (1964) (× × ×); after Knohl (1972).

defined in (7.5.10). If the lattice equations are given by (7.5.11), then $\mathscr{G}(z)$ is defined by

$$\mathscr{G}(z) = (\mathscr{D}^0 + \Delta - z)^{-1}. \tag{7.6.129}$$

With $\mathscr{G}_0(z) := (\mathscr{D}^0 - z)^{-1}$ this expression can be written

$$\mathscr{G} = (\mathscr{G}_0^{-1} + \Delta)^{-1} = [\mathscr{G}_0^{-1}(1 + \mathscr{G}_0\Delta)]^{-1} = (1 + \mathscr{G}_0\Delta)^{-1}\mathscr{G}_0 \tag{7.6.130}$$

and by an expansion in powers of Δ the expression

$$\mathscr{G} = \mathscr{G}_0 - \mathscr{G}_0\Delta\mathscr{G}_0 + \mathscr{G}_0\Delta\mathscr{G}_0\Delta\mathscr{G}_0 - \ldots = \mathscr{G}_0 - \mathscr{G}_0\mathscr{T}\mathscr{G}_0 \tag{7.6.131}$$

results with

$$\mathscr{T}(z) := \Delta[1 + \mathscr{G}_0(z)\Delta]^{-1}. \tag{7.6.132}$$

According to (7.5.13), the Green function of the ideal lattice is given by

$$\mathscr{G}_0(l\kappa, l'\kappa'|z) = (M_\kappa M_{\kappa'})^{-1/2} \sum_{j\mathbf{k}} \frac{\mathbf{e}(\kappa|j,\mathbf{k}) \otimes \mathbf{e}(\kappa'|j,\mathbf{k})^\times}{\omega(j,\mathbf{k})^2 - z} \exp\,[i\mathbf{k} \cdot (\mathbf{R}^0_{l\kappa} - \mathbf{R}^0_{l'\kappa'})]. \tag{7.6.133}$$

Hence, if we substitute (7.6.131) and (7.6.133) into (7.6.103) we obtain with the frequencies $\{\omega(tr,\lambda,\mathbf{k})\}$ of the transverse optical branches of the lattice for the absorption rate

$$\Delta B(\lambda,\mathbf{k}) = b(\lambda,\mathbf{k})\,\frac{4\pi\omega(\lambda,\mathbf{k})}{V}\,\frac{\gamma\hbar^{-1}}{[\omega(tr,\lambda,\mathbf{k})^2 - \omega(\lambda,\mathbf{k})^2]^2 + \gamma^2\hbar^{-2}} \tag{7.6.134}$$

$$+ b(\lambda,\mathbf{k})\,\frac{4\pi\omega(\lambda,\mathbf{k})}{V}\,\mathrm{Im}\left[\frac{1}{[\omega(tr,\lambda,\mathbf{k})^2 - z]^2}\sum_{jj'}\zeta_j(\lambda,\mathbf{k})\cdot\mathscr{T}_{jj'}(z)\cdot\zeta_{j'}(\lambda,\mathbf{k})\right]_{/z=\omega(\lambda,\mathbf{k})^2+i\gamma/\hbar}$$

Calculations with this kind of absorption formula were made by Martin (1971),

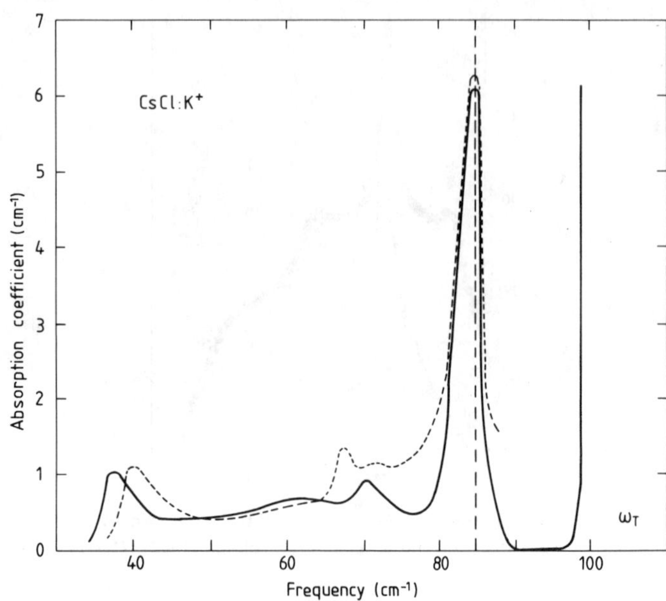

Figure 7.10:
Experimental (broken curve) and calculated (full curve) absorption coefficient at 4 K of 0.1 mol % K^+ in CsCl; after Martin (1971).

Figure 7.10, and by MacDonald, Klein and Martin (1969), Figure 7.11.
The calculation of infrared absorption constants, dielectric constants, dielectric susceptibilities, etc., is an involved problem. Although the majority of authors in this field use the formulae given here, there exist differences with respect to multiplicative constants, the use of imaginary or real parts, etc. This stems from the fact that for the derivation of such formulae, numerous methods with different suppositions have been applied, for instance, classical, semiclassical, and phenomenological approaches as well as response function theory, reversible quantum mechanics, and irreversible quantum statistics. Since our presentation is based on irreversible quantum statistics, our formulae rest on the general theory of absorption- and emission bands given in Section 3.7.
Infrared absorption has been treated in numerous books and review articles of which a few will be cited now. Born and Huang (1954) used in their book a semiclassical method for the derivation of dielectric constants and evaluated the corresponding formulae with respect to pure states of reversible quantum mechanics by performing a damping calculation. Maradudin (1966) in his review article based the treatment of infrared absorption on formulae of the type (7.6.121) which were derived by a semiclassical method. Klein (1968) based his review article on formulae for the absorption constants of the type (7.6.103) and (7.6.134), which he derived by classical considerations. The book of Birman (1974) contains a comprehensive treatment of the

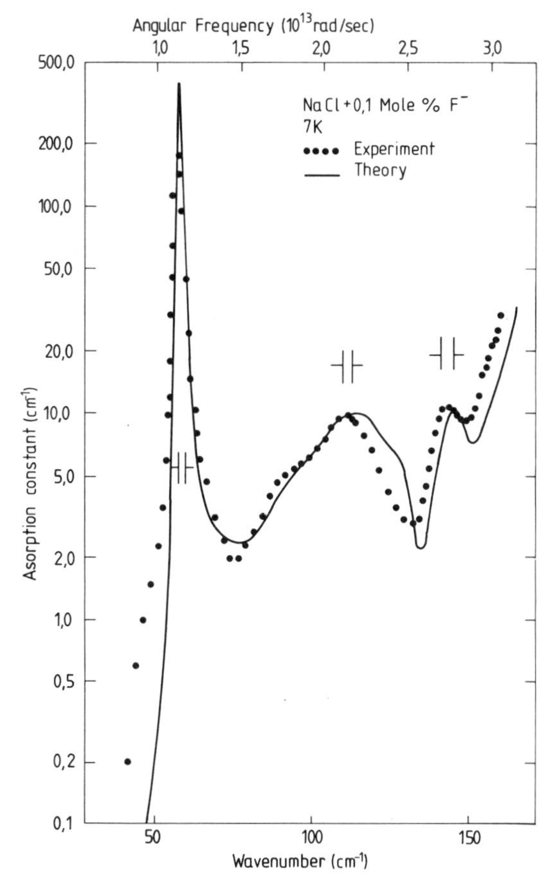

Figure 7.11:
Comparison of theory and experiment of the absorption constant for NaCl : F^-; after McDonald, Klein and Martin (1969).

theory of crystal space groups and its relation to infrared and Raman lattice processes. Balkanski (1972) published a review article where emphasis is laid on the evaluation of the phenomenological properties of dielectric crystals based on microscopic phonon-photon processes. In their book about phonons and resonances in solids, Di Bartolo and Powell (1976) devoted one chapter to the discussion of infrared absorption and Raman scattering. In this book further references can be found.

Concerning original papers we give some references from the last two decades, while for earlier work in this field we refer to books and review articles. The following topics were treated: Derivation of infrared active local lattice modes due to impurities, Wallis and Maradudin (1960). Investigation of higher order effects in the intrinsic infrared absorption of polar crystals by taking into account anharmonic lattice terms and higher order dipole moments, and by means of an absorption formula derived from the Pauli-Master equation, Szigeti (1960). Study of lattice anharmonicity and its influence on optical absorption in polar crystals by a formula of type (7.6.121), Maradudin and Wallis (1961). Infrared dispersion of alkali halogenides based on Born-Huang's damping calculation, Bilz, Genzel and Happ (1960), Bilz and Genzel (1962). Selection rules of infrared and Raman processes in ZnS-type crystals, Birman (1963). Classification of optically active lattice vibrations of zincblende and Wurtzite type crystals, Tsuboi (1964). General space group selection rules for two-phonon processes in infrared absorption and Raman scattering, Loudon (1965). Selection rules for second order infrared and Raman processes for crystals of caesium chloride structure, Ganesan, Burstein, Karo and Hardy (1965). Higher order dipole moments of harmonic and anharmonic crystals and their contribution to the second order moment of two-phonon infrared absorption, Keating and Rupprecht (1965). Calculation of linewidths for infrared absorption by third order anharmonic phonon interactions, Vredevoe (1965). Infrared lattice absorption by gap modes and resonance modes in KI based on local mode analysis, Sievers, Maradudin and Jaswal (1965). A note on infrared resonant-mode sidebands in alkali halides, Benedek and Nardelli (1966). Theory of infrared absorption and dispersion of crystals with inclusion of anharmonicities based on formula (7.6.121) and Green function technique, Wehner (1966). Sidebands of impurity-induced infrared lattice absorption spectra of alkali halide crystals with U-centers based on (7.6.121), Xinh (1966). Theory of the infrared absorption band width of U-centers, Zavt, Kristofel and Khizhnyakov (1966). Temperature dependence of the absorption line width of local oscillations of H^- and D^- ions in alkali halide crystals based on the evaluation of formula (7.6.121), Ipatova and Klotchichin (1966). Sidebands of infrared U-center modes in KBr based on (7.6.103), Timusk and Klein (1966), and the same topic based on (7.6.121), Page and Dick (1967). Discussion of infrared active local phonon modes and calculation of the corresponding impurity induced infrared absorption in alkali halide crystals by means of the formulae of the types (7.6.103) (7.6.134), Takeno (1967). Theory of infrared optical properties of fluorite crystals by damping calculations, Zernik (1967). Selection rules for two-phonon absorption processes, Hulin (1967). Impurity induced second-harmonic infrared absorption and Raman scattering by U-center localized modes in alkali halide crystals, Maradudin and Peretti (1967). Vibrational analysis of the sideband of the U-center local mode absorption based on formula (7.6.103), Kühner and Wagner (1967). Temperature dependence of the width of the fundamental lattice-vibration absorption peak in ionic crystals by means of a damping calculation with anharmonic terms, Ipatova, Maradudin and Wallis (1966) (1967). Infrared absorption spectra of Nitrate ions in alkali halides, Metselaar and van der Elsken (1968). Shell model treatment of the vibrations of crystals containing substitutional defects and defect induced one-phonon absorption, Strauch (1968). Far-infrared absorption induced by isotopes in NaCl and LiF by means of formulae of the types (7.6.103) (7.6.134), Klein and McDonald (1968). Theory of impurity induced infrared absorption by a formula of the type (7.6.121) and field theoretic methods, Wilson King and Kim (1968). Temperature dependence of the fundamental lattice vibration absorption by localized modes using a formula of type (7.6.121) and a damping calculation, Ipatova, Subashiev and Maradudin (1969). Deformable shell model calculation of infrared absorption frequencies of ionic crystals, Roy, Basu and

Sengupta (1969). Experimental and theoretical study of the far-infrared spectra of monovalent impurities in NaCl based on a formula of type (7.6.103), McDonald, Klein and Martin (1969). Calculated infrared absorption induced by impurities in CsCl and CsI based on (7.6.103), Martin (1971). Lattice dynamics and infrared absorption of cesium halides by the breathing shell model, Mahler and Engelhardt (1971). Infrared absorption in CsI containing impurity ions based on (7.6.103), Ram and Agrawal (1972). Infrared absorption in ionic crystals based on the breathing shell model and on (7.6.121), Knohl (1972). Theory of the multiplet structure of the far-infrared absorption by off-centered impurities, Pandey and Shukla (1971), Pandey, Shukla and Pandey (1973). Infrared properties of H^- ions in potassium halides based on quantum mechanical calculation of the binding forces and on (7.6.103), Wood and Ganguly (1973). Discussion of the far-infrared absorption by alkali halide crystals based on damping calculations with inclusion of selfenergy corrections, Bruce (1973). Theory of infrared light absorption by multiphonon processes with Green function technique, Bendow, Ying and Yukon (1973). Frequency and temperature dependence of anharmonicity induced multiphonon absorption, Bendow (1973). Infrared absorption of localized longitudinal-optical phonons with a phenomenological model, Barker (1973). Theory of impurity-induced infrared absorption in non-cubic crystals based on formula (7.6.121), Benson (1973). Multiphonon absorption in alkali halides, Rosenstock (1974). Theory of impurity induced infrared absorption in cubic crystals, Jaswal and Wahedra (1975). Quasiselection rules for multiphonon absorption in alkali halides, Duthler (1976), Harrington, Duthler, Patten and Hass (1976). Optical spectra of phonons bound to impurity centers, Rashba (1976). Shell model calculations of the far-infrared properties of CdI, Eldridge, Howard and Staal (1977). Impurity induced infrared spectra of CsBr based on the breathing shell model and on formulae (7.6.103), and (7.6.134), Haque and Strauch (1977). Infrared absorption of various isolated impurities in GaAs, Bellomonte (1977). Infrared absorption of Ag^+ in RbCl and RbBr based on (7.6.103), Mokross and Dick (1977). Far-infrared quartic damping in KI and the quasi-selection rule for gap compounds, Eldridge and Staal (1977). Far-infrared lattice absorption of Be in ZnS based on formula (7.6.103), Talwar and Agrawal (1978).

d) Raman scattering

Raman scattering inside a crystal is the inelastic scattering of light by the lattice vibrations of the crystal. Because energy must be conserved in the scattering processes the difference between the frequency of the incident light and the frequency of the scattered light must be equal to the frequency of the phonon responsible for the scattering. This conservation law holds for both imperfect and perfect crystals. In addition, the invariance of a perfect crystal against a displacement through any of its translation vectors leads to a conservation law of wave vectors: the wave vector of the incident light must be equal the sum of the wave vectors of the scattered light and of the phonon responsible for the scattering. In practice the smallness of the magnitude of the wave vector of the light in the crystal compared with the magnitude of the maximum phonon wave vector has the consequence that the incident light interacts only with

phonons of essentially zero wave vector. Because only the optical branches of the phonon spectrum have nonvanishing frequencies for $\mathbf{k}=0$, the first order Raman spectrum of a perfect crystal consists of a set of discrete lines on both sides of the line at the frequency of the incident light. These lines are shifted from the position of the latter line by the frequencies $\pm\omega(j0)$ of the optical modes. But the scattering of light by acoustic phonons is also possible as the acoustic phonons are equivalent to elastic deformations of the medium which are connected with electric moments. However, the energy of these phonons is much smaller as compared with the Raman effect for optical phonons. This kind of scattering processes is called Brillouin scattering. If impurities are present in a crystal its translational symmetry is destroyed and the selection rule $\mathbf{k}=0$ is relaxed. It is then possible to induce continuous first order Raman spectra. The line spectrum of the ideal crystal is then replaced by a continuous spectrum which reflects the singularities of the frequency distribution of the perfect host crystal and resonance and (or) localized modes if they are Raman active. The imperfections may also alter drastically some of the modes of the perfect crystal and so also change that part of the Raman spectrum belonging to the perfect crystal. In addition, Raman scattering of light may take place at lattice polaritons, excitonic polaritons, excitons, plasmons, etc. With respect to these processes we give no explicit treatment but refer to the literature. In principle, both coherent scattering and incoherent scattering may occur. While the latter is due to spontaneous emission, the coherent scattering depends on stimulated emission. Thus the theoretical approaches to Raman scattering are concerned with the calculation of corresponding coherent pure quantum states or with the calculation of irreversible quantum states. Only the latter shall be treated here.
Various treatments of irreversible quantum states have been developed. We give a treatment which fits into our general approach of irreversible quantum statistics and which partly makes use of the treatment of Placzek (1934). The connection with other approaches will be discussed later.
For Raman scattering the initial state is given by

$$|n,m,v\rangle=|n,m,\ldots v(j_1,\mathbf{k}_1)\ldots v(j_2,\mathbf{k}_2)\ldots\rangle \tag{7.6.135}$$

where $v(j_1,\mathbf{k}_1)$ and $v(j_2,\mathbf{k}_2)$ are the occupation numbers of photon modes involved in the scattering process. In the final state the photon mode $j_1,\mathbf{k}_1$ has lost one photon, while the photon mode $j_2,\mathbf{k}_2$ has gained one photon and at least one phonon has been created or annihilated. Thus the final state becomes

$$|n,m',v'\rangle=|n,m',\ldots v(j_1,\mathbf{k}_1)-1,\ldots v(j_2,\mathbf{k}_2)+1,\ldots\rangle. \tag{7.6.136}$$

For a first order transition, it can be seen that the general expression (2.3.14) for the interaction energy H^s between light and matter allows one or two photons to change their states. The one-photon transitions are connected with the linear terms of H^s in $\mathbf{A}$ while the two-photon transitions depend on the quadratic terms in $\mathbf{A}$. Birman (1974) has given arguments as to why the quadratic terms in $\mathbf{A}$ give no essential contributions to the Raman scattering. Although these arguments could be further improved, we neglect the quadratic terms of $\mathbf{A}$ in H^s and consider only the linear approximation of

H^s in $\mathbf{A}$. In this case only one photon state can change in a transition. Hence, in this approximation to reach the final state (7.6.136) from an initial state (7.6.135), an intermediate transition is required. The general form of such intermediate states reads

$$|n'',m'',v''\rangle = |n'',m'',\ldots v(j_1,\mathbf{k}_1)-1,\ldots v(j_2,\mathbf{k}_2)\ldots\rangle \tag{7.6.137}$$

or

$$|n'',m'',v''\rangle = |n'',m'',\ldots v(j_1,\mathbf{k}_1)\ldots v(j_2,\mathbf{k}_2)+1\ldots\rangle \tag{7.6.138}$$

i.e., for the intermediate states also a change in the electronic quantum number is admitted although in the final state, by definition, the original electronic state has to be restored. Transitions with intermediate states first occur in the second order transition probability (1.5.31). Hence these expressions have to be used for the calculation of Raman scattering rates. If we assume that the electrons are in their ground state, then any virtual transition from n to n'' can only be an excitation of the electrons. If further a photon has to be simultaneously created in the transition to the virtual intermediate state (7.6.138), then these states are very improbable, as they require a large deviation from the conserved energy. Hence we neglect these kinds of intermediate states. With the remaining intermediate states (7.6.137) where an energy compensation between crystal and electromagnetic field is possible, the second order transition probability (1.5.31) reads

$$W_{nm'v',nmv} = \left| \sum_{n''m''} \frac{\langle nm'v'|H^s|n''m''v''\rangle \, \langle n''m''v''|H^s|nmv\rangle}{[E_m^n - E_{m''}^{n''} + \hbar\omega(j_1,\mathbf{k}_1)]} \right|^2$$
$$\times \delta_\gamma [E_{m'}^n - E_m^n - \hbar\omega(j_1,\mathbf{k}_1) + \hbar\omega(j_2,\mathbf{k}_2)] \hbar^{-1}(2\pi) \tag{7.6.139}$$

if we immediately extend (1.5.31) to include resonance behavior according to Theorem 1.18.

This transition probability can only be evaluated if appropriate crystal models are used. We consider the simplest case consisting of an ideal crystal which has a filled valence band and which we describe by Bloch orbitals in the tight binding approximation.

Theorem 7.24: For an ideal insulating crystal in its ground state only the longitudinal optical modes for $\mathbf{h}=0$ are Raman active in the first order Raman effect, i.e., only one longitudinal optical phonon with $\mathbf{h}=0$ can be created or annihilated for Raman scattering, provided that no additional selection rules prevent this process.

Proof: In the linear approximation of H^s with respect to $\mathbf{A}$, the matrix elements of H^s between the states (7.6.135) and (7.6.137) read

$$\langle \ldots v(\lambda_1,\mathbf{k}_1)-1\ldots|H^s|\ldots v(\lambda_1,\mathbf{k}_1)\ldots\rangle = \mathfrak{P}^+(\lambda_1,\mathbf{k}_1)\,[v(\lambda_1,\mathbf{k}_1)]^{1/2} \tag{7.6.140}$$

$$\langle \ldots v(\lambda_1,\mathbf{k}_1)-1,\ldots,v(\lambda_2,\mathbf{k}_2)+1,\ldots|H^s|\ldots v(\lambda_1,\mathbf{k}_1)-1\ldots v(\lambda_2,\mathbf{k}_2)\ldots\rangle$$
$$= \mathfrak{P}^-(\lambda_2,\mathbf{k}_2)\,[v(\lambda_2,\mathbf{k}_2)+1]^{1/2}$$

where (7.6.105) is used. If the sum over $n''m''$ in (7.6.139) is split into two parts where the first contains only transitions for fixed $n'' \equiv n$, while the second contains all other transitions, then we obtain from (7.6.139) with (7.6.140) and the energy expression $\Delta E := E^n_{m'} - E^n_m - \hbar\omega(j_1, \mathbf{k}_1) + \hbar\omega(j_2, \mathbf{k}_2)$

$$W^{(2)}_{nm'v',nmv} = v(\lambda_1, \mathbf{k}_1)\,[v(\lambda_2, \mathbf{k}_2) + 1] \times \left| \sum_{m''} \frac{\langle nm'|\mathfrak{P}^-(\lambda_2, \mathbf{k}_2)|nm''\rangle \langle nm''|\mathfrak{P}^+(\lambda_1, \mathbf{k}_1)|nm\rangle}{E^n_m - E^n_{m''} + \hbar\omega(\lambda_1, \mathbf{k}_1)} + \sum_{\substack{n''m'' \\ n'' \neq n}} \frac{\langle nm'|\mathfrak{P}^-(\lambda_2, \mathbf{k}_2)|n''m''\rangle \langle n''m''|\mathfrak{P}^+(\lambda_1, \mathbf{k}_1)|nm\rangle}{E^n_m - E^{n''}_{m''} + \hbar\omega(\lambda_1, \mathbf{k}_1)} \right|^2 \delta_\gamma(\Delta E)\,\hbar^{-1}. \tag{7.6.141}$$

If (7.6.105) is approximated by (7.6.106), then from (7.6.107) we see that in the first term on the right-hand side of (7.6.141) we have $\langle n|\mathbf{p}_i|n\rangle = 0$, whereas for the second term the equation holds

$$\langle n|\langle m'|\mathbf{P}_j|m''\rangle|n''\rangle = 0 \tag{7.6.142}$$

i.e., the first term contains only pure phonon-photon interactions. The matrix elements of this term are all of the kind $\langle m'|\mathbf{P}_j|m''\rangle \langle m''|\mathbf{P}_h|m\rangle$. This implies, that this term only leads to processes where two phonons are involved. By definition, in the first order Raman process only one phonon participates. Hence to this order the first term contributes nothing and drops out. Then the transition probability for the first order effect becomes

$$W^{(2)}_{nm'v',nmv} = v(\lambda_1, \mathbf{k}_1)\,[v(\lambda_2, \mathbf{k}_2) + 1] \left(\frac{2\pi c\hbar}{k_1 V}\right) \left(\frac{2\pi c\hbar}{k_2 V}\right) \delta_\gamma(\Delta E)\,\hbar^{-1} \tag{7.6.143}$$
$$\left| \sum_{\substack{n''m'' \\ n'' \neq n}} \langle nm'|\sum_i \frac{e}{mc}\,\mathbf{p}_i \cdot \mathbf{e}(\lambda_2, \mathbf{k}_2) \exp(-i\mathbf{k}_2 \cdot \mathbf{R}^0_i)|n''m''\rangle \times \langle n''m''|\sum_l \frac{e}{mc}\,\mathbf{p}_l \cdot \mathbf{e}(\lambda_1, \mathbf{k}_1) \exp(i\mathbf{k}_1 \cdot \mathbf{R}^0_l)|nm\rangle\,[E^n_m - E^{n''}_{m''} + \hbar\omega(\lambda_1, \mathbf{k}_1)]^{-1} \right|^2$$

If (7.6.107) is applied to nondiagonal matrix elements, (7.6.143) takes the form

$$W^{(2)}_{nm'v',nmv} = v(\lambda_1, \mathbf{k}_1)\,[v(\lambda_2, \mathbf{k}_2) + 1] \left(\frac{2\pi c\hbar}{k_1 V}\right) \left(\frac{2\pi c\hbar}{k_2 V}\right) \delta_\gamma(\Delta E)\,\hbar^{-1} \tag{7.6.144}$$
$$\left| \sum_{\substack{n''m'' \\ n'' \neq n}} \frac{\langle m'|\mathbf{m}(\mathbf{R}, -\mathbf{k}_2)_{nn''}|m''\rangle \langle m''|\mathbf{m}(\mathbf{R}, \mathbf{k}_1)_{n''n}|m\rangle \cdot\cdot\, \mathbf{e}(\lambda_2, \mathbf{k}_2)\mathbf{e}(\lambda_1, \mathbf{k}_1)\omega^2_{nn''}}{E^n_m - E^{n''}_{m''} + \hbar\omega(\lambda_1, \mathbf{k}_1)} \right|^2$$

with

$$\mathbf{m}(\mathbf{R}, \mathbf{k})_{nn''} := \langle n|\sum_i \frac{e}{c}\,\mathbf{r}_i \exp(i\mathbf{k} \cdot \mathbf{R}^0_i)|n''\rangle. \tag{7.6.145}$$

The dependence of $\mathbf{m}_{nn''}$ on $\mathbf{R}$ results from the dependence of $\langle n|$ and $|n''\rangle$ on $\mathbf{R}$. An expansion of these states about $\mathbf{R} = \mathbf{R}^0$ gives with (7.2.66) and $l := j, \mathbf{h}$, $-l := j, -\mathbf{h}$

$$|n\rangle \equiv \chi_n(\mathbf{r}, \mathbf{R}) = \chi_n(\mathbf{r}, \mathbf{R}^0) + \sum_{il} \nabla_i \chi(\mathbf{r}, \mathbf{R})_{/\mathbf{R}=\mathbf{R}^0} \cdot \hat{\mathbf{a}}_i(l)\,(b_l - b^+_{-l}) + \dots \tag{7.6.146}$$

and from that it follows

$$\mathbf{m}(\mathbf{R},\mathbf{k})_{nn''}=\mathbf{m}(\mathbf{R}^0,\mathbf{k})_{nn''}+\sum_l \mathbf{m}(\mathbf{R}^0,\mathbf{k})^l_{nn''}(b_l-b^+_{-l})+\dots \tag{7.6.147}$$

if the normal coordinates are referred to $\mathbf{R}^n \approx \mathbf{R}^0$. In the lowest order where only the terms $\mathbf{m}(\mathbf{R},\mathbf{k})_{nm''}$ are taken into account no phonon participates in the process, i.e., we have $m=m'$ and owing to δ_γ in (7.6.139) or (7.6.144), resp. this contribution vanishes. Hence the first nonvanishing contributions come from the linear terms in q_l. In this case one phonon participates and (7.6.144) reads

$$\begin{aligned} W^{(2)}_{nm'v',nmv} &\overset{\text{first order}}{=} v(\lambda_1,\mathbf{k}_1)\,[v(\lambda_2,\mathbf{k}_2)+1]\left(\frac{2\pi c\hbar}{k_1 V}\right)\left(\frac{2\pi c\hbar}{k_2 V}\right)(m_l+1) \\ &\times \left|\sum_{n''}{}' \frac{\mathbf{m}(\mathbf{R}^0,-\mathbf{k}_2)_{nn''}\mathbf{m}(\mathbf{R}^0,\mathbf{k}_1)^l_{n''n}\cdot\cdot\,\mathbf{e}(\lambda_2,\mathbf{k}_2)\mathbf{e}(\lambda_1,\mathbf{k}_1)\omega^2_{nn''}}{E^n_0-E^{n''}_0-\hbar\omega_l+\hbar\omega(\lambda_1,\mathbf{k}_1)}\right. \\ &\left.+\sum_{n''}{}' \frac{\mathbf{m}(\mathbf{R}^0,-\mathbf{k}_2)^l_{nn''}\mathbf{m}(\mathbf{R}^0,\mathbf{k}_1)_{n''n}\cdot\cdot\,\mathbf{e}(\lambda_2,\mathbf{k}_2)\mathbf{e}(\lambda_1,\mathbf{k}_1)\omega^2_{nn''}}{E^n_0-E^{n''}_0+\hbar\omega(\lambda_1,\mathbf{k}_1)}\right|^2 \\ &\times \delta_\gamma[\hbar\omega_l-\hbar\omega(\lambda_1,\mathbf{k}_1)+\hbar\omega(\lambda_2,\mathbf{k}_2)]\,\hbar^{-1} \end{aligned} \tag{7.6.148}$$

for phonon creation, i.e., $m'=m_1\dots m_l+1\dots m_\infty$. A similar expression holds for phonon annihilation, i.e., $m'=m_1\dots m_l-1\dots m_\infty$. We will explicitly discuss only the creation processes and for annihilation we will give only the final formula. Since for polar crystals $|E^n_0-E^{n''}_0|\gg\hbar\omega_l$, we may neglect $\hbar\omega_l$ in the denominator of (7.6.148) and obtain with $\Delta E=\hbar\omega_l-\hbar\omega(\lambda_1,\mathbf{k}_1)+\hbar\omega(\lambda_2,\mathbf{k}_2)$

$$\begin{aligned} W^{(2)}_{nm'v',nmv} &= v(\lambda_1,\mathbf{k}_1)\,[v(\lambda_2,\mathbf{k}_2)+1]\left(\frac{2\pi c\hbar}{k_1 V}\right)\left(\frac{2\pi c\hbar}{k_2 V}\right)(m_l+1)\delta_\gamma(\Delta E)\,\hbar^{-1} \\ &\times \left|\sum_{n''}{}' \frac{\omega^2_{nn''}\mathbf{e}(\lambda_2,\mathbf{k}_2)\mathbf{e}(\lambda_1,\mathbf{k}_1)}{E^n_0-E^{n''}_0+\hbar\omega(\lambda_1,\mathbf{k}_1)}\cdot\cdot\,[\mathbf{m}(\mathbf{R}^0,-\mathbf{k}_2)_{nn''}\mathbf{m}(\mathbf{R}^0,\mathbf{k}_1)^l_{n''n}\right. \\ &\left.+\mathbf{m}(\mathbf{R}^0,-\mathbf{k}_2)^l_{nn''}\mathbf{m}(\mathbf{R}^0,\mathbf{k}_1)_{n''n}]\right|^2. \end{aligned} \tag{7.6.149}$$

The linear term in the expansion (7.6.147) reads explicitly

$$\begin{aligned} \mathbf{m}(\mathbf{R}^0,\mathbf{k})^l_{nn''} &= \sum_{ki}\{\langle \nabla_k\chi^0_n|\,\frac{e}{c}\,\mathbf{r}_i\,\exp\,(i\mathbf{k}\cdot\mathbf{R}^0_i)|\chi^0_{n''}\rangle \\ &+\langle\chi^0_n|\,\frac{e}{c}\,\mathbf{r}_i\,\exp\,(i\mathbf{k}\cdot\mathbf{R}^0_i)|\nabla_k\chi^0_{n''}\rangle\}\cdot\hat{\mathbf{a}}_k(l) \end{aligned} \tag{7.6.150}$$

where the label 0 indicates that these functions have the value at $\mathbf{R}=\mathbf{R}^0$. We now assume that in the ground state χ_n the electrons are so tightly bound that the polarization induced by the displacement of the ion itself and (or) by the displacements of the ions in the vicinity do not alter the electronic state χ_n to any degree. Hence we put

$\nabla_k \chi_n \approx 0$ and (7.6.150) goes over into

$$\mathbf{m}(\mathbf{R}^0,\mathbf{k})^l_{nn''} = \sum_{ki} \langle \chi_n^0 | \frac{e}{c} \, \mathbf{r}_i \exp(i\mathbf{k} \cdot \mathbf{R}_i^0) | \nabla_k \chi_{n''} \rangle \cdot \hat{\mathbf{a}}_k(l). \tag{7.6.151}$$

The calculation of (7.6.145) for $\mathbf{R}=\mathbf{R}^0$ and (7.6.151) depends on the states which are taken into account and on the state representations. We use a very simple model. We consider the ideal crystal which has in its ground state χ_n a filled valence band described by a determinant of Bloch orbitals, whereas the excited states arise from a substitution of one valence band orbital by a conduction band orbital. In the tight binding approximation these orbitals are given by

$$\psi(\mathbf{K},x|\mathbf{r}) = N^{-1/2} \sum_{s\lambda} \exp(i\mathbf{K} \cdot \mathbf{R}_{s\lambda}^0) a_x(\mathbf{r} - \mathbf{R}_{s\lambda}^0), \qquad x = v, c \tag{7.6.152}$$

where $a_x(\mathbf{r})$ are the corresponding atomic orbitals. These Bloch orbitals are referred to the ideal lattice positions $\mathbf{R}=\mathbf{R}^0$. If the lattice is moved from its ideal position, displacements $\{\mathbf{M}_{k\kappa} = (\mathbf{R}_{k\kappa} - \mathbf{R}_{k\kappa}^0)\}$ can be defined which change the potential energy of the lattice and the wave functions (7.6.152) are no longer appropriate solutions of the corresponding Schrödinger equation. In Section 5.2 a general formalism was developed to calculate the wave functions of single electrons in the presence of all other lattice electrons for a distorted lattice. This formalism can be applied to impurity center electrons as well as to valence and conduction band electrons in a distorted ideal crystal. If we neglect the contribution of the exchange forces, then from (5.2.56) we obtain the following expectation value of the linear electron lattice energy

$$\langle \psi | H_{\mathrm{el}} | \psi \rangle = (1+\alpha^e)^{-1} e \sum_{k\varrho} \int \varrho(\mathbf{r}) a_{k\varrho} \mathbf{M}_{k\varrho} \cdot \nabla_{k\varrho} C(\mathbf{r}, k\varrho) d^3 r. \tag{7.6.153}$$

Hence, if we calculate the conduction band wave functions for a distorted lattice by first order perturbation theory we have

$$\psi(\mathbf{K},c|\mathbf{M},\mathbf{r}) = \psi(\mathbf{K},c|\mathbf{r}) + \sum_{\mathbf{K}'}{}' \frac{\langle \mathbf{K}'c | H_{\mathrm{el}} | \mathbf{K}c \rangle}{E(\mathbf{K},c) - E(\mathbf{K}',c)} \psi(\mathbf{K}',c|\mathbf{r}) \tag{7.6.154}$$

and

$$\nabla_{l\kappa} \psi(\mathbf{K},c|\mathbf{M},\mathbf{r}) = \sum_{\mathbf{K}'}{}' \frac{\langle \mathbf{K}'c | a_{l\kappa} \nabla_{l\kappa} C(\mathbf{r}, l\kappa) | \mathbf{K}c \rangle}{E(\mathbf{K},c) - E(\mathbf{K}',c)} \psi(\mathbf{K}',c|\mathbf{r}) e (1+\alpha^e)^{-1}. \tag{7.6.155}$$

According to our definition the complete wave functions χ_n and $\chi_{n''}$ are represented by

$$\chi_n := \det |\psi(\mathbf{K}_1, v|\mathbf{r}) \ldots \psi(\mathbf{K}_N, v|\mathbf{r}_N)| (N!)^{-1/2} \tag{7.6.156}$$

and

$$\chi_{n''} = \chi(\mathbf{K}'', v|\mathbf{K}, c) = \det |\psi(\mathbf{K}_1, v|\mathbf{r}) \ldots \psi(\mathbf{K}, c|\mathbf{r}_s) \ldots \psi(\mathbf{K}_N, v|\mathbf{r}_N)| (N!)^{-1/2} \tag{7.6.157}$$

and due to our tight binding assumption

$$\nabla_{l\kappa} \chi_{n''} = \det |\psi(\mathbf{K}_1, v|\mathbf{r}_1) \ldots \nabla_{l\kappa} \psi(\mathbf{K}, c|\mathbf{r}_s) \ldots \psi(\mathbf{K}_N, v|\mathbf{r}_N)| (N!)^{-1/2} \tag{7.6.158}$$

If, in (7.6.145), we replace the term $i\mathbf{k}\cdot\mathbf{R}_i^0$ by its original value $i\mathbf{k}\cdot\mathbf{r}_i$ which stems from (3.7.6), we obtain by applying the calculation technique of (3.2.22) with $\mathbf{R}=\mathbf{R}^0$,

$$\mathbf{m}(\mathbf{R}^0,\mathbf{k})_{nn''}=\langle\chi_n|\sum_i \frac{e}{c}\,\mathbf{r}_i \exp(i\mathbf{k}\cdot\mathbf{r}_i)|\chi(\mathbf{K}'',v|\mathbf{K},c)\rangle$$
$$=\int \psi(\mathbf{K}'',v|\mathbf{r})\left[\frac{e}{c}\,\mathbf{r}\exp(i\mathbf{k}\cdot\mathbf{r})\right]\psi(\mathbf{K},c|\mathbf{r})d^3r. \tag{7.6.159}$$

If further in a rough approximation all overlap terms between different unit cells are neglected, then from (7.6.159) we have with (7.2.43)

$$\mathbf{m}(\mathbf{R}^0,\mathbf{k})_{nn''}=\Delta(\mathbf{K}''-\mathbf{K}+\mathbf{k})\mathbf{m}(v,c) \tag{7.6.160}$$

with

$$\mathbf{m}(v,c)=\sum_{\lambda\lambda'}\int a_v(\mathbf{r}-\mathbf{R}_{s\lambda}^0)\exp(i\mathbf{K}''\cdot\mathbf{x}(\lambda))\frac{e}{c}\,\mathbf{r}$$
$$\times\exp[i\mathbf{k}\cdot(\mathbf{r}-\mathbf{R}_s^0)]a_c(\mathbf{r}-\mathbf{R}_{s\lambda'}^0)\exp(-i\mathbf{K}\cdot\mathbf{x}(\lambda'))d^3r \tag{7.6.161}$$

being a translational invariant quantity. Hence $\mathbf{K}=\mathbf{K}''+\mathbf{k}+\mathbf{g}$ and by the same technique we obtain for (7.6.151) with (7.6.158) and (7.6.155)

$$\mathbf{m}(\mathbf{R}^0,\mathbf{k})_{n''n}^l=(1+\alpha^e)^{-1}e\sum_{h\kappa}\sum_{\mathbf{K}'}{}'\frac{\langle\mathbf{K}'c|a_{h\kappa}V_{h\kappa}C(\mathbf{r},h\kappa)|\mathbf{K}''+\mathbf{k}+\mathbf{g},c\rangle}{E(\mathbf{K}''+\mathbf{k}+\mathbf{g},c)-E(\mathbf{K}',c)}$$
$$\times\langle\mathbf{K}''c|\frac{e}{c}\,\mathbf{r}\exp(i\mathbf{k}\cdot\mathbf{r})|\mathbf{K}'v\rangle\hat{\mathbf{a}}_{h\kappa}(l). \tag{7.6.162}$$

If in order to start the calculation in the first Brillouin zone we put $\mathbf{g}=0$, then with (7.6.160)

$$\mathbf{m}(\mathbf{R}^0,\mathbf{k})_{n''n}^l=$$
$$e(1+\alpha^e)^{-1}\sum_{h\kappa}\frac{\langle\mathbf{K}''+\mathbf{k}+\mathbf{g}'|a_{h\kappa}V_{h\kappa}C(\mathbf{r},h\kappa)|\mathbf{K}''+\mathbf{k},c\rangle}{E(\mathbf{K}''+\mathbf{k},c)-E(\mathbf{K}''+\mathbf{k}+\mathbf{g}',c)}\hat{\mathbf{a}}_{h\kappa}(l)\mathbf{m}(v,c) \tag{7.6.163}$$

follows, where for simplicity only one nontrivial $\mathbf{g}'$ has been taken into account. The remaining integral in (7.6.163) can be evaluated if it is assumed that the atomic orbitals are sufficiently concentrated.

$$\sum_{hv}\sum_{h'v'}\int\exp[i(\mathbf{K}''+\mathbf{k}+\mathbf{g}')\cdot\mathbf{R}_{hv}^0]a_c(\mathbf{r}-\mathbf{R}_{hv}^0)\frac{a_{j\kappa}(\mathbf{r}-\mathbf{R}_{j\kappa}^0)}{|\mathbf{r}-\mathbf{R}_{j\kappa}^0|^3}$$
$$\times\exp[-i(\mathbf{K}''+\mathbf{k})\cdot\mathbf{R}_{h'v'}^0]a_c(\mathbf{r}-\mathbf{R}_{h'v'}^0)d^3r \tag{7.6.164}$$
$$\approx\sum_{hv}\exp(i\mathbf{g}'\cdot\mathbf{R}_{hv}^0)\frac{a_{j\kappa}(\mathbf{R}_{hv}^0-\mathbf{R}_{j\kappa}^0)}{|\mathbf{R}_{hv}^0-\mathbf{R}_{j\kappa}^0|^3}\approx 4\pi i\,\frac{\mathbf{g}'}{|\mathbf{g}'|^2}\exp(i\mathbf{g}'\cdot\mathbf{R}_{j\kappa}^0).$$

We now use the explicit form of $\hat{\mathbf{a}}_{h\kappa}(l)$ which follows from (7.2.66) and substitute (7.6.164) into (7.6.163). With $\Delta E' := E(\mathbf{k}''+\mathbf{k},c) - E(\mathbf{k}''+\mathbf{k}+\mathbf{g}',c)$ this gives the expression

$$\begin{aligned}\mathbf{m}(\mathbf{R}^0,\mathbf{k})^{\mathbf{h}j}_{n''n} &= (1+\alpha^e)^{-1} e(\Delta E')^{-1} \sum_{l\kappa} 4\pi i \frac{\mathbf{g}'}{|\mathbf{g}'|^2} N^{-1} \\ &\quad \times \exp\,[i(\mathbf{g}'+\mathbf{h})\cdot \mathbf{R}^0_{l\kappa}]\, a_{l\kappa} M_\kappa^{-1/2}\, \mathbf{e}(\kappa|\mathbf{h},j)\,\eta(\kappa|\mathbf{h},j)\,\mathbf{m}(v,c) \\ &= \frac{4\pi i e}{(1+\alpha^e)\Delta E'} \sum_\kappa \Delta(\mathbf{g}'+\mathbf{h}) \frac{a_{l\kappa}}{M_\kappa^{1/2}} \frac{\mathbf{g}'}{|\mathbf{g}'|^2}\, \mathbf{e}(\kappa|\mathbf{h},j)\,\eta(\kappa|\mathbf{h},j) \\ &\quad \times \exp\,[i(\mathbf{g}'+\mathbf{h})\cdot \mathbf{x}(\kappa)]\,\mathbf{m}(v,c).\end{aligned} \tag{7.6.165}$$

As $\mathbf{h}=\mathbf{g}'+\mathbf{g}''$ holds, the term (7.6.165) is $\neq 0$ if $\mathbf{e}(\kappa)\|\mathbf{h}$, i.e., if a longitudinal mode is considered. Further, $\mathbf{h}=\mathbf{g}'+\mathbf{g}''$ is equivalent to $\mathbf{h}=0$. In this case, according to (7.2.30), a normalization constant results and (7.6.165) takes the form

$$\mathbf{m}(\mathbf{R}^0,\mathbf{k})^{\mathbf{h}j}_{n''n} = \frac{4\pi i e}{(1+\alpha^e)\Delta E'}\, \Delta(\mathbf{h})\, \frac{1}{|\mathbf{g}'|}\, \mathbf{m}(v,c)\,\eta(0,l)\,\delta_{jl}. \tag{7.6.166}$$

Then (7.6.149) reads

$$\begin{aligned}W^{(2)}_{nm'v',nmv} &= v(\lambda_1,\mathbf{k}_1)\,[v(\lambda_2,\mathbf{k}_2)+1] \left(\frac{2\pi c\hbar}{k_1 V}\right)\left(\frac{2\pi c\hbar}{k_2 V}\right) [m(0,l)+1]\,\eta(0,l)^2 \\ &\quad \times \left|\frac{4\pi i e}{(1+\alpha^e)\Delta E'}\, \Delta(\mathbf{h})\,\delta_{jl}\, \frac{\omega^2_{nn''}}{E^n_0 - E^{n''}_0 + \hbar\omega(\lambda_1,\mathbf{k}_1)}\, \mathbf{m}(v,c)\mathbf{e}(\lambda_1,\mathbf{k}_1)\,\mathbf{m}(v,c)\mathbf{e}(\lambda_2,\mathbf{k}_2)\right|^2 \\ &\quad \times \delta_\gamma[\hbar\omega(0,l) - \hbar\omega(\lambda_1,\mathbf{k}_1) + \hbar\omega(\lambda_2,\mathbf{k}_2)]\,\hbar^{-1}.\end{aligned} \tag{7.6.167}$$

Similar considerations can be made for the absorption of a photon. For $\Delta E'=0$ a resonance catastrophe occurs. But this effect can be avoided by the use of a resonance calculation for (7.6.154) instead of a perturbation calculation Q.E.D.

Concerning this theorem and its proof it has to be noted that it represents the only consistent treatment of the first order Raman effect in the adiabatic coupling scheme. In the literature, cf. for instance Birman (1974), Section 120, di Bartolo and Powell (1976), Section 12.4, the operator (7.6.106) is replaced by the time derivative of the dipole moment operator $\mathbf{M}_j$ of the whole ion or atom, resp.:

$$\mathfrak{P}^{\mp}(\lambda,\mathbf{k}) = \sum_j \frac{d}{dt}\mathbf{M}_j \exp\,(\mp i\mathbf{k}\cdot\mathbf{R}^0_j)\,\mathbf{e}(\lambda,\mathbf{k})\left(\frac{2\pi c\hbar}{kV}\right)^{1/2} \tag{7.6.168}$$

and with $d/dt\,\mathbf{M}_j = [\mathscr{H},\mathbf{M}_j]$ the matrix elements $\langle nm'|\mathfrak{P}^{\mp}(\lambda,\mathbf{k})|nm''\rangle$ etc. are evaluated by means of $\mathscr{H}|nm\rangle = E^m_n|nm\rangle$. The latter relation, however, does not hold in the adiabatic coupling scheme since the states $|\mathrm{nm}\rangle$ are not eigenstates of $\mathscr{H}$. Hence a further evaluation of the Raman effect based on this formula is not correct. A similar error is made in Placzek's approximation. In the literature the Raman effect has so far been only correctly treated in the field theoretic version, where light scattering is calculated by means of a coupled electron, phonon and photon system with Hamiltonians in Fock space representation. Higher order Raman effects arise by taking into account higher powers of the lattice coordinates in (7.6.146). The resonant Raman scattering occurs if the denominator in (7.6.139) vanishes or is small. In this case the higher resonant transition probabilities have to be calculated beyond the diagonal approximation of Theorem 1.18.

The problem of calculating the scattering intensity due to the basic process (7.6.167) has been tackled in a great variety of ways. It would exceed the scope of this book to follow the various ramifications of the very special formulae. In principle, these formulae are derived either from Born and Huang's semiclassical scattering formula, or from field theoretic cross sections. As the former formula depends on the incorrect derivation of the dipole moment operator in the transition probabilities, we prefer the field theoretic cross sections. Nevertheless we will also present the semiclassical formula.

In the semiclassical approach the incident light is considered to be a classical quantity. Then the interaction Hamiltonian of crystal and field is given by $H' := \sum_j \mathbf{M}_j \cdot \mathbf{E}(t)$, where $\mathbf{E}(t) = \mathbf{E}^+ \exp(i\omega t) + \mathbf{E}^- \exp(-i\omega t)$ represents a spatially constant electric field. The energy scattered per unit time into the solid angle $d\Omega$ at the distance r from the scatterer is then given by

$$\bar{S}r^2\, d\Omega = \frac{(\omega + \omega_{nn'})^4}{2\pi c^3} \sum_{i=1}^{2} \sum_{\substack{\alpha\gamma \\ \beta\delta}} n_\alpha^i n_\gamma^i P_{\alpha\beta}^{nn'}(\omega) P_{\gamma\delta}^{nn'}(\omega) E_\beta^+ E_\delta^-\, d\Omega \tag{7.6.169}$$

where $\mathbf{n}^1$ and $\mathbf{n}^2$ are two unit vectors perpendicular to the direction of $\mathbf{E}$ and the Raman scattering tensor $P_{\alpha\beta}^{nn'}(\omega)$ has a close relation to the transition probability (7.6.141). If we write (7.6.141) in the form

$$W_{nm'v',nmv}^{(2)} = v(\lambda_1, \mathbf{k}_1)[v(\lambda_2, \mathbf{k}_2) + 1] \mathscr{P}_{nn}^{mm'}(\omega) \mathscr{P}_{nn}^{mm'}(\omega) \delta_\gamma(\Delta E) \hbar^{-1} \tag{7.6.170}$$

where $\mathscr{P}_{nn}^{mm'}(\omega)$ contains the matrix elements of $\mathfrak{P}^-$ and $\mathfrak{P}^+$, i.e., $\mathscr{P}_{nn}^{mm'}(\omega) = \mathscr{P}_{nn}^{mm'}[\omega, \mathfrak{P}^-, \mathfrak{P}^+]$, then the Raman scattering tensor can be obtained from $\mathscr{P}_{nn}^{mm'}(\omega)$ by replacing $\mathfrak{P}^-$ and $\mathfrak{P}^+$ by $\sum_j \mathbf{M}_j$. As we want to work with a totally quantized system and not with a semiclassical version, we do not go into details here, but refer to the original literature where this topic is extensively treated, for instance by Born and Huang (1954) or by di Bartolo and Powell (1976).

The field theoretic cross section consists of counting the outgoing particles in any direction and dividing by the intensity of the initial beam. Furthermore it depends on the distribution of the initial beam with respect to frequency and direction. If the Raman transition probabilities (7.6.139) are weighted with these initial beam distributions, then one obtains the differential cross section for the scattering of a light quantum $\mathbf{k}$ into an element of the solid angle $d\Omega$ by multiplication of this formula with $d\Omega$. As the result depends strongly on the analytic form of the initial distribution which can be varied and modified by the experimental arrangements we do not give any further discussion.

The Raman effect has been extensively discussed in numerous books and review articles. We refer for instance to the books of Born and Huang (1954), Birman (1974), die Bartolo and Powell (1976) and to the review articles of Loudon (1964) and Yacoby (1975). In Loudon's article a comprehensive list of early papers is given. Hence we restrict ourselves to referring to original papers of the last two decades including the Raman effect at lattice polaritons, excitonic polaritons, etc. The following topics were treated: The theory of the first-order Raman effect in crystals by means of a Fockspace representation of the coupled phonon-electron-photon system, Loudon (1963). Theory of stimulated Raman scattering from lattice vibrations, Loudon (1963). Note on the possibility of electronic Raman transition in crystals, Elliot and Loudon (1963). Theory of Raman scattering from crystals by evaluation of SCSF (≡semiclassical scattering formula) with the shell model of phonons, Cowley (1964). Review of non-linear optical effects in crystals including Raman processes, Ovander (1965). Note on Raman scattering by lattice polaritons, Henry and Hopfield (1965). Theory of stimulated Brillouin and Raman scattering, Shen and Bloembergen (1965). Theory of stimulated Raman effect by integration of the Heisenberg equations of motion for the coupled phonon-electron-photon field, Grob (1965). Raman effect for local modes and crystal modes by SCSF, Trifonov and Peuker (1965). Discussion of the second order Raman spectra of caesium halides by investigation of optically active phonon modes, Karo, Hardy and Morrison (1965). Impurity induced first order Raman scattering by alkali halide crystals with SCSF, Xinh, Maradudin and Coldwell-Horsfall (1965). Raman scattering from H^- centers in CaF_2 by SCSF, Ashkin (1965). Relation between second order Raman spectra and phonon dispersion curves, Karo and Hardy (1966), Hardy and Karo (1968). Theory of lattice Raman scattering in insulators by a Fockspace representation of the phonon-electron-photon system taking into account exciton states, Ganguly and Birman (1967). Raman scattering by color centers with SCSF, Benedek and Nardelli (1967). Theory of non-linear optical properties of ionic crystals, Genkin, Fain and Yashchin (1967). Raman scattering of polaritons, Burstein, Ushioda and Pinczuk (1968). Investigation of lattice dynamics and Raman spectra of caesium halides and RbCl based on the shell model, Haridasan and Krishnamurthy (1968). Theory of Raman scattering of excitonic polaritons by phonons based on field theoretic Green functions techniques, Mills and Burstein (1969). Resonant Raman scattering from longitudinal optical phonons in polar semiconductors by means of an interaction representation in Fock space, Hamilton (1969). Raman effect by localized modes with SCSF and application to U-centers in CaF_2, Maradudin and Wallis (1970). Quantum theory of higher order Raman processes by a Fock space representation, Wallis (1970). Theory of Raman scattering of photons by localized polaritons, Mills and Maradudin (1970). Theory of polariton Raman scattering by phonons based on a Hamiltonian in Fock space representation and evaluation of quantum mechanical scattering cross-sections, Bendow and Birman (1970) and Bendow (1970) (1971). Theory of resonant Raman scattering of photons in crystals by phonons via bare excitons, Bendow and Birman (1971). Shell model description of Raman spectra of crystals, Bruce and Cowley (1971). Raman scattering intensity of polaritons for general wave vectors based on a phenomenological model, Unger and Schaack (1971). Raman scattering of photons by plasmons coupled to phonons, Scott, Damen, Ruvalds and Zawadowski (1971). Theory of the one-phonon resonance Raman effect with FTSF (≡field theoretic scattering formula) and Fock space representation, Martin (1971). Theory of inelastic scattering of monochromatic infrared phonons by long-wave length acoustic modes, Humphreys and Maradudin (1972). Impurity induced Raman scattering in CsCl based on SCSF and further development by Xinh et al., Martin (1972). Two-phonon Raman spectra of alkali halide crystals based on SCSF and the breathing shell model, Bruce and Cowley (1972). Raman scattering from polar phonons based on SCSF, Shapiro and Axe (1972). Second order Raman spectrum of NaCl based on SCSF and non-linear shell model, Bruce (1972). First and second order ionic Raman effect based on SCSF and Fockspace representation, Humphreys (1972). Fluctuation-dissipation theorem of polariton Raman scattering in the vicinity of a phonon resonance, Strizhevskii and Obukhovskii (1972). Theory of stimulated Raman scattering by polaritons, Strizhevskii, Obukhovskii and Ponath (1972), Strizhevskii (1972). Biphonons, Fermi resonance and polariton effects in the theory of Raman scattering of light in crystals, Agranovich and Lalov (1972). Second order Raman spectra of fluorides based on shell model calculations, Krishnamurthy and Soots (1972). Theory of multipole-dipole resonance Raman scattering by photons, Birman (1973). Symmetry in the Raman scattering of light by cubic crystals with localized centers, Kaplyanski and Negodyiko (1973). Phonon dispersion relations and two-phonon Raman spectra of CsF, Gupta and Singh (1974). Impurity induced Raman scattering in CsBr and CsI based on SCSF and further development by Xinh et al., Buchanan, Bauhofer and Martin (1974). Second order Raman scattering in alkali fluorite crystals based on SCSF, Cunningham, Sharma, Jaswal, Hass and Hardy (1974). Second order Raman effect in alkaline earth oxides based on SCSF and the shell model, Buchanan, Haberkorn and Bilz (1974). Calculation of second order Raman scattering for KBr, NaCl and MgO crystals based on SCSF and the breathing shell model, Pasternak, Cohen and Gilat (1974). Peculiarities of Raman scattering by polaritons and polar optical phonons in anisotropic crystals, Strizhevskii and Yashkir (1974). Theory of Raman scattering from excitons, Peuker and Enderlein (1974). Raman scattering by polaritons under non-stationary conditions, Marchevskii and Strizhevskii (1975). Theory of multiphonon Raman spectra above the energy gap in semiconductors, Zeyher (1975). Impurity induced first order Raman spectra for alkali halides by means of the deformation dipole model, Karo and Hardy (1975). Multiple-order Raman scattering by a localized mode by a Fockspace representation, Martin (1976). Raman scattering by F-centers in KI, Buisson, Lefrant, Sadoc, Taurel and Billardon (1976). Local approach to polarizabilities and trends in the Raman spectra of semiconductors, Kunc and Bilz (1976). Discussion of the ω^4-law in microscopic theories of inelastic light scattering, Zeyher, Bilz and Cardona (1976). Double resonance in the two-phonon Raman scattering by longitudinal optical phonons with field theoretic Green functions, Abdumalikov and Klochikhin (1976). Time dependent description of resonance Raman scattering by density matrix formalism, Toyozawa (1976). Review of resonant Raman scattering in semiconductors, Richter (1976) and Richter and Zeyher (1976). Second order Raman spectra of RbCl, RbBr and RbI crystals based on SCSF and three-body force shell model, Goel and Dayal (1976). Stimulated Raman scattering from longitudinal and transverse polariton modes in a damped medium Reinisch, Biraud-Laval and Paraire (1976). Theoretical treatment of non-stationary light scattering by phonons and polaritons, Ponath and Schubert (1976). Raman scattering by coupled phonon-plasmon modes in n-GaAs, Katayama and Murase (1977) and Pinczuk, Abstreiter, Trommer and Cardona (1977). Review of the theory of Brillouin scattering of light in piezoelectric semiconductors with acoustic phonon-conduction-electron-interaction, Keller (1976). Hot luminescence and resonant Raman effect of impurity centers, Hizhnyakov and Tehver (1977). First order Raman scattering from off-center Ag^+ in RbCl based on SCSF, Mokross, Dick and Page (1977). Raman process in the spin-lattice relaxation of F-centers in alkali halides, Terrilè, Panepucci and Carvalho (1977). Second order Raman spectra and phonon spectrum of CsF based on SCSF, Kumar, Mahesh and Sharma (1977). Second order Raman scattering in MgO, Jaswal and Dilly (1977). First order Raman scattering induced by NO_2^- in KI, Rebane, Zavt and Haller (1977). Intensity of electronic and vibrational resonant Raman scattering, Tenan and Miranda (1977). Line shape changes in the first order resonance Raman spectra of *F*-centers in KCl as a function of incident wave length, Robbins and Page (1977). Resonant Raman scattering in InP, Sinyukov, Trommer and Cardona (1978). Theory of second order resonance Raman scattering in the case of strong excitonic effects, Bechstedt and Haus (1978). Theory of lineshapes for normal incident Brillouin scattering by acoustic phonons by field theoretic Green function technique, Loudon (1978). Quantum field theoretic formulation of Raman spectral line shapes, McKenzie and Stedman (1978). Multiphonon cascade processes and secondary radiation bands in polar semiconductors, Lang, Pavlov and Yashin (1978). Temperature dependence of second order Raman spectra of RbI based on SCSF, Goel and Dayal (1978). Two-phonon resonant Brillouin scattering in CdS, Yu and Evangilisti (1978). Resonant Raman scattering in GaAs, Trommer and Cardona (1978). First order Raman off- and on-resonance scattering induced by *F*-centers in CsCl and CsBr, Buisson, Lefrant, Ghomi and Taurel (1978). Two-phonon resonant Raman scattering by longitudinal optical phonons, Ivchenko, Lang and Pavlov (1978). Raman scattering from impurities in semiconductors based on SCSF, Barrie and Sharpe (1978). Time reversal symmetry in light-scattering theory, Loudon (1978).

8 External field effects

8.1 State representations with fields

The interactions of photons with electrons and phonons were already extensively discussed in the preceding chapters. These interactions are special cases of the general interactions of a crystal with quantized and nonquantized external fields, i.e., forces. The latter are given by static or quasistatic electric, magnetic, and electromagnetic fields, mechanical strain and pressure, thermal gradients etc. If several of these forces are simultaneously applied to the crystal, the reactions of the crystal with respect to these forces are correlated. We describe these correlations in the framework of irreversible quantum statistics which includes the special case of irreversible linear thermodynamics. The calculations are performed for homogeneous crystals and are developed in such a way that an extension to heterogeneous materials which are used for technical applications and which are not treated here can be performed along the same lines.
Among all interactions of a polar crystal with nonquantized external forces, theoretically and practically the most important ones are the interactions with static (quasistatic) electric and magnetic fields. Under the influence of such fields the electron states as well as the phonon states are deformed and transitions between the deformed states occur. In particular, the electron states show for local levels the Stark effect and the Zeeman effect, while for band states the spectrum is changed by the occurrence of Wannier levels (Stark ladder), Wannier (1959) (1960) (1962), and Landau levels, Landau (1930). Additionally, electrons and holes tunnel between the various states. The theoretical treatment is difficult and still controversial as the interaction energies are described by unbounded or only partially bounded operators which do not satisfy appropriate boundary conditions. Thus it is both of technical and theoretical interest to treat these forces in detail, in particular the case of homogeneous fields, which is the simplest one.
If an external force is switched on, the basic theoretical problem is to decide which part of the interaction is needed for the description of the formation (or deformation) of states and which part can be considered to be responsible for transitions. In ordinary quantum mechanics the influence of an external static electric and (or) magnetic field upon bound states is treated by time-independent perturbation theory, leading to the Stark effect and the Zeeman effect. But this treatment disguises the fact that the action of such a field upon the system causes drastic changes in the spectrum. For the occurrence of bound states, the corresponding potentials have to satisfy boundary conditions which are not fulfilled by the homogeneous external fields. Therefore, by applying such a field to a system with bound states, the spectrum is changed from a partly discrete one into a completely continuous one if no artificial boundary

conditions are imposed on the external field. It is just this effect which makes the use of a simple time-independent perturbation treatment dubious. Thus, in order to get an appropriate description we have to look for another theoretical treatment. As experimentally the field is switched on at a definite time $t=t_0$, it is obvious that a time-dependent description of the process is required.

Considering the set of crystal electrons, we know that at the time $t=t_0$ these electrons have to be localized in the crystal volume. But this means that the wave function $\chi(\mathbf{r},\mathbf{R})$ of these electrons must be normalizable. On the other hand, if we perform a time-dependent description of the process we have to represent $\chi(\mathbf{r},\mathbf{R})$ in terms of stationary eigenfunctions of the crystal with field which in general belong to the continuous spectrum. Denoting these eigenfunctions by the set $\{\chi(\mathbf{r},\mathbf{R},E)\}$ we therefore have for $t=t_0$

$$\chi(\mathbf{r},\mathbf{R})_{t=t_0}=\int c(E,t_0)\chi(\mathbf{r},\mathbf{R},E)\,dE \tag{8.1.1}$$

i.e., $\chi(\mathbf{r},\mathbf{R})$ is a wavepacket. This wavepacket is nontrivial. Since we are required for physical reasons to impose the condition

$$\int\chi(\mathbf{r},\mathbf{R})^{\times}\chi(\mathbf{r},\mathbf{R})\,d^3r=1 \tag{8.1.2}$$

and since the set $\{\chi(\mathbf{r},\mathbf{R},E)\}$ cannot be normalized in this way, the coefficients $c(E,t_0)$ must be spread out over the energy spectrum. Therefore, (8.1.1) cannot be a stationary eigensolution. From this it follows that imposing the normalization condition and requiring $\chi(\mathbf{r},\mathbf{R})$ to be an eigensolution are incompatible postulates. Thus we must choose one or the other.

For the physical reasons mentioned above we decide to choose the former. In this case a general wave function of the crystal $\Psi(t)$ can be represented by an expansion

$$\Psi(t)=\sum_{nm}c_{nm}(t)\chi_n(\mathbf{r},\mathbf{R})\varphi_m^n(\mathbf{R}) \tag{8.1.3}$$

where $\{\chi_n\}$ is a set of orthonormalized wavefunctions. These functions cannot be eigenfunctions of the adiabatic electron energy operator including the external field. However, we can form the expectation values of these functions with respect to this electronic adiabatic energy operator. Then in this case these values can replace the adiabatic lattice potential $U_n(\mathbf{R})$ and the lattice functions $\varphi_m^n(\mathbf{R})$ can be defined with respect to these modified potentials.

From this consideration it follows that for both a reversible and an irreversible evolution of the system with field, non-diagonal elements of the energy will appear. The transitions between various configurations $\chi_n\varphi_m^n$ which are induced by these nondiagonal elements are just those responsible for tunnel effects. In order to use a system of normalizable wavefunctions which enables us to fulfill the timescale postulates of the irreversible kinetics, it is necessary to look for a system which is (under the subsidiary condition of normalizability) optimally adapted to stationary states. In the framework of the theory developed so far it is obvious that we have to minimize the corresponding energy expectation values. This leads to the bound state field effects, while as mentioned above the nondiagonal energy elements lead to tunneling. Thus, under the assumption of normalizable states one gets a unique

description of all types of effects which ordinarily are not brought together. This program will be performed in detail for polar crystals in the following.
Concerning the minimization of energy expectation values, one has to distinguish between the minimization of the energy of bound states (local impurity center states) and of quasibound states (valence band and conduction band states). For bound states the inclusion of the external fields leads to Stark levels and Zeeman levels. For band states, calculations with simple one-electron Hamiltonians show that exactly in the approximation which corresponds to the minimization of the energy expectation value, Wannier levels and Landau levels occur, cf. Callaway (1974). If such improved valence band and conduction band states are used for the calculation of transition probabilities, the tunneling rate just exhibits the appropriate field dependence, namely $\exp(-c/E)$ for electric field ionization, i.e., interband transitions, Callaway (1974). Thus it seems to be desirable to use the improved state functions both for the bound states and for band states. On the other hand, it will be shown in the following section that the use of simple Bloch functions instead of field improved band functions leads to the Boltzmann equation with the ordinary drift term which is basic for nearly all phenomenological transport calculations. In order to connect our formalism with these phenomenological calculations, we therefore use Bloch functions instead of improved band functions.
If the improved band functions are used, the drift term is changed and the conventional Boltzmann equation with respect to Bloch functions has to be reformulated. Such a modified Boltzmann equation for external magnetic and electric fields which is split up into a set of Boltzmann equations for various Landau levels has been derived by Argyres (1958). Magnusson (1972) (1973) has used such equations for the calculation of electronic transport properties in polar semiconductors. Tausendfreund (1974) has shown that an exact integration of the drift term of the conventional Boltzmann equation for the case of an external electric field leads to the Wannier levels (Stark ladder) but the transition probabilities do, of course, not show the right field dependence if they are calculated with Bloch functions in first order perturbation theory. On the other hand, in our approach we use not the perturbative transition probabilities, but the resonance transition probabilities. A detailed investigation has not yet been made as to whether the resonance transition probabilities in the Bloch representation do give the correct field dependence or not. But apart from that it has to be emphasized that in our approach the question of the use of Bloch functions or of improved band functions is only a technical one and is not basic, since the calculations with both kinds of functions run in principle along the same lines. We will use the Bloch functions since they have the advantage of not depending on the special external forces. Then we have to consider only the modifications of the bound states. We discuss this for the sake of simplicity only for electric fields.
In order to obtain a simple system of rate equations we use for the description of field effects for the electronic states the H.F.-representation. Of course, more complicated representations such as configuration interaction are not excluded but they lead to a much more complicated reaction kinetics which we do not want to consider here.
The H.F.-representation is characterized by an appropriate set of one-electron states. The appropriate choice of these one-electron functions is determined by the minimum

principle of the H.F.-representation and by the crystal model to be used. As in the preceding chapters we use also in this chapter the simplest model of a defect crystal as defined in Section 3.1. The corresponding set of one-electron states in the field-free case is given by (3.1.1). In this set we have to distinguish between two types of states, namely bound states belonging to the discrete part of the spectrum and quasi-free states belonging to the quasi-continuous part of the spectrum. The valence band and conduction band states belong to the latter part, while the impurity center states etc. belong to the former part. The notation "quasi-continuous" is used since that part of the spectrum depends upon the boundary conditions which are used. If we consider an infinite crystal, the band levels are really continuous, in other cases like periodicity or even large but finite crystal volumes these levels are very close together. Switching on the static external field, a drastic change in the spectrum of the total energy operator takes place. The mixed discrete-quasicontinuous spectrum is changed into a pure quasi-continuous spectrum. According to our program for an appropriate description of the field effects in the framework of statistical quantum theory we have, in spite of this drastic change in the spectrum, to maintain the fieldfree classification of states, with the only difference that now the normalizable states are calculated by a minimum condition of the H.F.-energy including the external field.
To obtain nontrivial results we have, of course, to use the complete model of the perturbed crystal where s anion vacancies and s impurity center or conduction band electrons are present. However, with respect to the effective calculation of bound states we have suppressed the interaction between different centers. Thus, concerning state calculations (but not reaction kinetics) we work with the simpler model of one vacancy present in a practically infinite crystal colume (microblock model). In this section we consider a practically infinite crystal with one F-center, i.e., a microblock which is under the influence of a homogeneous static electric field. For this case, Renn (1975) has derived the following theorem:

Theorem 8.1 : If the total electronic wave functions $|\chi_n\rangle$ are represented by H.F.-states, the corresponding static one-electron states $\psi_s=\psi_n$ for F-centers under the influence of a homogeneous static electric field $\mathbf{E}$ have to be obtained by minimizing the expression

$$\begin{aligned} U_n[\psi,\mathbf{E}] &= U_0+\frac{1}{8\pi}\int \mathbf{E}^2 d^3r+U_n[\psi] \\ &\quad +e\int \varrho(\mathbf{r})V(\mathbf{r})d^3r+(\hat{\boldsymbol{\varrho}}-\mathbf{E})\cdot\mathfrak{K}\cdot(\hat{\boldsymbol{\varrho}}-\mathbf{E}) \\ &\quad +(\hat{\boldsymbol{\varrho}}-\mathbf{E})\cdot[\mathfrak{A}\cdot\mathfrak{G}-\mathfrak{g}\cdot\mathfrak{C}\cdot\mathfrak{A}\cdot\mathfrak{G}]\cdot\hat{\mathbf{v}}-\frac{1}{2}\,\hat{\mathbf{v}}\cdot\mathfrak{C}\cdot\hat{\mathbf{v}} \end{aligned} \tag{8.1.4}$$

with the charge tensor $\mathfrak{A}:=a_\mu\delta_{l\mu k\varrho}$ and with

$$\begin{aligned} \mathfrak{K} &:= -\frac{1}{2}\,\mathfrak{g}-\frac{1}{2}\,\mathfrak{A}\cdot\mathfrak{G}\cdot\mathfrak{A}+\mathfrak{A}\cdot\mathfrak{G}\cdot\mathfrak{A}\cdot\mathfrak{C}\cdot\mathfrak{g} \\ &\quad -\frac{1}{2}\,\mathfrak{g}\cdot\mathfrak{C}\cdot\mathfrak{A}\cdot\mathfrak{G}\cdot\mathfrak{A}\cdot\mathfrak{C}\cdot\mathfrak{g} \end{aligned} \tag{8.1.5}$$

where the definitions of the operators are given in the proof.

Proof: The proof runs analogously to that of Theorem 5.2. The only difference occurs due to the modified adiabatic energy which in the field case is given by

$$h(\mathbf{r},\mathbf{R},\mathbf{E}) := h(\mathbf{r},\mathbf{R}) + \sum_{i=1}^{M+1} eV(\mathbf{r}_i) + \sum_{\{l\mu\}} e_{l\mu} V(\mathbf{R}_{l\mu}) + \frac{1}{8\pi}\int \mathbf{E}^2 d^3r \tag{8.1.6}$$

with $\mathbf{E} = -\nabla V$. If we now define

$$U_n[\mathbf{R},\psi,\mathbf{E}] := \langle \chi_n | h(\mathbf{r},\mathbf{R},\mathbf{E}) | \chi_n \rangle \tag{8.1.7}$$

we obtain by introducing electronic and ionic dipole moments

$$\begin{aligned} U_n[\mathbf{M},\mathbf{m},\psi_n,\mathbf{E}] = U_0 + \frac{1}{8\pi}\int \mathbf{E}^2 d^3r + U_n[\psi_n] \\ + e\int \varrho(\mathbf{r}) V(\mathbf{r}) d^3r + \frac{1}{2}\sum_{\{l\mu\}} (\alpha_\mu^{-1}\mathbf{m}_{l\mu}^2 + \tau_\mu^{-1}\mathbf{M}_{l\mu}^2) \\ + \sum_{\{l\mu\}} \{(\mathbf{m}_{l\mu} + a_\mu \mathbf{M}_{l\mu}) \cdot \nabla_{l\mu}[\int \hat{\varrho}(\mathbf{r}) C(\mathbf{r},l\mu) d^3r + V(\mathbf{R}_{l\mu}^0)] \\ - \mathbf{M}_{l\mu} \cdot \nabla_{l\mu} \hat{V}^{\mathrm{ex}}(s,l\mu)\} \\ + \frac{1}{2}\sum_{\substack{l\mu \\ k\varrho}}' \{(\mathbf{m}_{l\mu} + a_\mu \mathbf{M}_{l\mu}) \cdot \nabla_{l\mu}\nabla_{k\varrho} C(l\mu,k\varrho) \cdot (\mathbf{m}_{k\varrho} + a_\varrho \mathbf{M}_{k\varrho}) \\ + \mathbf{M}_{l\mu} \cdot \nabla_{l\mu}\nabla_{k\varrho} C^{\mathrm{ex}}(l\mu,k\varrho) \cdot \mathbf{M}_{k\varrho}\} \end{aligned} \tag{8.1.8}$$

with

$$U_n[\psi] := -\frac{\hbar^2}{2m}\int \psi(\mathbf{r})^\times \Delta\psi(\mathbf{r}) d^3r + e\int \varrho(\mathbf{r}) \left[\sum_{l\mu} a_\mu C(\mathbf{r},l\mu)\right] d^3r \tag{8.1.9}$$

and

$$\hat{\varrho}(\mathbf{r}) = e\varrho(\mathbf{r}) - a_{l^i\mu^i}\delta(\mathbf{r} - \mathbf{R}_{l^i\mu^i}^0). \tag{8.1.10}$$

From the minimum condition with respect to the electronic dipoles

$$\frac{\partial}{\partial \mathbf{m}_{l\mu}} U_n[\mathbf{M},\mathbf{m},\psi_n,\mathbf{E}] = 0 \tag{8.1.11}$$

we obtain for these dipoles the values

$$\mathbf{m}_{k\varrho} = -\sum_{l\mu} \mathscr{g}_{k\varrho l\mu} \cdot \{\nabla_{l\mu}[\int \hat{\varrho}(\mathbf{r}) C(\mathbf{r},l\mu) d^3r + V(\mathbf{R}_{l\mu}^0)] + \sum_{n\kappa} \mathscr{C}_{l\mu n\kappa} a_\kappa \cdot \mathbf{M}_{n\kappa}\} \tag{8.1.12}$$

For brevity we use a matrix notation in (8.1.4) and (8.1.5)

$$\mathfrak{g}^{-1} := \mathscr{g}_{l\mu k\varrho}^{-1} := \alpha_\mu^{-1}\delta_{l\mu k\varrho} + \mathscr{C}_{l\mu k\varrho} \tag{8.1.13}$$

$$\mathfrak{C} := \mathscr{C}_{l\mu k\varrho} := \nabla_{l\mu}\nabla_{k\varrho} C(l\mu,k\varrho)\,(1 - \delta_{l\mu k\varrho}) \tag{8.1.14}$$

and a vector notation $\mathbf{m} := \mathbf{m}_{k\varrho}, \mathbf{M} := \mathbf{M}_{k\varrho}$ with corresponding multiplication formulae. Eliminating now the electronic dipoles from the total energy expression (8.1.8) by means of (8.1.12), we get the formula

$$U_n[\mathbf{M},\psi,\mathbf{E}] = U_0 + \frac{1}{8\pi}\int \mathbf{E}^2 d^3r + U_n[\psi]$$

$$+ e\int \varrho(\mathbf{r})V(\mathbf{r})d^3r - \frac{1}{2}\sum_{\substack{l\mu\\hv}}(\hat{\boldsymbol{\varrho}}_{l\mu} - \mathbf{E}_{l\mu})\cdot g_{l\mu hv}\cdot(\hat{\boldsymbol{\varrho}}_{hv} - \mathbf{E}_{hv})$$

$$+ \sum_{\substack{l\mu\\hv}} a_\mu \mathbf{M}_{l\mu}\cdot\left(\delta_{l\mu hv} - \sum_{g\kappa}\mathscr{C}_{l\mu g\kappa}\cdot g_{g\kappa hv}\right)\cdot(\hat{\boldsymbol{\varrho}}_{hv} - \mathbf{E}_{hv}) \tag{8.1.15}$$

$$+ \frac{1}{2}\sum_{\substack{l\mu\\k\varrho}}{}' \mathbf{M}_{l\mu}\cdot\mathscr{G}^{-1}_{l\mu k\varrho}\cdot\mathbf{M}_{k\varrho} - \sum_{l\mu}\mathbf{M}_{l\mu}\cdot\hat{\mathbf{v}}_{l\mu}$$

with the definitions

$$(\hat{\boldsymbol{\varrho}}_{l\mu} - \mathbf{E}_{l\mu}) := \nabla_{l\mu}\left[\int \hat{\varrho}(\mathbf{r})C(\mathbf{r},l\mu)d^3r + V(\mathbf{R}^0_{l\mu})\right];$$

$$\hat{\mathbf{v}}_{l\mu} = \nabla_{l\mu}V^{\mathrm{ex}}(s,l\mu) \tag{8.1.16}$$

$$\mathfrak{G}^{-1} := \mathscr{G}^{-1}_{l\mu hv} :=$$

$$\left\{\tau_\mu^{-1}\delta_{l\mu hv} + a_\mu\mathscr{C}_{l\mu hv}a_v - \sum_{\substack{k\varrho\\g\kappa}} a_\mu\mathscr{C}_{l\mu k\varrho}\cdot g_{k\varrho g\kappa}\cdot\mathscr{C}_{g\kappa hv}a_v + \nabla_{l\mu}\nabla_{hv}C^{\mathrm{ex}}(l\mu,hv)\right\}. \tag{8.1.17}$$

The minimum condition with respect to the lattice coordinates

$$\frac{\partial}{\partial\mathbf{M}_{l\mu}}U_n[\mathbf{M},\psi,\mathbf{E}] = 0 \tag{8.1.18}$$

leads to the values

$$\mathbf{M}_{hv} = -\sum_{\substack{l\mu\\j\alpha}}\mathscr{G}_{hvl\mu}a_\mu\cdot\left[\delta_{l\mu j\alpha} + \sum_{k\varrho}\mathscr{C}_{l\mu k\varrho}\cdot g_{k\varrho j\alpha}\right]\cdot(\hat{\boldsymbol{\varrho}}_{j\alpha} - \mathbf{E}_{j\alpha}) + \sum_{l\mu}\mathscr{G}_{hvl\mu}\cdot\hat{\mathbf{v}}_{l\mu} \tag{8.1.19}$$

of the displacement dipoles $\mathbf{M} = \mathbf{M}^n$. Eliminating these dipoles from (8.1.15) gives (8.1.4) with definition (8.1.5), Q.E.D.

This theorem shows that the derivation of effective energy expressions for one-electron states, as given in Theorem 5.2, can be extended to also include the case of external fields. However, it should be noted that the representation of an effective energy expression of the kind (8.1.4) is purely formal as long as the operators $\mathfrak{K}$, $\mathfrak{G}$, g, etc., are not explicitly calculated. Their calculation has to be performed by means of the techniques applied in Theorem 5.2. Since such calculations have been discussed in detail in Theorem 5.2, we do not repeat these procedures here.

We now proceed to the nondiagonal elements of the energy. Using a definite state representation we define the interactions by the nondiagonal elements of the total energy given in (2.1.6). Considering now a state system which is not completely adiabatic as is the case with the nonadiabatic Bloch functions and the adiabatic bound states, the meaning of the operator K^t has to be modified. According to Stumpf and Kleih (1980) the following theorem holds:

Theorem 8.2 : If in an H.F.-representation of the electronic state system $\{\chi_n\}$ the one-electron state functions of the discrete part of the spectrum are calculated with adiabatic electron-lattice coupling for the perturbed crystal, while the one-electron states of the continuous spectrum are chosen to be Bloch functions of the ideal crystal, then

$$K|\chi_n\rangle|\varphi_m^n\rangle=(E_m^n+K^t+K^s)|\chi_n\rangle|\varphi_m^n\rangle \tag{8.1.20}$$

is valid where

$$\begin{aligned}
&\langle\lambda_1\ldots\lambda_\alpha\ldots\lambda_z,m|K^t|\lambda_1\ldots\lambda'_\alpha\ldots\lambda_z,m'\rangle\equiv(3.3.13),\ \lambda'_\alpha\equiv\text{bound state}\\
&\qquad\qquad\qquad\qquad\qquad\qquad\equiv\ \ 0\quad,\ \lambda'_\alpha\equiv\text{cont. state}\\
&\langle\lambda_1\ldots\lambda_\alpha\ldots\lambda_z,m|K^s|\lambda_1\ldots\lambda'_\alpha\ldots\lambda_z,m'\rangle\equiv(3.3.18),\ \lambda'_\alpha\equiv\text{bound state}\\
&\equiv\langle\lambda_1\ldots\lambda_\alpha\ldots\lambda_z,m|\frac{\delta}{\delta\psi_{\lambda_\alpha}(\mathbf{r})^\times}\frac{\delta}{\delta\psi_{\lambda_\alpha}(\mathbf{r})}U_n[\mathbf{M},\psi_{\lambda_1}\ldots\psi_{\lambda_s}]|\lambda_1\ldots\lambda'_\alpha\ldots\lambda_z,m'\rangle,\\
&\lambda'_\alpha\equiv\text{cont. state}
\end{aligned} \tag{8.1.21}$$

with definition (8.1.27) and

$$\langle\lambda_1\ldots\lambda_\alpha\ldots\lambda_\beta\ldots\lambda_z,m|K^t|\lambda_1\ldots\lambda'_\alpha\ldots\lambda'_\beta\ldots\lambda_z,m'\rangle\equiv0,\ \lambda'_\alpha\equiv\text{cont. state or }\lambda'_\beta\equiv\text{cont. state}$$

while the two-particle transitions for K^s are unchanged.

Proof: To obtain the matrix elements of the interaction operators K^t and K^s we have to apply the crystal operator K to the state under consideration and afterwards to form the scalar product. Since the matrix elements for $\lambda'_\alpha\equiv$bound state are given by (3.3.13) (3.3.18), we have only to consider the case $\lambda'_\alpha\equiv$cont. state. Writing the definition (2.2.5) in the following way

$$K:=h(\mathbf{r},\mathbf{R})-\sum_{k=1}^{N}\frac{\hbar^2}{2M_k}\Delta_k \tag{8.1.22}$$

we observe that the interaction operator K^t results from the kinetic energy of the nuclei. The one-electron transition elements of this operator are given by (3.3.8). If the function $\psi^{n'}_{\lambda'_\alpha}(\mathbf{r},q^{n'})$ is a Bloch function of the ideal crystal then it does not depend on $q^{n'}$ and, therefore, this element vanishes. Thus we only have to consider the action of $h(\mathbf{r},\mathbf{R})$ on the state under consideration. The corresponding matrix elements are given by (3.3.18). To evaluate them we have to rewrite this expression. The adiabatic energy (3.3.15) is given equivalently by

$$U_n(\mathbf{R}) = \langle \chi_n | h(\mathbf{r},\mathbf{R}) | \chi_n \rangle_{\{\psi\}\,\mathrm{Min}} \tag{8.1.23}$$

i.e., the expectation value has to be taken at its minimum achieved by variation of the wave functions. Using the explicit expression (3.3.16) it can easily be seen that (3.3.18) can be written in the form

$$\begin{aligned} &\langle \lambda_1 \ldots \lambda_\alpha \ldots \lambda_z | h(\mathbf{r},\mathbf{R}) | \lambda_1 \ldots \lambda'_\alpha \ldots \lambda_z \rangle \\ &= \int \psi^n_{\lambda_\alpha}(\mathbf{r})^\times \left\{ \frac{\delta}{\delta \psi^n_{\lambda_\alpha}(\mathbf{r})^\times} \frac{\delta}{\delta \psi^n_{\lambda_\alpha}(\mathbf{r})} \langle \chi_n | h(\mathbf{r},\mathbf{R}) | \chi_n \rangle \right\}_{\{\psi\}\mathrm{Min}} \psi^{n'}_{\lambda_\alpha}(\mathbf{r})\, d^3r. \end{aligned} \tag{8.1.24}$$

The functional derivation of $\langle \chi_n | h | \chi_n \rangle$ is taken only with respect to the states involved in reaction processes, i.e., not with respect to ion shell functions. Thus we can commute differentiation and substitution of minimal ion shell functions. The expression $\langle \chi_n | h | \chi_n \rangle$, where only the minimum of the ion shell functions has been substituted, was discussed in detail in Chapter 5. It was formally denoted by

$$\langle \chi_n | h | \chi_n \rangle_{\substack{\{\psi\}\,\mathrm{Min} \\ \mathrm{ion\,shell}}} =: U_n[\mathbf{M}, \psi^n_{\lambda_1} \ldots \psi^n_{\lambda_s}] \tag{8.1.25}$$

where s electrons are in states $\{\psi^n_{\lambda_i}, 1 \leqslant i \leqslant s\}$ which are actively involved in reactions. Thus we may suppress all other electron quantum numbers and may write $|\lambda_1 \ldots \lambda_z\rangle \equiv |\lambda_1 \ldots \lambda_s\rangle$. For $s=1$ the expression (8.1.25) is given in the dipole approximation by (5.2.56). For $s>1$, the generalization of (5.2.56) was discussed in Theorem 5.5. According to this theorem, (8.1.25) is given by

$$\begin{aligned} U_n[\mathbf{M}, \psi^n_{\lambda_1} \ldots \psi^n_{\lambda_s}] :&= U_s(\mathbf{M}) + H(s) \\ &+ \sum_{\{l\mu\}} \left[\int \varrho_s(\mathbf{r}) e a_{l\mu} C(\mathbf{r}, l\mu)\, d^3r + V^{\mathrm{ex}}(s, \mathbf{R}_{l\mu}) \right] \\ &- \frac{1}{2} \frac{\alpha^e}{1+\alpha^e} \int \hat{\varrho}_s(\mathbf{r}) F(\mathbf{r},\mathbf{r}') \hat{\varrho}_s(\mathbf{r})\, d^3r\, d^3r' \\ &+ \frac{1}{1+\alpha^e} \sum_{\{l\mu\}} \int \hat{\varrho}_s(\mathbf{r}) a_{l\mu} \mathbf{M}_{l\mu} \cdot \boldsymbol{\nabla}_{l\mu} C(\mathbf{r}, l\mu)\, d^3r \end{aligned} \tag{8.1.26}$$

where $U_s(\mathbf{M})$ does not depend on $\{\psi_{\lambda_i}, 1 \leqslant i \leqslant s\}$ and $H(s)$ is given by (5.2.26) with $\varrho \equiv \varrho_s$, while $\varrho_s(\mathbf{r})$ and $\varrho_s(\mathbf{r},\mathbf{r}')$ are defined by (5.4.26). Finally $\hat{\varrho}_s(\mathbf{r})$ is defined by (5.4.33). The derivation of (8.1.26) with respect to $\psi^{n\times}_{\lambda_\alpha}$ and $\psi^n_{\lambda_\alpha}$ gives then

$$\begin{aligned} &\frac{\delta}{\delta \psi^n_{\lambda_\alpha}(\mathbf{r})^\times} \frac{\delta}{\delta \psi^n_{\lambda_\alpha}(\mathbf{r})} U_n[\mathbf{M}, \psi^n_{\lambda_1} \ldots \psi^n_{\lambda_s}] = \\ &-\frac{\hbar^2}{2m} \Delta + e \sum_{\{l\mu\}} a_{l\mu} C(\mathbf{r}, l\mu) + \frac{e}{1+\alpha^e} \sum_{\{k\varrho\}} a_{k\varrho} (\mathbf{M}_{k\varrho} \cdot \boldsymbol{\nabla}_{k\varrho}) C(\mathbf{r}, k\varrho) \\ &+ \int e^2 C(\mathbf{r},\mathbf{r}') \varrho_s(\mathbf{r}')\, d^3r' - \frac{\alpha^e e}{1+\alpha^e} \int F(\mathbf{r},\mathbf{r}') \hat{\varrho}_s(\mathbf{r}')\, d^3r' + \text{exch. terms.} \end{aligned} \tag{8.1.27}$$

By definition, the Bloch states are given by the wave functions of an excess electron in the ideal crystal. They therefore satisfy the equation

$$\left[-\frac{\hbar^2}{2m}\Delta+e\sum_{l\mu}^{M}a_{l\mu}C(\mathbf{r},l\mu)+\text{exch. terms}\right]\psi(\mathbf{r},\mathbf{K})=E(\mathbf{K})\psi(\mathbf{r},\mathbf{K}). \tag{8.1.28}$$

Identifying $\psi(\mathbf{r},\mathbf{K})$ with $\psi^{n'}_{\lambda'_\alpha}(\mathbf{r})$ and observing (8.1.27) and (8.1.28) the matrix element (8.1.24) goes over into

$$\begin{aligned}
&\langle\lambda_1\ldots\lambda_\alpha\ldots\lambda_z|h(\mathbf{r},\mathbf{R})|\lambda_1\ldots\lambda'_\alpha\ldots\lambda_z\rangle\\
&=\int\psi^n_{\lambda_\alpha}(\mathbf{r})^\times E(\mathbf{K})\psi(\mathbf{r},\mathbf{K})d^3r\\
&-\int\psi^n_{\lambda_\alpha}(\mathbf{r})^\times\left\{\sum_{i=1}^{s}\left[ea_{l^i\mu^i}C(\mathbf{r},l^i\mu^i)-e^2\int C(\mathbf{r},\mathbf{r}')\varrho^n_{\lambda_i}(\mathbf{r}')d^3r'\right]\right\}\psi(\mathbf{r},\mathbf{K})d^3r \qquad (8.1.29)\\
&+\frac{e}{1+\alpha^e}\int\psi^n_{\lambda_\alpha}(\mathbf{r})^\times\sum_{\{k\varrho\}}a_{k\varrho}(\mathbf{M}_{k\varrho}\cdot\nabla_{k\varrho})C(\mathbf{r},k\varrho)\psi(\mathbf{r},\mathbf{K})d^3r\\
&-\frac{\alpha^e e}{1+\alpha^e}\int\psi^n_{\lambda_\alpha}(\mathbf{r})^\times F(\mathbf{r},\mathbf{r}')\hat{\varrho}_s(\mathbf{r}')d^3r'\psi(\mathbf{r},\mathbf{K})d^3r+\text{exch. terms.}
\end{aligned}$$

Due to the orthonormality which is assumed to hold approximately also in this case, the first term on the right-hand side of (8.1.29) vanishes, while the second term just gives the sum over the impurity lattice potential and the interaction with the other active electrons. The third term gives the interaction with the lattice deformations, while the fourth term describes the interaction with the selftrapping potential of the lattice (polaron-exciton-selftrapping). The exchange terms are only formally included as they are very complicated in calculations.
Arguing along the same lines for two-particle transitions it can be shown by means of the results of Chapter 4 that K^t transition elements vanish, while K^s transition elements remain unchanged, Q.E.D.

In addition to the modification of the transition probabilities caused by the partly modified set of one-electron functions, transition operators occur which are due to the non-diagonal elements of $h(\mathbf{r},\mathbf{R},\mathbf{E})$ with respect to the external fields. This calculation is straightforward, so that no further comment is needed. We will take such transitions into account in the reaction kinetic calculations in the following section.
Concerning the literature, besides the elementary treatment of this topic in textbooks an extensive treatment was given in original papers. It is, however, a striking fact that in none of these papers the program of a clear separation between the reversible and the irreversible parts of the interaction has been pursued consequently. Therefore, with respect to this basic problem, a confusion occurs and the topic is not treated systematically from a unique point of view. Here we cite papers which are concerned with the pure state calculations. The majority of these papers deals with very simple one-electron Hamiltonians, and only a minority takes into account more complicated Hamiltonians with inclusion of electron-phonon coupling etc. We cite only papers of

the last two decades and refer for previous papers to the literature. The following topics have been treated:
Theory of solution of a one-electron equation with periodic potential (Bloch electron) in a magnetic field, Kohn (1959). Theory of internal field emission by means of a many-electron Hamiltonian in the rigid lattice and H.F.-states built up from Bloch electron states in an electric field, Kamphusmann (1959). Discussion of Bloch electron solutions in electric and magnetic fields, Wannier (1962). Improvement of Kohn's approach to Bloch electrons, Blount (1962). Derivation of an effective Hamiltonian for Bloch electrons in a magnetic field, Roth (1962). Derivation of effective Hamiltonians and discussion of general properties of the wave functions for Bloch electrons in homogeneous fields, Wannier and Fredkin (1962). Discussion of energy degeneracy of Bloch electrons in a magnetic field, Fischbeck (1963). Approximate treatment of Bloch electrons in external fields, Schnakenberg (1963). Lattice polarons in an electric field by performing a time-dependent perturbation theory of the electron-lattice interaction based on Bloch electron states with field, Larsen (1964). Definition of the magnetic translation group, investigation of its representations and group theoretical analysis of the influence of a periodic potential on the Landau levels, Zak (1964). Solution procedure for Bloch electrons in an external homogeneous magnetic field, Janussis (1964). Quantum levels of Bloch electrons in a semiconductor for a strong electric field, Bychkov and Dykhne (1965). Bloch electrons in external electric and magnetic fields, Praddaude (1965). Investigation of the interaction of optical phonons with magnetoplasma waves in ionic semiconductors based on phenomenological field equations, Casselman and Spector (1965). The wave function of a Bloch electron in a magnetic field, Chambers (1966). Motion of Bloch electrons in an electric field and calculation of the current, Kümmel (1966). Theory of weakly bound Bloch electrons in a magnetic field, Fischbeck (1967). Exact solution of the effective one-electron Hamiltonian of Luttinger and Kohn (1955) for the description of magnetic breakdown, Ruvalds and McClure (1967). Calculation of the linear Stark effect of the R-center (≡three *F*-centers forming an equilateral triangle) in a molecular model, Lanzl (1967). Theory of the Zeeman effect for rare-earth ions in a crystal field with c_{3h} symmetry, Syme, Haas, Spedding and Good (1968). General theory of the Bloch electron in a magnetic field, Fischbeck (1968). Lattice polaron in a magnetic field, Bajaj (1968). Theory of shallow *F*-centers in magnetic or electric fields by means of Pekar's continuum model, Perlin and Gifeisman (1968). Theory of excitons in high magnetic fields based on an effective two-particle Hamiltonian, Fritsche (1969). Stark effect for Bloch electrons in electric fields and related optical absorption with investigation of the combined action of Stark shift and tunneling for pure states, Enderlein, Keiper and Tausendfreund (1969). A perturbation approach to Bloch electrons in a magnetic field, Morris (1969). Symmetries of the Zeeman effect for ions in crystals, Earney (1969). Review of the calculation and properties of electronic band structures in external magnetic fields, Langbein (1969). State calculation for semiconductor electrons in interaction with a strong classical electromagnetic wave based on a Fock space representation, Galitskii, Goreslavskii and Elesin (1970). Review of the theory of Bloch electrons in a magnetic field, Fischbeck (1970). Calculation of groundstate energies and effective masses for strong coupled optical and piezoelectric

lattice polarons in a magnetic field, Porsch (1970). Interaction between optical phonons and electrons in a quantizing magnetic field by fieldtheoretic Green functions technique, Demikhovskii and Protogenov (1970). Coupling of helicons (≡magneto-plasma waves propagating parallel to an applied static magnetic field in a degenerate polar semiconductor) to transversal optical phonons by phenomenological equations, Katayama and Yokota (1970). Investigation of small gap semiconductors in crossed magnetic and electric field based on Bloch electron equations, Zawadzki and Kowalski (1971). Groundstate of an exciton in a magnetic field by a two-particle Hamiltonian, Cabib, Fabri and Fiorio (1971). A perturbation approach to Bloch electrons in crossed electric and magnetic fields, Godley and Morris (1971). Review of band structure calculations and properties for Bloch electrons in magnetic fields, Langbein (1971). Treatment of Bloch electrons in a finite-range constant electric field with the result that no Stark ladder appears, Rabinovitch and Zak (1971). Representations of the invariance group for a Bloch electron in a magnetic field, Boon (1972). Bound electron and phonon states in a strong magnetic field due to non-linear electron-phonon interaction, Levinson and Rashba (1972). Effects of the Coulomb interaction on excitonic energy levels in a magnetic field for InSb based on a two-particle model, Rees (1972). Derivation of an effective Schrödinger equation for an exciton in the rigid lattice with external magnetic fields in InSb-type crystals, Zhilich (1972). Calculation of phonon states of an ionic crystal in a strong static magnetic field, Holz (1972). Note on bound states of electrons, holes and phonons in a strong magnetic field, Levinson (1972). Investigation of the spectra and states of slow conduction electrons interacting with lattice vibrations in a strong magnetic field, Kukushkin (1973). Ground state energy and longitudinal effective mass of the lattice polaron in a magnetic field, Kartheuser and Negrete (1973). Derivation of an effective Hamiltonian for a lattice polaron in an external magnetic field and investigation of its properties, Röseler, Henneberger and Fischbeck (1973). Rigorous perturbation theory for Bloch electrons in a uniform electric field, Moyer (1973). Stark effect of free and bound excitons in a CdSe crystal based on a two-particle Hamiltonian, Razbirin, Uraltsev and Bogdanov (1973). Exciton levels in a magnetic field based on a two-particle Hamiltonian, Lee, Larsen and Lax (1973). Theory of the dynamic Stark effect and Franz-Keldysh effect in a resonant electromagnetic radiation field by means of field theoretic Green functions technique, Becker, Enderlein and Peuker (1973). A selfconsistent treatment of Bloch electrons in an electric field, Henneberger and Röseler (1973). Semiconductor impurity and exciton levels in a magnetic field based on a one-particle model, Roussel and O'Connell (1974). The fine structure of Wannier-Mott excitons in a cubic crystal in an electric field based on a two-particle Hamiltonian, Agekyan, Monozon and Shiryapov (1974). The effect of high magnetic fields on degenerate band excitons in InSb based on a two-particle Hamiltonian, Martin and Wallis (1974). Exciton states of semiconductors in a high magnetic field based on a two-particle Hamiltonian, Altarelli and Lipari (1974). Discussion of the compatibility of the momentum representation with infinite extended crystals for pure states, Rabinovitch (1974). Influence of magnetic fields on polariton states in uniaxial polar semiconductors, Gurevich and Tarkhanian (1974). Excitons in semiconductors with non-spherical bands in an external magnetic field based on a two-particle model, Zhilich and

Monozon (1975). Groundstate energy and wave function of an exciton in arbitrary homogeneous magnetic fields based on a two-particle model, Ekardt (1975). 1*s* and 2*s* states of a direct exciton in a magnetic field based on two-particle model, Swierkowski (1975). Theory of Bloch electrons in a homogeneous electric field, Fiddicke and Enderlein (1975). Phenomenological model for the Stark effect of the *F*-center in KCl, Bonciani and Grassano (1975). Localized states and effective Hamiltonians for Bloch electrons in external fields, Zak (1976). Groundstate energy of a lattice exciton in a magnetic field, Behnke, Büttner and Pollmann (1976). Bound states of large radius excitons in crossed fields based on a two-particle Hamiltonian, Monozon (1976). Energy levels and wave functions of Bloch electrons in magnetic fields, Hofstadter (1976). Exchange splitting and Zeeman effect in direct excitons based on a two-particle model, Suffczynski and Swierkowski (1976). Localized electron states at paramagnetic impurity centers in quantizing magnetic fields, Savvinykh (1976). Polaron mechanism of formation of a narrow band in quantizing magnetic fields, Osipov (1976). Groundstate energy and effective mass of a lattice polaron with an arbitrary coupling force in a magnetic field, Sheka, Khazan and Mozdor (1976). Calculation of the state density for an insulator in an external inhomogeneous electric field in a one-particle model, Zaiko (1976). True width of electron levels in a crystal in a static electric field based on a one-particle model, Berezhkovskii and Ovchinnikov (1976). Variational calculation of the groundstate energy of the lattice polaron Hamiltonian with external magnetic field, Lepine and Matz (1976). Theory of excitonic correlation in a strong magnetic field based on an electron-hole Hamiltonian in Fockspace representation and Green functions techniques, Tsuzuki (1976). Theory of excitons in zincblende structure semiconductors in intermediate magnetic fields based on a two-particle model, Ekardt (1977). Note on the calculation of free exciton levels in InP in high magnetic fields, Bimberg, Hess, Lipari, Fischbach and Altarelli (1977). Collective excitation modes in an excitonic phase under strong magnetic fields by means of field theoretic Green function technique, Kuramoto (1977). Review of the Stark effect of localized centers, Grassano (1977). Vibronic theory of the Stark effect on the relaxed excited state of the *F*-center based on a molecular model, Imanaka, Iida and Ohkura (1977). Derivation of Wannier functions for Bloch electrons in uniform magnetic fields, Zak (1977). Helicon wave instability in a two-valley semiconductor, Hsieh (1977). Phenomenological model of the propagation of high power helicons in *n*-InSb crystals, Guha and Ghosh (1977). Theory of Wannier-Mott excitonic polaritons in magnetic fields based on a two-particle exciton model, Bimberg (1977). Vibronic theory of magnetic properties of the relaxed excited state of the *F*-center in a molecular model, Imanaka, Iida and Ohkura (1978). Investigation of the energy spectrum of electrons and holes in a semiconductor in a strong electromagnetic field, Oleinik, Abakarov and Belousov (1978). Closure of bands for Bloch electrons in a magnetic field, Claro and Wannier (1978). Investigation of the groundstate properties of an electron-hole system in a strong magnetic field by means of a variational procedure, Kuramoto and Morimoto (1978).

8.2 Rate equations with fields

In phenomenological transport theory the motion of conduction electrons in a crystal is described by the Boltzmann equation, where the drift term contains the classical part of the statistical motion in an external field, while the collision term contains the quantized interactions of the electrons with phonons, impurity centers, etc. It is the task of the theory to derive this equation from first principles. To simplify the deduction we first consider the case of an external homogeneous electric field and proceed according to the program of Section 8.1. Stumpf and Kleih (1980) derived the following theorem:

Theorem 8.3: If an H.F.-representation with rigid Bloch functions is used and if an adiabatic coupling between conduction band reactions and impurity center reactions is assumed, then the reactions of the conduction band electrons under the influence of an external homogeneous electric field are governed by the Boltzmann equation

$$\begin{aligned}\left(\frac{\partial}{\partial t}+\frac{e}{\hbar}\mathbf{E}\cdot\nabla_{\mathbf{K}}\right)f(\mathbf{K},t)=\sum_{\mathbf{K}'}\{&W(\mathbf{K},\mathbf{K}')f(\mathbf{K}',t)[1-f(\mathbf{K},t)]\\ &-W(\mathbf{K}',\mathbf{K})f(\mathbf{K},t)[1-f(\mathbf{K}',t)]\}+[\quad]\end{aligned} \tag{8.2.1}$$

where the transition probabilities are given by $W(\mathbf{K},\mathbf{K}')=W^s(\mathbf{K},\mathbf{K}')+W^{k0}(\mathbf{K},\mathbf{K}')$ with W^{k0} defined by (3.3.4), while [ ] contains the contribution due to Auger processes. The equations for electronic impurity center reactions are given by (8.2.36) with (8.2.37).

Proof: We start with the equations (1.3.8) and observe (1.3.9). This gives

$$\begin{aligned}\dot{f}_\Pi(t)&=-iL_{\Pi K}f_K(t)\\ \dot{f}_K(t)&=-iL_{K\Pi}f_\Pi(t)-iL_{KK}f_K(t)\end{aligned} \tag{8.2.2}$$

i) To evaluate these equations we use the division of H into $H_0+\lambda H_1$. If no static electric field is present, H_1 is given by (2.2.32). If such a field is present we obtain

$$H_1:=K^s+K^t+H^s+H^w+V=:V_0+V \tag{8.2.3}$$

where V is the interaction energy of the crystal and the static field. The division of H into $H_0+\lambda H_1$ leads to a division of $\mathscr{L}$ into $\mathscr{L}_0+\lambda\mathscr{L}_1$. For this division (1.5.11) and (1.5.12) are valid and for $\lambda=1$ (8.2.2) goes over into

$$\begin{aligned}\dot{f}_\Pi(t)&=-iL_{1\Pi K}f_K(t)\\ \dot{f}_K(t)&=-iL_{1K\Pi}f_\Pi(t)-iL_{KK}f_K(t).\end{aligned} \tag{8.2.4}$$

For the further treatment we assume that the interaction of the external field with the nuclei or ions, resp. is already incorporated in H_0. Then the potential V acts only on the electronic coordinates and only the interaction of the external field with the active electrons or holes, resp. outside the closed shells is taken into account explicitly.

The projectors $\mathbb{\Pi}$ and $\mathbb{K}$ are referred to the general base set $\{|i\rangle\}$ which is assumed to be a complete set of eigenstates of H_0. In the case of a polar crystal in interaction with its surroundings and for the special model which we use, this set is given by (2.1.2) and (3.1.3), leading to the notation

$$\{|i\rangle\}:=\{|\lambda_1\ldots\lambda_k,\mu_1\ldots\mu_l,\nu_1\ldots\nu_m,\mathbf{K}_1\ldots\mathbf{K}_n,m,\nu,\varrho\rangle\}. \tag{8.2.5}$$

According to our model the total number of active electrons involved in the reactions is s. If in the conduction band n electrons are present, then in the impurity centers $n-s$ electrons are present. For brevity we write

$$\begin{aligned}&\Lambda(s-n):=\lambda_1\ldots\lambda_k,\mu_1\ldots\mu_l,\nu_1\ldots\nu_m\\&\mathbf{K}(n):=\mathbf{K}_1\ldots\mathbf{K}_n\\&\zeta:=m,\nu,\varrho.\end{aligned} \tag{8.2.6}$$

Then a division of (8.2.5) into subsets is given by the number n of conduction band electrons. We therefore have

$$\{|i\rangle\}=\bigcup_{n=0}^{s}\{|\Lambda(n-s),\mathbf{K}(n),\zeta\rangle\}. \tag{8.2.7}$$

For fixed n the quantum numbers $\Lambda(n-s)$, $\mathbf{K}(n)$, ζ can still be varied. This gives

$$\{|i\rangle\}=\bigcup_{n=0}^{s}\bigcup_{\Lambda(n-s)}\bigcup_{\mathbf{K}(n)}\bigcup_{\zeta}\{|\Lambda(n-s),\mathbf{K}(n),\zeta\rangle\} \tag{8.2.8}$$

and the projector $\mathbb{\Pi}$ can be written

$$\mathbb{\Pi}=\sum_{n=0}^{s}\sum_{\Lambda(n-s)}\sum_{\zeta}\mathbb{\Pi}(\Lambda(n-s),\zeta) \tag{8.2.9}$$

with $\Lambda\equiv\Lambda(n-s)$, $\mathbf{K}=\mathbf{K}(n)$ and

$$\mathbb{\Pi}(\Lambda(n-s),\zeta):=\sum_{\mathbf{K}(n)}|\Lambda,\mathbf{K},\zeta\rangle\langle\Lambda,\mathbf{K},\zeta|\otimes|\Lambda,\mathbf{K},\zeta\rangle\langle\Lambda,\mathbf{K},\zeta|. \tag{8.2.10}$$

In an analogous way the projector $\mathbb{K}$ can be decomposed into

$$\mathbb{K}=\mathbb{K}_0+\tilde{\mathbb{K}} \tag{8.2.11}$$

with

$$\mathbb{K}_0:=\sum_{n=0}^{s}\sum_{\Lambda(n-s)}\sum_{\zeta\zeta'}\sum_{\mathbf{K}(n)}\sum_{\mathbf{K}'(n)}|\Lambda,\mathbf{K},\zeta\rangle\langle\Lambda,\mathbf{K},\zeta|\otimes|\Lambda,\mathbf{K}',\zeta'\rangle\langle\Lambda,\mathbf{K}',\zeta'| \tag{8.2.12}$$

and

$$\tilde{\mathbb{K}}:=\mathbb{K}-\mathbb{K}_0. \tag{8.2.13}$$

We now split $\mathscr{L}_1$ into

$$\mathscr{L}_1=\mathscr{L}_1^0+\mathscr{L}_1^v \tag{8.2.14}$$

where $\mathscr{L}_1^v$ depends on the interaction V and introduce the definitions

$$\tilde{\mathscr{L}}_1^v := \mathscr{L}_1^v - \Pi\, \mathscr{L}_1^v \mathbb{K}_0 - \mathbb{K}_0 \mathscr{L}_1^v \mathbb{K}_0 - \mathbb{K}_0 \mathscr{L}_1^v \Pi \tag{8.2.15}$$

$$\tilde{\mathscr{L}}_1 := \mathscr{L}_1^0 + \tilde{\mathscr{L}}_1^v; \quad \tilde{\mathscr{L}} := \mathscr{L}_0 + \tilde{\mathscr{L}}_1. \tag{8.2.16}$$

Observing the various orthonormality relations between the projectors we obtain from (8.2.4) with (8.2.9) (8.2.11) (8.2.16)

$$\begin{aligned} \dot{f}_\Pi(t) &= -i\Pi\, \mathscr{L}_1^v \mathbb{K}_0 f_K(t) - i\tilde{L}_{1\Pi K} f_K(t) \\ \dot{f}_K(t) &= -i\mathbb{K}_0 \mathscr{L}_1^v \Pi f_\Pi(t) - i\mathbb{K}_0 \mathscr{L}_1^v \mathbb{K}_0 f_K(t) \\ &\quad - i\tilde{L}_{1K\Pi} f_\Pi(t) - i\tilde{L}_{KK} f_K(t). \end{aligned} \tag{8.2.17}$$

ii) For the further evaluation of equations (8.2.17) we consider the terms containing $\mathscr{L}_1^v$. By direct calculation we obtain

$$\begin{aligned} &\langle \Lambda(n-s), \mathbf{K}(n), \zeta \,|\, \Pi\, \mathscr{L}_1^v \mathbb{K}_0 f_K(t) \,|\, \Lambda(n-s), \mathbf{K}(n), \zeta \rangle \\ &= \hbar^{-1} \sum_{\mathbf{K}'(n)} \langle \Lambda(n-s), \mathbf{K}(n), \zeta \,|\, V \,|\, \Lambda(n-s), \mathbf{K}'(n), \zeta \rangle\; f(\Lambda, \mathbf{K}', \zeta \,|\, \Lambda, \mathbf{K}, \zeta) \\ &\quad - \hbar^{-1} \sum_{\mathbf{K}'(n)} \langle \Lambda(n-s), \mathbf{K}'(n), \zeta \,|\, V \,|\, \Lambda(n-s), \mathbf{K}(n), \zeta \rangle\; f(\Lambda, \mathbf{K}, \zeta \,|\, \Lambda, \mathbf{K}', \zeta) \end{aligned} \tag{8.2.18}$$

since V acts only on electronic coordinates.
To evaluate (8.2.18) we first consider the pure case. In this case the density operator is given by $f_{ij} = c_i^\times c_j$. As V does not depend on lattice coordinates, the phonon part ζ drops out. Furthermore, as V is a sum of one-particle operators we obtain in the H.F.-representation

$$\begin{aligned} &\sum_{\mathbf{K}'(n)} \langle \Lambda(n-s), \mathbf{K}'(n), \zeta \,|\, V \,|\, \Lambda(n-s), \mathbf{K}(n), \zeta \rangle\; c^\times(\Lambda, \mathbf{K}, \zeta)\, c(\Lambda, \mathbf{K}', \zeta) \\ &= \left[\sum_{i=1}^{n} \int_{\mathbf{K}_i'} \langle \mathbf{K}_i' | V | \mathbf{K}_i \rangle\; c^\times(\Lambda, \mathbf{K}_1 \ldots \mathbf{K}_i' \ldots \mathbf{K}_n, \zeta)\, d^3K_i' \right] c(\Lambda, \mathbf{K}_1 \ldots \mathbf{K}_n, \zeta) \end{aligned} \tag{8.2.19}$$

and the complex conjugate expression for the other term. Now according to Haug (1964), p. 101, the formula

$$\begin{aligned} &\int_{\mathbf{K}_i'} \langle \mathbf{K}_i' | V | \mathbf{K}_i \rangle\, c(\mathbf{K}_1 \ldots \mathbf{K}_i' \ldots \mathbf{K}_n)\, d^3K_i' \\ &= -ie\mathbf{E} \cdot [\nabla_{\mathbf{K}_i} c(\mathbf{K}_1 \ldots \mathbf{K}_n) + \langle u_{\mathbf{K}} | \nabla | u_{\mathbf{K}} \rangle\, c(\mathbf{K}_1 \ldots \mathbf{K}_n)] \end{aligned} \tag{8.2.20}$$

is valid. Without loss of generality it can be assumed that $u_{\mathbf{K}}$ is real. Then from the normalization condition it follows that the second term on the right-hand side of (8.2.20) vanishes. Observing this and (8.2.19), we obtain for (8.2.18)

$$\begin{aligned} &-i\langle \Lambda(n-s), \mathbf{K}(n), \zeta \,|\, \Pi\, \mathscr{L}_1^v \mathbb{K}_0 f_K(t) \,|\, \Lambda(n-s), \mathbf{K}(n), \zeta \rangle \\ &= -\frac{e}{\hbar} \mathbf{E} \cdot \left[\sum_i \nabla_{\mathbf{K}_i} \right] f(\Lambda(n-s), \mathbf{K}(n), \zeta \,|\, \Lambda(n-s), \mathbf{K}(n), \zeta). \end{aligned} \tag{8.2.21}$$

Defining the operator

$$\mathbb{V} := \frac{e}{\hbar}\mathbf{E}\cdot\sum_{n=0}^{s}\sum_{\Lambda(n-s)}\sum_{\zeta}\mathfrak{B}(\Lambda(n-s),\zeta) \tag{8.2.22}$$

with

$$\mathfrak{B}(\Lambda(n-s),\zeta) = \sum_{\mathbf{K}(n)}\left[\sum_{i}\nabla_{\mathbf{K}_i}\right]|\Lambda,\mathbf{K},\zeta\rangle\langle\Lambda,\mathbf{K},\zeta|\otimes|\Lambda,\mathbf{K},\zeta\rangle\langle\Lambda,\mathbf{K},\zeta| \tag{8.2.23}$$

the first equation of (8.2.17) can be written

$$\left(\frac{\partial}{\partial t}+\mathbb{V}\right)f_{\Pi}(t) = -i\tilde{L}_{1\Pi K}f_K(t) \tag{8.2.24}$$

for pure states. But as mixed states are a superposition of pure states of the same system and (8.2.24) is a linear equation it holds also in the case of mixed states.
In a similar way we treat the second equation of (8.2.17). Performing this treatment we have to observe that for pure states $f_{ij}=f_{ji}^{\times}$ is valid and that if the system is enclosed in a large box all electronic states can be assumed to be real. From this it follows that $f_{ji}=f_{ij}$ and that we can use the symmetric operator $f_{ij}=\frac{1}{2}(f_{ij}+f_{ji})$. But for this operator the drift terms in the second equation of (8.2.17) vanish. For details of this calculation we refer to Stumpf and Kleih (1980). If we treat the resulting system similar to Theorem 1.11, we obtain

$$\left(\frac{\partial}{\partial t}+\mathbb{V}\right)f_{\Pi}(t) = -\int_0^t \tilde{L}_{1\Pi K}e^{-iL_{KK}t'}\tilde{L}_{1K\Pi}f_{\Pi}(t-t')dt'. \tag{8.2.25}$$

iii) Starting from equation (8.2.25) we can proceed as in Theorem 1.16 or Theorem 1.18, resp. to derive the reaction equations.
It has only to be observed that the meaning of the transition probabilities has changed according to (8.1.21) and (8.2.15) and (8.2.16). We start with equation (8.2.25). The relation (1.4.11) reads in this case

$$P(\Lambda(n-s),\mathbf{K}(n),m,\nu,\varrho,t) := (\Pi(\Lambda(n-s),\ \mathbf{K}(n),m,\nu,\varrho),f_{\Pi}(t)) \tag{8.2.26}$$

and the equation (8.2.25) goes over by projection with $\Pi(\Lambda,\mathbf{K},\zeta)$ into the system of equations

$$\begin{aligned}
&\left(\frac{\partial}{\partial t}+\frac{e}{\hbar}\mathbf{E}\cdot\sum_{i=1}^{n}\nabla_{\mathbf{K}_i}\right)P(\Lambda,\mathbf{K},m,\nu,\varrho,t)\\
&\quad=\sum_{\Lambda'\mathbf{K}'n'm'} W^k(\Lambda(n-s),\mathbf{K}(n),m,\Lambda'(n'-s),\mathbf{K}'(n'),m')\\
&\qquad\times[P(\Lambda'(n'-s),\mathbf{K}'(n'),m',\nu,\varrho,t)-P(\Lambda,\mathbf{K},m,\nu,\varrho,t)]\\
&\quad+\sum_{\Lambda'\mathbf{K}'m'\nu'n'} W^s(\Lambda(n-s),\mathbf{K}(n),m,\nu,\Lambda'(n'-s),\mathbf{K}'(n'),m',\nu')\\
&\qquad\times[P(\Lambda'(n'-s),\ \mathbf{K}'(n'),m',\nu',\varrho,t)-P(\Lambda,\mathbf{K},m,\nu,\varrho,t)]\\
&\quad+\sum_{m'\varrho'} W^w(m,\varrho,m',\varrho')\,[P(\Lambda(n-s),\mathbf{K}(n),m',\nu,\varrho',t)-P(\Lambda,\mathbf{K},m,\nu,\varrho,t)]
\end{aligned} \tag{8.2.27}$$

where the transition probabilities W^k, W^s and W^w are evaluated in Lemma 8.1. By definition, the heat bath is in thermal equilibrium. If we exclude decay solutions of the phonon system which we have treated in Chapter 2, then the phonon system is in thermal equilibrium with respect to the initial states and we may make the ansatz

$$P\big(\Lambda(n-s),\mathbf{K}(n),m,v,\varrho;t\big)=P\big(\Lambda(n-s),\mathbf{K}(n),v,t\big)f_m(T)f_\varrho(T). \tag{8.2.28}$$

Then equations (8.2.27) go over into the set of equations

$$\begin{aligned}&\left(\frac{\partial}{\partial t}+\frac{e}{\hbar}\mathbf{E}\cdot\sum_{i=1}^{n}\nabla_{\mathbf{K}_i}\right)P(\Lambda,\mathbf{K},v,t)\\ &\quad=\sum_{\Lambda',\mathbf{K}',n'}[W^k(\Lambda,\mathbf{K},\Lambda',\mathbf{K}',T)P(\Lambda',\mathbf{K}',v,t)\\ &\quad-W^k(\Lambda',\mathbf{K}',\Lambda,\mathbf{K},T)P(\Lambda,\mathbf{K},v,t)]\\ &\quad+\sum_{\Lambda',\mathbf{K}',n',v'}[W^s(\Lambda,\mathbf{K},v,\Lambda',\mathbf{K}',v',T)P(\Lambda',\mathbf{K}',v',t)\\ &\quad-W^s(\Lambda',\mathbf{K}',v',\Lambda,\mathbf{K},v,T)P(\Lambda,\mathbf{K},v,t)]\end{aligned} \tag{8.2.29}$$

where we have used the definitions

$$\begin{aligned}&W^k(\Lambda,\mathbf{K},\Lambda',\mathbf{K}',T):=\sum_{m,m'}W^k\big(\Lambda(n-s),\mathbf{K}(n),m,\Lambda'(n'-s),\mathbf{K}'(n'),m'\big)f_{m'}(T)\\ &W^s(\Lambda,\mathbf{K},v,\Lambda',\mathbf{K}',v',T):=\sum_{m,m'}W^s\big(\Lambda(n-s),\mathbf{K}(n),m,v,\Lambda'(n'-s),\mathbf{K}'(n'),m',v'\big)f_{m'}(T).\end{aligned} \tag{8.2.30}$$

Furthermore, neglecting the direct correlations between electrons and photons, we may put

$$P\big(\Lambda(n-s),\mathbf{K}(n),v,t)\big)=P\big(\Lambda(n-s),\mathbf{K}(n),t\big)p_v(\sigma) \tag{8.2.31}$$

and obtain from (8.2.29)

$$\begin{aligned}&\left(\frac{\partial}{\partial t}+\frac{e}{\hbar}\mathbf{E}\cdot\sum_{i=1}^{n}\nabla_{\mathbf{K}_i}\right)P(\Lambda,\mathbf{K},t)\\ &\quad=\sum_{\Lambda',\mathbf{K}',n'}[W^k(\Lambda,\mathbf{K},\Lambda',\mathbf{K}',T)P(\Lambda',\mathbf{K}',t)-W^k(\Lambda'\mathbf{K}',\Lambda,\mathbf{K},T)P(\Lambda,\mathbf{K},t)]\\ &\quad+\sum_{\Lambda',\mathbf{K}',n'}[W^s(\Lambda,\mathbf{K},\Lambda',\mathbf{K}',T,\sigma)P(\Lambda',\mathbf{K}',t)-W^s(\Lambda',\mathbf{K}',\Lambda,\mathbf{K},T,\sigma)P(\Lambda,\mathbf{K},t)]\end{aligned} \tag{8.2.32}$$

with

$$W^s(\Lambda,\mathbf{K},\Lambda',\mathbf{K}',T,\sigma):=\sum_{vv'}W^s\big(\Lambda(n-s),\mathbf{K}(n),v,\Lambda'(n'-s),\mathbf{K}'(n'),v',T\big)p_{v'}(\sigma). \tag{8.2.33}$$

Finally, we separate the reactions of conduction electrons and impurity center electrons by the ansatz

$$P\big(\Lambda(n-s),\mathbf{K}(n),t\big)=P\big(\Lambda(n-s),n,t\big)f(\mathbf{K}_1\ldots\mathbf{K}_n,t). \tag{8.2.34}$$

This leads to the equations

$$\left(\frac{\partial}{\partial t}+\frac{e}{\hbar}\mathbf{E}\cdot\sum_{i=1}^{n}\nabla_{\mathbf{K}_i}\right)f(\mathbf{K}_1\ldots\mathbf{K}_n,t) \tag{8.2.35}$$

$$=\sum_{\mathbf{K}'_1\ldots\mathbf{K}'_n}\{[W^k(\Lambda,\mathbf{K},\Lambda,\mathbf{K}',T)+W^s(\Lambda,\mathbf{K},\Lambda,\mathbf{K}',T)]f(\mathbf{K}'_1\ldots\mathbf{K}'_n,t)$$

$$-[W^k(\Lambda,\mathbf{K}',\Lambda,\mathbf{K},T)+W^s(\Lambda,\mathbf{K}',\Lambda,\mathbf{K},T)]f(\mathbf{K}_1\ldots\mathbf{K}_n,t)\}$$

and to

$$\frac{\partial}{\partial t}P(\Lambda,n,t)=\sum_{\Lambda'n'}[W^k(\Lambda,\Lambda',T,f)P(\Lambda',n',t)-W^k(\Lambda',\Lambda,T,f)P(\Lambda,n,t)]$$

$$+\sum_{\Lambda'n'}[W^s(\Lambda,\Lambda',T,\sigma,f)P(\Lambda',n',t)-W^s(\Lambda',\Lambda,T,\sigma,f)P(\Lambda,n,t)] \tag{8.2.36}$$

if direct radiative excitations of the conduction band electrons are excluded. The transition probabilities occurring in (8.2.36) are defined by

$$W^k(\Lambda,\Lambda',T,f):=\sum_{\mathbf{K}(n),\mathbf{K}'(n')}W^k(\Lambda(n-s),\mathbf{K}(n),\Lambda'(n'-s),\mathbf{K}'(n'),T)f(\mathbf{K}'_1\ldots\mathbf{K}'_{n'},T)$$

$$W^s(\Lambda,\Lambda',T,\sigma,f):=\sum_{\mathbf{K}(n),\mathbf{K}'(n')}W^s(\Lambda(n-s),\mathbf{K}(n),\Lambda'(n'-s),\mathbf{K}'(n'),T,\sigma) \tag{8.2.37}$$

$$\times f(\mathbf{K}'_1\ldots\mathbf{K}'_{n'},T).$$

For the further evaluation of equations (8.2.35) we apply the techniques of Theorems 3.6 and 4.3, i.e., we introduce the occupation number representation and the formation of corresponding mean values. By these procedures, (8.2.35) goes over into

$$\dot{\bar{n}}_l+\sum_{n_1\ldots n_N}\left(\frac{e}{\hbar}\mathbf{E}\cdot\sum_{i=1}^{N}n_i\nabla_{\mathbf{K}_i}\right)n_lf(n_1\ldots n_N,t)$$

$$=\sum_{\substack{n_1\ldots n_N\\ n'_1\ldots n'_N}}n_l[W(n_1\ldots n_N,n'_1\ldots n'_N)f(n'_1\ldots n'_N,t) \tag{8.2.38}$$

$$-W(n'_1\ldots n'_N,n_1\ldots n_N)f(n_1\ldots n_N,t)].$$

The right-hand side of (8.2.38) can be treated according to the Theorems 3.6 and 4.3 and gives the right-hand side of equation (8.2.1). Thus we have only to evaluate the drift term of (8.2.38). We rewrite this term in the following way

$$\sum_{n_1\ldots n_N}\left(\frac{e}{\hbar}\mathbf{E}\cdot\sum_{i=1}^{N}n_i\nabla_{\mathbf{K}_i}\right)n_lf(n_1\ldots n_N,t)$$

$$=\sum_{n_1\ldots n_N}\frac{e}{\hbar}\mathbf{E}\cdot\left(\nabla_{\mathbf{K}_l}n_l^2+\sum_{\substack{i=1\\ \neq l}}^{N}n_i\nabla_{\mathbf{K}_i}n_l\right)f(n_1\ldots n_N,t). \tag{8.2.39}$$

As n_i, $1\leq i\leq N$ takes only the values 0 or 1, we may replace n_l^2 by n_l, while for the other terms we apply the mean value approximation $\overline{n_in_l}=\bar{n}_i\bar{n}_l$. Then, due to (3.6.14), i.e.,

$\bar{n}_l = :f(\mathbf{K}_l, t)$, the first term on the right-hand side of (8.2.39) gives $e\hbar^{-1}\mathbf{E}\cdot\nabla_{\mathbf{K}} f(\mathbf{K}, t)$ (with $\mathbf{K} = \mathbf{K}_l$), while due to $i \neq l$, the second term vanishes. Hence we obtain from (8.2.39) the drift term in (8.2.1), and thus the complete Boltzmann equation.
The set of equations (8.2.36) is concerned with impurity-electron reactions including those which lead to band-center transitions and vice versa. These equations (8.2.36) can be reduced to rate equations derived in Chapter 3 and 4. Evaluating the transition probabilities (8.2.37), it has to be observed that the various matrix elements are different with respect to the range where they are calculated, as has been pointed out in Section 8.1. In addition to the transition operators of the electric field occurring in W^k, due to (8.2.37) the reaction kinetics depends on the modified equilibrium distribution f under the influence of the electric field, Q.E.D.

To obtain the most general Boltzmann equation two extensions of equation (8.2.1) must be carried out.

i) The forces exerted by magnetic fields have to be included;
ii) the macroscopic variation of the distribution function has to be taken into account.

The inclusion of i) runs along the same lines as the foregoing proof for the electric field and leads instead of the Coulomb force in (8.2.1) to the Lorentz force. The inclusion of ii) runs along the same lines as the proof for the extension of the phonon Boltzmann equation to macroscopic variations given in Theorem 7.21 and leads to the gradient term with respect to the macroscopic position. Hence the most general Boltzmann equation reads

$$\left[\frac{\partial}{\partial t} + \mathbf{v}\cdot\nabla_{\mathbf{r}} + \frac{e}{\hbar}\left(\mathbf{E} + \frac{1}{c}\,\mathbf{v}\times\mathbf{B}\right)\cdot\nabla_{\mathbf{K}}\right] f(\mathbf{K},\mathbf{r},t)$$
$$= \sum_{\mathbf{K}'} \{W(\mathbf{K},\mathbf{K}') f(\mathbf{K}',\mathbf{r},t)\,[1 - f(\mathbf{K},\mathbf{r},t)] \tag{8.2.40}$$
$$- W(\mathbf{K}',\mathbf{K}) f(\mathbf{K},\mathbf{r},t)\,[1 - f(\mathbf{K}',\mathbf{r},t)]\} + [\quad]$$

where [ ] means the Auger term and $\mathbf{v} := \hbar^{-1}\nabla_{\mathbf{K}} E(\mathbf{K})$. In the following, we omit for brevity the Auger term [ ] and consider only stationary solutions for which f does not depend on t. Even if the Auger terms are neglected, equation (8.2.40) is still a very complicated nonlinear integro-differential equation. Hence we try to simplify it by appropriate approximations. For weak fields a linearized version can be derived from (8.2.40) by the following representation of f

$$f = f_0 - \frac{\partial f_0}{\partial E}\,\Phi \tag{8.2.41}$$

where f_0 is the equilibrium distribution (3.6.21). If each microblock or mosaicblock has its own specified temperature $T = T_i$ this manifests itself in the macroscopic description by a position dependence of T on $\mathbf{r}$, i.e., $T \equiv T(\mathbf{r})$, and thus f_0 depends on position. The following theorem holds:

Theorem 8.4 : If the temperature $T = T(\mathbf{r})$ is a function of position, then the linearized stationary version of (8.2.40) reads

$$\frac{\partial f_0}{\partial E}\mathbf{v}\cdot\left[\mathbf{A}-\frac{e}{\hbar c}(\mathbf{B}\times\nabla_{\mathbf{K}}\Phi(\mathbf{K}))\right]=\frac{1}{k_B T}\sum_{\mathbf{K}'} W_0(\mathbf{K}',\mathbf{K})\,[\Phi(\mathbf{K}')-\Phi(\mathbf{K})] \tag{8.2.42}$$

with

$$\mathbf{A}:=e\mathbf{E}-\nabla_{\mathbf{r}}\zeta-\frac{E(\mathbf{K})-\zeta}{T}\nabla_{\mathbf{r}}T \tag{8.2.43}$$

and

$$W_0(\mathbf{K}',\mathbf{K}):=W(\mathbf{K}',\mathbf{K})f_0(\mathbf{K})\,[1-f_0(\mathbf{K}')]. \tag{8.2.44}$$

Proof: By direct calculation it follows that

$$\begin{aligned}\nabla_{\mathbf{K}}f&=\frac{\partial f_0}{\partial E}\hbar\mathbf{v}(\mathbf{K})-\nabla_{\mathbf{K}}\frac{\partial f_0}{\partial E}\Phi\\ \nabla_{\mathbf{r}}f&=-\frac{\partial f_0}{\partial E}\left(\frac{E-\zeta}{T}+\frac{\partial\zeta}{\partial T}\right)\nabla_{\mathbf{r}}T+\dots\end{aligned} \tag{8.2.45}$$

In the linear approximation we take into account from each term only the lowest non-vanishing contribution. Hence we have

$$\begin{aligned}-\mathbf{v}\cdot\nabla_{\mathbf{r}}f&\approx\mathbf{v}\cdot\frac{\partial f_0}{\partial E}\left(\frac{E-\zeta}{T}+\frac{\partial\zeta}{\partial T}\right)\nabla_{\mathbf{r}}T\\ -\frac{e}{\hbar}\mathbf{E}\cdot\nabla_{\mathbf{K}}f&\approx-e\mathbf{E}\cdot\mathbf{v}\frac{\partial f_0}{\partial E}.\end{aligned} \tag{8.2.46}$$

In the magnetic field term, due to $\mathbf{v}\cdot(\mathbf{v}\times\mathbf{B})=0$ we obtain

$$\begin{aligned}-\frac{e}{\hbar}\frac{1}{c}(\mathbf{v}\times\mathbf{B})\cdot\nabla_{\mathbf{K}}f&=\frac{e}{\hbar}\frac{1}{c}(\mathbf{v}\times\mathbf{B})\cdot\left[\frac{\partial f_0}{\partial E}\nabla_{\mathbf{K}}\Phi+\Phi\frac{\partial^2 f_0}{\partial E^2}\hbar\mathbf{v}\right]\\ &=\frac{e}{\hbar}\frac{1}{c}\frac{\partial f_0}{\partial E}\mathbf{v}\cdot(\mathbf{B}\times\nabla_{\mathbf{K}}\Phi).\end{aligned} \tag{8.2.47}$$

Collecting these terms, the left-hand side of equation (8.2.42) results. For the collision term we get with (8.2.41) and the abbreviation $\varphi=(\partial f_0/\partial E)\Phi$

$$\begin{aligned}W(\mathbf{K}',\mathbf{K})f(\mathbf{K})\,[1-f(\mathbf{K}')]&=W(\mathbf{K}',\mathbf{K})f_0(\mathbf{K})\,[1-f_0(\mathbf{K}')]\left(1+\frac{\varphi}{f_0}\right)\left(1-\frac{\varphi'}{1-f_0'}\right)\\ &\approx W_0(\mathbf{K}',\mathbf{K})\left(1+\frac{\varphi}{f_0}-\frac{\varphi'}{1-f_0'}\right)\end{aligned} \tag{8.2.48}$$

where definition (8.2.44) is used. Further we obtain

$$W(\mathbf{K},\mathbf{K}')f(\mathbf{K}')\,[1-f(\mathbf{K})]\approx W_0(\mathbf{K},\mathbf{K}')\left(1+\frac{\varphi'}{f_0'}-\frac{\varphi}{1-f_0}\right) \tag{8.2.49}$$

with a similar definition for $W_0(\mathbf{K},\mathbf{K}')$. From Theorem 3.7 it follows that $W_0(\mathbf{K},\mathbf{K}') = W_0(\mathbf{K}',\mathbf{K})$ and if we observe that

$$\frac{\partial f_0}{\partial E} = \frac{1}{k_B T} f_0(1-f_0) \tag{8.2.50}$$

holds, then the terms (8.2.48) (8.2.49) can be rewritten to give the right-hand side of equation (8.2.42), Q.E.D.

If no magnetic field is present, the solution of the linearized equation (8.2.42) can be found or approximated, resp. by applying the variational principle of Kohler (1948). As this principle was discussed for the phonon Boltzmann equation in Chapter 7, we do not give further comments here but refer also for the electron Boltzmann equation to this section. Another solution procedure is achieved by the so-called relaxation time formalism which introduces a further simplification of the linearized collision term. In this method it is assumed that the collision term can be represented by

$$\frac{1}{k_B T} \sum_{\mathbf{K}'} W_0(\mathbf{K}',\mathbf{K}) \, [\Phi(\mathbf{K}') - \Phi(\mathbf{K})] = \frac{1}{\tau} \frac{\partial f_0}{\partial E} \Phi(\mathbf{K}) \tag{8.2.51}$$

where τ is the so-called relaxation time. This name stems from the fact that for the time-dependent Boltzmann equation (8.2.40) without external fields, etc., in the linearized version with the substitution of (8.2.51), the solutions are simply given by $f(t) = f_0 + [f(0) - f_0] \exp(-t/\tau)$. Although it is not possible to derive for all kinds of interactions a corresponding relaxation time, this method is of great interest as it allows a general physical discussion of the solutions obtained in this way. In the subsequent sections we shall discuss the calculation of relaxation times in detail. In this and the following section we first draw some conclusions concerning the physical consequences of this approach. A general solution is given by the following theorem, cf. Madelung (1970), Haug (1970), Seeger (1973).

Theorem 8.5: If it is assumed that the conduction band energy is parabolic, i.e., $E(\mathbf{K}) = \hbar^2 \mathbf{K}^2 / 2m^*$ and that the relaxation time depends only on E, i.e., $\tau = \tau(E)$, then the general solution of the linearized Boltzmann equation (8.2.42) in the relaxation time representation (8.2.51) is given by

$$\Phi(\mathbf{K}) = \frac{\tau \mathbf{v}}{1 + \tau^2 \mathbf{B}^{*2}} \cdot [\mathbf{A} - \tau(\mathbf{B}^* \times \mathbf{A}) + \tau^2 \mathbf{B}^*(\mathbf{B}^* \mathbf{A})] \tag{8.2.52}$$

with $\mathbf{v} = \hbar \mathbf{K}/m^*$, $\mathbf{B}^* = e\mathbf{B}/m^* c$ and $\mathbf{A}$ given by (8.2.43).

Proof: We combine equation (8.2.51) with equation (8.2.42). This gives

$$\Phi(\mathbf{K}) = \tau \mathbf{v} \cdot \left[\mathbf{A} - \frac{e}{\hbar c} (\mathbf{B} \times \nabla_{\mathbf{K}} \Phi(\mathbf{K})) \right]. \tag{8.2.53}$$

We try to solve (8.2.53) using the function defined by

$$\Phi(\mathbf{K}) := \mathbf{v} \cdot \mathbf{C}(E). \tag{8.2.54}$$

Owing to the assumption of the parabolic band we obtain

$$\nabla_{\mathbf{K}}\Phi(\mathbf{K})=\frac{\hbar}{m^*}\mathbf{C}(E)+\hbar\mathbf{v}\left(\mathbf{v}\cdot\frac{d\mathbf{C}(E)}{dE}\right). \tag{8.2.55}$$

Substitution of (8.2.54) and (8.2.55) in (8.2.53) leads to

$$\Phi(\mathbf{K})=\tau\mathbf{v}\cdot\left[\mathbf{A}-\frac{e}{m^*c}(\mathbf{B}\times\mathbf{C}(E))\right]=\mathbf{v}\cdot\mathbf{C} \tag{8.2.56}$$

and from this it follows that

$$\mathbf{C}+\tau(\mathbf{B}^*\times\mathbf{C})=\tau\mathbf{A}. \tag{8.2.57}$$

To derive $\mathbf{C}$ explicitly we may form the scalar and vector product of $\mathbf{B}^*$ with (8.2.55)

$$\mathbf{B}^*\cdot\mathbf{C}=\tau\mathbf{B}^*\cdot\mathbf{A} \tag{8.2.58}$$

$$\mathbf{B}^*\times\mathbf{C}+\tau\mathbf{B}^*\times(\mathbf{B}^*\times\mathbf{C})=\tau(\mathbf{B}^*\times\mathbf{A}). \tag{8.2.59}$$

From (8.2.59) it follows that

$$\mathbf{B}^*\times\mathbf{C}=\tau[\mathbf{B}^*\times\mathbf{A}-\mathbf{B}^*(\mathbf{B}^*\cdot\mathbf{C})+\mathbf{C}(\mathbf{B}^*\cdot\mathbf{B}^*)]. \tag{8.2.60}$$

Substitution of (8.2.58) in (8.2.60) and of (8.2.60) in (8.2.57) finally gives

$$\mathbf{C}(1+\tau^2\mathbf{B}^{*2})=\tau\mathbf{A}-\tau^2(\mathbf{B}^*\times\mathbf{A})+\tau^3\mathbf{B}^*(\mathbf{B}^*\cdot\mathbf{A}) \tag{8.2.61}$$

from which (8.2.52) follows, Q.E.D.

For the practical application of the coupled system of rate equations (8.2.35) and (8.2.36) the transition probabilities appearing in the general equation (8.2.27) have to be evaluated. The result of this evaluation is given by the following lemma due to Stumpf and Kleih (1980):

Lemma 8.1: Suppose that the set of base states $\{|nm\rangle\}$ for a crystal in an external (electric) field is chosen according to the prescription of Section 8.1, then the first order resonance transition probabilities W^s and W^w of equations (8.2.27) are given by (2.4.3) and (2.4.4), while W^k is given by

$$W^k_{nmv\varrho,n'm'v'\varrho'}=\frac{2\pi}{\hbar}|\langle nm|K^s+K^t|n'm'\rangle|^2\delta_{vv'}\delta_{\varrho\varrho'}\delta_\gamma(E^n_m-E^{n'}_{m'}) \tag{8.2.62}$$

for transitions within the conduction band, and by

$$W^k_{nmv\varrho,n'm'v'\varrho'}=\frac{2\pi}{\hbar}|\langle nm|K^s+K^t+V|n'm'\rangle|^2\delta_{vv'}\delta_{\varrho\varrho'}\delta_\gamma(E^n_m-E^{n'}_{m'}) \tag{8.2.63}$$

for all other transitions.

Proof: For simplicity we consider first order perturbation theoretic expressions, as the proof for resonance transitions is more complicated but runs along the same lines.

Instead of (1.3.13) we have now to start from (8.2.25). For the derivation of the generalized Pauli-Master equation (8.2.27) from (8.2.25) the same procedure can be applied as in Theorem 1.13, 1.14 and 1.16. Hence the transition probabilities read in analogy to (1.5.18) for the present case

$$W_{ij}^{k} = -\langle i| \left\{ \tilde{L}_{1\Pi K} \int_0^{\tau} e^{-i\tilde{L}^0_{KK}t} dt \tilde{L}_{1K\Pi} |j\rangle \langle j| \right\} |i\rangle. \tag{8.2.64}$$

We observe that owing to $\mathbb{\Pi}\mathbb{K} = \mathbb{K}\mathbb{\Pi} = 0$, $\mathbb{K}\mathbb{K}_0 = \mathbb{K}_0\mathbb{K} = \mathbb{K}_0$ and $\mathbb{\Pi}\mathbb{\Pi} = \mathbb{\Pi}$ for $\tilde{\mathscr{L}}_1^v$, given by (8.2.15), the following relation holds

$$\begin{aligned}\tilde{L}_{1\Pi K}^{v} &= L_{1\Pi K}^{v} - \mathbb{\Pi}\,(L_{1\Pi K_0}^{v} - L_{1K_0K_0}^{v} - L_{1K_0\Pi}^{v})\,\mathbb{K} \\ &= \mathbb{\Pi}\,\mathscr{L}_1^{v}\mathbb{K} - \mathbb{\Pi}\,\mathscr{L}_1^{v}\mathbb{K}_0 .\end{aligned} \tag{8.2.65}$$

Hence with $\tilde{L}_1$, given by (8.2.16), we obtain using (8.2.14) the relation

$$\tilde{L}_{1\Pi K} = \mathbb{\Pi}\,\mathscr{L}_1\mathbb{K} - \mathbb{\Pi}\,\mathscr{L}_1^{v}\mathbb{K}_0 \tag{8.2.66}$$

and the corresponding expression for $\tilde{L}_{1K\Pi}$. If these expressions are substituted into (8.2.64) and if $\mathbb{\Pi}|j\rangle\langle j| = |j\rangle\langle j|$ is observed, (8.2.64) yields

$$\begin{aligned}W_{ij}^{k} = &\langle i| \left\{ \mathscr{L}_1^{v}\mathbb{K}_0 \int_0^{\tau} e^{-iL^0_{KK}t} dt (\mathbb{K}\mathscr{L}_1 - \mathbb{K}_0\mathscr{L}_1^{v}) |j\rangle\langle j| \right\} |i\rangle \\ &- \langle i| \left\{ \mathscr{L}_1\mathbb{K} \int_0^{\tau} e^{-iL^0_{KK}t} dt (\mathbb{K}\mathscr{L}_1 - \mathbb{K}_0\mathscr{L}_1^{v}) |j\rangle\langle j| \right\} |i\rangle .\end{aligned} \tag{8.2.67}$$

The expression (8.2.67) can be evaluated directly by means of the techniques applied in Theorem 1.16. For states $|i\rangle$, $|j\rangle$ which differ only in quantum numbers of the conduction band electrons (irrespective of the behaviour of phonons and photons), the contributions of $\mathscr{L}_1^v$ cancel, and (8.2.62) remains, while for states $|i\rangle$, $|j\rangle$ which differ not only in quantum numbers of the conduction band electrons, all terms containing $\mathscr{L}_1^v\mathbb{K}_0$ or $\mathbb{K}_0\mathscr{L}_1^v$ lead to vanishing contributions so that (8.2.63) results. For details of this evaluation we refer to the original paper. Q.E.D.

In order to illustrate our equations for the reaction kinetics of a system with a mixed electronic bound-state-band-state spectrum in external fields, we discuss a special model, namely the model of Chapter 3 and 4. It consists of a polar crystal with s anion vacancies and s electrons where the crystal is coupled to the radiation field and to a heat reservoir, and it is assumed that the electrons, phonons and photons of this system perform rate processes under the action of external and internal forces. In addition we specify the external forces by assuming that an external homogeneous electric field acts on the system, while the photon field which, of course, is also an external force acting on the system, is incorporated in the total system from the beginning by definition. If we ignore phonon drag and photon drag and assume an adiabatic coupling between the conduction band occupation probability and the occupation probability of the discrete levels, then the general rate system (8.2.27) is reduced to the two sets of equations (8.2.35) and (8.2.36) by means of the product representations

(8.2.28) (8.2.31) and (8.2.34), which reflect our assumptions. We assume furthermore that photon absorption- and photon emission processes by conduction band electrons as well as Auger- and impact processes can be neglected. All these assumptions can, in principle, be abandoned. They are only made in order to give a transparent treatment. From (8.2.35) and Theorem 8.3 using the assumptions introduced above, the Boltzmann equation

$$\left(\frac{\partial}{\partial t} - e\mathbf{E}\cdot\nabla_k\right) f(\mathbf{K},t) = \sum_{\mathbf{K}'} \{W^{k0}(\mathbf{K},\mathbf{K}') f(\mathbf{K}',t)\,[1-f(\mathbf{K},t)] - W^{k0}(\mathbf{K}',\mathbf{K}) f(\mathbf{K},t)\,[1-f(\mathbf{K}',t)]\} \tag{8.2.68}$$

results, while according to Section 3.5 and Theorem 3.3 from (8.2.36) for the given model the rate equations

$$\begin{aligned}
\dot{\bar{n}}_F &= W(F,F^*)\bar{n}_{F^*} - W(F^*,F)\bar{n}_F - W(F',c)\bar{n}_F + W(c,F')\bar{n}_{F'}\\
\dot{\bar{n}}_{F^*} &= -W(F,F^*)\bar{n}_{F^*} + W(F^*,F)\bar{n}_F - W(c,F^*)\bar{n}_{F^*} + W(F^*,c)(\bar{n}_{F^*}+\bar{n}_c)\\
\dot{\bar{n}}_{F'} &= W(F',c)\bar{n}_F - W(c,F')\bar{n}_{F'}\\
\dot{\bar{n}}_c &= -W(F',c)\bar{n}_F + W(c,F')\bar{n}_{F'} + W(c,F^*)\bar{n}_{F^*} - W(F^*,c)(\bar{n}_{F^*}+\bar{n}_c)
\end{aligned} \tag{8.2.69}$$

can be derived. In the latter system of equations the transition probabilities are defined by

$$\begin{aligned}
W(c,a) &:= \sum_{\mathbf{K}} W(\mathbf{K},a)\\
W(a,c) &:= \sum_{\mathbf{K}} W(a,\mathbf{K}) f(\mathbf{K})\\
W(a,b) &:= W^s(a,b) + W^{k0}(a,b)
\end{aligned} \tag{8.2.70}$$

and

$$\begin{aligned}
W^{k0}(a,b) &:= \sum_{\substack{m,m'\\ \varrho,\varrho'}} W^{k0}_{nm\nu\varrho,n'm'\nu'\varrho'} f^{n'}_{m'}(\beta) f_{\varrho'}(\beta)\\
W^s(a,b) &:= \sum_{\substack{m,m'\\ \nu,\nu'\\ \varrho,\varrho'}} W^s_{nm\nu\varrho,n'm'\nu'\varrho'}\, p_{\nu'}(\sigma) f^{n'}_{m'}(\beta) f_{\varrho'}(\beta)
\end{aligned} \tag{8.2.71}$$

where the phonon distribution is given by

$$f^n_m(\beta) = Z^{-1}\exp(-\beta E^n_m) \tag{8.2.72}$$

and $p_{\nu'}(\sigma)$ is the parametric distribution of the photons and $f_\varrho(\beta)$ is the heat bath distribution which drops out due to the special form of W^k and W^s. Since only W^k of (8.2.63) differs for our model from W^k given by (2.4.2) for the model of Chapter 3, the technique of the derivation of rate equations from a given set of reaction equations (8.2.36) can be taken over from Section 3.5 without any change and we will not repeat this here. Due to (3.6.19)

$$\bar{n}_c(t) = \sum_{\mathbf{K}} f(\mathbf{K}, t) \tag{8.2.73}$$

holds and hence (8.2.68) and (8.2.69) are coupled systems of rate equations. These equations and their transition probabilities can be evaluated in order to derive a current-voltage characteristic for a doped polar insulator by ab initio calculations. The current of the conduction band electrons is given by (8.3.2). Since the solution of the Boltzmann equation (8.2.68) must be a functional of the external field $f \equiv f(\mathbf{K}, \mathbf{E}, t)$ we obtain from (8.3.2)

$$\bar{j} = \sum_{\mathbf{K}} e\mathbf{v} f(\mathbf{K}, \mathbf{E}, T) \equiv \bar{j}(T, \mathbf{E}) \tag{8.2.74}$$

for a stationary solution of temperature T and this expression gives the (generally non-linear) current-voltage characteristic. The simplest stationary approximate solution of the Boltzmann equation (8.2.68) in the relaxation time approximation is a drifted Maxwellian distribution. With the normalization according to (8.2.73) it reads

$$f(\mathbf{K}, \mathbf{E}, T) = \hat{n}_0^{-1} \bar{n}_c \exp\left[-\beta(\hbar\mathbf{K} - m^*\mathbf{v}_d)^2 (2m)^{-1}\right] = : \hat{f}(\mathbf{K}, \mathbf{E}, T)\bar{n}_c \tag{8.2.75}$$

with

$$\mathbf{v}_d = \frac{|e|}{m}\bar{\tau}\mathbf{E};\; n_0 = \frac{1}{4}\left(\frac{2m^*}{\pi\hbar^2\beta}\right)^{3/2};\; \hat{n}_0 := n_0\left(\frac{\pi}{2\beta m^*}\right)^{3/2} v_d^{-1}\Phi\left(\beta v_d\sqrt{\tfrac{m}{2}}\right) \tag{8.2.76}$$

where $\bar{\tau}$ is the mean relaxation time and m^* the effective electron mass in the conduction band while $\Phi(x)$ is the error function. Substitution of (8.2.75) into (8.2.70) leads to the transition probabilities $W(a, c)$ for stationary states

$$W(a, c) := \sum_{\mathbf{K}} W(a, \mathbf{K})\hat{f}(\mathbf{K}, \mathbf{E}, T)\bar{n}_c = : \bar{n}_c \hat{W}(a, c). \tag{8.2.77}$$

In this case the rate equations (8.2.69) achieve the following final form, Stumpf and Kleih (1980)

$$\begin{aligned}
\dot{\bar{n}}_F &= W(F, F^*)\bar{n}_{F^*} - W(F^*, F)\bar{n}_F - \hat{W}(F', c)\bar{n}_{F'}\bar{n}_c + W(c, F')\bar{n}_{F'} \\
\dot{\bar{n}}_{F^*} &= -W(F, F^*)\bar{n}_{F^*} + W(F^*, F)\bar{n}_F - W(c, F^*)\bar{n}_{F^*} \\
&\quad + \hat{W}(F^*, c)\bar{n}_c(\bar{n}_{F^*} + \bar{n}_c) \\
\dot{\bar{n}}_{F'} &= \hat{W}(F', c)\bar{n}_F\bar{n}_c - W(c, F')\bar{n}_{F'} \\
\dot{\bar{n}}_c &= -\hat{W}(F', c)\bar{n}_F\bar{n}_c + W(c, F')\bar{n}_{F'} + W(c, F^*)\bar{n}_{F^*} \\
&\quad - \hat{W}(F^*, c)\bar{n}_c(\bar{n}_{F^*} + \bar{n}_c).
\end{aligned} \tag{8.2.78}$$

The evaluation of these equations was performed by Kleih (1980) (1982). The mean value of $\bar{n}_c$ can be calculated directly from the system (8.2.78). According to Kleih an approximate stationary solution of (8.2.78) leads to the value

$$\bar{n}_c = \left[\frac{W(F^*, F)\, W(c, F^*)\, W(c, F')}{W(F, F^*)\, \hat{W}(F^*, c)\, \hat{W}(F', c)}\right]^{1/2} \tag{8.2.79}$$

If the values of $\bar{\tau}$, m^* and $\bar{n}_c$ are given, the current-voltage characteristic can be calculated. The values of $\bar{\tau}$ and m^* were (for simplicity) taken over from Ahrenkiel and Brown (1964) and from Redfield (1954), while the values of all interesting transition probabilities were calculated ab initio. Since these calculations run in principle along the same lines as the calculations described in Chapter 3 which are due to Rojas, and since the same techniques which were developed in Chapter 3 have been applied to this case, we do not present the calculation procedure. We rather give only the numerical results and refer for details to the original paper of Kleih.
In order to calculate the static electronic wave functions which are needed for the evaluation of the transition probabilities, Kleih applied a variational method to the energy expectation value (8.1.4) of Theorem 8.1. The electronic wave functions which were used as testfunctions are given by the formulas (5.6.42), (5.6.43) and (5.6.44). The first calculation of the static electronic wave functions in an external electric field by means of the energy operator (8.1.4) were done by Sieber (1973). The calculations of Kleih are an improved version of Sieber's calculations. In addition, Kleih performed the calculation of dynamical electron-lattice coupling for these states, cf. Section 5.6. The calculations were done for NaCl and the following crystal parameters were used:

Table 8.1:
Crystal parameters of NaCl which were used as input data for ab initio calculations of electronic wave functions and transition probabilities, after Kleih (1980).

Lattice parameter	d	2.82 Å
Effective electron mass	m^*	0.61 m_0
Phonon energy of the opt. long. branch	$\hbar\omega$	0.021 eV
Lattice polarizability	α^g	1.886
Electronic polarizability of the ions	α^e	0.902
Repulsion exponent	n	8.
Refractive index	n^*	1.54

For details of the wave function calculations we refer to the original work of Sieber (1973) and Kleih (1980). By means of such wave functions the transition probabilities

$$W(F,F^*)=W^k(F,F^*)+W^s(F,F^*);\ \hat{W}(F^*,c)=\hat{W}^k(F^*,c);\ \hat{W}(F',c)=\hat{W}^k(F',c);$$
$$W(c,F^*)=W^k(c,F^*);\ W(c,F')=W^k(c,F')$$

were calculated for an external electric field according to Lemma 8.1. The remaining optical transition probabilities $W(F^*,F)$ and $W(F',c)$ were taken over from Rojas, cf. Table 3.4. By these calculations all numerical values of the coefficients appearing in the rate equations (8.2.78) were determined. The following numerical results were obtained.

Table 8.2.a:
Nonradiative transition probability $W^k(F,F^*)$ for NaCl in dependence on the external electric field. Due to the weak field dependence the values – were not calculated; after Kleih (1980) (1982).

T \ E(kV/cm)	5	10	20	50
50	3.4 10^7	—	—	1.9 10^7
100	6.8 10^7	5.9 10^7	4.5 10^7	4.3 10^7
150	1.5 10^8	—	—	1.1 10^8
200	3.3 10^8	—	—	2.5 10^8
250	6.6 10^8	—	—	5.3 10^8
300	1.9 10^9	—	—	9.8 10^8

$W^k(F,F^*)$ [sec^{-1}]

Table 8.2.b:
Transition probability $W^s(F,F^*)$ of spontaneous optical emission for NaCl in dependence on the external electric field; after Kleih (1980) (1982).

T \ E(kV/cm)	0	50	100
50	4.2 10^7	6.3 10^7	9.2 10^7
100	4.4 10^7	6.7 10^7	9.7 10^7
150	4.6 10^7	7.0 10^7	1.0 10^8
200	4.7 10^7	7.2 10^7	1.0 10^8
250	4.8 10^7	7.3 10^7	1.1 10^8
300	4.9 10^7	7.4 10^7	1.1 10^8

$W^s(F,F^*)$ [sec^{-1}]

Table 8.2.c:
Transition probability $\hat{W}^k(F^*,c)$ of nonradiative capture of conduction band electrons into F^*-states for NaCl in dependence on the external electric field; after Kleih (1980) (1982).

T \ E(kV/cm)	0.001	5	10	20	30
50	1.3 10^{-11}	7.9 10^{-11}	9.3 10^{-11}	1.4 10^{-12}	2.1 10^{-16}
100	4.1 10^{-10}	6.5 10^{-10}	6.2 10^{-10}	2.7 10^{-11}	1.4 10^{-14}
150	1.5 10^{-9}	1.9 10^{-9}	1.6 10^{-9}	1.7 10^{-10}	1.0 10^{-11}
200	3.2 10^{-9}	3.5 10^{-9}	2.8 10^{-9}	6.2 10^{-10}	7.6 10^{-11}
250	5.0 10^{-9}	5.2 10^{-9}	4.2 10^{-9}	1.3 10^{-9}	2.6 10^{-10}
300	6.9 10^{-9}	6.9 10^{-9}	5.5 10^{-9}	2.3 10^{-9}	6.1 10^{-10}

$\hat{W}^k(F^*,c)$ [$\text{cm}^3\ \text{sec}^{-1}$]

Table 8.2.d:
Transition probability $\hat{W}^k(F',c)$ of nonradiative capture of conduction band electrons into F'-states for NaCl in dependence on the external electric field; after Kleih (1980) (1982).

T \ E(kV/cm)	5	10	20	40
50	$9.0\ 10^{-10}$	$1.7\ 10^{-9}$	$2.4\ 10^{-12}$	$3.8\ 10^{-12}$
100	$2.1\ 10^{-9}$	$2.3\ 10^{-9}$	$2.8\ 10^{-9}$	$1.2\ 10^{-10}$
150	$3.9\ 10^{-9}$	$3.9\ 10^{-9}$	$3.9\ 10^{-9}$	$4.1\ 10^{-9}$
200	$6.5\ 10^{-9}$	$6.5\ 10^{-9}$	$6.5\ 10^{-9}$	$6.6\ 10^{-9}$
250	$9.9\ 10^{-9}$	$9.9\ 10^{-9}$	$9.9\ 10^{-9}$	$9.9\ 10^{-9}$
300	$1.3\ 10^{-8}$	$1.3\ 10^{-8}$	$1.3\ 10^{-8}$	$1.3\ 10^{-8}$

$\hat{W}^k(F',c)$ [$cm^3\ sec^{-1}$]

Table 8.2.e:
Transition probability $W^k(c,F^*)$ of nonradiative (thermal) ionization of F^*-states for NaCl in dependence on the external electric field; after Kleih (1980) (1982).

T \ E(kV/cm)	5	10	20	30	40	50
50	$1.2\ 10^{3}$	$1.5\ 10^{5}$	$3.5\ 10^{6}$	$1.2\ 10^{7}$	$6.9\ 10^{8}$	$1.6\ 10^{9}$
100	$4.3\ 10^{7}$	$9.2\ 10^{8}$	$1.3\ 10^{10}$	$4.6\ 10^{10}$	$3.3\ 10^{11}$	$7.3\ 10^{11}$
150	$2.0\ 10^{9}$	$3.2\ 10^{10}$	$3.6\ 10^{11}$	$1.1\ 10^{12}$	$5.0\ 10^{12}$	$1.0\ 10^{13}$
200	$1.5\ 10^{10}$	$2.2\ 10^{11}$	$2.1\ 10^{12}$	$6.7\ 10^{12}$	$2.3\ 10^{13}$	$4.6\ 10^{13}$
250	$5.8\ 10^{10}$	$7.8\ 10^{11}$	$6.7\ 10^{12}$	$2.0\ 10^{13}$	$6.2\ 10^{13}$	$1.2\ 10^{14}$
300	$1.4\ 10^{11}$	$1.8\ 10^{12}$	$1.5\ 10^{13}$	$4.4\ 10^{13}$	$2.3\ 10^{14}$	

$W^k(c,F^*)$ [sec^{-1}]

Table 8.2.f:
Transition probability $W^k(c,F')$ of nonradiative (thermal) ionization of F'-states for NaCl in dependence on the external electric field; after Kleih (1980) (1982).

T \ E(kV/cm)	5	50
50	$3.8\ 10^{-79}$	$3.8\ 10^{-79}$
100	$2.1\ 10^{-30}$	$2.1\ 10^{-30}$
150	$1.9\ 10^{-14}$	$1.9\ 10^{-14}$
200	$1.0\ 10^{-6}$	$1.0\ 10^{-6}$
250	$1.3\ 10^{-1}$	$1.3\ 10^{-1}$
300	$2.2\ 10^{2}$	$2.2\ 10^{2}$

$W^k(c,F')$ [sec^{-1}]

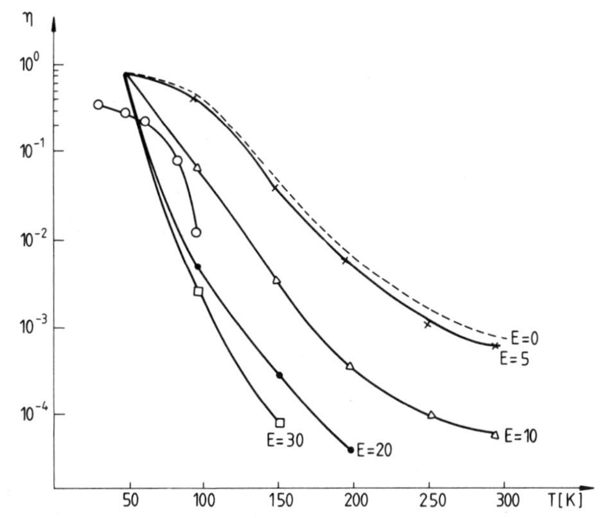

Figure 8.1:
The calculated quantum yield

$$\eta := W^s(F,F^*)[W^s(F,F^*)+W^k(F,F^*)+W^k(F^*,F)]^{-1}$$

in dependence on temperature and external electric field E in kV cm^{-1}. The symbols o represent the experimental values of η measured by Honda and Tomura (1972); after Kleih (1980) (1982).

Table 8.3.a:
Steady state conduction band electron concentration calculated from (8.2.77) for NaCl with an irradiation of $I_0=10^{12}$ cm^{-2} sec^{-1} into the F-band. The values of $\bar{n}_c$ are not sensitive to variations of the F-center concentration within the range of standard values, cf. Table 3.2, as $\bar{n}_c \ll \bar{n}_F$ holds. Therefore, the F-center concentration is not explicitly used as an additional parameter; after Kleih (1980) (1982).

T \ E(kV/cm)	5	10	20	30
50	$3.8\ 10^{-34}$	$1.7\ 10^{-33}$	$1.9\ 10^{-30}$	$3.4\ 10^{-28}$
100	$1.7\ 10^{-8}$	$6.8\ 10^{-8}$	$3.1\ 10^{-7}$	$3.7\ 10^{-4}$
150	$2.9\ 10^{0}$	$1.7\ 10^{1}$	$2.4\ 10^{2}$	$2.8\ 10^{3}$
200	$6.8\ 10^{4}$	$1.0\ 10^{5}$	$9.4\ 10^{5}$	$6.1\ 10^{6}$
250	$5.4\ 10^{6}$	$2.8\ 10^{7}$	$1.9\ 10^{8}$	$1.2\ 10^{9}$
300	$1.9\ 10^{8}$	$8.7\ 10^{8}$	$6.9\ 10^{9}$	$3.3\ 10^{10}$

$\bar{n}_c=\bar{n}_c(T,E)$ [cm^{-3}]

Table 8.3.b:
Mobility $\bar{\mu}$ of the conduction band electrons in dependence on the temperature; taken from Ahrenkiel and Brown (1964), and Redfield (1954).

T	50	100	150	200	250	300
$\bar{\mu}$ cm^2 V^{-1} s^{-1}	280	150	45	25	15	10

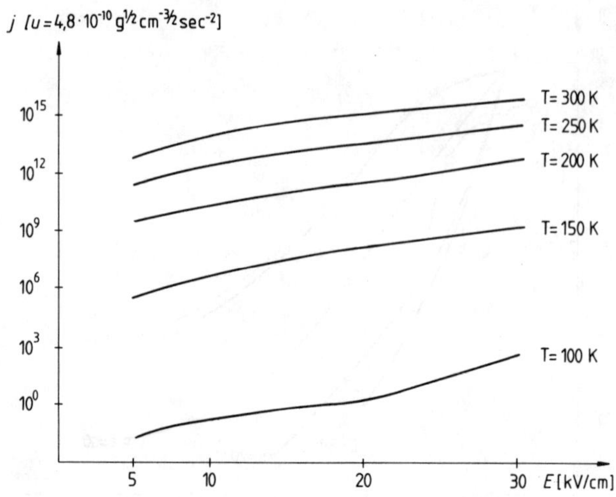

Figure 8.2:
Current voltage characteristic calculated from $|j|=\bar{n}_c e v_d = e\bar{n}_c \mu |E|$ with the values of Tables 8.3.a, 8.3.b for NaCl with an irradiation of $I_0 = 10^{12}$ cm^{-2} sec^{-1} into the *F*-band for various temperatures in units of $u=4{,}8 \cdot 10^{-10}$ g$^{1/2}$ cm$^{-3/2}$ sec^{-2}. No experimental data are available for comparison; after Kleih (1980) (1982).

For a discussion of these results we also refer to the original paper.
The electron transport processes which have been described in this example and which lead to a current-voltage characteristic are not complete from a physical point of view. First of all, in the general case the motion of holes has to be taken into account. This can be incorporated into the formalism by the derivation of an appropriate Boltzmann equation for non-localized holes and corresponding rate equations for localized hole occupation probabilities of acceptor states, Furthermore, in addition to transport processes via the conduction band and the valence band other transport mechanisms occur, namely the hopping motion of electrons, holes and ions, and the electron- and hole-impurity conduction by means of tunnel processes between the impurity centers. Hopping motion was already discussed in Section 5.5 with respect to localized excitons and holes and, naturally, it also occurs if localized polarons are present. Formally, localized excitons, holes and polarons can be considered to be a special type of impurity. Hence their levels have to be included in that part of the reaction or rate equations, resp. which contains the impurity processes, and the hopping motion is then a type of transition between impurity levels which leads to a change in the position of the electrons, etc. Such transitions are caused by the same interactions (including those with external fields) which are responsible for other types of transitions. Hence in principle hopping motion poses no new problems, and one can treat it with the methods developed so far for the treatment of rate equations. The same holds for impurity induced conduction. It has only to be observed that from these parts of the electronic processes a contribution to the various transport quantities has to be taken into account. Hence the transport processes via the conduction band and the valence band and via hopping motion and (or) impurity centers are competing mechanisms which are all contained in our reaction kinetic equations, provided they

are carefully examined. In the literature, however, these processes are usually treated separately. This stems from the fact that a combined reaction kinetics of conduction band states and impurity center states has so far been treated only by a few authors at a quantum statistical level, while the majority deal with this problem at a phenomenological level.

The quantum statistics of transport processes, in particular with respect to electronic transport processes, has been developed from the beginning of general quantum theory. However, in the early approaches no unified derivation of reversible and irreversible processes was performed based on the basic equations of quantum theory. Rather the irreversible processes were introduced in a phenomenological way. Such a presentation of transport theory can be found in almost any theoretical textbook on solid state theory, cf. for instance Ziman (1960). A review article of quantum transport theory which is more profound and which contains a comprehensive list of references of earlier work was published by Dresden (1961). Here we are not interested in the phenomenological introduction of irreversible transport theory. Rather we want to cite only those original papers which attempt to give a derivation of transport processes from first principles. Such attempts have been made mainly in the past two decades. Thus we cite literature from this period and only with respect to the derivation of Boltzmann-type equations or related kinetic equations while the hopping motion and impurity conduction will be surveyed in Section 8.3.e. Since transport processes are also of interest in metal physics, plasma physics, etc., the literature concerning transport equations is large and we cannot aim at completeness in any way.

In the literature three approaches to a theoretical description of transport processes are generally pursued, namely

i) derivation of transport equations from the equation for the density matrix,
ii) application of linear response formulae,
iii) use of test functions for approximate solutions of the Boltzmann-equation such as displaced Maxwellians, etc.

While in the approach i) the authors attempt to derive Boltzmann-type equations, the approach ii) provides only a general framework which has to be completed by special calculations. Usually the response formulae are transformed to field theoretic Green functions which are to be evaluated by perturbation theory or other many-body techniques, for instance Bethe-Salpeter equations. A review of these techniques is given by Ambegaokar (1962). There it is noted that the Bethe-Salpeter equation for the relevant response function can be rewritten as a Boltzmann equation, so that in certain approximations this approach is equivalent to the use of the Boltzmann equation. In addition, it was shown by several authors, for instance Montroll and Ward (1959), Chester and Thellung (1959) and Hubermann and Chester (1975), that the use of the response theory requires the summation of an infinite series of divergent terms in order to obtain the correct results of the Boltzmann equation solution for small scattering rates of the charge carriers. Furthermore the approaches i) and ii) have in common that a transition between reversible and irreversible motion cannot be seen clearly. Rather irreversibility appears by approximations which produce a sort of information loss and by averaging procedures. So i) and ii) cannot be considered to

be satisfactory deductions of irreversible transport theory. On the other hand, the application of the Boltzmann equation is often criticized although it is used in the majority of practical calculations. We prefer to work with the Boltzmann equation as in our way of derivation the transition from reversible to irreversible statistics can be clearly seen, the derivation allows the treatment of mixed band spectra and impurity spectra, and the derivation can be generalized to other basis systems, higher order transitions, resonance transitions, memory effects etc. leading to generalized Boltzmann equations etc. In particular with respect to the criticism the latter point is of interest, since most of the criticism is concerned with such extensions which are missed in the original form of the Boltzmann equation. Furthermore it should be noticed that the competing calculation methods so far have been applied only to very idealized models which do not reflect correctly the crystal electron situation, whereas our approach is concerned with the crystal model from the beginning.
Concerning the literature, the following topics were treated: Kohn and Luttinger (1957) developed a quantum theory of electric transport phenomena based on the density operator equations for the deviation p of the statistical operator from its equilibrium value. They showed that for the diagonal elements of the density operator an approximate Boltzmann equation for the conduction band electrons results by expanding the density matrix in ascending powers of the strength of the electronic scattering potential. Argyres (1958) presented a quantum theory of galvanomagnetic effects in semiconductors and semimetals, which is based on the density operator equation and a derivation of the Boltzmann equation for the conduction band electrons from the diagonal elements of the density operator and additional statistical averaging. Luttinger and Kohn (1958) improved their theory by an expansion in powers of the density of electronic scattering centers. Montroll and Ward (1959) analyzed the Kubo formula by performing a systematic cluster integral expansion for time relaxed momentum correlation functions. Prigogine and Ono (1959) studied transport equations of quantum gases in terms of Wigner functions which are derived from the density matrix. Argyres (1960) discussed the quantum theory of electric transport in a magnetic field along the lines of Luttinger and Kohn's approach. Gurevich and Firsov (1961) and Firsov and Gurevich (1962) based a theory of the electric conductivity of semiconductors in a magnetic field on Kubo's response function formalism and evaluated the corresponding expressions by field theoretic diagram techniques. Konstantinov and Perel (1961) defined mean electronic occupation numbers being averaged over the deviation of the statistical operator from its equilibrium value and derived by means of field theoretic diagram techniques kinetic equations for these quantities. This approach was extended by Gurevich and Nedlin (1961) to include also mean phonon occupation numbers and corresponding kinetic equations. Stinchcombe (1961) applied Kohn and Luttinger's approach to discuss the quantum theory of magnetoresistance in low fields. Baumann and Ranninger (1962) evaluated the response functions for electric conductivity of an electron-hole system in interaction with phonons by means of field theoretic Green function techniques. Klinger (1962) based a theory of kinetic effects for small current carrier mobility on an approximate evaluation of the Kubo formula. Kadanoff (1963) discussed a Boltzmann equation for polarons derived by phenomenological arguments. For the description of

the Hall effect in the polaron band regime Friedman (1963) derived a Boltzmann equation for the polaron motion by thermal averaging of the equation for the motion of polaronic wave packets. In analogy to the classical statistical many-particle correlation functions Ron (1964) defined many-particle quantum correlation functions $tr[d\varrho_n(\mathbf{r}_1 \ldots \mathbf{r}_n, \mathbf{r}'_1 \ldots \mathbf{r}'_n, t)]$ for electrons by means of the time-independent density matrix d of the system and the many-electron density operators $\varrho_n := \psi^+(\mathbf{r}_1, t) \ldots \psi^+(\mathbf{r}_n, t)\ \psi(\mathbf{r}'_1, t) \ldots \psi(\mathbf{r}'_n, t)$, $1 \leqslant n < \infty$ and derived a set of equations for these quantities which he treated for the case of an external magnetic field by a selfconsistent field approximation, i.e., a product representation of the higher point functions. Keldysh (1965) developed a diagram technique for the calculation of field theoretic Green functions of a system of particles in external fields and showed that for an appropriate one-particle distribution function derived from the two-point Green function a Boltzmann-like equation can be obtained. Dworin (1965) calculated the transverse conductivity of a system of independent electrons weakly interacting with optical phonons in an external magnetic field by means of Kubo's response formula. The Kubo formula was also used by Weijland (1966) for the calculation of the drift mobility of large polarons in the intermediate coupling region. Mahan (1966) treated this topic along the same lines. Wolman and Ron (1966) derived a kinetic equation for the quantum transport theory of electronic conduction in a strong magnetic field by approximating the two-particle density functions by a product of one-particle density functions and subsequent substitution of this expression in the equation for the one-particle density function, i.e., the first equation of the hierarchy of equations for all density functions. Argyres and Kirkpatrick (1967) considered a system of dynamically independent electrons interacting with a set of randomly distributed scattering centers under the influence of an external field. By defining the deviation p of the density operator from its equilibrium value they derived a kinetic equation for an appropriate mean value $\bar{p}$ by means of projection operator techniques and investigated the relation of this equation to the classical Boltzmann equation. Starting from the equations for the density operator, Hajdu and Keiter (1967) derived transport equations for electrons in an external magnetic field interacting with phonons and impurities. In the Born approximation these equations are generalizations of the corresponding Boltzmann equation. Holstein and Friedman (1968) treated the Hall mobility of the small polaron based on the evaluation of Kubo's formula. For a system of free electrons and randomly fixed impurities in a homogeneous magnetic field Bangert (1968) discussed the transverse electric conductivity by means of the Kubo formula. In a second paper Bangert (1968) derived transport equations for the one-electron density operator by means of field theoretic methods. Kalashnikov (1970) used a power series expansion of the statistical operator in terms of the external electric potential, the electron-electron interaction and the electron-phonon interaction in order to derive a set of transport equations for macroscopic variables in hot electron theory. Cunningham and Gruber (1970) carried out a quantum mechanical calculation of the electric conductivity for polar optical mode scattering of electrons in InSb by application of the Kubo formula. Levinson (1970) showed that the density matrix of an electron in uniform electric and magnetic fields is invariant with respect to translations which are applied simultaneously with appropriate gauge transformations of the field potentials, and derived the equation for the density matrix in the Wigner representation with inclusion of electron-phonon interaction in the weak coupling limit. Davies and Blum (1971) discussed a relaxation time ansatz for quantum transport theory by approximate treatment of the equation of motion for the single particle density matrix. Further references with respect to this approach are given there. Enderlein and Peuker (1971) based a theory of the electric conductivity of solids in a strong electric field on the evaluation of linear response formulae. Aldea (1971) performed a field theoretic calculation of polaronic conductivity by response functions. Also Gerhardts and Hajdu (1971) used response function formulae to calculate the transverse and longitudinal electric conductivity in strong magnetic fields in InSb by field theoretic evaluation techniques. Barker (1972) investigated the magnetophonon effect (≡oscillatory variation of electrical resistance with high magnetic field) in polar semiconductors by means of the Kubo formula and field theoretic techniques. Bryksin and Firsov (1972) (1973) derived expressions for the current in a semiconductor with a strong electric field and for the mobility of polarons in such fields. Klinger (1973) discussed the mechanism of small polaron transport based on response function formulae. Kümmel and Roldan (1973) performed a direct calculation of the current for a system of interacting electrons and phonons in strong electric fields. Barker (1973) developed a theory of nonlinear response of a system and applied it to the quantum transport theory of high field conduction in semiconductors. In evaluating the corresponding expressions he derived kinetic transport equations. Henneberger, Strehlow and Wünsche (1974) evaluated the Kubo formula for an electron-phonon system under the influence of an external electric field in order to calculate the tunnel conductivity of intraband transitions. Emin (1974) studied the conditions for the existence of free and selftrapped carriers in insulators, i.e., the occurrence of ordinary conduction and of hopping motion. Based on a linear response formula Bryksin and Firsov (1974) treated the quantum theory of galvano-magnetic effects in semiconductors, the transport in arbitrary electric and magnetic fields in absence of electron-electron interactions and the current in weak magnetic fields in the small polaron model. Bergman and Entin-Wohlmann (1974) discussed the appropriate choice of base systems for the representation of kinetic equations for electrons in an external electric field in interaction with phonons. Huberman and Chester (1975) investigated the use of the calculation of the electrical resistivity instead of the conductivity in order to avoid the summation of series of divergent graphs in the evaluation of the Kubo formula; they showed that a correct calculation offers no advantage of using one or the other expressions. Gerhardts (1975) applied field theoretic Green function techniques to the evaluation of the conductivity of an electron system in interaction with impurities and magnetic field by means of the Kubo formula. This formalism leads to "transport" equations derived for the calculation of the required Green functions. Jensen (1975) introduced the deviation p from the equilibrium density matrix and derived kinetic equations for p in the case of electrons which interact inelastically with phonons and infrared photons. He showed that by means of these generalized Boltzmann equations the results of second order time-dependent perturbation theory for electrons interacting with phonons and photons can be reproduced. Arora and Peterson (1975) developed a quantum theory of Ohmic

galvano- and thermomagnetic effects in semiconductors by direct evaluation of the density operator of a coupled electron-phonon system by means of a perturbation expansion. On the same lines Arora and Gomber (1976) treated electric conductivity, Aroa and Qureshi (1976) microwave conductivity, and Arora (1976) magnetomicrowave conductivity. Sawaki and Nishinaga (1977) used the deviation p from the equilibrium density matrix for a direct calculation of the electric current in an electron-phonon impurity system under the influence of a high electric field. Van Royen, de Sitter, Lemmens and Devreese (1977) applied the Kubo formula for the calculation of the frequency dependent conductivity tensor of polarons in a magnetic field by using the Fröhlich Hamiltonian. Anh and van Trong (1977) studied the excitation of electromagnetic vibrations of an electron-phonon system using the equation of motion for the electron distribution function coupled to the complete system of Maxwell's equations. Gurevich and Travnikov (1977) showed by means of an analysis of the Boltzmann equation coupled with electromagnetic wave equations that in a semiconductor in a strong magnetic field a strong helical wave can excite secondary weak helical waves. Ballini (1978) introduced the deviation p of the density operator from its value for the field free case and evaluated the corresponding equations with external electric field. He obtained an equation for the diagonal elements of p and derived from it by averaging an equation of the Boltzmann type. Beleznay and Serenyi (1978) used Feynman's path integral method for the formulation of a high field quantum magneto transport theory of polar semiconductors. Using the Wigner distribution function, van Weert and de Boer (1978) derived quantum kinetic equations for a system of interacting particles with a Hamiltonian in Fock space representation which in the lowest order approximation lead to the Boltzmann equation.
Concerning the reaction kinetics of a system with band states as well as with bound states, in the majority of papers this topic is treated at a phenomenological level. At this level the effect of electron-phonon transitions in the conduction band is described by the mobility of the charge carriers leading to a phenomenological conductivity and diffusion, while the other transitions are described by phenomenological rate equations, i.e., equations whose analytical form with corresponding values of the coefficients are not derived from microscopic quantum statistics. This approach has widely been applied to the calculation of reactions of the "classical" covalent semiconductors and to the photoconductivity, thermoluminescence, etc., of polar insulators. Hence we also cite some papers which deal with reactions of covalent materials. The phenomenological version of the reaction kinetics is discussed in many textbooks and monographies, for instance in the books of Adirowitsch (1953), Stasiw (1959), Blakemore (1962) and Haug (1970) and the review articles of Schön (1958) and of Stöckmann (1959). A general review of inorganic crystal phosphor reactions is given by Kröger (1956). In original papers the following topics were treated in this field at the phenomenological level: Rate equations at this level were first introduced by Schön (1937), Blochinzeff (1937) and de Groot (1939). By means of such equations Möglich and Rompe (1940) discussed energy conversion in crystal phosphors. Schön (1948) investigated the kinetics of luminescence of ZnS phosphors with several activators by solving the corresponding rate equations for stationary states. Stöckmann (1950) developed a theory of photoelectric conductivity in polar insulators and semicon-

ductors. Broser and Warminsky (1950) used the model of Riehl and Schön for the theoretical explanation of luminescence and electric conductivity in CdS crystals. In order to explain electronic conductivity and electric breakdown of insulators, Franz (1952) derived an elementary kinetic equation for the conduction band electrons and calculated the rate of generation of conduction band electrons under the influence of the external field by means of phenomenological rate equations for the discrete level occupation numbers. Shockley and Read (1952) treated the statistics of the recombinations of electrons and holes on the basis of a model in which recombination occurs through the mechanism of trapping. Isay (1953) gave a contribution to the theory of conductivity of semiconductors by an analysis of the solutions of the rate equations for electrons, holes and traps and of those for electrons, holes and activators. Böer and Vogel (1955) analyzed transient solutions of photoconductivity in order to obtain information about the impurity levels. Hornbeck and Haynes (1955) discussed the trapping kinetics of minority carriers in Si, and Rose (1955) gave an analysis of the recombination processes in insulators and semiconductors. Stöckmann (1955) (1957) discussed negative photoeffects in semiconductors and the dependence of photoelectric currents on the strength of the external electric field. Van Roosbroeck (1956) investigated the photomagnetoelectric effect in semiconductors by adding relaxation terms for the electron and hole numbers in the phenomenological equations of electron and hole currents. Sandiford (1957) calculated the carrier lifetime in semiconductors for transient conditions. Okada (1957) studied the recombination of excess carriers in semiconductors. Wertheim (1958) studied the transient recombination of excess carriers in semiconductors. Nomura and Blakemore (1958) (1961) calculated the decay rates of excess carriers in semiconductors. Haering and Adams (1960) studied thermally stimulated currents in photoconductors. Van Roosbroeck (1960) gave an extended version of his earlier approach including a discussion of the phenomenological rate equations for electron and hole recombination at traps and applied it to current carrier transport and photoconductivity in semiconductors. Conwell and Zucker (1961) discussed the recombination of hot carriers by solving a system of corresponding rate equations. Adirovich (1961) performed an analysis of the impurity-photoconductivity kinetics based on phenomenological rate equations. Using the model of Schön and Adirowitsch, the corresponding rate equations were solved by Bräunlich (1963) who studied by means of these solutions thermoluminescence and thermally stimulated conductivity. Kogan (1963) studied the photoconductivity with inclusion of variations of the carrier mobility. The effect of capture levels on the kinetics of impurity photoconductivity in semiconductors was investigated by Arkadeva (1963). Nicholas and Woods (1964) evaluated electron trapping parameters from conductivity glow curves in CdS. Bräunlich and Scharmann (1964) approximately solved the rate equations of a simple model in order to explain thermoluminescence and thermally stimulated conductivity in alkali halogenides. Richter (1964) developed a theory of the stationary photomagnetoelectric effect in InSb by means of an approximate solution of the Boltzmann equation with inclusion of recombination rates in the collision term. Stocker and Kaplan (1966) studied the oscillatory photoconductivity in semiconductors using the Boltzmann equation in the relaxation time approximation and adding relaxation terms due to electron recombination processes. Streetman (1966) investigated the effects of carrier recombination and trapping on transient photoconductive decay. Bräunlich and Scharmann (1966) gave an approximate solution of Schön's balance equations for the thermoluminescence and the thermally stimulated conductivity of inorganic photoconducting crystals. Vinetskii and Kholodar (1967) calculated the electric conductivity of semiconductors caused by the ionization of thermal lattice defects. Dussel and Bube (1967) gave a theory of thermally stimulated conductivity in a previously photoexcited crystal. Lyaguschenko and Yassievich (1968) discussed the photomagnetic effect on heated electrons by means of the Boltzmann equation supplemented by collision terms responsible for recombination and ionization. Bonch-Bruevich and Landsberg (1968) gave a review of various mechanisms of the charge carrier recombination in semiconductors. Saunders (1969) analyzed thermally stimulated luminescence and conductivity of insulators. Klassmann (1969) treated the isothermal Hall effect in impurity doped semiconductors. Stöckmann (1969) reviewed the effects of electric instabilities in semiconductors and photoconductors and their explanation. Folland (1970) derived an exact solution of the Boltzmann equation for the Stocker-Kaplan model of oscillatory photoconductivity. Gardavsky (1970) considered the decay of excess charge carriers in semiconductors. Kelly and Bräunlich (1970) developed a phenomenological theory of thermoluminescence. Crandall (1971) performed a calculation of the photocarrier density in a semiconductor that is undergoing impact ionization of shallow impurity states at low temperatures. Blakemore (1971) reviewed the phenomenological recombination rate theory in semiconductors. Simmons and Taylor (1972) developed a theory for the isothermal and thermally stimulated currents which flow in optically excited insulators subjected to high fields. Dneprovskaya (1972) studied the effects of light pumping on the conductivity of a multivalley n-type semiconductor. Kastalskii (1973) calculated the current-voltage characteristic of the low temperature breakdown of shallow impurity states. The influence of double optical transitions on the impurity photoconductivity spectrum of a compensated semiconductor was analyzed by Lebedev (1973). Osipov (1974) developed a phenomenological theory of the electroluminescence of heavily doped semiconductors. Chusov (1975) analyzed the photoconductivity of a compensated semiconductor. Parmon, Khairutdinov and Zamaraev (1975) analyzed the kinetics of tunneling electron transfer in solids. Yao, Inagaki and Maekawa (1975) studied the energy relaxation process in n-type InSb under strong longitudinal magnetic fields. Kachlishvili (1976) gave a review of galvanomagnetic and recombination effects in semiconductors in a strong electric field based on approximate solutions of the Boltzmann equation with inclusion of recombination collision terms. Weaver, Shultis and Faw (1977) derived analytic solutions of phenomenological rate equations for a model of radiation induced conductivity in insulators. Look (1977) used phenomenological current equations and elementary rate equations for an extension of the treatment of the photomagnetoelectric and photoconductivity effects in semiconductors with application to GaAs. Von Roos (1978) extended the Shockley-Read-Hall theory of free carrier recombination and generation via traps to degenerate semiconductors. Lhermitte, Carles and Viger (1978) studied non-equilibrium photoconductivity effects in semiconductors and insulators with recombination and trapping levels.

At the microscopic quantum level the band state-bound state reaction kinetics was treated by only a few authors. In connection with plasma problems the kinetics of bound states was considered by Klimontovich (1967) (1968). Peletminskii (1971) obtained kinetic equations in the presence of bound states of particles. Tsukanov (1975) described the kinetics of electron-impurity systems when bound states of electrons can be formed in impurities. His method is based on a direct evaluation for the density operator of the system in Fock space representation and is developed so far only at a formal level. So no decision can be made about its practical applicability to semiconductor reaction kinetics. Zehe and Röpke (1975) directly calculated the time derivatives of average values of characteristic observables of coupled electron-phonon-photon systems in Fock space representation by means of linear response theory for radiative recombination from non-equilibrium states in cathodo-excited semiconductors. Although in this case direct calculations can be performed, the disadvantage is the absence of kinetic rate equations which are required for the treatment of more complicated models and phenomena and the a priori knowledge required for the definition of the Fock representation which is too simple to cover more complicated models.

Finally the assumptions (8.2.28) and (8.2.31) may be abandoned. In this case the phonon and photon distributions are directly correlated to the non-equilibrium electron distributions and the mutual influence of these distributions leads to the so-called photon drag and phonon drag. The latter was introduced by Gurnevich (1945) (1946), Herring (1954), Price (1955) and Appel (1957) (1958); the coupled system of rate equations for electrons, phonons and photons was derived from quantum statistics by Stumpf (1957) (1961) (1971). We do not discuss this topic explicitly, but cite only some recent publications in this field. Gurevich and Nedlin (1961) derived a set of coupled kinetic equations for the phonon distribution functions and the electron density operator in order to calculate the thermoelectric transport coefficients in the presence of external magnetic fields. Plavitu (1965) considered the acoustic and optical phonon drag for the case of a pure ionic *n*-type semiconductor and evaluated the kinetic coefficients of transport theory. Spector (1965) formulated a quantum approach to the amplification of optical phonons in semiconductors. Yamashita and Nakamura (1966) treated electron-phonon interactions in strong electric fields by approximate solution of coupled Boltzmann equations for the electron and the phonon distribution functions. Georgescu (1967) treated the thermodynamics of the phonon drag in ionic semiconductors. Lee and Tzoar (1969) discussed a phonon instability due to electron-phonon interaction in a magnetic field. Alba and Kocevar (1969) calculated the perturbation of the phonon distribution by hot electrons in semiconductors. Plavitu (1969) calculated the contribution of the optical phonon drag to the electric conductivity of a polar semiconductor. Caccamo and Ferrante (1971) studied phonon amplifications in an electron-phonon system by the effect of electron drifting against the lattice. Kubalkova and Sakalas (1972) investigated the optical phonon drag effect in single CdSe crystals. Perrin and Budd (1972) studied the solutions of the coupled equations for electron and phonon distributions in semiconductors in a strong electric field. Yee (1972) developed a theory of the photon drag effect in polar crystals based on the solutions of a generalized Boltzmann equation with corresponding transition probabilities derived from a microscopic model of electron-phonon-photon interaction. Kocevar (1972) used a displaced Maxwellian distribution for hot electrons and studied its influence on the electron-phonon collision term in the corresponding phonon Boltzmann equation. By means of the momentum and energy balance relations for the electrons he established a feedback between phonon- and electron distributions and solved approximately the phonon Boltzmann equation for phonon amplification processes. Kocevar (1973) applied this model for an investigation of polar optical phonon and acoustic phonon disturbances and instabilities in *n*-InSb. Lang and Pavlov (1973) developed a theory of electron dragging by phonons in a magnetic field. Grigorev (1973) considered two coupled Boltzmann equations for two kinds of charge carriers in polar semiconductors and studied the drag effect being exerted from one kind on the other kind of charge carriers. Plavitu (1974) investigated the contribution of the phonon drag effect to the galvanomagnetic coefficients in polar semiconductors. Arora and Miller (1974) studied the effect of electron-phonon drag on the magneto-conductivity tensor of semiconductors by an evaluation of the corresponding density operator equation. Jay-Gerin (1975) calculated the effect of the quantization of the electron energy levels in a strong magnetic field on the phonon drag contribution to the transverse thermoelectric power of *n*-type GaS in the extreme quantum limit. Sharma and Agarwal (1975) discussed the effect of the modification of the phonon distribution in semiconductors by high electric fields on the transport coefficients. The effect of the mutual electron-phonon drag on the electric conductivity in a quantizing magnetic field and on the heating of electrons was studied by Zlobin (1976). Bagaev, Keldysh, Sibeldin and Tsvetkov (1976) investigated the dragging of excitons and electron-hole drops by non-equilibrium short-wave phonons (phonon wind) at a phenomenological level. Epshtein (1977) studied the amplification of optical lattice vibrations in semiconductors in the field of an electromagnetic wave. Tarasenko and Chumak (1977) discussed the non-equilibrium fluctuations of electrons and phonons in semiconductors and used for their description a coupled system of equations for the electron and phonon fluctuation operators, i.e., the distribution operators minus their average values ($\equiv$distribution functions). Belorusets, Grinberg and Imamov (1977) studied the photon drag of carriers in the photoionization of neutral deep impurity centers in semiconductors with a many-valley energy spectrum. Imamov, Kramer and Sagdullaeva (1977) treated the drag of electrons in photoionization of deep impurity centers, and Samoilovich and Buda (1976) considered the drag of electrons by phonons. Pinchuk and Korolyuk (1977) used a variational method for the calculation of the electron-phonon drag in semiconductors in a longitudinal quantizing magnetic field. Catalano, Cingolani, Cali and Riva-Sanseverino (1978) reported about the evidence of the photon drag effect in polar crystals.

We now turn to the literature with respect to the calculation of transition probabilities. The calculations of transition probabilities performed above show that these quantities are modified by the influence of external fields. This is due to the following reasons:

i) The occurrence of additional transition probabilities due to the interaction between particles and external fields;
ii) the occurrence of symmetry breaking as the external fields destroy the original symmetry;
iii) the modification of all transition probabilities as the crystal states are modified by the field;
iv) the modification of the thermal weighting factors in the transition probabilities as the thermal distributions are changed by the field.

As a result of these modifications we have to expect a great variety of field effects. Concerning the literature in this field, the treatment of these effects is not very systematic with respect to a clear application of irreversible quantum statistics. Rather approaches in various directions and with partly very simple models have been made. We mainly cite papers of the last two decades and refer for earlier attempts to the references given in the papers cited here. The following topics were treated in this field: Calculation of ionization rates of valence band electrons and impurity electrons by electric fields and by impact ionization with a one-electron and with a many-electron model in a rigid lattice by time-dependent perturbation theory, Franz and Tewordt (1956). Theory of electron multiplication in semiconductors by solution of the Boltzmann equation for the motion of electrons (or holes) in a high field, taking into account the effect of electron-phonon and pair-producing collisions on the distribution functions, Wolff (1954). Evaluation of Franz' theory of electric breakdown in alkali halogenides in a one-electron model with electron-phonon coupling, Veelken (1955). Phenomenological discussion of electroluminescence, Piper and Williams (1955), Thornton (1956). Review of the theory of dielectric breakdown, Franz (1956). Influence of an electric field on the optical absorption edge of electrons in insulators calculated by means of OEMRL (one-electron model in a rigid lattice), Franz (1958), Keldysh (1958). Calculation of electron-hole pair production probability in crystals with strong electric fields based on a one-electron Hamiltonian with electron-phonon coupling and the application of Houston functions and perturbation theory, Keldysh (1958). Calculation of the efficiency of electroluminescence in ZnS on the basis of a one-electron model with impact ionization in a barrier. Theory of the effect of a magnetic field on the optical absorption in semiconductors on the basis of a one-electron model with effective mass approximation, Roth, Lax and Zwerdling (1959). Evaluation of the non-linear dependence of the cyclotron resonance frequency on the magnetic field strength due to electron-phonon coupling in polar crystals by means of an electron-phonon Hamiltonian in Fock space representation, Tulub (1959). Note on the theory of the Zeeman effect in the photo-ionization of impurities, Wallis and Bowlden (1959). Treatment of Zener tunneling in semiconductors based on OEMRL, Kane (1959). Theory of line-shape of interband magneto-optical absorption in semiconductors based on an electron-phonon-photon Hamiltonian in Fockspace representation and an evaluation in the interaction representation, Ohta, Nagae and Miyakawa (1960). Theory of tunneling based on OEMRL, Kane (1961). Group theoretical analysis of the influence of an electric field on the quadrupole lines in the exciton absorption spectrum, Cherepanov, Druzhinin, Kargapolov and Nikoforov

(1962). Review of the magneto-optical effects, namely cyclotron resonance, magneto-reflection, magneto-absorption, Faraday effect and Voigt effect in semiconductors, Moss (1962). Review and treatment of the Faraday effect in semiconductors based on OEMRL, Boswarva, Howard and Lidiard (1962). Theory of interband tunneling due to a constant electric field and its dependence on a longitudinal magnetic field based on OEMRL by time-dependent perturbation theory, Argyres (1962). Calculation of optical absorption of semiconductors in the presence of a uniform electric field based on OEMRL, Tharmalingam (1963), Callaway (1963) (1964). Review of electroluminescence of zinc- and cadmium sulfid crystals, Matossi and Gutjahr (1963). Calculation of impact ionization probability of centers in polar crystals by electrons which have been accelerated in an electric field using time-dependent perturbation theory on the basis of OEMRL, Cardon (1963). Theory of field ionization of local levels based on OEMRL, Perlin and Cheban (1963). Theory of the Faraday effect in solids based on OEMRL, Roth (1964). Discussion of the effect of applied fields on the optical properties of color centers, Henry, Schnatterly and Slichter (1964). Electron tunneling and Schottky ionization from excited F-centers in an electric field based on OEMRL, Euwema and Smoluchowsky (1964). Calculation of the optical absorption coefficient for phonon assisted interband transitions in an electric field based on a one-electron model with electron-phonon coupling, Penchina (1965), Chester and Fritsche (1965). Calculation of transition rates for direct and indirect optical transitions in insulators and semiconductors subject to a strong electric field and the fine structure of the Stark splitting based on OEMRL and time-dependent wave functions, Yacoby (1965). Note on the magnetic field inversion effect in the exciton absorption range of CdS, Zhilich (1965). Treatment of the Faraday effect in the exciton absorption range based on an effective two-particle exciton Hamiltonian, Zhilich and Makarov (1965). Derivation of expressions for the weak field Faraday effect in solids based on OEMRL, Bennett and Stern (1965). Phonon assisted optical absorption in an electric field based on a one-electron model with electron-phonon coupling and time-dependent perturbation theory, Fritsche (1965). Investigation of the influence of a high magnetic field on the radiative recombination of electrons between the conduction band and an acceptor based on OEMRL and perturbation theory, Beleznay and Pataki (1965). Theory of phonon assisted interband tunneling in semiconductors in a homogeneous electric field based on a one-electron model with electron-phonon coupling and time-dependent perturbation theory, Tiemann and Fritzsche (1965). Extension of Euwema's and Smoluchowski's theory of the electric-field ionization of the excited F-center in KCl, Spinolo and Fowler (1965). Note on the optical absorption by excitons in a strong electric field, Duke (1965). Discussion of magneto-dispersion effects in semiconductors in crossed fields by means of an evaluation of the photon propagator with contributions of photon carrier interactions, Uritsky and Shuster (1965). Thermal ionization of impurity centers in polar semiconductors in an external electric field based on a one-electron model with electron-phonon coupling and time-dependent perturbation theory, Cheban (1965). Electric-field effects on optical absorption in the vicinity of the absorption edge in insulators or semiconductors based on OEMRL, Aspnes (1966). Generalization of Huang's phenomenological theory of lattice polarization in ionic crystals to include the interaction with an external static electric field and discussion of the related electro-optical effects in ionic crystals with zincblende structure, Kelly (1966). Inclusion of damping by impurity scattering in the quantum theory of interband magneto-optical absorption by means of the calculation of corrections to the one-particle density matrix, Korovin and Kharitonov (1966). Note on the polaron induced anomalies in the interband magneto-absorption of InSb, Johnson and Larsen (1966). Theory of electron tunneling in semiconductors with degenerate band structure based on OEMRL, Krieger (1966). Calculation of many-phonon thermal ionization rates of excited *F*-center states in an external field by means of a one-electron model with electron-phonon coupling and time-dependent perturbation theory, Rozneritsa and Cheban (1966). Phonon-assisted magnetoabsorption in direct-band semiconductors, Batra and Haering (1967). Electric-field effects on interband transition at the critical points in semiconductors and insulators based on OEMRL, Aymerich and Bassani (1967). Theory of impact ionization in ionic semiconductors in a strong electric field by means of the solution of the corresponding Boltzmann equation in the relaxation time approximation, Chuenkov (1967). Discussion of magneto-optical experiments on broad absorption bands in solids for the example of the *F*-center, Brown and Laramore (1967). Theory of the excitation of excitons and phonons by a strong electromagnetic field based on a Fockspace representation of the quasiparticle Hamiltonian, Fain (1968). Effect of electric field and energy dispersion law on plasma oscillations in semiconductors based on the solution of the Boltzmann equation with relaxation time approximation, Dykman and Tomchuk (1968). Calculation of direct multiphoton interband transitions induced by a strong oscillating electric field in semiconductors in the presence of a longitudinal or transverse magnetic field based on OEMRL, Weiler, Reine and Lax (1968). Theory of electron paramagnetic resonance line shape in external constant and alternating electric fields by means of the density matrix formalism and a one-electron spin Hamiltonian as zero order system, Glinchuk, Deigen and Korobko (1968). Investigation of intraband and of donor-band transitions of crystal electrons in high electric fields based on OEMRL, Döhler and Hacker (1968). Treatment of the effect of an electric field on the absorption edges of insulating crystals based on a one-electron model and taking into account electron-hole and carrier-lattice interactions, Rees (1968). High frequency Franz-Keldysh effect based on OEMRL, Yacoby (1968). Contribution to the theory of the Franz-Keldysh effect based on OEMRL, Ralph (1968). Review of the theory of *F*-center absorption spectra with inclusion of magneto-optical effects and stress effects, Henry and Slichter (1968). Theory of magnetic circular dichroism and stress dichroism in the absorption spectrum of impurity centers in alkali halide phosphors based on a one-electron-lattice molecular model, Cho (1969). Note on the stimulation of exciton lines in a strong magnetic field, Dyakonov, Mitchell and Efros (1969). Phonon-assisted absorption in crossed electric and magnetic fields in direct-band semiconductors, Batra (1969). Discussion of electron-hole pair generation phenomena in semiconductors based on OEMRL, Valton (1969). A review of electron tunneling processes mainly devoted to the electronic contact problem of materials and theoretically based on the evaluation of response functions by many-body techniques, Duke (1969). Optical absorption by bound and continuum states of Wannier excitons in an electric field based on OEMRL in the effective mass approximation to the exciton, Blossey (1970) (1971). Two photon transitions in solids in a magnetic field based on OEMRL, Bassani and Girlanda (1970). Electric dipole selection rules for band edge optical transitions arising from weakly bound two- or three carrier complexes in a uniaxially strained direct zincblende semiconductor by group theoretical analysis, Bailey (1970). Franz-Keldysh effect in an alternating electric field based on OEMRL, Subashiev and Chalikyan (1970). Discussion of the Franz-Keldysh effect and hot electron effects in the interband absorption of semiconductors in an external field by means of the Kubo formula and a Fockspace representation of the Hamiltonian of the coupled electron-phonon system, Enderlein (1970). Investigation of electric field-dependent plasma oscillations in semiconductors based on the solution of the corresponding Boltzmann equation, Krishnamurti and Paranjape (1970). Excitonic effects on dielectric properties of solids in a uniform electric field based on the solution of OEMRL in the effective mass approximation of the exciton, Yang (1971). Two-photon absorption in semiconductors in a magnetic field based on OEMRL, Girlanda (1971). Theory of impact ionization in semiconductors in crossed fields by means of the solution of the corresponding Boltzmann equation in the relaxation time approximation, Reuter and Hübner (1971). Review of the effects of external electric fields on paramagnetic resonance, Roitsin (1972). Optical properties of the small polaron in high electric fields based on a one-electron model with electron-phonon coupling to one breathing mode, Austin (1972). Calculation of the Stark effect of impurity centers by the method of average moments, Gardavsky (1972). Theory for impurity assisted tunneling by means of the direct evaluation of the current operator using a many-electron Hamiltonian in Fock space representation, Schattke and Birkner (1972). Theory of exciton-phonon complexes in a strong magnetic field by means of field theoretic Green function techniques, Rashba and Edelshtein (1972). The sprectrum of optical absorption by Mott excitons in an electric field based on OEMRL, Monozon, Seysyan and Shelekhin (1972). Indirect many-photon transitions in quantizing magnetic fields, Chaikovskii (1972). Paraelectric resonance of excitons in semiconductors, Zheru and Shmiglyuk (1973). Interband magneto-optical transitions based on OEMRL, Tanaka and Shinada (1973). Effect of internal excitations of polarons on magneto-optical absorption and cyclotron resonance in polar crystals based on a one-electron model with electron-phonon coupling, Devreese, Kartheuser, Evrard and Baldereschi (1973). Theory of avalanche ionization in transparent dielectrics under the action of a strong electromagnetic field based on the solution of the linearized Boltzmann equation, Rubinshtein and Fain (1973). Variation of luminescence band shape in a strong electric field based on OEMRL, Timashev (1973). Impact ionization of deep centers in an intense electric field based on a one-electron model with electron-phonon coupling, Timashev (1973). Effect of electric fields on the absorption of light by deep impurity centers based on OEMRL, Vinogradov (1973). Theory of cyclotron resonance absorption based on OEMRL, Argyres and Sigel (1974). Group theoretical selection rules for two-phonon processes in rocksalt type crystals under electrostatic fields, Hayashi and Kanamori (1974). Calculation of transition probabilities for impact ionization in the presence of an electric field in a many-electron model using time-dependent perturbation theory, Takeshima (1975). Investigation of a current transport mechanism due to interband tunneling without and with photon assistance by means of a Bloch electron Hamiltonian with electric field in Fock space representation and electron-photon coupling, Henneberger and Enderlein (1975). Study of the first order effects of an external electric field on optical transitions between localized electronic states of substitutional impurities in wurtzite type crystals based on the electron Hamiltonian of the impurity atom, Boyn and Gardavsky (1975), Gardavsky and Boyn (1975). Lineshape analysis of free-electron to bound-hole transitions in electric fields, Wagner and Bludau (1975). Theory of electric field effects on electronic spectra and electronic relaxation with applications to *F*-centers using a vibronic molecular model, Lin (1975). Non-uniform electric field effect on phonon assisted tunneling in semiconductors based on a one-electron model with electron-phonon coupling and time-dependent perturbation theory, Yang (1975). Intraband Raman scattering in semiconductors with non-parabolic bands in applied fields, Monozon and Zhilich (1975). Note on the optical absorption spectrum of diamagnetic excitons, Zhilich and Maksimov (1975). Influence of phonon dispersion on a threshold singularity in impurity magneto-optic absorption calculated by means of field theoretic Green function technique, Osipov and Yakovlev (1975). A note on photon thermal conductivity of semiconductors in strong electric fields, Dzhaksimov (1976). A note on far infrared Raman scattering by Landau levels in semiconductors, Vlasov (1976). Calculation of Auger and impact ionization probabilities for semiconductors in an external electric field based on a many-electron Hamiltonian in the rigid lattice and time-dependent perturbation theory, Hill (1976). Theory of photon-assisted Zener tunneling by means of an electron-photon Hamiltonian in the rigid lattice in Fock space representation and time-dependent perturbation theory, Henneberger (1976). Investigation of resonance Raman scattering in a strong electric field based on the evaluation of linear response functions, Bechstedt, Enderlein and Peuker (1976). Discussion of the parallel and perpendicular Stark effect of the *F*-center in RbBr and RbI, Banaii and Jacobs (1976). Calculation of the influence of the quadratic Stark effect of impurity centers on the absorption spectra based on a molecular model and the evaluation of Kubo's formula, Shekhtman (1976) (1977). Calculation of impact ionization probabilities for semiconductors in strong electric fields based on a many-electron Hamiltonian in the rigid lattice and time-dependent perturbation theory, Kyuregyan (1976). Note on the influence of impurities on Raman scattering by Landau levels, Genkin and Zilberberg (1976). Note on possible Raman scattering of light under Stark quantization conditions, Kryuchkov and Yakovlev (1976). Note on the theory of absorption of light in an intrinsic semiconductor in a homogeneous magnetic field, Zenchenko and Sinyavskii (1976). Note on many-phonon impurity absorption of light in the presence of an external electric field, Prodan and Rozneritsa (1976). Calculation of the effect of a magnetic field on the phonon-assisted two-photon absorption in semiconductors based on a one-electron model with electron-phonon coupling and time-dependent perturbation theory, Hassan (1976). Note on the influence of an electric field on the photoionization of deep centers as a result of electric-vibrational transitions, Baleinikov, Bulyarskii, Grushko and Gutkin (1976). Ionization of deep impurity centers in semiconductors by an electric field based on OEMRL, Perelman (1976). Calculation of the influence of an electric field on the

Auger recombination rates in semiconductors based on a many-electron model in a rigid lattice and time-dependent perturbation theory, Gebranzig, Haug and Rosenthal (1976). Calculation of indirect two-photon transition rates in semiconductors with a magnetic field based on a one-electron model with electron-phonon coupling, Hassan (1976). Effect of polaron coupling on free-electron cyclotron-resonance transitions between $n=1$ and $n=2$ Landau levels in InSb by application of perturbation theory, Koteles and Datars (1976). One- and two-photon magneto-optical transitions in InSb based on OEMRL, Zawadzki and Wlasak (1976). Note on the dependence of impact ionization of deep levels in semiconductors on the electric field, Kuzmin, Kryukova and Kyuregyan (1976). Infrared absorption by an optical polaron in a magnetic field, Huybrechts (1977). Ionization of impurity states in semiconductors by an electric field based on OEMRL, Korol (1977). Investigation of magnetic quantum oscillations of the Auger transition rate in narrow gap semiconductors based on a many-electron model in the rigid lattice, Gerhardts (1977). Intraband magneto-absorption in mixed semiconductors and in imperfect semiconductors based on OEMRL, Mycielski, Bastard and Rigaux (1977). Many-photon magneto-optical absorption in semiconductors with non-spherical bands based on OEMRL, Monozon and Zhilich (1977). Many-photon Franz-Keldysh effect in semiconductors with non-parabolic bands in a longitudinal magnetic field based on OEMRL, Monozon and Zhilich (1977). Theory of allowed resonance Raman scattering of first and second order in a magnetic field, Enderlein and Bechstedt (1977). Many-photon absorption by intraband transition in a magnetic field based on a one-electron model with electron scattering at impurities, phonons, etc., Muntyan, Pokatilov and Fomin (1977). Photoionization of deep impurity centers in a quantizing magnetic field based on OEMRL, Bulyanitsa and Grinberg (1977). Discussion of the Stark effect in F-center emission in Kbr, KI, RbBr and RbI crystals, Ohkura, Imanaka, Kamada, Mori and Iida (1977). Two-photon absorption in crossed and magnetic fields based on OEMRL, Hassan and Moussa (1977). Note on the capture of electrons by positively charged traps, El-Waheidy (1977). Phonon-assisted two-photon interband transitions in a magnetic field, Moussa and Hassan (1977). Theory of the Faraday and Cotton-Mouton effects for acoustic phonons in paramagnetic rare earth compounds, Thalmeier and Fulde (1978). Study of interband tunneling in a semiconductor in constant electric and magnetic fields based on an evaluation of the current operator in Fock space representation, Oleinik and Sinyak (1978). Effect of an electric field on exciton absorption based on a OEMRL of the exciton, Aronov and Ioselevich (1978). Multiphonon assisted electronic tunnel processes in a homogeneous field based on a one-electron model with electron-phonon coupling, Dalidchik (1978). Study of matrix elements of electronic dipole transitions of impurity centers in crystals in an external electric field and analysis of the effective field acting on the center, Geifman, Glinchuk, Deigen and Krulikovskii (1978). Thermal ionization of impurities in strong electric fields based on OEMRL, Abakumov, Kreshchuk and Yassievich (1978). Theory of magneto-Raman effect for rare earth ions in crystals, Chamberlain (1978).

8.3 Transport theory

Transport theory describes the macroscopic reactions of crystals under the influence of external forces. It is formulated by means of quantum statistics and the thermodynamics of irreversible processes. In application to semiconductors, it leads to the explanation of many important effects. As transport theory is basic for semiconductor physics and techniques it is treated extensively in the literature. Hence we give only a short account of these processes for homogeneous materials. Since we have a transport equation, we are able to derive the most important results directly without using irreversible thermodynamics.
The measurable quantities which play a central role for semiconductors are the average electric current and the average heat current. Leaving out the energy transport by phonons, the carriers of this charge and energy transport are electrons and holes if the motion of ions is excluded. Whether both kinds of carriers contribute to these currents depends on the physical preparation of the semiconductor crystal. If we have intrinsic conductivity, a simultaneous activation of electrons and holes takes place by raising electrons from the valence band into the conduction band, leaving behind holes in the valence band. For extrinsic conductivity, the charge carriers are generated by the ionization of impurity centers which leads either to holes in the valence band or to electrons in the conduction band. In practice, intrinsic or both kinds of extrinsic conductivity can be only approximately achieved. For the theoretical discussion it is, however, convenient to treat the pure cases separately. In the basic quantum statistical description of these processes the formalism of rate equations can cover all these cases. However, for mixed conductivity the system of rate equations is considerably more complicated than for the pure cases. As all techniques for establishing rate equations and for treating them can already be studied for the pure case, we used a semiconductor model in the foregoing chapters which leads to extrinsic electronic conductivity. Hence, we first treat in detail the extrinsic conductivity of this model and consider the most important facts about intrinsic conductivity afterwards.

a) Transport quantities

The treatment of extrinsic conductivity for electrons and holes runs along the same lines. Owing to our model we consider extrinsic electronic conductivity. In this case the carriers which contribute to the charge and energy current are the n electrons of the conduction band. According to (3.5.13), the probability distribution $f(\mathbf{K}_1 \dots \mathbf{K}_n, t)$ for the n electrons in the conduction band is normalized to unity. The individual current of a single electron in a Bloch state of the conduction band with wave vector $\mathbf{K}$ is given by $j = e\mathbf{v}(\mathbf{K})$. Thus the average current of the n conduction band electrons reads

$$\mathbf{J} := \sum_{i=1}^{n} \int e\mathbf{v}(\mathbf{K}_i) f(\mathbf{K}_1 \dots \mathbf{K}_n, t)\, d^3K_1 \dots d^3K_n. \tag{8.3.1}$$

Owing to Theorem 3.5 and equation (3.6.15) this can be equivalently written

$$\mathbf{J} = n \int e\mathbf{v}(\mathbf{K}) \bar{f}(\mathbf{K}, t)\, d^3K \tag{8.3.2}$$

and by using the solution of the Boltzmann equation $f(\mathbf{K},t)$ which is connected with $\tilde{f}(\mathbf{K},t)$ and $f(\mathbf{K}_1\ldots\mathbf{K}_n,t)$ by (3.6.19) (3.6.20) we obtain

$$\mathbf{J}=\int e\mathbf{v}(\mathbf{K})f(\mathbf{K},t)d^3K. \tag{8.3.3}$$

In contrast to $\tilde{f}(\mathbf{K},t)$, the solution of the Boltzmann equation is normalized to n and not to unity as follows from (3.6.19). Now in a stationary state of a system the number n is not independent of the processes occurring outside of the conduction band. Rather its mean value $\bar{n}$ is a stationary solution of rate equations derived for processes going on at impurities. Such solutions obviously depend on temperature, external fields, etc. Thus we have

$$\bar{n}:=n(T,\mathbf{E},\ldots) \tag{8.3.4}$$

and the number of conduction band electrons therefore depends on the physical state of the "underground" processes. Hence it is obvious that we have to replace (3.6.19) by the condition

$$\int f(\mathbf{K},t)d^3K=n(T,\mathbf{E},\ldots). \tag{8.3.5}$$

Thus the dependence of the current on the external forces does not only come about by the conduction band kinetics themselves but also by the normalization.
A further generalization of (8.3.5) is necessary if for a homogeneous medium the external forces depend on $\mathbf{r}$. In this case the statistical unit cells of the ensemble (microblocks, mosaicblocks) have to be correlated and f depends on the number i of the unit cell or in macroscopic coarse-graining on $\mathbf{r}$, i.e., $f\equiv f(\mathbf{K},\mathbf{r},t)$. In this case also $\bar{n}$ depends on i or $\mathbf{r}$, resp., i.e., $\bar{n}=n(\mathbf{r},T,\mathbf{E}\ldots)$ and (8.3.5) can be written as

$$\int f(\mathbf{K},\mathbf{r},t)d^3K=n(\mathbf{r},T,\mathbf{E},\ldots) \tag{8.3.6}$$

where for nonstationary solutions $\bar{n}$ depends also on t, i.e., $\bar{n}=n(\mathbf{r},t,\ldots)$. In the following we suppress the functional dependence of $\bar{n}$ on the external forces and write simply $\bar{n}=n(\mathbf{r},t)$.
If the impurity reactions are also influenced by the spatial inhomogenities of the forces, the system consisting of the position dependent Boltzmann equation and the position dependent rate equations, should be treated simultaneously, but no results are available for this most general case. Hence we confine ourselves to the discussion of the Boltzmann equation and assume that the changes in the concentration $n(\mathbf{r},t)$ with respect to $\mathbf{r}$ arise only from processes in the conduction band and are not influenced by the underground processes.
As the electric current does not explicitly depend on the kind of normalization, its expression (8.3.3) formally holds for the most general experimental arrangements if only electrons are involved. In a similar way, the heat current can be treated if its microscopic operator expression is known. To derive it we have to make use of some relations following from the Boltzmann equation and from elementary thermodynamic formulae.

Lemma 8.2: If the conduction band is a closed system with respect to the electron number, then for symmetric transition probabilities the equation of continuity

$$\frac{\partial}{\partial t} n(\mathbf{r},t) + \nabla \cdot [n(\mathbf{r},t)\bar{\mathbf{v}}(\mathbf{r},t)] = 0 \tag{8.3.7}$$

can be derived from the Boltzmann equation (8.2.40).

Proof: We integrate or sum, resp. equation (8.2.40) over **K**. Then because of $W(\mathbf{K}',\mathbf{K}) = W(\mathbf{K},\mathbf{K}')$ etc. it follows immediately that the collision term vanishes, i.e., the equation

$$\frac{\partial}{\partial t} n(\mathbf{r},t) + \int \mathbf{v} \cdot \nabla_{\mathbf{r}} f d^3K + \frac{e}{\hbar} \int \left(\mathbf{E} + \frac{1}{c} \mathbf{v} \times \mathbf{B}\right) \cdot \nabla_{\mathbf{K}} f d^3K = 0 \tag{8.3.8}$$

results. By partial integration it can be shown that the last term of (8.3.8) vanishes, while the second term can be written as

$$\int \mathbf{v} \cdot \nabla_{\mathbf{r}} f d^3K = \nabla_{\mathbf{r}} \cdot [n(\mathbf{r},t) \int \mathbf{v} \bar{f} d^3K] \tag{8.3.9}$$

where $\bar{f}$ is normalized to unity, Q.E.D.

A similar equation can be derived for the energy density u, where its microscopic expression is given by the one-electron energy $E(\mathbf{K})$.

Lemma 8.3: If the suppositions of Lemma 8.2 are satisfied and if in the collision term only elastic processes are taken into account, then for the energy density $u(\mathbf{r},t)$ the equation

$$\frac{\partial}{\partial t} u(\mathbf{r},t) + \nabla \cdot [\overline{u(\mathbf{r},t)\mathbf{v}(\mathbf{r},t)}] + \mathbf{E} \cdot \mathbf{J}(\mathbf{r},t) = 0 \tag{8.3.10}$$

can be derived from the Boltzmann equation (8.2.40).

Proof: We multiply (8.2.40) by $E(\mathbf{K})$ and integrate over **K**. Then due to the elastic collisions it follows immediately that the collision term vanishes, i.e., the equation

$$\frac{\partial}{\partial t} u(\mathbf{r},t) + \int E(\mathbf{K})\mathbf{v} \cdot \nabla_{\mathbf{r}} f d^3K + \frac{e}{\hbar} \int E(\mathbf{K}) \left(\mathbf{E} + \frac{1}{c} \mathbf{v} \times \mathbf{B}\right) \cdot \nabla_{\mathbf{K}} f d^3K = 0 \tag{8.3.11}$$

results. By partial integration it can be shown that the **B**-term of (8.3.11) vanishes, while the second term can be written as

$$\int E(\mathbf{K})\mathbf{v} \cdot \nabla_{\mathbf{r}} f d^3K = \nabla_{\mathbf{r}} \cdot [\overline{u(\mathbf{r},t)\mathbf{v}(\mathbf{r},t)}]. \tag{8.3.12}$$

The **E**-term gives by partial integration $e\, \mathbf{E} \cdot n(\mathbf{r},t)\bar{\mathbf{v}}(\mathbf{r},t)$, i.e., just $\mathbf{E} \cdot \mathbf{J}(\mathbf{r},t)$, *Q.E.D.*

These lemmas are special cases of a treatment of the Boltzmann equation by Spitzer (1962) and Seeger (1973). In addition to these lemmas for the derivation of the heat current, thermodynamic relations are required. A rigorous deduction of irreversible thermodynamics on the quantum statistical basis of Chapter 1 has been performed by Rieckers and Stumpf (1977). Such a deduction also includes the definition of heat currents, etc. If one is only interested in the definition of the heat current, it is sufficient to apply two basic suppositions of irreversible thermodynamics.

i) Irreversible thermodynamic processes occur between the subsystems Σ_α, $1 \leqslant \alpha \leqslant l$ of an aggregate system $\Sigma = (\Sigma_1, \ldots, \Sigma_l)$ where the entropy S of Σ is given by the sum over the entropies S_α of the subsystems Σ_α, $1 \leqslant \alpha \leqslant l$, i.e., the total entropy S is an additive quantity.
ii) The entropies S_α of the subsystems Σ_α, $1 \leqslant \alpha \leqslant l$ are given by the functional expressions of S_α for the thermostatic case. In the limiting case of very small subsystems a density formulation is used where the index α can be replaced by a continuous variable $\mathbf{r}$.

We now proceed to derive the heat current according to Madelung (1972). We consider the entropy of an aggregate system Σ which is composed of equivalent subsystems Σ_α, i.e., subsystems which have the same entropy functional $S_\alpha \equiv S_1$ but which are allowed to have different values of the independent variables $\{U_\alpha, V_\alpha, N_\alpha\}$. Then according to i) we obtain

$$S = \sum_{\alpha=1}^{l} S_1(U_\alpha, V_\alpha, N_\alpha) \tag{8.3.13}$$

and according to ii) for the differential dS it follows from ordinary thermodynamics that

$$dS = \sum_{\alpha=1}^{l} dS_1(U_\alpha, V_\alpha, N_\alpha) = \sum_{\alpha=1}^{l} \frac{1}{T_\alpha}(dU_\alpha - P_\alpha dV_\alpha - \zeta_\alpha dN_\alpha) \tag{8.3.14}$$

In the limiting case of continuous subsystems (8.3.14) can be written

$$dS = \int ds(\mathbf{r}, t) d^3r \tag{8.3.15}$$

where $s(\mathbf{r}, t)$ is the specific entropy. By comparison with (8.3.14) it follows for constant volume V, i.e., $dV_\alpha = 0$ that

$$\frac{\partial s}{\partial t} = \frac{1}{T(\mathbf{r})}\left(\frac{\partial u}{\partial t} - \zeta \frac{\partial n}{\partial t}\right) \tag{8.3.16}$$

where u and n are the specific internal energy and the electron concentration, resp. Substitution of (8.3.7) and (8.3.10) in (8.3.16) gives

$$\frac{\partial s}{\partial t} = -\frac{1}{T(\mathbf{r})} \nabla \cdot \left[\overline{u(\mathbf{r}, t)\mathbf{v}(\mathbf{r}, t) - \zeta n(\mathbf{r}, t)\bar{\mathbf{v}}(\mathbf{r}, t)}\right] - \frac{1}{T(\mathbf{r})} \mathbf{E} \cdot \boldsymbol{j}(\mathbf{r}, t). \tag{8.3.17}$$

If we compare this formula with $dS = dQ/T$, then we have

$$\frac{dq}{dt} = -\nabla \cdot [\overline{u\mathbf{v}} - \zeta n \bar{\mathbf{v}}] - \mathbf{E} \cdot \boldsymbol{j} \tag{8.3.18}$$

where the first term on the right-hand side of (8.3.18) is the divergence of the heat flow, while the second term is the local heat production. If the relations (8.3.9) and (8.3.12) are observed, the heat current reads

$$\mathbf{Q} := \int [E(\mathbf{K}) - \zeta]\mathbf{v}(\mathbf{K}) f(\mathbf{K}, \mathbf{r}, t) d^3K. \tag{8.3.19}$$

As in the equilibrium state f_0 of the conduction band electrons, no electric and thermal currents occur, the equations (8.3.3) and (8.3.19) take, with the aid of (8.2.41), the form

$$\mathbf{J}(\mathbf{r}, t) = \int e\mathbf{v}(\mathbf{K}) \frac{\partial f_0}{\partial E} \Phi(\mathbf{K}, \mathbf{r}, t) d^3K \tag{8.3.20}$$

$$\mathbf{Q}(\mathbf{r}, t) = \int [E(\mathbf{K}) - \zeta]\mathbf{v}(\mathbf{K}) \frac{\partial f_0}{\partial E} \Phi(\mathbf{K}, \mathbf{r}, t) d^3K. \tag{8.3.21}$$

The evaluation of these expressions in the relaxation time formalism was discussed, for instance by Madelung (1970) (1972) and Haug (1970). We give only a short account following the presentation of Madelung (1970) (1972).

b) Extrinsic conductivity

As only electrons (or holes) contribute to extrinsic conductivity, we can directly apply the formulae of Section a) since these formulae are concerned with one kind of particle in the conduction band. We use the following definitions:

$$\mathbf{b} := \frac{e\tau}{cm^*}\mathbf{B}; \quad \eta := \zeta - eV; \quad \mathbf{E} = -\nabla V \tag{8.3.22}$$

and assume that the suppositions of Theorem 8.5 hold. Then by substitution of the general solution (8.2.52) in (8.3.20) and (8.3.21) with $\nabla_{\mathbf{r}} \equiv \nabla$ the expressions

$$\begin{aligned}\mathbf{J} = {} & M_{00}\nabla\eta e^{-1} + M_{10}\mathbf{B}\times\nabla\eta e^{-1} + M_{20}\mathbf{B}(\nabla\eta e^{-1}\cdot\mathbf{B}) \\ & + M_{01}T^{-1}\nabla T + M_{11}\mathbf{B}\times T^{-1}\nabla T + M_{21}\mathbf{B}(T^{-1}\nabla T\cdot\mathbf{B})\end{aligned} \tag{8.3.23}$$

$$\begin{aligned}\mathbf{Q} = {} & M_{01}\nabla\eta e^{-1} + M_{11}\mathbf{B}\times\nabla\eta e^{-1} + M_{21}\mathbf{B}(\nabla\eta e^{-1}\cdot\mathbf{B}) \\ & + M_{02}T^{-1}\nabla T + M_{12}\mathbf{B}\times T^{-1}\nabla T + M_{22}\mathbf{B}(T^{-1}\nabla T\cdot\mathbf{B})\end{aligned} \tag{8.3.24}$$

result, where the symbols M_{ik} are given by

$$M_{ik} = -\frac{e}{3\pi^2}\left(\frac{2m^*}{\hbar^2}\right)^{3/2} \int_0^\infty \frac{E^{3/2}}{1+\mathbf{b}^2} \frac{\partial f_0}{\partial E} \left(\frac{e\tau(E)}{m^*}\right)^{i+1} \left(\frac{E-\zeta}{e}\right)^k dE. \tag{8.3.25}$$

For nondegenerate semiconductors, i.e., small electron numbers n in the conduction band, the Fermi distribution f_0 in (8.3.25) can be replaced by a Boltzmann distribution and in this case the integrals (8.3.25) go over into

$$M_{ik}=\frac{e}{3\pi^2}\left(\frac{2m^*}{\hbar}\right)^{3/2}\frac{n}{n_0k_BT}\int_0^\infty\frac{E^{3/2}}{1+\mathbf{b}^2}e^{-\frac{E}{k_BT}}\left(\frac{e\tau}{m^*}\right)^{i+1}\left(\frac{E-\zeta}{e}\right)^k dE \tag{8.3.26}$$

with $n_0=2/h^3(2\pi m_c k_BT)^{3/2}$. If, furthermore, for the relaxation time τ a power law in E is assumed, ($\tau=\tau_0E^r$) and an expansion of $(1+\mathbf{b}^2)^{-1}$ is made, then all integrals occurring in (8.3.26) can be reduced to integrals of the kind

$$\int_0^\infty E^\alpha e^{-\frac{E}{k_BT}}dE=(k_BT)^{\alpha+1}\Gamma(\alpha+1). \tag{8.3.27}$$

We first study (8.3.23) for a vanishing magnetic field. Then these equations take the form

$$\begin{aligned}\mathbf{J}&=M_{00}\nabla\eta e^{-1}+M_{01}T^{-1}\nabla T\\ \mathbf{Q}&=M_{01}\nabla\eta e^{-1}+M_{02}T^{-1}\nabla T.\end{aligned} \tag{8.3.28}$$

To connect the numbers M_{ik} in (8.3.29) with experimental quantities we rewrite (8.3.28) in the form

$$\nabla\eta e^{-1}=\sigma^{-1}\mathbf{J}+\varepsilon\nabla T \tag{8.3.29}$$

$$\mathbf{Q}=\Pi\mathbf{J}-\kappa\nabla T \tag{8.3.30}$$

with

$$\begin{aligned}&\sigma:=M_{00}\,;\quad \varepsilon=M_{01}/TM_{00}\\ &\kappa=(M_{00}M_{02}-M_{01}^2)/M_{02}\,;\quad \Pi=-M_{01}/M_{02}.\end{aligned} \tag{8.3.31}$$

According to (3.8.55), for extrinsic conduction ζ depends on $\mathbf{r}$ only indirectly by $T=T(\mathbf{r})$. Hence from $\nabla T=0$ the equation $\nabla\zeta=0$ follows. In an isothermal homogeneous extrinsic semiconductor, equation (8.3.29) therefore reads

$$\nabla\frac{\eta}{e}\equiv-\mathbf{E}=\sigma^{-1}\mathbf{J} \tag{8.3.32}$$

i.e., we obtain Ohm's law, while the second equation (8.3.30) goes over into

$$\mathbf{Q}=\Pi\mathbf{J} \tag{8.3.33}$$

i.e., the electric current is accompanied by a heat production. To analyze this heat production we replace the energy current $\overline{u\mathbf{v}}$ in (8.3.10) according to (8.3.19) by $\overline{u\mathbf{v}}=\mathbf{Q}+\zeta\mathbf{j}$, where $\mathbf{j}$ is the particle current. This gives

$$\begin{aligned}\frac{\partial u}{\partial t}&=-\nabla\cdot\mathbf{Q}-\nabla\cdot(\zeta\mathbf{j})-\mathbf{E}\cdot\mathbf{J}=\mathbf{E}\cdot\mathbf{J}-\nabla\cdot(\Pi\mathbf{J})-\nabla\cdot\left(\frac{\zeta}{e}\mathbf{J}\right)\\ &=\mathbf{J}\cdot\nabla\eta e^{-1}-\mathbf{J}\cdot\nabla\Pi=\sigma^{-1}\mathbf{J}^2-\mathbf{J}\cdot\nabla\Pi.\end{aligned} \tag{8.3.34}$$

The first term on the right-hand side of (8.3.34) is the Joule heat, the second term occurs only for spatial inhomogenities, where Π is discontinuous. This leads to the Peltier effect.

If we assume, on the other hand, $\nabla T \neq 0$, but $\mathbf{J}=0$, then (8.3.29) and (8.3.30) takes the form

$$\nabla \eta e^{-1} = \varepsilon \nabla T$$
$$\mathbf{Q} = -\kappa \nabla T. \tag{8.3.35}$$

The coefficient κ is the specific heat conductivity which connects the temperature gradient with the corresponding heat flow. The first equation, however, shows that a temperature gradient is always connected to a gradient of the electrochemical potential; and in an open circuit this leads to a thermovoltage. This is the so-called Seebeck effect.
For $\nabla T \neq 0$ also $\nabla \zeta \neq 0$ holds, and by evaluating $\nabla(\zeta \mathbf{j})$ and applying definition (8.3.22) for $\zeta = \eta + eV$, we may rewrite equation (8.3.34) in the form

$$\frac{\partial u}{\partial t} = -\nabla \cdot \mathbf{Q} + \mathbf{J} \cdot \nabla \eta e^{-1} - \zeta \nabla \cdot \mathbf{j}. \tag{8.3.36}$$

Owing to (8.3.6) and (8.3.7) for stationary solutions of the Boltzmann equation $\nabla \cdot \mathbf{j}$ must vanish. Hence, with substitution of (8.3.29) (8.3.30), equation (8.3.36) takes the form

$$\frac{\partial u}{\partial t} = \sigma^{-1} \mathbf{J}^2 + \varepsilon \mathbf{J} \cdot \nabla T - \nabla \cdot (\Pi \mathbf{J}) + \nabla \cdot (\kappa \nabla T) \tag{8.3.37}$$

and this equation reads for a homogeneous crystal, where Π depends only on T, $\Pi = \Pi(T)$

$$\frac{\partial u}{\partial t} = \sigma^{-1} \mathbf{J}^2 + \nabla \cdot (\kappa \nabla T) - \mu_E \mathbf{J} \cdot \nabla T \tag{8.3.38}$$

with $\mu_E := \partial \Pi / \partial T - \varepsilon$. The first term on the right-hand side of (8.3.38) describes the Joule heat, the second term the net result of heat conduction, and the third term a heat production which occurs additionally if electric current and heat current appear simultaneously (Thomson heating).
From (8.3.31) we obtain

$$\Pi = \varepsilon T \tag{8.3.39}$$

and from the definition of μ_E,

$$\mu_E = T \frac{\partial \varepsilon}{\partial T}. \tag{8.3.40}$$

These relations are the so-called Thomson relations.
We now consider (8.3.23) for a non-vanishing magnetic field. To simplify matters, we neglect the dependence of ζ on the temperature. Then $\nabla \eta e^{-1} = -\mathbf{E}$ and equation

(8.3.23) can be written in the general form

$$\begin{aligned}\mathbf{J} &= \alpha_{11}\mathbf{E} + \beta_{11}\mathbf{B}\times\mathbf{E} + \gamma_{11}\mathbf{B}(\mathbf{B}\cdot\mathbf{E}) + \alpha_{12}\nabla T\\ &\quad + \beta_{12}\mathbf{B}\times\nabla T + \gamma_{12}\mathbf{B}(\mathbf{B}\cdot\nabla T)\\ \mathbf{Q} &= \alpha_{21}\mathbf{E} + \beta_{21}\mathbf{B}\times\mathbf{E} + \gamma_{21}\mathbf{B}(\mathbf{B}\cdot\mathbf{E}) + \alpha_{22}\nabla T\\ &\quad + \beta_{22}\mathbf{B}\times\nabla T + \gamma_{22}\mathbf{B}(\mathbf{B}\cdot\nabla T).\end{aligned} \tag{8.3.41}$$

For a systematic discussion of equations (8.3.41), it is convenient to distinguish between galvano-magnetic and thermomagnetic effects, where the first kind of effect is produced by the combined action of an external magnetic and electric field, while the second kind of effect is produced by a magnetic field and a temperature gradient. If the directions of these fields are parallel, the effects are called longitudinal, while for orthogonal fields the effects are called transversal.
We consider only transversal effects. Without loss of generality we assume that the magnetic field has only an $\mathbf{e}_3$ component. The orthonormal electric field or temperature gradient then may have components in the $\mathbf{e}_1$ and $\mathbf{e}_2$ direction. In this case equations (8.3.41) read for the components of $\mathbf{J}$

$$\begin{aligned}J_1 &= \alpha_{11}E_1 - \beta_{11}B_3E_2 + \alpha_{12}\partial_1 T - \beta_{12}B_3\partial_2 T\\ J_2 &= \alpha_{11}E_2 - \beta_{11}B_3E_1 + \alpha_{12}\partial_2 T + \beta_{12}B_3\partial_1 T\\ J_3 &= 0\end{aligned} \tag{8.3.42}$$

and similar equations for the components of $\mathbf{Q}$, namely

$$\begin{aligned}q_1 &= \alpha_{21}E_1 - \beta_{21}B_3E_2 + \alpha_{22}\partial_1 T - \beta_{22}B_3\partial_2 T\\ q_2 &= \alpha_{21}E_2 - \beta_{21}B_3E_1 + \alpha_{22}\partial_2 T + \beta_{22}B_3\partial_1 T\\ q_3 &= 0.\end{aligned} \tag{8.3.43}$$

The various effects depend on the conditions which are imposed by the experimental arrangements. For instance, the isothermal case is defined by the conditions $\partial_2 T = I_2 = 0$, while the adiabatic case is defined by $q_2 = I_2 = 0$. If such conditions are imposed, this means that the role of the independent variables is changed. While for fixed $\mathbf{B} = B_3\mathbf{e}_3$ in (8.3.42) and (8.3.43) the independent variables are $(E_1, E_2, \partial_1 T, \partial_2 T)$, in the isothermal case the independent variables are $(I_1, I_2, \partial_1 T, \partial_2 T)$, and in the adiabatic case the independent variables are (I_1, I_2, q_1, q_2). For an appropriate treatment of these arrangements, equations (8.3.42) (8.3.43) have to be solved with respect to these variables. We do not perform this explicitly, but mention only some important effects.
The appearance of a voltage E_2 orthonormal to $I_1 \neq 0$ for $I_2 = 0$ in the adiabatic and isothermal case is called the Hall effect, and the factor

$$R = \frac{E_2}{I_1 B_3} \tag{8.3.44}$$

is called the Hall coefficient.

The appearance of a voltage E_2 orthonormal to a temperature gradient $\partial_1 T \neq 0$ for $I_2 = 0$ in the adiabatic and isothermal case is called the Nernst effect and the factor

$$Q = \frac{E_2}{B_3 \partial_1 T} \tag{8.3.45}$$

is called the Nernst factor.
The appearance of a temperature gradient $\partial_2 T$ orthonormal to $I_1 \neq 0$, or $\partial_1 T \neq 0$ or $q_1 \neq 0$ for $q_2 = 0$, resp. is called the Ettinghausen effect or the Righi-Leduc effect, resp. The corresponding coefficients

$$P = \frac{\partial_2 T}{I_1 B_3} \tag{8.3.46}$$

are called the Ettinghausen coefficient and

$$S = \frac{\partial_2 T}{\partial_1 T B_3} \tag{8.3.47}$$

the Righi-Leduc coefficient.
For further discussions of such effects we refer to the literature, for instance Seeger (1973).

c) Intrinsic conductivity

Intrinsic conductivity means that electrons and holes contribute simultaneously to electric and heat currents and that these carriers are generated by excitations of electrons from the valence band to the conduction band. An essential supposition for the treatment of intrinsic conductivity is the assumption that the motions of electrons and holes are not correlated. If this assumption is satisfied, it is unimportant for the simultaneous treatment of electrons and holes whether they are generated by valence band-conduction band excitations or by appropriate impurity excitations. Hence we may assume that a definite number n of electrons and a definite number p of holes are present in the conduction band or valence band, resp. If we work, to simplify matters, with the same model as in the preceding sections and if we only admit additional excitations of the valence band, then the set of states $\{|a\rangle\}$ characterizing the complete system can be denoted by

$$\{|a\rangle\} := \{|A(s-n'), \mathbf{K}_1 \ldots \mathbf{K}_n, \mathbf{K}'_1 \ldots \mathbf{K}'_p, m, v, \varrho\rangle\} \tag{8.3.48}$$

which is a generalization of the states (8.2.5), and where $\mathbf{K}'_1 \ldots \mathbf{K}'_p$ mean the p unoccupied electron states of the valence band.
In (8.2.5) the number of impurity center electrons $s-n'$ and the number of conduction band electrons n have always to satisfy $(s-n')+n=s$, i.e., the conservation of the electron number with $n'=n$. For the generalized states (8.3.48) this conservation reads $(s-n')+n-p=s$, from which $n=n'+p$ follows, i.e., the number of conduction band electrons equals the number of ionized impurity electrons plus the excited valence band

electrons. In this model the vacancies or F- and F'-centers, resp. act both as donors and acceptors. If these properties are shared between different centers, the conservation law has to be modified. But this extension is of no importance for our further discussion and hence we omit it.
In addition to equations for (8.2.5), rate equations for the states (8.3.48) can be derived. In the latter case not only the conduction band electron processes are adiabatically decoupled from the processes of the impurity center electrons but also the hole processes. This leads to Boltzmann equations both for electrons and holes. Formally, these equations are completely equivalent, the only differences arise from different numerical values of charges, energy surfaces and transition probabilities. Hence, keeping in mind these differences, both equations can be treated along the same lines. We do not perform these lengthy deductions explicitly. Rather, we consider the solution (8.2.52) of the Boltzmann equation for external fields which lead to the linear relations (8.3.23) and (8.3.24). These linear relations depend upon the equilibrium distribution f_0 for electrons. The corresponding equilibrium distribution for holes is given by $f_0^* := (1 - f_0)$ as this distribution just gives the unoccupied levels. Hence for holes all calculations have to be made with f_0^* instead of f_0 and for energy levels

$$E = E_v - \frac{\hbar \mathbf{K}^2}{2m_v} \tag{8.3.49}$$

where a parabolic valence band is assumed and E_v is the upper edge of the valence band, while m_v is the effective hole mass. By direct calculation it follows that f_0^* has the form

$$f_0^* = \left[\exp\left(E_v + \frac{\hbar \mathbf{K}^2}{2m_v} + \bar{\zeta} \right) (k_B T)^{-1} + 1 \right]^{-1} \tag{8.3.50}$$

with

$$\bar{\zeta} = -\zeta - \Delta E; \quad \Delta E = E_v - E_c. \tag{8.3.51}$$

Hence all calculations for electrons and holes are formally the same if only E_c is replaced by E_v, ζ by $\bar{\zeta}$, m_c by m_v and $-e$ by e.
The electric current and the heat current are then the superposition of the corresponding currents for electrons $\mathbf{J}_n$, $\mathbf{Q}_n$, and holes $\mathbf{J}_p$, $\mathbf{Q}_p$. This gives the total currents

$$\begin{aligned} \mathbf{J} &= \mathbf{J}_p + \mathbf{J}_n \\ \mathbf{Q} &= \mathbf{Q}_p + \mathbf{Q}_n . \end{aligned} \tag{8.3.52}$$

A detailed discussion of the various effects in external fields with both kinds of charge carriers contributing to currents can be found in Haug (1970). As we are more interested in the basic processes of statistical quantum theory we do not give further details here.

d) Literature

In order to obtain a clear description of the thermodynamic transport effects we have employed the relaxation time approximation for the solution of the linearized version of the Boltzmann equation. Naturally many attempts have been made to go beyond this approximation scheme. Since, however, even the explicit calculation of relaxation times needs a considerable amount of work, it would exceed the scope of this book to present the more sophisticated solution procedures of the Boltzmann equation or related quantum statistical approaches. In addition, in contrast to the relaxation time formalism, the approaches for the evaluation of the thermodynamic transport properties beyond this approximation scheme are not sufficiently well elaborated to allow a general presentation. Hence we restrict ourselves to the citation of some literature in this field.

Before referring to original papers we cite some books and review articles. A comprehensive treatment of semiconductor transport phenomena based on phenomenological theory, i.e., elementary statistical assumptions without quantum statistical foundations is given in the books of Madelung (1970) and Seeger (1973). Madelung's book contains an extensive list of references with respect to books, conference reports etc. in this field. A theoretical treatment of galvanomagnetic effects in semiconductors at a phenomenological level was given by Beer (1963). Busch and Winkler (1956) published a review article in which they described the determination of the basic quantities of the phenomenological theory by electric, optical and magnetic measurements. This article contains a comprehensive list of references of early papers in this field. Hence we restrict ourselves to mainly citing original papers of the last two decades, where we also include papers about scattering mechanism and relaxation time calculations as the latter cannot, in general, be clearly separated from the discussion of the other topics in most papers. The following topics were treated: Theory of impurity scattering in semiconductors, Mansfield (1956). Theory of transport effects in semiconductors with calculation of the Nernst coefficient and investigation of its relation to thermoelectric power based on PMTE (≡phenomenological macroscopic transport equations), Price (1956). Theory of the photomagnetoelectric effect in semiconductors based on PMTE, van Roosbroeck (1956). Theory of transport effects in semiconductors with respect to thermoelectricity based on PMTE, Price (1956). Calculation of relaxation times for scattering of electrons in semiconductors with spherically symmetric band structure, Barrie (1956). Theory of electric conductivity of semiconductors for high frequencies based on BERTA (≡Boltzmann equation with relaxation time approximation), Stolz (1957). Quantum theory of electronic conduction in crossed electric and magnetic fields for the limit of weak scattering based on a perturbation calculation for the density operator, Adams and Holstein (1959). Theory of electronic conduction in high magnetic fields based on an evaluation of the density operator in the Landau representation in powers of the external electric field, Argyres and Roth (1959). Selfenergy, mass and mobility of the lattice polaron, Schultz (1959). Theory of thermoelectric power of ionic crystals based on PMTE, Haga (1958) (1959). Theory of the effects of carrier-carrier scattering on the mobility of carriers in semiconductors based on variational solutions of the Boltzmann equation, McLean and Paige (1960). Calculation of electric current and heat flux for arbitrary energy band structure based on BERTA, Kolodziejczak (1961). Electron transport at high temperatures in the presence of impurities with BERTA, Frisch and Lebowitz (1961). Quantum theory of free carrier infrared absorption by means of second order perturbation calculation, Dumke (1961). Theory of the Faraday effect in anisotropic semiconductors based on PMTE and Maxwell's equations, Donovan and Webster (1962). Interband electron-electron scattering and transport phenomena in semiconductors treated by Kohler's variational principle for the linearized Boltzmann equation, cf. Section 8.5.b, Appel (1962). Discussion of thermomagnetic effects in semiconductors based on PMTE following from Kolodziejczak's theory, Zawadzki (1962). Photomagnetothermal effect in semiconductors based on PMTE, Gärtner, Loscoe and Mette (1962). Mobility of slow electrons in polar crystals, Feynman, Hellwarth, Iddings and Platzman (1962). Higher order relaxation times in electronic transport theory based on the Boltzmann equation, Schottky (1962). Investigation of galvanomagnetic effects by solution of Bloch's integral equation resulting from an approximate treatment of the Boltzmann equation, Langbein (1962). Calculation of the electric conductivity of polar crystals based on the evaluation of the states of lattice polarons and scattering cross sections caused by non-diagonalized parts of the Hamiltonian, Dogonadze and Chizmadzhev (1962). Photo-Hall effect and photoconductivity in a magnetic field based on BERTA, Dobrovolskii (1962). Longitudinal thermomagnetic effects in semiconductors in a strong longitudinal magnetic field based on BERTA, Anselm and Askerov (1962). Calculation of magnetoconductivity by variational solution of the Boltzmann equation with electron-polar-phonon scattering in ionic crystals, Garcia-Moliner (1963). Extensive discussion of transport phenomena in semiconductors in the presence of non-equilibrium concentrations based on BERTA, Zawadzki (1963). Approach for a general solution of the full Boltzmann equation in terms of a power series in the external perturbations and derivation of PMTE by means of this solution procedure, Fogarassi (1963). Discussion of the validity of the relaxation time approximation, Fogarassi (1963). Investigation of the conductivity of crystals with strong electron-phonon interaction based on the calculation of states of the lattice polaron Hamiltonian and a formal approximate solution of the Boltzmann equation, Nagaev (1963). Discussion of thermomagnetic phenomena in metals and semiconductors in quantizing magnetic fields, Obraztsov (1964) (1965). Spherical harmonics expansion of the electron distribution for an approximate solution of the Boltzmann equation in the case of high anisotropy applied to high field transport in semiconductors, Baraff (1964). Calculation of magnetoconductivity of hot electrons by solving the Boltzmann equation with partial use of relaxation times, Budd (1965). Calculation of the Hall constant in weak magnetic fields based on an approximate solution of the Boltzmann equation with electron-impurity collision terms, Zvyagin (1965). Treatment of the Hall effect and magnetoresistance for weak and intense magnetic fields based on the Boltzmann equation, Erezhepov (1964) (1966). Theory of mobility, Hall effect and magnetoresistance in semiconductors with charged defects, Pekar (1966). Note on the interaction of polarons with acoustic phonons, Zaitsev and Melnikova (1966). Quantum theory of thermomagnetic currents in semiconductors and metals based on approximate solutions of the equation for the density operator of OEMRL, Anselm, Obraztsov and Tarkhanyan (1966). Quantum theory of galvanomagnetic effects at extremely strong magnetic fields based on an evaluation of Kubo's response formula, Kubo, Miyake and Hashitsume (1965). Investigation of the thermoelectric properties of the small polaron by evaluation of Kubo's response formulae in lowest order perturbation theory, and use of their relation to PMTE, Schotte (1966). General solution procedure for the Boltzmann equation for crystal electrons by means of an expansion in powers of the external forces, Pfleiderer (1966). Theory of oscillatory photoconductivity in semiconductors based on BERTA with inclusion of electron generation due to photon absorption and electron decay due to recombination, Stocker and Kaplan (1966). Analysis of scattering processes of electrons in semiconductors by calculation of the collision term of the Boltzmann equation for various types of scattering, Kolodziejczak (1967). Theory of impact ionization of electrons in polar semiconductors by strong electric fields by means of a solution of the Boltzmann equation by an expansion of the distribution function into spherical harmonics, Chuenkov (1967). Review of high field transport in semiconductors based on the discussion of coupled electron-phonon rate equations and an analysis of scattering processes, Conwell (1967). Theory of spin-lattice relaxation of conduction electrons in a strong electric field based on BERTA, Kalashnikov (1967). Note on the theory of the electric conductivity of semiconductors in strong electric and magnetic fields, Pomortsev (1967). Effects of the thermoelectric current on the stability of the electric field and charge distribution in a semiconductor treated by means of phenomenological equations, Zvyagin (1967). Quantum theory of the Nernst-Ettinghausen effect in semiconductors based on the Anselm-Obraztsov-Tarkhanyan method, Anselm and Askerov (1967). Nonlinear excitations of electrons in semiconductors based on BERTA, Kolodziejczak (1968). Magneto-conductivity tensor in the presence of strong alternating current electric field based on BERTA, Kolodziejczak (1968). Calculation of the transverse conductivity for electrons interacting with waves in strong electric fields based on BERTA, Spector (1968). Theory of the photomagnetic effect on heated electrons based on an approximate solution of the Boltzmann equation taking into account electron generation and recombination, Lyagushchenko and Yassievich (1968). Theory of non-linear magneto-optical phenomena in semiconductors based on a solution procedure of the full Boltzmann equation by an expansion in powers of the external forces, Kolodziejczak (1968). Thermodynamics of hot electrons and electron density fluctuations in semiconductors based on PMTE of irreversible processes, Sato (1969). Investigation of inelastic scattering of an electron by a non-magnetic impurity in a crystal, Takeno (1969). Calculation of tunneling mobility of a small radius polaron, Bryksin and Firsow (1969). Relaxation time calculation for electron-phonon interaction of crystal electrons in high electric fields by means of perturbation theory, Hacker (1969). Theory of hot electrons in a strong magnetic field based on a free electron model with coupling to external forces, phonons and impurities and the evaluation of the equations for the corresponding density operator, Inoue and Yamashita (1969). Treatment of the Hall effect in hopping transport by evaluation of Kubo's response formula for the lattice polariton Hamiltonian, Klinger (1969). Note on the Hall mobility and the drift mobility of the polaron, Schober (1969). Discussion of the Hall effect in polaron semiconductors based on the evaluation of Kubo's response formula for the lattice polaron Hamiltonian, Firsov (1969). Review of electronic transport theory (in metals) based on the evaluation of response functions by field theoretic Green functions, Luttinger (1969). Derivation of non-linear optical effects of conduction electrons in semiconductors by means of a solution procedure of the Boltzmann equation by an expansion in powers of the external forces, Wang and Ressler (1969). Note on the theory of transport phenomena in semiconductors with low mobility, Efros (1969). Calculation of transport coefficients with inclusion of electron-electron collisions and the cooling of hot electrons at optical phonons by means of a solution of the Boltzmann equation by an expansion of the distribution function into spherical harmonics, Rabinovich (1970). Exact steady state solutions of the Boltzmann equation for the Stocker-Kaplan model for oscillatory photoconductivity, Folland (1970). Numerical study of deformation potential scattering of electrons by optical phonons in a longitudinal magnetic field, Peterson (1970). Discussion of the anomalous Hall effect in semiconductors based on linear response theory, Bastin, Lewiner and Fayet (1970). Theory of impurity scattering in semiconductors based on a one-electron model, Mattis and Sinha (1970). Note on the Hall effect in cubic polaron semiconductors, Bryksin and Firsov (1970). Calculation of high-field electron distribution functions in semiconductors by means of an iterative solution procedure of the Boltzmann equation, Vassell (1970). Theory of photoconductivity for hot photoelectrons in semiconductors based on BERTA, Ladyzhinskii (1970). Note on optical properties of small polarons at high electric fields, Reik (1970). Excitonic spectrum of a non-equilibrium electronic solid state plasma treated by phenomenological equations, Martinov (1971). Calculation of some transport coefficients in simple semiconductors at low temperatures and high electric fields by means of drifted Maxwellian distributions, Kamal and Sharma (1971). Note on plasmon contribution to the free carrier absorption, von Baltz (1971). Calculation of the electric conductivity tensor of polarons in a magnetic field based on the field theoretic evaluation of linear response formulae, Klyukanov and Pokatilov (1971). Study of impact ionization of electron-hole pairs in crossed fields in semiconductors based on BERTA for electrons and holes, Reuter and Hübner (1971). Study of hopping conductivity by small polarons in large electric fields based on Holstein's polaron model and direct evaluation of the current expectation value, Emtage (1971). Quantum transport theory of impurity-scattering-limited mobility in *n*-type semiconductors with inclusion of electron-electron scattering by evaluation of Kubo's response formula for a many-electron system in a rigid lattice, Luong and Shaw (1971). Calculation of carrier transport quantities in non-degenerate semiconductors at low fields by an iterative analytical solution of the Boltzmann equation based on an expansion in a power series of the electric field, Law and Kao (1971). Theory of the electric conductivity of solids in a strong electric field based on a many-electron model with electron-phonon coupling in Fockspace representation and evaluation of the linear response formulae for the conductivity tensor, Enderlein and Peuker (1971). Calculation of the mobility of conduction electrons due to the interaction with transverse optical vibrations in ionic crystals by means of a variational method of Ziman (1960) for the solution of the Boltzmann equation, Vinetskii, Itskovskii and

Kukushkin (1971). Derivation of the electron energy distribution function for piezoelectric scattering at high electric fields based on BERTA and an expansion of the distribution function into spherical harmonics, Sodha, Phadke and Chakravarti (1971). Note on a general expression for the thermoelectric power, Fritzsche (1971). Discussion of the predominance of ionized impurity scattering of warm carriers in semiconductors, Seeger (1971). Calculation of the electronic mobility due to inelastic impurity scattering of hot electrons in semiconductors based on the evaluation of an average energy balance equation, Kachlishvili (1971). Theory of photon absorption by free carriers in semiconductors taking into account scattering by impurities, phonons and plasmons and based on Dumke's approach, von Baltz and Escher (1972). Calculation of the distribution function and energy losses of hot electrons interacting with optical phonons by means of a direct solution of the Boltzmann equation, Gelmont, Lyagushchenko and Yassievich (1972). Effect of paramagnetic resonance on electric conductivity of semiconductors in strong electric fields, Zaitsev (1973). Note on scattering of slow polarons by *F*-centers, Shmelev and Zilberman (1972). Theory of impurity conduction in low compensated semiconductors based on a discussion by means of equilibrium distributions, Efros, Shklovskii and Yanchev (1972). Treatment of the photovoltaic effect by investigation of excess-carrier transport in anisotropic semiconductors based on PMTE and elementary rate equations, Shah and Schetzina (1972). General theory of high field transport in semiconductors based on the assumption of a drifted Maxwellian electronic distribution function, Stokoe and Cornwell (1972). Discussion of the anomalous Hall effect for polarized electrons in semiconductors based on a solution of the Boltzmann equation by an expansion in a power series of the external forces and the corresponding linear approximation, Abakumov and Yassievich (1972). Calculation of the high temperature Hall mobility in cubic polaron semiconductors based on the evaluation of Kubo's formula, Bryksin and Firsov (1972). Note on low temperature conductivity by hot electrons, Kachlishvili (1972). Theory of polaron conduction based on a modified Kubo formula and its evaluation by field theoretic Green functions, Sumi (1972). Discussion of polaron transport properties, von Baltz and Birkholz (1972). Theory of diffusion of electrons in semiconductors in high electric fields based on the method of moments applied to the Boltzmann equation, Cheung and Hearn (1972). Investigation of the influence of the finite width of the conduction band on the heating of electrons in an electric field by means of the solution of the corresponding Boltzmann equation, Levinson and Yasevichyute (1972). Magnetoresistance of semiconductors in strong electric fields, Kachlishvili (1972). Galvanomagnetic effects in semiconductors with several types of carriers, Borblik (1972). Calculation of the conductivity of semiconductors in strong electric fields based on Bryksin's and Firsov's approach, Bryksin (1973). Review of the effects of longitudinal optical lattice vibrations on electronic excitations of solids, in particular, photoconductivity, magnetoresistance, tunneling, etc., Harper, Hodby and Stradling (1973). Treatment of the anomalous Hall effect in semiconductors based on the use of an effective Hamiltonian and the evaluation of the linearized version of the Boltzmann equation, Nozieres and Lewiner (1973). Theory of avalanche ionization in transparent dielectrics under the action of a strong electromagnetic field by evaluation of elementary rate equations, Rubinshtein and

Fain (1973). Scattering probabilities for holes by polar optical scattering mechanism, Costato and Reggiani (1973). Calculation of relaxation times for scattering of conduction electrons by tunneling states of defects in ionic crystals, Deigen, Glinchuk and Suslin (1973). Note on the mobility of the polaron with strong coupling, Volovik, Melnikov and Edelshtein (1973). Analysis of the static conductivity expression of hot electrons interacting with phonons by the method of field theoretic double time Green functions starting from the Kubo-Kalashnikov formula, Licea (1973). Theory of thermogalvano magnetic effects in semiconductors at low temperatures based on BERTA, Yasevichyute (1973). Theory of the transport coefficients for inelastic scattering of electrons by optical phonons based on perturbation solutions of the Boltzmann equation, Polovinkin and Skok (1974). Theory of the conductivity of semiconductors in strong crossed electric and magnetic fields for inelastic electron scattering based on a phenomenological analysis of the distribution function, Ryzhii (1974). Review of magneto-phonon resonance in semiconductors, Parfenev, Kharus, Tsidilkovskii and Shalyt (1974). Theory of intraband infrared absorption in anisotropic ionic semiconductors, Gurevich , Lang and Parshin (1975). Calculation of the Ohmic hole mobility in cubic semiconductors based on the assumption of a drifted Maxwellian distribution, Costato, Gagliani, Jacoboni and Reggiani (1974). Calculation of the thermoelectric power of the extrinsic Mott semiconductor by use of linear response formulae, Bari (1974). Study of the thermoelectric power of semiconductors in the extreme quantum limit based on PMTE, Jay-Gerin (1974). Derivation of the polaron mobility at finite temperature in the case of finite coupling by means of the evaluation of a polaron Boltzmann equation, Okamoto and Takeda (1974). Note on electron scattering by local magnetic moments in semiconductors, Polnikov (1974). Calculation of the relaxation time for the interaction of an optical polaron with zero vibration of the lattice in high electric fields, Shmelev and Sibirskii (1974). Note on the calculation of ionized impurity scattering in semiconductors, Gerlach (1974). Temperature dependence of the Hall mobility in the small polaron model for moderate and high temperatures, Bryksin and Firsov (1975). Study on the absence of local field corrections in small polaron hopping conduction, Saglam and Friedman (1975). Note on the anomalous magnetoresistence of ionic semiconductors at low temperatures, Epshtein (1975). Investigation of the behaviour of transport coefficients in the presence of inelastic scattering by optical phonons based on the linearized version of the Boltzmann equation, Kravchenko, Kubalkova, Morozov, Polovinkin and Skok (1975). Evaluation of the conductivity tensor and scattering of carriers in isotropic crystals with parabolic bands based on BERTA, Shogenji and Okazaki (1975). Study of stimulated electron-phonon-photon interactions in non-degenerate semiconductors with energy-dependent carrier relaxation time based on the Boltzmann transfer equation, Agarwal, Sharma, Singh and Virmani (1975). Note on resonant scattering of polarons, Shmelev (1975). Excitonic polarization and polarization scattering of carriers in a highly excited semiconductor by a perturbation approach, Otsuka (1975). Note on the application of a variational principle to the solution of transport equations, Strekalov (1975). Quasiclassical theory of magneto-phonon resonance, Yasevichyute (1975). Discussion of the transport properties of small polarons, Emin (1975). Theory of hot electrons in a polar crystal based on an electron-phonon Hamiltonian in Fock space representation and its evaluation by field theoretic Green functions techniques, Hede and McMullen (1976). Calculation of hot electron microwave conductivity of wide bandgap semiconductors by means of drifted Maxwellian electron distribution functions, Das and Ferry (1976). Determination of the carrier distribution function in semiconductors with external electric fields, Wassef (1976). Damping effects in the calculation of polaron mobility based on the lattice polaron model in Fock space representation, Lai, Massicot and Nettel (1976). Influence of acoustic scattering on the breakdown of ionic crystals based on BERTA, Vershinin and Zotov (1976). Note on non-linear galvanomagnetic effects in semiconductors in longitudinal quantizing magnetic fields, Mikheev and Pomortsev (1976). Application of the variational method to the solution of the linearized Boltzmann equation in the theory of longitudinal transport phenomena in quantizing magnetic fields, Korolyuk and Pinchuk (1976). Discussion of transport phenomena and eigenoscillations of hot electrons in semiconductors with periodic impurity distribution in external electric fields based on PMTE, Dykman and Tomchuk (1976). Derivation of current expressions and transport coefficients from BERTA, including the spatial dependence of the band edge, the quasi Fermi level, the chemical potential and the temperature, van Vliet and Marshak (1976). Calculation of electron mobility for the discussion of the longitudinal magnetophonon effect in semiconductors containing magnetic impurities, Kossut and Walukiewicz (1976). Review of photoelectric phenomena in inhomogeneous semiconductors, Sheinkman and Shik (1976). Note on the iterative calculation of the polar optical mobility from the Boltzmann equation in semiconductors in the extreme quantum limit, Basu and Nag (1976). Application of a variational principle to the solution of the linearized Boltzmann equation, Strekalov (1976). Review of galvanomagnetic and recombination effects in semiconductors in a strong electric field based on BERTA and on calculation results of elementary rate processes for generation and recombination of charge carriers, Kachlishvili (1976). Non-equilibrium spin magnetization due to Coulomb attraction of electrons and holes in different bands in a semiconductor excited by a strong electromagnetic wave based on the evaluation of field theoretic Green functions, Popovkin (1976). Calculation of the mobility of hot electrons in a magnetic field based on BERTA, Shur (1976). Contribution to the theory of absorption of a strong electromagnetic field in semiconductors, Elesin (1976). Note on the inelastic scattering of electrons with bound excitons, Hönerlage and Rössler (1976). Absorption of a strong electromagnetic wave in a semiconductor near the optical phonon creation threshold, Gurevich and Parshin (1977). Note on the effect of non-parabolicity on transverse magnetoresistance in non-degenerate semiconductors, Wu and Chen (1977). Theory of scattering of holes from impurity potentials based on a one-particle effective mass Hamiltonian, Ralph (1977). Discussion of the quantum limit magnetoresistance for acoustic phonon scattering based on the Kubo formula, Arora, Cassiday and Spector (1977). Investigation of the thermopower and the transverse Nernst-Ettinghausen effect in semiconductors in a strong electric field based on BERTA with inclusion of the phonon drag of electrons and the heating of the phonons, Babaev and Gasymov (1977). Note on the relation between the resistivity formula and the Boltzmann equation, Christoph and Röpke (1977). Derivation of the quantum energy distribution function of hot electrons in crossed electric and magnetic field by means of an evaluation of solutions of the Pauli-Master equation for a free electron gas in external fields coupled to phonons, Calecki, Lewiner and Nozieres (1977). Use of the effective electron temperature method to study the effect of an electric field and a longitudinal quantizing magnetic field on the electron gas in a non-degenerate semiconductor, Pinchuk (1977). Note on the photoconductivity due to electron scattering accompanied by the emission of optical phonons, Vasko (1977). Derivation of a transport equation for electrons with an arbitrary dispersion law in an electromagnetic field based on the evaluation of the average one-electron density operator, Strekalov (1977). Exact calculation of the mobility due to scattering by point defects based on a one-electron model, Yartsev (1977). Analytic solutions of a model for radiation-induced conductivity in insulators based on phenomenological rate equations, Weaver, Shultis and Faw (1977). Investigation of the impurity breakdown in a magnetic field based on the discussion of the breakdown conditions derived from stationary non-equilibrium solutions of the Boltzmann equation, Bhattacharya and Kachlishvili (1977).Reconciliation of the Conwell-Weisskopf and Brooks-Herring formulae for charged impurity scattering in semiconductors by inclusion of third-body interference, Ridley (1977). Review of the theory of scattering of electrons at impurities in crystals mainly based on one-particle Hamiltonians, Rennie (1977). Note on the theory of galvanomagnetic effects in a strong electric field, Bhattacharya and Kachlishvili (1977). Note on transient injection currents in semiconductors, Arkhipov and Rudenko (1977). Calculation of the relaxation time due to electron-hole scattering, Lyapilin (1977). Derivation of current voltage characteristics for hot electrons in a quantizing magnetic field with inclusion of impact ionization of impurities in addition to common scattering processes, based on an evaluation of the diagonal elements of the density operator, Zlobin (1977). Investigation of the negative electron diffusion in high electric fields based on BERTA, Nag and Chattopadhyay (1978). High-field electronic conduction in insulators based on an evaluation of the Boltzmann equation, Thornber (1978). Review of hot electron quantum magneto transport theory based on an evaluation of the Boltzmann equation, Barker (1978). Derivation of steady state carrier distribution functions of electrons in semiconductors in external fields by variational solution of the Boltzmann equation, Zukotynski and Howlett (1978). Survey of calculations of hot electron phenomena based on the solution of the Boltzmann equation, Price (1978). Discussion of a modification of the relaxation time approximation, Polder and Weiss (1978). Analysis of the photogalvanic effect due to free carriers in non-centrosymmetric crystals by means of field theoretic Green functions, Belinicher (1978). Theory of nonlinear electric conductivity based on the Boltzmann equation with an inversion of the drift term by a corresponding Green function, Boffi, Molinari and Wonneberger (1978). Calculation of hot electron distributions in strong magnetic fields by means of a modified diffusion equation, Partl, Müller, Kohl and Gornik (1978). Investigation of the effect of electrons capture by negatively charged traps on the energy distribution of hot electrons in semiconductors based on BERTA, supplied by generation terms and application of a variational method, El-Waheidy and Farahat (1978). Calculation of elastic and inelastic scattering rates of electrons by vibrating impurities in semiconductors by means of a one-electron model with electron-phonon coupling, Suzuki and Frood (1978). Calculation of electric current and relaxation times of electrons in a lattice in a strong electric field leading to a Stark ladder, Kümmel, Rauh and Bangert (1978). Study of the paramagnetic relaxation of the spin system of the conduction electrons in semiconductors owing to the Coulomb interactions between the electrons, Timerkaev and Khabibullin (1978). Discussion of hot electron and phonon phenomena under high intensity photoexcitation of semiconductors, Shah (1978). Investigation of nonlinear current densities in semiconductors based on BERTA, Stramska (1978).

e) Electronic hopping motion

The physical concept of this type of electron or electronic quasiparticle motion and charge transport (which includes also impurity conduction) has already been explained in Section 8.2. It depends on the existence of local electron or electronic quasiparticle states. In this subsection we discuss its theoretical formulation.
Theoretically, hopping processes are special types of non-radiative processes, where for electrons or electronic quasiparticles the corresponding interaction energies are given by the non-adiabatic terms, the Coulomb interactions and the potentials of external fields. The hopping process is then considered to be a transition produced by these interaction terms, where the initial state describes an electron or electronic quasiparticle with a wave function centered at a point $\mathbf{R}_i$ in the lattice, while the final state describes such a particle with a wave function centered at a point $\mathbf{R}_f$. In the adiabatic coupling scheme such selftrapped electron states are accompanied by static lattice deformations around $\mathbf{R}_i$ or $\mathbf{R}_f$, resp. and by corresponding local lattice modes. Hence the corresponding transition matrix elements contain a phonon part with overlap integrals between discrete phonon modes centered at different lattice positions. As the electron levels in the selftrapped states are discrete, whether the transitions take place in the discrete spectrum or in the continuous spectrum depends on the spectral decomposition of the lattice equilibrium shifts $(\mathbf{R}_k^i - \mathbf{R}_k^f)$ with respect to the lattice modes. In the former case the spectral decomposition contains only local modes with discrete frequencies while in the latter case also non-localized acoustic and optical modes with continuous frequencies participate. Accordingly, resonance transitions or ordinary perturbation transitions, resp. may be used. As it is easier to calculate the overlap integrals for a small number of local modes than for a great number of non-local modes, resonance transition probabilities should be used whenever possible. In addition, the amount of linebreadth in the resonance transition probabilities allows an estimate of the validity of the hopping motion since large linebreadths indicate a wrong choice of the base system. For the further theoretical treatment of these transitions the full formalism for the treatment of irreversible rate processes which was developed in the preceding chapters can also be applied to these processes. However, a complete systematic treatment in this way has not yet been given in the literature, although the underlying physical idea of hopping motion given here has been widely accepted.

As in ordinary conductivity theory also in this field numerous theoretical approaches have been developed. Concerning the formulation of hopping processes in irreversible quantum statistics, except for details, the following approaches have been used. Hopping theory can be based on

i) rate equations of type (3.6.13) or Pauli-Master equations;
ii) direct evaluation of the density operator equations;
iii) evaluation of linear response theory formulae;
iv) treatment by stochastic models and techniques.

For the sake of brevity we can only give a short survey of the literature which is concerned with this comprehensive field.
The majority of authors used the approach i) which is completely embedded in our general formalism of rate equations and no essentially new theoretical developments are required to cover this topic. The approaches ii) and iii) will be dicussed shortly in citing the literature. The approach iv) will be omitted as it does not fit into a deductive treatment based on quantum theory.
In recent years the applicability and the consistency of the approach i) was doubted although it is obvious from our discussion that i) must be correct, provided it is carefully evaluated. A clarification was given by Barker (1976) who showed that the controversy is resolved by the observation that the antirate-equation theories have neglected an infinite sequence of terms appearing in perturbation theory which restore the disputed chemical potential terms to the rate equation formulation. Further clarification has been contributed by Zvyagin (1978) by a correct definition and evaluation of the current density within the rate equation formalism which was also the subject of controversy before.
Impurity conduction was first proposed by Gudden and Schottky (1935). The hopping process was suggested by Conwell (1956) and Mott (1956). Mott and Twose (1961) gave a comprehensive review of the theory of impurity conduction and based their calculations of the conductivity on a rate equation of the type (3.6.13). A review of hopping conductivity in ordered and disordered solids which contains a comprehensive list of references in this field, was given by Böttger and Bryksin (1976). These authors based their analytical discussion mainly on the equations for density operators from which they derived rate equations of the type (3.6.13) and more generalized ones with memory effects. Concerning the original literature we refer to some papers of the last two decades. Kasuya and Koide (1958) formulated a theory of electronic impurity conduction with phonon assistance for very low impurity concentrations based on rate equations of type (3.6.13), while Kasuya (1958) considered the phononless conduction processes for higher impurity concentrations. Holstein (1959) used a wave function of a coupled one-electron-lattice system in the tight binding approximation and calculated by time-dependent perturbation theory the transition probabilities for electronic hopping motion in the deformable lattice. While the diagonal elements of the perturbation lead to the formation of a polaron band, the non-diagonal elements cause a slowing down of the directed polaron motion. Miller and Abrahams (1960) based an investigation of impurity conduction at low concentrations on a rate equation

of the type (3.6.13) for steady state currents and calculated the transition probabilities in an effective mass model for the one-electron motion together with an electron-deformation potential interaction. Yamashita and Kurosawa (1960) and Kurosawa (1960) discussed the hopping motion of electrons and holes based on a Pauli-Master equation with inclusion of lattice deformation in polar substances and adiabatic electron lattice states. With the same type of equation Pollak and Geballe (1961) studied low frequency conductivity due to hopping processes in silicon. By means of linear response theory Sewell (1963) investigated a model of thermally activated electronic hopping motion in solids. Mikoshiba (1963) calculated the transition probability for an electron transition from a neutral donor to an ionized donor in a weak magnetic field. A remark on the hopping transition probability of polaron holes in NiO was made by Appel (1966). Efros (1967) based a theory of kinetic phenomena in semiconductors on the evaluation of a Pauli-Master equation for hopping motion of polarons. Schnakenberg (1968) investigated polaronic hopping motion and conduction between impurities by response functions and field theoretic evaluation techniques. Lang and Firsov (1968) calculated the activation probability for a jump of a small-radius polaron. Lang (1969) discussed the theory of the mobility of small polarons along the lines of Lang and Firsov's paper. Fukuyama, Saitoh, Uemura and Shiba (1970) studied the transport properties of electrons in impurity bands with external magnetic fields. Hacker and Obermair (1970) directly diagonalized the Hamiltonian for a one-band hopping model in the rigid lattice with static homogeneous electric field and proved the existence of the Wannier Stark ladder. Bagley (1970) applied rate equations to the calculation of the field dependent mobility of localized electronic carriers in semiconductors. Sussmann (1971) discussed the quantum theory of diffusion based on hopping processes and reviewed the preceding approaches. Brenig, Döhler and Wölfle (1971) developed a theory of thermally assisted electron hopping in amorphous solids based on rate equations of the type (3.6.13) and determined the d.c. (≡direct current) hopping conductivity. Emin (1971) used the three-dimensional analogue of Holstein's one-dimensional hopping model to study the relaxation of the lattice after a small polaron hop and showed that the polaron drift mobility is significantly affected by lattice relaxation effects. Butcher (1972) discussed hopping motion and the a.c. (≡alternating current) conductivity by means of the evaluation of a system of rate equations of the type (3.6.13). Jones and Schaich (1972) developed an extension of Miller and Abraham's theory of impurity conduction. Shklovskii (1972) (1973) developed a theory of hopping conductivity of semiconductors, in particular, in strong magnetic fields and used the percolation theoretic approach (cf. the review of Böttger and Bryksin). Klinger (1973) discussed the mechanism of small polaron transport. Based on Kubo's formula, Scher and Lax (1973) developed a theory of stochastic transport in disordered solids and applied it to impurity conduction in semiconductors. Brenig, Döhler and Wölfle (1973) discussed thermally assisted hopping processes in a disordered system on the basis of the Kubo formula. Toyabe and Asai (1973) investigated the phonon assisted electronic hopping conduction in polar piezoelectric semiconductors based on Miller and Abraham's theory. Butcher (1974) gave a stochastic interpretation of the rate equation formulation of hopping transport theory in the d.c. limit and discussed in addition diffusion and thermopower. Emin (1974) studied the conditions of the existence of free and selftrapped charge carriers in insulators, i.e., the occurrence of ordinary conduction and of hopping conduction by means of the three-dimensional analogue of Holstein's model. Entin-Wohlman and Bergmann (1974) applied a procedure similar to that of Nakajima-Zwanzig to derive equations for the occupation probability of states and discussed by means of this formalism the transition from the polaron tunneling motion to polaron hopping motion in dependence on the temperature. Emin (1975) calculated phonon-assisted transition rates based on Holstein's model, and by means of rate equations of the type (3.6.13) he developed a theory of the thermoelectric power due to electronic hopping motion. Tsu and Döhler (1975) considered hopping conduction in a superlattice. Abram and Silbey (1975) studied the connection between electronic energy transfer and the spectral line shapes of impurities in crystals by means of linear response theory. Skal, Shklovskii and Efros (1975) calculated the activation energy of hopping conduction by means of the percolation theoretic approach. Munn (1975) investigated the local field effects on carrier hopping mobilities. Capek (1975) derived exact expression for the d.c. hopping conductivity by means of a Hamiltonian in Fock space representation with electron phonon interaction and linear response theory. Kosarev (1975) performed a phenomenological analysis of the thermoelectric power of lightly doped semiconductors in the hopping conduction region. Starting with the Kubo formula for a.c. and d.c. conductivity of a hopping electron-phonon system, Barker (1976) derived by direct evaluation of this formula a system of electronic rate equations equivalent to those of Butcher's approach and gave a critical comparison of various theoretical approaches to hopping conductivity. Dieterich, Peschel and Schneider (1977) investigated the hopping conductivity of interacting particles based on the solution of a corresponding Pauli-Master equation. Capek (1977) studied phonon-assisted electronic d.c. hopping conductivity at weak electron-phonon coupling by means of an evaluation of the Kubo formula. Banyai and Aldea (1977) analyzed hopping conduction based on linear response theory. Butcher, Hayden and McInnes (1977) and Butcher and Hayden (1977) derived an approximate analytical formula for the d.c. hopping conductivity by using equations for the equivalent conductance network based on Miller's and Abraham's paper. Capek (1977) analyzed the validity of approaches for an approximate solution of rate equations of type (3.6.13) in the d.c. limit. Pollak (1977) discussed the connection between the non-markovian transport theory of Scher and Montroll (1975) and the markovian transport theory as provided, for instance, by the Boltzmann equation for the case of hopping transport with inclusion of trapping. Holstein (1978) improved his former calculations of the polaron hopping motion by a detailed investigation of the transition probabilities derived from perturbation theory. Dykman and Tarasov (1978) derived Pauli-Master equations for tunneling processes between impurity centers by means of the density matrix equations. Zvyagin (1978) generalized rate equations of the type (3.6.13) to take into account the effect of strong external electromagnetic waves and analyzed corresponding solutions for photon-phonon-assisted electronic hopping conductivity. Thomas and Wuertz (1978) calculated d.c. hopping conductivity and thermopower by percolation theory.

f) Ionic conduction

In addition to electrons and electronic quasiparticles also imperfections and regular lattice particles may undergo a hopping motion which leads to defect formation, ionic conduction etc. The underlying physical idea is the same as for electrons and rests on the localizability of the lattice particles, which is all the better satisfied the heavier the particles are. Theoretically this type of motion can be considered as a special type of non-radiative transition, where in contrast to electronic hopping motion the interaction energies are now due to the anharmonic forces between the lattice particles, the forces between the imperfections and the forces owing to external fields and where now in the transitions only different phonon states are involved. Then the transitions take place between local phonon mode states centered at the different positions at which the lattice particles or imperfections may be located. This formulation of lattice hopping motion was proposed by Stumpf (1961) and fits in our treatment of irreversible quantum statistics. However, a complete elaboration of ionic hopping motion in this way has not been performed so far in the literature.

The majority of papers in this field is concerned with a phenomenological treatment by diffusion equations etc. Reviews of this topic have been given in the article of Lidiard (1957), the book of Stasiw (1959) and the articles of Friauf (1977), Beniere, Beniere, Catlow, Shukla and Rao (1977) and of Slifkin (1978). A review of the theoretical aspects and applications of high ionic conductivity in solids was given by Holzapfel and Rickert (1975). Most studies of imperfect solids concentrate on the treatment of a sample of individual non-interacting defects. However, many physical phenomena are determined by defect processes, in which defects interact or the defect state of the lattice evolves and for which the motion of defects etc. plays an essential role. A phenomenological review of these more general processes was given by Stoneham (1978).

With respect to quantum statistical calculations based on microscopic models the theory of ionic hopping motion, etc., is not very well developed. The first approaches to a treatment at microscopic level were made by means of classical lattice models. In these theories expressions for the jump rates of ions, vacancies, etc., are derived which are based on the harmonic approximation and the canonical thermal distribution of the lattice energy. The various approaches are then distinguished by different evaluation techniques of the jump rate formulae. A review of these theories was given by Glyde (1967). By employing the same formalism, later on, these theories were generalized to the use of quantized lattice energies, cf. for instance Weiner (1968). Zhukov, Kuzin and Fedorov (1976) criticized the omission of anharmonic lattice energy terms and included such terms in the formalism. A review of these and related approaches is given in the papers of Sussmann (1971) and Kehr (1973). A characteristic feature of these "rate" theories is that they have nothing in common with the formalism of irreversible quantum statistics pursued here. Another theoretical approach to the treatment of ionic hopping motion at the microscopic level is the transfer of Holstein's small polaron model. Although in this model in contrast to Stumpf's proposal only the harmonic parts of the lattice energies are taken into account, it fits in principle in our formalism. However, this model was mainly

employed for the investigation of the diffusion of hydrogen in metals, cf. for instance Kehr (1973).
In addition to hopping motion also tunneling, reorientation, etc., of ions and imperfections may occur in the lattice. Theoretically, all these types of motions are described by non-radiative transitions, although in the literature a distinction is made. A review of tunneling was given by Narayanamurti and Pohl (1970) and with respect to reorientation by Stoneham (1975). We cite only some recent papers which work with microscopic models in quantum theoretical description: Theory of one- and two-phonon reorientation rates of paraelectric defects in ionic crystals, Dick and Strauch (1970). Migration of V_k centers in alkali halides, Song (1970). Calculation of reorientation rates of OH^- defects in RbBr, Shore and Sander (1972). Model for OH^- in alkali halide matrices, Pandey, Shukla and Pandey (1973). Renormalized tunneling of off-center impurities in alkali halides, Shore and Sander (1975). Effects of in-band defect induced phonon resonant modes on phonon-assisted defect tunneling, Dick (1977).
Analogously to electronic motion there are irreversible thermodynamic effects connected with the ionic motion. Allnatt and Jacobs (1960) treated the thermoelectric power of ionic conduction in ionic crystals, while Girvin (1978) considered the thermoelectric power of superionic conductors.

g) Energy transport

An example of energy transfer without charge transfer was already discussed in Chapter 4. In this case the Coulomb interaction between electrons located at different impurities is responsible for the energy transfer process between impurities, where one impurity loses its electronic excitation energy, while the other impurity gains this energy. This transfer mechanism rests on a distant action and is not comparable with general conduction phenomena treated in this chapter. Another mechanism of energy transfer, however, which is a genuine conduction phenomenon, is provided by the motion of excitons. This motion or migration of excitons takes place if either wavepackets of non-localized excitons or the hopping of local excitons is considered. Both processes are in complete analogy to the two different forms of polaron motion and can be treated along the same lines. In the literature nearly all papers are concerned with non-localized exciton motion. Calculations of the mean free path of non-localized excitons were performed by Anselm and Firsov (1955), (1956), Haken (1956), Tulub (1959), Genkin (1962) and Porsch (1965). Scattering processes and decay processes which limit the mean free path were considered in detail by Haken (1959), Goodman and Oen (1959), Korolyuk (1962), Demidenko (1963), Gross, Permogorov and Razbirin (1966) for interactions with lattice modes, by Trlifaj (1959) for interactions with F-centers and other defects, by Kachlishvili (1961) (1962) for scattering at F-centers, F'-centers and U-centers, by Lipnik (1962) for binding and decay due to phonons and impurities, by Kachlishvili (1962) for scattering at impurities, by Efremov and Kozhushner (1972) for exciton-exciton scattering, by Onipko and Sugakov (1973) for exciton-exciton interaction, by Zgierski (1973) for the influence of vibronic coupling, by Bobrysheva and Vybornov (1976) for biexciton formation, by Ivchenko, Pikus, Razbirin and Starukhin (1977) for resonant scattering of light, by Giner, Korovin and Pavlov (1977) for interaction with acoustic phonons, by Aristova, Giner, Lang and Pavlov (1978) for interaction with longitudinal optical phonons. A systematic survey of exciton annihilation processes was given by Trlifaj (1959). A qualitative discussion of exciton transport phenomena is given in the book of Knox (1963).

8.4 Relaxation time calculations

By introducing the concept of a relaxation time we were able to find general solutions of the linearized Boltzmann equation with arbitrary external fields. We did not, however, justify this concept in the foregoing section. In this section this problem will be discussed and in addition methods for the calculation of relaxation times will be given. The following theorem holds, cf. Seeger (1973), Madelung (1972).

Theorem 8.6: If we assume that for the linearized Boltzmann equation (8.2.42) the following suppositions hold:

i) the solutions $f(\mathbf{K})$ are in the non-degenerate region;
ii) the collisons are elastic and bimolecular;
iii) the energy surface is parabolic.

then the relaxation time is given by the expression

$$\tau^{-1} := \frac{m^* k}{2\pi\hbar^3} \int_0^{\pi} \tilde{W}(\mathbf{K},\mathbf{K}')(1 - \cos\Theta)\sin\Theta \, d\Theta \tag{8.4.1}$$

where the integration has to be performed in polar coordinates (8.4.5) and the transition probability $\tilde{W}(\mathbf{K},\mathbf{K}')$ is defined by (8.4.12).

Proof: We first consider the collision term on the right-hand side of (8.2.40) in its non-linearized form. For the non-degenerate case we have $f(\mathbf{K}) \ll 1$ in the energy region of interest. Hence we may replace $(1-f(\mathbf{K}))$ by 1 etc. If we further observe equation (3.6.25) we obtain

$$\begin{aligned}\sum_{\mathbf{K}'}\{\ \}_{\text{Coll}} &\approx \int \{W(\mathbf{K},\mathbf{K}')f(\mathbf{K}) - W(\mathbf{K}',\mathbf{K})f(\mathbf{K}')\} d^3K' \\ &= \int W(\mathbf{K},\mathbf{K}') \left[f(\mathbf{K}) - f(\mathbf{K}') \frac{f_0(\mathbf{K})}{f_0(\mathbf{K}')} \right] d^3K'.\end{aligned} \tag{8.4.2}$$

For elastic collisions we have $E(\mathbf{K}) = E(\mathbf{K}')$ and therefore $f_0(\mathbf{K}) = f_0(\mathbf{K}')$, so that (8.4.2) takes the form

$$\sum_{\mathbf{K}'}\{\ \}_{\text{Coll}} \approx \int W(\mathbf{K},\mathbf{K}')[f(\mathbf{K}) - f(\mathbf{K}')] d^3K'. \tag{8.4.3}$$

In the linearized version of the Boltzmann equation the solutions are given by (8.2.41) (8.2.52) where Φ can be represented by $\Phi(\mathbf{K}) \equiv \mathbf{v} \cdot \mathbf{G}$. Hence, due to energy conservation we have

$$f(\mathbf{K})-f(\mathbf{K}')=f_0(\mathbf{K})-f_0(\mathbf{K}')-\frac{\partial f_0}{\partial E}\frac{\hbar}{m}\mathbf{K}\cdot\mathbf{G}+\frac{\partial f_0}{\partial E'}\frac{\hbar}{m}\mathbf{K}'\cdot\mathbf{G}$$

$$=-\frac{\partial f_0}{\partial E}\frac{\hbar}{m}[\mathbf{K}\cdot\mathbf{G}-\mathbf{K}'\cdot\mathbf{G}]. \tag{8.4.4}$$

Introducing polar coordinates relative to the **K**-direction we have with $|\mathbf{K}|=k$

$$\begin{aligned} d^3K' &= k'^2 dk' \sin\Theta\, d\Theta\, d\varphi \\ \mathbf{K}\cdot\mathbf{G} &= kG\cos\vartheta \\ \mathbf{K}'\cdot\mathbf{G} &= k'G(\cos\vartheta\cos\Theta+\sin\vartheta\sin\Theta\cos\varphi). \end{aligned} \tag{8.4.5}$$

If we substitute these expressions in (8.4.4) and (8.4.4) in (8.4.3) the term with $\cos\varphi$ drops out by integration over φ. The remaining terms read

$$\sum_{\mathbf{K}'}\{\ \}_{\mathrm{Coll}} \approx -2\pi\int k'^2 W(\mathbf{K},\mathbf{K}')\frac{\hbar}{m}\frac{\partial f_0}{\partial E}kG\cos\vartheta(1-\cos\Theta)\sin\Theta\, dk'd\Theta$$

$$=2\pi[f(\mathbf{K})-f_0(\mathbf{K})]\int k'^2 W(\mathbf{K},\mathbf{K}')(1-\cos\Theta)\sin\Theta d\Theta dk' \tag{8.4.6}$$

and by comparison with the definition of a relaxation time

$$\sum_{\mathbf{K}'}\{\ \}_{\mathrm{Coll}} =: \frac{1}{\tau}[f(\mathbf{K})-f_0(\mathbf{K})] \tag{8.4.7}$$

it follows that

$$\tau^{-1} := 2\pi\int k'^2 dk' \int_0^\pi W(\mathbf{K},\mathbf{K}')(1-\cos\Theta)\sin\Theta d\Theta. \tag{8.4.8}$$

The transition probabilities $W(\mathbf{K},\mathbf{K}')$ for non-radiative electronic transitions are given according to Section 2.4 in the general form

$$W(\mathbf{K},\mathbf{K}') := \frac{2\pi}{\hbar}\sum_{mm'}|\langle nm|H_{\mathrm{p}}|n'm'\rangle|^2 f_{m'}\delta_\gamma(E_m^n-E_{m'}^{n'}) \tag{8.4.9}$$

where H_{p} is the perturbation operator. If the transitions are assumed to be (approximately) elastic and if they take place between two Bloch states with wave vectors **K** and **K**′ then we may write approximately

$$\delta_\gamma(E_m^n-E_{m'}^{n'}) \approx \delta\left(\frac{\hbar^2k^2}{2m^*}-\frac{\hbar^2k'^2}{2m^*}\right)=\frac{m^*}{\hbar^2k'}\delta(k-k'). \tag{8.4.10}$$

Substitution of (8.4.10) in (8.4.9) and of (8.4.9) in (8.4.8) gives

$$\tau^{-1}=\frac{m^*k}{2\pi\hbar^3}\int_0^\pi\sum_{mm'}|\langle nm|H_{\mathrm{p}}|n'm'\rangle|^2 f_{m'}(1-\cos\Theta)\sin\Theta d\Theta. \tag{8.4.11}$$

Since the quantum numbers m and m' only occur in the summation we may define

$$\tilde{W}(\mathbf{K},\mathbf{K}') := \sum_{mm'} |\langle nm|H_p|n'm'\rangle|^2 f_{m'} \tag{8.4.12}$$

and therefore obtain (8.4.1), Q.E.D.

This theorem shows that a relaxation time can be defined only under very restrictive suppositions, which are generally not satisfied. Nevertheless, we will discuss some important processes where the concept of relaxation times is valid at least to a certain degree of approximation.
The perturbation operators for Bloch electrons which lead to collisions are given by Theorem 8.2. In keeping with the formalism of this chapter we concentrate on the one-particle transitions. In this case, according to Theorem 8.2, for such transitions the matrix elements (8.1.21) have to be calculated. If the initial states are Bloch-states, the matrix elements of the operator K^t vanish and only the matrix elements of K^s remain. In the proof of Theorem 8.2 it is shown that the matrix elements (8.1.21) of K^s can be expressed by (8.1.29).
Hence only the matrix elements (8.1.29) with the corresponding transitions have to be discussed. Equation (8.1.29) contains interactions with impurities, lattice deformations, selftrapping potentials, exchange forces etc. Only a small part of these interactions has been treated in detail in the literature, often using phenomenological arguments. We briefly give some of the most important results and discuss separately the various transition mechanisms. The separate treatment of these processes is possible as long as the corresponding transition probabilities do not interfere, i.e., as long as they are additive. In this case the splitting of the total transition probability into several parts and their separate treatment is consistent with the general definition of the collision operator in (8.2.40), since there the transition probability enters linearly. If it is split up, any part can be treated individually, so that the final expression is a sum over the single parts, i.e.,

$$\sum_{\mathbf{K}'} \{ \quad \}_{\text{Coll}} = \sum_{\alpha} \sum_{\mathbf{K}'} \{ \quad \}_{\text{Coll}}^{\alpha} \tag{8.4.13}$$

where α denotes the various contributions. The transition probabilities for static potentials, i.e., without phonon participation and those for dynamical potentials, i.e., with phonon participation do not interfere. Hence we can treat them separately. We first consider the main static contribution.

a) Impurity scattering

According to (8.1.29) the interaction of a Bloch electron with an impurity depends on its electron states. In our simple model these defects may be vacancies, F-centers or F'-centers. But from the derivation it follows that other impurities can be treated equally well along the same lines. The simplest case is a single vacancy, for which electron scattering has first been treated by Conwell and Weisskopf (1950) and Brooks and Herring (1955) and Brooks (1951) who used plane waves instead of Bloch waves. A straighforward calculation of τ_i with Coulomb potentials and plane waves for an infinite crystal, however, runs into difficulties. The integral (8.4.1) has no finite value. It has to be expected that this divergence can be avoided if finite crystals and/or genuine Bloch states calculated with pseudopotentials are applied. As both improvements require a considerable calculation effort, a phenomenological way out was proposed. It is assumed that the impurity charges are screened by the conduction band electrons (or valence band holes) and that the effect of the screening can be taken into account by a change of the Coulomb potential $C(\mathbf{r}, l\mu)$ into a Yukawa potential

$$V(\mathbf{r}, l\mu) := a_{l\mu} C(\mathbf{r}, l\mu) \exp\left(-\beta|\mathbf{r}-\mathbf{R}_{l\mu}^0|\right) \tag{8.4.14}$$

where β is a suitable cut-off parameter. The quantum mechanical calculation of the cross section based on (8.4.14) has an analytical result even if plane waves are used. We assume that the vacancy is located at $\mathbf{R}_{l\mu}^0 = 0$ and that only one vacancy is present. Then the matrix element (8.1.29) for the scattering of a Bloch wave $\psi(\mathbf{r}, \mathbf{K}')$ into a Bloch wave $\psi(\mathbf{r}, \mathbf{K})$ by a single vacancy yields the following value in the plane wave approximation

$$\begin{aligned} &\langle \lambda_1 \ldots \mathbf{K} \ldots \lambda_z, m | eV(\mathbf{r}) | \lambda_1 \ldots \mathbf{K}' \ldots \lambda_z, m' \rangle \\ &= \frac{ea_0}{\varepsilon} \int C(\mathbf{r}) \exp\left[-\beta r + i(\mathbf{K}-\mathbf{K}')\mathbf{r}\right] d^3 r \, \delta_{mm'} \\ &= \frac{ea_0}{\varepsilon} 4\pi \left(|\mathbf{K}-\mathbf{K}'|^2 + \beta^2\right)^{-1} \delta_{mm'} . \end{aligned} \tag{8.4.15}$$

where only the contribution of the static potential in (8.1.29) was taken into account which corresponds to the second term on the right-hand side of (8.1.29) with $\varrho_{\lambda_i}(\mathbf{r}) \equiv 0$, i.e., a vacancy.

As no phonons are involved in this process, the phonon part reduces simply to $\delta_{mm'}$. For elastic collision $|\mathbf{K}| = |\mathbf{K}'|$ holds and therefore we obtain $|\mathbf{K}-\mathbf{K}'| = 2k \sin \Theta/2$. Then with (8.4.15) equation (8.4.12) takes the form

$$\tilde{W}(\mathbf{K}, \mathbf{K}') = \left(\frac{\pi e a_0}{\varepsilon k^2}\right)^2 \left(\sin^2 \Theta/2 + \frac{\beta^2}{4k^2}\right)^{-2} \tag{8.4.16}$$

and the relaxation time follows from (8.4.1) with (8.4.16)

$$\tau_i^{-1} = \frac{2\pi m^* e^2 a_0^2}{\varepsilon^2 \hbar^3 k^3} \int_0^\pi \frac{\sin^3 \Theta/2 \cos \Theta/2}{\left(\sin^2 \Theta/2 + \frac{\beta^2}{4k^2}\right)^2} d\Theta . \tag{8.4.17}$$

Direct calculation leads to

$$\tau_i^{-1} = A\left[\ln(1+b) - \frac{b}{1+b}\right] \tag{8.4.18}$$

with

$$A = \pi \varepsilon^{-2} e^2 a_0^2 (2m^*)^{-1/2} E(\mathbf{K})^{-3/2}; \quad b = 8m^* E(\mathbf{K}) (\hbar\beta)^{-2} . \tag{8.4.19}$$

The screening parameter β depends on the temperature

$$\beta = \left(\frac{4\pi e^2 n}{\varepsilon k_B T}\right)^{1/2} \tag{8.4.20}$$

cf. Haug (1970), where n is the average charge density of conduction electrons. Using (8.4.20) in (8.4.19) we obtain the equation

$$b = \frac{2m\varepsilon k_B T E(\mathbf{K})}{\pi \hbar^2 e^2 n}. \tag{8.4.21}$$

As $\ln b(E(\mathbf{K}))$ increases more weakly than any power of $E(\mathbf{K})$, τ_i is assumed to be proportional to $E(\mathbf{K})^{3/2}$.

b) Acoustic phonon scattering

The lattice deformations in the perturbation operator (8.1.29) are characterized by terms with the ionic displacements $\{\mathbf{M}_{l\mu}\}$. These displacements are referred to the ideal lattice positions. With (2.2.14) they can be split up into a static displacement $\mathbf{M}_{l\mu}^n$ and a dynamic displacement $\mathbf{u}_{l\mu}^n$

$$\mathbf{R}_{l\mu} - \mathbf{R}_{l\mu}^0 =: \mathbf{M}_{l\mu} = (\mathbf{R}_{l\mu}^n - \mathbf{R}_{l\mu}^0) + (\mathbf{R}_{l\mu} - \mathbf{R}_{l\mu}^n) =: \mathbf{M}_{l\mu}^n + \mathbf{u}_{l\mu}^n \tag{8.4.22}$$

where $\{\mathbf{R}_{l\mu}^n\}$ are the equilibrium positions in the electron state n. Hence the conduction band electrons are scattered at static deformation potentials and at dynamic deformation potentials, where the latter give rise to electron-phonon scattering. As the displacements $\{\mathbf{M}_{l\mu}^n\}$ can be calculated by the methods of Chapter 5, the static lattice deformation potentials are well defined. They are potentials which are similar to the impurity potentials of subsection a) and can be treated in a similar way. Hence this kind of interaction brings about no new general information. Thus we treat only electron-phonon scattering. These scattering processes have been extensively discussed in the literature. For diatomic polar lattices optical and acoustic phonon-electron scattering occur. Only the latter admits the definition of a relaxation time for all energy ranges. We first treat this type of scattering. The terms which are responsible for electron-phonon scattering in (8.1.29) are given by

$$V_P := e(1+\alpha^e)^{-1} \sum_{\{k\varrho\}} a_{k\varrho}(\mathbf{M}_{k\varrho} \cdot \nabla_{k\varrho}) C(\mathbf{r}, k\varrho) + \text{exch. terms.} \tag{8.4.23}$$

If we substitute (8.4.22) in (8.4.23), only the terms with $\mathbf{u}_{k\varrho}^n$ contribute to electron phonon scattering. In order to simplify matters we make two approximations:

i) the sum in (8.4.23) is extended over all lattice points, i.e., the imperfect lattice is replaced by a perfect lattice;
ii) the dynamic displacements $\mathbf{u}_{k\varrho}^n$ and their expansion in phonon coordinates are replaced by the displacements of the ideal crystal $\mathbf{u}_{k\varrho}^0$ and their expansion in ideal crystal phonon modes.

With these two approximations we replace the interaction terms (8.4.23) by

$$V_P=\sum_{k\varrho} \mathbf{u}_{k\varrho}^0 \cdot \nabla_{k\varrho} V_\varrho(\mathbf{r}-\mathbf{R}_{k\varrho}^0) \tag{8.4.24}$$

where the ionic potential energy $V_\varrho(\mathbf{r}-\mathbf{R}_{k\varrho}^0)$ is a suitable combination of the Coulomb interaction and the exchange interaction of an ion with its surroundings.
According to equations (7.2.66), (7.2.67) the expansion of $\mathbf{u}_{k\varrho}^0$ in phonon modes reads

$$\mathbf{u}(k\varrho)=\sum_{\mathbf{q}j}\left(\frac{\hbar}{2M_\varrho\omega(j,\mathbf{q})}\right)^{1/2}\mathbf{e}_\varrho(j,\mathbf{q}) \exp(i\mathbf{q}\cdot\mathbf{R}_k^0)N^{-1/2}[b(j,\mathbf{q})-b^+(j,-\mathbf{q})]. \tag{8.4.25}$$

If we now substitute (8.4.25) in (8.4.24) and if we apply the common calculation rules for phonon matrix elements of Chapter 3 and Chapter 7, specialized to ideal crystal modes, we obtain for the corresponding matrix element of (8.4.24)

$$\langle\lambda_1\ldots\mathbf{K}\ldots\lambda_z,m|\sum_{k\varrho}\mathbf{u}_{k\varrho}^0\cdot\nabla_{k\varrho}V_\varrho(\mathbf{r}-\mathbf{R}_{k\varrho}^0)|\lambda_1\ldots\mathbf{K}'\ldots\lambda_z,m'\rangle$$

$$=\sum_{j\mathbf{q}}\left(\frac{\hbar}{2M\omega(j,\mathbf{q})}\right)^{1/2}[m(j,\mathbf{q})+\tfrac{1}{2}\mp\tfrac{1}{2}]^{1/2}\delta_{m'(j,\mathbf{q}),m(j,\mathbf{q})\mp 1} \tag{8.4.26}$$

$$\times\prod_{\substack{\mathbf{q}'j'\\ \neq \mathbf{q}j}}\delta_{m'(j',\mathbf{q}'),m(j',\mathbf{q}')}I^{\pm}$$

where the $\{m(j,\mathbf{q})\}$ are the occupation numbers of the phonon modes $\{j,\mathbf{q}\}$ and where the electronic part $I^{\pm}$ is defined by

$$I^{\pm}:=\mathbf{e}(j,\mathbf{q})\cdot\int\psi(\mathbf{r},\mathbf{K})^{\times}\sum_{k\varrho}e^{\pm i\mathbf{q}\cdot\mathbf{R}_k^0}\nabla_{k\varrho}V_\varrho(\mathbf{r}-\mathbf{R}_{k\varrho}^0)\psi(\mathbf{r},\mathbf{K}')d^3r. \tag{8.4.27}$$

The vector

$$\mathbf{V}^{\pm}(\mathbf{r}):=\sum_{k\varrho}e^{\pm i\mathbf{q}\cdot\mathbf{R}_k^0}\nabla_{k\varrho}V_\varrho(\mathbf{r}-\mathbf{R}_{k\varrho}^0) \tag{8.4.28}$$

is an eigenstate of the translation operator $T(\mathbf{r})$ as

$$T(\mathbf{m})\mathbf{V}^{\pm}(\mathbf{r})\equiv\mathbf{V}^{\pm}(\mathbf{r}+\mathbf{m})=\exp(\pm i\mathbf{q}\mathbf{m})\mathbf{V}^{\pm}(\mathbf{r}) \tag{8.4.29}$$

holds if $\mathbf{m}$ is assumed to be a lattice vector.
From this it follows that (8.4.28) can be rewritten in the form

$$\mathbf{V}^{\pm}(\mathbf{r})=e^{\pm i\mathbf{q}\cdot\mathbf{r}}\mathbf{v}^{\pm}(\mathbf{r}) \tag{8.4.30}$$

where

$$\mathbf{v}^{\pm}(\mathbf{r})=\sum_{k\varrho}\exp[\pm i\mathbf{q}\cdot(\mathbf{R}_k^0-\mathbf{r})]\nabla_{k\varrho}V_\varrho(\mathbf{r}-\mathbf{R}_{k\varrho}^0) \tag{8.4.31}$$

is a periodic function $\mathbf{v}^{\pm}(\mathbf{r}+\mathbf{m})=\mathbf{v}^{\pm}(\mathbf{r})$. Using this representation of $\mathbf{V}$ we may write the integral (8.4.27) in the form

$$I^{\pm}=N^{-1}\mathbf{e}(j,\mathbf{q})\cdot\int_V\exp[i(\mathbf{K}-\mathbf{K}'\pm\mathbf{q})\cdot\mathbf{r}]u(\mathbf{r},\mathbf{K})^{\times}\mathbf{v}^{\pm}(\mathbf{r})u(\mathbf{r},\mathbf{K}')d^3r, \tag{8.4.32}$$

if we observe the definition of a Bloch function (6.6.6). Using the periodicity of u and $\mathbf{v}$ we can evaluate (8.4.32) and obtain

$$I^{\pm} = \Delta(\mathbf{K}-\mathbf{K}'\pm\mathbf{q})\mathbf{e}(j,\mathbf{q}) \cdot \int_{V_0} \exp\,[i(\mathbf{K}-\mathbf{K}'\pm\mathbf{q}) \cdot \mathbf{r}]u(\mathbf{r},\mathbf{K})^{\times}\mathbf{v}^{\pm}(\mathbf{r})u(\mathbf{r},\mathbf{K}')d^3r \tag{8.4.33}$$

where V_0 is the volume of the unit cell. According to (7.2.43), the Δ-function does not vanish if

$$\mathbf{K}' = \mathbf{K} \pm \mathbf{q} + \mathbf{g} \tag{8.4.34}$$

i.e., in analogy to the phonon-phonon interaction also the electron-phonon interaction allows normal processes ($\mathbf{g}=0$) and Umklapp-processes ($\mathbf{g}\neq 0$). The further evaluation of (8.4.33) depends on the special model and the physical situation which is of interest. We first evaluate the general transition probability (8.4.9). The phonon equilibrium distribution $f_{m'}$ is given by a product over the equilibrium distributions for the single phonon modes

$$f_m := \prod_{j\mathbf{q}} f[m(j,\mathbf{q}), T] \tag{8.4.35}$$

where any single mode distribution satisfies the conservation of probability, i.e., its sum over $m(j,\mathbf{q})=1,2\ldots$ is normalized to unity. As in (8.4.9) the sum runs over all quantum number configurations m and m' we obtain with (8.4.26)

$$\begin{aligned} &\frac{2\pi}{\hbar} \sum_{mm'} |\langle nm| \sum_{k\varrho} \mathbf{u}^0_{k\varrho} \cdot \nabla_{k\varrho} V_\varrho(\mathbf{r}-\mathbf{R}^0_{k\varrho})|n'm'\rangle|^2 f_{m'}\delta_\gamma(E^n_m - E^{n'}_{m'}) \\ &= \sum_{j\mathbf{q}} \frac{2\pi|I^{\pm}|^2}{2M\omega(j,\mathbf{q})} \left[\bar{m}(j,\mathbf{q}) + \frac{1}{2} \mp \frac{1}{2}\right] \delta_\gamma[E(\mathbf{K}) - E(\mathbf{K}') \mp \hbar\omega(j,\mathbf{q})] \end{aligned} \tag{8.4.36}$$

as all mixed terms for $j,\mathbf{q} \neq j',\mathbf{q}'$ drop out. From (8.4.36) it can be seen that if phonons participate in the collision processes, these processes are never elastic in the strict sense. If the temperature, however, is sufficiently high, then the average electron energy $\bar{E}(\mathbf{K})$ can be assumed to be considerably larger than the energy of the phonons of the acoustic branch. Thus approximately we can neglect the phonon energies in the energy balance and may treat the contribution of the acoustic branches to the transition probability (8.4.36) by the relaxation time formalism.
According to the additivity of the contributions of the various phonon branches in (8.4.36), the corresponding transition probabilities do not interfere with those of other processes and this leads to the definition of a special relaxation time according to (8.4.13). The corresponding reduced probability (8.4.12) is then given, with the aid of (8.4.36), by

$$\tilde{W}(\mathbf{K},\mathbf{K}')_a = \sum_{\substack{\mathbf{q} \\ \{j\}=\text{acoust.}}} \frac{\pi|I^{\pm}|^2}{M\omega(j,\mathbf{q})} [\bar{m}(j,\mathbf{q}) + \tfrac{1}{2} \mp \tfrac{1}{2}] \tag{8.4.37}$$

where the label a denotes the acoustic contribution. For a rough estimate of the corresponding relaxation time we evaluate $I^{\pm}$ by considering its long wave limit and by admitting only normal processes. For normal processes with $\mathbf{g}\equiv 0$, we obtain from (8.4.33)

$$I^{\pm}=\Delta(\mathbf{K}-\mathbf{K}'\pm\mathbf{q})\mathbf{e}(j,\mathbf{q})\cdot\int_{V_0} u(\mathbf{r},\mathbf{K}'\mp\mathbf{q})^{\times}\mathbf{v}^{\pm}(\mathbf{r})u(\mathbf{r},\mathbf{K}')d^3r, \tag{8.4.38}$$

and the long wave limit corresponds to an expansion of the integral in (8.4.38) in powers of $\mathbf{q}$, where only the first nontrivial term is taken into account. If we write

$$\begin{aligned}&\int_{V_0} u(\mathbf{r},\mathbf{K}'\mp\mathbf{q})^{\times}\mathbf{v}^{\pm}(\mathbf{r})u(\mathbf{r},\mathbf{K}')d^3r\\&=\mathfrak{C}_0(\mathbf{K}')\pm i\mathbf{q}\cdot\mathfrak{C}_1(\mathbf{K}')+\tfrac{1}{2}i\mathbf{q}\cdot\mathfrak{C}_2(\mathbf{K}')\cdot i\mathbf{q}\pm\ldots\end{aligned} \tag{8.4.39}$$

the integral $\mathfrak{C}_j(\mathbf{K}')$ is a tensor of rank $j+1$. It can be shown, cf. Haug (1970), that $\mathfrak{C}_0(\mathbf{K}')$ vanishes. Hence the first nontrivial term is $\pm i\mathbf{q}\cdot\mathfrak{C}_1(\mathbf{K}')$. For isotropic materials $\mathfrak{C}_1(\mathbf{K}') = C(\mathbf{K}')\mathbb{1}$ holds, and (8.4.38) takes the form

$$I^{\pm}=\pm i\Delta(\mathbf{K}-\mathbf{K}'\pm\mathbf{q})\mathbf{e}(j,\mathbf{q})\cdot\mathbf{q}C(\mathbf{K}'). \tag{8.4.40}$$

According to Section 7.2 in the long wave limit $\mathbf{e}(j,\mathbf{q})$ is either orthogonal or parallel to $\mathbf{q}$. Hence only for longitudinal acoustic modes $(\mathbf{q}\parallel\mathbf{e}(j,\mathbf{q}))$ does (8.4.40) give a non-vanishing contribution.
If (8.4.40) is substituted in (8.4.37) the sum in (8.4.37) can be evaluated immediately by means of the Δ-function if we confine ourselves to normal processes. In this case (8.4.37) goes over into

$$\tilde{W}(\mathbf{K},\mathbf{K}')_a=\sum_{\{j\}=\mathrm{ac}}\frac{\pi[\mathbf{e}(j,\mathbf{K}-\mathbf{K}')\cdot(\mathbf{K}-\mathbf{K}')]^2C(\mathbf{K}')^2}{M\omega(j,\mathbf{K}-\mathbf{K}')}[\bar{m}(j,\mathbf{K}-\mathbf{K}')+\tfrac{1}{2}\mp\tfrac{1}{2}]. \tag{8.4.41}$$

For the temperature range where the collisions can be considered to be elastic the mean value of $\bar{m}$ is approximately given by

$$\bar{m}(j,\mathbf{q})\approx\frac{k_BT}{\hbar\omega(j,\mathbf{q})}\gg 1;\qquad \mathbf{q}=\mathbf{K}-\mathbf{K}'. \tag{8.4.42}$$

Hence the terms $1/2\pm 1/2$ can be neglected in (8.4.41). We further use the long wave limit for ω which is given by $\omega(j,\mathbf{q})\approx v_j q$. Furthermore the relaxation time (8.4.1) equivalently reads

$$\tau_a^{-1}=\frac{m^*k}{2\pi\hbar^3}\int_0^{\pi}\tilde{W}(\mathbf{K},\mathbf{K}')_a\sin^3\Theta/2\cos\Theta/2\,d\Theta. \tag{8.4.43}$$

For elastic collisions the relation $|\mathbf{K}'-\mathbf{K}|=2k\sin\Theta/2$ holds and with the new variable $z=2k\sin\Theta/2$ and the relation, cf. Haug (1970)

$$\sum_{j=1}^{3}\frac{[\mathbf{e}(j,\mathbf{q})\cdot\mathbf{q}]^2}{[\omega(j,\mathbf{q})]^2}=\frac{1}{v_s^2} \tag{8.4.44}$$

where v_s is the average value of the velocities of sound, we obtain with the aid of (8.4.41) and (8.4.42)

$$\tau_a^{-1} = \frac{m^*}{2\pi\hbar^4} \frac{k_B T}{Mk^3} C^2 \frac{1}{v_s^2} \int_0^{z_0} z^3 dz \tag{8.4.45}$$

$$= \frac{m^*}{2\pi\hbar^4} \frac{k_B T}{M} \frac{C^2}{v_s^2} \frac{z_0^4}{4} E(\mathbf{K}')^{-3/2} \left(\frac{2m}{\hbar^2}\right)^{-3/2} \tag{8.4.46}$$

and from this it follows that

$$\tau_a \sim E(\mathbf{K}')^{3/2} T^{-1}. \tag{8.4.47}$$

For the detailed discussion of the factors occurring in (8.4.46) cf. Haug (1970).

c) Optical phonon scattering

For the optical branches in (8.4.36) the energy contributions of the phonons to the energy conservation cannot be ignored in the long wave length limit, because in contrast to the acoustic phonons their frequencies have finite values everywhere. In this case relaxation times can be derived only for certain energy and temperature ranges while in other ranges the Boltzmann equation has to be solved with inclusion of the collision term. For the treatment of the optical branches usually the interaction potential (8.4.23) is calculated in the long wave limit, since for this limit the interaction between conduction band electron and optical phonons is strongest. Furthermore, as the electric interaction is essential for this kind of collision process, we neglect the exchange terms in (8.4.23). With the effective charge

$$a_{k\varrho}^* \equiv a_\varrho^* := (1+\alpha^e)^{-1} a_{k\varrho} \tag{8.4.48}$$

(8.4.23) reads in this case

$$V_P := \sum_{\{k\varrho\}} e a_\varrho^* (\mathbf{M}_{k\varrho} \cdot \nabla_{k\varrho}) C(\mathbf{r}, k\varrho). \tag{8.4.49}$$

We now assume that the same suppositions i) ii) hold for (8.4.49) as for (8.4.23) in b). Then for two ions in the unit cell, (8.4.49) can be written using suppositions i) and ii)

$$V_P = -ee^* \sum_k \left[\mathbf{M}_{k1}^0 \cdot \frac{(\mathbf{r}-\mathbf{R}_{k1}^0)}{|\mathbf{r}-\mathbf{R}_{k1}^0|^3} - \mathbf{M}_{k2}^0 \cdot \frac{(\mathbf{r}-\mathbf{R}_{k2}^0)}{|\mathbf{r}-\mathbf{R}_{k2}^0|^3} \right] \tag{8.4.50}$$

if the two effective charges are assumed to have the same absolute value $|a_\varrho^*| = e^*$, $\varrho = 1,2$. If $\mathbf{R}_k^0$ is the center of mass of the k-th until cell, then V_P can be approximately replaced by the first term of a Taylor expansion of any cell potential about $\mathbf{R}_k^0$

$$V_P \approx -ee^* \sum_k (\mathbf{M}_{k1}^0 - \mathbf{M}_{k2}^0) \cdot \frac{(\mathbf{r}-\mathbf{R}_k^0)}{|\mathbf{r}-\mathbf{R}_k^0|^3}. \tag{8.4.51}$$

With $\mathbf{M}_{k\varrho}^0 = \mathbf{u}_{k\varrho}^0$ where $\mathbf{u}_{k\varrho}^0$ is given by (8.4.25), we obtain

$$(\mathbf{M}_{k1}^0 - \mathbf{M}_{k2}^0) = N^{-1/2} \sum_{\substack{\mathbf{q} \\ \{j\}=\mathrm{op}}} \left(\frac{\hbar}{2\omega(j,\mathbf{q})}\right)^{1/2} \times \left[\frac{\mathbf{e}_1(j,\mathbf{q})}{M_1^{1/2}} - \frac{\mathbf{e}_2(j,\mathbf{q})}{M_2^{1/2}}\right] [b(j,\mathbf{q}) e^{i\mathbf{q}\cdot\mathbf{R}_k^0} + b^+(j,\mathbf{q}) e^{-i\mathbf{q}\cdot\mathbf{R}_k^0}]. \tag{8.4.52}$$

In the long wave limit, according to (7.2.30), the relations

$$\mathbf{e}_1 = \mathbf{e}\left(\frac{M_2}{M_1+M_2}\right)^{1/2}; \quad \mathbf{e}_2 = -\mathbf{e}\left(\frac{M_1}{M_1+M_2}\right)^{1/2} \tag{8.4.53}$$

and

$$M_0^{-1} := M_1^{-1} + M_2^{-1} \tag{8.4.54}$$

hold. In this limit (8.4.52) goes over into

$$(\mathbf{M}_{k1}^0 - \mathbf{M}_{k2}^0) = \left(\frac{\hbar}{2NM_0}\right)^{1/2} \sum_{\substack{\mathbf{q} \\ \{j\}=\mathrm{op}}} \frac{\mathbf{e}(j,\mathbf{q})}{\omega(j,\mathbf{q})^{1/2}} \times [b(j,\mathbf{q}) \mathbf{e}^{i\mathbf{q}\cdot\mathbf{R}_k^0} + b^+(j,\mathbf{q}) e^{-i\mathbf{q}\cdot\mathbf{R}_k^0}] \tag{8.4.55}$$

and by substitution into (8.4.51) this expression takes the form

$$V_P \approx -ee^* \sum_k \sum_{\substack{\mathbf{q} \\ \{j\}=\mathrm{op}}} \left(\frac{\hbar}{2NM_0\omega(j,\mathbf{q})}\right)^{1/2} \mathbf{e}(j,\mathbf{q}) \times [b(j,\mathbf{q}) e^{i\mathbf{q}\cdot\mathbf{R}_k^0} + b^+(j,\mathbf{q}) e^{-i\mathbf{q}\cdot\mathbf{R}_k^0}] \frac{(\mathbf{r}-\mathbf{R}_k^0)}{|\mathbf{r}-\mathbf{R}_k^0|^3}. \tag{8.4.56}$$

The further evaluation runs along the same lines as in b). In particular the matrix element corresponding to (8.4.26) now reads

$$\langle \lambda_1 \ldots \mathbf{K} \ldots \lambda_z, m | V_P | \lambda_1 \ldots \mathbf{K}' \ldots \lambda_z, m' \rangle = \sum_{\substack{\mathbf{q} \\ \{j\}=\mathrm{op}}} \left(\frac{\hbar}{2M\omega(j,\mathbf{q})}\right)^{1/2} [m(j,\mathbf{q}) + \tfrac{1}{2} \mp \tfrac{1}{2}]^{1/2} \delta_{m'(j,\mathbf{q}), m(j,\mathbf{q}) \mp 1} \prod_{\substack{j'\mathbf{q}' \\ \neq j\mathbf{q}}} \delta_{m'(j',\mathbf{q}'), m(j',\mathbf{q}')} I^{\pm} \tag{8.4.57}$$

with

$$I^{\pm} := \mathbf{e}(j,\mathbf{q}) \cdot \int \psi(\mathbf{r},\mathbf{K})^{\times} \sum_k e^{\pm i\mathbf{q}\cdot\mathbf{R}_k^0} (-ee^*) \frac{(\mathbf{r}-\mathbf{R}_k^0)}{|\mathbf{r}-\mathbf{R}_k^0|^3} \psi(\mathbf{r},\mathbf{K}') d^3r. \tag{8.4.58}$$

The potential energy corresponding to (8.4.31) is therefore defined by

$$\mathbf{v}^{\pm}(\mathbf{r}) = \sum_k e^{\pm i\mathbf{q}\cdot(\mathbf{R}_k^0 - \mathbf{r})} (-ee^*) \frac{(\mathbf{r}-\mathbf{R}_k^0)}{|\mathbf{r}-\mathbf{R}_k^0|^3} \tag{8.4.59}$$

and if the sum in (8.4.59) is evaluated in the integral approximation we obtain

$$\mathbf{v}^{\pm}(\mathbf{r}) \approx \pm i(-ee^*)\frac{4\pi}{V_0}\frac{\mathbf{q}}{q^2} \tag{8.4.60}$$

where V_0 is the volume of the unit cell. For transversal optical modes $\mathbf{e}(j,\mathbf{q})$ is orthogonal to $\mathbf{q}$. Hence, if (8.4.60) is substituted into (8.4.33) we obtain for longitudinal modes with $\mathbf{e}(j,\mathbf{q})$ parallel to $\mathbf{q}$ and $\mathbf{e}(j,\mathbf{q}) \cdot \mathbf{q} = q$

$$\begin{aligned} I^{\pm} &= \mp \Delta(\mathbf{K}-\mathbf{K}' \pm \mathbf{q}) i(ee^*) \frac{4\pi}{V_0} \\ &\times \frac{1}{q} \int_{V_0} \exp\left[i[\mathbf{K}-\mathbf{K}' \pm \mathbf{q}) \cdot \mathbf{r}\right] u(\mathbf{r},\mathbf{K})^{\times} u(\mathbf{r},\mathbf{K}') d^3r. \end{aligned} \tag{8.4.61}$$

As in the long wave limit no Umklapp-processes occur, we have $(\mathbf{K}-\mathbf{K}' \pm \mathbf{q}) = 0$ and hence

$$I^{\pm} = \mp i(ee^*)\frac{4\pi}{V_0}\frac{1}{q}\Delta(\mathbf{K}-\mathbf{K}' \pm \mathbf{q}) \int_{V_0} u(\mathbf{r},\mathbf{K})^{\times} u(\mathbf{r},\mathbf{K}') d^3r. \tag{8.4.62}$$

If for simplicity free electron states are used, the overlap integral gives just V_0/V and with $\Delta \equiv \Delta^2$ the transition probability for V_P reads

$$\begin{aligned} &\frac{2\pi}{\hbar} \sum_{mm'} |\langle nm| V_P |n'm'\rangle|^2 f_{m'} \delta_\gamma (E_m^n - E_{m'}^{n'}) \\ &= \frac{1}{V^2} \sum_{\mathbf{q}} \frac{(2\pi)^2 (ee^* 4\pi)^2}{2\hbar M_0 \omega(j,\mathbf{q}) q^2} \Delta(\mathbf{K}-\mathbf{K}' \pm \mathbf{q}) [\bar{m}(j,\mathbf{q}) + \tfrac{1}{2} \mp \tfrac{1}{2}] \\ &\times \delta_\gamma [E(\mathbf{K}) - E(\mathbf{K}') \mp \hbar\omega(j,\mathbf{q})]. \end{aligned} \tag{8.4.63}$$

By reducing the sum over $\mathbf{K}'$ by means of $\Delta(\mathbf{K}-\mathbf{K}' \pm \mathbf{q})$ we obtain for the optical part of the collision term of (8.2.40) the expression

$$\begin{aligned} \sum_{\mathbf{K}'} \{\ \}_{\text{Coll}}^{\text{op}} = \sum_{\mathbf{q}} \{ & W_-^*(\mathbf{K},\mathbf{K}-\mathbf{q}) f(\mathbf{K}-\mathbf{q}) [1 - f(\mathbf{K})] \\ & + W_+^*(\mathbf{K},\mathbf{K}+\mathbf{q}) f(\mathbf{K}+\mathbf{q}) [1 - f(\mathbf{K})] \\ & - W_-^*(\mathbf{K}-\mathbf{q},\mathbf{K}) f(\mathbf{K}) [1 - f(\mathbf{K}-\mathbf{q})] \\ & - W_+^*(\mathbf{K}+\mathbf{q},\mathbf{K}) f(\mathbf{K}) [1 - f(\mathbf{K}+\mathbf{q})] \} \end{aligned} \tag{8.4.64}$$

where $W_{\pm}^*$ mean the transition probabilities (8.4.63) from which Δ has been removed. In the linear approximation of Theorem 8.4 the collision term becomes with (8.2.42)

$$\begin{aligned} \sum_{\mathbf{K}'} \{\ \}_{\text{Coll}}^{\text{op}} &\approx -\frac{1}{k_B T} \sum_{\mathbf{K}'} W_0(\mathbf{K}',\mathbf{K}) [\Phi(\mathbf{K}) - \Phi(\mathbf{K}')] \\ &= \sum_{\mathbf{q}} \{ W_{0-}^*(\mathbf{K}-\mathbf{q},\mathbf{K}) [\Phi(\mathbf{K}) - \Phi(\mathbf{K}-\mathbf{q})] \\ &\quad + W_{0+}^*(\mathbf{K}+\mathbf{q},\mathbf{K}) [\Phi(\mathbf{K}) - \Phi(\mathbf{K}+\mathbf{q})] \} \end{aligned} \tag{8.4.65}$$

with

$$W^*_{0\pm}(\mathbf{K}\pm\mathbf{q},\mathbf{K}) := W^*_{\pm}(\mathbf{K}\pm\mathbf{q},\mathbf{K}) f_0(\mathbf{K}\pm\mathbf{q})[1-f_0(\mathbf{K})]. \tag{8.4.66}$$

If in the long wave limit $\omega(j,\mathbf{q})\approx\omega_0$ is assumed, the transition probabilities W^* are given by

$$W^*_{\pm}(\mathbf{K}\pm\mathbf{q},\mathbf{K}) := \frac{1}{V^2}\frac{(2\pi)^2(ee^*4\pi)^2}{2\hbar M_0\omega_0 q^2}[\bar{m}(j,\mathbf{q})+\tfrac{1}{2}\mp\tfrac{1}{2}] \tag{8.4.67}$$
$$\times\,\delta_\gamma[E(\mathbf{K}\pm\mathbf{q})-E(\mathbf{K})\mp\hbar\omega_0]$$

and since the phonon thermal equilibrium distribution depends only on $\omega(j,\mathbf{q})\approx\omega_0$ the average values $\bar{m}(j,\mathbf{q})$ do not depend on $\mathbf{q}$, i.e., $\{\bar{m}(j,\mathbf{q})\approx\bar{m}\forall j,\mathbf{q}=\text{opt.}\}$. In (8.4.65) the summation over $\mathbf{q}$ is taken over the first Brillouin zone. Due to the symmetry of this zone a summation over $\mathbf{q}$ or $-\mathbf{q}$, resp. is completely equivalent. Hence $\sum_{\mathbf{q}}$ can be replaced by $\sum_{-\mathbf{q}}$ and if this is done for the W_- term in (8.4.65), by substitution of (8.4.67) into (8.4.65) the following expression results

$$\sum_{\mathbf{K}'}\{\ \}^{\text{op}}_{\text{Coll}} = -\frac{(8\pi^2 ee^*)^2}{V^2 2\hbar\omega_0 M_0}\frac{1}{k_B T}\sum_{\mathbf{q}}\frac{1}{q^2}f_0(\mathbf{K}+\mathbf{q})\,[1-f_0(\mathbf{K})]$$
$$\times\{(\bar{m}+1)\delta_\gamma[E(\mathbf{K})-E(\mathbf{K}+\mathbf{q})+\hbar\omega_0] \tag{8.4.68}$$
$$+\bar{m}\delta_\gamma[E(\mathbf{K})-E(\mathbf{K}+\mathbf{q})-\hbar\omega_0]\}[\Phi(\mathbf{K})-\Phi(\mathbf{K}+\mathbf{q})].$$

We now replace the summation over $\mathbf{q}$ by an integration which removes the factor V^{-1} from (8.4.68) and make use of the formula (8.2.50). Since f_0 depends only on $E(\mathbf{K})$ the following integrals result from (8.4.68)

$$\frac{(8\pi^2 ee^*)^2}{V2\hbar\omega_0 M_0}\frac{\partial f_0}{\partial E}\int\frac{1}{q^2}\frac{f_0(E\pm\hbar\omega_0)}{f_0(E)}(\bar{m}+\tfrac{1}{2}\pm\tfrac{1}{2})$$
$$\times\delta_\gamma[E(\mathbf{K})-E(\mathbf{K}+\mathbf{q})\pm\hbar\omega_0][\Phi(\mathbf{K})-\Phi(\mathbf{K}+\mathbf{q})]d^3q. \tag{8.4.69}$$

According to Wilson (1953), the integrals can be evaluated if the following suppositions are made:

i) the conduction band is parabolic;
ii) the first Brillouin zone is approximated by a sphere of radius q_0;
iii) only an electric field is admitted as external force.

If we assume that a relaxation time exists then due to (8.2.52) we have $\Phi(\mathbf{K})=\mathbf{K}\cdot\mathbf{G}$ where the vector $\mathbf{G}$ is parallel to $\mathbf{E}$. Without loss of generality, it can be assumed that $\mathbf{E}$ points in the $\mathbf{e}_1$ direction. Then $\Phi(\mathbf{K})$ can be written

$$\Phi(\mathbf{K})=\mathbf{K}\cdot\mathbf{G}=k\cos\Theta c(E) \tag{8.4.70}$$

where Θ is the angle between $\mathbf{K}$ and $\mathbf{e}_1$. Further we obtain

$$\Phi(\mathbf{K}+\mathbf{q})=(\mathbf{K}+\mathbf{q})\cdot\mathbf{G}=(k\cos\Theta+q\cos\Theta_1)c(E+\hbar\omega_0) \tag{8.4.71}$$

where Θ_1 is the angle between $\mathbf{q}$ and $\mathbf{e}_1$. For parabolic energy bands we have $E(\mathbf{K}) = \mathbf{K}^2/2m$ and hence

$$E(\mathbf{K}+\mathbf{q})-E(\mathbf{K})\pm\hbar\omega_0 = q\frac{dE}{dk}\cos\vartheta+\frac{1}{2}q^2\frac{d^2E}{dk^2}\pm\hbar\omega_0 \tag{8.4.72}$$

where ϑ is the angle between $\mathbf{K}$ and $\mathbf{q}$. We now introduce polar coordinates q, ϑ, φ in $\mathbf{q}$-space, the polar axis being in the direction of $\mathbf{K}$, and $d^3q = q^2 \sin\vartheta\, dq d\vartheta\, d\varphi$. In this coordinate system one has

$$q\cos\Theta_1 = q\cos\Theta\cos\vartheta + q\sin\Theta\sin\vartheta\cos\varphi \tag{8.4.73}$$

and if (8.4.73) is substituted into (8.4.71) and (8.4.71) (8.4.70) are substituted into (8.4.69), by integration over φ the last term of (8.4.73) vanishes, while all other terms give 2π. Hence the integrals (8.4.69) go over into

$$\begin{aligned}&\frac{(2\pi)^3(ee^*)^2}{V2\hbar\omega_0 M_0}\frac{df_0}{dE}\int\frac{f_0(E\pm\hbar\omega_0)}{f_0(E)}(\bar{m}+\tfrac{1}{2}\pm\tfrac{1}{2})\\ &\times\delta_\gamma\left(q\frac{dE}{dk}\cos\vartheta+\frac{1}{2}q^2\frac{d^2E}{dk^2}\pm\hbar\omega_0\right)\\ &\times\cos\Theta\,[kc(E)-(k+q\cos\vartheta)c(E+\hbar\omega_0)]\sin\vartheta\, dq d\vartheta.\end{aligned} \tag{8.4.74}$$

With $\cos\vartheta = x$ the integral over ϑ can be evaluated and gives, due to the δ-distribution

$$\cos\vartheta = -\left[\frac{1}{2}q^2\frac{d^2E}{dk^2}\pm\hbar\omega_0\right]\left(q\frac{dE}{dk}\right)^{-1} \tag{8.4.75}$$

and due to $\cos\Theta = k_1/k$, we obtain for the collision term the expression

$$\begin{aligned}\sum_{\mathbf{K}'}\{\ \}^{\text{op}}_{\text{Coll}} = &-\frac{(2\pi)^3(ee^*)^2k_1}{V2\hbar\omega_0 M_0(dE/dk)}\frac{df_0}{dE}\\ &\times\int_{q_1}^{q_2}\left\{(\bar{m}+1)\frac{f_0(E+\hbar\omega_0)}{f_0(E)}\left[\left(1+\frac{\hbar\omega_0}{2E}-\frac{q^2}{2k^2}\right)c(E+\hbar\omega_0)-c(E)\right]\right.\\ &\left.+\bar{m}\frac{f_0(E-\hbar\omega_0)}{f_0(E)}\left[\left(1-\frac{\hbar\omega_0}{2E}-\frac{q^2}{2k^2}\right)c(E-\hbar\omega_0)-c(E)\right]\right\}\frac{dq}{q}.\end{aligned} \tag{8.4.76}$$

From this expression Howarth and Sondheimer (1953) derived a relaxation time by the following procedure.

First the limits of integration q_1, q_2 of the integral in (8.4.76) have to be considered. The limits of integration q_1, q_2 in (8.4.76) are determined by the requirement that the angle ϑ for which $E(\mathbf{K})-E(\mathbf{K}+\mathbf{q})\pm\hbar\omega_0 = 0$ must be a real angle. This leads to the condition

$$|\cos\vartheta| = \left|-\frac{q}{2k}\pm\frac{m\hbar\omega_0}{\hbar^2 kq}\right| \leqslant 1 \tag{8.4.77}$$

where the upper sign refers to the first set of terms in (8.4.76) and the lower sign to the second set. The values of q_1 and q_2 are thus obtained by solving a quadratic equation

For the first set of terms it is found that

$$\begin{aligned} q_1 &= (2m/\hbar^2)^{1/2}[(E+\hbar\omega_0)^{1/2}-E^{1/2}]; \\ q_2 &= (2m/\hbar^2)^{1/2}[(E+\hbar\omega_0)^{1/2}+E^{1/2}] \end{aligned} \tag{8.4.78}$$

while, for the second set the values

$$\begin{aligned} q_1 &= (2m/\hbar^2)^{1/2}[E^{1/2}-(E-\hbar\omega_0)^{1/2}]; \\ q_2 &= (2m/\hbar^2)^{1/2}[E^{1/2}+(E-\hbar\omega_0)^{1/2}] \end{aligned} \tag{8.4.79}$$

result, if $E \geqq \hbar\omega_0$, whereas there are no real values of q_1 and q_2 when $E < \hbar\omega_0$. The second set of terms is therefore entirely absent when $E < \hbar\omega_0$, corresponding to the fact that an electron with energy $E < \hbar\omega_0$ cannot emit a lattice quantum. In addition, values of q greater than q_0 are not allowed. For all except very high energies, however, this condition is less restrictive than the limits given by (8.4.78) (8.4.79), and the latter are therefore to be used as the limits of the effective lattice vibrations for polar crystals. Combining (8.4.76) and (8.4.78) (8.4.79) and performing the integration over q, the final expression for the collision operator is found to be

$$\sum_{\mathbf{K}'}\{\ \}^{\mathrm{op}}_{\mathrm{Coll}} = -\frac{(2\pi)^3(ee^*)k_1}{V2\hbar\omega_0 M_0(dE/dk)}\frac{df_0}{dE}L[c] \tag{8.4.80}$$

where

$$\begin{aligned} L[c] = {} & (\bar{m}+1)\frac{f_0(E+\hbar\omega_0)}{f_0(E)}\{c(E+\hbar\omega_0)[(2E+\hbar\omega_0)\sinh^{-1}(E/\hbar\omega_0)^{1/2} \\ & -(E^2+E\hbar\omega_0)^{1/2}]-2Ec(E)\sinh^{-1}(E/\hbar\omega_0)^{1/2}\} \\ & +h(E-\hbar\omega_0)\bar{m}\frac{f_0(E-\hbar\omega_0)}{f_0(E)}\{c(E-\hbar\omega_0)[(2E-\hbar\omega_0)\cosh^{-1}(E/\hbar\omega_0)^{1/2} \\ & -(E^2-E\hbar\omega_0)^{1/2}]-2Ec(E)\cosh^{-1}(E/\hbar\omega_0)^{1/2}\}. \end{aligned} \tag{8.4.81}$$

$h(x)$ being Heaviside's unit function $h(x)=0$ if $x<0$, $h(x)=1$, $x>0$ and the positive value of $\cosh^{-1}(E/\hbar\omega_0)^{1/2}$ being understood.

It will be shown that a universal time of relaxation can be uniquely defined provided that the temperature is sufficiently high. When $kT > \hbar\omega_0$ the collision operator can be expanded in powers of $\hbar\omega_0/kT$ and $\hbar\omega_0/E$ (except for very slow electrons for which $E<\hbar\omega_0$ which, at high temperatures, make a negligible contribution to the conductivity). Since

$$\begin{aligned} & c(E\pm\hbar\omega_0)[(2E\pm\hbar\omega_0)\begin{matrix}\sinh^{-1}\\ \cosh^{-1}\end{matrix}(E/\hbar\omega_0)^{1/2}-(E^2\pm E\hbar\omega_0)^{1/2}] \\ & -2Ec(E)\begin{matrix}\sinh^{-1}\\ \cosh^{-1}\end{matrix}(E/\hbar\omega_0)^{1/2} = -Ec(E)\left[1+0\left(\frac{\hbar\omega_0}{E}\ln\frac{\hbar\omega_0}{E}\right)\right] \end{aligned} \tag{8.4.82}$$

and

$$(\bar{m}+\tfrac{1}{2}\pm\tfrac{1}{2})\frac{f_0(E\pm\hbar\omega_0)}{f_0(E)} = \frac{k_B T}{\hbar\omega_0}\left[1+0\left(\frac{\hbar\omega_0}{k_B T}\right)\right] \tag{8.4.83}$$

it follows that (8.4.81) reduces to

$$L[c] = -2Ek_BTc(E)/\hbar\omega_0 \tag{8.4.84}$$

if only the leading terms are kept. Thus

$$\sum_{\mathbf{K}'} \{\ \}^{\text{op}}_{\text{Coll}} = -(f-f_0)\frac{2\pi(2m)^{1/2}(ee^*)^2 k_B T}{VM_0(\hbar\omega_0)^2 E^{1/2}}. \tag{8.4.85}$$

A time of relaxation therefore exists, and is given by

$$\tau_{\text{op}} = \frac{VM_0(\hbar\omega_0)^2 E^{1/2}}{2\pi(2m)^{1/2}(ee^*)^2 k_B T}. \tag{8.4.86}$$

The corresponding mean free path is $l = \tau v = \tau(2E/m)^{1/2}$, and is proportional to E/T. For very high energies such that $2k > q_0$, where the condition $q \leqslant q_0$ is more restrictive than the limits given by (8.4.78) (8.4.79), it is easily seen that (8.4.86) must be multiplied by $(2k/q_0)^2$, giving

$$\tau^*_{\text{op}} = (2k/q_0)^2 \tau_{\text{op}} \tag{8.4.87}$$

Since a time of relaxation exists at high temperatures, the formal theory of conductivity applies, and the expressions for the various transport phenomena may be obtained at once for arbitrary degrees of degeneracy of the electron gas, by substituting (8.4.86) or (8.4.87) into the standard formulae of the formal theory.

The foregoing results hold whatever the degree of degeneracy of the electrons, provided that the temperature is sufficiently high. In the particular case of classical statistics, however, we have $f_0(E) = \exp\left[-(E-\zeta)/k_BT\right]$, and in this case (8.4.80) reduces to the more general expression

$$\sum_{\mathbf{K}'} \{\ \}^{\text{op}}_{\text{Coll}} = -(f-f_0)\frac{\pi(2m)^{1/2}(ee^*)^2}{VM_0\hbar\omega_0 E^{1/2}}(2\bar{m}+1) \tag{8.4.88}$$

at all temperatures, provided only that $E \gg \hbar\omega_0$. The result (8.4.88) (and the corresponding expression for $2k > q_0$) was obtained by Fröhlich (1937) in developing his theory of dielectric breakdown in polar crystals.

The use of (8.4.88) to discuss the conductivity of degenerate semiconductors at low temperatures is not justified. At low temperatures the theory becomes much more difficult, since a universal time of relaxation can no longer be defined for the conduction phenomena, and much more refined mathematical methods are required to obtain a solution.

8.5 Conductivity of polar semiconductors

Among the partially covalent semiconductors, the so-called III–V-compounds are of special technical interest. In a narrow sense these substances are the so-called polar semiconductors and their state representations can be obtained by the methods described in Section 5.7. Hence they can be treated in the theoretical framework developed here. Of particular interest are the electronic transport phenomena which we will treat in this section. A characteristic feature of the III–V-compounds are their complicated energy bands which influence the transport phenomena decisively. These energy bands are described in detail in Section 6.6. The energetic distances are shown in Figure 8.3, for the six most interesting substances.

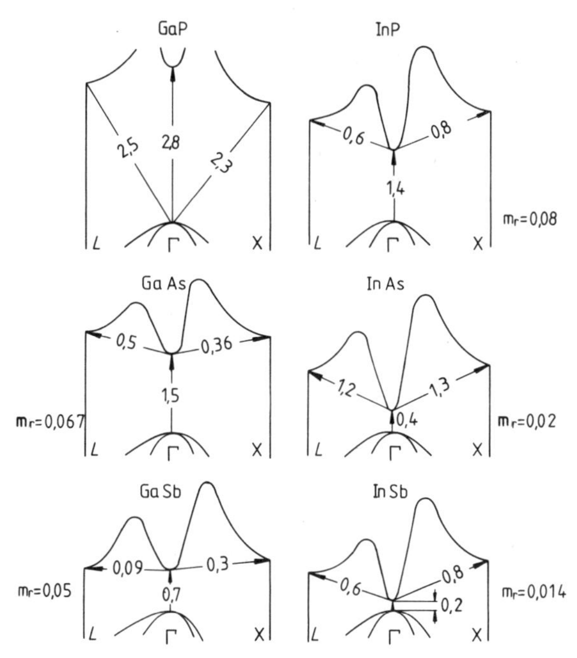

Figure 8.3:
Energy band structure of some III–V compounds. The maximum of the valence band at the center of the Brillouin zone (Γ-point) and the minima of the conduction band at the Γ-point, at the boundary of the Brillouin zone in (1,0,0)-direction (X-point) and at the boundary of the Brillouin zone in (1,1,1)-direction (L-point) are given. The arrows indicate the energy differences im eV. The effective mass m^* is given in units of the free electron mass m_0. The compounds InAs and InSb are usually called as one-valley-semiconductors, as the energy gaps between the various conduction band valleys are considerably larger than the energy gap between the lowest conduction band valley and the valence band. In this case the higher conduction band valleys play no significant role for the conduction process in contrast to the other compounds which are usually called many-valley semiconductors where several conduction band valleys participate in the conduction process; after Kranzer, Hillbrand, Pötzl and Zimmerl (1972).

The maxima of the valence bands lie at the Γ-point (center of the Brillouin-zone), while the deepest minima of the conduction band lie at the X-point (boundary of the Brillouin-zone in 100-direction) and the L-point (boundary of the Brillouin-zone in the 111-direction). Except for GaP the deepest valley is at the Γ-point with an especially strong curvature. This causes an especially small effective mass and an especially high electron mobility. Usually InAs and InSb are called one-valley semiconductors as the energy differences between higher valleys and the Γ-valley are larger than the energetic distance of the Γ-valley to the valence band. In this case electron excitations in the lowest conduction band are primarily transferred into electron-hole production processes and not into transition processes into higher valleys. In contrast to these substances, GaSb, for instance, exhibits such small differences between the conduction band minima that all valleys contribute to the conductivity, i.e., are populated even for ordinary temperatures. For a successful theoretical treatment of the transport phenomena in these substances we have to extend the one-valley formalism of the preceding section to a many-valley formalism. The theoretical transition from the one-valley case to the general many-valley case can be subdivided into several steps of increasing complexity:

i) one valley with non-parabolic, non-spherical form;
ii) many valleys with non-parabolic, non-spherical form, but with no intraband transitions;
iii) many valleys with non-parabolic, non-spherical form including intraband transitions.

We will chiefly study case i), which is realized for InSb and related substances. The band structure of InSb has been described by Kane (1957). For not very high energies the effect of higher bands is negligible and the conduction and valence band energies are described by the following secular equation

$$E'(E'+E_0)(E'+E_0+\Delta)-k^2P^2(E'+E_0+\tfrac{2}{3}\Delta)=0 \tag{8.5.1}$$

where the definitions $E':=E-\hbar^2k^2/2m_0$, $m_0:=$ free electron mass, $E_0:=$ energy gap, $\Delta:=$ spin-orbit splitting; $P:=$ conduction band-valence band interaction are used. We concentrate on the solutions of (8.5.1) for the conduction band. For $E' \ll E_0+\frac{2}{3}\Delta$, the resulting equation becomes approximately

$$E'(1+E'/E_0)=\hbar^2k^2/2m_e \tag{8.5.2}$$

with

$$m_e^{-1}:=(4P^2/3\hbar^2E_0)(\tfrac{2}{3}\Delta+E_0)(\Delta+E_0)^{-1}. \tag{8.5.3}$$

For InSb at low temperature $m_e=0{,}0145\ m_0$, $E_0=0{,}236$ eV, $\Delta\approx 0{,}9$ eV, Pidgeon, Brown (1966), so that the assumption made above is satisfied to a good approximation. Kane's theoretical description is essentially valid for the conduction bands of other III–V-compounds, although the relative influence of other bands may be stronger than in InSb.

a) Relaxation time calculations

The example of InSb shows that for this interesting substance it is sufficient to consider one non-parabolic but spherical valley. With a slight generalization we write for a non-parabolic and ellipsoidal energy band the formula

$$\gamma(E)=a_{\alpha\beta}k_\alpha k_\beta. \tag{8.5.4}$$

Zukotynski and Kolodziejczak (1963). The arbitrary function $\gamma(E)$ describes the non-parabolicity of the band. Here and in all subsequent formulas the summation convention is used. The evaluation of the Boltzmann equation and the transport quantities has now to be performed with such generalized energy bands. Concerning this evaluation we follow the approach of Zawadzki (1974). According to (8.2.40) the stationary Boltzmann equation reads

$$\hbar^{-1}\mathbf{F}\cdot\nabla_{\mathbf{K}}f+\mathbf{v}\cdot\nabla_{\mathbf{r}}f+\{\quad\}=0 \tag{8.5.5}$$

where $\{\quad\}$ denotes the collision term and where $\mathbf{F}$ is the Lorentz force

$$\mathbf{F}:=e\mathbf{E}+\frac{e}{c}(\mathbf{v}\times\mathbf{B}) \tag{8.5.6}$$

and for electrons $e=-|e|$. A solution of (8.5.5) is assumed to be of the form (8.2.41). If temperature and carrier concentration depend upon position $\mathbf{r}$ we obtain

$$\partial_l f_0=-\frac{\partial f_0}{\partial E}\left[\frac{E-\zeta}{T}\partial_l T+\left(\frac{\partial\zeta}{\partial T}\right)_n\partial_l T+\left(\frac{\partial\zeta}{\partial n}\right)_T\partial_l n\right] \tag{8.5.7}$$

and with (8.2.41), (8.2.42) and the linearized version of the collision term in the relaxation time approximation (8.2.49), equation (8.5.5) can be approximated by the equation

$$\frac{e}{\hbar c}(\mathbf{v}\times\mathbf{B})\cdot\nabla_{\mathbf{K}}\Phi+\left[e\mathbf{E}'-C\nabla_{\mathbf{r}}T-\left(\frac{\partial\zeta}{\partial n}\right)_T\nabla_{\mathbf{r}}n\right]\cdot\mathbf{v}+\Phi\tau^{-1}=0 \tag{8.5.8}$$

with

$$C:=\frac{(E-\zeta)}{T};\ \mathbf{E}':=\mathbf{E}-\frac{1}{e}\nabla_{\mathbf{r}}\zeta. \tag{8.5.9}$$

Equation (8.5.8) can be rewritten to give

$$\frac{e}{\hbar c}\varepsilon_{\alpha\beta\gamma}B_\alpha\frac{\partial\Phi}{\partial k_\beta}v_\gamma+\tau^{-1}\Phi+W_\gamma v_\gamma=0 \tag{8.5.10}$$

where $\varepsilon_{\alpha\beta\gamma}$ is the completely antisymmetric tensor, and

$$W_\gamma:=eE'_\gamma-C\partial_\gamma T-\left(\frac{\partial\zeta}{\partial n}\right)_T\partial_\gamma n. \tag{8.5.11}$$

The solution of equation (8.5.10) is assumed to be linear in W_γ and to have the form

$$\Phi=-\tau g_{\lambda\mu}v_\lambda W_\mu \tag{8.5.12}$$

where $g_{\lambda\mu}$ is an unknown tensor depending only on the energy, so that

$$\frac{\partial\Phi}{\partial k_\beta}=-\tau g_{\eta\mu}W_\mu\frac{\partial v_\eta}{\partial k_\beta}-\hbar v_\eta v_\beta\frac{d}{dE}(\tau g_{\eta\mu}W_\mu). \tag{8.5.13}$$

The components of the vector W_μ are functions of energy only. Inserting equation (8.5.13) into equation (8.5.10) and observing that $\varepsilon_{\alpha\beta\gamma} v_\beta v_\gamma = 0$ one obtains

$$\left(\frac{e\tau}{\hbar c}\varepsilon_{\alpha\beta\gamma} B_\alpha g_{\eta\mu}\frac{\partial v_\eta}{\partial k_\beta} + g_{\gamma\mu}\right) v_\gamma W_\mu = \delta_{\gamma\mu} v_\gamma W_\mu. \tag{8.5.14}$$

By definition we have $\mathbf{v} = \hbar^{-1}\nabla_{\mathbf{K}} E$ and from this it follows with (8.5.4) that

$$\hbar^{-1}\frac{\partial v_\eta}{\partial k_\beta} = \hbar^{-2}\frac{\partial^2 E}{\partial k_\eta \partial k_\beta} = m_{\eta\beta}^{-1} \tag{8.5.15}$$

with

$$m_{ij}^{-1} = 2\hbar^{-2}\left(\frac{d\gamma^{-1}(a_{\alpha\beta}k_\alpha k_\beta)}{d(a_{\alpha\beta}k_\alpha k_\beta)}\right) a_{ij}. \tag{8.5.16}$$

The expression (8.5.15) is the mass tensor. Introducting the mobility μ as

$$\mu := 2\frac{e\tau}{\hbar^2}\left(\frac{d\gamma^{-1}(a_{\alpha\beta}k_\alpha k_\beta)}{d(a_{\alpha\beta}k_\alpha k_\beta)}\right) \tag{8.5.17}$$

one obtains from equation (8.5.14)

$$\frac{\mu}{c}\varepsilon_{\alpha\beta\gamma} B_\alpha a_{\beta\eta} g_{\eta\mu} + g_{\gamma\mu} = \delta_{\gamma\mu}. \tag{8.5.18}$$

The solution of this equation is given by

$$g_{\gamma\mu} = \frac{1}{\Delta}\left(\delta_{\gamma\mu} - \frac{\mu}{c}\varepsilon_{\alpha\beta\gamma} B_\alpha a_{\beta\mu} + \frac{\mu^2}{c^2} A B_\alpha B_\mu a_{\gamma\alpha}^{-1}\right) \tag{8.5.19}$$

with

$$\Delta := 1 + \frac{\mu^2}{c^2} A B_\alpha B_\gamma a_{\gamma\alpha}^{-1} \tag{8.5.20}$$

and $A = \det|a_{\alpha\beta}|$. Knowing the tensor $g_{\gamma\mu}$ and hence from equation (8.5.12) the function Φ, one can calculate the currents. According to (8.3.3) the electric current is

$$J_\nu = \frac{2}{8\pi^3} e \int v_\nu f(\mathbf{K},\mathbf{r}) d^3k. \tag{8.5.21}$$

If we add the lattice thermal conductivity to the electronic thermal conductivity, according to (8.3.19) and (7.6.47) the heat current reads

$$Q_\nu = \frac{2}{8\pi^3}\int v_\nu (E-\zeta) f(\mathbf{K},\mathbf{r}) d^3k - \kappa_{\nu\mu}\partial_\mu T. \tag{8.5.22}$$

Taking into account that the equilibrium distribution f_0 does not contribute to the currents we get from equations (8.2.41) and (8.5.12)

$$J_\nu = \frac{2e}{8\pi^3}\int -\frac{\partial f_0}{\partial E}\tau g_{\eta\mu} v_\eta v_\nu W_\mu d^3k. \tag{8.5.23}$$

From the definition of $\mathbf{v}$ we obtain

$$v_\eta v_\nu = \hbar^2 m_{\eta\alpha}^{-1} m_{\nu\beta}^{-1} k_\alpha k_\beta \tag{8.5.24}$$

and since all other quantities in (8.5.23) depend only on E, it is convenient to introduce a spherical coordinate system with E, Θ, φ for integration. If the heat flux is treated along the same lines one obtains finally

$$J_\nu = \sigma_{\nu\mu} E'_\mu - \Theta_{\gamma\nu} \partial_\mu T - e D_{\nu\mu} \partial_\mu n \tag{8.5.25}$$

$$Q_\nu = \chi_{\nu\mu} E'_\mu - (\xi_{\nu\mu} + \kappa_{\nu\mu}) \partial_\mu T - e B_{\nu\mu} \partial_\mu n \tag{8.5.26}$$

where the three basic tensors $\sigma_{\nu\mu}$, $\Theta_{\nu\mu}$ and $\xi_{\nu\mu}$ are given by

$$\sigma_{\nu\mu} := \frac{e}{4\pi^3} a_{\nu\beta} I_{\alpha\beta} \langle G_{\alpha\mu} \rangle \tag{8.5.27}$$

$$\Theta_{\nu\mu} := \frac{1}{4\pi^3} T^{-1} a_{\nu\beta} I_{\alpha\beta} \langle (E-\zeta) G_{\alpha\mu} \rangle \tag{8.5.28}$$

$$\xi_{\nu\mu} := \frac{1}{4\pi^3} e T a_{\nu\beta} I_{\alpha\beta} \langle (E-\zeta)^2 G_{\alpha\mu} \rangle \tag{8.5.29}$$

with

$$I_{\alpha\beta} := \int_0^\pi \int_0^{2\pi} \frac{k_\alpha k_\beta}{k^2} \left(\frac{a_{ij} k_i k_j}{k^2} \right)^{-5/2} \sin \Theta d\Theta d\varphi \tag{8.5.30}$$

and

$$\langle A \rangle := \int_0^\infty -\frac{\partial f_0}{\partial E} A \gamma(E)^{3/2} dE \tag{8.5.31}$$

and

$$G_{\alpha\mu} = \mu a_{\eta\alpha} g_{\eta\mu}. \tag{8.5.32}$$

It can be shown that

$$a_{\nu\beta} I_{\alpha\beta} = \tfrac{1}{3} I \delta_{\nu\beta} \tag{8.5.33}$$

with

$$I = \int_0^\pi \int_0^{2\pi} \left(\frac{a_{\alpha\beta} k_\alpha k_\beta}{k^2} \right)^{-3/2} \sin \Theta d\Theta d\varphi. \tag{8.5.34}$$

Equations (8.5.25) (8.5.26) contain the description of the electric resistance, transverse and planar Halls effects, magnetoresistance, galvanomagnetic effects, thermoelectric power, longitudinal, planar and transverse Nernst-Ettinghausen effects, thermal resistivity, longitudinal, planar and transverse Maggi-Righi-Leduc effects, Nernst effect, longitudinal, planar and transverse Ettinghausen-Nernst effects. These effects were already partly discussed in Section 8.3, for parabolic energy bands. A full discussion of these effects on the phenomenological basis of equations (8.5.25) (8.5.26) would exceed the scope of this book and we refer to the literature, cf. for

instance Zawadzki (1973) and Seeger (1973). Here we discuss only the quantum statistical foundations of these formulas. Obviously the derivation of the general formulas (8.5.25) (8.5.26) and of a solution of (8.5.12) of the linearized Boltzmann equation depends upon the possibility of defining relaxation times.
Conduction band electrons can be scattered by

i) localized imperfections of the lattice (charged impurities, neutral impurities, dislocations);
ii) lattice vibrations (polar optical, non-polar optical, piezoacoustic, acoustic interactions);
iii) other free carriers (electrons, holes, quasiparticles).

One must distinguish between elastic and inelastic collisions. If the effective mass of heavy holes is much larger than that of electrons and if other quasiparticles are neglected, the only important inelastic processes are due to optical phonon-electron and electron-electron scattering processes. Strictly speaking for the latter processes relaxation times cannot be introduced. It is possible, however, to introduce a relaxation time at least approximately for some temperature ranges. Such calculations have been made by Zawadzki and Szymanska (1971) and we give a short review of this approach. Some results, which do not use the relaxation time concept, will be given later.
In the preceding section we embedded the relaxation time formalism in the many-particle state formulation of quantum statistical processes. In this section we content ourselves with the pure one-particle state formalism as it results from a naive interpretation of the Boltzmann equation. Concerning the many-particle aspects of the following calculations we thus refer to Section 8.4. The calculations are performed for InSb and we give a brief summary of the results of Kane's model which are relevant for these calculations. We will consider the energy range of formula (8.5.2) with $E' \ll E_0 + \frac{2}{3}\Delta$. Furthermore, we neglect the free electron term, since the effective mass in small-gap materials is just a small fraction of the free electron mass. Under these assumptions, equation (8.5.1) gives for the conduction band (for both spin orientations)

$$E(\mathbf{K}) = -\frac{E_0}{2} + \left[\left(\frac{E_0}{2}\right)^2 + E_0 \frac{\hbar^2 \mathbf{K}^2}{2m_0^*}\right]^{1/2} \tag{8.5.35}$$

where the effective mass at the bottom of the band is

$$m_0^{*\,-1} = \frac{4P^2}{3\hbar^2 E_0}(\Delta + \tfrac{3}{2}E_0)(\Delta + E_0)^{-1}. \tag{8.5.36}$$

The effective mass, as defined in (8.5.16) becomes

$$m^* = m_0^*\left(1 + 2\frac{E}{E_0}\right). \tag{8.5.37}$$

The initial unperturbed Hamiltonian for an electron is

$$H_0 = \frac{\mathbf{p}^2}{2m} + V_0(\mathbf{r}) + \frac{\hbar}{4m^2c^2}(\boldsymbol{\sigma} \times \nabla V_0) \cdot \mathbf{p} \tag{8.5.38}$$

where $V_0(\mathbf{r})$ is the periodic potential of the lattice and the last term represents the spin-orbit interaction in the standard notation. The eigenvalue problem

$$H_0 \psi(n, \mathbf{K}, j_3, \mathbf{r}) = E_n(\mathbf{K}, j) \psi(n, \mathbf{K}, j_3, \mathbf{r}) \tag{8.5.39}$$

is satisfied by the Bloch wave functions, normalized over the crystal volume V

$$\psi(n, \mathbf{K}, j_3, \mathbf{r}) = V^{-1/2} \exp(i\mathbf{K} \cdot \mathbf{r}) u(n, \mathbf{K}, j_3, \mathbf{r}). \tag{8.5.40}$$

Here n denotes the band (being interested in the conduction band alone we shall leave out the band index in the following), $\mathbf{K}$ is the wave vector, j_3 the projection of the total angular momentum on the $\mathbf{e}_3$-direction. The eigenvalue problem (8.5.39) is solved by the $\mathbf{K} \cdot \mathbf{p}$ method expanding $\mathbf{K}$-dependent Bloch amplitudes in terms of $\mathbf{K}$-independent Luttinger-Kohn amplitudes taken at the band's extremum, i.e., at $\mathbf{K}=0$ in our case. In the simplified model one takes four bands (eight including spin) neglecting all other bands. Thus we have at $\mathbf{K}=0$ an s-like conduction band ($j=1/2$) separated by the energy gap E_0 from degenerate light- and heavy-hole valence bands ($j=3/2$) and the p-like band ($j=1/2$) separated from the $j=3/2$ bands by the energy of the spin-orbit interaction Δ. Apart from E_0 and Δ, there is one more parameter in the theory, $P = -(i\hbar/m)\langle S|p_3|Z\rangle$ which characterizes the interband interaction. The functions S, X, Y, Z denote the Luttinger-Kohn amplitudes, which transform under the tetrahedral group like s, p_1, p_2, and p_3 atomic wave functions, respectively. For the Bloch functions in the form (8.5.40), they are normalized to $v_0^{-1}\langle S|S\rangle = v_0^{-1}\langle Z|Z\rangle = 1$ where the integration is over the volume of the unit cell v_0. The value of P remains almost constant in different III–V-compounds whereas the value of E_0/Δ varies widely.
In order to calculate the scattering transition probability one also needs to know the periodic components of the electron Bloch functions for the same range of energies. Under the above assumptions they are found to be, cf. Bowers and Yafet (1959)

$$\begin{aligned} u(\mathbf{K}, 1/2) &= \left(iaS - \frac{b - c2^{1/2}}{2} \frac{k_-}{k} R_- + \frac{b + c2^{1/2}}{2} \frac{k_-}{k} R_+ + c\frac{k_3}{k} Z \right) \eta_+ \\ &\quad - b\left(\frac{k_3}{k} R_+ - \frac{k_+}{k} 2^{-1/2} Z \right) \eta_- \\ u(\mathbf{K}, -1/2) &= \left(iaS + \frac{b + c2^{1/2}}{2} \frac{k_-}{k} R_- - \frac{b - c2^{1/2}}{2} \frac{k_-}{k} R_+ + c\frac{k_3}{k} Z \right) \eta_- \\ &\quad + b\left(\frac{k_3}{k} R_- - \frac{k_-}{k} 2^{-1/2} Z \right) \eta_+ \end{aligned} \tag{8.5.41}$$

where $k_\pm := k_1 \pm ik_2$ and $R_\pm = (X \pm iY)2^{-1/2}$ and η_+ and η_- denote the spin-up and spin-down functions, respectively.

The total angular momentum $j=1/2$ is quantized along the $\mathbf{e}_3$-direction (in the following we supress the subscript 3). The coefficients a, b and c are

$$a^2=\frac{E_0+E}{E_0+2E-\alpha E};\; b^2=\frac{1}{3}\frac{E}{E_0+2E-\alpha E}\beta^2$$
$$c^2=\frac{2}{3}\frac{E}{E_0+2E-\alpha E}\gamma^2 \tag{8.5.42}$$

where

$$\beta^2=\Delta^2(\Delta+E_0)^{-1}(\Delta+\tfrac{3}{2}E_0)^{-1}$$
$$\gamma^2=(\Delta+\tfrac{3}{2}E_0)(\Delta+E_0)^{-1} \tag{8.5.43}$$

and

$$\alpha=\Delta E_0 2^{-1}(\Delta+\tfrac{3}{2}E_0)^{-1}(\Delta+E_0)^{-1}.$$

Since $\alpha \lesssim 1/10$ for all values of the (E_0/Δ)-ratio, the term αE can be neglected in the above expressions. Hence, to a very good approximation

$$a^2=1-L;\quad b^2=\tfrac{1}{3}L\beta^2;\quad c^2=\tfrac{2}{3}L\gamma^2 \tag{8.5.44}$$

where $L=E/(E_0+2E)$. Normalization of the Bloch amplitudes (8.5.41) is equivalent to the condition $a^2+b^2+c^2=1$. It can be seen that both Bloch amplitudes are not pure spin functions due to the non-vanishing spin-orbit interaction ($b\neq 0$).

When calculating matrix elements of perturbing potentials we shall deal with integrals over the crystal volume of slowly varying envelope functions φ and quickly varying periodic amplitudes u. According to common procedure these integrals will be broken into

$$\int_V \varphi(\mathbf{r})u(\mathbf{r})d^3r=\frac{1}{v_0}\int_V \varphi(\mathbf{r})d^3r\int_{v_0} u(\mathbf{r})d^3r. \tag{8.5.45}$$

Transition probabilities for all scattering modes are calculated in the first Born approximation.

A systematic calculation of relaxation times for the main scattering processes in the nonparabolic conduction bands of InSb-type materials as described by the Kane model has been carried out first by Korenblit and Sherstobitov (1968), the improved calculation on this basis was made by Zawadzki and Szymanska (1971) which we present here.

α) Charged impurity scattering

We shall consider electron scattering by the screened Coulomb potential of a single ion:

$$V(\mathbf{r})=\frac{e^2}{\varepsilon_0}C(\mathbf{r})\exp(-r/\lambda) \tag{8.5.46}$$

where ε_0 is the static dielectric constant and λ the screening length. We have to calculate the matrix elements of this potential between initial and final electron states. Since in the Bloch amplitudes of (8.5.41), the spin variables are mixed with the coordinate variables, the spin-flip scattering transitions are also possible. There are four matrix elements

$$\langle \mathbf{K}',j'|V|\mathbf{K},j\rangle = V^{-1}\int_V V(\mathbf{r})\exp\left[i(\mathbf{K}-\mathbf{K}')\cdot\mathbf{r}\right]u(\mathbf{K}',j')^{\times}u(\mathbf{K},j)d^3r, \tag{8.5.47}$$

$\mathbf{K}',j'$ denoting a final electron state. Following equation (8.5.45) we shall break this integral into two, since the scattering potential extends over many unit cells. The resulting scalar product of appropriate Bloch amplitudes gives

$$\langle \mathbf{K}',\tfrac{1}{2}|V|\mathbf{K},\tfrac{1}{2}\rangle = V^{-1}[a^2+(b^2+c^2)\cos\Theta]\,\tilde{V}(|\mathbf{K}'-\mathbf{K}|) \tag{8.5.48}$$

$$\langle \mathbf{K}',-\tfrac{1}{2}|V|\mathbf{K},\tfrac{1}{2}\rangle = V^{-1}b(\tfrac{1}{2}b-c2^{1/2})\sin\Theta\exp(i\varphi)\,\tilde{V}(|\mathbf{K}'-\mathbf{K}|) \tag{8.5.49}$$

where Θ and φ are the polar angles of $\mathbf{K}'$ in the spherical coordinate system with the $\mathbf{K}$-direction taken as the polar axis. The slowly varying part is just the standard Fourier transform of the screened Coulomb potential,

$$\tilde{V}(|\mathbf{K}'-\mathbf{K}|) = \frac{4\pi e^2}{\varepsilon_0}\left[4k^2\sin^2\frac{\Theta}{2}+\lambda^{-2}\right]^{-1} \tag{8.5.50}$$

where we have put $\kappa = k'$. The matrix elements with the initial value of $j_3 = -1/2$ are the complex conjugates of those given by equations (8.5.48) and (8.5.49). Thus the transition probability from either $j_3 = 1/2$ or $j_3 = -1/2$ state is

$$\begin{aligned} W(\mathbf{K},\mathbf{K}') &= \frac{2\pi}{\hbar}\left(|\langle \mathbf{K}',\tfrac{1}{2}|V|\mathbf{K},\tfrac{1}{2}\rangle|^2 + |\langle \mathbf{K}',-\tfrac{1}{2}|V|\mathbf{K},\tfrac{1}{2}\rangle|^2\right)\delta[E(\mathbf{K}')-E(\mathbf{K})] \\ &= \frac{2\pi}{\hbar}\left(\frac{4\pi e^2}{\varepsilon_0}\right)^2 V^{-2}\left(4k^2\sin^2\frac{\Theta}{2}+\lambda^{-2}\right)^{-2}\delta[E(\mathbf{K}')-E(\mathbf{K})] \\ &\quad\times\left\{1-4a^2(1-a^2)\sin^4\frac{\Theta}{2}+[2b^2(2c-b2^{-1/2})^2-4(1-a^2)]\sin^2\frac{\Theta}{2}\cos^2\frac{\Theta}{2}\right\} \end{aligned} \tag{8.5.51}$$

where the δ-function indicates that the scattering is elastic. The total scattering probability is obtained by averaging over the initial states, and summing over all possible final states. Since, as we have already observed, the scattering probability from the $j_3 = 1/2$ state is equal to that from $j_3 = -1/2$, the average over the two initial states is equivalent to using just equation (8.5.51) without any change. The relaxation time is given by (8.4.8)

$$\tau(\mathbf{K})^{-1} = \frac{VN}{8\pi^3}\int W(\mathbf{K},\mathbf{K}')(1-\cos\Theta)d^3K' \tag{8.5.52}$$

where N is a number of charged ions in the volume V. We observe that

$$\delta[E(\mathbf{K}')-E(\mathbf{K})] = \frac{dk}{dE}\delta(k'-k) \tag{8.5.53}$$

which is directly related to the density of states $\varrho = \pi^{-2}k^2 dk/dE$. In the spherical coordinate system the only nontrivial integration is over Θ. The final result is, Zawadzki and Szymanska (1971)

$$\tau_i(E) = \frac{1}{2\pi}\frac{\varepsilon_0^2\hbar}{e^4 N_i} F_i^{-1}\frac{dE}{dk}k^2 \tag{8.5.54}$$

where $N_i = N/V$ is the concentration of charged ions. F_i is given by

$$\begin{aligned}F_i = &\ln(\xi+1) - \xi(\xi+1)^{-1} - (4L - fL^2)[1 + (\xi+1)^{-1} - 2\xi^{-1}\ln(\xi+1)] \\ &+ \tfrac{1}{2}L^2(4-f)[1 - 4\xi^{-1} + 6\xi^{-2}\ln(\xi+1) - 2\xi^{-1}(\xi+1)^{-1}]\end{aligned} \tag{8.5.55}$$

where

$$\xi = (2k\lambda)^2; \quad f = \frac{\beta^2}{9}(16\gamma^2 - 8\beta\gamma + \beta^2) \tag{8.5.56}$$

and L, β and γ are defined in equations (8.5.43) and (8.5.44). For $\Delta \gg E_0$, one has $f \approx 1$ and from this for the case of InSb where $E_0 = 0{,}2$ eV and $\Delta \approx 0{,}8$ eV, it follows that $f \approx 0.8$. For small electron energies, i.e., for $E \ll E_0$ we have $L \approx 0$ and only the first two terms are left in (8.5.55). This represents the result for a parabolic band. In the nonparabolic region of energies, i.e., for higher electron concentrations, one has $\xi \gg 1$ (in InSb at room temperature it varies from $\xi \approx 20$ for intrinsic samples to $\xi \approx 30$ for $n \approx 10^{19}$ cm^{-3}). Hence, to a good approximation equation (8.5.55), may be simplified to the following form

$$\begin{aligned}F_i = &\ln(\xi+1) - \xi(\xi+1)^{-1} - (4L - fL^2)[1 - 2\xi^{-1}\ln(\xi+1)] \\ &+ \tfrac{1}{2}L^2(4-f)(1 - 4\xi^{-1}).\end{aligned} \tag{8.5.57}$$

In InSb at $n \approx 10^{19}$ cm^{-3}, $L \approx 0{,}4$ and the last two terms resulting from the mixing of p-like functions reduce the value of F_i by about 30% increasing the corresponding mobility by the same proportion

β) Polar optical scattering

In III–V-compounds which have two different ions in a unit cell and no inversion symmetry, an electron can interact with lattice vibrations in a number of different ways. We first consider polar optical scattering. The perturbing potential can be derived from (8.4.50) in a continuum approximation, cf. Haug (1970), and reads

$$V_P = i\frac{4\pi e e^*}{v_0}\left(\frac{\hbar}{2NM\omega_0}\right)^{1/2}\sum_{\mathbf{q}}\frac{1}{q}[b(\mathbf{q})e^{i\mathbf{q}\cdot\mathbf{r}} - b^+(\mathbf{q})e^{-i\mathbf{q}\cdot\mathbf{r}}] \tag{8.5.58}$$

where $\mathbf{q}$ is the phonon wave vector, ω_0 the frequency of longitudinal phonons, N the number of unit cells in the volume V, M the reduced mass of the ions $M^{-1} := M_1^{-1} + M_2^{-1}$ and e^* is the effective ionic charge defined according to Callen (1949) as

$$e^{*2} := \frac{v_0 M\omega_0^2}{4\pi}(\varepsilon_\infty^{-1} - \varepsilon_0^{-1}) \tag{8.5.59}$$

where ε_∞ and ε_0 are the high frequency and the static dielectric constants, respectively. According to Ehrenreich (1959) the screening of the initial interaction (8.5.58) by other free carriers in the band weakens the initial interaction according to the relation

$$V'_P = V_P[1+(q\lambda)^{-2}]^{-1} \tag{8.5.60}$$

where λ is again the screening length, the same which appeared in the Coulomb interaction. In III–V-compounds the effect of screening in the phonon dispersion can be neglected.

Next we have to calculate the matrix elements between the initial and final states $|\mathbf{K},j,m(\mathbf{q})\rangle$ and $|\mathbf{K}',j',m'(\mathbf{q})\rangle$, where $m(\mathbf{q})$ is the number of phonons characterized by $\mathbf{q}$. There are two non-vanishing matrix elements of the potential between phonon states

$$\begin{aligned} \langle m(\mathbf{q})-1|b(\mathbf{q})|m(\mathbf{q})\rangle &= m(\mathbf{q})^{1/2}; \quad m'(\mathbf{q})=m(\mathbf{q})-1 \\ \langle m(\mathbf{q})+1|b^+(\mathbf{q})|m(\mathbf{q})\rangle &= [m(\mathbf{q})+1]^{1/2}; \quad m'(\mathbf{q})=m(\mathbf{q})+1 \end{aligned} \tag{8.5.61}$$

corresponding to phonon absorption and emission. Using the representation (8.5.40) for the initial and final electron states, the matrix element is obtained in the form

$$\begin{aligned} \langle \mathbf{K}',j',m'(\mathbf{q})|V'_P|\mathbf{K},j,m(\mathbf{q})\rangle &= i\frac{4\pi e^* e}{v_0}\left(\frac{\hbar}{2NM\omega_0}\right)^{1/2}\frac{1}{q} \\ \times [\Delta(\mathbf{K}'-\mathbf{K}-\mathbf{q})m(\mathbf{q})^{1/2} &- \Delta(\mathbf{K}'-\mathbf{K}+\mathbf{q})\,(m(\mathbf{q})+1)^{1/2}]\,[1+(q\lambda)^{-2}]^{-1}I_{jj'}(\mathbf{q}) \end{aligned} \tag{8.5.62}$$

where the integral over the crystal volume has been broken into the sum of integrals over unit cells and the relation

$$\Delta(\mathbf{K}'-\mathbf{K}\pm\mathbf{q}) = N^{-1}\sum_{n=1}^{N} \exp\,[i(\mathbf{K}-\mathbf{K}'\pm\mathbf{q})\cdot\mathbf{R}_n^0] \tag{8.5.63}$$

has been used. Choosing $\mathbf{K}=(0,0,\,k)$ which gives $\mathbf{K}\pm\mathbf{q}=(\pm q_1,\ \pm q_2,\ k\pm q_3)$, one obtains

$$I_{\frac{1}{2},\frac{1}{2}}(\mathbf{q}) = \frac{1}{v_0}\int_{v_0} u(\mathbf{K}\pm\mathbf{q},\tfrac{1}{2})^\times u(\mathbf{K},\tfrac{1}{2})d^3r = 1-(b^2+c^2)\frac{q^2}{2k^2} \tag{8.5.64}$$

and

$$\begin{aligned} I_{\frac{1}{2},-\frac{1}{2}}(\mathbf{q}) &= \frac{1}{v_0}\int_{v_0} u(\mathbf{K}\pm\mathbf{q},\,-\tfrac{1}{2})^\times u(\mathbf{K},\tfrac{1}{2})d^3r \\ &= \pm b\left(\frac{b}{2}-c2^{1/2}\right)\frac{q}{k}\left(1-\frac{q^2}{4k^2}\right)^{1/2} e^{i\varphi}. \end{aligned} \tag{8.5.65}$$

Hence the transition probability for both processes is

$$W(\mathbf{K},\mathbf{K}+\mathbf{q}) = w(k,q)\bar{m}(\mathbf{q})\delta[E(\mathbf{K}+\mathbf{q})-E(\mathbf{K})-\hbar\omega_0]$$

and (8.5.66)

$$W(\mathbf{K},\mathbf{K}-\mathbf{q}) = w(k,q)[\bar{m}(\mathbf{q})+1]\delta[E(\mathbf{K}-\mathbf{q})-E(\mathbf{K})+\hbar\omega_0]$$

where

$$w(k,q)=\frac{2\pi}{\hbar}\left(\frac{4\pi ee^*}{v_0}\right)^2 \frac{\hbar}{2NM\omega_0}\left(\frac{\lambda^2 q}{1+\lambda^2 q^2}\right)^2 (|I_{\frac{1}{2},\frac{1}{2}}|^2+|I_{\frac{1}{2},-\frac{1}{2}}|^2). \tag{8.5.67}$$

The relaxation time may be introduced for the region of temperatures where $k_B T \gg \hbar\omega_0 =: k_B\Theta_1$. Since the electron energy is much larger than the phonon energy, the collisions may be regarded as quasi-elastic, and the term $\hbar\omega_0$ may be neglected in the argument of the δ-functions. Using a reasoning analogous to that of equation (8.5.53), we get for any spherical energy band

$$\delta[E(\mathbf{K}\pm\mathbf{q})-E(\mathbf{K})]=\frac{1}{q}\frac{dk}{dE}\,\delta\left(\frac{q}{2k}\pm\cos\Theta\right) \tag{8.5.68}$$

where Θ is the angle between **K**- and **q**-vectors. Next the transport scattering transition rate between **K**- and **K**′-states is to be found and summed over all possible final states, which is equivalent to summation over **q**. According to Zawadzki and Szymanska (1971) this procedure gives in the case of a spherical band the following general result for the relaxation time:

$$\tau(k)^{-1}=\frac{V}{8\pi}\frac{1}{k^2}\frac{dk}{dE}\int_0^{2k} w(k,q)\,(2\bar{m}(\mathbf{q})+1)\,q^3\,dq. \tag{8.5.69}$$

For $T \gg \Theta_1$ there is $2\bar{m}(\mathbf{q})+1 \approx 2k_B T/\hbar\omega_0$ and after the integration we obtain, Zawadzki and Szymanska (1971)

$$\tau_{\text{op}}(E)=\frac{1}{8\pi}\frac{v_0 M\hbar\omega_0^2}{(ee^*)^2 k_B T}F_{\text{op}}^{-1}\frac{dE}{dk} \tag{8.5.70}$$

where

$$F_{\text{op}}:=A-\tfrac{1}{2}(4L-fL^2)B+\tfrac{1}{3}(4-f)L^2 C \tag{8.5.71}$$

with

$$\begin{aligned} A&:=1-2\xi^{-1}\ln(\xi+1)+(\xi+1)^{-1}\\ B&:=1-4\xi^{-1}+6\xi^{-2}\ln(\xi+1)-2\xi^{-1}(\xi+1)^{-1}\\ C&:=1-3\xi^{-1}+9\xi^{-2}-12\xi^{-3}\ln(\xi+1)+3\xi^{-2}(\xi+1)^{-1}. \end{aligned} \tag{8.5.72}$$

f and ξ are defined in (8.5.56). Using the previous approximations ($\xi \gg 1$), F_{op} can be simplified to the form

$$\begin{aligned} F_{\text{op}}=1&-2\xi^{-1}\ln(\xi+1)-\tfrac{1}{2}(4L-fL^2)(1-4\xi^{-1})\\ &+\tfrac{1}{3}(4-f)L^2(1-3\xi^{-1}). \end{aligned} \tag{8.5.73}$$

As before, the terms containing L are directly due to the mixing of p-like functions into the conduction band wave function, so that for very small electron energies only the first two terms are left.

γ) Acoustic scattering

Using the initial Hamiltonian (8.5.38), the interaction between electrons and acoustic phonons can be written as

$$V_a = \delta V_0 + \frac{\hbar}{4m_0^2c^2}(\sigma \times \nabla \delta V_0) \cdot \mathbf{p} \tag{8.5.74}$$

where δV_0 is the change of the periodic potential of the crystal due to acoustic wave propagation. This change is given by

$$\begin{aligned}\delta V_0 &= V - V_0 = -\mathbf{u} \cdot \nabla V_0 \\ &= -\sum_{\mathbf{q},j}\left(\frac{\hbar}{2NM\omega(j,\mathbf{q})}\right)^{1/2}(\mathbf{e}(j,\mathbf{q}) \cdot \nabla V_0)\,[b(j,\mathbf{q})e^{i\mathbf{q}\cdot\mathbf{r}} + b^+(j,\mathbf{q})e^{-i\mathbf{q}\cdot\mathbf{r}}]\end{aligned} \tag{8.5.75}$$

where $\mathbf{u}$ is the displacement, $\mathbf{e}(j,\mathbf{q})$ denotes the polarization of the wave, and $M = M_1 + M_2$. The summation is over the wave vectors $\mathbf{q}$ and the three possible polarizations j (one longitudinal and two transverse branches). Again, we have to calculate the matrix elements of V_a between initial and final electron and phonon states. Using a directional derivative $\mathbf{e}(j,\mathbf{q}) \cdot \nabla V_0 = \partial V_0/\partial \mathbf{s}_j$ and equations (8.5.40) and (8.5.61) the matrix elements are obtained in the form

$$\begin{aligned}\langle \mathbf{K}',j',m'(\mathbf{q})|V_a|\mathbf{K},j,m(\mathbf{q})\rangle &= -\sum_{\mathbf{q},h} I_{jj'}^h(\mathbf{q})\left(\frac{\hbar}{2NM\omega(h,\mathbf{q})}\right)^{1/2} \\ &\times [\Delta(\mathbf{K}'-\mathbf{K}-\mathbf{q})m(h,\mathbf{q})^{1/2} + \Delta(\mathbf{K}'-\mathbf{K}+\mathbf{q})\,(m(h,\mathbf{q})+1)^{1/2}]\end{aligned} \tag{8.5.76}$$

where

$$\begin{aligned}&I_{jj'}^h(\mathbf{q}) := \\ &v_0^{-1}\int_{v_0} u(\mathbf{K}\pm\mathbf{q},j')^\times\left[\frac{\partial V_0}{\partial \mathbf{s}_h} + \frac{\hbar}{4m^2c^2}\left(\sigma \times \nabla\frac{\partial V_0}{\partial \mathbf{s}_h}\right) \cdot (\mathbf{p}+\hbar\mathbf{K})\right]u(\mathbf{K},j)d^3r.\end{aligned} \tag{8.5.77}$$

We shall now transform equation (8.5.77) according to the well-known procedure. Using the unperturbed Hamiltonian H_0 of equation (8.5.38), the last equation can be written as

$$I_{jj'}^h(\mathbf{q}) = v_0^{-1}\int_{v_0} u(\mathbf{K}\pm\mathbf{q},j')^\times\left[\frac{\partial}{\partial \mathbf{s}_h}, \left(H_0 + \frac{\hbar}{m}\mathbf{K}\cdot\pi\right)\right]_- u(\mathbf{K},j)d^3r \tag{8.5.78}$$

where $\pi := \mathbf{p} + (\hbar/4mc^2)(\sigma \times \nabla V_0)$ and $[A,B]_-$ denotes the commutator. Observing that $[H_0 + (\hbar/m)\mathbf{K}\cdot\pi]u(\mathbf{K},j) = [E(\mathbf{K}) - \hbar^2\mathbf{K}^2/2m]u(\mathbf{K},j)$ and neglecting the value of $E(\mathbf{K}) - E(\mathbf{K}+\mathbf{q}) = (\mathbf{K}^2 - |\mathbf{K}+\mathbf{q}|^2)\hbar^2/2m$ as compared to $(\hbar/m)\,(\mathbf{q}\cdot\tau)$, equation (8.5.77) is finally transformed into

$$I_{jj'}^h(\mathbf{q}) = \frac{\hbar}{m}\frac{i}{v_0}\int_{v_0}[\mathbf{q}\cdot\nabla u(\mathbf{K}\pm\mathbf{q},j')]\,[\mathbf{e}(\mathbf{q},h)\cdot\nabla u(\mathbf{K},j)]d^3r \tag{8.5.79}$$

where we have approximated the operator π by $\mathbf{p}$.

Regarding periodic parts of the Bloch functions as given by (8.5.41), it can be seen that there are five different non-vanishing integrals (and their cyclic equivalents) in the matrix element (8.5.79), namely

$$\varepsilon_0 := \frac{\hbar^2}{m} \frac{1}{3v_0} \int_{v_0} (\nabla S)^2 d^3r; \quad \varepsilon_1 := \frac{\hbar^2}{m} \frac{1}{v_0} \int_{v_0} \left(\frac{\partial X}{\partial x_1}\right) d^3r \tag{8.5.80}$$

$$\varepsilon_2 := \frac{\hbar^2}{m} \frac{1}{v_0} \int_{v_0} \left(\frac{\partial X}{\partial x_2}\right) d^3r; \quad D := \frac{\hbar^2}{m} \frac{1}{v_0} \int_{v_0} \frac{\partial X}{\partial x_1} \frac{\partial Y}{\partial x_2} d^3r;$$

$$E := \frac{\hbar^2}{m} \frac{1}{v_0} \int_{v_0} \frac{\partial X}{\partial x_2} \frac{\partial Y}{\partial x_1} d^3r. \tag{8.5.81}$$

Values of these integrals cannot be directly calculated since the explicit form of S, X, Y and Z-functions is not known. These values can be bracketed between the atomic limit at one end (i.e., taking for S, X, Y, and Z the hydrogen functions of s- and p-states), and at the other end by the empty-lattice approximation. We shall use in the following estimations the averages of these values which are given by $\varepsilon_1/\varepsilon_0 = 29/40$ and $\varepsilon_2/\varepsilon_0 = 3/40$.

Polarizations of the three phonon branches were chosen in the following way. We denote the directional cosines of $\mathbf{q}$ as $\mathbf{e}$. Thus $\mathbf{q} = q(e_1, e_2, e_3)$. Hence for the longitudinal acoustic wave one has $\mathbf{e}(\mathbf{q}, l) = (e_1 e_2 e_3)$. The first transverse polarization is chosen to be perpendicular to both the $\mathbf{q}$ and $\mathbf{e}_3$-directions. This gives $\mathbf{e}(\mathbf{q}, tr_1) = (e_2^1 + e_2^2)^{1/2} \times (-e_2, e_1, 0)$. The second transverse mode must be perpendicular to the above two. This gives for the polarization vector

$$\mathbf{e}(\mathbf{q}, tr_2) = (e_1^2 + e_2^2)^{-1/2} (-e_3 e_1, \ -e_3 e_2, \ e_1^2 + e_2^2). \tag{8.5.82}$$

There are six different matrix elements of equation (8.5.79), corresponding to spin-conserving and spin-flip transitions for the three branches. The main contribution to acoustic scattering is given by the longitudinal spin-conserving transitions, although the others are not negligible. After choosing, as before, the initial $\mathbf{K}$-vector in the $\mathbf{e}_3$-direction and calculating the matrix elements (8.5.79) it turns out that in five cases, namely in longitudinal spin-flip and four transverse processes there appear terms which in the spherical coordinate system depend on the azimuthal angle φ. In principle this fact would prevent a rigourous introduction of the relaxation time; fortunately, however, the troublesome terms are very small. Even for the highest achievable energies, they do not exceed the value of 0.01, which can be neglected as compared to the terms of the order of unity. Furthermore, in order to simplify the calculations, we have assumed $D = \varepsilon_1$ and $E = \varepsilon_2$, which are good approximations since the quantities involved represent the same types of integrals.

With the above simplifications, the squares of the matrix elements for the longitudinal mode are, Zawadzki and Szymanska (1971)

$$|I^{l}_{\frac{1}{2},\frac{1}{2}}(\mathbf{q})|^2 = \varepsilon_0^2 q^2 \left\{ a^2 + \frac{\varepsilon_1}{\varepsilon_0}\frac{b^2}{2} + \frac{\varepsilon_2}{\varepsilon_0}\left(c^2 + \frac{b^2}{2}\right) - \frac{q^2}{4k^2}\left[\frac{\varepsilon_1}{\varepsilon_0}\left(c^2+\frac{b^2}{2}\right) + \frac{\varepsilon_2}{\varepsilon_0}\left(5c^2+\frac{b^2}{2}\right)\right] + \frac{q^4}{4k^4}\frac{\varepsilon_2}{\varepsilon_0}\left(c^2 - \frac{b^2}{2}\right)\right\}^2 \tag{8.5.83}$$

$$|I^{l}_{\frac{1}{2},-\frac{1}{2}}(\mathbf{q})|^2 = \varepsilon_1^2 \frac{q^4}{4k^2}\left(1-\frac{q^2}{4k^2}\right) b^2 c^2 \tag{8.5.84}$$

where in equation (8.5.83), after squaring, the terms proportional to $\varepsilon_1 \varepsilon_2$ and ε_2^2 are to be neglected. For the transverse modes we have, Zawadzki and Szymanska (1971)

$$|I^{tr_1}_{\frac{1}{2},\frac{1}{2}}(\mathbf{q})|^2 = \varepsilon_1^2 \frac{q^2}{4}\left(1-\frac{q^2}{4k^2}\right)\left[b^2 - \frac{q^2}{k^2}\left(\frac{bc}{2^{1/2}}+\frac{b^2}{2}\right)\right]^2 \tag{8.5.85}$$

$$|I^{tr_1}_{\frac{1}{2},-\frac{1}{2}}(\mathbf{q})|^2 = \varepsilon_1^2 \frac{q^4}{4k^2}\left(1-\frac{q^2}{4k^2}\right)\left[\left(b^2+\frac{bc}{2^{1/2}}\right) - \frac{q^2}{2k^2}\left(\frac{b^2}{2}+\frac{bc}{2^{1/2}}\right)\right]^2 \tag{8.5.86}$$

$$|I^{tr_2}_{\frac{1}{2},\frac{1}{2}}(\mathbf{q})|^2 = \varepsilon_1^2 \frac{q^4}{4k^2}\left(1-\frac{q^2}{4k^2}\right)\left(c^2+\frac{b^2}{2}\right)^2 \tag{8.5.87}$$

$$|I^{tr_2}_{\frac{1}{2},-\frac{1}{2}}(\mathbf{q})|^2 = \varepsilon_1^2 \frac{q^2}{2}\left(1-\frac{q^2}{2k^2}\right)^2 b^2 c^2. \tag{8.5.88}$$

In all the above equations we have put $e_3 = \cos\Theta = -q2k$ for phonon absorption (see equation (8.5.68)). Now the scattering probability with absorption and emission of a phonon can be calculated in analogy to equation (8.5.66), where now

$$w(k,q,h) := \frac{2\pi}{\hbar}\left(\frac{\hbar}{2NM\omega(h,\mathbf{q})}\right)\sum_{j'} |I^{h}_{jj'}(q)|^2. \tag{8.5.89}$$

For acoustic phonons of small energies we have $\omega = v_{\parallel} q$ and $\omega = v_{\perp} q$ i.e., one must make a distinction between the velocities of longitudinal and transverse modes.
It can be shown that the acoustic phonon energy in (8.5.66) is negligible compared to electron energies for all temperatures above few degrees Kelvin, so that equations (8.5.68) and (8.5.69) may be used without any change. For the same condition one has $2\bar{m}(h,\mathbf{q}) + 1 \approx 2k_B T/\hbar v_h q$. The integration indicated in (8.5.69) can now be performed to give finally for the longitudinal mode with the crystal density $\varrho = MN/V$ the expression

$$\tau_a(l) = \frac{\pi\hbar v_{\parallel}^2 \varrho}{k_B T \varepsilon_0^2} F_a(l)^{-1} \frac{dE}{dk} k^{-2} \tag{8.5.90}$$

where, to a very good approximation,

$$F_a(l) = \left[1 - L\left(1 + \frac{7}{18}\frac{\varepsilon_1}{\varepsilon_0}a_1 + \frac{5}{18}\frac{\varepsilon_2}{\varepsilon_0}a_2\right)\right]^2 \tag{8.5.91}$$

with

$$a_1 = \frac{1}{7}(8\gamma^2 - \beta^2), \quad a_2 = \frac{1}{5}(4\gamma^2 + \beta^2). \tag{8.5.92}$$

The relaxation time $\tau_a(tr)$ for both transverse modes can also be expressed by (8.5.90) with $v_\parallel^2$ replaced by $v_\perp^2$, and $F_a(l)$ replaced by $F_a(tr)$ where

$$F_a(tr) = \frac{17}{12}\left(\frac{\varepsilon_1}{\varepsilon_0}\right)^2 L^2 a_3 \tag{8.5.93}$$

with

$$a_3 = \frac{1}{51}(3\beta^4 + 32\beta^2\gamma^2 + 16\gamma^4). \tag{8.5.94}$$

For $\Delta \gg E_0$ it follows that $a_1 \approx a_2 \approx a_3 \approx 1$.

δ) Piezo-acoustic scattering

In III–V-compounds this mode is possible due to the lack of inversion symmetry and it becomes of importance at low temperatures where in pure samples it competes with the regular acoustic scattering. It has been shown by Harrison (1956) that the two modes can be considered independently. In calculating piezo-acoustic scattering we shall follow in general the approach of Hutson (1961).
First we have to determine the scattering potential due to the piezo-effect associated with acoustic wave deformation. In an ideal dielectric (no movable charge) the electric induction is zero, i.e.,

$$4\pi\mathfrak{E} \cdot \mathfrak{S} + \kappa_0 \mathbf{E} = 0 \tag{8.5.95}$$

where $\mathfrak{E}$ is the piezo-electric tensor of the third order, $\mathfrak{S}$ the deformation tensor $S_{ij} = 1/2\,(\partial u_i/\partial x_j + \partial u_j/\partial x_i)$, and $\mathbf{u}$, as before, denotes the displacement connected with the acoustic wave. Using equation (8.5.75) we have

$$S_{ij} = \frac{i}{2}\sum_{\mathbf{q},h}\left(\frac{\hbar}{2NM\omega(h,\mathbf{q})}\right)^{1/2}[e(h,\mathbf{q})_i q_j + e(h,\mathbf{q})_j q_i] \times [b(h,\mathbf{q})e^{i\mathbf{q}\cdot\mathbf{r}} - b^+(h,\mathbf{q})e^{-i\mathbf{q}\cdot\mathbf{r}}]. \tag{8.5.96}$$

In cubic crystals of InSb type there is one non-vanishing component of the piezo-electric tensor: $e_{123} = e_{132} = e_{231} = e_{213} = e_{312} = e_{321} = P/2$. Hence, using equations (8.5.95) and (8.5.96) to obtain components of the electric field and next employing the Poisson equation, we get for the unscreened perturbing potential energy

$$V_{pa} = -\frac{4\pi Pe}{\varepsilon_0}\sum_{\mathbf{q}h}\left(\frac{\hbar}{2NM\omega(h,\mathbf{q})}\right)^{1/2} K_h[b(h,\mathbf{q})e^{i\mathbf{q}\cdot\mathbf{r}} - b^+(h,\mathbf{q})e^{-i\mathbf{q}\cdot\mathbf{r}}] \tag{8.5.97}$$

with

$$K_h = e(h,\mathbf{q})_1 e_2 e_3 + e(h,\mathbf{q})_2 e_1 e_3 + e(h,\mathbf{q})_3 e_1 e_2 \tag{8.5.98}$$

where, as before, $\mathbf{e}(h,\mathbf{q})$ is the unit polarization vector of the acoustic wave and $\mathbf{e}$ the unit vector along its propagation direction. It can be seen that V_{pa} is strongly anisotropic. For instance, it vanishes for $\mathbf{e}=(1,0,0)$ for both longitudinal and transverse modes, and for $\mathbf{e}=(1,1,0)$ for the longitudinal mode. If there are movable carriers in the band, one is not dealing with a perfect dielectric and the screening effects come into play in this long-range interaction. According to Hutson (1961) the screened piezo-acoustic interaction takes the form

$$V'_{pa}=V_{pa}[1+(q\lambda)^{-2}]^{-1}. \tag{8.5.99}$$

That means, the effect of screening is the same as for polar optical interaction. The matrix elements between initial and final states are

$$\langle \mathbf{K}',j',m'(\mathbf{q})\,|V'_{pa}|\mathbf{K},j,m(\mathbf{q})\rangle=-\frac{4\pi Pe}{\varepsilon_0}\left(\frac{\hbar^2}{2NM\omega(h,\mathbf{q})}\right)^{1/2} \tag{8.5.100}$$
$$\times[\Delta(\mathbf{K}'-\mathbf{K}-\mathbf{q})m(\mathbf{q})^{1/2}-\Delta(\mathbf{K}'-\mathbf{K}+\mathbf{q})\,(m(\mathbf{q})+1)^{1/2}]K_h[1+(q\lambda)^{-2}]^{-1}I_{jj'}(q)$$

where $I_{jj'}(q)$ are defined in (8.5.64) and (8.5.65). In principle, due to the strong anisotropy of the scattering, the usual relaxation time cannot be introduced in a rigorous way. However, following the common procedure, we shall introduce a relaxation time by averaging over the angles. Making use of $\omega(h,\mathbf{q})=v_h q$, where v_h is the sound velocity depending on the direction of propagation and the mode in question, we have $v_h^2=c/\varrho$ with c representing an appropriate combination of elastic constants. Following equation (8.5.69) and performing the integration as in the case of polar optical scattering, the final formula for the relaxation time is obtained in the form

$$\tau_{pa}^{-1}=8\pi\,\frac{P^2e^2}{\varepsilon_0^2}\,\frac{k_BT}{\hbar}\left(\overline{\frac{K_h}{c}}\right)\frac{dk}{dE}\,F_{pa} \tag{8.5.101}$$

where F_{pa} is equal to F_{op} of equation (8.5.73). We do not elaborate on the averaging procedure, since it has been done by Meijer and Polder (1953), Hutson (1961) and Zook (1964). In order to avoid all misunderstanding, we note that equation (8.5.101) was obtained by simply taking out (K_h/c) in front of the integral sign, since it does not depend on $\mathbf{q}$- or $\mathbf{e}(h,\mathbf{q})$-directions.
According to (8.4.13) the reciprocal relaxation times τ_α^{-1} are additive. Hence the total relaxation time is given by

$$\tau^{-1}=\sum_\alpha \tau_\alpha^{-1} \tag{8.5.102}$$

where α runs over the contributions of various scattering processes. If the partial mobilities μ_α are defined by inserting τ_α into (8.5.17), then from (8.5.102) it follows that the total mobility is given by

$$\mu^{-1}=\sum_\alpha \mu_\alpha^{-1}. \tag{8.5.103}$$

For InSb materials one obtains with (8.5.17) and $m^{-1}=(\hbar^2k)^{-1}dE/dk$ the expression

$$\mu_\alpha=\frac{e\tau_\alpha}{m}. \tag{8.5.104}$$

If these values are inserted in (8.5.32), and (8.5.32) is inserted in (8.5.27), then for $H_\alpha \equiv 0$ the average Ohmic mobility can be calculated directly. The results are presented in Figure 8.4 for various electron concentrations at the temperature $T = 300°$ K.

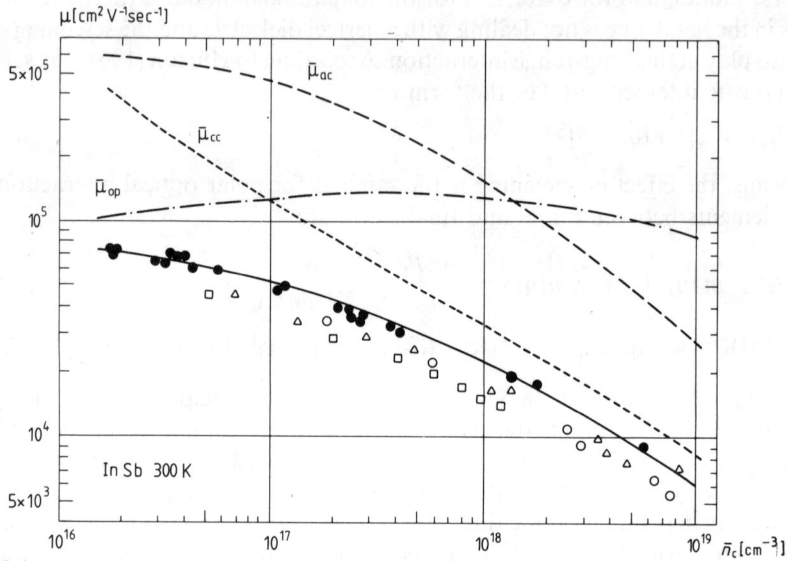

Figure 8.4:
Electron mobility in n-InSb at room temperature versus free electron concentration. Dashed lines denote the theoretical mobilities for charged center, polar optical, and acoustic scattering mode. Experimental data: △ Barrie and Edmond (1955), ● Rupprecht, Weber and Weiss (1960), □ Kessler and Sutter (1964), ○ Galavanov, Nasledov and Filipchenko (1965); after Zawadzki (1974).

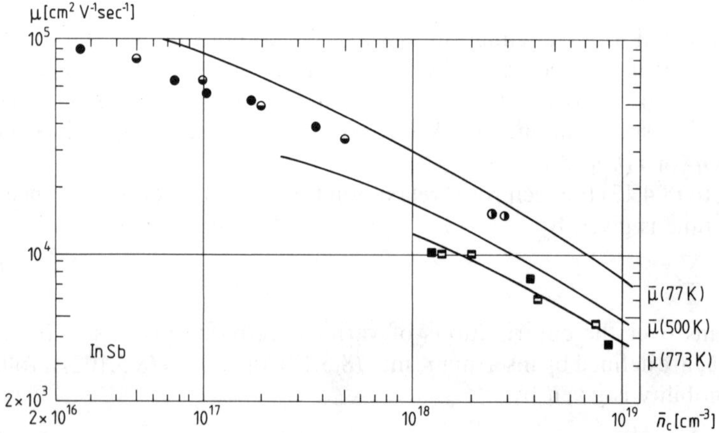

Figure 8.5:
Electron mobility in n-InSb versus free electron concentration at various temperatures. Solid lines show the total theoretical mobilities calculated for the corresponding temperatures (the theory for 77 K does not take into account polar optical scattering). Experimental data: ◑ Champness (1958), ⊟ Putley (1959), ◒ Strauss (1959), ● Rupprecht, Weber and Weiss (1960), ■ (773 K) Galavanov, Nasledov and Filipchenko (1965); after Zawadzki (1974).

If also the temperature is varied, the Figure 8.5 results.
The discrepancy between theory and experiment at 77° K for low electron concentrations is due to the fact that the polar optical mode has not been included in the theory for this temperature since the relaxation time approximation cannot be used for $T<\Theta$.

b) General solution procedures

For temperatures below the Debye temperature the calculation of the mobility has to be performed directly by means of the stationary Boltzmann equation. For simplicity, we consider only an external electric field **E**, and apply the linearized version (8.2.42) which reads for this case

$$e\frac{df_0}{dE}\mathbf{v}\cdot\mathbf{E}=\frac{1}{k_BT}\sum_{\mathbf{K}'}W_0(\mathbf{K},\mathbf{K}')\,[\Phi(\mathbf{K}')-\Phi(\mathbf{K})] \tag{8.5.105}$$

with the definition (8.2.44). Analoguous to the relaxation time approach we try to find a solution of the type $\Phi(\mathbf{K})=\mathbf{K}\cdot\mathbf{G}$, where now the vector **G** is not parallel to **E** and $\Phi(\mathbf{K})$ can be written

$$\Phi(\mathbf{K})=f_1(k)\cos\vartheta \tag{8.5.106}$$

with ϑ being the angle between **K** and **E**. The general transition probability for phonon electron collisions which is also valid for optical phonons is given by (8.4.36). If the spectrum of optical phonons is assumed to be degenerate then (8.4.36) reads for the optical branches

$$W(\mathbf{K},\mathbf{K}')_{\mathrm{op}}=\sum_{\substack{\mathbf{q}\\ \{j\}\mathrm{op}}}\frac{2\pi|I^{\pm}|}{2M\omega_0}[\bar{m}(j,\mathbf{q})+\tfrac{1}{2}\mp\tfrac{1}{2}]\,\delta_\gamma\big(E(\mathbf{K})-E(\mathbf{K}')\mp\hbar\omega_0\big). \tag{8.5.107}$$

If we now split the collision operator in (8.5.105) into a part containing the elastic collisions and a part containing the inelastic collisions, we may introduce for the first part a common relaxation time, while the second part has to be treated explicitly. Formally we write

$$e\frac{df_0}{dE}\mathbf{v}\cdot\mathbf{E}=\frac{1}{\tau}\,\Phi(\mathbf{K})+\frac{1}{k_BT}\sum_{\mathbf{K}'}W_0(\mathbf{K},\mathbf{K}')_{\mathrm{op}}[\Phi(\mathbf{K}')-\Phi(\mathbf{K})]. \tag{8.5.108}$$

With spherical coordinates E, ϑ, φ we have $\Phi(\mathbf{K})=\bar{f}_1(E)\cos\vartheta$ and the δ-distribution in (8.5.107) depends only on E and E', while the summation over $\mathbf{K}'$ is changed into an integration over E', ϑ', φ'.
Due to the δ-distribution the integration over E' can immediately be performed and if equation (8.5.105) is integrated over ϑ and φ one obtains an equation for $\bar{f}_1(E)$ which

reads

$$-\frac{e|\mathbf{E}|}{\hbar}\frac{df_0}{dE}\left(\frac{dk(E)}{dE}\right)^{-1}=a(E)\tilde{f}_1(E-\hbar\omega_0)-\lambda(E)\tilde{f}_1(E)+c(E)\tilde{f}_1(E+\hbar\omega_0) \tag{8.5.109}$$

where $\lambda(E)$ contains the total collision loss of elastic and inelastic collisions. With the continuum approximation for the optical phonon-electron interaction, such an equation was derived by Grigorev, Dykman, Tomchuk (1968). But it should be emphasized that also the collision term (8.4.76) of Wilson (1953), Howarth and Sondheimer (1953) has this general structure.

Equation (8.5.109) can be considered to generate a system of linear equations for the values of the distribution function $f_1(E)$ at the points $\{E'=E+n\hbar\omega_0,\ n \text{ integer}\}$. The boundary conditions of this system are given by $f_1(E')=0$ for $E'<0$ and $\lim_{E\to\infty} f_1(E')=0$. Kranzer, Hillbrand, Pötzl and Zimmerl (1972) solved such a system by truncating it for finite energies. For temperatures between 40 K and 300 K they took into account the following scattering mechanisms: polar-optical scattering, acoustic scattering, charged impurity scattering and piezoelectric scattering. They obtained results which are given in the Figures 8.6a and 8.6b.

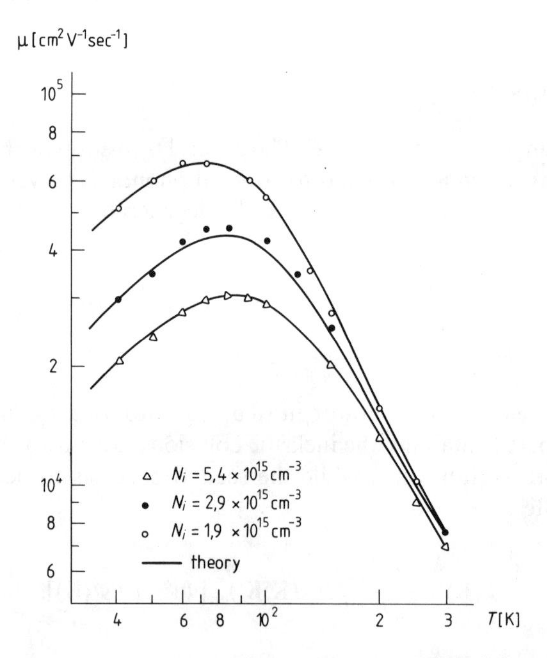

Figure 8.6.a:
Calculated temperature dependence of the electronic mobility $\mu=-|j|/e\bar{n}_c|\mathbf{E}|$ in n-GaAs for various concentrations of foreign ions N_i. For one-valley semiconductors with InSb band structure or with simpler band structures the definition of μ coincides with the value (8.5.104) which follows from the more general definition (8.5.17). For more general band structures the definition given above is only an average value which in general is different from (8.5.17). Experimental values are given by the symbols ○, ●, △; after Kranzer, Hillbrand, Pötzl, and Zimmerl (1972).

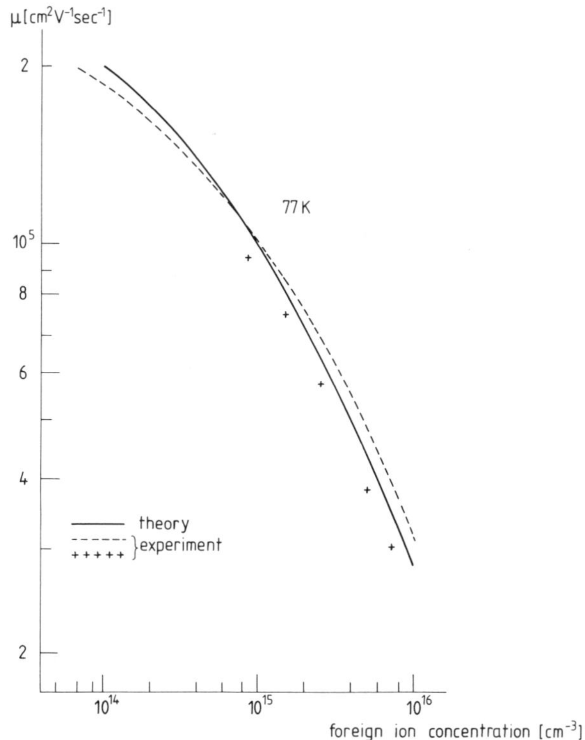

Figure 8.6.b:
Calculated dependence of the mobility $\mu = -|j|/e\bar{n}_c|\mathbf{E}|$ on the concentration of foreign ions N_i at $T=77$ K. Experimental values are given by Wolfe, Stillman and Lindley (1970), Wolfe, Stillman and Dimmrock (1970); after Kranzer, Hillbrand, Pötzl, and Zimmerl (1972).

The same authors extended the calculation to "quasistatic" electromagnetic fields in the microwave region using the same calculation technique.
Another approach is the variational principle of Kohler (1948) (1949) which may not only be applied to the phonon Boltzmann equation but also to the electron Boltzmann equation. We follow the version given by Wilson (1953). In the linear approximation and with vanishing magnetic field the Boltzmann equation reads according to (8.2.42)

$$\frac{df_0}{dE}\mathbf{v}\cdot\left[e\mathbf{E}-\nabla_{\mathbf{r}}\zeta-\frac{E(\mathbf{K})-\zeta}{T}\nabla_{\mathbf{r}}T\right]=L[\Phi] \tag{8.5.110}$$

with

$$L[\Phi]:=\frac{1}{k_B T}\sum_{\mathbf{K}'} W_0(\mathbf{K},\mathbf{K}')\,[\Phi(\mathbf{K})-\Phi(\mathbf{K}')]. \tag{8.5.111}$$

Making the assumption

$$\Phi=(e\mathbf{E}-\nabla_{\mathbf{r}}\zeta)\cdot\mathbf{a}-\frac{1}{T}\nabla_{\mathbf{r}}T\cdot\mathbf{b} \tag{8.5.112}$$

then with $\eta(\mathbf{K}) := E(\mathbf{K}) - \zeta \equiv \eta$ from (8.5.110) we obtain the equations

$$L[\mathbf{a}] = \mathbf{v}\frac{df_0}{dE} \tag{8.5.113}$$

$$L[\mathbf{b}] = \mathbf{v}\eta\frac{df_0}{dE}. \tag{8.5.114}$$

The solution of these equations can be achieved by the following theorem:

Theorem 8.7: Let F be the solution of the equation

$$L[F] = v_i E^n \frac{\partial f_0}{\partial E} \tag{8.5.115}$$

and let G be any function satisfying the relation

$$(G,G) = \frac{1}{4\pi^3}\int G L[G] d^3K = \frac{1}{4\pi^3}\int G v_i E^n \frac{\partial f_0}{\partial E} d^3K. \tag{8.5.116}$$

Then of all the functions G satisfying the above relation, F is the one which makes (G,G) a maximum.

Proof: The proof of this theorem depends upon the relations $(F,G) = (G,F)$ and $(G,G) \geqslant 0$ which are consequences of the symmetry property

$$W_0(\mathbf{K},\mathbf{K}') = W_0(\mathbf{K}',\mathbf{K}) \tag{8.5.117}$$

and of the fact that $W_0(\mathbf{K},\mathbf{K}')$ being a transition probability, is essentially positive. To establish these relations we write

$$\begin{aligned}(F,G) &= \frac{1}{4\pi^3 k_B T}\iint W_0(\mathbf{K},\mathbf{K}')F(\mathbf{K})[G(\mathbf{K}) - G(\mathbf{K}')]d^3K d^3K' \\ &= \frac{1}{4\pi^3 k_B T}\iint W_0(\mathbf{K}',\mathbf{K})F(\mathbf{K}')[G(\mathbf{K}') - G(\mathbf{K})]d^3K d^3K'\end{aligned} \tag{8.5.118}$$

since (8.5.117) holds. Hence, by addition we have

$$(F,G) = \frac{1}{4\pi^3 k_B T}\iint W_0(\mathbf{K},\mathbf{K}')[F(\mathbf{K}) - F(\mathbf{K}')][G(\mathbf{K}) - G(\mathbf{K}')]d^3K d^3K'. \tag{8.5.119}$$

Since this is symmetrical in F and G, it establishes the relation

$$(F,G) = (G,F). \tag{8.5.120}$$

The inequality

$$(G,G) \geqslant 0 \tag{8.5.121}$$

follows at once from (8.5.119) by putting $F = G$, since the integrand is then essentially positive.

To prove the variation principle we proceed as follows. If we multiply the equation (8.5.115) by G and integrate with respect to $\mathbf{K}$ we find that

$$(G,F)=\frac{1}{4\pi^3}\int Gv_i E^n \frac{\partial f_0}{\partial E} d^3K=(G,G) \tag{8.5.122}$$

by (8.5.116). Now $(F-G,F-G)\geqslant 0$ and so $(F,F)-2(F,G)+(G,G)\geqslant 0$. Hence, on using (8.5.122), we have

$$(G,G)\leqslant(F,F) \tag{8.5.123}$$

which proves the maximum principle, Q.E.D.

Theorem 8.8 : Let G be any function for which (G,G) exists. Then of all such functions the solution F of equation (8.5.115) is the one which makes

$$(G,G)\left[\frac{1}{4\pi^3}\int Gv_i E^n \frac{\partial f_0}{\partial E} d^3K\right]^{-2} \tag{8.5.124}$$

a minimum.

Proof: To prove this we note that

$$(\lambda F+\mu G,\lambda F+\mu G)=\lambda^2(F,F)+2\lambda\mu(F,G)+\mu^2(G,G)\geqslant 0$$

for all real λ and μ. The discriminant is therefore positive, i.e.,

$$(F,F)(G,G)\geqslant(F,G)^2. \tag{8.5.125}$$

Now, if F satisfies equation (8.5.115), we have

$$\begin{aligned}(F,F)&=\frac{1}{4\pi^3}\int Fv_i E^n \frac{\partial f_0}{\partial E} d^3K;\\ (G,F)&=\frac{1}{4\pi^3}\int Gv_i E^n \frac{\partial f_0}{\partial E} d^3K.\end{aligned} \tag{8.5.126}$$

Hence dividing the inequality (8.5.125) by $(F,F)(F,G)^2$ which is positive, we have

$$\frac{(G,G)}{(F,G)^2}\geqslant\frac{1}{(F,F)} \tag{8.5.127}$$

which by (8.5.126) can be written as

$$(G,G)\left[\frac{1}{4\pi^3}\int Gv_i E^n \frac{\partial f_0}{\partial E} d^3K\right]^{-2}\geqslant(F,F)\left[\frac{1}{4\pi^3}\int Fv_i E^n \frac{\partial f_0}{\partial E} d^3K\right]^{-2} \tag{8.5.128}$$

which proves the minimum principle, Q.E.D.

These theorems remain valid if E^n is replaced by η^n. Eqs. (8.5.113), (8.5.114) are then treated by the following ansatz, cf. Wilson (1953), Haug (1970):

$$\mathbf{a}(\mathbf{K})=\mathbf{v}(\mathbf{K})\sum_r \alpha_r[E(\mathbf{K})-\zeta]^r; \quad \mathbf{b}(\mathbf{K})=\mathbf{v}(\mathbf{K})\sum_r \beta_r[E(\mathbf{K})-\zeta]^r \tag{8.5.129}$$

Then we have for any component a_i, $i=1,2,3$ of $\mathbf{a}$

$$(a_i, L[a_i])=\sum_{rr'} \alpha_r\alpha_{r'} D^i_{rr'}\,; \quad \left(a_i, v_i \frac{df_0}{dE}\right)=\sum_r \alpha_r C^i_r \tag{8.5.130}$$

with the definitions

$$D^i_{rr'}:=\int v_i\eta^r L[v_i\eta^r]d^3K; \quad C^i_r:=-\int v_i^2\eta^r \frac{df_0}{dE}\,d^3K \tag{8.5.131}$$

and $(a_i, L[a_i])$ must be a maximum subject to the condition $(a_i, L[a_i])=(a_i, v_i df_0/dE)$. With a variational parameter λ this leads to the equations

$$\frac{\partial}{\partial\alpha_s}\left(\sum_{rr'}\alpha_r\alpha_{r'}D^i_{rr'}+\lambda\sum_r \alpha_r C^i_r\right)=0. \tag{8.5.132}$$

From (8.5.132) the value $\lambda=-2$ results and by substitution the linear system

$$\sum_r \alpha_r\, D^i_{sr}=C^i_s, \qquad s=0,1,\ldots \tag{8.5.133}$$

emerges for the determination of the α_r. In the same way eq. (8.5.114) can be treated. With regard to an improved treatment, see Langbein (1961), Klein (1961).
Howarth and Sondheimer (1953) used this variational principle to calculate the conductivity, etc., in the low temperature limit for polar semiconductors with parabolic energy bands. Ehrenreich (1957) (1959) extended this approach to polar semiconductors with non-parabolic bands, in particular to InSb. In this case the overlap integral in (8.4.61) plays an important role for the calculation if Kane band functions are used.
Physically, Kohler's variational method can be interpreted as an extremum principle for the entropy production of the electron or phonon system, resp. due to collision processes. Dorn (1957) developed a generalization of Kohler's variational principle which is aimed at the simultaneous determination of the electron and phonon non-equilibrium distributions for interacting electron-phonon systems. Starting from the coupled system of an electron Boltzmann equation and a phonon Boltzmann equation, he showed that under certain subsidiary conditions the entropy production of the electron-phonon system takes an extremum value for transport processes. This principle can be used for the calculation of transport quantities. Another approach which can rigourously be justified is an iteration procedure for solving the Boltzmann equation introduced by Rode (1970) (1971). Using the contraction mapping principle, Rode (1973) showed how interated solutions of the Boltzmann equation with Fermi statistics are obtained. The mathematical formalism allows a manipulation of the Boltzmann equation separately from considerations of particular electron scattering mechanisms and proves existence, uniqueness, and convergence properties of the solutions. This technique applies potentially to a broad range of electronic transport

problems. Rode, in particular, considered Hall effects and magnetoresistance in III–V and II–VI compounds.
Concerning the literature, the general discussion given in Section 8.3.d) is valid also for the case of polar semiconductors. Hence we do not repeat it here. We rather cite original papers in this field, where we shall apply the abbreviations which we already used in Section 8.3.d). In addition, in most cases we omit the explicit specification of the class of substances, namely polar semiconductors, as the papers which we cite here deal in particular with these materials. An extensive review of electron transport phenomena in small gap semiconductors was given by Zawadski (1974), and of electron scattering and transport phenomena in small gap zincblende semiconductors by Szymanska and Dietl (1978).
In the original literature the following topics were treated: Theory of electronic conduction based on variational solutions of the Boltzmann equation and discussion of the validity of the relaxation time approach, Howarth and Sondheimer (1953). Extension of the Howarth-Sondheimer approach to the treatment of magnetoresistance effects, Lewis and Sondheimer (1955). Discussion of various calculations of the mobility of electrons and holes in PbS, Petritz and Scanlon (1955). Review of infrared absorption in semiconductors, Fan (1956). Calculation of mobility of electrons and the thermoelectric power in intrinsic InSb based on a one-electron model with electron-phonon and electron-hole coupling and an extension of Howarth and Sondheimer's method to non-parabolic bands, Ehrenreich (1957). Theory of thermomagnetic effects of non-polar isotropic semiconductors with simultaneous determination of electron and phonon non-equilibrium distributions for a mutually coupled electron and phonon Boltzmann equation, Appel (1958). Theory of carrier heating in non-degenerate non-polar semiconductors under the influence of both ion scattering and lattice scattering based on BERTA, Seeger (1959). Derivation of transport coefficients based on the Boltzmann equation with the collison operator of Howarth and Sondheimer and direct numerical integration of this equation, Delves (1959). Calculation of transport coefficients with an improved version of the previous theory, Ehrenreich (1959). Inclusion of screening effects in the calculation of transport coefficients, Ehrenreich (1959). Solution of the Boltzmann equation by a series expansion for the Howarth-Sondheimer model and calculation of transport coefficients, Durney (1961). Current-carrier transport with space charge in semiconductors based on PMTE, von Roosbroeck (1961). Transport of current carriers in n-type InSb at low temperatures based on BERTA, Kolodziejczak (1961). Transport phenomena in semiconductors with arbitrary spherical energy bands based on BERTA, Kolodziejczak (1961), and application of this method to thermomagnetic effects, Zawadzki (1962). Magneto-optical effects, thermoelectromotive force and Nernst-Ettinghausen effect in InSb based on BERTA, Kolodziejczak (1962), Kolodziejczak and Sosnowski (1962). Calculation of mobility and thermoelectric power in many-valley polar semiconductors by variational solution of the Boltzmann equation, Olechna and Ehrenreich (1962). Development of a variational solution procedure for the Boltzmann equation with direct evaluation of collision terms and calculation of the electric conductivity of polar semiconductors, Dykman and Tomchuk (1960) (1961) (1962) (1963), Tomchuk (1962). Investigation of hot electron-phonon scattering, Kogan (1963). Theory of

transport phenomena in semiconductors with non-spherical and non-quadratic energy bands based on BERTA, Zukotynski and Kolodziejczak (1963). Theory and analysis of free carrier infrared absorption in III–V semiconductors, Haga and Kimura (1963) (1964). Diffusion coefficient of electrons in InSb based on previous calculations, Zawadzki (1963), and magnetic susceptibility of semiconductors with non-parabolic energy bands, Zawadzki (1963). Thermodynamic properties of free carriers in semiconductors with arbitrary energy bands, Zawadzki and Kolodziejczak (1964). Non-linear theory of current saturation in piezoelectric semiconductors based on PMTE, Abe (1964). Galvano- and thermomagnetic effects in semiconductors with non-spherical and non-parabolic energy bands based on BERTA, Kolodziejczak and Zukotynski (1964). Review of intraband transitions by light absorption of free charge carriers in semiconductors, Kessler (1964). Treatment of the Boltzmann equation with inclusion of a recombination term for the description of the stationary photo-magnetoelectric effect in InSb, Richter (1964). Calculation of the relaxation time of electron-ionized impurity scattering in degenerate many-valley semiconductors, Robinson and Rodriguez (1964). Transport properties of cubic semiconductors with non-parabolic energy bands based on BERTA, Zawadski and Kolodziejczak (1964). Note on the theory of hot electrons in semiconductors, Kurosawa (1965). Calculation of relaxation times for electron-phonon scattering processes in polar semiconductors, Krishnamurthy and Sinha (1965). Calculation of temperature and mobility of hot electrons in polar semiconductors by means of Dykman's and Tomchuk's variational method, Grigorev, Dykman and Tomchuk (1966). Note on hot electron scattering in polar semiconductors with non-parabolic bands, Matz (1966). Calculation of stationary distributions of hot electrons in an anisotropic semiconductor in a strong electric field with electron-acoustic phonon scattering by approximate solution of the Boltzmann equation, Gurevich and Katilius (1966). Theory of free carrier electro-magneto-optical phenomena based on BERTA and a power series expansion of the distribution function in terms of the static electric field, Kolodziejczak and Stramska (1966). Note on the variation of the electron temperature by an electric field, Licea (1966). Calculation of mobility and temperature of electrons in polar semiconductors in electric fields based on the approximate solution of the Boltzmann equation by a displaced Maxwellian distribution, Blötekjaer (1966). Free carrier electro-magneto-optical phenomena in semiconductors treated by the formalism of the previous paper, Kolodziejczak and Kierzek-Pecold (1967), Kolodziejczak (1967). Intervalley-scattering selection rules in III–V semiconductors, Birman, Lax and Loudon (1966). Note on resonant electron-phonon scattering, Sher and Thornber (1967). Theory of electrical conductivity of a many-valley semiconductor in a strong electric field based on Gantsevich's solution method of the Boltzmann equation, Gantsevich (1967). Calculation of hot carrier distribution functions for non-parabolic energy bands by means of an expansion of the distribution functions in terms of Legendre functions and BERTA for an iterative solution, Matz (1967). Note on hot optical phonons in polar semiconductors, Korenblit (1967). Theory of avalanche breakdown in InSb and InAs based on PMTE and elementary rate equations, Dumke (1968). Effects of non-parabolicity on non-Ohmic transport in InSb and InAs based on Matz' method for the solution of the Boltzmann equation, Matz (1968). Analysis of the collision operator for electron-optical phonon scattering for a semiconductor with an arbitrary non-parabolic energy band in a strong electric field using the electron temperature approximation for the distribution function, Licea (1968). Calculation of the distribution function and of the temperature dependence of the mobility in III–V compounds by direct solution of the Boltzmann equation, Grigorev, Dykman and Tomchuk (1968). Calculation of hot-electron galvanomagnetic coefficients by means of shifted Maxwellian distribution etc., based on the Boltzmann equation, Das and Sahni (1968). Modified variational calculation of the zero field mobility based on Howarth's and Sondheimer's method, Heinle (1968). Calculation of hot electron mobility based on Licea's paper with displaced Maxwellian distributions, Licea (1968). Calculation of electron mobility at intermediate and high electric fields based on the evaluation of the Boltzmann equation with displaced Maxwellian distributions, de Alba and Warman (1968). Analysis of plasma states in semiconductors with anisotropic energy bands based on BERTA and the Maxwell equations, Zukotynski and Saleh (1969). Investigation of the Faraday ellipticity of hot electrons in many-valley semiconductors, Gupta (1970). Discussion of the Faraday effect of hot electrons in many-valley semiconductors by means of an approximate solution of the Boltzmann equation for the case of an external microwave field together with a static magnetic and electric field, Asche (1970). Investigation of the high frequency behaviour of hot electrons in one-valley polar semiconductors by means of drifted Maxwellian distributions with time-dependent temperature and drift momentum, Bonek, Pötzl and Richter (1970). Analysis of electron-ionized impurity scattering processes in semiconductors with an arbitrary non-parabolic spherical conduction band in a strong electric field by means of a drifted Maxwellian distribution, Licea (1970). Note on the anisotropy of the electron temperature in polar single-valley semiconductors, Hillbrand and Kranzer (1970). Angular dependence of matrix elements for scattering in III–V compounds using Ehrenreich's model, Vassell, Ganguly and Conwell (1970). Calculation of the mobility of holes in III–V compounds by means of relaxation times, Wiley and di Domenico (1970). Calculation of the electron drift mobility in direct gap III–V semiconductors by means of an approximate solution of the corresponding Boltzmann equation, Rode (1970), and along the same lines for II–VI semiconductors, Rode (1970). Analysis of polar optical scattering of electrons in GaAs by means of an approximate solution of the corresponding Boltzmann equation, Fortini, Diguet and Lugand (1970). Calculation of galvano-magnetic phenomena in semiconductors with non-parabolic dispersion law and non-elastic scattering of carriers in the diffusion approximation of the Boltzmann equation, Tolpygo (1970). Investigation of the magneto-conductivity of hot electrons in semiconductors with arbitrary non-parabolic, spherically symmetric conduction band by evaluation of the Boltzmann equation in the electron temperature approximation of the distribution function, Licea (1970). Theory of electron-impurity scattering in semiconductors with multi-valley conduction bands based on a one-electron model, Mattis and Sinha (1970). Effect on non-parabolic bands upon free carrier absorption, Demidenko (1970). Calculation of galvano- and thermomagnetic transport coefficients for semiconductors with non-parabolic conduction band, in particular InSb, at high electric fields based on BERTA, Dubey and Chakravarti (1971). Study of hot-electron galvano-thermomagnetic transport properties of III–V semiconductors at low temperatures based on BERTA and a special evaluation of the collision term for electron-optical phonon processes, Sharma and Dubey (1971). Phenomenological analysis of the two-band Hall effect and magnetoresistance, Kwan, Basinski and Woolley (1971). Calculation of longitudinal galvanomagnetic effects in the non-Ohmic regime with magnetophonon resonances based on the evaluation of a corresponding Pauli-Master equation and averaged transport equations for a displaced Maxwellian distribution function, applied to InSb, Peterson, Magnusson and Weissglas (1971). Calculation of the mobility of holes in III–V compounds based on an extension of Ehrenreich's method, Wiley (1971). Calculation of relaxation times for elastic electron scattering in InSb-type semiconductors, Zawadski and Szymanska (1971). Calculation of the electron drift mobility in GaAs by evaluation of the corresponding collision terms, Rode and Knight (1971). Solution of the Boltzmann equation in the diffusion approximation and calculation of electron-transport properties for InSb, InAs and InP, Rode (1971). Note on high frequency conductivity of polar semiconductors, Kranzer and König (1971). Treatment of high frequency conductivity using the difference equation approach (8.5.109) for the solution of the Boltzmann equation, Kranzer (1971). Hot carrier conduction in a degenerate many-valley semiconductor based on BERTA, Sodha, Tripathi and Kamal (1971). Treatment of the non-linear Faraday effect in non-parabolic semiconductors based on BERTA, Chakravarti (1971). Effect of non-parabolic band structure on longitudinal magneto-acoustic phenomena based on an evaluation of the density operator equation, Wu and Spector (1971). Review of scattering of current carriers and transport phenomena in PbTe and PbS, Ravich, Efimova and Tamarchenko (1971). High frequency conductivity of hot carriers in semiconductors with non-parabolic conduction band in external electric and magnetic fields based on BERTA, Almasov (1972). Study of non-linear electric transport phenomena of hot electrons in *n*-type InSb under magnetic quantum conditions by means of Kurosawa's method, Kotera, Yamada and Komatsubara (1972). Investigation of transport properties of electrons in a quantizing magnetic field by direct solution of the Boltzmann equation for a one-electron model with electron-phonon coupling, Magnusson (1972). Analysis of the longitudinal magnetoresistance using displaced Maxwellian distribution functions, Peterson (1972). Investigation of the convergence of a Legendre polynomial expansion of the distribution function for hot carriers at high temperatures based on BERTA, Sodha, Phadke and Chakravarti (1972). Study of galvanomagnetic effects in many-valley semiconductors for strong magnetic fields and heating electric fields based on BERTA, Mitin (1972). Treatment of transport properties of semiconductors with non-parabolic bands in strong electric and magnetic fields based on the solution of the Boltzmann equation in the effective temperature approximation, Pogrebnyak (1972). An exact solution of the linearized Boltzmann equation for arbitrary magnetic field strengths and polar mode and elastic scattering with applications to Hall mobility and Hall factor of *n*-type GaAs, Fletcher and Butcher (1972). Investigation of transport properties of holes in cubic semiconductors with polar mode scattering using Monte Carlo techniques for the solution of the Boltzmann equation, Costato, Jacoboni and Reggiani (1972). Study of methods for the solution of the Boltzmann equation for III–V compound semiconductors,

Kranzer, Hillbrand, Pötzl and Zimmerl (1972). Study of degeneracy and screening effects in the transport coefficients of n-type InSb at high temperatures based on previous work of the authors, Sodha, Phadke and Chakravarti (1972). Calculation of low field mobility of electrons based on a solution of the Boltzmann equation by an expansion in Legendre polynomials, Hammar and Magnusson (1972). Note on the polar mobility of electrons and holes, Kranzer (1972). Theory of light absorption by free carriers in semiconductors with a non-parabolic band, Aliev and Gashimzade (1972). Theory of electric conductivity of many-valley semiconductors in strong alternating electric fields based on a solution procedure of the Boltzmann equation, Chuenkov (1973). Calculation of the mobility of hot electrons in non-parabolic bands with acoustic and polar optical phonon scattering by means of a drifted Maxwellian distribution function and average conservation equations, Harris and Ridley (1973). Theory of hot electrons in strong magnetic fields based on Kurosawa's approach of Boltzmann like transport equations in energy space applied to n-type InSb, Yamada and Kurosawa (1973). Derivation of low-field transport properties in a quantizing parallel magnetic field by means of direct solution of the corresponding linearized Boltzmann equation, Magnusson (1973). Calculation of the Hall coefficient factor for semiconductors with non-parabolic conduction band by means of direct integration of the corresponding Boltzmann equation, Kranzer (1973). Theory of electron galvanomagnetic effects by application of the contraction mapping principle to iterative solutions of the Boltzmann equation which allows the proof of existence, uniqueness and convergence of the iteration procedure, Rode (1973). Calculation of the Hall and drift mobility of polar n-type semiconductors by direct solution of the Boltzmann equation including light to heavy hole band intervalley scattering as well as the common scattering mechanism, Kranzer (1973). Calculation of collision terms, i.e., integrals for small band gap semiconductors with respect to polar optical, acoustic and intervalley phonon scattering, ionized impurity scattering and carrier-carrier scattering, Stokoe and Cornwell (1973). Calculation of scattering probabilities for holes with respect to deformation potential and ionized impurity scattering mechanism, Costato and Reggiani (1973). Influence of resonance interactions between carriers on the transport coefficients, Ravich and Tamarchenko (1973). Investigation of electron-phonon interaction and intervalley scattering in semiconductors by means of field theoretic Green function techniques, Herbert (1973). Calculation of scattering processes of hot electrons using Kurosawa's energy diffusion equation and taking into account the disturbance of the polar optical phonon distribution due to carrier flow, Ferry (1973). Extensive treatment of the quantum theory of free carrier absorption in polar semiconductors based on a rate equation approach and perturbation theory, Jensen (1973). Estimate of the energy loss mechanism of hot electrons by two-phonon emission, Baumann (1973). Note on the hot electron conductivity in polar semiconductors, Licea (1974). Theory of free carrier absorption in polar semiconductors due to electron-longitudinal optical phonon interaction, Mycielski, Aziza, Mycielski and Balkanski (1974). Discussion of high field transport in GaAs and InP, Fawcett and Herbert (1974), Herbert, Fawcett and Hilsum (1976). Review of mobility calculations of holes for zincblende III–V and II–VI compounds, Kranzer (1974). Pseudopotential calculation of the piezoelectric constant of GaAs using the deformable ion model,

Zeller (1974). Calculation of electron transport quantities for polar semiconductors in external electric fields based on the solution of the Boltzmann equation by the Monte Carlo method, Aas and Blotekjaer (1974). Calculation of the electron mobility limited by two-phonon mode polar optical scattering by means of an iteration method for the solution of the Boltzmann equation, Nag (1974). Note on Monte Carlo calculation of hot electron mobility in PbS, Chattopadhyay (1974). Calculation of galvano- and thermomagnetic phenomena in *n*-InSb-type semiconductors with inelastic optical scattering based on the solution of the Boltzmann equation by Grigorev, Dykman and Tomchuk's method, Tolpygo (1974). Note on the electric conductivity of many-valley semiconductors with charged point defects, Kornyushin (1974). Note on the diffusion thermoelectric power in *n*-type InSb, Kosarev and Tamarin (1974). Theory of nonlinear galvanomagnetic effects in anisotropic many-valley semiconductors based on BERTA, Chuenkov (1974) (1975). Note on the interaction between hot electrons and acoustic phonons in InSb-type semiconductors, Akulinichev, Volkov and Kumekov (1975). Discussion of the phonon drag effect in InSb, Kosarev and Tamarin (1975). Calculation of electron transition probabilities and high temperature relaxation times for non-polar optical phonon scattering in small gap semiconductors, Boguslawski (1975). Investigation of the influence of electron-electron scattering on the transport coefficients in many-valley semiconductors by approximate solution of the Boltzmann equation, Mitin and Tolpygo (1975). Extension of the method of Fletcher and Butcher for solving the Boltzmann equation to calculate the mobility of electrons due to polar optical scattering to the case of non-parabolic bands, Basinski and Woolley (1975). Theory of electron diffusion under a temperature gradient in degenerate semiconductors with non-parabolic energy bands using BERTA, Chakravarti (1975). Note on the quantum theory of the Nernst-Ettinghausen effect in semiconductors with Kane's dispersion law, Agaeva, Askerov and Eminov (1975). Calculation of the relaxation time and the mobility for scattering of conduction electrons by magnetic impurities in semiconductors of InSb-type and HgTe-type band structure, Kossut (1975). Note on the influence of multiple scattering on the pseudo-resonance structure in the magnetophonon effect, Barker and Magnusson (1975). Longitudinal magneto-resistance in the extreme quantum limit in non-parabolic semiconductors based on BERTA, Phadke and Sharma (1975). Free carrier absorption in semiconductors with non-parabolic and ellipsoidal energy band structures, Das and Nag (1975). Density operator approach to the calculation of linear magneto-resistance in parabolic semiconductors, Arora (1975), its application to Ohmic galvano and thermomagnetic effects, Arora and Peterson (1975), and the effect of non-parabolicity on Ohmic magneto-resistance, Arora and Jaafarian (1976). Asymptotic solutions to the Boltzmann equation for degenerate polar semiconductors with electron-optical phonon mode scattering, Christie and McGill (1976). Contribution of the phonon drag effect to the calculation of transport coefficients in polar semiconductors, Plavitu (1976). Calculation of the warm electron coefficient in polar semiconductors based on an iterative solution of the Boltzmann equation, Chattopadhyay and Nag (1976). Influence of scattering of electrons by magnetic impurities on transport phenomena in small gap and zero gap semiconductors, Kossut (1976). Derivation of the electron distribution for high electric fields at low temperatures and carrier concentrations by direct solution of the Boltzmann equation, Devreese and Evrard (1976). Calculation of the hot electron mobility and electron temperature in n-type InSb by means of a displaced Maxwellian distribution function and the average conservation equations of energy and momentum, Komissarov (1976). Derivation of matrix elements of electron-phonon non-polar interactions in PbSe-like semiconductors, Szymanski (1976). Note on scattering on short-range potentials in semiconductors with small energy gap, Litwin-Staszewska and Szymanska (1976). Monte Carlo calculations of hot electron transient response to a short pulse of a spatially uniform electric field for CdTe and GaAs and of electron heating in a non-parabolic band semiconductor, Matulenis, Pozhela and Reklaitis (1976). Investigation of the influence of the recombination of carriers on the appearance of a negative differential conductance of hot electrons in a magnetic field, Asratyan and Kachlishvili (1976). Calculation of relaxation times of the conductivity of n-type GaAs in the warm electron range by means of the electron temperature approximation, Kagan, Landsberg and Elenkrig (1976). Effects of energy band non-parabolicity on the free carrier absorption in n-type GaP, Das and Nag (1976). Calculation of the mobility of holes in p-type InSb based on BERTA, Filipchenko and Bolshakov (1976). Note on the mechanism of intervalley electron transitions involving quasidiscrete levels in the Gunn effect, Timashev (1976). Phenomenological theory of intraband photoconductivity of compensated n-type InSb, Gulyaev, Listvin, Potapov, Chusov and Yaremenko (1976). Ohmic magneto-resistance for inelastic acoustic phonon scattering in semiconductors, Arora (1976). Quantum theory of non-linear galvanomagnetic effects in Kane semiconductors by evaluation of Adam and Holstein's current expression, Agaeva, Askerov and Gashimzade (1976). Calculation of electron mobility in ternary III–V compounds, Harrison and Hauser (1976). Derivation of an effective mass equation for multi-valley semiconductors, Shindo and Nara (1976). Criterion of electric breakdown in many-valley semiconductors, Chuenkov (1977). Calculation of electron mobility in InP by iterative solution of the Boltzmann equation, Nag and Dutta (1978). Calculation of the optical conductivity of free carriers in GaAs with BERTA and frequency dependent relaxation times, Jensen (1977). Note on the average energy of conduction electrons in non-parabolic semiconductors, Chattopadhyay (1977). Extension of the theory of the photo-magnetoelectric and photoconductivity effects to include unequal excitation rates of electrons and holes due to impurity photo excitation based on PMTE and elementary rate equations with application to GaAs, Look (1977). Analysis of the scattering of electrons by the deformation potential in doped InSb, Demchuk and Tsidilkovskii (1977). Study of thermoelectric power and thermomagnetic effects of hot electrons in Kane semiconductors in heating electric and non-quantizing magnetic fields by approximate solution of the Boltzmann equation, Gasymov, Katanov and Babaev (1977). Current-voltage characteristics of many-valley semiconductors with multi-valued dependence of the current on the electric field, Mitin (1977). Note on transport phenomena in semiconductors with wurtzite structure in longitudinal quantizing magnetic fields, Pinchuk (1977). Note on photon transfer of excitation of non-equilibrium carriers in GaAs, Epifanov, Galkin, Bobrova and Vavilov (1977). Convective instability of an electromagnetic wave in n-InSb in the presence of crossed electric and magnetic fields based on PMTE, Guha and Gosh (1977). Analysis of the transferred-electron effect in III–V semiconductors based on the average energy- and momentum balance equations, Ridley (1977). Note on low-temperature electron mobility in InSb, Nag and Dutta (1977). Note on the calculation of the energy relaxation time of hot electrons in InSb, Kobayashi (1977). Note on the screening of the deformation potential by free electrons in the multivalley conduction band, Boguslawski (1977). Note on the mobility in polar semiconductors in the extreme quantum limit, Sarkar and Basu (1977). Inclusion of the medium range disorder in the spatial distribution of impurities in the theory of galvanomagnetic phenomena, Robert, Pistoulet, Raymond, Aulombard, Bernard and Bousquet (1978). Theory of transient high-field conductivity in polar semiconductors based on analytical solutions of the Boltzmann equation, Leburton and Evrard (1978). Derivation of expressions for electric conductivity, thermoelectric power and thermal conductivity of conduction band electrons in small gap zincblende semiconductors by means of a variational solution procedure of the Boltzmann equation with inclusion of phonon drag and an extensive analysis of the scattering processes and corresponding collision terms, Szymanska and Dietl (1978). Calculation of deformation potential acoustic scattering of electrons in a Kane semiconductor under quantizing magnetic fields in the extreme quantum limit, Sarkar, Basu and Chattopadhya (1978). Derivation of the effective conductivity tensor for electrons in narrow band gap semiconductors under the influence of external electromagnetic fields and static magnetic fields based on an evaluation of the equation for the density operator with an energy-dependent relaxation time, Sutherland and Spector (1978). Investigation of the influence of acoustic phonon disturbances on the conductivity of piezo-electric semiconductors by means of BERTA for phonons and a drifted Maxwellian distribution function approach for electrons, Kocevar and Fitz (1978). Note on the effect of the electron spin on the magneto-Seebeck coefficient of polar semiconductors at large magnetic fields, Sarkar and Chattopadhyay (1978). Derivation of transport coefficients for electrons in InSb-type semiconductors by means of a variational solution procedure of the Boltzmann equation with inclusion of inelastic scattering of electrons by optic phonons, Pfeffer and Zawadzki (1978). Calculation of high field collision rates of hot electrons based on generalized transport equation derived from the equations for the density operator, Barker (1978). Review of theory and experiments of hot electrons and phonon phenomena under high intensity photoexcitation of semiconductors, Shah (1978). Model of removal and scattering of charge carriers by defect clusters in semiconductors, Kolchenko and Lomako (1978). Determination of the mean energy of photoexcited hot electrons in high magnetic fields by means of a shifted Maxwellian distribution and BERTA, Lewiner and Calecki (1978). Determination of the hot electron distribution in GaAs by means of a Monte Carlo solution of the Boltzmann equation, Inoue, Takenaka, Shirafuji and Inuishi (1978). Infrared absorption spectrum of *n*-type GaAs, Jensen (1978). Parametric excitation of electromagnetic vibrations in narrow-gap semiconductors with non-parabolic energy bands, Anh and Dat (1978).

8.6 Non-equilibrium phase transitions

An open system which is coupled to large reservoirs such as energy sources or sinks, e.g., external pumps or fields, can be driven into a state far from thermal equilibrium and maintained in a region where linear thermodynamics of irreversible processes is no longer valid. In such cases the state of the system has to be described by the solutions of a Pauli-Master equation or by rate equations or by other appropriate equations of irreversible quantum statistics. If the system is maintained by external forces in a time-independent thermodynamic non-equilibrium state, this state must be a stationary solution of the corresponding equations. Thus the stationary solutions are the analogues and the generalizations of the stationary thermodynamic equilibrium states for closed systems, or systems which are simply coupled to a heat reservoir. At certain threshold values of the driving forces or "control parameters" of an open system a stationary state of such a system may become unstable and change. Because this behaviour is in close analogy to phase transitions of systems in thermal equilibrium, these phenomena are called non-equilibrium phase transitions. There are numerous examples of this in many different branches of physics, chemistry, biology, ecology etc., see the review by Haken (1977). Discussing this topic we restrict ourselves to semiconductor non-equilibrium phase transitions. In the preceding chapters the processes in semiconductors far from equilibrium were described by rate equations. Hence, to study non-equilibrium phase transitions we have to discuss the properties of these rate equations. The notion of non-equilibrium phase transitions can be explained by comparison with equilibrium phase transitions. Hence we first give a short review of the theorems (without proofs) which are basic for such a comparison, cf. Stumpf and Rieckers (1976).

The basic quantity of equilibrium thermodynamics is the entropy function $S = S(z_1 \ldots z_m)$, where $\{z_i, 1 \leqslant i \leqslant m\}$ is a complete set of independent extensive variables of the system under consideration. We suppose that the entropy function S of a thermodynamic system Σ is known; then the following theorem holds:

Theorem 8.9 : A necessary condition for the stability of the thermodynamic system Σ in a state $z^0 \equiv \{z_i^0, 1 \leqslant i \leqslant m\}$ is the validity of the equation

$$\Delta^2_{(m-1)} S := \frac{1}{2} \sum_{ij=1}^{m-1} \frac{\partial^2 S(y_1 \ldots y_m)}{\partial y_i \partial y_j}\bigg|_{y_l = z_l^0} \Delta y_i \Delta y_j \leqslant 0 \tag{8.6.1}$$

for arbitrary values of the difference variables $\{\Delta y_i = (z_i - z_i^0), 1 \leqslant i \leqslant m-1\}$. If even $<$ holds, then (8.6.1) is a sufficient condition.

Equivalent to this theorem is the following theorem:

Theorem 8.10 : A necessary condition for the stability of Σ in z^0 is the negative-semidefiniteness of the matrix

$$S_{ik} := \left\{ \frac{\partial^2 S}{\partial y_i \partial y_k}\bigg|_{y_l = z_l^0} \right\}; \quad 1 \leqslant i,k \leqslant m-1 \tag{8.6.2}$$

A sufficient condition for stability is the negative definiteness of (8.6.2).

For a system with a given entropy function in general not the whole space $\mathbb{R}^m$, but only a subspace $G_s \subset \mathbb{R}^m$ is the set of stable states z.

Def. 8.1 : The boundary R_s of the region of stability G_s is called the stability boundary.

Def. 8.2 : A thermodynamic system is said to undergo a thermodynamic equilibrium phase transition if by means of a thermodynamic process the system reaches R_s or crosses R_s.

The equilibrium stability condition for S_{ik} can be formulated by means of the state equations for the conjugate intensive variables of the entropy representation. We define them by

$$X_i := \frac{\partial S}{\partial y_{i/y=z}} \equiv X_i(z_1 \ldots z_m), \quad 1 \leqslant i \leqslant m. \tag{8.6.3}$$

As the set $\{z\}$ contains the independent variables, the set $\{X\}$ contains dependent variables. It can be shown, however, in thermostatics that for the set $\{X\}$ an independent measurement prescription can be given, so that the validity of state equations (8.6.3) can be checked experimentally. The stability condition is then

$$S_{ik} = \left\{ \frac{\partial X_i(y_1 \ldots y_m)}{\partial y_k}{}_{/y=z^0} \right\}; \ 1 \leqslant i,k \leqslant m-1; \text{ neg. def.} \tag{8.6.4}$$

Instead of the entropy function S we may just as well use the internal energy function $U = U(z')$ with $z' = \{z'_i, \ 1 \leqslant i \leqslant m\}$ for the formulation of the stability condition. This leads then to the requirement

$$U_{ik} := \left\{ \frac{\partial^2 U}{\partial y'_i \partial y'_{k/y'_i = z'^0_i}} \right\}; \ 1 \leqslant i,k \leqslant m-1; \text{ pos. def.} \tag{8.6.5}$$

With the conjugate intensive variables of the energy representation

$$X'_i := \frac{\partial U}{\partial y'_{i/y'=z'}} = X'_i(z'_1 \ldots z'_m), \ 1 \leqslant i \leqslant m \tag{8.6.6}$$

(8.6.5) can be written

$$U_{ik} = \left\{ \frac{\partial X'_i(y'_1 \ldots y'_m)}{\partial y'_k}{}_{/y'=z'^0} \right\}; \ 1 \leqslant i,k \leqslant m-1; \text{ pos. def.} \tag{8.6.7}$$

It is supposed in thermostatics that all differentiations commute (Maxwell relations). Hence S_{ik} and U_{ik} are symmetric matrices and can be diagonalized by an orthogonal similarity transformation. The negative definiteness or positive definiteness, resp. of S_{ik}, or U_{ik}, can then be expressed by the properties of the eigenvalues of S_{ik}, or U_{ik}. If the eigenvalues of S_{ik} are denoted by $\{s_\lambda, 1 \leqslant \lambda \leqslant m-1\}$ and of U_{ik} by $\{u_\lambda, 1 \leqslant \lambda \leqslant m-1\}$, then for a stable point the conditions $\{s_\lambda < 0, \text{ real}, \ 1 \leqslant \lambda \leqslant m-1\}$, or $\{u_\lambda > 0, \text{ real}, \ 1 \leqslant \lambda \leqslant m-1\}$ resp. hold. The discussion of thermal equilibrium phase transitions is then partly done by means of the investigation of S_{ik}, or U_{ik} and partly by the direct use of the state equations (8.6.3), or (8.6.6).

Phase transitions of the first order are said to occur if two or more phases can coexist and if latent heats are involved. This includes that the specific entropies and internal energies of the coexisting phases have different values for common points in the coexistence region. It can be considered as a discontinuity of the entropy function, and internal energy function for a transition from one phase to another phase, i.e., $s^{\mathrm{I}} \neq s^{\mathrm{II}}$, $u^{\mathrm{I}} \neq u^{\mathrm{II}}$ at the same point. Higher order phase transitions are said to occur if for the transition from one phase to another phase the specific entropy and internal energy are continuous functions, i.e., $s^{\mathrm{I}} - s^{\mathrm{II}} = 0$, $u^{\mathrm{I}} - u^{\mathrm{II}} = 0$ at the same point, while some of the derivatives are discontinuous. In particular, if all derivates of order $k < n$ are continuous, while the n-th derivative is discontinuous, the transitions are said to be n-th order phase transitions. This classification can be justified by a systematic evaluation of the stability conditions of the matrix (8.2.6), cf. Stumpf and Rieckers (1976).

We now turn to the non-equilibrium phase transitions. They have to be derived from the equations of irreversible quantum statistics. As our deductions are concerned with rate equations we concentrate on these equations and investigate their stability properties. It will turn out that non-equilibrium phase transitions are closely connected with dynamic stability of the solutions of rate equations. The dynamic stability can be introduced by the following considerations.

If a physical system is excited by an interaction with external forces, two responses may generally occur after switching off the excitation

i) the system returns to a stationary equilibrium state under energy dissipation;
ii) the system does not return to a stationary state.

In the latter case, for instance (if the system did not explode) such a system shows periodic motions and thus many physical observables will not converge as time approaches infinity. Mathematically the various situations which lead to a different time asymptotic behaviour of the system can be characterized by the variation of suitably chosen parameters incorporated in the system of rate equations. These parameters are called "driving forces" or "control parameters" in complete analogy to the parameters by which equilibrium phase transitions are influenced. Before we discuss the situation for semiconductors, we give without proof some mathematical theorems concerning stability. In a general form the rate equations which have been derived in the preceding chapters can be written as

$$\frac{dn_i}{dt} = Y_i(n_1 \ldots n_r, t); \quad 1 \leqslant i \leqslant r \tag{8.6.8}$$

where $\mathfrak{n} = \{n_i, 1 \leqslant i \leqslant r\}$ are the mean occupation numbers, for instance of electrons. But likewise, other occupation numbers of various kinds can be considered. In vector notation (8.6.8) reads

$$\frac{d\mathfrak{n}}{dt} = \mathfrak{Y}(\mathfrak{n}, t) \tag{8.6.9}$$

with $\mathfrak{n} := (n_1 \ldots n_r)$, $\mathfrak{Y} := (Y_1 \ldots Y_r)$.

Def. 8.3: The point $\mathfrak{n} \equiv \mathfrak{n}^l$ is called a singular point of the system (8.6.9), if for all possible t the condition $\mathfrak{Y}(\mathfrak{n}^l, t) = 0$ is satisfied.

According to Definition 8.3, a singular point $\mathfrak{n}^l$ of (8.6.9) is a stationary solution of the system (8.6.9). Concerning the asymptotic behaviour of timedependent solutions $\mathfrak{n} = \mathfrak{n}(t)$, the following definition can be used

Def. 8.4: Assume that $\mathfrak{n}^l$ is a singular point. Then

i) $\mathfrak{n}^l$ is called locally stable if for all $\varepsilon > 0$ and all $t_0 \geq \tau$ there exists a function $\eta(\varepsilon, t_0) > 0$ such that from $\|\mathfrak{n}(t_0) - \mathfrak{n}^l\| < \eta$ it follows that $\|\mathfrak{n}(t) - \mathfrak{n}^l\| < \varepsilon$ for $t > t_0$.

ii) $\mathfrak{n}^l$ is called locally asymptotically stable if for all $t_0 \geq \tau$ there exists a function $\eta(t_0)$ such that from $\|\mathfrak{n}(t_0) - \mathfrak{n}^l\| < \eta$ it follows that $\lim_{t \to \infty} \|\mathfrak{n}(t) - \mathfrak{n}^l\| = 0$.

iii) $\mathfrak{n}^l$ is called unstable if for any $\varepsilon, \eta > 0$ and arbitrary t_0 there exists a time $t_1 > t_0$ such that from $\|\mathfrak{n}(t_0) - \mathfrak{n}^l\| \leq \eta$ it follows that $\|\mathfrak{n}(t_1) - \mathfrak{n}^l\| = \varepsilon$.

The physical interpretation of this definition is obvious. To derive statements about stability one may use the theorems of Ljapunov which connect such properties with the properties of the functions $Y_i(\mathfrak{n}, t)$ ocurring in (8.6.8).

Def. 8.5: i) A function $W(\mathfrak{n})$ is positive definite in an open neighbourhood Ω of $\mathfrak{n}^l$ if it has continuous first derivatives in Ω, vanishes in $\mathfrak{n}^l$ and is strictly positive in $\Omega / \{\mathfrak{n}^l\}$.

ii) A function $V(\mathfrak{n}, t)$ is positive definite in an open neighbourhood Ω of $\mathfrak{n}^l$ if

a) $V(\mathfrak{n}, t)$ is defined on Ω for all $t \geq 0$,

b) $V(\mathfrak{n}^l, t) = 0$ for all $t \geq 0$,

c) a positive definite function $W(\mathfrak{n})$ exists on Ω with $W(\mathfrak{n}) \leq V(\mathfrak{n}, t)$ for all $\mathfrak{n} \in \Omega$ and for all $t \geq 0$.

iii) $V(\mathfrak{n}, t)$ is a Ljapunov-function (L-function) around $\mathfrak{n}^l$ if it is positive definite on an open neighbourhood Ω of $\mathfrak{n}^l$, if it has continuous first derivatives and if

$$\frac{dV}{dt} = \frac{\partial V}{\partial t} + \dot{\mathfrak{n}} \cdot \nabla_{\mathfrak{n}} V \tag{8.6.10}$$

is negative on Ω.

We first consider the special case $\mathfrak{n}^l = 0$. We call $\mathfrak{n}^l = 0$ the origin. The following theorems of Ljapunow hold:

Theorem 8.11: If in a neighbourhood of the origin there exists an L-function, then the origin is stable.

Theorem 8.12: If in Ω the L-function $V(\mathfrak{n}, t)$ is dominated by a positive definite function $W(\mathfrak{n})$ and if dV/dt is negative definite, then the origin is asymptotically stable.

Theorem 8.13: If $V(\mathfrak{n}, t)$ has continuous first partial derivatives in Ω, if $V(0, t) = 0$, if dV/dt is positive definite and if V can assume positive values in any region containing the origin, then the origin is unstable.

Additionally it can be shown that the existence of such functions is necessary and sufficient for the stability or instability, resp. of the origin of the system.
The general case is $\mathfrak{n}^l \neq 0$. This case can be reduced to the special case of $\mathfrak{n}^l = 0$ by performing the transformation

$$\mathfrak{n}' = \mathfrak{n} - \mathfrak{n}^l. \tag{8.6.11}$$

Then, if $\mathfrak{n}^l \neq 0$ is the singular point of the $\mathfrak{n}$-basis, it follows that $\mathfrak{n}'^l = 0$ i.e., the origin is the singular point of the $\mathfrak{n}'$-basis. By substitution of (8.6.11) into (8.6.9), the transformed system reads

$$\frac{d\mathfrak{n}'}{dt} = \mathfrak{Y}'(\mathfrak{n}', t) \equiv \mathfrak{Y}(\mathfrak{n}' + \mathfrak{n}^l, t). \tag{8.6.12}$$

A Taylor expansion of (8.6.12) about the point $\mathfrak{n}' = \mathfrak{n}^l$ gives

$$\frac{d\mathfrak{n}'}{dt} = \mathfrak{Y}(\mathfrak{n}^l, t) + \left.\frac{\partial \mathfrak{Y}(\mathfrak{n}' + \mathfrak{n}^l, t)}{\partial \mathfrak{n}'}\right|_{\mathfrak{n}'=0} \cdot \mathfrak{n}' + \mathfrak{R} \tag{8.6.13}$$

where $\mathfrak{R}$ contains the higher order terms. As $\mathfrak{n}^l$ is assumed to be a singular point of (8.6.9), we have $\mathfrak{Y}(\mathfrak{n}^l, t) = 0$ for all possible t. Therefore, we obtain from (8.6.13)

$$\frac{d\mathfrak{n}'}{dt} = \left.\frac{\partial \mathfrak{Y}(\mathfrak{n}' + \mathfrak{n}^l)}{\partial \mathfrak{n}'}\right|_{\mathfrak{n}'=0} \cdot \mathfrak{n}' + \mathfrak{R}. \tag{8.6.14}$$

Under certain assumptions about $\mathfrak{R}$, Ljapunow derived stability criteria for (8.6.14). They can be summarized by

Theorem 8.14: The solutions of (8.6.14) are locally asymptotically stable with respect to $\mathfrak{n}^l = 0$, if all eigenvalues of the Jacobian matrix

$$A_{ik} := \left\{ \left.\frac{\partial Y_i(\mathfrak{n}' + \mathfrak{n}^l)}{\partial n'_k}\right|_{n'_i=0} \right\}; \quad 1 \leqslant i, k \leqslant r \tag{8.6.15}$$

have strictly negative real parts. The solutions are unstable if at least one eigenvalue has a strictly positive real part.

By this theorem it is possible to reduce local stability investigations to the linear part of the system which can be treated much more easily than the total expression. The negative definiteness conditions for the eigenvalues of $\mathscr{A}$ can be expressed in various forms. As the eigenvalues can be calculated from det $|\mathscr{A} - \lambda| = 0$ leading to a polynomial in λ the roots of which are the eigenvalues, the following theorem can be used:

Theorem 8.15: A necessary and sufficient condition for the negative definiteness of the real parts of the roots of $f(z) = \sum_{\nu=0}^{n} a_\nu z^{n-\nu}$ is $a_i/a_0 > 0$ and $\Delta_i > 0$, $1 \leqslant i \leqslant n$ where Δ_i are the Hurwitz determinants

$$\Delta_i := \det \begin{vmatrix} a_1 & a_0 & 0 & 0 & \dots & \dots \\ a_3 & a_2 & a_1 & a_0 & 0 \dots & \dots & \dots \\ a_5 & a_4 & a_3 & a_2 & a_1 & \dots & \dots \\ a_7 & a_6 & a_5 & a_4 & a_3 & \dots & \dots \\ \cdot \\ \cdot \\ \cdot \\ a_{2i-1} & \dots & \dots & \dots & \dots & a_i \end{vmatrix}, \quad 1 \leqslant i \leqslant n$$

with $a_j = 0$ if $j > n$

In the case $n=2$, the conditions of Theorem 8.15 reduce to $-a_1 \equiv \mathrm{tr}\,\mathscr{A} < 0$, $a_2 \equiv \det \mathscr{A} > 0$.

As has already been emphasized, stability is of interest if the system considered is under the influence of external forces or of internal forces resulting from the interaction of the various subsystems. For brevity we assume that the effect of all these influences and interactions can be described by one vectorial parameter $\mathfrak{k} = (k_1 \dots k_s)$. Therefore, we consider the rate equations

$$\frac{d\mathfrak{n}}{dt} = \mathfrak{Y}(\mathfrak{n}, \mathfrak{k}, t) \tag{8.6.16}$$

with $\mathfrak{k}$ real. Then the singular point of the system $\mathfrak{n}^l = \mathfrak{x}^l(\mathfrak{k})$ depends on $\mathfrak{k}$ for $a_i \leqslant k_i \leqslant b_i$, $1 \leqslant i \leqslant s$, if k_i is allowed to vary in this interval. It is now possible that for $k_i \leqslant k_i^0$, the singular point is asymptotically stable, but that for $k_i > k_i^0$ it is unstable. That means that for any solution $\mathfrak{n}(t)$ in the range $k_i \leqslant k_i^0$, $\lim \mathfrak{n}(t) = \mathfrak{n}^l(\mathfrak{k})$, while for any solution $\mathfrak{n}(t)$ in the range $k_i > k_i^0$, $\mathfrak{n}(t, \mathfrak{k})$ may converge towards a new stationary state or a periodical solution. This behaviour is called a bifurcation. In the case of instability obviously at least one of the roots of (8.6.15) has non-negative real parts. If at certain threshold values of the external control parameters (critical points) stable singular points turn unstable, one has to expect a drastic qualitative change in the nature of the stationary non-equilibrium states of the system. These qualitative changes are called non-equilibrium phase transitions of first order if $\mathfrak{n}(\mathfrak{k})$ is discontinuous and of second order if $\mathfrak{n}(\mathfrak{k})$ is continuous but $\frac{\partial}{\partial \mathfrak{k}} \mathfrak{n}(\mathfrak{k})$ is discontinuous.

The analogy between equilibrium and non-equilibrium phase transitions is obvious if the conditions for A_{ik} of (8.6.15) are compared with those for U_{ik}, or S_{ik} of (8.6.4), or (8.6.7). In analogy to the case of equilibrium phase transitions the conditions for A_{ik} may be studied or the transitions may be derived directly from a discussion of the state equation which is given by $\mathfrak{Y}(\mathfrak{n}) = 0$ in the non-equilibrium case.

Investigations of non-equilibrium phase transitions for chemical reaction systems by means of rate equations which are very similar to semiconductor rate equations were made by Schlögl (1972). Landsberg and Pimpale (1976) found recombination induced non-equilibrium phase transitions in simple semiconductor models. Schöll (1978) made an extensive investigation of phase transitions in intrinsic and extrinsic semiconductors, see also Landsberg, Robbins and Schöll (1978), Schöll (1979), Schöll and Landsberg (1979), Robbins, Landsberg and Schöll (1981), Schöll (1982). Here we give a short review of the results obtained by Schöll (1978).

a) Models for second-order phase transitions

α) Band-trap impact ionization

It can be shown by an analysis of the stability condition (8.6.15) for semiconductor rate equations that (band-band or band-trap) impact ionization is a key process for the occurrence of non-equilibrium phase transitions, Schöll (1979). The simplest model of a second-order transition with impact ionization of traps

$$e+N_1 \xrightarrow{W_1^a} 2e+N_0 \tag{8.6.17}$$

involves conduction electrons e of concentration n, occupied traps N_1 of concentration n_t and empty traps N_0 of concentration p_t. The other processes to be considered between these species are

$$N_1 \xrightarrow{W_1^t} e+N_0 \quad \text{(thermal ionization)} \tag{8.6.18}$$

$$e+N_0 \xrightarrow{W_2^t} N_1 \qquad \text{(electron capture by traps)} \tag{8.6.19}$$

$$2e+N_0 \xrightarrow{W_2^a} e+N_1 \qquad \text{(Auger recombination).} \tag{8.6.20}$$

Conservation of the total number of traps N_t requires

$$n_t+p_t=N_t \tag{8.6.21}$$

while charge conservation gives

$$n_t+n=N_D \tag{8.6.22}$$

where we assume the semiconductor to be partially compensated by N_D additional, fully ionized shallow donors to provide for free carriers; $N_D<N_t$, cf. McCombs and Milnes (1972).

From (8.6.21) and (8.6.22) two of the three variables can be eliminated:

$$\begin{aligned} n_t&=N_D-n \\ p_t&=(N_t-N_D)+n \end{aligned} \tag{8.6.23}$$

so that we are effectively dealing with a single-variable system.

If thermal ionization (8.6.18) is neglected, the steady state for n, using the reactions (8.6.17) (8.6.19) and the equation (8.6.23), may be written

$$\begin{aligned}\dot{n} &= W_1^a n n_t - W_2^t n p_t = 0 \\ &= n[W_1^a N_D - W_2^t(N_t - N_D)] - n^2[W_2^t + W_1^a].\end{aligned} \tag{8.6.24}$$

The above equation has always the trivial solution $n=0$, and for

$$W_1^a N_D - W_2^t(N_t - N_D) \geqslant 0 \tag{8.6.25}$$

an additional positive solution

$$n = \frac{W_1^a N_D - W_2^t(N_t - N_D)}{W_1^a + W_2^t}. \tag{8.6.26}$$

The solution (8.6.26) is stable whenever it exists, and $n=0$ is stable otherwise. The stable steady states of (8.6.24) are therefore with (8.6.23)

$$\begin{gathered} n = \begin{cases} 0 & , \quad W_1^a \leqslant W_2^t \dfrac{N_t - N_D}{N_D} \\[2ex] \dfrac{W_1^a N_D - W_2^t(N_t - N_D)}{W_1^a + W_2^t}, & \quad W_1^a > W_2^t \dfrac{N_t - N_D}{N_D} \end{cases} \\[2ex] n_t = \begin{cases} N_D \\ \dfrac{W_2^t N_t}{W_1^a + W_2^t} \end{cases}, \quad p_t = \begin{cases} N_t - N_D \\ \dfrac{W_1^a N_t}{W_1^a + W_2^t} \end{cases}. \end{gathered} \tag{8.6.27}$$

A plot of the steady state concentrations (order parameter) against the impact ionization coefficient W_1^a, taken as a control parameter (Fig. 8.7) shows the typical picture of a second-order phase transition. The critical point where the order parameter $n(W_1^a)$ is continuous but has a discontinuous derivative is given by

$$W_1^a = W_2^t \frac{N_t - N_D}{N_D}. \tag{8.6.28}$$

As W_1^a depends on the applied electric field, the transition can be brought about by increasing this. Physically, this corresponds to the transition from a non-conducting ($n=0$) to a conducting phase ($n>0$) when the impact ionization is strong enough to compensate the electron losses by trapping.

The critical point and the qualitative behaviour of the steady state concentrations vs. the control parameter is not affected if Auger recombination (8.6.20) is also taken into account.

Note that the threshold condition (8.6.25) can be attained all the easier, the larger N_D is in proportion to the trap concentration N_t. If, however, the semiconductor is fully compensated ($N_D = N_t$), or even overcompensated ($N_D > N_t$), the phase transition disappears, because (8.6.25) is then always satisfied, and the positive, conducting

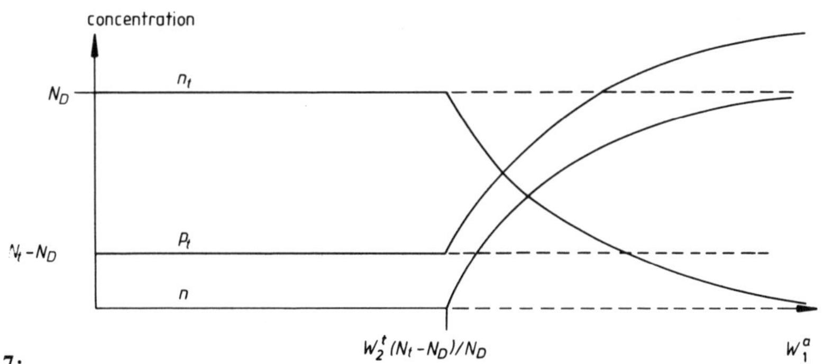

Figure 8.7:
Steady state concentrations (8.6.27) versus impact ionization coefficient W_1^a (schematic); after Schöll (1978).

steady state is unique and stable throughout the whole range of the impact ionization coefficients. In this case $n=0$ no longer corresponds to a physical state, since this would imply $p_t<0$ by (8.6.23). Physically, this means that the fully ionized donors provide more conduction electrons than can be captured by traps.
On the other hand, $N_D=0$ would render $n=0$ the only physical steady state by (8.6.25); there would be no free electrons at all available.

β) Dielectric breakdown

The second-order phase transition (8.6.27) has a remarkable connection to dielectric breakdown in solids. This phenomenon is characterized by the fact that the carrier concentration, and thus the current, in a semiconductor suddenly increases by many orders of magnitude at a certain strength of the applied static electric field, due to the impact ionization of traps, McCombs and Milnes (1972), Bannaya et al. (1973), Cohen and Landsberg (1967). At still higher fields (10^5–10^6 V/cm) the carriers are hot (i.e. energetic) enough to impact ionize electrons from the valence band, and the so-called avalanche breakdown eventually destroys the whole semiconductor device, cf. O'Dwyer (1958).
The breakdown criterion for band-trap impact ionization is usually derived from the steady state equation of processes (8.6.17)–(8.6.19) (Auger recombination (8.6.20) is generally neglected)

$$0=W_1^t n_t - W_2^t n p_t + W_1^a n n_t \tag{8.6.29}$$

in the following way, cf. McCombs and Milnes (1972), Koenig et al. (1962): For small n, (8.6.23) becomes approximately

$$n_t \approx N_D, \quad p_t \approx N_t - N_D \tag{8.6.30}$$

Substitution into (8.6.29) yields the electron concentration

$$n \approx \frac{W_1^t N_D}{W_2^t(N_t-N_D)-W_1^a N_D} \tag{8.6.31}$$

which becomes infinite for

$$W_1^a N_D = W_2^t (N_t - N_D). \tag{8.6.32}$$

Eq. (8.6.32) is the breakdown criterion which determines the breakdown field via W_1^a. Of course, approximation (8.6.30) is no longer valid in the breakdown region, and the exact solution of the quadratic equation given by (8.6.29) with (8.6.23)

$$0 = W_1^t N_D - n[W_1^t + W_2^t(N_t - N_D) - W_1^a N_D] - n^2[W_2^t + W_1^a] \tag{8.6.33}$$

yields a finite value of n.

The threshold (8.6.32) equals precisely the critical point (8.6.28) of the second-order transition, which is marked by a sharp rise from zero to nonzero concentration n, if the thermal ionization W_1^t is neglected in (8.6.29). Although arising from quite different approximations, (8.6.27) and (8.6.31) describe the same physical phenomenon: an abrupt increase in the carrier concentration.

The approximations (8.6.31) and (8.6.27) and the exact solution of (8.6.33) are plotted together in Fig. 8.8. A finite W_1^t, however small it may be, smoothes out the second-order phase transition, just as does a magnetic field in the case of a ferromagnet. In the limit $W_1^t \to 0$ dielectric breakdown caused by impact ionization of traps thus represents a second-order nonequilibrium phase transition.

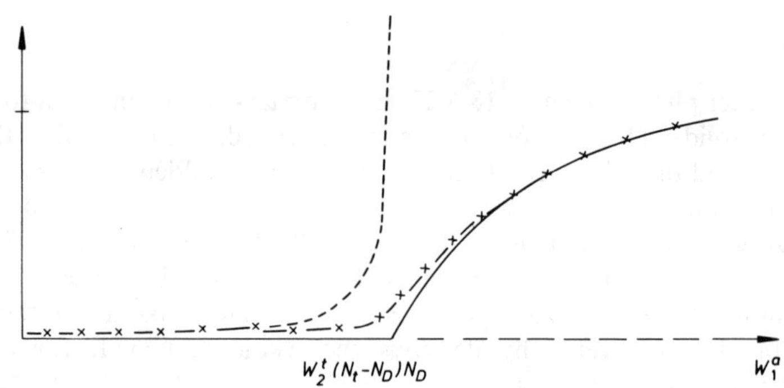

Figure 8.8:
Solution of (8.6.33) (—×—), approximate solution (8.6.31) (— — —), and solution (8.6.27) which follows from (8.6.33) for $W_1^t = 0$ (———) (schematic); after Schöll (1978).

Typical numerical values for gold in silicon (acceptor level $E_c - 0.54$ eV) are, McCombs and Milnes (1972):

$$N_D \approx 10^{15}\text{–}10^{16}\ \text{cm}^{-3}$$
$$N_t - N_D \approx 10^{14}\text{–}10^{15}\ \text{cm}^{-3}$$
$$W_2^t \approx 10^{-8}\text{–}10^{-9}\ \text{cm}^3\ \text{s}^{-1}\ (77\ \text{K–}300\ \text{K})$$
$$W_1^a \approx 10^{-7}\ \text{cm}^3\ \text{s}^{-1}\ (\text{for breakdown fields } 10^3\text{–}10^4\ \text{V/cm}).$$

The neglect of W_1^t is a good approximation at temperatures $k_B T \ll 0.54$ eV.

γ) Phosphor with colour centers

Single electron processes

As a practical application of the previous chapters we investigate the possibility of phase transitions in a well-known complex multivariable system with multivalent flaws and excited levels. As a model for photoconducting phosphors ionic crystals with colour centres have been studied, Fowler (1968). These impurities consist of anion vacancies, usually in an alkali halide crystal, which can be empty (α-centres) or occupied with one (F-centre) or two (F'-centre) bound electrons. Hence such a centre forms a divalent flaw with charge conditions $s=0$, 1, 2. Speaking in terms of a photoconducting phosphor, the F-center acts as an activator, the F'-center as a trap. Sometimes the first excited state of the F-center (F^*) is also considered. The single-electron processes have been discussed by means of reaction kinetics, Lüty (1961), Hoffmann, Stöckmann and Tödheide-Haupt (1973).

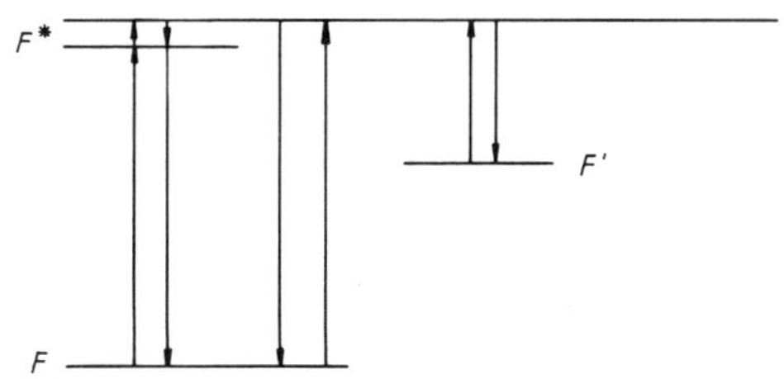

Figure 8.9:
Energy level diagram of an $F-F'$-system.

The following processes are usually considered (Fig. 8.9):

$$F \xrightarrow{W_1^s} F^* \tag{8.6.34}$$

(optical excitation by irradiation into the F-band)

$$F^* \xrightarrow{W_1} F \tag{8.6.35}$$

(non-radiative or radiative relaxation (fluorescence))

$$F^* \xrightarrow{W_5} e+\alpha \tag{8.6.36}$$

(thermal or optical ionization of an excited F-centre)

$$e+\alpha \xrightarrow{W_2} F^* \tag{8.6.37}$$

(electron capture from the conduction band by an α-centre into the excited state F^*)

$$F' \xrightarrow{W_4} e+F \tag{8.6.38}$$

(thermal or optical ionization of an F'-centre)

$$e+F \xrightarrow{W_3} F' \tag{8.6.39}$$

(electron capture by an F-centre)

$$F \xrightarrow{W_2^s} e+\alpha \tag{8.6.40}$$

(optical ionization of an F-centre)

$$e+\alpha \xrightarrow{W_6} F \tag{8.6.41}$$

(direct capture of an electron by an α-centre into the ground state F).

W_1^s is the incident photon current density multiplied by the optical absorption cross section of the F-centre; W_1 is the inverse radiative and non-radiative decay time of the F^*-centre; W_2, W_6 and W_3 are recombination coefficients; W_4 and W_5 are ionization coefficients. Direct capture (8.6.41) predominates the recombination via the F^* state (8.6.37) plus subsequent relaxation (8.6.35) above the temperature range where F'-centres become thermally unstable, Hoffmann et al. (1973); e.g. $T \geqslant 200$ K for KCl. Reactions involving the valence band can be neglected since they are hampered by the large energy gap in alkali halides (several eV).

Tunnel and Auger processes

Tunnel processes between neighbouring impurities have been deduced from experiments, DeMartini, Grassano and Simoni (1974).

$$2F^* \rightarrow F' + \alpha. \tag{8.6.42}$$

Of the same type are the reactions

$$2F \rightarrow F' + \alpha \tag{8.6.43}$$

$$F + F^* \rightarrow F' + \alpha \tag{8.6.44}$$

and the reverse reactions should also be possible if the energy is provided by photons or phonons:

$$\begin{aligned} &F' + \alpha \rightarrow 2F^* \\ &F' + \alpha \rightarrow 2F \\ &F' + \alpha \rightarrow F + F^* \end{aligned} \tag{8.6.45}$$

Auger transitions have also been found in crystals with colour centers, Fröhlich and Mahr (1966), Berezin (1969). The following Auger transitions and their corresponding reverse reactions are theoretically possible, Tröster (1976):

$$
\begin{array}{lll}
(a) & 2e+\alpha & \rightleftarrows e+F \\
(b) & 2e+\alpha & \underset{W_1^a}{\overset{W_2^a}{\rightleftarrows}} e+F^* \\
(c) & 2e+F & \rightleftarrows e+F' \\
(e) & e+F^* & \rightleftarrows e+F \\
(f) & 2F^* & \rightleftarrows e+F+\alpha \\
(g) & F^*+F' & \rightleftarrows e+2F \\
(h) & F^*+F & \rightleftarrows e+F+\alpha \\
(i) & e+F^*+\alpha & \rightleftarrows e+F+\alpha \\
(j) & e+F'+\alpha & \rightleftarrows e+2F \\
(h) & e+F'+\alpha & \rightleftarrows e+F+F^*.
\end{array}
\qquad (8.6.46)
$$

δ) Impact ionization of F^*-centers

It has been deduced from photoconducting experiments, Hoffmann et al. (1973) that alkali halide crystals contain additional α-centers (Schottky defects) which are frozen in at high temperatures with concentration $n_\alpha(0)$ when the crystal samples are quenched to room temperature. The charge of these additional α-centers is compensated by other charged defects.
In the simplest model F'-centers are not taken into regard. The conservation of charge and of the total impurity concentration then imposes the following conditions:

$$n-n_\alpha+n_\alpha(0)=0 \qquad (8.6.47)$$

and

$$n_F+n_{F^*}+n=n_0 \qquad (8.6.48)$$

where $n_F \gg n_{F^*}$, n_α, n_0 denote the concentration of F-, F^*-, α-centers and of anion vacancies, resp.
We use the impact ionization of F^*-centers (8.6.46) (b) as autocatalytic key reaction. As in α), a second-order phase transition can be expected if the generation of electrons

$$F^* \xrightarrow{W_5} e+\alpha \qquad (8.6.49)$$

is neglected. For low temperature, where thermalization of F^*-centers is negligible, $W_5=0$ is a reasonable approximation.
We will explicitly evaluate the steady states for the processes (8.6.34–37), (8.6.46) (b) with $W_5=0$. The rate equations for F-centers and electrons are:

$$\dot{n}_F=-W_1^s n_F+W_1 n_{F^*} \qquad (8.6.50)$$

$$\dot{n}=-W_2 n_\alpha n-W_2^a n^2 n_\alpha+W_1^a n_{F^*} n. \qquad (8.6.51)$$

Elimination of n_{F*}, n_α by (8.6.47) and (8.6.48) yields in the steady state equations

$$0 = -W_1^s n_F + W_1 n_0 - W_1 n_F - W_1 n \;\rightarrow\; n_F = \frac{W_1}{W_1^s + W_1}(n_0 - n) \tag{8.6.52}$$

$$\begin{aligned} 0 = & -W_2 n_\alpha(0) n - W_2 n^2 - W_2^a n_\alpha(0) n^2 - W_2^a n^3 \\ & + W_1^a n_0 n - W_1^a n_F n - W_1^a n^2 \end{aligned} \tag{8.6.53}$$

and substitution of (8.6.52) into (8.6.53) gives a cubic equation for n:

$$\begin{aligned} & \left[W_1^a \frac{W_1^s}{W_1^s + W_1} n_0 - W_2 n_\alpha(0) \right] n \\ & - \left[W_2 + W_2^a n_\alpha(0) + W_1^a \frac{W_1^s}{W_1^s + W_1} \right] n^2 - W_2^a n^3 = 0 \end{aligned} \tag{8.6.54}$$

which has the trivial solution $n=0$ and an additional positive solution

$$\begin{aligned} n' = & \frac{W_2 + W_2^a n_\alpha(0) + W_1^a \dfrac{W_1^s}{W_1^s + W_1}}{2\, W_2^a} \\ & \times \left\{ \left[1 + \frac{4 W_2^a \left(W_1^a \dfrac{W_1^s}{W_1^s + W_1} n_0 - W_2 n_\alpha(0) \right)}{\left(W_2 + W_2^a n_\alpha(0) + W_1^a \dfrac{W_1^s}{W_1^s + W_1} \right)^2} \right]^{\frac{1}{2}} - 1 \right\} \end{aligned} \tag{8.6.55}$$

for

$$W_1^a n_0 > W_2 n_\alpha(0) \frac{W_1^s + W_1}{W_1^s}. \tag{8.6.56}$$

Inserting $n=0$ into (8.6.52), (8.6.47), (8.6.48) yields the concentration of the other species in the "zero steady state":

$$n_F = \frac{W_1}{W_1^s + W_1} n_0; \quad n_{F*} = \frac{W_1^s}{W_1^s + W_1} n_0; \quad n_\alpha = n_\alpha(0). \tag{8.6.57}$$

The Jacobian Matrix $A \equiv \begin{pmatrix} \partial \dot{n}_F / \partial n_F & \partial \dot{n}_F / \partial n \\ \partial \dot{n} \;/ \partial n_F & \partial \dot{n} \;/ \partial n \end{pmatrix}$, calculated from (8.6.52) (8.6.53), taken at the zero steady state (8.6.57), is

$$A = \begin{pmatrix} -(W_1^s + W_1), & -W_1 \\ 0 \quad , & W_1^a n_0 \dfrac{W_1^s}{W_1^s + W_1} - W_2 n_\alpha(0) \end{pmatrix}. \tag{8.6.58}$$

The zero steady state (8.6.57) is asymptotically stable if both eigenvalues of A are negative, i.e.,

$$W_1^a n_0 \frac{W_1^s}{W_1^s + W_1} < W_2 n_\alpha(0). \tag{8.6.59}$$

The Jacobian matrix A' at the positive steady state n' (8.6.55) is, using (8.6.53) with $n \neq 0$

$$A' = \begin{pmatrix} -(W_1^s + W_1), & -W_1 \\ -W_1^a n' \quad , & -(W_2 n' + W_2^a n_\alpha(0) n' + 2 W_2^a n'^2 + W_1^a n') \end{pmatrix}. \tag{8.6.60}$$

Obviously,

$$\operatorname{tr} A' < 0$$

and

$$\det A' = (W_1^s + W_1)\,(W_2 n' + W_2^a n_\alpha(0) n' + 2 W_2^a n'^2 + W_1^s W_1^a n') > 0$$

therefore the positive solution n' of (8.6.54) is stable whenever it exists by Theorem 8.15.

Thus, just as in model α) a second-order phase transition is induced by impact ionization and recombination of traps in the absence of thermal ionization ($W_1^t = 0$ and $W_5 = 0$), but the critical point

$$W_1^a = W_2 \frac{n_\alpha(0)}{n_0}\left(1 + \frac{W_1}{W_1^s}\right) \tag{8.6.61}$$

differs from (8.6.28) by a term W_1/W_1^s, i.e., the ratio of the rate constants for F^*-center relaxation and excitation, due to the influence of the excited trap level.

Since both photon current density W_1^s and impact ionization coefficient W_1^a can be driven externally, either may be taken as a control parameter. The qualitative behaviour of the system is retained if the Auger recombination is neglected ($W_2^a = 0$), or if reactions (8.6.38) (8.6.39) are added.

The interesting feature of the phase transition of this model is that – in contrast to other semiconductor models, Landsberg and Pimpale (1976) – it is enhanced by irradiation. At temperatures where normal photoconductivity is quenched because the F^*-centers are not thermalized, free carriers ($n > 0$) can spontaneously be created by application of a sufficiently strong electric field and simultaneous irradiation into the F-band. Neither of these alone can produce the effect.

The stronger the irradiation W_1^s and the smaller the number of frozen-in α-centers $n_\alpha(0)$, the lower is the required field strength.

Avalanche breakdown in alkali halides caused by band-band impact ionization has been investigated by Vorobev et al. (1976), Yasojima et al. (1971, 1975) and Okumura et al. (1976) at fields $\approx 10^6$ V/cm, but no impurity breakdown of the above kind has been observed so far.

Typical numerical values for KCl lie within the following ranges:

$n_0 = 10^{16}$–10^{18} cm^{-3} Lüty (1961), Hoffmann (1973)

$n_\alpha(0) = 10^{11}$–10^{12} cm^{-3} Hoffmann (1973)

$W_2 = 5\times 10^{-3}$–10^{-5} cm^3 s^{-1} Hoffmann (1973) for temperatures $T=50$–100 K

$W_1 \approx 10^6$ s^{-1} Swank and Brown (1963) Honda and Tomura (1972) for $T=4$–130 K

Normal irradiation, Hoffmann (1973), yields $W_1^s = 10^{-6}$–10^{-2} s^{-1}, laser irradiation with 100 kW cm^{-2}, DeMartini et al. (1974), reaches up to $W_1^s = 10^8$ s^{-1} (calculated with an absorption cross section of 10^{-16} cm^2, Bosi (1974)). The resulting critical values of the impact ionization coefficient $W_1^{a,\mathrm{crit.}}$ according to (8.6.61) are given for some combinations of the above values in Table 8.4:

Table 8.4:
Numerical values for the calculation of the impact ionization coefficient (8.6.61), after Schöll (1978)

$n_\alpha(0)/n_0$	10^{-7}	10^{-7}	10^{-4}	10^{-4}
W_2/cm^3 s^{-1}	10^{-5}	10^{-5}	10^{-4}	10^{-4}
W_1/s^{-1}	10^6	10^6	10^6	10^6
W_1^s/s^{-1}	10^{-4}	10^8	10^{-4}	10^8
$W_1^{a,\mathrm{crit.}}$/cm^3 s^{-1} by (8.6.61)	10^{-2}	10^{-12}	10^2	10^{-8}

Impact ionization data for KCl are not available, but the values for traps of similar depth in Ge and Si are of the order of

$W_1^a \approx 10^{-9}$ cm^3 s^{-1} for S ($E_c - 0.275$ eV) in Ge at 82 K and fields $\approx 10^4$ V/cm, Rosier and Sah (1971)

and

$W_1^a \approx 10^{-7}$–10^{-9} cm^3 s^{-1} for In, Ni, Au in Si at 65–200 K and at fields $\approx 10^3$–10^4 V/cm, McCombs and Milnes (1972).

Similar considerations are possible for the impact ionization of F-centers and also lead to a second order phase transition, Schöll (1978).

b) Models for first order phase transitions

α) Extrinsic semiconductor model

A model exhibiting a first order transition in addition to a second order one can be realized by an extrinsic semiconductor with band-band recombination and impact ionization of both trapped electrons N_1 and trapped holes N_0:

$$\begin{aligned} &e+N_1 \xrightarrow{W_1^a} 2e+N_0 \\ &h+N_0 \xrightarrow{W_2^a} 2h+N_1 \\ &e+h \xrightarrow{W_1} \text{photon.} \end{aligned} \tag{8.6.62}$$

Charge and impurity conservation give two linear relations between the concentrations n, p, n_t, p_t of free electrons, holes, and trapped electrons, holes:

$$\begin{aligned} &n_t+p_t=N_t \\ &n_t+n-p=N_D \end{aligned} \tag{8.6.63}$$

or

$$\begin{aligned} &n_t=N_D-n+p \\ &p_t=P_D-p+n \end{aligned} \tag{8.6.64}$$

setting

$$\begin{aligned} &P_D:=N_t-N_D \\ &(\text{assume } N_t>N_D). \end{aligned} \tag{8.6.65}$$

The steady state equations from (8.6.62) for the two independent concentrations n, p are:

$$\begin{aligned} &\dot{n}=W_1^a n n_t - W_1 np=0 \\ &\dot{p}=W_2^a p p_t - W_1 np=0 \end{aligned} \tag{8.6.66}$$

or, by (8.6.64):

$$0=n[W_1^a N_D - W_1^a n - (W_1 - W_1^a)p] \tag{8.6.67}$$

$$0=p[W_2^a P_D - W_2^a p - (W_1 - W_2^a)n] \tag{8.6.68}$$

with the solutions

$$\begin{aligned} &(a)\quad n=0, \quad p=0 \\ &(b)\quad n=0, \quad p=P_D \\ &(c)\quad n=N_D, \quad p=0 \\ &(d)\quad n=\frac{W_2^a(W_1 P_D - W_1^a N_t)}{W_1(W_1 - W_1^a - W_2^a)}; \quad p=\frac{W_1^a(W_1 N_D - W_2^a N_t)}{W_1(W_1 - W_1^a - W_2^a)}. \end{aligned} \tag{8.6.69}$$

The stability of (8.6.69) follows from the Jacobian of (8.6.67) (8.6.68):

$$A=\begin{pmatrix} W_1^a N_D - 2W_1^a n - (W_1 - W_1^a)p, & -(W_1 - W_1^a)n \\ -(W_1 - W_2^a)p & , \quad W_2^a P_D - 2W_2^a p - (W_1 - W_2^a)n \end{pmatrix}. \tag{8.6.70}$$

Substitution of (8.6.69) (a) into (8.6.70) yields

$$A^{(a)}=\begin{pmatrix} W_1^a N_D, & 0 \\ 0 & , \quad W_2^a P_D \end{pmatrix} \tag{8.6.71}$$

therefore, (8.6.69) (a) is always unstable by Theorem 8.14.
Substitution of (8.6.69) (b) into (8.6.70) yields

$$A^{(b)}=\begin{pmatrix} W_1^a N_t - W_1 P_D \ , & 0 \\ -(W_1 - W_2^a)P_D, & -W_2^a P_D \end{pmatrix} \tag{8.6.72}$$

which gives stability by Theorem 8.14 for

$$W_1^a N_t < W_1 P_D. \tag{8.6.73}$$

Similarly, substitution of (8.6.69) (c) into (8.6.70) yields

$$A^{(c)}=\begin{pmatrix} -W_1^a N_D, & -(W_1 - W_1^a)N_D \\ 0 & , \quad W_2^a N_t - W_1 N_D \end{pmatrix} \tag{8.6.74}$$

and thus stability for

$$W_2^a N_t < W_1 N_D \tag{8.6.75}$$

Using (8.6.67) (8.6.68) with $n, p \neq 0$ in (8.6.70) yields

$$A^{(d)}=\begin{pmatrix} -W_1^a n & , \quad -(W_1 - W_1^a)n \\ -(W_1 - W_1^a)p, & -W_2^a p \end{pmatrix} \tag{8.6.76}$$

where n, p satisfy (8.6.69) (d). We find

$$\operatorname{tr} A^{(d)} = -W_1^a n - W_2^a p < 0 \text{ always}$$

and (8.6.77)

$$\det A^{(d)} = [W_1^a + W_2^a - W_1]\, W_1 np > 0 \quad \text{if} \quad W_1^a + W_2^a > W_1$$

which gives the stability condition for (8.6.69) (d) by Theorem 8.15.
In its stability domain (8.6.77), solution (8.6.69) (d) exists as a positive, i.e., physical steady state if and only if

$$W_1^a N_t > W_1 P_D; \quad W_2^a N_t > W_1 N_D. \tag{8.6.78}$$

If (8.6.73) and (8.6.75) are satisfied simultaneously, the model (8.6.62) has two stable steady states (use (8.6.69) (8.6.64)):

$$\begin{array}{llll}
\text{(b) } n=0 & & \text{and} & \text{(c) } n=N_D \\
p=N_t-N_D & & & p=0 \\
n_t=N_t & & & n_t=0 \\
p_t=0 & & & p_t=N_t \\
(p\text{-type semiconductor}) & & & (n\text{-type semiconductor})
\end{array} \tag{8.6.79}$$

Thus, under conditions where the band-band recombination of electrons and holes is stronger than the carrier generation by impact ionization, a steady state can be maintained in two different ways:

(8.6.79) (b): All traps are occupied. Since the donors cannot supply enough electrons ($N_D < N_t$), the additionally required $N_t - N_D$ electrons are raised from the valence band by trap-band impact ionization of holes.

(8.6.79) (c): All traps are empty, and all electrons from the ionized donors are in the conduction band.

First-order phase transitions should be possible between these two steady states (8.6.79).
It is remarkable that the steady state (8.6.69) (a)

$$n=p=0,\; n_t=N_D,\; p_t=N_t-N_D \tag{8.6.80}$$

in which the traps are occupied by electrons only to that extent of which the donors can supply them, is unstable.
The domains of stability for each of the solutions (8.6.69) (b)–(d) are best illustrated in a plot of the steady state concentrations n, p against the control parameter W_1^a (Fig. 8.10). Since both impact ionization coefficients W_1^a and W_2^a depend on the external electric field, they cannot be driven independently. We assume $W_1^a = W_2^a$; this is reasonable for trap levels approximately in the middle of the bandgap. If (for $P_D > N_D$) by suitable initial conditions a system is prepared in the n-type state (8.6.69) (c), and if the electric field is slowly increased, at the threshold

$$W_1^a = W_1 \frac{N_D}{N_t} \tag{8.6.81}$$

this state turns unstable, and the system is expected to jump into the p-type state (8.6.69) (b) in a typical first-order phase transition. If the electric field is further increased, at the critical point

$$W_1^a = W_1 \frac{N_t - N_D}{N_t} \tag{8.6.82}$$

a second-order phase transition occurs after which both carriers have a non-zero concentration which eventually increases asymptotically linearly with the impact ionization coefficient $W_1^a = W_2^a$.

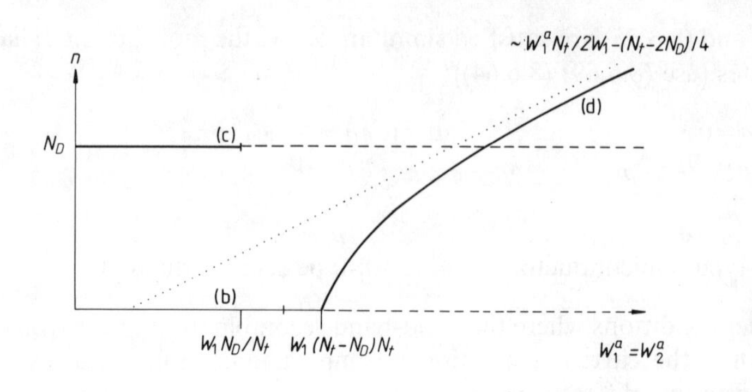

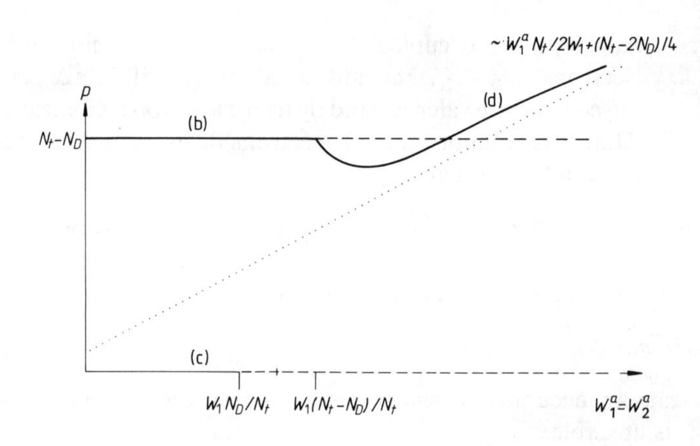

Figure 8.10:
Steady states (8.6.69) of model (8.6.62) against impact ionization coefficient $W_1^a = W_2^a$ (schematic). $P_D := N_t - N_D > N_D$ is assumed. For $P_D < N_D$ exchange $n \leftrightarrow p$, $N_D \leftrightarrow N_t - N_D$. (b), (c), (d) are referred to the cases given in (8.6.69); after Schöll (1978).

For $W_1^a = W_2^a$, $P_D = N_D$ (i.e. $N_t = 2N_D$; Fig. 8.) n and p are completely symmetric, and the steady states (8.6.69) read

$$\begin{aligned}
&(a)\ n = p = 0\\
&(b)\ n = 0,\ p = N_D\\
&(c)\ n = N_D,\ p = 0\\
&(d)\ n = p = N_D W_1^a / W_1 .
\end{aligned} \tag{8.6.83}$$

The first and the second order transitions merge at the degenerate critical point

$$W_1^a = \tfrac{1}{2} W_1 . \tag{8.6.84}$$

β) Model with band-trap recombination

As an elaboration of model (8.6.62) we take into account also the direct band-trap recombination

$$\begin{aligned} e + N_0 &\xrightarrow{W_2^t} N_1 \\ h + N_1 &\xrightarrow{W_3^t} N_0 . \end{aligned} \tag{8.6.85}$$

The steady state equations for model (8.6.62) combined with (8.6.85) read:

$$\begin{aligned} \dot{n} &= W_1^a n n_t - W_1 np - W_2^t np_t = 0 \\ \dot{p} &= W_2^a pp_t - W_1 np - W_3^t pn_t = 0 \end{aligned} \tag{8.6.86}$$

or, by (8.6.64)

$$\begin{aligned} & n[W_1^a N_D - W_2^t P_D - (W_1^a + W_2^t)n - (W_1 - W_1^a - W_2^t)p] = 0 \\ & p[W_2^a P_D - W_3^t N_D - (W_2^a + W_3^t)p - (W_1 - W_2^a - W_3^t)n] = 0. \end{aligned} \tag{8.6.87}$$

The solutions of (8.6.87) are,

$$\begin{aligned} &\text{(a) } n=0;\ p=0;\ n_t = N_D;\ p_t = P_D \\ &\text{(b) } n=0;\ p=\frac{(W_2^a P_D - W_3^t N_D)}{W_2^a + W_3^t};\ n_t = N_t \frac{W_2^a}{W_2^a + W_3^t};\ p_t = N_t \frac{W_3^t}{W_2^a + W_3^t} \\ &\text{(c) } n=\frac{W_1^a N_D - W_2^t P_D}{W_1^a + W_2^t};\ p=0;\ n_t = N_t \frac{W_2^t}{W_1^a + W_2^t};\ p_t = N_t \frac{W_1^a}{W_1^a + W_2^t} \\ &\text{(d) } n = [C - W_1(W_2^a P_D - W_3^t N_D)]D^{-1} \\ &\qquad p = [C - W_1(W_1^a N_D - W_2^t P_D)]D^{-1} \end{aligned} \tag{8.6.88}$$

with

$$D := W_1(W_1^a + W_2^a + W_2^t + W_3^t - W_1)$$

and

$$C := (W_1^a W_2^a - W_2^t W_3^t) N_t .$$

The solution (8.6.88) (a) is stable for

$$W_1^a < W_2^t \frac{P_D}{N_D};\ W_2^a < W_3^t \frac{N_D}{P_D} \tag{8.6.89}$$

(8.6.88) (b) is stable for

$$W_2^a > W_3^t \frac{N_D}{P_D} \tag{8.6.90}$$

and

$$W_1(W_2^a P_D - W_3^t N_D) > C \tag{8.6.91}$$

(8.6.88) (c) is stable for

$$W_1^a > W_2^t \frac{P_D}{N_D} \tag{8.6.92}$$

and

$$W_1(W_1^a N_D - W_2^t P_D) > C \tag{8.6.93}$$

(8.6.88) (d) is stable for

$$W_1^a + W_2^a + W_2^t + W_3^t > W_1 \tag{8.6.94}$$

in which case it exists as a physical state if and if only (8.6.91) and (8.6.93) are violated. Whenever (8.6.90)–(8.6.93) are satisfied simultaneously, two stable steady states (8.6.88) (b), (c) exist, and a first-order phase transition should be possible.
Besides, a second-order phase transition of precisely the same kind as in Section 8.6 a), (8.6.27) is possible from the zero steady state (8.6.88) (a) to the n-type steady state (8.6.88) (c) or to the p-type state (8.6.88) (b), depending on which of these two emerges first as a stable physical state with increasing impact ionization. For $W_1^a = W_2^a$, $W_2^t = W_3^t$ this depends on the compensation ratio by (8.6.90), (8.6.92): $P_D < N_D$ favours the n-type (8.6.88) (c), the converse the p-type (8.6.88) (b). Note that the threshold condition given by (8.6.92) agrees with the critical point (8.6.28).
In order to give an idea of the domains of stability of (8.6.88) (b), (c) we consider the symmetric case

$$W_1^a = W_2^a; \; W_2^t = W_3^t; \; N_D = P_D. \tag{8.6.95}$$

In this case the domain of simultaneous stability is smaller than in model (8.6.62) and exists only for

$$W_2^t < \frac{1}{4} W_1 . \tag{8.6.96}$$

γ) Model with band-band generation

If band-band generation

$$0 \xrightarrow{W_g} e + h \tag{8.6.97}$$

is taken into regard in addition to (8.6.62), the steady state equation

$$\dot{n} = W_g + n[W_1^a N_D - W_1^a n - (W_1 - W_1^a)p] = 0 \tag{8.6.98}$$

$$\dot{p} = W_g + p[W_2^a P_D - W_2^a p - (W_1 - W_2^a)n] = 0 \tag{8.6.99}$$

no longer factorizes. The easiest way to obtain a qualitative picture of the steady states and their stability character is a plot of the two curves (8.6.98) (8.6.99) in the $n-p$

phase plane. For the case $W_1^a = W_2^a > W_1$ these curves can intersect in one non-negative point only which is then stable, but for $W_1^a = W_2^a < W_1$ either one or three physical states are possible. The latter situation allows first-order phase transitions between a stable (n-type) state with many free electrons and few holes (corresponding to (8.6.69) (c)), and one with few electrons and many holes (corresponding to (8.6.69) (b)).

In the symmetric case $W_1^a = W_2^a$, $N_D = P_D$, an analytical solution of (8.6.98) (8.6.99) can readily be obtained by subtraction of (8.6.99) from (8.6.98):

$$(n-p)\, W_1^a (N_D - n - p) = 0 \tag{8.6.100}$$

which gives

$$n = p \quad \text{or} \quad n + p = N_D.$$

Substitution of $n = p$ into (8.6.98) yields the steady state (1)

$$n_1 = p_1 = \frac{W_1^a N_D}{2 W_1} \left\{ 1 + \left(1 + \frac{4 W_g W_1}{(W_1^a N_D)^2} \right)^{1/2} \right\} \tag{8.6.101}$$

Substitution of $n + p = N_D$ into (8.6.98) yields the steady states (2), (3)

$$n_{2,3} = p_{3,2} = \frac{N_D}{2} \left\{ 1 \pm \left(1 - \frac{4 W_g}{(W_1 - 2 W_1^a) N_D^2} \right)^{1/2} \right\} \tag{8.6.102}$$

with both n and p positive if and only if

$$W_1^a < \tfrac{1}{2} W_1 - \frac{2 W_g}{N_D^2}. \tag{8.6.103}$$

A plot of (8.6.101) (8.6.102) against W_1^a then shows that two stable states are possible only if

$$W_g < \tfrac{1}{4} W_1 N_D^2. \tag{8.6.104}$$

Conclusion

In all three models the band-band recombination W_1 plays an important role. If it is neglected, then two stable steady states cannot coexist. Therefore, in the search for materials which might display first-order phase transitions, semiconductors with a strong band-band recombination are desirable. Band-trap recombination $W_2^t (\approx W_3^t)$ should satisfy $W_2^t < \frac{1}{4} W_1$ (8.6.96), band-band generation $W_g < \frac{1}{4} W_1 N_D^2$ (8.6.104). In the limit

$$\begin{aligned} &W_2^t, W_3^t \ll \tfrac{1}{4} W_1 \\ &W_g \ll \tfrac{1}{4} W_1 N_D^2 \quad \text{(low or no irradiation)} \end{aligned} \tag{8.6.105}$$

and low temperature (thermalization of traps negligible) the simple model (8.6.62) is a reasonable approximation, and transitions from n- to p-type states can be expected for impact ionization of the order

$$W_1^a \approx W_2^a \approx c W_1 \quad \text{in case of } c < \tfrac{1}{2} \text{ by (8.6.75) (8.6.81)}$$

and from p- to n-type for

$$W_1^a \approx W_2^a \approx (1-c) W_1 \quad \text{in case of } c > \tfrac{1}{2} \text{ by (8.6.73)}$$

where $c := \frac{N_D}{N_t}$ is the compensation ratio.
$N_D = 0$ or $N_D = N_t$ both destroy the phase transition by (8.6.73) (8.6.75).
Experimental data on capture coefficients of various traps, cf. McCombs and Milnes (1972), Bonch-Bruevich and Landsberg (1968) suggest that (8.6.96) or (8.6.110) can be satisfied for repulsive centres,

$$W_2^t,\ W_3^t \approx 10^{-14} - 10^{-12} \text{ cm}^3 \text{ s}^{-1}$$

with band-band recombination, Landsberg and Adams (1973)

$$W_1 \approx 10^{-10} - 10^{-8} \text{ cm}^3 \text{ s}^{-1}.$$

An application of model β) to the switching effect in semiconductors has been given by Landsberg, Robbins and Schöll (1978). For further elaborations of models $\alpha) - \beta$) including Auger recombination and for a detailed analysis of the critical behaviour see Schöll (1980).
Robbins, Landsberg and Schöll (1981) considered the dynamics of the non-equilibrium phase transition in model β) and calculated switching and delay times.
A different model for first and second order phase transitions based upon coupled impact ionization of ground state and excited donors was analysed by Schöll (1982). The homogeneous steady state was shown to become unstable against the formation of high and low current layers or filaments, and equal area rules for their spatial coexistence were established, similar to Maxwell's rule in equilibrium thermodynamics. The filament formation can be interpreted as a new kind of non-equilibrium phase transition.
This model has been extended to embrace the effect of a magnetic field and used to explain the observation of impact ionization induced negative photoconductivity in n-GaAs, see Schöll, Heisel and Prettl (1982).

References

In the reference index the names of authors with prefixes (de, di, van, von, etc.) are alphabetically arranged corresponding to the initials of the surname.

Aas, E.J., Bløtekjaer, K., J. Phys. Chem. Solids **35**, 1053 (1974)

Abakumov, V.N., Yassievich, I.N., Sov. Phys. JETP **34**, 1375 (1972)

Abakumov, V.N., Kreshchuk, L.N., Yassievich, I.N., Sov. Phys. JETP **47**, 535 (1978)

Abarenkov, I.V., Antonova, I.M., Phys. Stat. Sol. **20**, 643 (1967)

Abarenkov, I.V., Phys. Stat. Sol. (b) **61**, 757 (1974)

Abarenkov, I.V., Antonova, I.M., Phys. Stat. Sol. (b) **64**, 747 (1974)

Abarenkov, I.V., Antonova, I.M., Phys. Stat. Sol. (b) **65**, 325 (1974)

Abarenkov, I.V., Antonova, I.M., Sov. Phys. Solid State **20**, 326 (1978)

Abdumalikov, A.A., Klochikhin, A.A., Sov. Phys. Solid State **18**, 1028 (1976)

Abe, R., Progr. Theor. Phys. **31**, 957 (1964)

Abeles, F. (Ed.), Optical Properties of Solids, Amsterdam: North-Holland 1972

Abou-Ghantous, M., Bates, C.A., Chandler P.E., Stevens, K.W.H., J. Phys. C: Solid State Phys. 7, 309 (1974)

Abou-Ghantous, M., Bates, C.A., Stevens, K.W.H., J. Phys. C: Solid State Phys. **7**, 325 (1974)

Abragam, A., Pryce, M.H.L., Proc. Phys. Soc. (London) **A63**, 409 (1950)

Abragam, A., Bleaney, B., Electron Paramagnetic Resonance of Transition Ions, Oxford: Clarendon Press 1970

Abram, I.I., Silbey, R., J. Chem. Phys. **63**, 2317 (1975)

Abram, I.I., Chem. Phys. **25**, 87 (1977)

Achar, B.N.N., Barsch, G.R., Phys. Stat. Sol. (a) **6**, 247 (1971)

Achieser, A.I., Berestezki, W.B., Quantenelektrodynamik, Frankfurt/M.: Verlag H. Deutsch, 1962

Adamov, M.N., Ledovskaya, E.M., Rebane, T.K., Sov. Phys. Solid State **8**, 2541 (1967)

Adamowski, J., Bednarek, S., Suffczyński, M., Solid State Commun. **20**, 785 (1976)

Adamowski, J., Bednarek, S., Suffczyński, M., Solid State Commun. **25**, 89 (1978)

Adams, D.J., McDonald, I.R., J. Phys. C: Solid State Phys. **7**, 2761 (1974)

Adams, E.N., Holstein, T.D., J. Phys. Chem. Solids **10**, 254 (1959)

Adams, W.H., J. Chem. Phys. **34**, 89 (1961)

Adams, W.H., J. Chem. Phys. **37**, 2009 (1962)

Adirovich, E.I., Dokl. Akad. Nauk. SSSR **60**, 361 (1948)

Adirowitsch, E.I., Einige Fragen zur Theorie der Lumineszenz der Kristalle, Berlin: Akademie Verlag 1953

Adirovich, E.I., Sov. Phys. Solid State **2**, 2123 (1961)

Adler, S.L., Phys. Rev. **126**, 118 (1962)

Adrian, F.J., Jette, A.N., Phys. Rev. B **9**, 3587 (1974)

Agabekyan, A.S., Opt. Spectrosc. **27**, 348 (1969)

Agabekyan, A.S., Melikyan, A.O., Opt. Spectrosc. **32**, 153 (1972)

Agaeva, R.G., Askerov, B.M., Eminov, R.F., Phys. Stat. Sol. (b) **69**, K63 (1975)

Agaeva, R.G., Askerov, B.M., Gashimzade, F.M., Sov. Phys. Semicond. **10**, 1268 (1976)

Agarwal, A.K., Sharma, S.K., Singh, S.P., Virmani, S.K., J. Appl. Phys. **46**, 846 (1975)

Agarwal, L.D., Shanker, J., Indian J. Pure Appl. Phys. **11**, 734 (1973)

Agarwal, L.D., Shanker, J., Indian J. Pure Appl. Phys. **12**, 85 (1974)

Agarwal, S.K., Phys. Rev. B **13**, 1817 (1976)

Agarwal, S.K., J. Phys. Chem. Solids **38**, 199 (1977)

Agekyan, V.T., Monozon, B.S., Shiryapov, I.P., Phys. Stat. Sol. (b) **66**, 359 (1974)

Aggarwal, K.G., Pathak, K.N., Phys. Rev. **B 7**, 4449 (1973)

Agranovich, V.M., Konobeev, Yu.V., Sov. Phys. Solid State **3**, 260 (1961)

Agranovich, V.M., Dubovskii, O.A., Sov. Phys. Solid State **12**, 1631 (1971)

Agranovich, V.M., Lalov, I.I., Sov Phys. JETP **34**, 350 (1972)

Agrawal, B.K., Ram, P.N., Phys. Rev. B **4**, 2774 (1971)
Agrawal, B.S., Hardy, J.R., Solid State Commun. **14**, 239 (1974)
Agrawal, G.G., Shanker, J., J. Chem. Phys. **68**, 5244 (1978)
Agrawal, S.C., Sharma, L.P., Shanker, J., Indian J. Pure Appl. Phys. **16**, 438 (1978)
Agrawal, V.K., Garg, H.C., Phys. Rev. B **8**, 843 (1973)
Agrawal, V.K., Garg, H.C., Indian J. Pure Appl. Phys. **12**, 168 (1974)
Ahrenkiel, R.K., Brown, F.C., Phys. Rev. **136**, A223 (1964)
Akimoto, O., Hanamura, E., Solid State Commun. **10**, 253 (1972)
Akulinichev, V.V., Volkov, A.S., Kumekov, S.E., Sov. Phys. Semicond. **9**, 666 (1975)
de Alba, E., Warman, J., J. Phys. Chem. Solids **29**, 69 (1968)
de Alba, E., Kocevar, P., Acta Phys. Austr. **29**, 257 (1969)
Alben, R., Burns, G., Phys. Rev. B **16**, 3746 (1977)
Albert, J.P., Jouanin, C., Gout, C., Phys. Rev. B **16**, 925 (1977)
Albert, J.P., Jouanin, C., Gout, C., Phys. Rev. B **16**, 4619 (1977)
Albert, J.P., Jouanin, C., Gout, C., Solid State Commun. **22**, 199 (1977)
Aldea, A., Z. Physik **244**, 206 (1971)
Alder, B., Fernbach, S., Rotenberg, M. (Eds.), Energy Bands of Solids; Methods in Computational Physics, v. 8, New York: Academic Press 1968
Aldrich, C., Bajaj, K.K., Solid State Commun. **22**, 157 (1977)
Al-Edani, M.C., Dubey, K.S., Phys. Stat. Sol. (b) **86**, 741 (1978)
Al-Edani, M.C., Dubey, K.S., Phys. Stat. Sol. (b) **87**, K47 (1978)
Aleksandrov, I.V., Theor. Exper. Chem. **2**, 51 (1966)
Aliev, T.A., Gashimzade, F.M., Sov. Phys. Semicond. **6**, 395 (1972)
Alig, R.C., Phys. Rev. B **2**, 2108 (1970)
Alig, R.C., Phys. Rev. B **3**, 536 (1971)
Allcock, G.R., Adv. Phys. **5**, 412 (1956)
Allen, J.W., J. Phys. C:Solid State Phys. **1**, 1136 (1968)
Allen, J.W., J. Phys. C: Solid State Phys. **2**, 1077 (1969)
Allen, J.W., J. Phys. C: Solid State Phys. **4**, 1936 (1971)
Allnatt, A.R., Jacobs, P.W.M., Proc. Roy. Soc. **A** (London) **260**, 350 (1961)
Allnatt, A.R., Jacobs, P.W.M., Proc. Roy Soc. **A** (London) **267**, 31 (1962)
Allnatt, A.R., Cohen, M.H., J. Chem. Phys. **40**, 1860 (1964)
Allnatt, A.R., Cohen, M.H., J. Chem. Phys. **40**, 1871 (1964)
Allnatt, A.R., Loftus, E., J. Chem Phys. **59**, 2541 (1973)
Allnatt, A.R., Loftus, E., J. Chem. Phys. **59**, 2550 (1973)
Allnatt, A.R., Yuen, P.S., J. Phys. C: Solid State Phys. **8**, 2199 (1975)
Allnatt, A.R., Yuen, P.S., J. Phys. C: Solid State Phys. **9**, 431 (1976)
Almairac, R., Benoit, C., J. Phys. C: Solid State Phys. **7**, 2614 (1974)
Almasov, L.A., Phys. Stat. Sol. (b) **54**, 87 (1972)
Altarelli, M., Iadonisi, G., Nuovo Cim. **5 B**, 21 (1971)
Altarelli, M., Iadonisi, G., Nuovo Cim. **5B**, 36 (1971)
Altarelli, M., Bassani, F., J. Phys. C: Solid State Phys. **4**, L328 (1971)
Altarelli, M., Lipari, N.O., Phys. Rev. B **9**, 1733 (1974)
Altarelli, M., Sabatini, R.A., Lipari, N.O., Solid State Commun. **25**, 1101 (1978)
Altenberger-Siczek, A., Phys. Rev. B **15**, 64 (1977)
Altschuler, S.A., Kosyrew, B.M., Paramagnetische Elektronenresonanz, Zürich: Verlag H. Deutsch 1964
Altshuler, A., Vekilov, Yu.Kh., Kadyshevich, A.E., Rusakov, A.P., Sov. Phys. Solid State **16**, 1852 (1975)
Altshuler, A.M., Vekilov, Yu.Kh., Izotov, A.D., Phys. Stat. Sol. (b) **68**, 165 (1975)
Altshuler, A.M., Vekilov, Yu.Kh., Umarov, G.R., Phys. Stat. Sol. (b) **69**, 661 (1975)
Altshuler, A.M., Vekilov, Yu.Kh., Izotov, A.D., Phys. Stat. Sol. (b) **70**, 347 (1975)
Altshuler, A.M., Vekilov, Yu.Kh., Izotov, A.D., Sov. Phys. Solid State **17**, 1389 (1976)
Altshuler, A.M., Vekilov, Yu.Kh., Kacherets, T.I., Sov. Phys. Solid State **20**, 9 (1978)
Altukhov, V.I., Phys. Stat. Sol. (b) **64**, 403 (1974)
Altukhov, V.I., Zavt, G.S., Phys. Stat. Sol. (b) **65**, 83 (1974)
Altukhov, V.I., Zavt, G.S., Sov. Phys. Solid State **19**, 615 (1977)
Altymyshov, L., Sov Phys. Solid State **19**, 1372 (1977)
de Alvarez, C.V., Walter, J.P., Boyd, R.W., Cohen M.L., J. Phys. Chem. Solids **34**, 337 (1973)
Ambegaokar, V., in: Astrophysics and the Many-Body Problem, ed. Ford, Brandeis Summer Institute 1962, v. 2, New York: Benjamin 1963, p. 321
Aminov, L.K., Sov. Phys. JETP **15**, 547 (1962)
Aminov, L.K., Malkin, B.Z., Sov. Phys. Solid Stat **9**, 1030 (1967)
Anderson, C.L., Crowell, C.R., Phys. Rev. B **5**, 2267 (1972)
Anderson, P.W., Phys. Rev. Lett. **21**, 13 (1968)
Andriesh, I.S., Gamurar, V.Y., Perlin, Y.E., Cryst Latt. Defects **5**, 35 (1974)
Anh, Vo H., Van Trong, N., Phys. Stat. Sol. (b) 83, 395 (1977)

Anh, Vo H., Dat, N.N., Phys. Stat. Sol. (b) **86**, 585 (1978)
Ansbacher, F., Z. Naturforsch. **14a**, 889 (1959)
Ansel'm, A.I., Firsov, Iu.A., Sov. Phys. JETP **1**, 139 (1955)
Ansel'm, A.I., Firsov, Iu.A., Sov. Phys. JETP **3**, 564 (1956)
Ansel'm, A.I., Askerov, M.B., Sov. Phys. Solid State **4**, 1154 (1962)
Ansel'm, A.I., Obraztsov, Yu.N., Tarkhanyan, R.G., Sov. Phys. Solid State **7**, 2293 (1966)
Ansel'm, A.I., Askerov, B.M., Sov. Phys. Solid State **9**, 22 (1967)
Antončik, E., J. Phys. Chem. Solids **10**, 314 (1959)
Antonov-Romanovskii, V.V., Dokl. Akad. Nauk. SSSR **85**, 517 (1952)
Antonov-Romanovskii, V.V., Opt. Spectrosc. **8**, 35 (1960)
Antonov-Romanovskii, V.V., Optika i Spektrosk. **8**, 73 (1960)
Antonov-Romanovskii, V.V., Phys. Stat. Sol. **19**, 417 (1967)
Antonov-Romanovskii, V.V., J. Luminescence **17**, 201 (1978)
Appel, J., Z. Naturforsch. **12a**, 410 (1957)
Appel, J., Z. Naturforsch. **13a**, 386 (1958)
Appel, J., Phys. Rev. **125**, 1815 (1962)
Appel, J., Phys. Rev. **141**, 506 (1966)
Arenstein, M., Hatcher, R.D., Neuberger, J., Phys. Rev. **132**, 73 (1963)
Areshev, I.P., Sov. Phys. Semicond. **11**, 567 (1977)
Argyres, P.N., Phys. Rev. **109**, 1115 (1958)
Argyres, P.N., Roth, L.M., J. Phys. Chem. Solids **12**, 89 (1959)
Argyres, P.N., Phys. Rev. **117**, 315 (1960)
Argyres, P.N., Phys. Rev. **126**, 1386 (1962)
Argyres, P.N., Kelley, P.L., Phys. Rev. **143**, A 98 (1964)
Argyres, P.N., Kirkpatrick, E.S., Ann. Physics **42**, 513 (1967)
Argyres, P.N., Sigel, J.L., Phys. Rev. B. **10**, 1139 (1974)
Aristova, K.A., Giner, C.T., Lang, I.G., Pavlov, S.T., Phys. Stat. Sol. (b) **85**, 351 (1978)
Arkad'eva, E.N., Sov. Phys. Solid State **4**, 2233 (1963)
Arkhipov, S.M., Malkin, B.Z., Sov. Phys. Solid State **17**, 806 (1975)
Arkhipov, V.I., Rudenko, A.I., Sov. Phys. Semicond. **11**, 811 (1977)
Aronov, A.G., Ioselevich, A.S., Sov. Phys. JETP **47**, 548 (1978)
Aronov, D.A., Shamasov, R.G., Sov. Phys. Solid State **8**, 1318 (1966)
Arora, H.L., Kim, S., Wang, S., Phys. Stat. Sol. **25**, 223 (1968)
Arora, H.L., Wang, S., J. Phys. Chem. Solids **30**, 1649 (1969)
Arora, H.L., Mahutte, C.K., Wang, S., J. Phys. Chem. Solids **30**, 2623 (1969)
Arora, V.K., Miller, S.C., Phys. Rev. B **10**, 688 (1974)
Arora, V.K., Peterson, R.L., Phys. Rev. B **12**, 2285 (1975)
Arora, V.K., Phys. Stat. Sol. (b) **71**, 293 (1975)
Arora, V.K., Gomber, K.L., Phys. Stat. Sol. (b) **74**, K111 (1976)
Arora, V.K., Qureshi, E., Phys. Stat. Sol. (b) **77**, 77 (1976)
Arora, V.K., Phys. Rev. B **13**, 2532 (1976)
Arora, V.K., Jaafarian, M., Phys. Rev. B **13**, 4457 (1976)
Arora, V.K., Phys. Rev. B **14**, 679 (1976)
Arora, V.K., Cassiday, D.R., Spector, H.N., Phys. Rev. B **15**, 5996 (1977)
Arya, K., Hassan, A.R., Solid State Commun. **21**, 301 (1977)
Asahi, T., Prog. Theor. Phys. Suppl. No. **23**, 59 (1962)
Asano, S., Tomishima, Y., J. Phys. Soc. Japan **13**, 1119 (1958)
Asano, S., Tomishima, Y., J. Phys. Soc. Japan **13**, 1126 (1958)
Asche, M., Phys. Stat. Sol. (b) **41**, 67 (1970)
Ashkin, M., J. Phys. (Paris) **26**, 709 (1965)
Aspnes, D.E., Phys. Rev. **147**, 554 (1966)
Asratyan, K.Kh., Kachlishvili, Z.S., Sov. Phys. Semicond. **10**, 1292 (1976)
Aten, A.C., Haanstra, J.H., de Vries, H., Philips Research Reports **20**, 395 (1965)
Augst, G.R., Phys. Stat. Sol. (b) **83**, 55 (1977)
Austin, B.J., Heine, V., Sham, L.J., Phys. Rev. **127**, 276 (1962)
Austin, I.G., J. Phys. C:Solid State Phys. **5**, 1687 (1972)
Auvergne, D., Camassel, J., Phys. Stat. Sol. (b) **44**, 687 (1971)
Aven, M., Prener, J.S. (Eds.), Physics and Chemistry of II-VI Compounds, Amsterdam: North-Holland 1967
Avery, J.S., Proc. Phys. Soc. **89**, 677 (1966)
Avilov, V.V., Sov. Phys. Solid State **14**, 2209 (1973)
Avvakumov, V.I., Sov. Phys. JETP **37**, 723 (1960)
Axe, J.D., Phys. Rev. **139**, A 1215 (1965)
Aymerich, F., Bassani, F., Nuovo Cim. **48**, 358 (1967)

Babaev, M.M., Gasymov, T.M., Phys. Stat. Sol. (b) **84**, 473 (1977)
Bacci, M., Phys. Letters **57A**, 475 (1976)
Bacci, M., Phys. Stat. Sol. (b) **82**, 169 (1977)
Bäuerle, D., Fritz, B., Solid State Commun. **6**, 453 (1968)
Bäuerle, D., Hübner, R., Phys. Rev. B **2**, 4252 (1970)
Bagaev, V.S., Keldysh, L.V., Sibel'din, N.N., Tsvetkov, V.A., Sov. Phys. JETP **43**, 362 (1976)
Bagley, B.G., Solid State Commun. **8**, 345 (1970)

Bailey, P.T., Phys. Rev. B **1**, 588 (1970)
Bailly, F., J. Phys. Radium **27**, 335 (1966)
Bajaj, K.K., Phys. Rev. **170**, 694 (1968)
Bajaj, K.K., Nuovo Cim. **55B**, 244 (1968)
Bajaj, K.K., Clark, T.D., Phys. Stat. Sol. (b) **52**, 195 (1972)
Bajaj, K.K., Phys. Stat. Sol. (b) **64**, K107 (1974)
Baker, J.M., Davies, E.R., J. Phys. C: Solid State Phys. **8**, 1869 (1975)
Bakhshi, P.S., Goyal, S.C., Shanker, J., Indian J. Pure Appl. Phys. **14**, 598 (1976)
Balabanyan, G.O., Phys. Lett. **45A**, 303 (1973)
Balabanyan, G.O., Phys. Lett. **48A**, 275 (1974)
Balcar, E., Acta Phys. Austr. **23**, 234 (1966)
Baldereschi, A., Diaz, M.G., Nuovo Cim. **68**, 217 (1970)
Baldereschi, A., Lipari, N.O., Phys. Rev. B **3**, 439 (1971)
Baldereschi, A., Hopfield, J.J., Phys. Rev. Lett. **28**, 171 (1972)
Baldereschi, A., Phys. Stat. Sol. (b) **59**, 629 (1973)
Baldereschi, A., J. Luminescence **7**, 79 (1973)
Baldereschi, A., Lipari, N.O., J. Luminescence **12/13**, 489 (1976)
Baldini, G., Guzzi, M., Phys. Stat. Sol. **30**, 601 (1968)
Baleinikov, L.A., Bulyarskii, S.V., Grushko, N.S., Gutkin, A.A., Sov. Phys. Semicond. **10**, 216 (1976)
Balescu, R., Prigogine, I., Physica **25**, 281 (1959)
Balescu, R., Prigogine, I., Physica **25**, 302 (1959)
Balescu, R., J. Math. Phys. **4**, 1009 (1963)
Balkanski, M., des Cloiseaux, J., J. Phys. Radium (Paris) **21**, 825 (1960)
Balkanski, M., des Cloizeaux, J., J. Phys. Radium (Paris) **22**, 41 (1961)
Balkanski, M., Optical Properties of Solids, ed. F. Abeles, Amsterdam: North-Holland 1972, p. 529
Balkanski, M., Leite, R.C.C., Porto, S.P.S. (Eds.), Light Scattering in Solids, Paris: Flammarion Sciences 1976
Balkanski, M. (Ed.), International Conference on Lattice Dynamics, Paris 1977, Paris: Flammarion 1978
Ballhausen, C.J., Introduction to Ligand Field Theory, New York: McGraw-Hill 1962
Ballinger, R.A., Major, K.G., Mallinson, J.R., J. Phys. C: Solid State Phys. **6**, 2573 (1973)
Ballini, Y., J. Phys. C: Solid State Phys. **11**, 2039 (1978)
von Baltz, R., Phys. Stat. Sol. (b) **43**, K133 (1971)
von Baltz, R., Birkholz, U., Festkörperprobleme **12**, 233 (1972) (Ed. O. Madelung, Braunschweig: Vieweg)
von Baltz, R., Escher, W., Phys. Stat. Sol. (b) **51**, 499 (1972)
Banaii, N., Jacobs, G., Phys. Stat. Sol. (b) **77**, 707 (1976)
Bandura, A.V., Evarestov, R.A., Phys. Stat. Sol. (b) **64**, 635 (1974)
Banerjee, R., Varshni, Y.P., Can. J. Phys. **47**, 451 (1969)
Bangert, E., Z. Physik **215**, 177 (1968)
Bangert, E., Z. Physik **215**, 192 (1968)
Bangert, E., Kästner, P., Phys. Stat. Sol. (b) **54**, 173 (1972)
Bannaya, V.F., Velesova, L.I., Gerzhenzon, E.M., Chuenkov, V.A., Sov. Phys. Semicond, **7**, 1315 (1974)
Di Bartolo, B., Optical Interactions in Solids, New York: J. Wiley & Sons 1968
Di Bartolo, B., Powell, Phonons and Resonances in Solids, New York: Wiley 1976
Bartram, R.H., Swenberg, C.E., La, S.Y., Phys. Rev. **162**, 759 (1967)
Bartram, R.H., Stoneham, A.M., Gash, P., Phys. Rev. **176**, 1014 (1968)
Banyai, L., Aldea, A., Phys. Stat. Sol. (b) **79**, 365 (1977)
Baraff, G.A., Phys. Rev. **133**, A26 (1964)
Barentzen, H., Phys. Stat. Sol. (b) **71**, 245 (1975)
Barentzen, H., Polansky, O.E., J. Chem. Phys. **68**, 4398 (1978)
Bari, R.A., Phys. Rev. B **10**, 1560 (1974)
Barker, A.S., Loudon, R., Rev. Mod. Phys. **44**, 18 (1972)
Barker, A.S., Phys. Rev. B **7**, 2507 (1973)
Barker, J.R., J. Phys. C: Solid State Phys. **5**, 1657 (1972)
Barker, J.R., J. Phys. C: Solid State Phys. **6**, 2663 (1973)
Barker, J.R., Magnusson, B.G., Phys. Letters **54**, 283 (1975)
Barker, J.R., J. Phys. C: Solid State Phys. **9**, 4397 (1976)
Barker, J.R., Solid-State Electronics **21**, 267 (1978)
Barker, J.R., Solid-State Electronics **21**, 197 (1978)
Barrie, R., Edmond, J.T., J. Electron. **1**, 161 (1955)
Barrie, R., Proc. Phys. Soc. B **69**, 553 (1956)
Barrie, R., Sharpe, I.W., Can. J. Phys. **50**, 222 (1972)
Barrie, R., Sharpe, I.W., Jones, B.L., Can. J. Phys **50**, 231 (1972)
Barrie, R., Chow, H.-C., Can. J. Phys. **56**, 526 (1978)
Barrie, R., Sharpe, I.W., Can. J. Phys. **56**, 550 (1978)
Barriol, J., Nikitine, S., Sieskind, M., Compt. Rend. **242**, 790 (1956)
Barron, T.H.K., Phys. Rev. **123**, 1995 (1961)
Barron, T.H.K., Batana, A., Phys. Rev. **167**, 814 (1968)
Barron, T.H.K., Klein, M.L., Perturbation Theory of Anharmonic Crystals, in: Dynamical Properties of Solids I, ed. G.K. Horton and A.A. Maradudin, Amsterdam: North-Holland 1974
Barsch, G.R., Achar, B.N.N., Phys. Stat. Sol. **35**, 881 (1969)

Bartram, R.H., Harmer, A.L., Hayes, W., J. Phys. C: Solid State Phys. **4**, 1665 (1971)
Bashenov, V.K., Alarashi, R.A., Foigel, M.G., Phys. Stat. Sol. (b) **63**, 403 (1974)
Bashenov, V.K., Soloshenko, V.I., Phys. Stat. Sol. (b) **67**, K73 (1975)
Basinski, J., Woolley, J.C., J. Phys. C: Solid State Phys. **8**, 1841 (1975)
Bassani, F., Fumi, F.G., Nuovo Cim. **XI**, 274 (1954)
Bassani, F., Yoshimine, M., Phys. Rev. **130**, 20 (1963)
Bassani, F., Knox, R.S., Fowler, W.B., Phys. Rev. **137**, A1217 (1965)
Bassani, F., Iadonisi, G., Preziosi, B., Phys. Rev. **186**, 735 (1969)
Bassani, F., Girlanda, R., Optics Commun. **1**, 359 (1970)
Bassani, F., Hassan, A.R., Optics Commun. **1**, 371 (1970)
Bassani, F., Theory of Impurity States, Trieste Lectures 1970, Wien: International Atom. Energy Agency 1971, p. 265
Bassani, F., Hassan, A.R., Nuovo Cim. B **7**, 313 (1972)
Bassani, F., Giuliano, E.S., Nuovo Cim. B **8**, 193 (1972)
Bassani, F., Forney, J.J., Quattropani, A., Phys. Stat. Sol. (b) **65**, 591 (1974)
Bassani, F., Iadonisi, G., Preziosi, B., Rep. Prog. Phys. **37**, 1099 (1974)
Bassani, F., Rovere, M., Solid State Commun. **19**, 887 (1976)
Bastin, A.J.F., Lewiner, C., Fayet, N., J. Phys. Chem. Solids **31**, 817 (1970)
Basu, A.N., Sengupta, S., Phys. Stat. Sol. **29**, 367 (1968)
Basu, A.N., Sengupta, S., J. Phys. C: Solid State Phys. **5**, 1158 (1972)
Basu, A.N., Sengupta, S., Phys. Rev. B **8**, 2982 (1973)
Basu, A.N., Roy, D., Sengupta, S., Phys. Stat. Sol. (a) **23**, 11 (1974)
Basu, A.N., Sengupta, S., Phys. Rev. B **14**, 2633 (1976)
Basu, C., Ghosh, U.S., Phys. Stat. Sol. (b) **60**, 97 (1973)
Basu, C., Ghosh, U.S., Phys. Stat. Sol. (b) **70**, K95 (1975)
Basu, P.K., Nag. B.R., Phys. Stat. Sol. (b) **73**, K15 (1976)
Bates, C.A., J. Phys. C: Solid State Phys. **1**, 877 (1968)
Bates, C.A., Chandler, P.E., J. Phys. C: Solid State Phys. **4**, 2713 (1971)
Bates, C.A., Chandler, P.E., J. Phys. C: Solid State Phys. **6**, 1975 (1973)
Bates, C.A., Chandler, P.E., Stevens, K.W.H., J. Phys. C: Solid State Phys. **7**, 3969 (1974)
Bates, C.A., Szymczak, H., Phys. Stat. Sol. (b) **74**, 225 (1976)
Bates, C.A., Steggles, P., Physica **86–88B**, 1130 (1977)
Batra, I.P., Haering, R.R., Can. J. Phys. **45**, 3401 (1967)
Batra, I.P., Can J. Phys. **47**, 521 (1969)
Bauer, R., Phys. Stat. Sol. (b) **50**, 491 (1972)
Bauer, R., Differt, K., Schwan, L., Semiconduct. Insulators **2**, 217 (1977)
Bauer, R., Leutz, R.K., Semicond. Insulators **2**, 227 (1977)
Baumann, K., Ranninger, J., Ann. Physics **20**, 157 (1962)
Baumann, K., Acta Phys. Austr. **37**, 350 (1973)
Baumann, K., Phys. Stat. Sol. (b) **63**, K71 (1974)
Bazhenov, V.K., Soloshenko, V.I., Timofeenko, V.V., Inorganic Materials **12**, 823 (1976)
Bazhenov, V.K., Mutal', A.M., Soloshenko, V.I., Sov. Phys. Semicond. **9**, 1247 (1976)
Beattie, A.R., Landsberg, P.T., Proc. Roy. Soc. London A **249**, 16 (1959)
Beattie, A.R., Smith, G., Phys. Stat. Sol. **19**, 577 (1967)
Bebb, H.B., Gold, A., Phys. Rev. **143**, 1 (1966)
Bechstedt, F., Enderlein, R., Peuker, K., Phys. Stat. Sol. (b) **78**, 711 (1976)
Bechstedt, F., Henneberger, F., Phys. Stat. Sol. (b) **81**, 211 (1977)
Bechstedt, F., Haus, D., Phys. Stat. Sol. (b) **88**, 163 (1978)
Becker, C.A.L., Meek, D.W., Dunn, T.M., J. Phys. Chem. **74**, 1568 (1970)
Becker, L., Enderlein, R., Peuker, K., Phys. Stat. Sol. (b) **60**, 579 (1973)
Becker, R., Theorie der Wärme, Berlin: Springer Verlag 1964
Beer, A.C., Galvanomagnetic Effects in Semiconductors, Solid State Physics Suppl. **4**, ed. F. Seitz and D. Turnbull, New York: Academic Press 1963
Behnke, G., Büttner, H., Pollmann, J., Solid State Commun. **20**, 873 (1976)
Beleznay, F., Pataki, G., Phys. Stat. Sol. **8**, 805 (1965)
Beleznay, F., Serényi, M., Solid-State Electronics **21**, 215 (1978)
Belinicher, V.I., Sov. Phys. JETP **48**, 322 (1978)
Bell, D.G., Hum, D.M., Pincherle, L., Sciama, D.W., Woodward, P.M., Proc. Roy. Soc. (London) **A217**, 71 (1953)
Bell, S., Warsop, P.A., J. Molec. Spectrosc. **20**, 425 (1966)
Bell, S., Warsop, P.A., J. Molec. Spectrosc. **22**, 360 (1967)
Bellomonte, L., J. Phys. Chem. Solids **38**, 59 (1977)
Bellomonte, L., J. Phys. Chem. Solids **38**, 1137 (1977)
Belorusets, E.D., Grinberg, A.A., Imamov, E.Z., Sov. Phys. Semicond. **11**, 786 (1977)
Belousov, A.V., Kovarskii, V.A., Sinyavesii, E.P., Sov. Phys. Solid State **20**, 113 (1978)

Belyavskii, V.I., Shalimov, V.V., Sov. Phys. Semicond. **11**, 884 (1977)
Belyavskii, V.I., Sov. Phys. Semicond. **11**, 1130 (1977)
Bendow, B., Birman, J.L., Phys. Rev. B **1**, 1678 (1970)
Bendow, B., Phys. Rev. B **2**, 5051 (1970)
Bendow, B., Phys. Rev. B **4**, 552 (1971)
Bendow, B., Birman, J.L., Phys. Rev. B **4**, 569 (1971)
Bendow, B., Ying, S.-Ch., Yukon, S.P., Phys. Rev. B **8**, 1679 (1973)
Bendow, B., Phys. Rev. B **8**, 5821 (1973)
Bendow, B., Ting, C.S., Birman, J.L., Solid State Commun. **15**, 1395 (1974)
Bendow, B., Ting, C.S., Phys. Rev. B **12**, 695 (1975)
Benedek, G., Nardelli, G.F., Phys. Rev. Letters **17**, 1136 (1966)
Benedek, G., Solid State Commun. **5**, 101 (1967)
Benedek, G., Nardelli, G.F., Phys. Rev. **154**, 872 (1967)
Benedek, G., Nardelli, G.F., J. Chem Phys. **48**, 5242 (1968)
Benedek, G., Mulazzi, E., Phys. Rev. **179**, 906 (1969)
Benedek, G., Phys. Stat. Sol. (b) **43**, 509 (1971)
Benedek, R., Ho, P.S., J. Phys. F: Metal Phys. **3**, 1285 (1973)
Beni, G., Rice, T.M., Phys. Rev. Letters **37**, 874 (1976)
Beni, G., Rice, T.M., Phys. Rev. B **15**, 840 (1977)
Beni, G., Rice, T.M., Solid State Commun. **23**, 871 (1977)
Beniere, F., J. Physique, Suppl., **37**, C7-260 (1976)
Beniere, F., J. Sci. Industrial Research **35**, 503 (1976)
Beniere, M., Chemla, M., Beniere, F., J. Phys. Chem. Solids **37**, 525 (1976)
Beniere, M., Beniere, F., Catlow, C.R.A., Shukla, A.K., Rao, C.N.R., J. Phys. Chem. Solids **38**, 521 (1977)
Benin, D., Phys. Rev. B **1**, 2777 (1970)
Benin, D., Phys. Rev. B **5**, 2344 (1971)
Bennemann, K.H., Phys. Rev. **130**, 1757 (1963)
Bennemann, K.H., Phys. Rev. **137**, A1497 (1965)
Bennett, H.S., Lidiard, A.B., Phys. Letters **18**, 253 (1965)
Bennett, H.S., Stern, E.A., Phys. Rev. **137**, A 448 (1965)
Bennett, H.S., Phys. Rev. **169**, 729 (1968)
Bennett, H.S., Phys. Rev. **184**, 918 (1969)
Bennett, H.S., Phys. Rev. B **1**, 1702 (1970)
Bennett, H.S., Phys. Rev. B **3**, 2763 (1971)
Bennett, H.S., Phys. Rev. B **4**, 1327 (1971)
Bennett, H.S., Phys. Rev. B **6**, 3936 (1972)
Benoit à la Guillaume, C., J. Luminescence **12/13**, 57 (1976)
Benson, G.C., Wyllie, G., Proc. Phys. Soc. A **64**, 276 (1951)
Benson, H.J., Mills, D.L., Phys. Rev. B **1**, 4835 (1970)
Benson, H.J., Can J. Phys. **51**, 1737 (1973)
Berezhkovskii, A.M., Ovchinnikov, A.A., Sov. Phys. Solid State **18**, 1908 (1976)
Berezin, A.A., Opt. Spectrosc. **25**, 610 (1968)
Berezin, A.A., Sov. Phys. Solid State **9**, 2170 (1968)
Berezin, A.A., Sov. Phys. Solid State **10**, 2280 (1969)
Berezin, A.A., Sov. Phys. Solid State **11**, 1285 (1969)
Berezin, A.A., Opt. Spectrosc. **29**, 430 (1970)
Berezin, A.A., Phys. Stat. Sol. (b) **49**, 51 (1972)
Berezin, A.A., Phys. Stat. Sol. (b) **49**, K201 (19
Berezin, A.A., Z. Naturforsch. **37a**, 613 (1982)
Bergmann, D.J., Entin-Wohlmann, O., Physica **78**, 31 (1974)
Bergmann, D.J., Entin-Wohlman, O., Physica **78**, 45 (1974)
Bergstresser, T.K., Cohen, M.L., Phys. Letters **23**, 8 (1966)
Bergstresser, T.K., Cohen, M.L., Phys. Rev. **164**, 1069 (1967)
Bersuker, I.B., Sov. Phys. JETP **16**, 933 (1963)
Bersuker, I.B., Sov. Phys. JETP **17**, 836 (1963)
Bersuker, I.B., Vekhter, B.G., Phys. Stat. Sol. **16**, 63 (1966)
Bersuker, I.B., Polinger, V.Z., Phys. Stat. Sol. (b **60**, 85 (1973)
Bersuker, I.B., Polinger, V.Z., Sov. Phys. JETP **39**, 1023 (1974)
Bess, L., Phys. Rev. **105**, 1469 (1957)
Bethe, H., Ann. Physik (Leipzig) **3**, 133 (1929)
Bhattacharya, D.P., Kachlishvili, Z.S., Sov. Phys Semicond. **11**, 1181 (1977)
Bhattacharya, D.P., Kachlishvili, Z.S., Phys. Stat Sol. (a) **39**, 49 (1977)
Bhattacharyya, B.D., Phys. Stat. Sol. (b) **48**, K87 (1971)
Bhattacharyya, B.D., J. Phys. Chem. Solids **32**, 2357 (1971)
Bhattacharyya, B.D., Phys. Stat. Sol. (b) **71**, K181 (1975)
Biem, W., Z. Physik **164**, 199 (1961)
Biernacki, S.W., Phys. Stat. Sol. (b) **51**, 829 (1972)
Biernacki, S.W., Phys. Stat. Sol. (b) **84**, 699 (1977)
Biernacki, S.W., Phys. Stat. Sol. (b) **87**, 607 (19
Bikbaev, N.Kh., Ivanov, A.I., Lomakin, G.S., Ponomarev, O.A., Sov. Phys. JETP **47**, 1121 (1978)
Bill, H., Silsbee, R.H., Phys. Rev. B **10**, 2697 (1974)
Bilz, H., Genzel, L., Happ. H., Z. Physik **160**, 535 (1960)
Bilz, H., Genzel, L., Z. Physik **169**, 53 (1962)
Bilz, H., Zeyher, R., Wehner, R.K., Phys. Stat. Sol. **20**, K167 (1967)

Bilz, H., Gliss, B., Hanke, W., Dynamical Properties of Solids I, ed. G.K. Horton and A.A. Maradudin, Amsterdam: North-Holland 1974, p. 343
Bilz, H., Kress, W., Phonon Dispersion Relation in Insulators, Heidelberg: Springer 1979
Bimberg, D., Hess, K., Lipari, N.O., Fischbach, J.U., Altarelli, M., Physica **89B**, 139 (1977)
Bimberg, D., Festkörperprobleme **17**, 195 (1977) (Ed. J. Treusch, Braunschweig: Vieweg Verlag)
Bimberg, D., Dean, P.J., Phys. Rev. B **15**, 3917 (1977)
Bir, G.L., Sov. Phys. JETP **24**, 372 (1967)
Bir, G.L., Razbirin, B.S., Ural'tsev, I.N., Sov Phys. Solid State **14**, 360 (1972)
Bir, G.L., Aronov, A.G., Pikus, G.E., Sov. Phys. JETP **42**, 705 (1976)
Bir, G.L., Sov. Phys. Solid State **18**, 946 (1976)
Birman, J.L., Phys. Rev. **109**, 810 (1958)
Birman, J.L., J. Phys. Chem. Solids **6**, 65 (1958)
Birman, J.L., Phys. Rev. **115**, 1493 (1959)
Birman, J.L., Phys. Rev. **131**, 1489 (1963)
Birman, J.L., Lax, M., Loudon, R., Phys. Rev. **145**, 620 (1966)
Birman, J.L., Solid State Commun. **13**, 1189 (1973)
Birman, J.L., Handbuch der Physik XXV/2b, ed. S. Flügge, L. Genzel, Heidelberg: Springer Verlag 1974, p. 1
Birman, J.L., Dynamical Properties of Solids I, ed. G.K. Horton and A.A. Maradudin, Amsterdam: North-Holland 1974, p. 83
Birtcher, R.C., Deutsch, P.W., Wendelken, J.F., Kunz, A.B., J. Phys. C: Solid State Phys. **5**, 562 (1972)
Bisti, V.E. Sov. Phys. Lebedev Inst. Rep. **1**, 31 (1977)
Bjorken, D., Drell, S.D., Relativistic Quantum Fields, New York: McGraw-Hill 1965
Blackman, M., Handbuch der Physik VII, pt. 1, ed. S. Flügge, Heidelberg: Springer 1955, p. 325
Blaesser, G., Peretti, J., Toth, G., Phys. Rev. **171**, 665 (1968)
Blakemore, J.S., Semiconductor Statistics, London: Pergamon Press 1962
Blakemore, J.S., Recombination Rate Theory in Semiconductors, in: Problems in Thermodynamics and Statistical Physics, ed. P.T. Landsberg, London: Pion Ltd. 1971
Blatt, F.J., Physics of Electronic Conduction in Solids, New York: Mc Graw-Hill 1968
Blatt, J.M., Weisskopf, V., Theoretical Nuclear Physics, New York 1966
Bloch, F., Z. Physik **52**, 555 (1928)
Blochinzew, D., Phys. Z. Sowj. Union **12**, 586 (1937)
Bloembergen, N., Nonlinear Optics, New York: Benjamin 1965
Blötekjær, K., Arkiv Fysik **33**, 105 (1966)
Bloom, S., Bergstresser, T.K., Solid State Commun. **6**, 465 (1968)
Bloom, S., Ortenburger, I., Phys. Stat. Sol. (b) **58**, 561 (1973)
Blossey, D.F., Phys. Rev. B **2**, 3976 (1970)
Blossey, D.F., Phys. Rev. B **3**, 1382 (1971)
Blount, E.I., Phys. Rev. **126**, 1636 (1962)
Blumberg, W.E., Das, T. P., Phys. Rev. **110**, 647 (1958)
Blumberg, W. E., Phys. Rev. **119**, 1842 (1960)
Bluthardt, W., Schneider, W., Wagner, M., Phys. Stat. Sol. (b) **56**, 453 (1973)
Boardman, A.D., Parker, M.R., Allos, T.I.Y., Phys. Letters **60A**, 437 (1977)
Bobrysheva, A.I., Phys. Stat. Sol. **16**, 337 (1966)
Bobrysheva, A.I., Moskalenko, S.A., Shmiglyuk, M.I., Bull. Acad. Sci. USSR, Phys. Ser. **31**, 2043 (1967)
Bobrysheva, A.I., Miglei, M.F., Shmiglyuk, M. I., Phys. Stat. Sol. (b) **53**, 71 (1972)
Bobrysheva, A.I., Vybornov, V.I., Phy. Stat. Sol. (b) **69**, 267 (1975)
Bobrysheva, A. I., Karp, I. A., Moskalenko, S. A., Sov. Phy. Semicond. **9**, 407 (1975)
Bobrysheva, A.I., Vybornov, V.I., Sov. Phys. Semicond. **10**, 268 (1976)
Böer, K.W., Vogel, H., Ann. Physik **17**, 10 (1955)
Böer, K.W., Voigt, J., Z. Naturforsch. **16a**, 873 (1961)
Böhm, M., Sharmann, A., Phys. Stat. Sol. (a) **4**, 99 (1971)
Boese, F.-K., Wagner, M., Z. Physik **235**, 140 (1970)
Böttger, H., Phys. Stat. Sol. **26**, 681 (1968)
Böttger, H., Bryksin, V.V., Phys. Stat. Sol. (b) **78**, 9 (1976)
Böttger, H., Bryksin, V.V., Phys. Stat. Sol. (b) **78**, 415 (1976)
Boffi, V.C., Molinari, V.G., Wonneberger, W., Nuovo Cim. **45B**, 109 (1978)
Boganov, A.G., Cheremisin, I.I., Rudenko, V. S., Sov. Phys. Solid State **8**, 1512 (1966)
Boguslawski, P., Phys. Stat. Sol. (b) **70**, 53 (1975)
Boguslawski, P., J. Phys. C: Solid State Phys. **10**, L417 (1977)
Bohm, D., Pines, D., Phys. Rev. **92**, 609 (1953)
Bollmann, W., Phys. Stat. Sol. (a) **24**, 181 (1974)
Bolonin, O.N., Tolpygo, K.B., Sov. Phys. Solid State **18**, 446 (1976)
Bolonin, O.N., Sov. Phys. Solid State **19**, 1088 1977)
Bonch-Bruevich, V.L., Tyablikov, S.V., The Greenfunction Method in Statistical Mechanics, Amsterdam: North-Holland 1962
Bonch-Bruevich, V.L., Drugova, A.A., Sov. Phys. Solid State **7**, 2491 (1966)
Bonch-Bruevich, V.L., Landsberg, E.G., Phys. Stat. Sol. **29**, 9 (1968)

Bonciani, M., Grassano, U.M., Phys. Stat. Sol. (b) **67**, 371 (1975)
Bonek, E., Pötzl, H.W., Richter, K., J. Phys. Chem. Solids **31**, 1151 (1970)
Bonfiglioli, G., Brovetto, P., Cortese, C., Phys. Rev. **114**, 951 (1959)
Boon, M.H., J. Math. Phys. **13**, 1268 (1972)
Borblik, V.L., Sov. Phys. Semicond. **6**, 719 (1972)
Born, M., Oppenheimer, R., Ann. Physik (Germany) **84**, 457 (1927)
Born, M. Huang, K., Dynamical Theory of Crystal Lattices, Oxford University Press, 1954, 1st ed.
Born, M., Huang, K., Dynamical Theory of Crystal Lattices, Oxford: Clarendon Press 1966, 2nd ed.
Borstel, G., Merten, L., Z. Naturforsch. **26a**, 653 (1971)
Bosacchi, B., Robinson, J.E., Solid State Commun. **10**, 797 (1972)
Bose, D., Basu, A.N., Phys. Stat. Sol. (b) **89**, K35 (1978)
Bosi, L., Phys. Stat. Sol. (b) **66**, 285 (1974)
Bosi, L., Fantola-Lazzarini, A.L., Lazzarini, E., Phys. Stat. Sol. (b) **66**, 285 (1974)
Bosi, L., Cova, S., Spinolo, G., Phys. Stat. Sol. (b) **68**, 603 (1975)
Boswarva, I.M., Howard, R.E., Lidiard, A.B., Proc. Roy. Soc. London A **269**, 125 (1962)
Boswarva, I.M., Franklin, A.D., Phil Mag. **11**, 335 (1965)
Boswarva, I.M., Lidiard, A.B., Phil. Mag. **16**, 805 (1967)
Boswarva, I.M., Phil. Mag. **16**, 827 (1967)
Boswarva, I.M., J. Phys. C: Solid State Phys. **5**, L5 (1972)
Boswarva, I.M., Simpson, J.H., Can. J. Phys. **51**, 1923 (1973)
Bouckaert, L., Smoluchowski, R., Wigner, E., Phys. Rev. **50**, 58 (1936)
Bourne, J., Jacobs, R.L., J. Phys. C: Solid State Phys. **5**, 3462 (1972)
Bowers, R., Yafet, Y., Phys. Rev. **115**, 1165 (1959)
Bowers, R.L., Mahan, G.D., Phys. Rev. **185**, 1073 (1969)
Bowlden, H.J., Phys. Rev. **106**, 427 (1957)
Bowman, R.C., J. Chem. Phys. **59**, 2215 (1973)
Boyer, L.L., Hardy, J.R., Phys. Rev. B **7**, 2886 (1973)
Boyer, L.L., Phys. Rev. B **9**, 2684 (1974)
Boyn, R., Gardavsky, J., Phys. Stat. Sol. (b) **68**, 275 (1975)
Bozic-Popovic, M., Lalovic, D.I., Tosic, B.S., Zakula, R.B., Can. J. Phys. **50**, 898 (1972)
Bräunlich, P., Ann. Physik **12**, 262 (1963)
Bräunlich, P., Scharmann, A., Z. Physik **177**, 320 (1964)
Bräunlich, P., Scharmann, A., Phys. Stat. Sol. **18**, 307 (1966)
Bräunlich, P., Phys. Stat. Sol. **22**, 391 (1967)
Bräunlich, P., Kelly, P., Phys. Rev. B **1**, 1596 (1970)
Bramanti, D., Mancini, M., Ranfagni, A., Phys. Rev. B **3**, 3670 (1971)
Brand, S., Jaros, M., Solid State Commun. **21**, 875 (1977)
Brauer, P., Z. Naturforsch. **6a**, 560 (1951)
Brauer, P., Z. Naturforsch. **7a**, 741 (1952)
Braunstein, R., Phys. Rev. **125**, 475 (1962)
Braunstein, R., Ockman, N., Phys. Rev. **134**, A499 (1964)
Brauwers, M., Evrard, R., Kartheuser, E., Phys. Rev. B **12**, 5864 (1975)
Brauwers, M., Evrard, R., Kartheuser, E., Solid State Commun. **19**, 857 (1976)
Brauwers, M., Vail, J.M., Solid State Commun. **21**, 709 (1977)
Breit, G., Phys. Rev. **34**, 553 (1929)
Brener, N.E., Fry, J.L., Phys. Rev. B **6**, 4016 (1972)
Brener, N.E., Phys. Rev. B **7**, 1721 (1973)
Brener, N.E., Phys. Rev. B **11**, 1600 (1975)
Brenig, W., Statistische Mechanik, Lecure Notes of the Physics Department, TU München 1968
Brenig, W., Döhler, G., Wölfle, P., Z. Physik **246**, 1 (1971)
Brenig, W., Döhler, G.H., Wölfle, P., Z. Physik **258**, 381 (1973)
Brini, J., Kamarinos, G., Viktorovitch, P., Revue Phys. Appl. **9**, 451 (1974)
Brinkman, W.F., Rice, T.M., Bell, B., Phys. Rev. B **8**, 1570 (1973)
Bron, W.E., Wagner, M., Phys. Rev. **139**, A 233 (1965)
Bron, W.E., Phys. Rev. **140**, A2005 (1965)
Bron, W.E., Wagner, M., Phys. Rev. **145**, 689 (1966)
Brooks, H., Phys. Rev. **83**, 879 (1951)
Brooks, H., Advances in Electronics and Electron Physics, Vol. 7, ed. L. Marton, New York: Academic Press 1955, p. 85
Broser, I., Warminsky, R., Ann. Physik **7** 289 (1950)
Broser, I., Broser-Warminsky, R., Ann. Physik **16**, 361 (1955)
Broser, I., Gumlich, H.E., Moser, R., Z. Naturforsch. **20a**, 1648 (1965)
Bross, H., Phys. Stat. Sol. **2**, 481 (1962)
Brout, R., Prigogine, I., Physica **22**, 621 (1956)
Brout, R., Phys. Rev. **113**, 43 (1959)
Brown, F.C., Laramore, G., Appl. Optics **6**, 669 (1967)
Brown, R.J., Vail, J.M., Phys. Stat. Sol. **40**, 737 (1970)
Bruce, A.D., Cowley, R.A., Indian J. Pure Appl. Phys. **9**, 877 (1971)

Bruce, A.D., Cowley, R.A., J. Phys. C: Solid State Phys. **5**, 595 (1972)
Bruce, A.D., J. Phys. C: Solid State Phys. **5**, 2909 (1972)
Bruce, A.D., J. Phys. C: Solid State Phys. **6**, 174 (1973)
Brühl, S., Sigmund, E., Z. Naturforsch. **32a**, 111 (1977)
Brühl, S., Sigmund, E., Wagner, M., Ann. Phys. (Leipzig) **34**, 73 (1977)
Brühl, S., Wagner, M., in: Proceedings of the Int. Conference on Lattice Dynamics, ed. M. Balkanski, Paris 1977, Paris: Flammarion Sciences 1978, p. 215
Brust, D., Solid State Commun. 8, 1725 (1970)
Brust, D., Phys. Lett. **35A**, 182 (1971)
Brust, D., Solid State Commun, 9, 481 (1971)
Brust, D., Solid State Commun. **12**, 125 (1973)
Bryksin, V.V., Firsov, Yu.A., Sov. Phys. JETP **29**, 457 (1969)
Bryksin, V.V., Firsov, Yu.A., Sov. Phys. Solid State **10**, 2049 (1969)
Bryksin, V.V., Firsov, Yu.A., Sov. Phys. Solid State **12**, 480 (1970)
Bryksin, V.V., Firsov, Yu.A., Sov. Phys. Solid State **12**, 809 (1970)
Bryksin, V.V., Firsov, Yu.A., Sov. Phys. JETP **31**, 551 (1970)
Bryksin, V.V., Firsov, Yu.A., Sov. Phys. Solid State **14**, 384 (1972)
Bryksin, V.V., Firsov, Yu.A., Sov. Phys. Solid State **13**, 2729 (1972)
Bryksin, V.V., Firsov, Yu.A., Sov. Phys. Solid State **14**, 3019 (1973)
Bryksin, V.V., Sov. Phys. Solid State **14**, 2505 (1973)
Bryksin, V.V., Firsov, Yu.A., Sov. Phys. Solid State **15**, 2158 (1974)
Bryksin, V.V., Firsov, Yu.A., Sov. Phys. Solid State **15**, 2224 (1974)
Bryksin, V.V., Firsov, Yu.A., Sov. Phys. Solid State **16**, 524 (1974)
Bryksin, V V., Firsov, Yu.A., Sov. Phys. Solid State **16**, 1266 (1975)
Bube, R.H., Photoconductivity of Solids, New York: J. Wiley & Sons 1960
Bube, R.H., Electronic Properties of Crystalline Solids, New York: Academic Press 1974
Buch, T., Gelineau, A., Phys. Rev. B **4**, 1444 (1971)
Buchanan, M., Woll, E.J., Can. J. Phys. **47**, 1757 (1969)
Buchanan, M., Haberkorn, R., Bilz, H., J. Phys. C: Solid State Phys. **7**, 439 (1974)
Buchanan, M., Bauhofer, W., Martin, T.P., Phys. Rev. B **10**, 4358 (1974)
Budd, H.F., Phys. Rev. **140**, A2170 (1965)
Büttner, H., Festkörperprobleme **13**, 145 (1973) (Ed. H.J. Queisser, Braunschweig: Vieweg Verlag)
Bufáiçal, R.F., Maffeo, B., Brandi H.S., Phys. Rev. B **15**, 4091 (1977)
Buhks, E., J. Phys C: Solid State Phys. **8**, 1601 (1975)
Buhrow, J., Z. Phys. Chemie (Leipzig) **225**, 249 (1964)
Buisson, J.P., Lefrant, S., Sadoc, A., Taurel, L., Billardon, M., Phys. Stat. Sol. (b) **78**, 779 (1976)
Buisson, J.P., Lefrant, S., Ghomi, M., Taurel, L., Phys. Rev. B **18**, 885 (1978)
Bulyanitsa, D.S., Grinberg, A.A., Sov. Phys. Semicond. **11**, 474 (1977)
Burstein, E., in: Phonons and Phonon Interactions, Aarhus Summer School Lectures, 1963, ed. T.A. Bak, New York: Benjamin 1964
Burstein, E., J. Phys. Chem. Solids Suppl. **1**, 315 (1965)
Burstein, E., Proceedings of the International Conference on Lattice Dynamics, Copenhagen, Denmark, ed. R.F. Wallis, Oxford: Pergamon Press 1965
Burstein, E., Ushioda, S., Pinczuk, A., Solid State Commun. **6**, 407 (1968)
Buryakovskii, G.Yu., Mashkevich, V.S., Uritskii, Z.I., Sov. Phys. Solid State **16**, 584 (1974)
Buryakovskii, G.Yu., Mashkevich, V.S., Sov. Phys. Semicond. **8**, 1313 (1975)
Busch, G., Z. angew. Math. Phys. **1**, 4 (1950)
Busch, G., Winkler, U., Ergebn. exakt. Naturwiss. **29**, 145 (1956) (Ed. S. Flügge, F. Trendelenburg, Berlin: Springer Verlag)
Butcher, P.N., J. Phys. C: Solid State Phys. **5**, 1817 (1972)
Butcher, P.N., J. Phys. C: Solid State Phys. **7**, 2645 (1974)
Butcher, P.N., J. Phys. C: Solid State Phys. **7**, 879 (1974)
Butcher, P.N., Hayden, K.J., McInnes, J.A., Phil. Mag. **36**, 19 (1977)
Butcher, P.N., Hayden, K.J., Phil. Mag. **36**, 657 (1977)
Buzano, C., Rasetti, M., Nuovo Cim. B **67**, 55 (1970)
Bychkov, Yu.A., Dykhne, A.M., Sov. Phys. JETP **21**, 783 (1965)
Bychkov, Yu.A., Dykhne, A.M., Sov. Phys. JETP **31**, 928 (1970)

Cabib, D.,Fabri, E., Fiorio, G., Solid State Commun. **9**, 1517 (1971)
Caccamo, C., Ferrante, G., Nuovo Cim. **2B**, 93 (1971)
Cahn, R.N., Cohen, M.L., Phys. Rev. B **1**, 2569 (1970)
Calabrese, E., Fowler, W.B., Phys. Stat. Sol. (b) **56**, 621 (1973)
Calabrese, E., Fowler, W.B., Phys. Stat. Sol. (b) **57**, 135 (1973)
Calais, J.L., Mansikka, K., Petterson, G., Vallin, J., Ark. Fysik **34**, 361 (1967)

Calecki, D., Lewiner, C., Nozières, P., J. Physique **38**, 169 (1977)
Callaway, J., Phys. Rev. **113**, 1046 (1959)
Callaway, J.,Phys. Rev. **120**, 731 (1960)
Callaway, J., Phys. Rev. **121**, 1351 (1961)
Callaway, J., Phys. Rev. **130**, 549 (1963)
Callaway, J., Energy Band Theory, New York: Academic Press 1964
Callaway, J., Phys. Rev. **134**, A 998 (1964)
Callaway, J., Quantum Theory of the Solid State, Part A, B. New York: Academic Press 1974
Callen, H.B., Phys. Rev. **76**, 1394 (1949)
Calvo, R., Passeggi, M.C.G., Tovar, M., Phys. Rev. B **4**, 2876 (1971)
Capek, V., Czech. J. Phys. B **25**, 1020 (1975)
Capek, V., Czech, J. Phys. B **27**, 449 (1977)
Capek, V., J. Phys. Chem. Solids **38**, 623 (1977)
Carabatos, C., Prevot, B., Phys. Stat. Sol. (b) **44**, 701 (1971)
Carabatos, C., Prevot, B., Can. J. Phys. **50**, 122 (1972)
Cardon, F., Phys. Stat. Sol. **3**, 339 (1963)
Cardon, F., Phys. Stat. Sol. **12**, 805 (1965)
Cardon, F., Phys. Stat. Sol. **13**, 219 (1966)
Care, C.M., Tucker, J.W., J. Phys C: Solid State Phys. **9**, 4237 (1976)
Care, C.M., Tucker, J.W., J. Phys. C: Solid State Phys. **10**, 2773 (1977)
Carruthers, P., Rev. Mod. Phys. **33**, 92 (1961)
Carruthers, P., Phys. Rev. **125**, 123 (1962)
Carruthers. P., Phys. Rev.. **126**, 1448 (1962)
Carter, A.C., Dean, P.J., Skolnick, M.S., Stradling, R.A., J. Phys. C: Solid State Phys. **10**, 5111 (1977)
Carvalho, R.A., Terrile, M.C., Panepucci, H., Phys. Rev. B **15**, 1116 (1977)
Casella, R.C., Phys. Rev. **104**, 1260 (1956)
Casella, R.C., Phys. Rev. Letters **5**, 371 (1960)
Casimir, H.B.G., Physica **5**, 495 (1938)
Casselman, T.N., Markham, J.J., J. Chem. Phys. **42**, 4178 (1965)
Casselman, T.N., Spector, H.N., Phys. kondens. Materie **4**, 179 (1965)
Castner, T.G., Känzig, W., J. Phys. Chem. Solids **3**, 178 (1957)
Catalano, I.M., Cingolani, A., Minafra, A., Phys. Rev. B **5**, 1629 (1972)
Catalano, I.M., Cingolani, A., Cali, C., Riva-Sanseverino, S., Solid State Commun. **25**, 1 (1978)
Catlow, C.R.A., Norgett, M.J., J. Phys. C: Solid State Phys. **6**, 1325 (1973)
Catlow, C.R.A., Chem. Phys. Letters **39**, 497 (1976)
Catlow, C.R.A., J. Phys. C: Solid State Phys. **9**, 1845 (1976)
Catlow, C.R.A., J. Phys. C: Solid State Phys. **9**, 1859 (1976)
Catlow, C.R.A., Faux, D., Norgett, M.J., J. Phys. C: Solid State Phys. **9**, 419 (1976)
Catlow, C.R.A., Diller, K.M., Norgett M.J., J. Phys. C: Solid State Phys. **10**, 1395 (1977)
Chadi, D.J., Cohen, M.L., Grobman, W.D., Phys. Rev. B **8**, 5587 (1973)
Chaikovskii, I.A., Sov. Phys. Semicond. **6**, 198 (1972)
Chakrabarti, S.K., Sarkar, S.K., Sengupta, S., Phys. Stat. Sol. (b) **77**, 329 (1976)
Chakraborty, B., Pan. Y.K., Appl. Spectrosc Rev. **7**, 283 (1973)
Chakravarti, A.N., J. Appl. Phys. **42**, 2875 (1971)
Chakravarti, A.N., Czech. J. Phys. B **25**, 778 (1975)
Chamberlain, J.R., J. Phys. C: Solid State Phys. **11**, 1223 (1978)
Chambers, R.G., Proc. Phys. Soc. **89**, 695 (1966)
Champness, C.H., J. Electron. **4**, 201 (1958)
Chandler, P.E., J. Phys. C: Solid State Phys. **8**, 316 (1975)
Chandra, S., Pandey, G.K., Agrawal, V.K., Phys. Rev. **144**, 738 (1966)
Chaney, R.C., Lafon, E.E., Lin, C.C., Phys. Rev. B **4**, 2734 (1971)
Chaney, R.C., Lin, C.C., Phys. Rev. B **13**, 843 (1976)
Chaney, R.C., Phys. Rev. B **14**, 4578 (1976)
Chatterjee, R., Dixon, J.M., Lacroix, R., Weber J., Phys. Letters **43A**, 393 (1973)
Chatterjee, S., Sinha, P., Phys. Stat. Sol. (b) **70**, 283 (1975)
Chattopadhyay, D., Solid State Commun. **15**, 325 (1974)
Chattopadhyay, D., Nag, B.R., J. Phys. C: Solid State Phys. **9**, 3095 (1976)
Chattopadhyay, D., Nag, B.R., Phys. Letters **57A**, 347 (1976)
Chattopadhyay, D., Phys. Stat. Sol. (b) **81**, K51 (1977)
Chaudhuri, S., Roy, D., Ghosh, A.K., Indian J. Phys. **48**, 1033 (1974)
Chaudhuri, S., Roy, D., Ghosh, A.K., Indian J. Phys. **49**, 928 (1975)
Chaudhuri, S., Roy, D., Ghosh, A.K., Phys. Stat. Sol. (b) **70**, K33 (1975)
Cheban, A.G., Opt. Spectrosc. **10**, 253 (1961)
Cheban, A.G., Opt. Spectrosc. **14**, 269 (1963)
Cheban, A.G., Sov. Phys. Solid State **7**, 1054 (1965)
Chelikowsky, J.R., Cohen, M.L., Phys. Rev. Letters **32**, 674 (1974)
Chelikowsky, J.R., Cohen, M.L., Phys. Rev. B **14**, 556 (1976)
Chelikowsky, J.R., Cohen, M.L., Phys. Rev. Letters **36**, 229 (1976)
Chelikowsky, J.R., Solid State Commun. **22**, 351 (1977)

Chen, A.-B., Sher, A., Phys. Rev. B **17**, 4726 (1978)
Cherepanov, V.I., Druzhinin, V.V., Kargapolov, Yu.A., Nikoforov, A.E., Sov. Phys. Solid State **3**, 2179 (1962)
Chester, G.V., Thellung, A., Proc. Phys. Soc. **73**, 745 (1959)
Chester, M., Fritsche, L., Phys. Rev. **139**, A 518 (1965)
Cheung, P.S., Hearn, C.J., J. Phys. C: Solid State Phys. **5**, 1563 (1972)
Cheung, T.H., Wang, S., Phys. Stat. Sol. **24**, 509 (1967)
Chiarotti, G., Riv. Nuovo Cim. I (Numero Speciale), 30 (1969)
Chiarotti, G., Spectroscopy of Localized States, Trieste Lectures 1970, Wien: Intern. Atom Energy Agency 1971, p. 313
Child, M.S., Longuet-Higgins, H.C., Phil. Trans. Roy. Soc. London A **254**, 259 (1962)
Child, M.S., Phil. Trans. Roy. Soc. London A **255**, 31 (1963)
Cho, K., Kamimura, H., Uemura, Y., J. Phys. Soc. Japan **21**, 2244 (1966)
Cho, K., J. Phys. Soc. Japan **23**, 1296 (1967)
Cho, K., J. Phys. Soc. Japan **27**, 646 (1969)
Cho, K., Toyozawa, Y., J. Phys. Soc. Japan **30**, 1555 (1971)
Cho, K., Phys. Rev. B **14**, 4463 (1976)
Choh, S.H., Yi, K.S., J. Phys. C: Solid State Phys. **11**, 725 (1978)
Choquard, P., The Anharmonic Crystal, New York: Benjamin 1967
Choudhury, B.J., Phys. Rev. B **8**, 4849 (1973)
Chow, H.C., Can J. Phys. **56**, 537 (1978)
Chowdhury, S., Sen, S.K., Roy, D., Phys. Stat. Sol. (b) **56**, 403 (1973)
Christie, J., McGill, N.C., Physica **84B**, 155 (1976)
Christoph, V., Röpke, G., Phys. Stat. Sol. (b) **80**, K117 (1977)
Christy, R.W., Am. J. Phys. **40**, 40 (1972)
Chuenkov, V.A., Sov. Phys. Solid State **9**, 35 (1967)
Chuenkov, V.A., Sov. Phys. Semicond. **6**, 1233 (1973)
Chuenkov, V.A., Sov. Phys. Semicond. **8**, 556 (1974)
Chuenkov, V.A., Sov. Phys. Semicond. **8**, 1210 (1975)
Chuenkov, V.A., Sov. Phys. Semicond. **11**, 624 (1977)
Chusov, I.I., Sov. Phys. Semicond. **8**, 1236 (1975)
Cianchi, L., Mancini, M., Moretti, P., Rivista Nuovo Cim. **5**, 187 (1975)
Clack, D.W., Williams, W.T., J. inorg. nucl. Chem. Lett. **8**, 367 (1972)
Clark, T.D., Kliewer, K.L., Phys. Letters **27A**, 167 (1968)
Clark, T.D., Bajaj, K.K., Phys. Stat. Sol. (b) **56**, 211 (1973)
Claro, F., Wannier, G.H., Phys. Stat. Sol. (b) **88**, K147 (1978)
Claus, R., Merten, L., Brandmüller, J., Light Scattering by Phonon-Polaritons, Springer Tracts in Modern Physics **75**, ed. G. Höhler, Berlin: Springer 1975
Cochran, W., Proc. Roy. Soc. London A **253**, 260 (1959)
Cochran, W., Phil Mag. **4**, 1082 (1959)
Cochran, W., Rep. Progr. Phys. **26**, 1 (1963)
Cochran, W., Cowley, R.A., Phonons in Perfect Crystals, in: Handbuch der Physik, Band XXV/2a, ed. S. Flügge, Berlin-Heidelberg-New York 1967, p. 59
Cochran, W., Crit. Rev. Solid State Sci. **2**, 1 (1971)
Cohen, M.H., Heine, V., Phys. Rev. **122**, 1821 (1961)
Cohen, M.L., Bergstresser, T.K., Phys. Rev. **141**, 789 (1966)
Cohen, M.E., Landsberg, P.T., Phys. Rev. **154**, 683 (1967)
Cohen, M.E., Landsberg, P.T., Phys. Stat. Sol. (b) **64**, 39 (1974)
Collins, T.C., Euwema, R.N., De Witt, J.S., Phys. Soc. Japan **21**, Suppl., 15 (1966)
Collins, T.C., Stukel, D.J., Euwema, R.N., Phys. Rev. B **1**, 724 (1970)
Compton, W., Rabin, D., Solid State Physics **16**, 121 (1964)
Comte, C., Optics Commun. **14**, 79 (1975)
Conan, A., Phys. Stat. Sol. (b) **50**, 217 (1972)
Conklin, J.B., Johnson, L.E., Pratt, G.W., Phys. Rev. **137**, A1282 (1965)
Conwell, E., Weisskopf, V.F., Phys. Rev **77**, 388 (1950)
Conwell, E.M, Phys. Rev. **103**, 51 (1956)
Conwell, E.M., Zucker, J., J. Phys. Chem. Solids **22**, 149 (1961)
Conwell, E.M., High Field Transport in Semiconductors, Solid State Physics Suppl. 9, ed. F. Seitz, D. Turnbull, H. Ehrenreich, New York: Academic Press 1967
Corish, J., Parker, B.M.C., Jacobs, P.W.M., Can. J. Chem. **54**, 3839 (1976)
Cornwell, J.F., Group Theory and Electronic Energy Bands in Solids, Amsterdam: North-Holland 1969
Costato, M., Jacoboni, C., Reggiani, L., Phys. Stat. Sol. (b) **52**, 461 (1972)
Costato, M., Reggiani, L., Phys. Stat. Sol. (b) **58**, 47 (1973)
Costato, M., Reggiani, L., Phys. Stat. Sol. (b) **58**, 471 (1973)
Costato, M., Gagliani, G., Jacoboni, C., Reggiani, L., J. Phys. Chem. Solids **35**, 1605 (1974)
Coulson, C.A., Rèdei, L.B., Stocker, D., Proc. Roy. Soc. (London) A **270**, 357 (1962)

Cowan, W.B., Zuckermann, M.J., Solid State Comm. **16**, 207 (1975)
Cowley, E.R., Cowley, R.A., Proc. Roy. Soc. London A **287**, 259 (1965)
Cowley, E.R., Cowley, R.A., Proc. Roy. Soc. London A **292**, 209 (1966)
Cowley, E.R., J. Phys. C: Solid State Phys. **4**, 988 (1971)
Cowley, E.R., Jacucci, G., Klein, M.L., McDonald, I.R., Phys. Rev. B **14**, 1758 (1976)
Cowley, R.A., Proc. Roy. Soc. London A **268**, 109 (1962)
Cowley, R.A., Proc. Roy. Soc. London A **268** , 121 (1962)
Cowley, R.A., Adv. Phys. **12**, 421 (1963)
Cowley, R.A., Cochran, W., Brockhouse, B.N., Woods, A.D.B., Phys. Rev. **131**, 1030 (1963)
Cowley, R.A., Proc. Phys. Soc. **84**, 281 (1964)
Cowley, R.A., J. Physique (France) **26**, 659 (1965)
Cran, G.C., Sangster, M.J.L., J. Phys. C: Solid State Phys. 7, 1937 (1974)
Crandall, R.S., Phys. Rev. **138A**, 1242 (1965)
Crandall, R.S., J. Appl. Phys. **42**, 3933 (1971)
Croitoru, M., Grecu, D., Phys. Stat. Sol. **42**, 137 (1970)
Culik, F., Czech. J. Phys. B **16**, 194 (1966)
Cunningham, R.W., Gruber, J.B., J. Phys. Chem. Solids **31**, 2017 (1970)
Cunningham, S.L., Sharma, T.P., Jaswal, S.S., Hass, M., Hardy, J.R., Phys. Rev. B **10**, 3500 (1974)
Curby, R.C., Ferry, D.K., Phys. Stat. Sol. (a) **15**, 319 (1973)
Cuthbert, J.D., Thomas, D.G., Phys. Rev. **154**, 763 (1967)
Czachor, A., Holas, A., Acta Phys. Pol. **A49**, 307 (1976)
Czaja, W., Phys. kondens. Materie **12**, 226 (1971)
Czaja, W., Festkörperprobleme **11**, 65 (1971) (Ed. O. Madelung, Braunschweig: Vieweg Verlag)

Dahl, J.P., Switendick, A.C., J. Phys. Chem. Solids **27**, 931 (1966)
Dalgaard, E., Proc. Roy. Soc. London A **361**, 487 (1978)
Dalidchik, F.I., Sov. Phys. JETP **47**, 247 (1978)
Damask, A.C., Dienes, G.J., Phys. Rev. **125**, 444 (1962)
Danemar, A., Royce, B.S., Welch, D.O., Phys. Stat. Sol. (b) **70**, 663 (1975)
Das, A.K., Nag, B.R., Phys. Stat. Sol. (b) **69**, 329 (1975)
Das, A.K., Nag, B.R., Phys. Rev. B **13**, 1857 (1976)
Das, C.D., Keer, H.V., Rao, R.V.G., Z. phys. Chemie (Leipzig) **224**, 377 (1963)
Das, P., Sahni, V., J. Phys. C (Proc. Phys. Soc.) **1**, 1398 (1968)
Das, P., Ferry, D.K., Solid-State Electronics **19**, 851 (1976)
Das, T.P., Jette, A.N., Knox, R.S., Phys. Rev. **134**, A1079 (1964)
Das, T.P., Phys. Rev. **140**, A 1957 (1965)
Dass, L., Saxena, S.C., J. Chem. Phys. **43**, 1747 (1965)
Daude, N., Jouanin, C., Gout, C., Phys. Rev. B **15**, 2399 (1977)
Davies, E.B., Commun. Math. Phys. **39**, 91 (1974)
Davies, J.A., J. Math. Phys. **13**, 1207 (1972)
Davies, J.A., Mainville, C.L., J. Math. Phys. **16**, 1156 (1975)
Davies, J.A., J. Math. Phys. **17**, 388 (1976)
Davies, R.W., Blum, F.A., Phys. Rev. B **3**, 3321 (1971)
Davydov, A.S., Lubchenko, A.F., Sov. Phys. Doklady **13**, 325 (1968)
Davydov, A.S., Phys. Stat. Sol. **27**, 51 (1968)
Davydov, A.S., Phys. Stat. Sol. **30**, 357 (1968)
Davydov, A.S., Serikov, A.A., Phys. Stat. Sol. (b) **51**, 57 (1972)
Davydov, A.S., Pestryakov, G.M., Phys. Stat. Sol. (b) **49**, 505 (1972)
Dawber, P.G., Elliott, R.J., Proc. Roy Soc. London A **273**, 222 (1963)
Dawber, P.G., Parker, I.M., J. Phys. C: Solid State Phys. **3**, 2186 (1970)
Dean, P., J. Phys. C (Proc. Phys. Soc.) **1**, 22 (1968)
Dean, P.J., Faulkner, R.A., Phys. Rev. **185**, 1064 (1969)
DeCicco, P.D., Johnson, F.A., Proc. Roy. Soc. London A **310**, 111 (1969)
Dederichs, P.H., Zeller, R., Phys. Rev. B **14**, 2314 (1976)
Deigen, M.F., J. Exp. Theor. Phys. USSR **33** 773 (1957)
Deigen, M.F., Roitsin, L.B., J. Exp. Theor. Phys. USSR **36**, 176 (1959)
Deigen, M.F., Glinchuk, M.D., Korobko, G.V., Sov. Phys. Solid State **12**, 391 (1970)
Deigen, M.F., Glinchuk, M.S., Suslin, L.A., Sov. Phys. Solid State **14**, 2096 (1973)
Delves, R.T., Proc. Phys. Soc. **73**, 572 (1959)
DeMartini, F., Grassano, U.M., Simoni, F., Opt. Commun. **11**, 8 (1974)
Demchuk, K.M., Tsidilkovskii, I.M., Phys. Stat Sol. (b) **82**, 59 (1977)
Demidenko, A.A., Sov. Phys. Solid State **3**, 869 (1960)
Demidenko, A.A., Sov. Phys. Solid State **5**, 357 (1963)
Demidenko, Z.A., Tolpygo, K.B., Sov. Phys. Solid State **3**, 2493 (1962)
Demidenko, Z.A., Solid State Commun. **8**, 533 (1970)

Demikhovskii, V.Ya., Protogenov, A.P., Sov. Phys. JETP **31**, 348 (1970)
Denisov, M.M., Makarov, V.P., Phys. Stat. Sol. (b) **56**, 9 (1973)
Desnica, U.V., Urli, N.B., Phys. Stat. Sol. (b) **83**, K41 (1977)
Dettmann, K., Ludwig, W., Phys. kondens. Materie **2**, 241 (1964)
Déverin, J.A., Nuovo Cim. **63B**, 1 (1969)
Déverin, J.A., Helv. Phys. Acta **42**, 397 (1969)
Devreese, J., Evrard, R., Kartheuser, E., Solid State Commun. **7**, 767 (1969)
Devreese, J., Huybrechts, W., Lemmens, L., Phys. Stat. Sol. (b) **48**, 77 (1971)
Devreese, J., De Sitter, J., Goovaerts, M., Phys. Rev. B **5**, 2367 (1972)
Devreese, J.T., Kunz, A.B., Collins, T.C., Solid State Commun. **11**, 673 (1972)
Devreese, J.T., Kartheuser, E.P., Evrard, R., Baldereschi, A., Phys. Stat. Sol. (b) **59**, 629 (1973)
Devreese J.T., Evrard, R., Phys. Stat. Sol. (b) **78**, 85 (1976)
Dexter, D.L., Phys. Rev. **83**, 435 (1951)
Dexter,D.L., J. Chem. Phys. **21**, 836 (1953)
Dexter, D.L., Phys. Rev. **108**, 707 (1957)
Dexter, D.L., Solid State Physics **6**, 355 (1963), (Ed. F. Seitz, D. Turnbull, New York: Academic Press 1963)
Dheer, J.D., Sharan, B., Proc. Phys. Soc. **91**, 225 (1967)
Dichtel, K., Z. Physik **190**, 414 (1966)
Dick, B.G., Overhauser, A.W., Phys. Rev. **112**, 90 (1958)
Dick, B.G., Phys. Stat. Sol. **28**, 223 (1968)
Dick, B.G., Strauch, D., Phys. Rev. B **2**, 2200 (1970)
Dick, B.G., Phys. Rev. B **16**, 3359 (1977)
Dickerson, R.E., Gray, H.B., Haight, G.P., Chemical Principles, Menlo Park: Benjamin 1974
Dienes, G.J., Damask, A.C., Phys. Rev. **125**, 447 (1962)
Dienes, G.J., Hatcher, R.D., Smoluchowski R., Phys. Rev. **157**, 692 (1967)
Dienes, G.J., Hatcher, R.D., Lazareth, O.W., Royce, B.S.H., Smoluchowski, R., Phys. Rev. B **7**, 5332 (1973)
Dieterich, W., Peschel, I., Schneider, W.R., Commun. Phys. **2**, 175 (1977)
Dimmock, J.O., Solid State Phys. **26**, 103 (1971)
Dixit, V.K., Sharma, M.N., Indian J. Phys. **46**, 489 (1972)
Dixon, J.M., Smith, R.M., J. Phys. C: Solid State Phys. **5**, 2941 (1972)
Djordjević, R., Tosić, B.S., Prog. Theor. Phys. **54**, 1299 (1975)
Dneprovskaya, T.S., Phys. Stat. Sol. (b) **52**, 39 (1972)
Dobrovol'skii, V.N., Sov. Phys. Solid State **4**, 236 (1962)
Dobrzynski, L., J. Phys. Chem. Solids **30**, 2395 (1969)
Döhler, G., Hacker, K., Phys. Stat. Sol. **26**, 551 (1968)
Döhler, G.H., Phys. Stat. Sol. (b) **45**, 705 (1971)
Dogonadze, R.R., Chizmadzhev, Yu.A., Sov. Phys. Solid State **3**, 2693 (1962)
Dogonadze, R.R., Kuznetsov, A.M., Vorotyntsev, M.A., Phys. Stat. Sol. (b) **54**, 125 (1972)
Doktorov, E.V., Malkin, I.A., Manko, V.I., J. Phys. B: Atom. Molec. Phys. **9**, 507 (1976)
Donato, E., Giuliano, E.S., Ruggeri, R., Nuovo Cim. **15B**, 77 (1973)
Doni, E., Parravicini, G.P., Girlanda, R., Solid State Commun. **14** 873 (1974)
Donovan, B., Webster, J., Proc. Phys. Soc. **79**, 46 (1962)
Dorn, D., Z. Naturforsch. **12a**, 739 (1957)
Dorner, B., von der Osten, W., Bührer, W., J. Phys. C: Solid State Phys. **9**, 723 (1976)
Dos, K., Haug, A., Rohner, P., Phys. Stat. Sol. **30**, 619 (1968)
Dow, J.D., Redfield, D., Phys. Rev. Letters **26**, 762 (1971)
Dow, J.D., Redfield, D., Phys. Rev. B **5**, 594 (1972)
Dow, J.D., Final-State Interactions in the Optical Spectra of Solids: Elements of Exciton Theory, in: Optical Properties of Solids, ed. B.O. Seraphin, Amsterdam: North-Holland 1975, p. 33
Dowson, A.E.K., Ter Haar, D., Z. Naturforsch. **22a**, 1300 (1967)
Drabble, J.R., Goldsmid, H.J., Thermal Conduction in Semiconductors, Oxford: Pergamon Press 1961
Dresden, M., Rev. Mod. Phys. **33**, 265 (1961)
Dresselhaus, G., Phys. Chem. Solids **1**, 14 (1955)
Dresselhaus, G., Phys. Rev. **100**, 580 (1955)
Dreybrodt, W., Silber, D., Phys. Stat. Sol. **34**, 559 (1969)
Drost, D.M., Fry, J.L., Phys. Rev. B **5**, 684 (1972)
Druzhinin, V.V., Cherepanov, V.I., Levin, V.S., Sov. Phys. Solid State **7**, 2023 (1966)
Druzhinin, V.V., Moskvin, A.S., Sov. Phys. Solid State **11**, 1088 (1969)
Dubey, K.S., Verma, G.S., J. Phys. Soc. Japan **32**, 1202 (1972)
Dubey, K.S., Indian J. Pure Appl. Phys. **12**, 720 (1974)
Dubey, K.S., Phys. Stat. Sol. (b) **63**, K35 (1974)
Dubey, K.S., Phys. Stat. Sol. (b) **81**, K83 (1977)
Dubey, K.S., Phys. Stat. Sol. (b) **82**, K63 (1977)

Dubey, K.S., Misho, R.H., Phys. Stat. Sol. (b) **84**, 69 (1977)
Dubey, P.K., Chakravarti, A.K., Phys. Stat. Sol. (b) **43**, 89 (1971)
Dubey, P.K., Paranjape, V.V., Anand, M.M.L., Phys. Rev. B 7, 1718 (1973)
DuBois, D.F., Goldman, M.V., Phys. Rev. Letters **40**, 257 (1978)
Duke, C.B., Phys. Rev. Letters **15**, 625 (1965)
Duke, C.B., Tunneling in Solids, Solid State Physics Suppl. 10, ed. F. Seitz, D. Turnbull, H. Ehrenreich, New York: Academic Press 1969
Dumke, W.P., Phys. Rev. **108**, 1419 (1957)
Dumke, W.P., Phys. Rev. **124**, 1813 (1961)
Dumke, W.P., Phys. Rev. **167**, 783 (1968)
Dunn, D., J. Phys. C: Solid State Phys. **8**, 814 (1975)
Duran, J., Semicond. Insulators **3**, 329 (1978)
Durney, B., Proc. Phys. Soc. **78** 1384 (1961)
Dussel, G.A., Bube, R., Phys. Rev. **155**, 764 (1967)
Duthler, C.J., Phys. Rev. B **14**, 4606 (1976)
Dworin, L., Phys. Rev. **140**, A1689 (1965)
D'yachenko, V.G., Tyutin, M.S., Sov. Phys. Solid State **13**, 775 (1971)
D'yakonov, M.I., Mitchell, D.L., Efros, A.L., Sov. Phys. Solid State **10**, 2021 (1969)
D'yakonov, M.I., Perel', V.I., Sov. Phys. JETP **33**, 1053 (1971)
Dykman, I.M., Pekar, S.J., Dokl. Akad. Nauk USSR **83**, 825 (1952)
Dykman, I.M., J. Exp. Theor. Phys. USSR **26**, 307 (1954)
Dykman, I.M., Tomchuk, P.M., Sov. Phys. Solid State **2**, 1988 (1960)
Dykman, I.M., Tomchuk, P.M., Sov. Phys. Solid State **3**, 1393 (1961)
Dykman, I.M., Tomchuk, P.M., Sov. Phys. Solid State **4**, 798 (1962)
Dykman, I.M., Tomchuk, P.M., Sov. Phys. Solid State **4**, 2600 (1963)
Dykman, I.M., Tomchuk, P.M., Sov. Phys. JETP **27**, 318 (1968)
Dykman, I.M., Tomchuk, P.M., Phys. Stat. Sol. (b) **76**, 385 (1976)
Dykman, M.I., Ivanov, M.A., Sov. Phys. Solid State **18**, 415 (1976)
Dykman, M.I., Tarasov, G.G., Sov. Phys. JETP **47**, 557 (1978)
Dymnikov, V.D., Sov. Phys. Semicond. **11**, 868 (1977)
Dzhaksimov, E., Sov. Phys. Solid State **17**, 1849 (1976)

Eagles, D.M., Proc. Phys. Soc. 78, 204 (1961)
Eagles, D.M., Phys. Rev. **145**, 645 (1966)
Earney, J.J., J. Phys. C: Solid State Phys. **2**, 457 (1969)
Eckelt, P., Madelung, O., Treusch, J., Phys. Rev. Letters **18**, 656 (1967)
Eckelt, P., Phys. Stat. Sol. **23**, 307 (1967)
Eckstein, B., Z. Phys. Chemie **103**, 311 (1976)
Ecsedy, D.J., Klemens, P.G., Phys. Rev. B **15**, 5957 (1977)
Edelshtein, V.M., Sov. Phys. Solid State **13**, 2519 (1972)
Edelstein, V.M., Solid State Commun. **23**, 499 (1977)
Efremov, N.A., Kozhushner, M.A., Sov. Phys. Solid State **13**, 2993 (1972)
Efros, A.L., Sov. Phys. Solid State **9**, 901 (1967)
Efros, A.L., Sov. Phys. Solid State **10**, 2245 (1969)
Efros, A.L., Shklovskii, B.I., Yanchev, I.Y., Phys. Stat. Sol. (b) **50**, 45 (1972)
Egler, W., Haken, H., Z. Physik B **28**, 51 (1977)
Ehrenreich, H., J. Phys. Chem. Solids **2**, 131 (1957)
Ehrenreich, H., J. Phys. Chem. Solids **8**, 130 (1959)
Ehrenreich, H., J. Phys. Chem. Solids **9**, 129 (1959)
Ehrenreich, H., Cohen, M.H., Phys. Rev. **115**, 786 (1959)
Eisenriegler, E., Z. Physik **222**, 341 (1969)
Ekardt, W., Solid State Commun. **16**, 233 (1975)
Ekardt, W., Phys. Stat. Sol. (b) **68**, 53 (1975)
Ekardt, W., Phys. Stat. Sol. (b) **68**, 491 (1975)
Ekardt, W., Sheboul, M.I., Phys. Stat. Sol. (b) **74**, 523 (1976)
Ekardt, W., Sheboul, M.I., Phys. Stat. Sol. (b) **76**, K89 (1976)
Ekardt, W., Sheboul, M.I., Phys. Stat. Sol. (b) **80**, 51 (1977)
Ekardt, W., Phys. Stat. Sol. (b) **84**, 293 (1977)
Ekardt, W., Sheboul, M.I., Phys. Stat. Sol. (b) **86**, 535 (1978)
Eldridge, J.E., Howard, R., Staal, P.R., Can. J. Phys. **55**, 227 (1977)
Eldridge, J.E., Staal, P.R., Phys. Rev. B **16**, 3834 (1977)
Elesin, V.F., Sov. Phys. JETP **42**, 291 (1976)
Elkomoss, S.G., Phys. Rev. B **4**, 3411 (1971)
Elkomoss, S.G., Phys. Rev. B **6**, 3913 (1972)
Elkomoss, S.G., Amer, A.S., Phys. Rev. B **11**, 2222 (1975)
Elliott, R.J., Phys. Rev. **108**, 1384 (1957)
Elliott, R.J., Loudon, R., Phys. Letters **3**, 189 (1963)
Elliott, R.J., Physica **86–88B**, 1118 (1977)
El-Waheidy, E.F., Indian J. Phys. **51A**, 278 (1977)
El-Waheidy, E.F., Farahat, M.F., Acta Phys. Polon. **A53**, 365 (1978)
Emch, G., Helv. Phys. Acta **37**, 532 (1964)
Emel'yanov, V.I., Klimontovich, Yu.L., Sov. Phys. JETP **35**, 411 (1972)

Emin, D., Phys. Rev. B **4**, 3639 (1971)
Emin, D., Phys. Rev. Letters **28**, 604 (1972)
Emin, D., Phys. Rev. Letters **32**, 303 (1974)
Emin, D., J. Solid State Chem. **12**, 246 (1975)
Emin, D., Phys. Rev. Letters **35**, 882 (1975)
Emin, D., Adv. Phys. **24**, 305 (1975)
Emtage, P.R., Phys. Rev. B **3**, 2685 (1971)
Enderlein, R., Keiper, R., Tausendfreund, W., Phys. Stat. Sol. **33**, 69 (1969)
Enderlein, R., Phys. Stat. Sol. **41**, 107 (1970)
Enderlein, R., Peuker, K., Phys. Stat. Sol. (b) **48**, 231 (1971)
Enderlein, R., Bechstedt, F., Phys. Stat. Sol. (b) **80**, 225 (1977)
Engelmann, F., Z. Physik **145**, 430 (1956)
Engineer, M., Tzoar, N., Phys. Rev. B **8**, 702 (1973)
Engineer, M., Solid State Commun. **14**, 123 (1974)
Englman, R., Jortner, J., Molecular Phys. **18**, 145 (1970)
Englman, R., Caner, M., Toaff, S., J. Phys. Soc. Japan **29**, 306 (1970)
Englman, R., The Jahn-Teller Effect in Molecules and Crystals, London/New York: Wiley Interscience 1972
Englman, R., Halperin, B., J. Phys. C: Solid State Phys. **6**, L219 (1973)
Englman, R., J. Chem. Phys. **66**, 2212 (1977)
Englman, R., Non-radiative decay of ions and molecules in solids; Amsterdam: North-Holland 1979
Entin-Wohlman, O., Bergman, D.J., Physica **74**, 559 (1974)
Epifanov, M.S., Galkin, G.N., Bobrova, E.A., Vavilov, V.S., Sov. Phys. Semicond. **11**, 41 (1977)
Epshtein, E.M., JETP Letters **22**, 79 (1975)
Epshtein, E.M., Sov. Phys. Semicond. **11**, 243 (1977)
Erdös, P., Kang, J.P., Phys. Rev. B **6**, 3393 (1972)
Erezhepov, M., Sov. Phys. Solid State **6**, 1952 (1964)
Erezhepov, M., Sov. Phys. Solid State **6**, 2794 (1964)
Erezhepov, M., Sov. Phys. Solid State **8**, 1436 (1966)
Ermakov, V.N., Nitsovich, V.M., Tkach, N.V., Sov. Phys. Solid State **19**, 1867 (1977)
Ermolovich, I.B., Matvievskaja, G.I., Sheinkman, M.K., J. Luminescence **10**, 58 (1975)
Ermoshkin, A.N., Evarestov, R.A., Opt. Spectrosc. **35**, 456 (1973)
Ermoshkin, A.N., Evarestov, R.A., Phys. Stat. Sol. (b) **66**, 687 (1974)
Ermoshkin, A.N., Kotomin, E.A., Evarestov, R.A., Phys. Stat. Sol. (b) **72**, 787 (1975)
Ermoshkin, A.N., Evarestov, R.A., Kotomin, E.A., Phys. Stat. Sol. (b) **73**, 81 (1976)
Eschrig, H., Phys. Stat. Sol. (b) **56**, 197 (1973)
Euwema, R.N., Smoluchowski, R., Phys. Rev. **133**, A 1724 (1964)
Euwema, R.N., Collins, T.C., Shankland, D.G., DeWitt, D.G., DeWitt, J.S., Phys. Rev. **162**, 710 (1967)
Euwema, R.N., Stukel, D.J., Phys. Rev. B **1**, 4692 (1970)
Euwema, R.N., Wepfer, G.G., Surratt, G.T., Wilhite, D.L., Phys. Rev. B **9**, 5249 (1974)
Evans, B.D., Kemp, J.C., Phys. Rev. B **2**, 4179 (1970)
Evans, D.A., Landsberg, P.T., J. Phys. Chem. Solids **26**, 315 (1965)
Evarestov, R.A., Optika i Spektrosk. **16**, 361 (1964)
Evarestov, R.A., Phys. Stat. Sol. **31**, 401 (1969)
Evarestov, R.A., Treiger, V.M., Phys. Stat. Sol. **33**, 873 (1969)
Evarestov, R.A., Phys. Stat. Sol. (b) **46**, K13 (1971)
Evarestov, R.A., Slonim, V.Z., Phys. Stat. Sol. (b) **47**, K59 (1971)
Evjen, H.M., Phys. Rev. **39**, 675 (1932)
Evrard, R., Kartheuser, E., Williams, F., J. Luminescence **14**, 81 (1976)
Evseev, Z.Ya., Tolpygo, K.B., Sov. Phys. Solid State **4**, 2665 (1963)
Evseev, Z.Ya., Tolpygo, K.B., Sov. Phys. Solid State **9**, 1 (1967)
Evseev, Z.Ya., Tolpygo, K.B., Sov. Phys. Solid State **10**, 947 (1968)
Evseev, Z.Ya., Tolpygo, K.B., Phys. Stat. Sol. **34**, 45 (1969)
Ewald, P.P., Ann. Phys. (Germany) **54**, 519 (1917)
Ewald, P.P., Ann. Phys. (Germany) **54**, 557 (1917)
Ewald, P.P., Z. Krist. **56**, 129 (1921)
Ewald, P.P., Ann. Phys. (Germany) **64**, 253 (1921)
Ewing, D.H., Seitz, F., Phys. Rev. **50**, 760 (1936)
Eyring, H., Walter, J., Kimball, G.E., Quantum Chemistry, J. Wiley, New York 1944

Fain, V.M., Sov. Phys. JETP **26**, 1171 (1968)
Falter, C., Zierau, W., Varotsos, P., Solid State Commun. **27**, 401 (1978)
Fan, H.Y., Rep. Prog. Phys. **19**, 107 (1956)
Fano, U., Phys. Rev. **103**, 1202 (1956)
Fano, U., Phys. Rev. **118**, 451 (1960)
Farge, Y., Parodi, O., Toulouse, G., J. Physique **28**, Suppl. au no 2, C 1–111 (1967)
Fáthy, F., Bukovszky, F., Acta Phys. Hung. **8**, 89 (1957)
Fáthy, F., Bukovszky, F., Acta Phys. Hung. **9**, 275 (1959)
Faulkner, R.A., Phys. Rev. **175**, 991 (1968)
Faux, I.D., J. Phys. C: Solid State Phys. **4**, L211 (1971)
Fawcett, W., Herbert, D.C., J. Phys. C: Solid State Phys. **7**, 1641 (1974)

Feder, J., Pytte, E., Phys. Rev. B **8**, 3978 (1973)
Fedoseev, V., Hizhnyakov, V., Phys. Stat. Sol. **27**, 751 (1968)
Fedoseev, V.G., Sov. Phys. Solid State **13**, 253 (1971)
Fedoseyev, V.G., Solid State Commun. **19**, 569 (1976)
Fedyanin, V.K., Makhankov, V.G., Yakushevich, L.V., Phys. Letters **61A**, 256 (1977)
Feldkamp, L.A., J. Phys. Chem. Solids **33**, 711 (1972)
Feltham, P., Andrews, I., Phys. Stat. Sol. **10**, 203 (1965)
Feltham, P., Phys. Stat. Sol. **20**, 675 (1967)
Ferreira, L.G., J. Phys. Chem. Solids **30**, 1113 (1969)
Ferreira, L.G., J. Phys. C: Solid State Phys. **4**, 3 (1971)
Ferry, D.K., Phys. Rev. B **8**, 1544 (1973)
Ferry, D.K., Phys. Rev. B **9**, 4277 (1974)
Feuchtwang, T.E., Phys. Rev. **126**, 1616 (1962)
Feuchtwang, T.E., Phys. Rev. **126**, 1628 (1962)
Feynman, R.P., Phys. Rev. **97**, 660 (1955)
Feynman, R.P., Hellwarth, R.W., Iddings, C.K., Platzman, P.M., Phys. Rev. **127**, 1004 (1962)
Fick, E., Joos, G., Kristallspektren, in: Handbuch der Physik, Bd. XXVIII, ed. S. Flügge, Berlin-Göttingen-Heidelberg: Springer 1957, p. 203
Fick, E., Z. Physik **169**, 100 (1962)
Fiddicke, J., Enderlein, R., Phys. Stat. Sol. (b) **71**, 497 (1975)
Fidler, F.B., Tucker, J.W., J. Phys. C: Solid State Phys. **4**, 2583 (1971)
Fieschi, R., Nardelli, G.F., Terzi, N., Phys. Rev. **138**, A203 (1965)
Filipchenko, A.S., Lang, I.G., Nasledov, D.N., Pavlov, S.T., Radaikina, L.N., Phys. Stat. Sol. (b) **66**, 417 (1974)
Filipchenko, A.S., Bolshakov, L.P., Phys. Stat. Sol. (b) **77**, 53 (1976)
Firsov, Yu.A., Gurevich, V.L., Sov. Phys. JETP **14**, 367 (1962)
Firsov, Yu.A., Sov. Phys. Solid State **10**, 1537 (1969)
Firsov, Yu.A., Sov. Phys. Solid State **10**, 2387 (1969)
Fischbeck, H.J., Phys. Stat. Sol. **3**, 1082 (1963)
Fischbeck, H.J., Phys. Stat. Sol. **22**, 235 (1967)
Fischbeck, H.J., Phys. Stat. Sol. **30**, 779 (1968)
Fischbeck, H.J., Phys. Stat. Sol. **38**, 11 (1970)
Fischer, K., Z. Physik **155**, 59 (1959)
Fischer, K., Bilz, H., Haberkorn, R., Weber, W., Phys. Stat. Sol. (b) **54**, 285 (1972)
Fischer, K., Phys. Stat. Sol. (b) **66**, 295 (1974)
Fischer, K., Phys. Stat. Sol. (b) **66**, 449 (1974)
Fischer, S., J. Chem. Phys. **53**, 3195 (1970)
Flato, M., J. Mol. Spectrosc. **17**, 300 (1965)
Fleming, R.J., Phys. Letters **21**, 256 (1966)
Fleming, R.J., Phys. Stat. Sol. **29**, K77 (1968)
Fletcher, G.C., Austral. J. Phys. **12**, 237 (1959)
Fletcher, G.C., Austral. J. Phys. **14**, 420 (1961)
Fletcher, J.R., Stevens, K.W.H., J. Phys. C: Solid State Phys. **2**, 444 (1969)
Fletcher, J.R., J. Phys. C: Solid State Phys. **5**, 852 (1972)
Fletcher, J.R., Solid State Commun. **11**, 601 (1972)
Fletcher, K., Butcher, P.N., J. Phys. C: Solid State Phys. **5**, 212 (1972)
Fleurov, V.N., Kikoin, K.A., J. Phys. C: Solid State Phys. **9**, 1673 (1976)
Flinn, P.A., Maradudin, A.A., Ann. Phys. (N.Y.) **22**, 223 (1963)
Flügge, S., Lehrbuch der Theoretischen Physik IV: Quantentheorie I, Heidelberg: Springer 1964
Fock, H., Kramer, B., Büttner, H., Phys. Stat. Sol. (b) **67**, 199 (1975)
Fock, H., Kramer, B., Büttner, H., Phys. Stat. Sol. (b) **72**, 155 (1975)
Fock, V., Petrashen, M., Phys. Z. Sowj. **8**, 547 (1935)
Förster, Th., Ann. Physik, 6. Folge, **2**, 55 (1948)
Fogarassy, B., Phys. Stat. Sol. **3**, 1646 (1963)
Fogarassy, B., Phys. Stat. Sol. **3**, 2347 (1963)
Foglio, M.E., Phys. Stat. Sol. (b) **86**, 459 (1978)
Foldy, L.L., Wouthuysen, S.A., Phys. Rev. **78**, 29 (1950)
Foldy, L.L., Phys. Rev. B **17**, 4889 (1978)
Folland, N.O., Phys. Rev. B **2**, 418 (1970)
Fong, C.Y., Phys. Rev. **165**, 462 (1968)
Fong, F.K., Miller, M.M., Chem. Phys. Letters **10**, 408 (1971)
Fong, F.K., Naberhuis, S.L., Miller, M.M., J. Chem. Phys. **56**, 4020 (1972)
Fong, F.K., Diestler, D.J., J. Chem. Phys. **56**, 2875 (1972)
Fong, F.K., Diestler, D.J., J. Chem. Phys. **57**, 4953 (1972)
Fong, F.K., Wassam, W.A., J. Chem. Phys. **58**, 956 (1973)
Fong, F.K., Theory of Molecular Relaxation, New York: Wiley Interscience 1975
Fong, F.K. (Ed.), Radiationless Processes in Molecules and Condensed Phases, Topics in Applied Physics 15, Berlin: Springer Verlag 1976
Fonger, W.H., Struck, C.W., J. Chem, Phys. **60**, 1994 (1974)
Fonger, W.H., Struck, C.W., J. Luminescence **8**, 452 (1974)
Foo, E.–N., Phys. Rev. B **14**, 4338 (1976)
Ford, R.L., Fong, F.K., J. Chem. Phys. **55**, 2532 (1971)
Forney, J., Quattropani, A., Bassani, F., Nuovo Cim. **22**, 153 (1974)

Fortini, A., Diguet, D., Lugand, J., J. Appl. Phys. **41**, 3121 (1970)
Fouchaux, R.D., J. Phys. Chem. Solids **31**, 1113 (1970)
Fowler, W.B., Dexter, D.L., Phys. Stat. Sol. **2**, 821 (1962)
Fowler, W.B., Phys. Rev. **135**, A 1725 (1964)
Fowler, W.B., Phys. Rev. **151**, 657 (1966)
Fowler, W.B., Calabrese, E., Smith, D.Y., Solid State Commun. **5**, 569 (1967)
Fowler, W.B. (Ed.), Physics of Color Centers, New York: Academic Press 1968
Fowler, W.B., Phys. Rev. **174**, 988 (1968)
Fowler, W.B., Kunz, A.B., Phys. Rev. **186**, 956 (1969)
Fowler, W.B., Cope, D.N., J. Phys. Chem. Solids **31**, 2477 (1970)
Fowler, W.B., Kunz, A.B., Phys. Stat. Sol. **40**, 249 (1970)
Fowler, W.B., Phys. Stat. Sol. (b) **52**, 591 (1972)
Fowler, W.B., Marrone, M.J., Kabler, M.N., Phys. Rev. B 8, 5909 (1973)
Franceschetti, D.R., J. Appl. Phys. **48**, 3439 (1977)
Franklin, A.D., J. Phys. Chem. Solids **29**, 823 (1968)
Franz, W., Z. Physik **132**, 285 (1952)
Franz, W., Tewordt, L., Halbleiterprobleme III, ed. W. Schottky, Braunschweig: Vieweg Verlag 1956, p.1
Franz, W., Handbuch der Physik XVII, ed. S. Flügge, Berlin: Springer Verlag 1956, p. 155
Franz, W., Z. Naturforsch. **13a**, 484 (1958)
Fraser, S., Phys. Cond. Matter **17**, 71 (1974)
Freed, K.F., Jortner, J., J. Chem. Phys. **52**, 6272 (1970)
Freed, K.F., Chem. Phys. Letters **29**, 143 (1974)
Freed, K.F., Energy Dependence of Electronic Relaxation Processes in Polyatomix Molecules, in: Radiationless Processes in Molecules and Condensed Phases, Topics in Applied Physics 15, ed. F.K. Fong, Berlin: Springer 1976, p. 23
Frei, V., Velický, B., Czech. J. Phys. B **15**, 43 (1965)
Frenkel, J., Phys. Rev. **37**, 17 (1931)
Frenkel, J., Phys. Rev. **37**, 1276 (1931)
Frey, A., Jüngst, K.-L., Z. Physik B **27**, 47 (1977)
Friauf, R.F., J. Physique **38**, 1077 (1977)
Friedmann, L., Phys. Rev. **131**, 2445 (1963)
Friedrich, H., Jungk, G., Phys. Stat. Sol. **27**, 237 (1968)
Frisch, H.L., Lebowitz, J.L., Phys. Rev. **123**, 1542 (1961)
Fritsche, L., Phys. Stat. Sol. **11**, 381 (1965)
Fritsche, L., Phys. Stat. Sol. **34**, 195 (1969)
Fritzsche, H., Solid State Commun. **9**, 1813 (1971)
Fröhlich, D., Z. Physik **169**, 114 (1962)
Fröhlich, D., Z. Physik **177**, 126 (1964)
Fröhlich, D., Mahr, H., Phys. Rev. **141**, 692 (1966)
Fröhlich, D., Mahr, H., Phys. Rev. **148**, 868 (1966)
Fröhlich, D., Mohler, E., Uihlein, Ch., Phys. Stat. Sol. (b) **55**, 175 (1973)
Fröhlich, H., Proc. Roy. Soc. London A. **160**, 280 (1937)
Fröhlich, H., O'Dwyer, J., Proc. Phys. Soc. A **63**, 81 (1950)
Fröhlich, H., Pelzer, H., Zienau, S., Phil. Mag. **41**, 221 (1951)
Fröhlich, H., Proc. Roy. Soc. London A **215**, 291 (1952)
Fröhlich, H., Adv. Phys. **3**, 325 (1954)
Fröman, A., Löwdin, P.-O., J. Phys. Chem. Solids **23**, 75 (1962)
Fues, E., Stumpf, H., Z. Naturforsch. **9a**, 897 (1954)
Fues, E., Stumpf, H., Z. Naturforsch. **10a**, 136 (1955)
Fues, E., Stumpf, H., Z. Naturforsch. **10a**, 1055 (1955)
Fues, E., Stumpf, H., Z. Naturforsch. **14a**, 142 (1959)
Fukuda, A., Matsushima, A., Masunaga, S., J. Luminescence **12/13**, 139 (1976)
Fukuyama, H., Saitoh, M., Uemura, Y., Shiba, H., J. Phys. Soc. Japan **28**, 842 (1970)
Fulinski, A., Kramarczyk, W.J., Physica **39**, 575 (1968)
Fuliński, A., Physica **92 A**, 198 (1978)
Fulton, T., Phys. Rev. **103**, 1712 (1956)
Fumi, F.G., Tosi, M.P., Phys. Rev. **117**, 1466 (1960)
Fumi, F.G., Tosi, M.P., J. Phys. Chem. Solids **25**, 31 (1964)
Furukawa, H., Prog. Theor. Phys. **50**, 332 (1973)

Gabriel, H., Z. Naturforsch. **19a**, 1591 (1964)
Gärtner, W.W., Loscoe, C., Mette, H., Phys. Rev. **126**, 1680 (1962)
Gaididei, Yu.B., Loktev, V.M., Theor. Math. Phys. **16**, 714 (1973)
Galavanov, V.V., Nasledov, D.N., Filipchenko, A.S., Phys. Stat. Sol. **8**, 671 (1965)
Galindo, H., Mejia, C.R., Phys. Stat. Sol. (a) **22**, K165 (1974)
Galitskii, V.M., Goreslavskii, S.P., Elesin, V.F., Sov. Phys. JETP **30**, 117 (1970)
Gallina, V., Omini, M., Phys. Stat. Sol. **7**, 29 (1964)
Gallina, V., Omini, M., Phys. Stat. Sol. **7**, 405 (1964)
Gamurar, V.Ya., Opt. Spectrosc. **27**, 524 (1969)
Gamurar, V.Y., Perlin, Y.E., Tsukerblat, B.S., Sov. Phys. Solid State **11**, 970 (1969)
Ganesan, S., Burstein, E., Karo, A.M., Hardy, J.R., J. Physique **26**, 639 (1965)

Ganguly, A.K., Birman, J.L., Phys. Rev. **162**, 806 (1967)
Gantsevich, S.V., Sov. Phys. Solid State **9**, 707 (1967)
Garbato, L., Manca, P., Mula, G., J. Phys. C: Solid State Phys. **6**, L441 (1973)
Garcia-Moliner, F., Phys. Rev. **130**, 2290 (1963)
Gardavsky, J., Phys. Stat. Sol. (a) **3**, 939 (1970)
Gardavsky, J., Phys. Stat. Sol. (b) **52**, 493 (1972)
Gardavsky, J., Boyn, R., Phys. Stat. Sol. (b) **68**, 575 (1975)
Gasanov, M.S., Sov. Phys. Solid State **17**, 754 (1975)
Gasymov, T.M., Katanov, A.A., Babaev, M.M., Sov. Phys. Semicond. **11**, 1161 (1977)
Gaur, N.K.S., Physica **82B**, 262 (1976)
Gaur, S.P., Vetelino, J.F., Mitra, S.S., J. Phys. Chem. Solids **32**, 2737 (1971)
Gauthier, N., Walker, M.B., Phys. Rev. Letters **31**, 1211 (1973)
Gautier, F., J. Physique **28**, supp. au fasc. 5–6 C3-3 (1967)
Gay, J.G., Smith, J.R., Phys. Rev. B **9**, 4151 (1974)
Gazzinelli, R., Reik, H.G., Solid State Commun. 8, 745 (1970)
Gebhardt, W., Festkörperprobleme **9**, 99 (1969) (Ed. O. Madelung, Braunschweig: Vieweg Verlag)
Gebranzig, U., Haug, A., Rosenthal, W., J. Luminescence **12/13**, 547 (1976)
Gehring, G.A., J. Phys. C: Solid State Phys. **7**, L379 (1974)
Gehring, G.A., Gehring, K.A., Rep. Prog. Phys. **38**, 1 (1975)
Gehring, G.A., Laugsch, J., Phys. Stat. Sol. (b) **78**, 211 (1976)
Geifman, I.N., Glinchuk, M.D., Deigen, M.F., Krulikovskii, B.K., Sov. Phys. JETP **47**, 84 (1978)
Gelbart, W.M., Freed, K.F., Rice, S.A., J. Chem. Phys. **52**, 2460 (1970)
Gel'mont, B.L., Lyagushchenko, R.I. Yassievich, I.N., Sov. Phys. Solid State **14**, 445 (1972)
Gel'mont, B.L., Sov. Phys. Semicond. **9**, 1257 (1976)
Gel'mont, B.L., Phys. Letters **66 A**, 323 (1978)
Genkin, G.M., Sov. Phys. Solid State **3**, 1523 (1962)
Genkin, G.M., Fain, V.M., Yashchin, E.G., Sov. Phys. JETP **25**, 592 (1967)
Genkin, V.M., Zilberberg, V.V., Sov. Phys. Semiconduct. **10**, 479 (1976)
Georgescu, L., Physica **35**, 107 (1967)
Gerhardts, R., Hajdu, J., Z. Physik **245**, 126 (1971)
Gerhardts, R.R., Z. Physik B **22**, 327 (1975)
Gerhardts, R.R., Solid State Commun. **23**, 137 (1977)
Gerlach, B., Phys. Stat. Sol. (b) **63**, 459 (1974)
Gerlach, B., Pollmann, J., Phys. Stat. Sol. (b) **67**, 93 (1975)
Gerlach, E., Phys. Stat. Sol. (b) **61**, K97 (1974)
Gethins, T., Timusk, T., Woll, E.J., Phys. Rev. **157** 744 (1967)
de Geus, J.W., Gijzeman, O.L.J., J. Phys. Chem. Solids **39**, 577 (1978)
Ghaem-Maghami, V., Paranjape, V.V., Hawton, M.H., Solid State Commun. **10**, 923 (1972)
Ghate, P.B., Phys. Rev. **139**, A 1666 (1965)
Ghosh, A., Basu, A.N., Sengupta, S., Proc. Roy. Soc. London A **340**, 199 (1974)
Ghosh, A., Sarkar, A.K., Basu, A.N., J. Phys. C.: Solid State Phys. **8**, 1332 (1975)
Ghosh, A., Basu, A.N., J. Phys. C: Solid State Phys. **9**, 4365 (1976)
Ghosh, A., Basu, A.N., Phys. Rev. B **14**, 4616 (1976)
Ghosh, A. Basu, A.N., Phys. Rev. B **17**, 4558 (1978)
Ghosh, A.K., Manna, A., Deb, S.K., J. Phys. Chem Solids **38**, 701 (1977)
Giallorenzi, T.G., Phys. Rev. B **5**, 2314 (1972)
Giaquinta, P.V., Parrinello, M., Tosatti, E., Tosi, M.P., J. Phys. C: Solid State Phys. **9**, 2031 (1976)
Giehler, M., Jahne, E., Phys. Stat. Sol. (b) **73**, 503 (1976)
Giesecke, P., von der Osten, W., Röder, U., Phys. Stat. Sol. (b) **51**, 723 (1972)
Gifeisman, Sh.N., Perlin, Yu.E., Sov. Phys. JETP **22**, 857 (1966)
Gifeisman, Sh.N., Sov. Phys. Solid State **8**, 942 (1966)
Gilbert, R.L., Markham, J.J., J. Phys. Chem. Solid **30**, 2699 (1969)
Gilbert, T.L., in: Molecular Orbitals in Chemistry Physics, and Biology, ed. P.O. Löwdin and B. Pullman, New York: Academic Press 1964, p. 405
Gilbert, T.L., in: Sigma Molecular Orbital Theory ed. O. Sinanoglu and K. Wiberg, Yale Universi Press 1970, p. 249
Gill, J.C., Rep. Prog. Phys. **38**, 91 (1975)
Gillis, N.S., Phys. Rev. B **3**, 1482 (1971)
Giner, R.T., Korovin, L.I., Pavlov, S.T., Sov. Phy Solid State **19**, 1438 (1977)
Ginzberg, V.L., Rukhadze, A.A., Silin, V.P., Sov Phys. Solid State **3**, 1337 (1961)
Ginzburg, V.L., Rukhadze, A.A., Silin, V.P., J. Phys. Chem. Solids **23**, 85 (1962)
Giorgianni, U., Mondio, G. Saitta, G., Vermiglio, G., Phys. Stat. Sol. (b) **74**, 317 (1976)
Girlanda, R., Nuovo Cim. **6 B**, 53 (1971)
Girvin, S.M., J. Solid State Chem. **25**, 65 (1978)
Giterman, M.Sh., Tolpygo, K.B., Sov. Phys. JET **5**, 713 (1957)
Glasstone, S., Laidler, K.J., Eyring, H., The The of Rate Processes, New York: MacGraw-Hill 1941
Glinchuk, M.D., Deigen, M.F., Sov. Phys. JETP 952 (1968)
Glinchuk, M.D., Deigen, M.F., Korobko, G.V., S Phys. Solid State **9**, 2519 (1968)

Glinski, R.G., J. Phys. C: Solid State Phys. **6**, 3689 (1973)
Gliss, B., Zeyher, R., Bilz, H., Phys. Stat. Sol. (b) **44**, 747 (1971)
Glodeanu, A., Phys. Stat. Sol. **35**, 481 (1969)
Glyde, H.R., Rev. Mod. Phys. **39**, 373 (1967)
Glyde, H.R., Klein, M.L., Crit. Rev. Solid State Sci. **2**, 181 (1971)
Gnutzmann, U., Z. Physik **243**, 274 (1971)
Godley, J.S., Morris, S.P., Nuovo Cim. **4B**, 341 (1971)
Goel, C.M., Dayal, B., Phys. Stat. Sol. (b) **75**, 697 (1976)
Goel, C.M., Dayal, B., Phys. Stat. Sol. (b) **81**, 403 (1977)
Goel, C.M., Dayal, B., Solid State Commun. **26**, 369 (1978)
Goeppert-Mayer, M. Ann. Physik (Leipzig) **9**, 273 (1931)
Götze, W., Michel, K.H., Phys. Rev. **157**, 738 (1967)
Götze, W., Michel, K.H., Selfconsistent Phonons, in: Dynamical Properties of Solids, Vol. 1, ed. G.K. Horton and A.A. Maradudin, Amsterdam: North-Holland 1974
Gold, A., J. Chem. Phys. **35**, 2180 (1961)
Goldberg, P. (Ed.), Luminescence of Inorganic Solids, New York: Academic Press 1966
Goldberger, M.L., Watson, K.M., Collision Theory, New York: J. Wiley & Sons 1964
Goldmann, A., Phys. Stat. Sol. (b) **81**, 9 (1977)
Goldschmidt, V.M., Chem. Berichte (Germany) **60**, 1288 (1927)
Golka, J., Mostowski, J., Solid State Commun. **18**, 991 (1976)
Gontier, Y., Trahin, M., Phys. Rev. **172**, 83 (1968)
Goodman, B., Phys. Rev. **110**, 888 (1958)
Goodman, B., Oen, O.S., J. Phys. Chem. Solids **8**, 291 (1959)
van Gool, W., Proc. Kon. Ned. Ac. Wet. B **66**, 311 (1964)
Goossens, P., Phariseau, P., Physica **45**, 587 (1970)
Goossens, P., Phariseau, P., Physica **45**, 575 (1970)
Gorbachenko, B.I., Tolpygo, K.B., Sov. Phys. Solid State **8**, 192 (1966)
Gorbenko, P.K., Opt. Spectrosc. **20**, 247 (1966)
Gorczyka, I., Miasek, M., Acta Phys. Polon. **A48**, 621 (1975)
Gorczyca, I., Phys. Stat. Sol. (b) **82**, 279 (1977)
Gorzkowski, W., Acta Phys. Polon. **A50**, 159 (1976)
Gorzkowski, W., Phys Stat. Sol. (b) **79** K151 (1977)
Gosar, P., J. Phys. C: Solid State Phys. **8**, 3584 (1975)
Gourary, B.S., Adrian, F.I., Phys. Rev. **105**, 1180 (1957)
Gourary, B.S., Phys. Rev. **112**, 337 (1958)
Gourary, B.S., Adrian, F.J., Solid State Physics **10**, 128 (1960) (Ed. F. Seitz and D. Turnbull, New York: Academic Press)
Gout, C., Frandon, R., Sadaca, J., Phys. Letters **29A**, 656 (1969)
Govindarajan, J., Haridasan, T.M., Phys. Letters **29A**, 387 (1969)
Govindarajan, J., Jacobs, P.W.M., Nerenberg, M.A., J. Phys. C: Solid State Phys. **9, 3911 (1976)**
Govindarajan, J., Jacobs, P.W.M., Nerenberg, M.A., J. Phys. C: Solid State Phys. **10**, 1809 (1977)
Goyal, S.C., Prakash, R., Tripathi, S.P., Phys. Stat. Sol. (b) **85**, 477 (1978)
Grabert, H., Weidlich, W., Z. Physik **268**, 139 (1974)
Grabert, H., Phys. Letters **57A**, 105 (1976)
Graf, C.J.F., Maffeo, B., Brandi, H.S., Phys Stat. Sol. (b) **76**, K137 (1976)
Grant, W.J.C., J. Phys. Chem. Solids **25**, 751 (1964)
Grant, W.J.C., Strandberg, M.W.P., Phys. Rev. **135**, A715 (1964)
Grant, W.J.C., Phys. Rev. B **4**, 648 (1971)
Grassano, U.M., Nuovo Cim. **39B** 368 (1977)
Gray, A.M., Phys. Stat. Sol. **37**, 11 (1970)
Grekhov, A.M., Roitsin, A.B., Phys. Stat. Sol. (b) **74**, 323 (1976)
Griffith, J.S., The Theory of Transition-Metal Ions, Cambridge University Press 1961, 1971
Grigor'ev, N.N., Dykman, I.M., Tomchuk, P.M., Sov. Phys. Solid State **7**, 2719 (1966)
Grigor'ev, N.N., Dykman, I.M., Tomchuk, P.M., Sov. Phys. Solid State **10**, 837 (1968)
Grigor'ev, N.N., Sov. Phys. Semicond. **6**, 1576 (1973)
Grimley, T.B., Proc. Phys. Soc. A **70**, 123 (1957)
Grimley, T.B., Proc. Phys. Soc. **71**, 749 (1958)
Grimm, A., Maradudin, A.A., Ipatova, I.P., Subashiev, A.V., J. Phys. Chem. Solids **33**, 775 (1972)
Grimmeiss, H.G., Ledebo, L.A., J. Phys. C: Solid State Phys. **8**, 2615 (1975)
Grinberg, A.A., Sov. Phys. Semicond. **10**, 1117 (1976)
Grob, K., Z. Physik **184**, 395 (1965)
de Groot, W., Physica **6**, 275 (1939)
Gross, E.F., Permogorov, S.A., Razbirin, B.S., Sov. Phys. Solid State **8**, 1180 (1966)
Gross, E.P., Ann. Physics **8**, 78 (1959)
Gross, E.P., Ann. Physics **99**, 1 (1976)
Gross, H., Wahl, F., Z. Naturforsch. **14a**, 285 (1959)
Grover, M., Silbey, R., J. Chem. Phys. **54** 4843 (1971)
Gschwind, G., Haberland, D.H., Nelkowski, H., Stais, A., Phys. Stat. Sol. (a) **21**, 167 (1974)
Gudden, B., Schottky, W., Z. techn. Physik **16**, 323 (1935)
von Guérard, B., Peisl, H., Waidelich, W., Phys. Stat. Sol. **29**, K59 (1968)
Guha, S., Ghosh, S., J. Phys. Chem. Solids **38**, 1377 (1977)
Guha, S., Ghosh, S., Phys. Stat. Sol. (a) **41**, 249 (1977)

Gulyaev, Yu. V., Listvin, V.N., Potapov, V.T., Chusov, I.I., Yaremenko, N.G., Sov. Phys. Semicond. **9**, 972 (1976)
Gunther, L., Phys. Rev. **138**, **A**1697 (1965)
Gupta, B.M., Phys. Stat. Sol. **37**, 115 (1970)
Gupta, H.N., Singh, R.K., Phys. Stat. Sol. (b) **61**, 681 (1974)
Gupta, R.K., Haridasan, T.M., Chem. Phys. Letters **27**, 385 (1974)
Gupta, R.K., Phys. Letters **50A**, 269 (1974)
Gupta, R.K., Mathur, P., Singh, A.K., J. Phys. Chem. Solids **38**, 809 (1977)
Gupta, R.K., Singh, A.K., Phys. Letters **59A**, 489 (1977)
Gupta, R.K.,Singh, A.K., Solid State Commun. **25** 47 (1978)
Gupta, S.K., Narchal, M.L., Phys. Rev. B **2**, 1405 (1970)
Gupta, S.N., Quantum Electrodynamics, New York: Gordon and Breach 1977
Gurari, M., Phil. Mag. **44**, 329 (1953)
Gurari, M.L., Kozhushner, M.A., Sov. Phys. JETP **31**, 1060 (1970)
Gurevich, L.E., Nedlin, G.M., Sov. Phys JETP **13**, 568 (1961)
Gurevich, L.E., Tarkhanian, R.G., Phys. Stat. Sol. (b) **66**, 69 (1974)
Gurevich, L.E., Travnikov, V.S., Sov. Phys. Solid State **19**, 769 (1977)
Gurevich, V.L., Firsov, Yu.A., Sov. Phys. JETP **13** 137 (1961)
Gurevich, V.L., Katilius, R., Sov. Phys. JETP **22**, 796 (1966)
Gurevich, V.L., Lang, I.G., Parshin, D.A., Sov. Phys. Solid State **16**, 2356 (1975)
Gurevich, V.L., Parshin, D.A., Sov. Phys. Solid State **19**, 1406 (1977)
Gurevich, V.L., Parshin, D.A., Sov. Phys. JETP **45**, 834 (1977)
Gurnevich, L., J. Phys. USSR **9**, 477 (1945)
Gurnevich, L., J. Phys. USSR **10**, 67 (1946)
Gurskij, B.A., Gurskij, Z.A., Ukr. Fiz. Zh. (USSR) **21**, 1609 (1976)
Gurskij, B.A., Gurskij, Z.A., Ukr. Fiz. Zh. (USSR) **21**, 1603 (1976)

Haake, F., in: Quantum Statistics in Optics and Solid State Physics, Springer Tracts in Modern Physics 66, ed. G. Höhler, Berlin: Springer 1973, p. 98
Hacker, K., Phys. Stat. Sol. **33**, 607 (1969)
Hacker, K., Obermair, G., Z. Physik **234**, 1 (1970)
Haering, R.R., Adams, E.N., Phys. Rev. **117**, 451 (1960)
Hafemeister, D.W., Flygare, W.H., J. Chem. Phys. **43**, 795 (1965)
Haga, E., J. Phys. Soc. Japan **13**, 1090 (1958)
Haga, E., J. Phys. Soc. Japan **14**, 992 (1959)
Haga, E., J. Phys. Soc. Japan **14**, 1176 (1959)
Haga, E., Kimura, H., J. Phys. Soc. Japan **18**, 777 (1963)
Haga, E., Kimura, H., J. Phys. Soc. Japan **19**, 471 (1964)
Haga, E., Kimura, H., J. Phys. Soc. Japan **19**, 658 (1964)
Haga, E., Kimura, H., J. Phys. Soc. Japan **19**, 1596 (1964)
Hagen, D.E., Van Zandt, L.L., Prohofsky, E.W., Phys. Rev. B **2**, 553 (1970)
Hagston, W.E., Proc. Phys. Soc. **92**, 1101 (1967)
Hagston, W.E., Mol. Phys. **19**, 593 (1970)
Hagston, W.E., Phys. Stat. Sol. **39**, 551 (1970)
Hagston, W.E., Phys. Stat. Sol. **42**, 879 (1970)
Hagston, W.E., Phys. Stat. Sol. (b) **47**, 281 (1971)
Hagston, W.E., J. Phys. C: Solid State. Phys. **5**, 69 (1972)
Hagston, W.E., Lowther, J.E., Physica **70**, 40 (1973)
Hagston, W.E., Lowther, J.E., J. Phys. Chem. Solids **34**, 1773 (1973)
Hajdu, J., Keiter, H., Z. Physik **201**, 507 (1967)
Haken, H., Z. Naturforsch. **9a**, 228 (1954)
Haken, H., Halbleiterprobleme I, 72 (1954) (Ed. W. Schottky, Braunschweig, Vieweg Verlag)
Haken, H., Z. Physik **139**, 66 (1954)
Haken, H., Schottky, W., Z. Physik **144**, 91 (1956
Haken, H., Z. Physik **146**, 527 (1956)
Haken, H., Nuovo Cim. **3**, 1230 (1956)
Haken, H., Z. Naturforsch. **11a** 875 (1956)
Haken, H., Halbleiterprobleme IV, 1 (1958) (Ed. W. Schottky, Braunschweig: Vieweg Verlag)
Haken, H., Fortschr. Phys. **6**, 271 (1958)
Haken, H., Schottky, W., Z. physik. Chem. **16**, 21 (1958)
Haken, H., J. Phys. Chem. Solids **8**, 166 (1959)
Haken, H., Z. Physik **155**, 223 (1959)
Haken, H., in: Polarons and Excitons, ed. C.G. Kuper and G.D. Whitfield, London: Oliver an Boyd 1962, p. 295
Haken, H., Strobl, G., in: The Triplet State, ed. A.B. Zahlan, Cambridge University Press 196
Haken, H., Licht und Materie Ic, Handbuch der Physik XXV/2c, ed. S. Flügge Berlin: Spring 1970
Haken, H., Reineker, P., Z. Physik **249**, 253 (19
Haken, H., Quantenfeldtheorie des Festkörpers, Stuttgart: Teubner Verlag 1973
Haken, H., Z. Physik **262**, 119 (1973)
Haken, H., Schenzle, A., Z. Physik **258**, 231 (1973)
Haken, H., Strobl, G., Z. Physik **262**, 135 (1973
Haken, H., Nonlinear Interaction between Excit and Coherent Light, in: Proceedings of the In ternational Summerschool on Physics "Enric Fermi", Course 54, Varenna 1975, Amsterda North-Holland 1977, p. 350
Haken, H., Synergetics, Berlin: Springer 1977
Haldane, F.D.M., Anderson, P.W., Phys. Rev. B 2553 (1976)
Halperin, B., Phys. Rev. B **7**, 894 (1973)
Halperin, B., Englman, R., Phys. Rev. Letters **3** 1052 (1973)

Halperin, B., Englman, R., Solid State Commun. **13**, 1185 (1973)
Halperin, B., Englman, R., Phys. Rev. B **9**, 2264 (1974)
Halperin, B., Phys. Letters **55A**, 301 (1975)
Halperin, B., Englman, R., Phys. Rev. B **12**, 388 (1975)
Halpern, V., J. Phys. C: Solid State Phys. **3**, 1900 (1970)
Ham, F.S., Phys. Rev. **138**, A1727 (1965)
Ham, F.S., Phys. Rev. **166**, 307 (1968)
Ham, F.S., Schwarz, W.M., O'Brien, M.C.M., Phys. Rev. **185**, 548 (1969)
Ham, F.S., Slack, G.A., Phys. Rev. B **4**, 777 (1971)
Ham, F.S., Phys. Rev. Letters **28**, 1048 (1972)
Ham, F.S., Phys. Rev. B **8**, 2926 (1973)
Ham, F.S., Grevsmühl, U., Phys. Rev. B **8**, 2945 (1973)
Ham, F.S., Leung, C.H., Kleiner, W.H., Solid State Commun. **18**, 757 (1976)
Ham, N.S., Spectrochim. Acta **18**, 775 (1962)
Hamera, M., Phys. Stat. Sol. (b) **69**, K45 (1975)
Hamilton, D.C., Phys. Rev. **188** 1221 (1969)
Hamilton, R.A.H., Parrott, J.E., Phys. Rev. **178**, 1284 (1969)
Hamilton, R.A.H., J. Phys. C: Solid State Phys. **6**, 2653 (1973)
Hammar, C., Magnusson, B., Phys. Scripta **6**, 206 (1972)
Hanamura, E., Inui, T., J. Phys. Soc. Japan **18**, 690 (1963)
Hanamura, E., J. Phys. Soc. Japan **29**, 50 (1970)
Hanamura, E., J. Phys. Soc. Japan **37**, 1545 (1974)
Hanamura, E., J. Phys. Soc. Japan **37**, 1553 (1974)
Hanamura, E., J. Phys. Soc. Japan **39**, 1506 (1975)
Hanamura, E., J. Phys. Soc. Japan **39**, 1516 (1975)
Hanamura, E., Excitonic Molecules, in: Optical Properties of Solids, ed. B.O. Seraphin, Amsterdam: North-Holland 1975, p. 81
Hanamura, E., Haug, H., Phys. Reports **33**, 209 (1977)
Handel, P.H., Phys. Rev. B **7**, 5183 (1973)
Hanisch, G., Phys. kondens. Materie **4**, 297 (1965)
Hanlon, J.E., Lawson, A.W., Phys. Rev. **113**, 472 (1959)
Haque, M.S., Strauch, D., Phys. Rev. B **15**, 5898 (1977)
Hardy, J.R., Phil. Mag. **4**, 1278 (1959)
Hardy, J.R., J. Phys. Chem. Solids **15**, 39 (1960)
Hardy, J.R., Phil. Mag. **6**, 27 (1961)
Hardy, J.R., J. Phys. Chem. Solids **23**, 113 (1962)
Hardy, J.R., Phil. Mag. **7**, 315 (1962)
Hardy, J.R., Lidiard, A.B., Phil. Mag. **15**, 825 (1967)
Hardy, J.R., Karo, A.M., Phys. Rev. **168**, 1054 (1968)
Hardy, J.R., in: Dynamical Properties of Solids I, ed. G.K. Horton and A.A. Maradudin, Amsterdam: North-Holland 1974, p 157
Hardy, R.J., Phys. Rev. **132**, 168 (1963)
Hardy, R.J., J. Math. Phys. **6**, 1749 (1965)
Hardy, R.J., J. Math. Phys. **7**, 1435 (1966)
Hardy, R.J., Karo, A.M., J. Chem. Phys. **48**, 3173 (1968)
Hardy, R.J., Karo, A.M., J. Appl. Phys. **41**, 5144 (1970)
Hardy, R.J., Karo, A.M., Phys. Rev. B **7**, 4696 (1973)
Haridasan, T.M., Krishnamurthy, N., Indian J. Pure Appl. Phys. **6**, 407 (1968)
Haridasan, T.M., Gupta, R.K., Ludwig, W., Solid State Commun. **12**, 1205 (1973)
Haridasan, T.M., Gupta, R.K., Ludwig, W., Chem. Phys. Letters **23**, 217 (1973)
Harker, A.H., J. Phys. C: Solid State Phys. **6**, 2993 (1973)
Harker, A.H., J. Phys. C: Solid State Phys. **7**, 3224 (1974)
Harker, A.H., J. Phys. C: Solid State Phys. **9**, 2273 (1976)
Harker, A.H., J. Phys. C: Solid State Phys. **9**, 3141 (1976)
Harker, A.H., Lyon, S.B., Wasiela, A., Solid State Commun. **21**, 1053 (1977)
Harker, A.H., J. Phys. C: Solid State Phys. **11**, 1059 (1978)
Harper, P.G., Hodby, J.W., Stradling, R.A., Rep. Prog. Phys. **36**, 1 (1973)
Harrington, J.A., Duthler, C.J., Patten, F.W., Hass, M., Solid State Commun. **18**, 1043 (1976)
Harris, J.J., Ridley, B.K., J. Phys. Chem. Solids **34**, 197 (1973)
Harris, S.M., Prohofsky, E.W., Phys. Rev. **170**, 749 (1968)
Harrison, J.W., Hauser, J.R., J. Appl. Phys. **47**, 292 (1976)
Harrison, W.A., Phys. Rev. **101**, 903 (1956)
Harrison, W.A., Phys. Rev. B **8**, 4487 (1973)
Harrison, W.A., Festkörperprobleme **17**, 135 (1977) (Ed. J. Treusch, Braunschweig: Vieweg Verlag)
Hartmann, W.M., Gilbert, T.L., Kaiser, K.A., Wahl, A.C., Phys. Rev. B **2**, 1140 (1970)
den Hartog, H.W., Arends, J., Phys. Stat. Sol. **23**, 713 (1967)
den Hartog, H.W., Phys. Stat. Sol. **38**, 457 (1970)
Hartree, D.R., Proc. Roy. Soc. London A **151**, 96 (1935)
Hartree, D.R., Hartree, W., Proc. Roy. Soc. London A **156**, 45 (1936)
Hartree, D.R., Hartree, W., Proc. Roy. Soc. London A **166**, 450 (1938)
Hashitsume, N., Shibata, F., Shingu, M., J. Stat. Phys. **17**, 155 (1977)
Hassan, A.R., Nuovo Cim. **LXX B**, 21 (1970)
Hassan, A.R., Nuovo Cim. **13B**, 19 (1973)
Hassan, A.R., Solid State Commun. **12**, 99 (1973)
Hassan, A.R., J. Phys. C: Solid State Phys. **9**, 2383 (1976)
Hassan, A.R., Solid State Commun. **18**, 437 (1976)
Hassan, A.R., Moussa, A.R., Nuovo Cim. **40 B**, 354 (1977)

Hassan, A.R., Phys. Stat. Sol. (b) **87**, 31 (1978)
Hassan, A.R., Solid State Commun. **25**, 817 (1978)
Hassan, S.S.A.Z., Proc. Phys. Soc. **85**, 783 (1965)
Hatcher, R.D., Dienes, G.J., Phys. Rev. **124**, 726 (1961)
Hatcher, R.D., Dienes, G.J., Phys. Rev. **134**, A214 (1964)
Hattori, K., J. Phys. Soc. Japan **37**, 1215 (1974)
Hattori, K., J. Phys. Soc. Japan **38**, 360 (1975)
Hattori, K., J. Phys. Soc. Japan **38**, 351 (1975)
Hattori, K., J. Phys. Soc. Japan **38**, 356 (1975)
Hattori, K., J. Phys. Soc. Japan **38**, 669 (1975)
Hattori, K., Phys. Stat. Sol. (b) **73**, 717 (1976)
Hattori, K., Phys. Stat. Sol. (b) **76**, 281 (1976)
Hauffe, K., Halbleiterprobleme I, 107 (1954) (Ed. W. Schottky, Braunschweig: Vieweg Verlag)
Hauffe, K., Ilschner, B., Z. Elektrochem. **58**, 467 (1954)
Haug, A., Halbleiterprobleme I, 227 (1954) (Ed. W. Schottky, Braunschweig: Vieweg Verlag)
Haug, A., Z. Physik **146**, 75 (1956)
Haug, A., Z. Physik **148**, 504 (1957)
Haug, A., Sauermann, G., Z. Physik **153**, 269 (1958)
Haug, A., Theoretische Festkörperphysik, Bd. 2, Wien: Fr. Deuticke Verlag 1970
Haug, A., Festkörperprobleme **12**, 411 (1972) (Ed. O. Madelung, Braunschweig: Vieweg Verlag)
Haug, A., Solid State Commun. **22**, 537 (1977)
Haug, A., Solid State Commun. **25**, 477 (1978)
Haug, A., Theoretische Festkörperphysik, Bd. 1, Wien: Fr. Deuticke Verlag 1964
Hawranek, J.P., Lowndes, R.P., Solid State Commun. **11**, 1473 (1972)
Hayashi, S., Kanamori, H., J. Phys. Soc. Japan **37**, 1399 (1974)
Hayes, W., Wiltshire, M.C.K., J. Phys. C: Solid State Phys. **6**, 1149 (1973)
Haynes, J.R., Phys. Rev. Lett. **4**, 361 (1960)
Hayns, M.R., Dissado, L., Theoret. Chim. Acta **37**, 147 (1975)
Heck, R.J., Woodruff, T.O., Phys. Rev. B **3**, 2056 (1971)
Hede, B., McMullen, T., Can. J. Phys. **54**, 2101 (1976)
Hede, B., McMullen, T., Can. J. Phys. **54**, 2093 (1976)
Heine, V., Solid State Physics **24**, 1 (1970) (Ed. F. Seitz, D. Turnbull and H. Ehrenreich, New York: Academic Press)
Heine, V., Henry, C.H., Phys. Rev. B **11**, 3795 (1975)
Heinle, W., Z. Physik **217**, 150 (1968)
Heinrichs, J., Kumar, N., Solid State Commun. **16**, 1035 (1975)
Heinzel, W., Zur Theorie der Temperaturverschiebung der Banden des F-Zentrums in Ionenkristallen, Thesis, University of Tübingen 1973
Heinzel, W., Phys. Cond. Matter **17**, 99 (1974)
Heitler, W., The Quantum Theory of Radiation, Oxford University Press 1944, 1949, 1954
Heller, D.F., Freed, K.F., Gelbart, W.M., Chem. Phys. Letters **23**, 56 (1973)
Hellmann, H., J. Chem. Phys. **3**, 61 (1935)
Hellmann, H., Kassatotschkin, W., Acta Physicochim. USSR **5**, 23 (1936)
Helmis, G., Ann. Physik (Leipzig), 6. Folge, **17**, 356 (1956)
Helmis, G., Ann. Physik (Leipzig) **19**, 41 (1956)
Hemstreet, L.A., Phys. Rev. B **12**, 1212 (1975)
Henderson, B., Garrison, A.K., Adv. Phys. **22**, 423 (1973)
Henderson, B., Contemp. Phys. **19**, 225 (1978)
Henderson, B., O'Connell, D.C., Semicond. Insulators **3**, 299 (1978)
Henisch, H.K., Electroluminescence, Oxford: Pergamon Press 1962
Henneberger, F., Phys. Stat. Sol. (b) **70**, K169 (1975)
Henneberger, F., Henneberger, K., Voigt, J., Phys. Stat. Sol. (b) **83**, 439 (1977)
Henneberger, K., Röseler, J., Phys. Stat. Sol. (b) **58**, 575 (1973)
Henneberger, K., Strehlow, R., Wünsche, H.-J., Phys. Stat. Sol. (b) **61**, 455 (1974)
Henneberger, K., Enderlein, R., Phys. Stat. Sol. (b) **72**, 547 (1975)
Henneberger, K., Phys. Stat. Sol. (b) **74**, 567 (1976)
Henry, C.H., Schnatterly, S.E., Slichter, C.P., Phy Rev. Lett. **13**, 130 (1964)
Henry, C.H., Schnatterly, S.E., Slichter, C.P., Phy Rev. **137**, A 583 (1965)
Henry, C.H., Hopfield, J.J., Phys. Rev. Letters **15** 964 (1965)
Henry, C.H., Slichter, C.P., Moments and Degeneracy in Optical Spectra, in: Physics of Color Centers, ed. W.B. Fowler, New York: Academ Press 1968, p. 351
Henry, C.H., Lang, D.V., Phys. Rev. B **15**, 989 (1977)
Herbert, D.C., J. Phys. C: Solid State Phys. **6**, 27 (1973)
Herbert, D.C., Fawcett, W., Hilsum, C., J. Phys. C Solid State Phys. **9**, 3969 (1976)
Herman, F., Phys. Rev. **88**, 1210 (1952)
Herman, F., Phys. Rev. **93**, 1214 (1954)
Herman, F., Rev. Mod. Phys. **30**, 102 (1958)
Herman, F., Skillman, S., Atomic Structure Calculation, Englewood Cliffs, N.J.: Prentice-Hall 1963
Herman, F., Kortum, R.L., Ortenburger, I.B., van Dyke, J.P., J. Physique **29**, Suppl. au no 11-1 C4–62 (1968)
Herman, R.C., Meyer, C.F., Hopfield, H.S., J. Op Soc. Am. **38**, 999 (1948)
Herman, R.C., Wallis, M.C., Wallis, R.F., Phys. R **103**, 87 (1956)

Hermanson, J., Phillips, J.C., Phys. Rev. **150**, 652 (1966)
Hermanson, J., Phys. Rev. B **6**, 2427 (1972)
Hermanson, J.C., Phys. Rev. B **2**, 5043 (1970)
Herring, C., Phys. Rev. **57**, 1169 (1940)
Herring, C., Phys. Rev. **96**, 1163 (1954)
Hersh, H.N., Phys. Rev. **148**, 928 (1966)
Herzfeld, C.M., Meijer, P.H., Solid State Physics **12**, 1 (1962) (Ed. F. Seitz and D. Turnbull, New York: Academic Press)
Heuer, K., Phys. Stat. Sol. **4**, 461 (1964)
van Heufelen, A., J. Chem. Phys. **46**, 4903 (1967)
Hewat, A.W., Solid State Commun. **8**, 187 (1970)
Heyna, H.J., Lokalisierte Gitterschwingungen in Ionenkristallen mit einer Leerstelle, Thesis, University of Tübingen 1974
Hill, D., Landsberg, P.T., Proc. Roy. Soc. London A **347**, 547 (1976)
Hill, D., Proc. Roy. Soc. London A **347**, 565 (1976)
Hill, D., J. Phys. C: Solid State Phys. **9**, 3527 (1976)
Hillbrand, H., Kranzer, D., Phys. Stat. Sol. **42**, K79 (1970)
Hilsum, C., Rose-Innes, A.C., Semiconducting III–V Compounds, Oxford: Pergamon Press 1961
Hirai, M., Matsuyama, T., J. Phys. Soc. Japan **28**, 1240 (1970)
Hizhnyakov, V., Tehver, I., Phys. Stat. Sol. **21**, 755 (1967)
Hizhnyakov, V., Tehver, I., Phys. Stat. Sol. (b) **82**, K89 (1977)
Hizhnyakov, V.V., Sherman, A.V., Phys. Stat. Sol. (b) **85**, 51 (1978)
Hizhnyakov, V., Zazubovich, S., Phys. Stat. Sol. (b) **86**, 733 (1978)
Hobey, W.D., McLachlan, A.D., J. Chem. Phys. **33**, 1695 (1960)
Höhler, G., Z. Naturforsch. **9a**, 801 (1954)
Höhler, G., Nuovo Cim. **2**, 691 (1955)
Höhler, G., Z. Physik **140**, 192 (1955)
Höhler, G., Z. Physik **146**, 372 (1956)
Höhler, G., Z. Physik **146**, 571 (1956)
Höhler, G., Müllensiefen, A., Z. Physik **157**, 159 (1959)
Hoenerlage, B., Wiesner, H., Z. Physik **242**, 406 (1971)
Hönerlage, B., Rössler, U., J. Luminescence **12/13**, 593 (1976)
Hönerlage, B., Klingshirn, C., Grun, J.B., Phys. Stat. Sol. (b) **78**, 599 (1976)
Hoffmann, H.J., Stöckmann, F., Tödheide-Haupt, U., Phys. Stat. Sol. (b) **56**, 549 (1973)
Hoffmann, H.J., Phys. Stat. Sol. (b) **57**, 123 (1973)
Hofstadter, D.R., Phys. Rev. B **14**, 2239 (1976)
Holland, B.W., Phil. Mag. **8**, 87 (1963)
Holstein, T., Ann. Physics **8**, 325 (1959)
Holstein, T., Ann. Physics **8**, 343 (1959)
Holstein, T., Friedman, L., Phys. Rev. **165**, 1019 (1968)
Holstein, T., Phil. Mag. B **37**, 49 (1978)
Holstein, T., Phil. Mag. B **37**, 499 (1978)
Holz, A., Nuovo Cim. **9B**, 83 (1972)
Holzapfel, G., Rickert, H., Festkörperprobleme **15**, 318 (1975) (Ed. H.J. Queisser, Braunschweig: Vieweg Verlag)
Honda, S., Tomura, M., J. Phys. Soc. Japan **33**, 1003 (1972)
Honig, J.M., Wahnsiedler, W.E., Dimmock, J.O., J. Solid State Chem. **5**, 452 (1972)
Honma, A., Sci. of Light (Japan) **16**, 212 (1967)
Honma, A., Sci. of Light (Japan) **16**, 229 (1967)
Honma, A., J. Phys. Soc. Japan **24**, 1082 (1968)
Honma, A., J. Phys. Soc. Japan **32**, 483 (1972)
Honma, A., Sci. of Light (Japan) **22**, 101 (1973)
Honma, A., Sci. of Light (Japan) **23**, 1 (1974)
Honma, A., Sci. of Light (Japan) **24**, 33 (1975)
Honma, A., Sci. of Light (Japan) **24**, 57 (1975)
Honma, A., Ooaku, S., J. Phys. Soc. Japan **41**, 152 (1976)
Hopfield, J.J., Phys. Rev. **112**, 1555 (1958)
Hopfield, J.J., J. Phys. Chem. Solids **10**, 110 (1959)
Hopfield, J.J., Thomas, D.G., J. Phys. Chem. Solids **12**, 276 (1960)
Hopfield, J.J., J. Appl. Phys. **32**, Suppl., 2277 (1961)
Hopfield, J.J., Thomas, D.G., Phys. Rev. **122**, 35 (1961)
Hopfield, J.J., Worlock, J.M., Phys. Rev. **137**, A 1455 (1965)
Hopfield, J.J., J. Phys. Soc. Japan **21**, Suppl., 77 (1966)
Hopfield, J.J., Thomas, D.G., Lynch, R.T., Phys. Rev. Lett. **17**, 312 (1966)
Hopfield, J.J., Phys. Rev. **182**, 945 (1969)
Horie, C., Prog. Theor. Phys. **21**, 113 (1959)
Horie, C., Krumhansl, J.A., Phys. Rev. **136**, A1397 (1964)
Hornbeck, J.A., Haynes, J.R., Phys. Rev. **97**, 311 (1955)
Horner, H., Strongly Anharmonic Crystals with Hard Core Interactions, in: Dynamical Properties of Solids I, ed. G.K. Horton and A.A. Maradudin, Amsterdam: North-Holland 1974
Horton, G.K., Maradudin, A.A. (Eds.), Dynamical Properties of Solids, Vol. 1, Amsterdam: North-Holland 1974
Hougen, J.T., J. Mol. Spectrosc. **13**, 149 (1964)
Houlier, B., J. Phys. C: Solid State Phys. **10**, 1419 (1977)
van Hove, L., Phys. Rev. **89**, 1189 (1953)
van Hove, L., Physica **21**, 517 (1955)
van Hove, L., Physica **23**, 441 (1957)
Howarth, D.J., Sondheimer, E.H., Proc. Roy. Soc. London A **219**, 53 (1953)
Howgate, D.W., Phys. Rev. **177**, 1358 (1969)
Howland, L.P., Phys. Rev. **109**, 1927 (1958)

Howlett, W., Zukotynski, S., Phys. Rev. B **16**, 3688 (1977)
Hrivnak, L., Czech. J. Phys. **9**, 439 (1959)
Hrivnak, L., Czech. J. Phys. B **26**, 670 (1976)
Hsieh, H.C., J. Appl. Phys. **48**, 1240 (1977)
Huang, C.Y., Phys. Rev. **158**, 280 (1967)
Huang, C.Y., Phys. Rev. **154**, 215 (1967)
Huang, C.Y., Phys. Rev. **168**, 334 (1968)
Huang, C.Y., Lue, J.T., Prog. Theor. Phys. **43**, 10 (1970)
Huang, K., Rhys, A., Proc. R. Soc. London A **204**, 406 (1950)
Huang, K., Proc. Roy. Soc. London A **208**, 352 (1951)
Huang, K., Nature **167**, 779 (1951)
Huang, K., Z. Physik **171**, 213 (1963)
Hubbard, J., Proc. Roy. Soc. London A **276**, 238 (1963)
Hubbard, J., Proc. Roy. Soc. London A **277**, 237 (1964)
Hubbard, J., Proc. Roy. Soc. London A **281**, 401 (1964)
Hubbard, J., Proc. Roy. Soc. London A **296**, 82 (1966)
Hubbard, J., Proc. Roy. Soc. London A **296**, 100 (1966)
Hubbard, J., Proc. Phys. Soc. **92**, 921 (1967)
Huberman, M., Chester, G.V., Adv. Phys. **24**, 489 (1975)
Hübner, K., Phys. Letters **31A**, 365 (1970)
Hübner, K., Bashenov, V.K., Phys. Stat. Sol. (b) **77**, 473 (1976)
Hübner, R., Z. Physik **222**, 380 (1969)
Hughes, A.E., J. Phys. C: Solid State Phys. **3**, 627 (1970)
Hughes, F., Allard, J., Phys. Rev. **125**, 173 (1962)
Hughes, F., Phys. Stat. Sol. **5**, 55 (1964)
Huldt, L., Phys. Stat. Sol. (a) **8**, 173 (1971)
Huldt, L., Phys. Stat. Sol. (a) **24**, 221 (1974)
Huldt, L., Phys. Stat. Sol. (a) **33**, 607 (1976)
Hulin, M., Phys. Stat. Sol. **21**, 607 (1967)
Humphreys, L.B., Maradudin, A.A., Phys. Rev. B **6**, 3868 (1972)
Humphreys, L.B., Phys. Rev. B **6**, 3886 (1972)
Hutson, A.R., J. Appl. Phys. Suppl. **32**, 2287 (1961)
Huybrechts, W.J., J. Phys. C: Solid State Phys. **10**, 3761 (1977)
Huybrechts, W.J., Phys. Stat. Sol. (b) **81**, 585 (1977)
Hylleraas, E.A., Z. Physik **63**, 771 (1930)

Iadonisi, G., Preziosi, B., Nuovo Cim. **48B**, 92 (1967)
Iadonisi, G., Zucchelli, G.P., Phys. Stat. Sol. (b) **62**, 625 (1974)
Ida, Y., Phys. Rev. B **1**, 2488 (1970)
Iida, T., Kurata, K., Muramatsu, S., J. Phys. Chem. Solids **33**, 1225 (1972)
Iida, T., J. Phys. Chem. Solids **33**, 1423 (1972)
Iida, T., Monnier, R., Phys. Stat. Sol. (b) **74**, 91 (1976)
Imanaka, K., Iida, T., Ohkura, H., J. Phys. Soc. Japan **43**, 519 (1977)
Imanaka, K., Iida, T., Ohkura, H., J. Phys. Soc. Japan **44**, 1632 (1978)
Imamóv, E.Z., Kramer, N.I., Sagdullaeva, S.A., Sov. Phys. Semiconduct. **11**, 1065 (1977)
Inglis, G.B., Williams, F., J. Luminescence **12/13**, 525 (1976)
Inkson, J.C., J. Phys. C: Solid State Phys. **6**, L181 (1973)
Inkson, J.C., J. Phys. C: Solid State Phys. **9**, 1177 (1976)
Inomata, H., Horie, C., Phys. Letters **31A**, 418 (1970)
Inoue, M., Sati, R., Wang, S., Can. J. Phys. **48**, 1694 (1970)
Inoue, M., Mahutte, C.K., Wang, S., Phys. Rev. B **2**, 539 (1970)
Inoue, M., Okazaki, M., J. Phys. Soc. Japan **31**, 1313 (1971)
Inoue, M., J. Phys. Soc. Japan **37**, 1560 (1974)
Inoue, M., Takenaka, N., Shirafuji, J., Inuishi, Y., Solid-State Electronics **21**, 29 (1978)
Inoue, S., Yamashita, J., Prog. Theor. Phys. **42**, 158 (1969)
Insepov, Z.A., Norman, G.E., Sov. Phys. JETP **46** 798 (1977)
Inui, T., Uemura, A., Prog. Theor. Phys. **5**, 252 (1950)
Iona, M., Phys. Rev. **60**, 822 (1941)
Ipatova, I.P., J. Techn. Phys. (russ.) **26**, 2787 (1956)
Ipatova, I.P., Klotchichin, A.A., Sov. Phys. JETP **23**, 1068 (1966)
Ipatova, I.P., Maradudin, A.A., Wallis, R.F., Sov. Phys. Solid State **8**, 850 (1966)
Ipatova, I.P., Maradudin, A.A., Wallis, R.F., Phys Rev. **155**, 882 (1967)
Ipatova, I.P., Klochinin, A.A., Subashiev, A.V., Phys. Stat. Sol. **23**, 467 (1967)
Ipatova, I.P., Subashiev, A.V., Maradudin, A.A., Ann. Physics **53**, 376 (1969)
Isay, W.–H., Ann. Physik **13**, 327 (1953)
Itoh, N., Crystal Latt. Def. **3**, 115 (1972)
Itoh, N., Goto, T., Radiation Eff. **37**, 45 (1978)
Ivanov, A.I., Ponomarev, O.A., Theor. Math. Phy **30**, 246 (1977)
Ivanov, M.A., Pogorelov, Yu. G., Sov. Phys. JET **45**, 1155 (1977)
Ivchenko, E.L., Pikus, G.E., Razbirin, B.S., Starukhin, A.I., Sov. Phys. JETP **45**, 1172 (1977)
Ivchenko, E.L., Lang, I.G., Pavlov, S.T., Sov. Ph Solid State **19**, 1610 (1977)
Ivchenko, E.L., Lang, I.G., Pavlov, S.T., Sov. Ph Solid State **19**, 718 (1977)
Ivchenko, E.L., Lang, I.G., Pavlov, S.T., Phys. S Sol. (b) **85**, 81 (1978)
Izmailov, S.V., Rozman, G.A., Opt. Spectrosc. **3** 507 (1975)

Jackson, H.E., Walker, C.T., Phys. Rev. B **3**, 1428 (1971)
Jacobs, P.W.M., Menon, A.K., J. Chem. Phys. **55**, 5357 (1971)
Jacucci, G., McDonald, I.R., Rahman, A., Phys. Rev. A **13**, 1581 (1976)
Jahn, H.A., Teller, E., Proc. Roy. Soc. London A **161**, 220 (1937)
Jahne, E., Gutsche, E., Phys. Stat. Sol. **21**, 57 (1967)
Jain, J.K., Shanker, J., Khandelwal, D.P., Phys. Rev. B **13**, 2692 (1976)
Jain, K.P., Prabhakaran, A.K., Phys. Letters **40A**, 61 (1972)
Jain, K.P., Prabhakaran, A.K., Phys. Rev. B **6**, 596 (1972)
Jain, S.C., Sharma, T.P., Arora, N.D., J. Phys. Chem. Solids **37**, 81 (1976)
Jain, V.C., Shanker, J., Physica **94 B**, 346 (1978)
Jain, V.C., Shanker, J., Phys. Stat. Sol. (b) **89**, 213 (1978)
James, H.M., Phys. Rev. **76**, 1611 (1949)
Janak, J.F., Solid State Commun. **20**, 151 (1976)
Janner, A., Helv. Phys. Acta **35**, 47 (1962)
Jannussis, A., Phys. Stat. Sol. **6**, 217 (1964)
Jansen, L., Phys. Rev. A **135**, 1292 (1964)
Jansen, L., Lombardi, E., Disc. Faraday Soc. **40**, 78 (1965)
Jaros, M., Ross, S.F., J. Phys. C: Solid State Phys. **6**, 1753 (1973)
Jaros, M., J. Phys. C: Solid State Phys. **8**, 2455 (1975)
Jaros, M., Brand, S., Phys. Rev. B **14**, 4494 (1976)
Jaros, M., Srivastava, G.P., J. Phys. Chem. Solids **38**, 1399 (1977)
Jaros, M., Phys. Rev. B **16**, 3694 (1977)
Jaros, M., Solid State Commun. **25**, 1071 (1978)
Jaswal, S.S., Montgomery, D.J., Phys. Rev. **135**, A1257 (1964)
Jaswal, S.S., Phys. Rev. **140**, A687 (1965)
Jaswal, S.S., Phys. Rev. **137**, A302 (1965)
Jaswal, S.S., Striefler, M.E., Solid State Commun. **6**, 351 (1968)
Jaswal, S.S., Sharma, T.P., J. Phys. Chem. Solids **34**, 509 (1973)
Jaswal, S.S., Wadehra, J.M., Phys. Rev. B **11**, 4055 (1975)
Jaswal, S.S., Phys. Rev. Letters **35**, 1600 (1975)
Jaswal, S.S., Dilly, V.D., Phys. Rev. B **15**, 2366 (1977)
Jaswal, S.S., Dilly, V.D., Solid State Commun. **24**, 577 (1977)
Jaswal, S.S., J. Phys. C: Solid State Phys. **11**, 3559 (1978)
Jay-Gerin, J.P., J. Phys. Chem. Solids **35**, 81 (1974)
Jay-Gerin, J.P., Phys. Rev. B **12**, 1418 (1975)
Jefferson, J.H., Hagston, W.E., Proc. Roy. Soc. London A **355**, 355 (1977)
Jenkins, H.D.B., Pratt, K.F., Proc. Roy. Soc. London A **356**, 115 (1977)
Jennison, D.R., Kunz, B.A., Phys. Rev. B **13**, 5597 (1976)
Jensen, B., Ann. Physics **80**, 284 (1973)
Jensen, B., Ann. Physics **95**, 229 (1975)
Jensen, B., Solid State Commun. **24**, 853 (1977)
Jensen, B., Phys. Stat. Sol. (b) **86**, 291 (1978)
Jette, A.N., Phys. Rev. **184**, 604 (1969)
Jette, A.N., Gilbert, T.L., Das, T.P., Phys. Rev. **184**, 884 (1969)
Jette, A.N., Das, T.P., Phys. Rev. **186**, 919 (1969)
Jex, H., Phys. Stat. Sol. (b) **62**, 393 (1974)
Johnson, E.J., Larsen, D.M., Phys. Rev. Letters **16**, 655 (1966)
Johnson, F.A., Proc. Roy. Soc. London A **310**, 79 (1969)
Johnson, F.A., Proc. Roy. Soc. London A **310**, 89 (1969)
Johnson, F.A., Proc. Roy. Soc. London A **310**, 101 (1969)
Johnson, K.H., Phys. Rev. **150**, 429 (1966)
Johnson, L.F., Guggenheim, H.J., Rich, T.C., Ostermayer, F.W., J. Appl. Phys. **43**, 1125 (1972)
Johri, G., Paranjape, V.V., Can. J. Phys. **54**, 2147 (1976)
Jones, R., Schaich, W., J. Phys. C: Solid State Phys. **5**, 43 (1972)
de Jong, C., Solid State Commun. **9**, 527 (1971)
Joshi, S.K., Gupta, R., Phys. Rev. **126**, 933 (1962)
Joshi, Y.P., Phys. Stat. Sol. (b) **68**, 145 (1975)
Joshi, V.P., Singh, D.P., Phys. Rev. B **14**, 1733 (1976)
Jost, R., Pais, A., Phys. Rev. **82**, 840 (1951)
Jouanin, C., Albert, J.P., Gout, C., C.R. Acad. Sc. Paris B **278**, 743 (1974)
Jouanin, C., Albert, J.P., Gout, C., Nuovo Cim. **28B**, 483 (1975)
Jouanin, C., Albert, J.P., Gout, C., J. Physique **37**, 595 (1976)
Judd, B.R., Can. J. Phys. **52**, 999 (1974)
Judd, B.R., Vogel, E.E., Phys. Rev. B **11**, 2427 (1975)
Judd, B.R., Lie Groups and the Jahn-Teller Effect for a Color Center, in: 4th International Colloquium on Group Theoretical Methods in Physics, Nijmegen 1975, ed. A. Janner, T. Janssen, M. Boon, Berlin: Springer Verlag 1976, p. 312
Jüngst, K.-L., Kuri, P., Crystal Lattice Defects **2**, 83 (1971)

Kachlishvili, Z.S., Sov. Phys. Solid State **3**, 361 (1961)
Kachlishvili, Z.S., Sov. Phys. Solid State **4**, 538 (1962)
Kachlishvili, Z.S., Sov. Phys. Solid State **3**, 2091 (1962)
Kachlishvili, Z.S., Sov. Phys. Solid State **3**, 1554 (1962)
Kachlishvili, Z.S., Phys. Stat. Sol. (b) **48**, 65 (1971)

Kachlishvili, Z.S., Sov. Phys. Solid State **13**, 1903 (1972)
Kachlishvili, Z.S., Sov. Phys. Semicond. **6**, 487 (1972)
Kachlishvili, Z.S., Phys. Stat. Sol. (a) **33**, 15 (1976)
Kadanoff, L.P., Phys. Rev. **130**, 1364 (1963)
Kagan, M.S., *Landsberg, E.G.*, *Elenkrig, B.B.*, Sov. Phys. Semicond. **10**, 1110 (1976)
Kagan, V.D., Sov. Phys. Solid State **17**, 1717 (1976)
Kahn, A., *Kittel, C.*, Phys. Rev. **89**, 315 (1953)
Kaiser, F., *Wagner, M.*, Z. Naturforsch. **32a**, 375 (1977)
Kaiser, F., Z. Naturforsch. **32a**, 697 (1977)
Kaiser, F., Z. Naturforsch. **32a**, 805 (1977)
Kalashnikov, V.P., Sov. Phys. JETP **24**, 956 (1967)
Kalashnikov, V.P., Sov. Phys. Solid State **8**, 1693 (1967)
Kalashnikov, V.P., Physica **48**, 93 (1970)
Kalok, L., *Treusch, J.*, Phys. Stat. Sol. (b) **52**, K125 (1972)
Kalyani, S., *Haridasan, T.M.*, Chem. Phys. Letters **44**, 184 (1976)
Kalyani, S., *Haridasan, T.M.*, J. Phys. Chem. Solids **38**, 735 (1977)
Kamal, J., *Sharma, S.*, Can. J. Phys. **49**, 876 (1971)
Kambara, T., *Haas, W.J.*, *Spedding, F.H.*, *Good, R.H.*, J. Chem. Phys. **56**, 4475 (1972)
Kamimura, H., *Yamaguchi, T.*, J. Phys. Soc. Japan **25**, 1138 (1968)
van Kampen, N.G., Physica **20**, 603 (1954)
Kamphusmann, J., Z. Naturforsch. **14a**, 165 (1959)
Kane, E.O., J. Phys. Chem. Solids **1**, 249 (1957)
Kane, E.O., J. Phys. Chem. Solids **12**, 181 (1959)
Kane, E.O., J. Appl. Phys. **32**, 83 (1961)
Kanzaki, H., J. Phys. Chem. Solids **2**, 24, 107 (1957)
Kanzaki, H., Semicond. Insulators **3**, 285 (1978)
Kaplan, H., *Sullivan, J.J.*, Phys. Rev. **130**, 120 (1963)
Kaplyanskii, A.A., *Negodyiko, V.K.*, Opt. Spectrosc. **35**, 269 (1973)
Kapoor, A.K., *Jain, J.K.*, Phil. Mag. B **37**, 545 (1978)
Kapor, D.V., *Stojanovic, S.D.*, *Skrinjar, M.J.*, *Tosic, B.S.*, Phys. Stat. Sol. (b) **74**, 103 (1976)
Karageorgy-Alkalaev, P.M., *Leidermann, Y.A.*, Phys. Stat. Sol. **26**, 419 (1968)
Karakhanyan, K.I., *Kazaryan, E.M.*, *Bezirganyan, P.A.*, Sov. Phys. Solid State **18**, 295 (1976)
Karo, A.M., J. Chem. Phys. **31**, 1489 (1959)
Karo, A.M., J. Chem. Phys. **33**, 7 (1960)
Karo, A.M., *Hardy, J.R.*, Phil. Mag. **5**, 859 (1960)
Karo, A.M., *Hardy, J.R.*, Phys. Rev. **129**, 2024 (1963)
Karo, A.M., *Hardy, J.R.*, *Morrison, I.*, J. Physique **26**, 668 (1965)
Karo, A.M., *Hardy, J.R.*, Phys. Rev. **141**, 696 (1966)
Karo, A.M., *Hardy, J.R.*, J. Chem. Phys. **48**, 3173 (1968)
Karo, A.M., *Hardy, J.R.*, Phys. Rev. **181**, 1272 (1969)
Karo, A.M., *Hardy, J.R.*, Phys. Rev. B **3**, 3418 (1971)
Karo, A.M., *Hardy, J.R.*, Phys. Rev. B **12**, 690 (1975)
Karp, I.A., *Moskalenko, S.A.*, Sov. Phys. Semicond. 8, 183 (1974)
Kartheuser, E.P., *Negrete, P.*, Phys. Stat. Sol. (b) **57**, 77 (1973)
Kassim, H.A., *Dubey, K.S.*, Phys. Stat. Sol. (b) **86**, K107 (1978)
Kastalskii, A.A., Phys. Stat. Sol. (a) **15**, 599 (1973
Kasuya, T., J. Phys. Soc. Japan **13**, 1096 (1958)
Kasuya, T., *Koide, S.*, J. Phys. Soc. Japan **13**, 128 (1958)
Katayama, S., *Yokota, I.*, J. Phys. Soc. Japan **29**, 667 (1970)
Katayama, S., *Murase, K.*, J. Phys. Soc. Japan **42**, 886 (1977)
Kayanuma, Y., J. Phys. Soc. Japan **40**, 363 (1976)
Kayanuma, Y., *Toyozawa, Y.*, J. Phys. Soc. Japan **40**, 355 (1976)
Kayanuma, Y., *Kondo, Y.*, Solid State Commun. **24**, 447 (1977)
Kayanuma, Y., *Kondo, Y.*, J. Phys. Soc. Japan **45**, 528 (1978)
Kazakov, A.E., *Nagaev, E.L.*, Sov. Phys. Solid Stat **8**, 2302 (1967)
Keating, P.N., *Rupprecht, G.*, Phys. Rev. **138**, A866 (1965)
Kehr, K.W., in: Hydrogen in Metals I, ed. G. Alefeld and J. Völkl, Berlin: Springer 1978, p. 197
Keil, T.H., Phys. Rev. **140**, A601 (1965)
Keil, T.H., Phys. Rev. **144**, 582 (1966)
Keiper, R., Phys. Stat. Sol. (b) **44**, 593 (1971)
Keldysh, L.V., Zh. exp. teor. fiz. **34**, 1138 (1958)
Keldysh, L.V., Sov. Phys. JETP **6**, 763 (1958)
Keldysh, L.V., Sov. Phys. JETP **7**, 665 (1958)
Keldysh, L.V., Sov. Phys. JETP **20**, 1018 (1965)
Keldysh, L.V., Sov. Phys. JETP **20**, 1307 (1965)
Keldysh, L.V., *Kozlov, A.N.*, Sov. Phys. JETP **27**, 521 (1968)
Keldysh, L.V., *Silin, A.P.*, Sov. Phys. Lebedev Ins Rep. **8**, 30 (1976)
Keldysh, L.V., *Silin, A.P.*, Sov. Phys. JETP **42**, 535 (1976)
Keldysh, L.V., *Tikhodeev, S.G.*, Sov. Phys. Lebedev Inst. Rep. **8**, 6 (1976)
Keller, O., Spectrosc. Letters **9**, 545 (1976)
Kellermann, E.W., Phil. Trans. Roy. Soc. London A **238**, 513 (1940)
Kelly, P., *Bräunlich, P.*, Phys. Rev. B **1**, 1587 (1970)
Kelly, P., *Bräunlich, P.*, Phys. Rev. B **3**, 2090 (1971)
Kelly, P., *Laubitz, M.J.*, *Bräunlich, P.*, Phys. Rev. B **4**, 1960 (1971)

Kelly, R.L., Phys. Rev. **151**, 721 (1966)
Kemp, J.C., Neeley, V.I., Phys. Rev. **132**, 215 (1963)
Kemp, J.C., Bull. Am. Phys. Soc. **8**, 484 (1964)
Kenkre, V.M., Knox, R.S., Phys. Rev. B **9**, 5269 (1974)
Kern-Bausch, L., Z. Naturforsch. **21a**, 798 (1966)
Kern-Bausch, L., Z. Naturforsch. **23a**, 393 (1968)
Kersten, R., Phys. Stat. Sol. **29**, 575 (1968)
Kessler, F.R., Phys. Stat. Sol. **5**, 3 (1964)
Kessler, F.R., Sutter, E., in: Proc. VII Int. Conf. Semiconductors, Paris: Dunod 1964, p. 175
Khadzhi, P.I., Moskalenko, S.A., Opt. Spectrosc. **37**, 544 (1974)
Khadzhi, P.I., Sov. Phys. Semicond. **9**, 264 (1975)
Khadzhi, P.I., Sov. Phys. Semicond. **9**, 1402 (1976)
Khan, M.A., J. Phys. Chem. Sol. **31**, 2309 (1970)
Khan, M.A., Solid State Commun. **11**, 587 (1972)
Khan, M.A., J. Physique **34**, 597 (1973)
Khan, M.A., Phys. Stat. Sol. (b) **60**, 641 (1973)
Khan, M.A., Phys. Letters **61A**, 421 (1977)
Khazan, L.S., Pekar, S.I., Sheka, V.I., Phys. Stat. Sol. (b) **65**, 263 (1974)
Khazan, L.S., Sheka, V.I., Mozdor, E.V., Phys. Stat. Sol. (b) **69**, 741 (1975)
Khuri, N.N., Phys. Rev. **107**, 1148 (1957)
Kikoin, K. A., Fleurov, V. N., J. Phys. C: Solid State Phys. **10**, 4295 (1977)
Kim, Y.S., Gordon, R.G., Phys. Rev. B **9**, 3548 (1974)
Kimmel, H., Z. Naturforsch. **20a**, 359 (1965)
Kip, A., Kittel, S., Levy, A., Portis, M., Phys. Rev. **91**, 1066 (1953)
Kiselev, A.A., Abarenkov, I.V., Optika i Spektrosk. **9**, 765 (1960)
Kiselev, V.A., Zhilich, A.G., Sov. Phys. Solid State **14**, 1233 (1972)
Klassmann, W., Z. Physik **218**, 237 (1969)
Klassmann, W., Z. Physik **226**, 409 (1969)
Kleefstra, M., J. Phys. Chem. Solids **24**, 1567 (1963)
Kleih, W., Zur Theorie eines ionischen Photohalbleiters im elektrischen Feld, Thesis, University of Tübingen, 1980
Kleih, W., Quantum Statistical Calculation of Current Voltage Characteristics for Impure Polar Crystals, Preprint, Institut für Theoretische Physik, Universität Tübingen 1982
Klein, A., Phys. Rev. **115**, 1136 (1959)
Klein, M.V., Phys. Rev. **131**, 1500 (1963)
Klein, M.V., Phys. Rev. **141**, 716 (1966)
Klein, M.V., Localized Modes and Resonance States in Alkali Halides, in: Physics of Color Centres, ed. W.B. Fowler, New York: Academic Press 1968, p. 430
Klein, M.V., Macdonald, H.F., Phys. Rev. Letters **20**, 1031 (1968)
Klein, M.V., Phys. Rev. **186**, 839 (1969)
Klein, R., Z. Naturforsch. **16a**, 116 (1961)
Klein, R., Wehner, R.K., Phys. kondens. Materie **8**, 141 (1968)
Klein, R., Wehner, R.K., Phys. kondens. Materie **10**, 1 (1969)
Kleinman, L., Shurtleff, R., Phys. Rev. **188**, 1111 (1969)
Klemens, P.G., Proc. Roy. Soc. London A **208**, 108 (1951)
Klemens, P.G., Proc. Phys. Soc. **68**, 1113 (1955)
Klemens, P.G., Thermal Conductivity of Solids at Low Temperatures, in: Handbuch der Physik, Band 14/1, ed. S. Flügge, Heidelberg: Springer Verlag 1956, p. 198
Klemens, P.G., Solid State Physics **7**, 1 (1958) (Ed. F. Seitz and D. Turnbull, New York: Academic Press)
Klemens, P.G., Phys. Rev. **122**, 443 (1961)
Klemens, P.G., Phys. Rev. **148**, 845 (1966)
Kliewer, K.L., Koehler, J.S., Phys. Rev. **140**, A1226 (1965)
Kliewer, K.L., Phys. Rev. **140**, A1241 (1965)
Klimontovich, Yu.E., Sov. Phys. JETP **25**, 820 (1967)
Klimontovich, Yu.E., Sov. Phys. JETP **27**, 75 (1968)
Klinger, M.I., Sov. Phys. Doklady **7**, 123 (1962)
Klinger, M.I., Blakher, E., Phys. Stat. Sol. **31**, 515 (1969)
Klinger, M.I., Phys. Stat. Sol. **31**, 545 (1969)
Klinger, M.I., Phys. Stat. Sol. (b) **58**, 831 (1973)
Klyukanov, A.A., Pokatilov, E.P., Sov. Phys. JETP **33**, 170 (1971)
Knauss, D.C., Wilson, R.S., Phys. Rev. B **10**, 4383 (1974)
Knohl, U., Phys. Stat. Sol. (b) **53**, 295 (1972)
Knowles, J., Morris, P., Blount, C.E., Phys. Letters **44A**, 257 (1973)
Knox, R.S., Theory of Excitons, Solid State Physics Suppl. 5, ed. F. Seitz and D. Turnbull, New York: Academic Press 1963
Knox, R.S., Gold, A., Symmetry in the Solid State, New York: Benjamin 1964
Kobayashi, T., J. Appl. Phys. **48**, 3154 (1977)
Kocevar, P., J. Phys. C: Solid State Phys. **5**, 3349 (1972)
Kocevar, P., Acta Phys. Austr. **37**, 270 (1973)
Kocevar, P., Acta Phys. Austr. **37**, 259 (1973)
Kocevar, P., Fitz, E., Phys. Stat. Sol. (b) **89**, 225 (1978)
Koch, B.P., Phys. Stat. Sol. (b) **68**, 193 (1975)
König, E., Schnakig, R., Chem. Phys. Letters **32**, 553 (1975)
König, E., Schnakig, R., Phys. Stat. Sol. (b) **77**, 657 (1976)
König, E., Schnakig, R., Phys. Stat. Sol. (b) **78**, 543 (1976)
Koenig, S.H., Brown, R.D., Schillinger, W., Phys. Rev. **128**, 1668 (1962)
Kogan, Sh.M., Sov. Phys. Solid State **4**, 1386 (1963)

Kogan, Sh.M., Sov. Phys. Solid State **4**, 1813 (1963)
Kohler, M., Z. Physik **124**, 772 (1948)
Kohler, M., Z. Physik **125**, 679 (1949)
Kohli, M., Liu, N.L.H., Phys. Rev. B **9**, 1008 (1974)
Kohn, W., Phys. Rev. **105**, 509 (1957)
Kohn, W., Luttinger, J.M., Phys. Rev. **108**, 590 (1957)
Kohn, W., Phys. Rev. **110**, 857 (1958)
Kohn, W., Phys. Rev. **115**, 1460 (1959)
Kohn, W., Phys. Rev. Letters **19**, 439 (1967)
Kohn, W., Onffroy, J., Bull. Am. Phys. Soc. **18**, 456 (1973)
Kohn, W., Onffroy, J., Phys. Rev. B **8**, 2485 (1973)
Koide, S., Z. Naturforsch. **15a**, 123 (1960)
Koidl, P., Phys. Stat. Sol. (b) **74**, 477 (1976)
Kojima, K., Nishimaki, N., Kojima, T., J. Phys. Soc. Japan **16**, 2033 (1961)
Kojima T., J. Phys. Soc. Japan **12**, 908 (1957)
Kolchenko, T.I., Lomako, V.M., Phys. Stat. Sol. (a) **48**, 263 (1978)
Kolodziejczak, J., Acta Phys. Polon. **20**, 289 (1961)
Kolodziejczak, J., Acta Phys. Polon. **20**, 379 (1961)
Kolodziejczak, J., Sosnowski, L., Acta Phys. Polon. **21**, 399 (1962)
Kolodziejczak, J., Acta Phys. Polon. **21**, 637 (1962)
Kolodziejczak, J., Zukotynski, S., Phys. Stat. Sol. **5**, 145 (1964)
Kolodziejczak, J., Stramska, H., Phys. Stat. Sol. **17**, 701 (1966)
Kolodziejczak, J., Phys. Stat. Sol. **24**, 323 (1967)
Kolodziejczak, J., Phys. Stat. Sol. **19**, 231 (1967)
Kolodziejczak, J., Kierzek-Pecold, E., Phys. Stat. Sol. **19**, 623 (1967)
Kolodziejczak, J., Kierzek-Pecold, E., Phys. Stat. Sol. **19**, K55 (1967)
Kolodziejczak, J., Phys. Stat. Sol. **29**, 645 (1968)
Kolodziejczak, J., Acta Phys. Polon. **33**, 585 (1968)
Kolodziejczak, J., Acta Phys. Polon. **33**, 183 (1968)
Komissarov, V.S., Phys. Stat. Sol. (b) **76**, 105 (1976)
Kondo, J., Prog. Theor. Phys. **28**, 1026 (1962)
Konstantinov, O.V., Perel', V.I., Sov. Phys. JETP **12**, 142 (1961)
Kopvillem, U.Kh., Samartsev, V.V., Sheibut, Yu.E., Phys. Stat. Sol. (b) **70**, 799 (1975)
Kopylov, A.A., Pikhtin, A.N., Sov. Phys. Semicond. **10**, 7 (1976)
Kopylov, A.A., Pikhtin, A.N., Sov. Phys. Semicond. **11**, 510 (1977)
Korenblit, I.Ya., JETP Letters **5**, 77 (1967)
Korenblit, L.L., Sherstobitov, V.E., Fiz. Tekh. Poluprov **2**, 675 (1968)
Kornyushin, Yu.V., Sov. Phys. Semicond. **7**, 947 (1974)
Korol', E.N., Tolpygo, K.B., Sov. Phys. Solid State **5**, 1597 (1964)
Korol', E.N., Tolpygo, K.B., Bull. Acad. Sci. USSR, Phys. Ser. **28**, 846 (1965)
Korol, E.N., Tolpygo, K.B., Phys. Stat. Sol. (b) **45**, 71 (1971)
Korol, E.N., Sov. Phys. Solid State **19**, 1327 (1977)
Korolyuk, S.L., Sov. Phys. Solid State **4**, 580 (1962)
Korolyuk, S.L., Pinchuk, I.I., Sov. Phys. Semicond. **10**, 1069 (1976)
Korovin, L.I., Kharitonov, E.V., Phys. Stat. Sol. **14**, 445 (1966)
Korsak, K.V., Krivoglaz, M.A., Sov. Phys. Solid State **10**, 1952 (1969)
Korzhevykh, E.I., Sov. Phys. Solid State **17**, 2174 (1976)
Kosarev, V.V., Tamarin, P.V., Sov. Phys. Semicond. **7**, 1079 (1974)
Kosarev, V.V., Sov. Phys. Semicond. **8**, 897 (1975)
Kosarev, V.V., Tamarin, P.V., Sov. Phys. Semicond. **9**, 172 (1975)
Koshino, S., Ando, T., J. Phys. Soc. Japan **16**, 1151 (1961)
Kossut, J., Phys. Stat. Sol. (b) **72**, 359 (1975)
Kossut, J., Phys. Stat. Sol. (b) **78**, 537 (1976)
Kossut, J., Walukiewicz, W., Solid State Commun. **18**, 343 (1976)
Koster, G.F., Slater, J.C., Phys. Rev. **95**, 1167 (1954)
Koster, G.F., Solid State Physics **5**, 174 (1957) (Ed. F. Seitz and D. Turnbull, New York: Academic Press)
Koteles, E.S., Datars, W.R., Phys. Rev. B **14**, 1571 (1976)
Kotelnikov, Yu.E., Kochelaev, B.I., Phys. Stat. Sol. (b) **81**, 747 (1977)
Kotera, N., Yamada, E., Komatsubara, K.F., J. Phys. Chem. Solids **33**, 1311 (1972)
Kotova, L.V., Vörös, T., Acta Phys. Acad. Sci. Hung. **36**, 297 (1974)
Kovarskii, V.A., Opt. i Spektrosk. **5**, 222 (1958)
Kovarskii, V.A., Sov. Phys. Solid State **4**, 1200 (1962)
Kovarskii, V.A., Sinyavskii, E.P., Sov. Phys. Solid State **4**, 2345 (1963)
Kovarskii, V.A., Vitin, E.V., Bull Acad. Sci. USSR (Phys. Ser.) **31**, 2046 (1967)
Kovarskii, V.A., Vitin, E.V., Sinyavskii, E.P., Sov. Phys. Solid State **12**, 543 (1970)
Kovarskii, V.A., Perlin, E.Yu., Phys. Stat. Sol. (b) **45**, 47 (1971)

Kovarskii, V.A., Popov, E.A., Chaikovskii, I.A., Phys. Stat. Sol. (b) **67**, 427 (1975)
Kozhushner, M.A., Sov. Phys. JETP **33**, 121 (1971)
Kozhushner, M.A., Sov. Phys. Solid State **13**, 2183 (1972)
Kozhushner, M.A., Sov. Phys. Solid State **16**, 895 (1974)
Kraeft, W.D., Kilimann, K., Kremp, D., Phys. Stat. Sol. (b) **72**, 461 (1975)
Kramarczyk, W.J., Voss, K., Ann. Physik (Germany), 7. Folge, **21**, 167 (1968)
Kranzer, D., König, W., Phys. Stat. Sol. (b) **48**, K133 (1971)
Kranzer, D., Phys. Stat. Sol. (b) **46**, 591 (1971)
Kranzer, D., Phys. Stat. Sol. (b) **50**, K109 (1972)
Kranzer, D., Hillbrand, H., Pötzl, H., Zimmerl, O., Acta Phys. Austr. **35**, 110 (1972)
Kranzer, D., J. Phys. C: Solid State Phys. **6**, 2967 (1973)
Kranzer, D., J. Phys. C: Solid State Phys. **6**, 2977 (1973)
Kranzer, D., J. Phys. Chem. Solids **34**, 9 (1973)
Kranzer, D., Phys. Stat. Sol. (a) **26**, 11 (1974)
Kravchenko, A.F., Kubalkova, S., Morozov, B.V., Polovinkin, V.G., Skok, E.M., Phys. Stat. Sol. (b) **72**, 221 (1975)
Kravchenko, V.Ya., Sov. Phys. Solid State **4**, 1319 (1963)
Kravchenko, V.Ya., Vinetskii, V.L., Sov. Phys. Solid State **7**, 1 (1965)
Kravchenko, V.Ya., Vinetskii, V.L., Sov. Phys. Solid State **6**, 1638 (1965)
Kremp, D., Ebeling, W., Kraeft, W.D., Phys. Stat. Sol. (b) **69**, K59 (1975)
Kress, W., Phys. Stat. Sol. (b) **62**, 403 (1974)
Krieger, J.B., Ann. Phys. (New York) **36**, 1 (1966)
Krishan, K., J. Phys. C: Solid State Phys. **4**, 1550 (1971)
Krishnamurthy, B.S., Sinha, K.P., J. Phys. Chem. Solids **26**, 1949 (1965)
Krishnamurthy, B.S., Paranjape, V.V., Phys. Stat. Sol. **40**, 221 (1970)
Krishnamurthy, N., Proc. Phys. Soc. 88, 1015 (1966)
Krishnamurthy, N., Haridasan, T.M., Indian J. Pure Appl. Phys. **4**, 255 (1966)
Krishnamurthy, N., Haridasan, T.M., Indian J. Pure Appl. Phys. **7**, 89 (1969)
Krishnamurthy, N., Soots, V., Can. J. Phys. **50**, 1350 (1972)
Krishnamurthy, N., Haridasan, T.M., J. Phys. Chem. Solids **39**, 69 (1978)
Krishnan, K.S., Roy, S.K., Proc. Roy. Soc. London A **207**, 447 (1951)
Krishnan, K.S., Roy, S.K., Proc. Roy. Soc. London A **210**, 481 (1952)
Kristofel, N.N., Zavt, G.S., Sov. Phys. Solid State **5**, 932 (1963)
Kristofel, N.N., Tyurkson, E.E., Sov. Phys. Solid State **6**, 969 (1964)
Kristofel, N.N., Zavt, G.S., Opt. Spectrosc. **20**, 373 (1966)
Kristofel, N.N., Bull. Acad. Sci. USSR, Phys. Ser. **30**, 1598 (1966)
Kristofel, N.N., Zavt, G.S., Sov. Phys. Solid State **9**, 1246 (1967)
Kristofel, N.N., Zavt, G.S., Bull. Acad. Sci USSR, Phys. Ser., **33**, 859 (1969)
Kristoffel, N.N, Opt. Spectrosc. **35**, 179 (1973)
Kristoffel, N.N., Sigmund, E., Wagner, M., Z. Naturforsch. **28a**, 1782 (1973)
Krivoglaz, M.A., Sov. Phys. Solid State **6**, 1340 (1964)
Krivoglaz, M.A., Sov. Phys. JETP **21**, 204 (1965)
Krivoglaz, M.A., Pinkevich, I.P., Sov. Phys. JETP **24**, 772 (1967)
Krivoglaz, M.A., Pinkevich, I.P., Sov. Phys. Solid State **11**, 69 (1969)
Krivoglaz, M.A., Levenson, G.F., Sov. Phys. Solid State **12**, 293 (1970)
Kröger, F.A., Ergebnisse der exakten Naturwissenschaften, Vol. 29, p. 61 (1956) (Ed. S. Flügge and F. Trendelenburg, Berlin: Springer Verlag)
Kröger, F.A., Vink, H.J., J. Phys. Chem. Solids **5**, 208 (1958)
Kröger, F.A., J. Phys. Chem. Solids **26**, 901 (1965)
Krumhansl, J.A., J. Appl. Phys., Suppl., **33**, 307 (1962)
Krumhansl, J.A., Matthew, J.A.D., Phys. Rev. **140**, A1812 (1965)
Krumhansl, J.A., Matthew, J.A.D., Phys. Rev. **166**, 856 (1968)
Krupka, D.C., Silsbee, R.H., Phys. Rev. **152**, 816 (1966)
Kryuchkov, S.V., Yakovlev, V.A., Sov. Phys. Semicond. **10**, 724 (1976)
Kubalkova, S., Sakalas, A., Phys. Stat. Sol. (b) **50**, 119 (1972)
Kubo, R., J. Phys. Soc. Japan **3**, 254 (1948)
Kubo, R., Phys. Rev. **86**, 929 (1952)
Kubo, R., Toyozawa, Y., Prog. Theor. Phys. **13**, 160 (1955)
Kubo, R., J. Phys. Soc. Japan **12**, 570 (1957)
Kubo, R., Miyake, S.J., Hashitsume, N., Solid State Physics **17**, 270 (1965) (Ed. F. Seitz and D. Turnbull, New York: Academic Press)
Kucher, T.I., Tolpygo, K.B., Sov. Phys. JETP **4**, 883 (1957)
Kucher, T.I., Sov. Phys. JETP **5**, 418 (1957)
Kucher, T.I., Sov. Phys. JETP **7**, 274 (1958)
Kucher, T.I., Tolpygo, K.B., Sov. Phys. Solid State **2**, 2052 (1961)
Kudinov, E.K., Sov. Phys. Solid State **16**, 323 (1974)
Kudykina, T.A., Tolpygo, K.B., Zubkova, S.M., Phys. Stat. Sol. **28**, 807 (1968)

Kudykina, T.A., Tolpygo, K.B., Sov. Phys. Solid State **13**, 921 (1971)
Kudykina, T.A., Tolpygo, K.B., Sov. Phys. State **13**, 921 (1971)
Kudykina, T.A., Tolpygo, K.B., Sov. Phys. Solid State **13**, 2402 (1972)
Kudykina, T.A., Tolpygo, K.B., Sov. Phys. Solid State **14**, 535 (1972)
Kudykina, T.A., Tolpygo, K.B., Sov. Phys. Solid State **14**, 2187 (1973)
Kudzmanskas, S.P., Batarunas, J.V., Phys. Stat. Sol. (b) **49**, 769 (1972)
Kübler, J.K., Friauf, R.J., Phys. Rev. **140**, A 1742 (1965)
Kühne, R., Reineker, P., Phys. Stat. Sol. (b) **89**, 131 (1978)
Kühner, D., Wagner, M., Z. Physik **207**, 111 (1967)
Kühner, D., Wagner, M., Phys. Stat. Sol. **40**, 517 (1970)
Kühner, D., Z. Physik **230**, 108 (1970)
Kühner, D., Wagner, M., Z. Physik **256**, 22 (1972)
Kühner, D.H., Phys. Rev. B **9**, 1792 (1974)
Kümmel, R., Z. Physik **196**, 123 (1966)
Kümmel, R., Roldan, J., Z. Physik **259**, 411 (1973)
Kümmel, R., Rauh, H., Bangert, E., Phys. Stat. Sol. (b) **87**, 99 (1978)
Kugel, G., Carabatos, C., Hennion, B., Prevot, B., Revcolevschi, A., Tocchetti, D., Phys. Rev. B **16**, 378 (1977)
Kukharskii, A.A., Solid State Commun. **13**, 1761 (1973)
Kukharskii, A.A., Sov. Phys. Semicond. **9**, 1165 (1975)
Kukushkin, L.S., Sov. Phys. JETP **17**, 476 (1963)
Kukushkin, L.S., Sov. Phys. Solid State **15**, 591 (1973)
Kumar, A., Srivastava, A.K., Verma, G.S., Phys. Rev. **178**, 1480 (1969)
Kumar, M., Mahesh, P.S., Sharma, T.P., Phys. Stat. Sol. (b) **83**, 155 (1977)
Kumar, S., Proc. Nat. Inst. Sci. India A **25**, 364 (1959)
Kunc, K., Czech. J. Phys. B **15**, 883 (1965)
Kunc, K., Phys. Stat. Sol. **15**, 683 (1966)
Kunc, K., Balkanski, M., Nusimovici, M.A., Phys. Rev. B **12**, 4346 (1975)
Kunc, K., Bilz, H., Solid State Commun. **19**, 1027 (1976)
Kung, A.Y.S., Vail, J.M., Phys. Stat. Sol. (b) **79**, 663 (1977)
Kunz, A.B., Phys. Rev. **151**, 620 (1966)
Kunz, A.B., Van Sciver, W.J., Phys. Rev. **142**, 462 (1966)
Kunz, A.B., Phys. Rev. **162**, 789 (1967)
Kunz, A.B., Phys. Rev. **159**, 738 (1967)
Kunz, A.B., Phys. Letters **25A**, 538 (1967)
Kunz, A.B., Phys. Stat. Sol. **29**, 115 (1968)
Kunz, A.B., Phys. Letters **27A**, 401 (1968)
Kunz, A.B., Phys. Rev. **175**, 1147 (1968)
Kunz, A.B., Phys. Rev. **180**, 934 (1969)
Kunz, A.B., Phys. Stat. Sol. **36**, 301 (1969)
Kunz, A.B., J. Phys. C: Solid State Phys. **3**, 1542 (1970)
Kunz, A.B., J. Phys. Chem. Solids **31**, 265 (1970)
Kunz, A.B., Phys. Rev. B **2**, 2224 (1970)
Kunz, A.B., Phys. Rev. B **2**, 5015 (1970)
Kunz, A.B., Lipari, N.O., Phys. Rev. B **4**, 1374 (1971)
Kunz, A.B., Lipari, N.O., J. Phys. Chem. Solids **32**, 1141 (1971)
Kunz, A.B., Phys. Stat. Sol. (b) **46**, 697 (1971)
Kunz, A.B., Phys. Rev. B **6**, 606 (1972)
Kunz, A.B., Phys. Rev. B **8**, 1690 (1973)
Kunz, A.B., Mickish, D.J., Phys. Rev. B **11**, 1700 (1975)
Kunz, B.A., Klein, D.L., Phys. Rev. B **17**, 4614 (1978)
Kuramoto, Y., J. Phys. Soc. Japan **42**, 1143 (1977)
Kuramoto, Y., Morimoto, M., J. Phys. Soc. Japan **44**, 1759 (1978)
Kurosawa, T., J. Phys. Soc. Japan **13**, 153 (1958)
Kurosawa, T., J. Phys. Soc. Japan **15**, 1211 (1960)
Kurosawa, T., J. Phys. Soc. Japan **16**, 1298 (1961)
Kurosawa, T., Prog. Theor. Phys. **29**, 159 (1963)
Kurosawa, T., J. Phys. Soc. Japan **20**, 937 (1965)
Kushida, T., J. Phys. Soc. Japan **34**, 1318 (1973)
Kushida, T., J. Phys. Soc. Japan **34**, 1327 (1973)
Kushida, T., J. Phys. Soc. Japan **34**, 1334 (1973)
Kushkulei, L.M., Perlin, Yu.E., Tsukerblat, B.S., Engel'gardt, G.R., Sov. Phys. JETP **43**, 1162 (1976)
Kustov, E.F., Opt. Spectrosc. **32**, 171 (1971)
Kuz'min, V.A., Kryukova, N.N., Kyuregyan, A.S., Sov. Phys. Semicond. **9**, 1136 (1976)
Kuz'min, Yu.M., Sov. Phys. Solid State **18**, 584 (1976)
Kuzuba, T., Era, K., J. Phys. Soc. Japan **40**, 134 (1976)
Kwan, C.C.Y., Basinski, J., Woolley, J.C., Phys. Stat. Sol. (b) **48**, 699 (1971)
Kwon, T.H., Henkel, J.H., Can. J. Phys. **47**, 325 (1969)
Kyuregyan, A.S., Sov. Phys. Semicond. **10**, 410 (1976)

La, S.Y., Bartram, R.H., Phys. Rev. **144**, 670 (1966)
Lacina, A., Czech. J. Phys. B **24**, 284 (1974)
Lacroix, R., Helv. Phys. Acta **47**, 689 (1974)

Ladyzhinskii, Yu.P., Sov. Phys. Solid State **11**, 1842 (1970)
von der Lage, F.C.,Bethe, H.A., Phys. Rev. **71**, 612 (1947)
Lagu, M., Dayal, B., Indian Pure Appl. Phys. **6**, 670 (1968)
Lagu, M., Dayal, B., J.Phys. C: Solid State Phys. **8**, 961 (1975)
Lai, P.K., Massicot, P.O., Nettel, S.J., Z. Physik B **23**, 97 (1976)
Lal, H.H., Verma, M.P., Indian J. Pure Appl. Phys. **8**, 380 (1970)
Lal, H.H., Verma, M.P., Phys. Stat. Sol. **38**, K19 (1970)
Lal, H.H., Verma, M.P., J. Phys. C: Solid State Phys. **5**, 543 (1972)
Lal, H.H., Verma, M.P., J. Phys. C: Solid State Phys. **5**, 1038 (1972)
Lamatsch, H., Rossel, J., Saurer, E., Phys. Stat. Sol. (b) **46**, 687 (1971)
Lamb, W.E., Phys. Rev. **55**, 190 (1939)
Lampert, M.A., Phys. Rev. Letters **1**, 450 (1958)
Lamprecht, G., Merten, L., Phys. Stat. Sol. **35**, 353 (1969)
Landau, L.D., Z. Physik (Germany) **64**, 629 (1930)
Landau, L.D., Phys. Z. Sowj. **3**, 644 (1933)
Landsberg, P.T., Proc. Phys. Soc. B **70**, 282 (1957)
Landsberg, P.T., Rhys-Roberts, C., Lal, P., Proc. Phys. Soc. **84**, 915 (1964)
Landsberg, P.T., Phys. Stat. Sol. **15**, 623 (1966)
Landsberg, P.T., Phys. Stat. Sol. **41**, 457 (1970)
Landsberg, P.T., Adams, M.J., J. Luminescence **7**, 3 (1973)
Landsberg, P.T., Adams, M.J., Proc. Roy. Soc. London A **334**, 523 (1973)
Landsberg, P.T., Pimpale, A., J. Phys. C: Solid State Phys. **9**, 1243 (1976)
Landsberg, P.T., Robbins, D.J., Schöll,E., Phys. Stat. Sol. (a) **50**, 423 (1978)
Landshoff, R., Z. Physik (Germany) **102**, 201 (1936)
Landshoff, R., Phys. Rev. **52**, 246 (1937)
Lang, I.G., Firsov, Yu.A., Sov. Phys. JETP **27**, 443 (1968)
Lang, I.G., Sov. Phys. Solid State **10**, 1873 (1969)
Lang, I.G., Pavlov, S.T., Sov. Phys. JETP **36**, 793 (1973)
Lang, I.G., Pavlov, S.T., Yashin, G.Yu., JETP Letters **26**, 305 (1977)
Lang, I.G., Sov. Phys. JETP **45**, 1130 (1977)
Lang, M., Phys. Rev. B **2**, 4022 (1970)
Langbein, D., Z. Physik **162**, 542 (1961)
Langbein, D., Z. Physik **167**, 83 (1962)
Langbein, D., Festkörperprobleme **9**, 255 (1969) (Ed. O. Madelung, Braunschweig: Vieweg)
Langbein, D., Festkörperprobleme **11**, 255 (1971), Braunschweig: Vieweg Verlag
Langer, D.W., Euwema, R.N., Era, K., Koda,T., Phys. Rev. B **2**, 4005 (1970)
Lanoo, M., Decarpigny, J.N., Phys. Rev. B **8**, 5704 (1973)
Lanoo, M., Phys. Rev. B **10**, 2544 (1974)
Lanz, L., Ramella, G., Physica **44**, 499 (1969)
Lanzl, F., Phys. Letters **25A**, 596 (1967)
Laplaze, D., Vergnoux, A.–M., Benoit, C., Ann. Phys. (Paris) **6**, 105 (1971)
Larsen, D.M., Phys. Rev. **133**, A860 (1964)
Larsen, D.M., Phys. Rev. **144**, 697 (1966)
Larsen, D.M., Phys. Rev. **174**, 1046 (1968)
Larsen, D.M., Phys. Rev. **180**, 919 (1969)
Larsen, D.M., Phys. Rev. **187**, 1147 (1969)
Larsen, D.M., Phys. Rev. B. **2**, 4209 (1970)
Lauer, H.V., Fong, F.K., J. Chem. Phys. **60**, 274 (1974)
Laughlin, C., Öpik, U., Solid State Commun. **2**, 309 (1964)
Laughlin, C., Solid State Commun. **3**, 55 (1965)
Law, H.C., Kao, K.C., Phys. Rev. B **4**, 2524 (1971)
Lawless, W.N., J. Phys. Chem. Solids **28**, 1755 (1967)
Lax, M., J. Chem. Phys. **20**, 1752 (1952)
Lax, M., Phys. Rev. **94**, 1391 (1954)
Lax, M., Symmetry Principles in Solid State and Molecular Physics, New York: J. Wiley & Sons 1974
Lebedev, A.A., Sov. Phys. Semicond. **7**, 373 (1973)
Leburton, J.P., Evrard, R., J. Low Temp. Phys. **32**, 323 (1978)
Ledovskaya, E.M., Phys. Stat. Sol. **31**, 507 (1969)
Lee, C.C., Fan, H.Y., Phys. Rev. B **9**, 3502 (1974)
Lee, N., Larsen, D.M., Lax, B., J. Phys. Chem. Solids **34**, 1059 (1973)
Lee, T.D., Low, F.E., Pines, D., Phys. Rev. **90**, 297 (1953)
Lee, Y.C., Tzoar, N., Phys. Rev. **178**, 1303 (1969)
Lefrant, S., Harker, A.H., Taurel, L., J. Phys. C: Solid State Phys. **8**, 1119 (1975)
Leibfried, G., Schlömann, E., Nachr. Akad. Wiss. Göttingen, Math. Physik Kl.IIa, 71 (1954)
Leibfried, G., Gittertheorie der mechanischen und thermischen Eigenschaften der Kristalle, in: Handbuch der Physik VII, pt. 1, ed. S. Flügge, Berlin: Springer 1957, p. 104
Leibfried, G., Ludwig, W., Z. Physik **169**, 80 (1960)
Leibfried, G., Ludwig, W., Solid State Physics **12**, 276 (1961) (Ed. F. Seitz and D. Turnbull, New York: Academic Press)

Leibler, K., Wilamowski, Z., Phys. Stat. Sol. (b) **55**, 811 (1973)
Lemmens, L.F., De Sitter, J., Devreese, J.T., Phys. Rev. B 8, 2717 (1973)
Lemmens, L.F., Devreese, J.T., Brosens, F., Phys. Stat. Sol. (b) **82**, 439 (1977)
Lemos, A.M., Markham, J.J., J. Phys. Chem. Solids **26**, 1837 (1965)
Lemos, A.M., Phys. Rev. **151**, 727 (1966)
Lemoyne, D., Duran, J., Billardon, M., Dang, Le Si, Phys. Rev. B **14**, 747 (1976)
Lemoyne, D., Duran, J., Badoz, J., J. Phys. C: Solid State Phys. **10**, 1255 (1977)
Lengeler, B., Ludwig, W., Z. Physik **171**, 273 (1963)
Lepine, Y., Matz, D., Can. J. Phys. **54**, 1979 (1976)
Lepine, Y., Bader, G., Matz, D., Phys. Stat. Sol. (b) **89**, 53 (1978)
Leung, C.H., Song, K.S., Phys. Rev. B **18**, 922 (1978)
Levanyuk, A.P., Osipov, V.V., Sov. Phys. Semicond. 7, 721 (1973)
Levanyuk, A.P., Osipov, V.V., Sov. Phys. Semicond. 7, 727 (1973)
Levine, B.F., Phys. Rev. B 7, 2591 (1973)
Levinson, I.B., Sov. Phys. JETP **30**, 362 (1970)
Levinson, I.B., JETP Letters **15**, 409 (1972)
Levinson, I.B., Rashba, E.I., Sov. Phys. JETP **35**, 788 (1972)
Levinson, I.B., Yasevichyute, Ya., Sov. Phys. JETP **35**, 991 (1972)
Levinson, Y.B., Rashba, E.I., Rep. Prog. Phys. **36**, 1499 (1973)
Lewiner, C., Calecki, D., Solid-State Electronics **21**, 185 (1978)
Lewis, B.F., Sondheimer, E.H., Proc. Roy. Soc. London A **227**, 241 (1955)
Lewis, W.B., Pretzel, F.E., J. Phys. Chem. Solids **19**, 139 (1961)
Lhermitte, C., Carles, D., Viger, C., C.R.Acad. Sci. Serie B (Paris) **286**, 73 (1978)
Licari, J.J., Whitfield, G., Phys. Rev. B **9**, 1432 (1974)
Licea, I., Phys. Stat. Sol. **18**, K9 (1966)
Licea, I., Phys. Stat. Sol. **25**, 461 (1968)
Licea, I., Phys. Stat. Sol. **26**, 115 (1968)
Licea, I., Phys. Stat. Sol. **38**, 841 (1970)
Licea, I., Phys. Stat. Sol. **39**, 255 (1970)
Licea, I., Physica **63**, 172 (1973)
Licea, I., J. Phys. Chem. Solids **35**, 1344 (1974)
Lidiard, A.B., Ionic Conductivity, in: Handbuch der Physik XX, ed. S. Flügge, Berlin: Springer 1957, p. 246
Lidiard, A.B., Phys. Rev. **112**, 54 (1958)
Lidiard, A.B., J. Appl. Phys. **33**, Suppl., 414 (1962)
Lidiard, A.B., Comments Solid State Phys. **2**, 76 (1969)
Lidiard, A.B., Thermodynamics and Kinetics of Point Defects, in: Trieste Lectures 1970, Wien: Intern. Atom Energy Agency 1971, p. 339
Liehr, A.D., Z.Naturforsch. **16a**, 641 (1961)
Liehr, A.D., J. Phys. Chem. **67**, 389 (1963)
Liehr, A.D., J. Phys. Chem. **67**, 1314 (1963)
Lifshitz, I.M., J. Phys. USSR **7**, 215 (1943)
Lifshitz, I.M., J. Phys. USSR **7**, 249 (1943)
Lifshitz, I.M., J. Phys. USSR **8**, 89 (1944)
Lifsic, M., Nuovo Cim. Suppl. **3** (Serie X), 716 (1956)
Lifshitz, I.M., Kosevich, A.M., Rep. Prog. Phys. **29**, 217 (1966)
Lin, P.J., Kleinman, L., Phys. Rev. **142**, 478 (1966)
Lin, S.H., J. Chem. Phys. **44**, 3759 (1966)
Lin, S.H., J. Chem. Phys. **61**, 3810 (1974)
Lin, S.H., J. Chem. Phys. **62**, 4500 (1975)
Lin, S.H., J. Chem. Phys. **65**, 1053 (1976)
Lipari, N.O., Kunz, A.B., Phys. Rev. B **4**, 4639 (1971)
Lipari, N.O., Kunz, A.B., Phys. Rev. B **3**, 491 (1971)
Lipari, N.O., Baldereschi, A., Phys. Rev. B **3**, 2497 (1971)
Lipari, N.O., Phys. Rev. B **4**, 4535 (1971)
Lipari, N.O., Nuovo Cim. **8B**, 406 (1972)
Lipari, N.O., Baldereschi, A., Phys. Rev. B **6**, 3764 (1972)
Lipnik, A.A., Sov. Phys. Solid State **2**, 1835 (1961)
Lipnik, A.A., Sov. Phys. Solid State **3**, 1683 (1962)
Litwin-Staszewska, E., Szymanska, W., Phys. Stat. Sol. (b) **74**, K89 (1976)
Litzmann, O., Czech. J. Phys. **8**, 633 (1958)
Litzmann, O., Rozsa, P., Proc. Phys. Soc. **85**, 285 (1965)
Litzmann,O., Phys. Stat. Sol. **13**, 71 (1966)
Litzmann, O., Bartusek, M., Zavadil, V., J. Phys. C: Solid State Phys. **5**, 287 (1972)
Litzmann, O., Rozsa, P., Phys. Stat. Sol. (b) **58**, 451 (1973)
Liu, S.W.W., Rabii, S., Phys. Rev. B **13**, 1675 (1976)
Lochmann, W., Phys. Stat. Sol. (a) **45**, 423 (1978)
Löffler, A., Z. Naturforsch. **22a**, 1771 (1967)
Löffler, A., Z. Naturforsch. **24a**, 516 (1969)
Löffler, A., Z. Naturforsch. **24a**, 530 (1969)
Löwdin, P.O., Arch. f. Math., Astr. o Fysik **35A**, No.9 (1947)
Löwdin, P.O., A Theoretical Investigation into Some Properties of Ionic Crystals, Thesis, Uppsala: Almquist and Wiksells 1948
Löwdin, P.O., J. Chem. Phys. **18**, 365 (1950)
Löwdin, P.O., J. Chem. Phys. **19**, 1570 (1951)
Löwdin, P.O., Advances in Physics **5**, 1 (1956)

Löwdin, P.O., Appel, K., Phys. Rev. **103**, 1746 (1956)
Löwdin, P.O., J. Appl. Phys. **33**, Suppl., 251 (1962)
Logatchov, Yu.A., Phys. Stat. Sol. **26**, 765 (1968)
Logatchov, Yu.A., Evarestov, R.A., Phys. Stat. Sol. **4o**, 493 (1970)
Loidl, A., Jex, H., Daubert, J., Müllner, M., Phys. Stat. Sol. (b) **76**, 581 (1976)
Loidl, A., Jex, H., Phys. Letters **61A**, 188 (1977)
Lombardi, E., Jansen, L., Phys. Rev. **136**, A1011 (1964)
Lombardi, E., Jansen, L., Phys. Rev. **140**, A275 (1965)
Lomont, J.S., Applications of Finite Groups, New York: Academic Press 1959
Long, D., J. Appl. Phys. **33**, 1682 (1962)
Longini, R.L., Greene, R.F., Phys. Rev. **102**, 992 (1956)
Longuet-Higgins, H.C., Öpik, U., Pryce, M.H.L., Sack, R.A., Proc. Roy. Soc. London A **244**, 1 (1958)
Longuet-Higgins, H.C., Adv. Spectrosc. **2**, 429 (1961)
Look, D.C., Phys. Rev. B **16**, 5460 (1977)
Loose, P., Wöhlecke, M., Phys. Stat. Sol. (b) **73**, 265 (1976)
Lord, N.W., Phys. Rev. **103**, 756 (1957)
Los, V.F., Sov. Phys. Solid State **15**, 1337 (1974)
Loudon, R., Proc. Phys. Soc. **80**, 952 (1962)
Loudon, R., Proc. Phys. Soc. **82**, 393 (1963)
Loudon, R., Proc, Roy. Soc. London A **275**, 218 (1963)
Loudon, R., Adv. Phys. **13**, 423 (1964)
Loudon, R., Phys. Rev. **137**, A1784 (1965)
Loudon, R., J. Phys. C: Solid State Phys. **11**, 403 (1978)
Loudon, R., J. Raman Spectrosc. 7, 10 (1978)
Louis, E., Yndurain, F., J. Phys. C: Solid State Phys. 7, L303 (1974)
Low, W., Paramagnetic Resonance in Solids, New York: Academic Press 1960
Lowndes, R.P., Martin, D.H., Proc. Roy. Soc. London A **308**, 473 (1969)
Lowther, J.E., Killingbeck, J.P., J. Phys. C: Solid State Phys. **3**, 1621 (1970)
Lowther, J.E., J. Phys. C: Solid State Phys. **5**, 676 (1972)
Lowther, J.E., Phys. Stat. Sol. (b) **50**, 287 (1972)
Lowther, J.E., van Wyk, J.A., J. Magn. Resonance **13**, 328 (1974)
Lowther, J.E., Chem. Phys. Letters **35**, 136 (1975)
Lowther, J.E., J. Phys. C: Solid State Phys. **10**, 1247 (1977)
Lubchenko, A.F., Phys. Stat. Sol. **6**, 319 (1964)
Lubchenko, A.F., Phys. Stat. Sol. **27**, K73 (1968)
Lubchenko, A.F., Zushman, I.M., Phys. Stat. Sol. **32**, 703 (1969)
Lubchenko, A.F., Dudkin, S.I., Phys. Stat. Sol. **40**, 43 (1970)
Lubchenko, A.F., Fishuk, I.I., Phys. Stat. Sol. (b) **48**, K11 (1971)
Ludwig, W., Recent Developments in Lattice Theory, Springer Tracts in Modern Physics 43, ed. G. Höhler, Berlin: Springer 1967
Ludwig, W., Festkörperphysik I, Frankfurt a.M.: Akademische Verlagsgesellschaft 1970
Lüty, F., Z. Physik **160**, 1 (1960)
Lüty, F., Halbleiterprobleme VI, 238 (1961) (Ed. F. Sauter, Braunschweig: Vieweg)
Lugiato, L.A., Physica **44**, 337 (1969)
Lugiato, L.A., Physica **81A**, 565 (1975)
Lukashin, A.V., Makshantsev, B.I., Sov. Phys. Solid State **16**, 533 (1974)
Lulek, T., Acta Phys. Pol. **A43**, 705 (1973)
Lulek, T., Acta Phys. Pol. **A48**, 657 (1975)
Lulek, T., Acta Phys. Pol. **A48**, 669 (1975)
Lundqvist, S.O., Fröman, P.O., Ark, Fysik **2**, 431 (1950)
Lundqvist, S.O., Ark, Fysik **6**, 25 (1952)
Lundqvist, S.O., Ark. Fysik 8, 177 (1954)
Lundqvist, S.O., Arkiv Fysik **9**, 435 (1955)
Lundqvist, S.O., Arkiv Fysik **12**, 263 (1957)
Lundqvist, S.O., Ark. Fysik **19**, 113 (1961)
Luong, M., Shaw, A.W., Phys. Rev. B **4**, 2436 (1971)
Lurie, D., Particles and Fields, New York: Interscience Publ. 1968
Lushchik, Ch.B., Vitol, I.K., Elango, M.A., Sov. Phys. Solid State **10**, 2166 (1969)
Luttinger, J.M., Kohn, W., Phys. Rev. **97**, 869 (1955)
Luttinger, J.M., Kohn, W., Phys. Rev. **109**, 1892 (1958)
Luttinger, J.M., Transport Theory, in: Mathematical Methods in Solid State and Superfluid Theory, 8th Scottish Univ. Summer School, St. Andrews 1967, ed. R.C.Clark and G.H. Derrick, Edinburgh: Oliver and Boyd, 1969, p. 157
Lyagushchenko, R.I., Yassievich, I.N., Sov. Phys. Solid State **9**, 2794 (1968)
Lyapilin, I.I., Sov. Phys. Semicond. **11**, 504 (1977)
Lyddane, R.H., Herzfeld, K.F., Phys. Rev. **54**, 846 (1938)
Lynch, D.W., Robinson, D.A., Phys. Rev. **174**, 1050 (1968)

Macdonald, H.F., Klein, M.V., Martin, T.P., Phys. Rev. **177**, 1292 (1969)
Macfarlane, R.M., Wong, J.Y., Sturge, M.D., Phys. Rev. **166**, 250 (1968)

Macfarlane, R.M., Morawitz, H., Phys. Rev. Lett. **27**, 151 (1971)
Maddox, R., Mills, D.L., Phys. Rev. B **11**, 2229 (1975)
Madelung, O., Physics of III-V Compounds, New York: J. Wiley & Sons, 1964
Madelung, O., Grundlagen der Halbleiterphysik, Berlin: Springer, 1970
Madelung, O., Festkörpertheorie I, Berlin: Springer 1972
Madelung, O., Festkörpertheorie II, Berlin: Springer, 1972
Madelung, O., Introduction to Solid-State Theory, Berlin: Springer, 1978
Magnusson, B., Phys. Stat. Sol. (b) **52**, 361 (1972)
Magnusson, B., Phys. Stat. Sol. (b) **56**, 269 (1973)
Mahan, G.D., Phys. Rev. **142**, 366 (1966)
Mahan, G.D., Phys. Rev. **170**, 825 (1968)
Mahan, G.D., Phys. Rev. B **15**, 4587 (1977)
Mahanti, S.D., Varma, C.M., Phys. Rev. B **6**, 2209 (1972)
Mahanty, J., Maradudin, A.A., Weiss, G.H., Prog. Theor. Phys. **20**, 369 (1958)
Mahanty, J., Maradudin, A.A., Weiss, G.H., Prog. Theor. Phys. **24**, 648 (1960)
Mahanty, J., Proc. Phys. Soc. **88**, 1011 (1966)
Mahanty, J., Phys. Letters **29A**, 583 (1969)
Mahanty, J., Sachdev, M., J. Phys. C: Solid State Phys. **3**, 773 (1970)
Mahanty, J., Paranjape, V.V., Phys. Rev. B **10**, 2596 (1974)
Mahanty, J., Paranjape, V.V., Solid State Commun. **16**, 733 (1975)
Mahanty, J., Paranjape, V.V., Phys. Rev. B **13**, 1830 (1976)
Mahler, G., Schröder, U., Phys. Rev. Letters **27**, 1358 (1971)
Mahler, G., Engelhardt, P. Phys. Stat. Sol. (b) **45**, 543 (1971)
Mahler, G., Schröder, U., Phys. Stat. Sol. (b) **61**, 629 (1974)
Maiman, T.H., Phys. Rev. **123**, 1145 (1961)
Majlis, N., Anda, E., J. Phys. C: Solid State Phys. **11**, 1607 (1978)
Malkin, B.Z., Sov. Phys. Solid State **5**, 2262 (1964)
Malkin, B.Z., Sov. Phys. JETP **21**, 1101 (1965)
Man, A., Bron, W.E., Phys. Rev. B **13**, 5591 (1976)
Mani, K.K., Singh, R.P., Phys. Stat. Sol. (b) **56**, 723 (1973)
Mani, K.K., Singh, R.P., Phys. Stat. Sol. (b) **57**, 289 (1973)
Mansfield, R., Proc. Phys. Soc. B **69**, 76 (1956)
Mansikka, K., Bystrand, F., J. Phys. Chem. Solids **27**, 1073 (1966)
Maradudin, A. A., Flinn, P. A., Coldwell-Horsfall, R. A., Ann. Physics **15**, 360 (1961)
Maradudin, A. A., Flinn, P. A., Coldwell-Horsfall, R. A., Ann. Physics **15**, 337 (1961)
Maradudin, A.A., Wallis, R.F., Phys. Rev. **123**, 777 (1961)
Maradudin, A.A., Weiss, G.H., Phys. Rev. **123**, 1968 (1961)
Maradudin, A.A., Phys. Stat. Sol. **2**, 1493 (1962)
Maradudin, A.A., Dynamic Theory of Imperfect Crystals, in: Astrophysics and the Many-Body Problem, Brandeis Summer Institute 1962, v. 2, ed. K.W. Ford, New York: Benjamin, 1963, p. 107
Maradudin, A.A., Ann. Phys. (New York) **30**, 371 (1964)
Maradudin, A.A., Rep. Prog. Phys. **28**, 331 (1965)
Maradudin, A.A., Solid State Physics **18**, 273 (1966) (Ed. F. Seitz and D. Turnbull, New York: Academic Press)
Maradudin, A.A., Solid State Physics **19**, 1 (1966) (Ed. F. Seitz and D. Turnbull, New York: Academic Press)
Maradudin, A.A., Peretti, J., Phys. Rev. **161**, 852 (1967)
Maradudin, A.A., Vosko, S.H., Rev. Mod. Phys. **40**, 1 (1968)
Maradudin, A.A., Nardelli, G.F. (Eds.), Elementary Excitations in Solids, New York: Plenum Press, 1969
Maradudin, A.A., Oitmaa, J., Solid State Commun. 7, 1143 (1969)
Maradudin, A.A., Wallis, R.F., Phys. Rev. B **2**, 4294 (1970)
Maradudin, A.A., Montroll, E.W., Weiss, G.H., Ipatova, I.P., Theory of Lattic Dynamics in the Harmonic Approximation, New York: Academic Press, 1971
Maradudin, A.A., in: Dynamical Properties of Solids, v. 1, ed. Horton, G.K. and Maradudin, A.A., Amsterdam: North-Holland 1974, p. 1
Maradudin, A.A., Phys. Rev. B **13**, 658 (1976)
Marchevskii, F.N., Strizhevskii, V.L., J. Raman Spectrosc. **3**, 7 (1975)
Marchevskii, F.N., Strizhevskii, V.L., J. Raman Spectrosc. **3**, 15 (1975)
Marcus, A., Phys. Rev. **135**, A527 (1964)
Margerie, J., Martin-Brunetiere, F., J. Phys. C: Solid State Phys. **10**, 2701 (1977)
Marinkovic, M.M., Tosic, B.S., Phys. Stat. Sol. (b) **67**, 435 (1975)
Marinkovic, M.M., Phys. Stat. Sol. (b) **69**, 291 (1975)
Markham, J.J., Phys. Rev. **103**, 588 (1956)
Markham, J.J., Rev. Mod. Phys. **31**, 956 (1959)
Markham, J.J., Z. Physik **188**, 139 (1965)
Markham, J.J., F-Centers in Alkali Halides, Solid State Physics, Suppl. 8, ed. F. Seitz and D. Turnbull, New York: Academic

Marshall, J.T., Stewart, B.U., Phys. Rev. B **2**, 4001 (1970)
Marshall, R.F., Blint, R.J., Kunz, A.B., Phys. Rev. B **13**, 3333 (1976)
Marshall, R.F., Blint, R.J., Kunz, A.B., Solid State Commun. **18**, 731 (1976)
Marston, R.L., Dick, B.G., Solid State Commun. **5**, 731 (1967)
Martin, B.G., Wallis, R.F., Solid State Commun. **15**, 361 (1974)
Martin, R.M., Phys. Rev. **186**, 871 (1969)
Martin, R.M., Phys. Rev. B **4**, 3676 (1971)
Martin, T.P., Fowler, W.B., Phys. Rev. B **2**, 4221 (1970)
Martin, T.P., J. Phys. C: Solid State Phys. **4**, 2269 (1971)
Martin, T.P., J. Phys. C: Solid State Phys. **5**, 493 (1972)
Martin, T.P., Phys. Rev. B **13**, 3617 (1976)
Martinez, G., Schlüter, M., Cohen, M.L., Phys. Rev. B **11**, 651 (1975)
Martinov, N.K., J. Phys. C: Solid State Phys. **4**, 998 (1971)
Maschke, K., Rössler, U., Phys. Stat. Sol. **28**, 577 (1968)
Mashkevich, V.S., Tolpygo, K.B., Sov. Phys. JETP **5**, 435 (1957)
Mashkevich, V.S., Shadchin, E.A., Sov. Phys. Solid State **14**, 716 (1972)
Matejec, R., Z. Physik **151**, 595 (1958)
Matossi, F., Gutjahr, H., Phys. Stat. Sol. **3**, 167 (1963)
Matsushima, A., Fukuda, A., Phys. Stat. Sol. (b) **66**, 663 (1974)
Matsushima, A., Fukuda, A., Phys. Rev. B **14**, 3664 (1976)
Matsuura, M., Wang, S., Solid State Commun. **15**, 613 (1974)
Matsuura, M., Mavroyannis, C., J. Low Temp. Phys. **28**, 129 (1977)
Matta, M.L., Sukheeja, B.D., Narchal, M.L., J. Magn. Resonance **9**, 121 (1973)
Matta, M.L., Sukheeja, B.D., Narchal, M.L., J. Phys. Chem. Solids **35**, 1339 (1974)
Mattausch, H.J., Uihlein, Ch., Solid State Commun. **25**, 447 (1978)
Matthew, J.A.D., Proc. Phys. Soc. **89**, 683 (1966)
Matthew, J.A.D., Hart-Davis, A., Phys. Rev. **168**, 936 (1968)
Matthew, J.A.D., Green, B., J. Phys. C: Solid State Phys. **4**, L101 (1971)
Mattis, D.C., Lieb, E.H., J. Math. Phys. **7**, 2045 (1966)
Mattis, D., Sinha, O., Ann. Physics **61**, 214 (1970)
Matulenis, A., Pozhela, Yu., Reklaitis, A., Sov. Phys. Semicond. **10**, 169 (1976)
Matulionis, A., Pozela, J., Reklaitis, A., Phys. Stat. Sol. (a) **35**, 43 (1976)
Matulis, A., Sov. Phys. Collect. **14**, 22 (1974)
Matz, D., Solid State Commun. **4**, 491 (1966)
Matz, D., J. Phys. Chem. Solids **28**, 373 (1967)
Matz, D., Phys. Rev. **168**, 843 (1968)
Matz, D., Can. J. Phys. **53**, 2665 (1975)
Mauser, K.E., Niesert, B., Winnacker, A., Z. Physik B **26**, 107 (1977)
Mavroyannis, C., J. Low Temp. Phys. **25**, 501 (1976)
Maxia, V., Phys. Rev. B **17**, 3262 (1978)
McCanny, J.V., Murray, R.B., J. Phys. C: Solid State Phys. **10**, 1211 (1977)
McCarty, M., Maycock, J.N., Can. J. Phys. **49**, 2005 (1971)
McClure, D.S., The Electric States and Spectra of Ions and Imperfections in Solids, in: Phonons in Perfect Lattices and Lattices with Point Imperfections, Scottish Univ. Summer School 1965, ed. R.W.H. Stevenson, Edinburgh: Oliver and Boyd, 1966, p. 314
McCombie, C.W., Matthew, J.A.D., Murray, A.M., J. Appl. Phys., Suppl., **33**, 359 (1962)
McCombie, C.W., Slater, J., Proc. Phys. Soc. **84**, 499 (1964)
McCombs, A.E., Milnes, A.G., Int. J. Electron. **32**, 361 (1971)
McConnel, H.M., McLachlan, A.D., J. Chem. Phys. **34**, 1 (1961)
McCumber, D.E., J. Math. Phys. **5**, 221 (1964)
McCumber, D.E., J. Math. Phys. **5**, 508 (1964)
McCumber, D.E., Phys. Rev. **135**, A 1676 (1964)
McCumber, D.E., Phys. Rev. **136**, A 954 (1964)
McKenzie, B.J., Stedman, G.E., J. Phys. C: Solid State Phys. **11**, 589 (1978)
McLean, T.P., Paige, E.G.S., J. Phys. Chem. Solids **16**, 220 (1960)
Medvedev, E.S., Opt. Spectrosc. **30**, 567 (1971)
Meijer, H.J.G., Polder, D., Physica **19**, 255 (1953)
Mejia, C.R., J. Phys. Soc. Japan **34**, 320 (1973)
Mel'nikov, V.I., Volovik, G.E., Sov. Phys. JETP **38**, 819 (1974)
Mel'nikov, V.I., Sov. Phys. JETP **47**, 404 (1978)
Melvin, J.S., Pirie, J.D., Smith, T., Phys. Rev. **175**, 1082 (1968)
Melvin, J.S., Smith, T., Solid State Commun. **11**, 1723 (1972)
Meneses, E.A., Luzzi, R., Solid State Commun. **12**, 447 (1973)
Menne, T.J., Phys. Rev. **180**, 350 (1969)
Menne, T.J., Phys. Rev. B **1**, 4496 (1970)
Merchant, H.D., Srivastava, K.K., Pandey, H.D., Crit. Rev. Solid State Sci. **3**, 451 (1973)
Merten, L., Z. Naturforsch. **13a**, 662 (1958)
Merten, L., Z. Naturforsch. **13a**, 1067 (1958)
Merten, L., Z. Naturforsch. **15a**, 47 (1960)
Merten, L., Z. Naturforsch. **17a**, 65 (1961)
Merten, L., Z. Naturforsch. **22a**, 359 (1967)
Merten, L., Phys. Stat. Sol. **30**, 449 (1968)

Merten, L., Festkörperprobleme **12**, 343 (1972) (Ed. O. Madelung, Braunschweig: Vieweg)
Merten, L., Borstel, G., Z. Naturforsch. **27a**, 1073 (1972)
Merten, L., Borstel, G., Z. Naturforsch. **27a**, 1792 (1972)
Merten, L., Phys. Stat. Sol. (b) **55**, K143 (1973)
Merz, J.L., Phys. Rev. **176**, 961 (1968)
Merz, J.L., Faulkner, R.A., Dean, P.J., Phys. Rev. **188**, 1228 (1969)
Messiah, A., Quantum Mechanics II, Amsterdam: North-Holland, 1962
Messiah, A., Quantum Mechanics, Amsterdam: North-Holland 1970
Metselaar, R., van der Elsken, J., Phys. Rev. **165**, 359 (1968)
Meyer, A., Wood, R.F., Phys. Rev. **133**, A1436 (1964)
Meyer, H. J. G., Physica **20**, 1016 (1954)
Meyer, H.J.G., Physica **22**, 109 (1956)
Meyer, H.J.G., Halbleiterprobleme III, p. 230 (1956) (ed. W. Schottky, Braunschweig: Vieweg)
Michel, H., Wagner, M., Phys. Stat. Sol. (b) **75**, 507 (1976)
Michel, H., Wagner, M., Phys. Stat. Sol. (b) **85**, 195 (1978)
Michielsen, J., Woerlee, P., van de Graaf, F., Ketelaar, J.A.A., J. Chem. Soc. Faraday Trans. II **71**, 1730 (1975)
Mickish, D.J., Kunz, A.B., J. Phys. C: Solid State Phys. **6**, 1723 (1973)
Mickish, D.J., Kunz, A.B., Collins, T.C., Phys. Rev. B **9**, 4461 (1974)
Mickish, D.J., Kunz, A.B., Phys. Stat. Sol. (b) **73**, 193 (1976)
Mikami, Y., Mizunoya, K., Nakajima, T., Chem. Phys. Letters **30**, 373 (1975)
Mikhailova, M.P., Rogachev, A.A., Yassievich, I.N., Sov. Phys. Semicond. **10**, 866 (1976)
Mikheev, V.M., Pomortsev, R.V., Sov. Phys. Semicond. **9**, 1177 (1976)
Mikoshiba, N., J. Phys. Chem. Solids **24**, 341 (1963)
Mikoshiba, N., J. Phys. Soc. Japan **20**, 2160 (1965)
Miller, A., Abrahams, E., Phys. Rev. **120**, 745 (1960)
Miller, A., J. Phys. Chem. Solids **35**, 641 (1974)
Miller, M.M., Fong, K.F., J. Chem. Phys. **59**, 1528 (1973)
Mills, D.L., Phys. Rev. **146**, 336 (1966)
Mills, D.L., Burstein, E., Phys. Rev. **188**, 1465 (1969)
Mills, D.L., Burstein, E., Comments Solid State Phys. **3**, 12 (1970)
Mills, D.L., Maradudin, A.A., Phys. Rev. B **1**, 903 (1970)
Milnes, A.G., Deep Impurities in Semiconductors, New York: J. Wiley & Sons, 1973
Mimura, H., Uemura, Y., J. Phys. Soc. Japan **14**, 1011 (1959)
Mishima, N., Petrosky, T.Y., Yamazaki, M., J. Statist. Phys. **14**, 359 (1976)
Misra, B.N., Faujdar, Kripal, R., Nuovo Cim. **42 B**, 77 (1977)
Misra, K.D., Dixit, V.K., Sharma, M.N., Z. Naturforsch. **29a**, 1601 (1974)
Mitani, Y., Takeno, S., Prog. Theor. Phys. **33**, 779 (1965)
Mitin, V.V., Phys. Stat. Sol. (b) **49**, 125 (1972)
Mitin, V.V., Tolpygo, E.I., Phys. Stat. Sol. (b) **72**, 51 (1975)
Mitin, V.V., Sov. Phys. Semicond. **11**, 727 (1977)
Mitra, S.S., Marshall, R., J. Chem. Phys. **41**, 3158 (1964)
Mitra, S.S., Singh, R.S., Brada, Y., Phys. Rev. **182**, 953 (1969)
Mitra, T.K., Phys. Letters **42A**, 223 (1972)
Mitskevich, V.V., Sov. Phys. Solid State **3**, 2202 (1962)
Mitskevich, V.V., Sov. Phys. Solid State **3**, 2211 (1962)
Mitskevich, V., Phys. Stat. Sol. (b) **61**, 675 (1974)
Mitskevich, V., Phys. Stat. Sol. (b) **67**, 677 (1975)
Miura, M., Murata, H., Shiro, Y., J. Phys. Chem. Solids **38**, 1071 (1977)
Miura, M., Murata, H., Shiro, Y., J. Phys. Chem. Solids **39**, 669 (1978)
Miyakawa, T., J. Phys. Soc. Japan **24**, 768 (1968)
Miyakawa, T., Oyama, S., J. Phys. Soc. Japan **24**, 996 (1968)
Miyakawa, T., Dexter, D.L., Phys. Rev. B **1**, 2961 (1970)
Miyakawa, T., Solid State Commun. **25**, 133 (1978)
Miyake, S.J., J. Phys. Soc. Japan **41**, 747 (1976)
Modine, F.A., Chen, Y., Major, R.W., Wilson, T.M., Phys. Rev. B **14**, 1739 (1976)
Möglich, F., Rompe, R., Physik. Zeitschr., Jg. 1940, S. 236
Moffitt, W., Liehr, A.D., Phys. Rev. **106**, 1195 (1957)
Moffitt, W., Thorson, W., Phys. Rev. **108**, 1251 (1957)
Mokhracheva, L.P., Tskhai, V.A., Geld, P.V., Phys. Stat. Sol. (b) **78**, 465 (1976)
Mokross, B.J., Dick, B.G., Phys. Rev. B **15**, 5938 (1977)
Mokross, B.J., Dick, B.G., Page, J.B., Phys. Rev. B **15**, 5946 (1977)
Mollenauer, L.F., Pan, S., Phys. Rev. B **6**, 772 (1972)

Monnier, R., Song, K.S., Stoneham, A.M., J. Phys. C: Solid State Phys. **10**, 4441 (1977)
Monozon, B.S., Seysyan, R.P., Shelekhin, Yu. L., Phys. Stat. Sol. (b) **54**, 719 (1972)
Monozon, B.S., Zhilich, A.G., Sov. Phys. Semicond. **8**, 950 (1975)
Monozon, B.S., Sov. Phys. Semicond. **10**, 107 (1976)
Monozon, B.S., Zhilich, A.G., Sov. Phys. Solid State **19**, 791 (1977)
Monozon, B.S., Zhilich, A.G., Sov. Phys. Solid State **19**, 2131 (1977)
Montroll, E.W., Potts, R.B., Phys. Rev. **100**, 525 (1955)
Montroll, E.W., Ward, J.C., Physica **25**, 423 (1959)
Montroll, E.W., Boulder Lectures in Theoretical Physis **3**, 221 (1960)
Montroll, E.W., in: Fundamental Problems in Statistical Mechanics, ed. E.G.D. Cohen, Amsterdam: North-Holland, 1962, p. 230
Moran, P.R., Phys. Rev. **137**, A 1016 (1965)
Moreno, M., Barriuso, M.T., Solid State Commun. **17**, 1035 (1975)
Morgan, G.J., Proc. Phys. Soc. **89**, 476 (1966)
Morgan, G.J., J. Physique (Paris) **33**, 49 (1972)
Morgan, G.J., Shahtahmasebi, N., J. Phys. C: Solid State Phys. **6**, 3385 (1973)
Morgan, T.N., Phys. Rev. Letters **24**, 887 (1970)
Morgan, T.N., Nuovo Cim. **39B**, 602 (1977)
Morita, A., Takahashi, K., Prog. Theor. Phys. **19**, 257 (1958)
Morita, A., J. Phys. Chem. Solids **8**, 363 (1959)
Morita, A., Azuma, M., Nara, H., J. Phys. Soc. Japan **17**, 1570 (1962)
Morita, A., J. Phys. Soc. Japan **18**, 1437 (1963)
Morokuma, K., Freed, K.F., J. Chem. Phys. **61**, 4342 (1974)
Morris, S.P., J. Phys. Chem. Solids **30**, 1873 (1969)
Moskalenko, S.A., Opt. i Spektrosk. **5**, 147 (1958)
Moskalenko, S.A., Tolpygo, K.B., Zh. Eksp. i Teoret. Fiz. **36**, 149 (1959)
Moskalenko, S.A., Shmiglyuk, M.I., Sov. Phys. Solid State **6**, 2831 (1965)
Moskalenko, S.A., Shmiglyuk, M.I., Chinik, B.I., Sov. Phys. Solid State **10**, 279 (1968)
Moskalenko, S.A., Miglei, M.F., Shmiglyuk, M.I., Khadzhi, P.I., Lelyakov, A.V., Sov. Phys. JETP **37**, 902 (1973)
Moskalenko, W.A., J. Exp. Theor. Phys. **30**, 959 (1956)
Moss, T.S., Phys. Stat. Sol. **2**, 601 (1962)
Mostoller, M., Ganguly, B.N., Wood, R.F., Phys. Rev. B **4**, 2015 (1971)
Mott, N.F., Littleton, M.I., Trans. Faraday Soc. **34**, 485 (1938)
Mott, N.F., Gurney, R.W., Electronic Processes in Ionic Crystals, Oxford University Press, 1940
Mott, N.F., Gurney, R.W., Electronic Processes in Ionic Crystals, Oxford: Clarendon Press, 1953, 2nd Ed.
Mott, N.F., Can. J. Phys. **34**, 1356 (1956)
Mott, N.F., Twose, W.D., Adv. Phys. **10**, 107 (1961)
Moussa, A.R., Hassan, A.R., Indian J. Pure Appl. Phys. **15**, 68 (1977)
Moyer, C.A., Phys. Rev. B **7**, 5025 (1973)
Müller, P., Phys. Stat. Sol. **21**, 693 (1967)
Müser, H., Z. Naturforsch. **7a**, 729 (1952)
Mulazzi, E., Nardelli, G.F., Terzi, N., Phys. Rev. **172**, 847 (1968)
Mulazzi, E., Nardelli, G.F., Terzi, N., Solid State Commun. **8**, 1875 (1970)
Mulazzi, E., Terzi, N., Phys. Rev. B **10**, 3552 (1974)
Munn, R.W., J. Phys. C: Solid State Phys. **8**, 2721 (1975)
Munn, R.W., Silbey, R., J. Chem. Phys. **68**, 2439 (1978)
Munschy, G., Phys. Stat. Sol. (b) **53**, 377 (1972)
Munschy, G., Stebe, B., Phys. Stat. Sol. (b) **72**, 135 (1975)
Muntyan, A.P., Pokatilov, E.P., Fomin, V.M., Sov. Phys. Semicond. **11**, 735 (1977)
Muramatsu, S., Phys. Stat. Sol. (b) **56**, 631 (1973)
Muramatsu, S., Phys. Stat. Sol. (b) **72**, K177 (1975)
Muravev, V.I., Yunusov, N.B., Phys. Stat. Sol. (b) **84**, 791 (1977)
Murthy, C.S.N., Murti, Y.V.G.S., J. Phys. C: Solid State Phys. **4**, 1108 (1971)
Murti, Y.V.G.S., Murthy, C.S.N., J. Phys. C: Solid State Phys. **5**, 401 (1972)
Murti, Y.V.G.S., Usha, V., Physica **83B**, 275 (1976)
Murti, Y.V.G.S., Usha, V., Physica **93B**, 219 (1978)
Muto, T., Prog. Theor. Phys. **IV**, 181 (1949)
Muto, T., Prog. Theor. Phys. **IV**, 243 (1949)
Muto, T., Okuno, H., J. Phys. Soc. Japan **11**, 633 (1956)
Muto, T., Progr. Theor. Phys. Suppl. **12**, 3 (1959)
Mycielski. J., Aziza, A., Mycielski, A., Balkanski, M., Phys. Stat. Sol. (b) **67**, 447 (1974)
Mycielski, J., Bastard, G., Rigaux, C., Phys. Rev. B **16**, 1675 (1977)

Naberhuis, S.L., Fong, F.K., J. Chem. Phys. **56**, 1174 (1972)
Nag, B.R., Phys. Stat. Sol. (b) **66**, 719 (1974)
Nag, B.R., Dutta, G.M., J. Appl. Phys. **48**, 3621 (1977)
Nag, B.R., Dutta, G.M., J. Phys. C: Solid State Phys. **11**, 119 (1978)

Nag, B.R., Chattopadhyay, D., Solid State Electr. **21**, 303 (1978)
Nag, B.R., Electron Transport in Compound Semiconductors, Berlin: Springer, 1980
Nagae, M., J. Phys. Soc. Japan **18**, 207 (1963)
Nagaev, E.L., Sov. Phys. Solid State **4**, 1611 (1963)
Nagaev, E.L., Theor. Math. Phys. **14**, 67 (1973)
Nagai, S., J. Phys. Soc. Japan **20**, 1366 (1965)
Nagamiya, T., Tatsuuma, N., J. Phys. Soc. Japan **9**, 307 (1954)
Nagamiya, T., Kojima, T., Kondoh, H., J. Phys. Soc. Japan **9**, 310 (1954)
Nagashima, T., Horie, C., J. Phys. Soc. Japan **37**, 614 (1974)
Nagel, S., Maschke, K., Baldereschi, A., Phys. Stat. Sol. (b) **76**, 629 (1976)
Nakajima, S., Prog. Theor. Phys. **20**, 948 (1958)
Nakamura, A., J. Chem. Phys. **64**, 185 (1976)
Namjoshi, K.V., Mitra, S.S., Vetelino, J.F., Solid State Commun. **9**, 185 (1971)
Namjoshi, K.V., Mitra, S.S., Vetelino, J.F., Phys. Rev. B **3**, 4398 (1971)
Nara, H., Kobayasi, T., Yamazaki, H., J. Phys. Soc. Japan **38**, 631 (1975)
Narayan, R., Ramaseshan, S., J. Phys. Chem. Solids **37**, 395 (1976)
Narayanamurti, V., Pohl, R.O., Rev. Mod. Phys. **42**, 201 (1970)
Nardelli, G.F., Tettamanzi, N., Phys. Rev. **126**, 1283 (1962)
Nascimento, O.R., Brandi, H.S., Ribeiro, S.C., Solid State Commun. **15**, 1153 (1974)
Nasu, K., Kojima, T., Prog. Theor. Phys. **51**, 26 (1974)
Nasu, K., Z. Naturforsch. **30a**, 1060 (1975)
Natadze, A.L., Ryskin, A.I., Solid State Commun. **24**, 147 (1977)
Natsume, Y., J. Phys. Soc. Japan **41**, 607 (1976)
Neckel, A., Vinek, G., Z. Phys. Chem. **42**, 129 (1964)
Nedoluha, A., Phys. Rev. B **1**, 864 (1970)
Neogy, C., Deb, S.K., Indian J. Pure Appl. Phys. **15**, 308 (1977)
Neogy, C., Deb, S.K., J. Phys. Chem. Solids **39**, 585 (1978)
Nettel, S.J., Phys. Rev. **121**, 425 (1961)
Nettel, S.J., Phys. Rev. **128**, 2573 (1962)
Neugebauer, Th., Acta Phys. Hung. **12**, 279 (1961)
Neumann, H., Hess, E., Topol, I., Czech. J. Phys. B **25**, 174 (1975)
Neumark, G.F., Phys. Rev. B **7**, 3802 (1973)
Nicholas, K.H., Woods, J., Brit. J. Appl. Phys. **15**, 783 (1964)
Niedermann, H.P., Wagner, M., Phys. Stat. Sol. (b) **78**, 615 (1976)
Nikitine, S., Compt. Rend. (Paris) **240**, 1415 (1955)
Nikitine, S., Prog. Semicond. **6**, 233 (1961)
Niklasson, G., Sjölander, A., Ann. Phys. (New York) **49**, 249 (1968)
Niklasson, G., Ann. Phys. (New York) **59**, 263 (1970)
Nishikawa, K., Barrie, R., Can. J. Phys. **41**, 1135 (1963)
Nishikawa, K., Aono, S., Prog. Theor. Phys. **50**, 345 (1973)
Nitzan, A., Jortner, J., J. Chem. Phys. **55**, 1355 (1971)
Nitzan, A., Jortner, J., Mol. Phys. **25**, 713 (1973)
Nitzan, A., Jortner, J., Theoret. Chim. Acta **29**, 97 (1973)
Nitzan, A., Jortner, J., Theor. Chim. Acta **30**, 217 (1973)
Nomura, K.C., Blakemore, J.S., Phys. Rev. **112**, 1607 (1958)
Nomura, K.C., Blakemore, J.S., Phys. Rev. **121**, 734 (1961)
Norgett, M.J., J. Phys. C: Solid State Phys. **4**, 298 (1971)
Norgett, M.J., J. Phys. C: Solid State Phys. **4**, 1284 (1971)
Norgett, M.J., J. Phys. C: Solid State Phys. **4**, 1289 (1971)
Norgett, M.J., Stoneham, A.M., J. Phys. C: Solid State Phys. **6**, 238 (1973)
Norgett, M.J., Stoneham, A.M., J. Phys. C: Solid State Phys. **6**, 229 (1973)
Norgett, M.J., Stoneham, A.M., Pathak, A.P., J. Phys. C: Solid State Phys. **10**, 555 (1977)
Norris, G.B., Whitfield, G.D., Phys. Rev. B **17**, 5040 (1978)
Novak, P., Stevens, K.W.H., J. Phys. C: Solid State Phys. **3**, 1703 (1970)
Novikov, V.S., Sov. Phys. Solid State **11**, 690 (1969)
Nozieres, P., Lewiner, C., J. Physique **34**, 901 (1973)
Nudelman, S., Mitra, S.S. (Eds.), Optical Properties of Solids, New York: Plenum Press, 1969
Nüsslein, V., Schröder, U., Phys. Stat. Sol. **21**, 309 (1967)
Nusimovici, M.A., Birman, J.L., Phys. Rev. **156**, 925 (1967)
Nusimovici, M.A., Ann. Phys. (Paris) **4**, 97 (1969)
Nusimovici, M.A., Balkanski, M., Birman, J.L., Phys. Rev. B **1**, 603 (1970)
Nusimovici, M.A., Balkanski, M., Birman, J.L., *Phys. Rev. B* **1**, 595 (1970)
Nusimovici, M.A. (Ed.), Phonons, Paris: Flammarion, 1971

Obata, Y., Sasaki, K., J. Phys. Soc. Japan **42**, 36 (1977)
Obraztsov, Yu.N., Sov. Phys. Solid State **6**, 331 (1964)
Obraztsov, Yu.N., Sov. Phys. Solid State 7, 455 (1965)
O'Brien, M.C.M., Proc. Roy. Soc. London A **281**, 323 (1964)
O'Brien, M.C.M., Proc. Phys. Soc. **86**, 847 (1965)
O'Brien, M.C.M., Phys. Rev. **187**, 407 (1969)
O'Brien, M.C.M., J. Phys. C: Solid State Phys. **4**, 2524 (1971)
O'Brien, M.C.M., J. Phys. C: Solid State Phys. **5**, 2045 (1972)
O'Brien, M.C.M., J. Phys. C: Solid State Phys. **9**, 3153 (1976)
O'Brien, W.P., Hernandez, J.P., Phys. Rev. B **9**, 3560 (1974)
O'Dwyer, J.J., Adv. Phys. **7**, 349 (1958)
O'Dwyer, J.J., Nickle, H.H., Phys. Rev. B **2**, 5063 (1970)
Öpik, U., Pryce, M.H.L., Proc. Roy. Soc. London A **238**, 425 (1957)
Öpik, U., Wood, R.F., Phys. Rev. **179**, 772 (1969)
Ohashi, K., Ohashi, Y.H., J. Phys. C: Solid State Phys. **9**, 733 (1976)
Ohkura, H., Imanaka, K., Kamada, O., Mori, Y., Iida, T., J. Phys. Soc. Japan **42**, 1942 (1977)
Ohta, T., Nagae, M., Miyakawa, T., Prog. Theor. Phys. **23**, 229 (1960)
Ohta, K., J. Phys. Soc. Japan **26**, 1196 (1969)
Ohtaka, K., Fujiwara, T., J. Phys. Soc. Japan **27**, 901 (1969)
Oitmaa, J., Maradudin, A.A., Solid State Commun. 7, 1371 (1969)
Okada, J., J. Phys. Soc. Japan **12**, 1338 (1957)
Okamoto, K., J. Phys. Soc. Japan **32**, 46 (1972)
Okamoto, K., J. Phys. Soc. Japan **36**, 657 (1974)
Okamoto, K., Takeda, S., J. Phys. Soc. Japan **37**, 333 (1974)
Okumura, N. et al., Jap. J. Appl. Phys. **15**, 2259 (1976)
Olechna, D.J., Ehrenreich, H., J. Phys. Chem. Solids **23**, 1513 (1962)
Oleinik, V.P., Sinyak, V.A., Phys. Stat. Sol. (b) **86**, 319 (1978)
Oleinik, V.P., Abakarov, D.I., Belousov, I.V., Sov. Phys. JETP **48**, 155 (1978)
Oliveira, L.E., Oliveira, P.M., Maffeo, B., Phys. Stat. Sol. (b) **87**, 25 (1978)
Olszewski, S., Dubejko, M., J. Chem. Phys. **48**, 5576 (1968)
Ong, C.K., Vail, J.M., Phys. Rev. B **8**, 1636 (1973)
Ong, C.K., Vail, J.M., Phys. Rev. B **15**, 3898 (1977)
Onipko, A.I., Sugakov, V.I., Opt. Spectrosc **35**, 108 (1973)
Onipko, A.I., Chem. Phys. Letters **38**, 594 (1976)
Onodera, Y., Okazaki, M., J. Phys. Soc. Japan **21**, 1273 (1966)
Onodera, Y., Okazaki, M., J. Phys. Soc. Japan **21**, 2229 (1966)
Onodera, Y., Okazaki, M., Inui, T., J. Phys. Soc. Japan **21**, 816 (1966)
Onodera, Y., Toyozawa, Y., J. Phys. Soc. Japan **22**, 833 (1967)
Opechowski, W., Physica 7, 552 (1940)
Opie, A.H., Phys. Rev. **172**, 640 (1968)
Opie, A.H., Phys. Rev. **187**, 1168 (1969)
Orbach, R., Proc. Roy. Soc. London A **264**, 458 (1961)
Orbach, R., Simanek, E., Phys. Rev. **158**, 310 (1967)
Ore, A., J. Chem. Phys. **33**, 31 (1960)
O'Rourke, R.C., Phys. Rev. **91**, 265 (1953)
Osad'ko, I.S., Sov. Phys. Solid State **14**, 2522 (1973)
Osad'ko, I.S., Sov. Phys. Solid State **17**, 2098 (1976)
Osad'ko, I.S., Sov. Phys. JETP **45**, 827 (1977)
Osadko, I.S., Phys. Stat. Sol. (b) **82**, K107 (1977)
Osaka, Y., Osaka, Y.S., Goto, F., J. Phys. Soc. Japan **17**, 1715 (1962)
Osherov, V.I., Sov. Phys. Solid State **10**, 21 (1968)
Osherov, V.I., Theor. Exp. Chem. **4**, 171 (1968)
Osherov, V.I., Medvedev, E.S., Phys. Stat. Sol. **36**, 69 (1969)
Osipov, E.B., Yakovlev, V.A., Sov. Phys. Semicond. **9**, 781 (1975)
Osipov, E.B., Yakovlev, V.A., Sov. Phys. Semicond. **10**, 447 (1976)
Osipov, E.B., Sov. Phys. Solid State **18**, 2085 (1976)
Osipov, V.V., Sov. Phys. Semicond. 7, 1405 (1974)
Osipov, V.V., Foigel, M.G., Sov. Phys. Semicond. **10**, 311 (1976)
Osipov, V.V., Soboleva, T.I., Foigel, M.G., Sov. Phys. Semicond. **11**, 752 (1977)
Ostrovskii, V.S., Kharkyanen, V.N., Sov. Phys. Solid State **19**, 903 (1977)
Otsuka, E., Prog. Theor. Phys., Suppl., **57**, 1 (1975)
Ovander, L.N., Sov. Phys. Uspekhi **86**, 337 (1965)
Overhauser, A.W., Phys. Rev. **101**, 1702 (1956)
Overhof, H., Rössler, U., Phys. Stat. Sol. **37**, 691 (1970)
Overhof, H., Phys. Stat. Sol. (b) **43**, 575 (1971)
Overhof, H., Treusch, J., Solid State Commun. **9**, 53 (1971)
Overton, J., Hernandez, J.P., Phys. Rev. B **7**, 778 (1973)

Owens, F.J., Phys. Stat. Sol. (b) **63**, 115 (1974)
Oyama, S., Miyakawa, T., J. Phys. Soc. Japan **21**, 868 (1966)

Paerschke, H., Süsse, K.E., Welsch, D.G., Ann. Physik (Leipzig) **34**, 405 (1977)
Pässler, R., Czech. J. Phys. B **24**, 322 (1974)
Pässler, R., Phys. Stat. Sol. (b) **65**, 561 (1974)
Pässler, R., Czech. J. Phys. B **25**, 219, (1975)
Pässler, R., Phys. Stat. Sol. (b) **68**, 69 (1975)
Pässler, R., Phys. Stat. Sol. (b) **76**, 647 (1976)
Pässler, R., Phys. Stat. Sol. (b) **78**, 625 (1976)
Page, J.B., Dick, B.G., Phys. Rev. **163**, 910 (1967)
Page, J.B., Strauch, D., Phys. Stat. Sol. **24**, 469 (1967)
Page, J.B., Phys. Rev. **184**, 905 (1969)
Page, J.B., Helliwell, K.G., Phys. Rev. B **12**, 718 (1975)
Page, L.J., Hygh, E.H., Phys. Rev. B **1**, 3472 (1970)
Palistrant, M.E., Opt. Spectrosc. **15**, 425 (1963)
Palistrant, M.E., Moskalenko, V.A., Opt. Spectrosc. **17**, 392 (1964)
Pan, D.S., Smith, D.L., McGill, T.C., Phys. Rev. B **17**, 3284 (1978)
Pan, D.S., Smith, D.L., McGill, T.C., Phys. Rev. B **17**, 3297 (1978)
Pandey, A., Shukla, A.K., J. Phys. Chem. Solids **38**, 15 (1977)
Pandey, A.N., Upadhyaya, K.S., Phys. Stat. Sol. (b) **78**, 399 (1976)
Pandey, B.P., Phys. Letters **50A**, 215 (1974)
Pandey, B.P., Dayal, B., Solid State Commun. **15**, 1667 (1974)
Pandey, G.K., Shukla, D.K., Phys. Rev. B **3**, 4391 (1971)
Pandey, G.K., Shukla, D.K., Pandey, A., J. Phys. C: Solid State Phys. **6**, 3514 (1973)
Pandey, G.K., Shukla, D.K., Pandey, A., J. Phys. Chem. Solids **34**, 1833 (1973)
Pandey, G.K., Shukla, D.K., Pandey, A., J. Phys. C: Solid State Phys. **7**, 1242 (1974)
Pandey, R.N., Dayal, B., Solid State Commun. **13**, 21 (1973)
Pandey, R.N., Sharma, T.P., Dayal, B., J. Phys. Chem. Solids **38**, 329 (1977)
Pankove, J.I. (Ed.), Electroluminescence, Berlin: Springer 1977
Pantelides, S.T., Mickish, D.J., Kunz, A.B., Phys. Rev. B **10**, 2602 (1974)
Pantelides, S.T., Mickish, D.J., Kunz, A.B., Solid State Commun. **15**, 203 (1974)
Pantelides, S.T., Mickish, D.J., Kunz, A.B., Phys. Rev. B **10**, 5203 (1974)
Pantelides. S.T., Phys. Rev. B **10**, 638 (1974)
Pantelides, S.T., Harrison, W.A., Phys. Rev. B **11**, 3006 (1975)
Papazian, H.A., Chem. Phys. Letters **49**, 330 (1977)
Pappert, R.A., Phys. Rev. **119**, 525 (1960)
Parada, N.J., Phys. Rev. B **3**, 2042 (1971)
Paranichev, V.N., Opt. Spectrosc. **9**, 62 (1960)
Paranjape, V.V., Krishnamurthy, B.S., J. Phys. C: Solid State Phys. **3**, 2098 (1970)
Parfen'ev, R.V., Kharus, G.I., Tsidil'kovskii, I.M., Shalyt, S.S., Sov. Phys. Usp. **17**, 1 (1974)
Parmenter, R.H., Phys. Rev. **100**, 573 (1955)
Parmon, V.N., Khairutdinov, R.F., Zamaraev, K.I., Sov. Phys. Solid State **16**, 1672 (1975)
Parrot, R., Naud, C., Porte, C., Fournier, D., Boccara, A.C., Rivoal, J.C., Phys. Rev. B **17**, 1057 (1978)
Partl, H., Müller, W., Kohl, F., Gornik, E., J. Phys. C: Solid State Phys. **11**, 1091 (1978)
Passeggi, M.C.G., Stevens, K.W.H., J. Phys. C: Solid State Phys. **6**, 98 (1973)
Passeggi, M.C.G., Stevens, K.W.H., Physica **71**, 141 (1974)
Pasternak, A., Cohen, E., Gilat, G., Phys. Rev. B **9**, 4584 (1974)
Pasternak, A., J. Phys. C: Solid State Phys. **9**, 2987 (1976)
Pathak, K.N., Phys. Rev. **139**, **A**1569 (1965)
Pathak, P.D., Pandya, N.M., Acta Cryst. **A 31**, 155 (1975)
Patzer, K., Phys. Stat. Sol. **32**, 11 (1969)
Paul, D., Takeno, S., Phys. Rev. B **5**, 2328 (1972)
Paul, S., Sarkar, A.K., Sengupta, S., Phys. Stat. Sol. (b) **54**, 321 (1972)
Paul, S., Sengupta, S., Phys. Stat. Sol. (b) **83**, 645 (1977)
Pauli, W., Festschrift zum 60. Geburtstag von A. Sommerfeld, Leipzig: Hirzel Verlag, 1928
Pauling, L., The Nature of the Chemical Bond, London: Oxford University Press, 1940
Pêcheur, P., Kauffer, E., Gerl, M., Phys. Rev. B **14**, 4521 (1976)
Pedrini, C., Phys. Stat. Sol. (b) **87**, 273 (1978)
Pedrotti, F.L., Reynolds, D.C., Phys. Rev. **127**, 1584 (1962)
Peier, W., Physica **57**, 565 (1972)
Peierls, R., Ann. Physik (Leipzig) **3**, 1055 (1929)
Peierls, R., Quantum Theory of Solids, New York: Oxford University Press, 1955
Pekar, S.I., Z. Exp. Theor. Physik **16**, 335 (1946)
Pekar, S.I., Z. Exp. Theor. Physik **16**, 341 (1946)
Pekar, S.I., Z. Exp. Theor. Physik **16**, 933 (1946)
Pekar, S.I., Zh. Eksp. Theor. Fiz. **19**, 796 (1949)
Pekar, S.I., J. Exp. Theor. Phys. **20**, 510 (1950)
Pekar, S.I., Untersuchungen über die Elektronentheorie der Kristalle, Berlin: Akademie Verlag, 1954

Pekar, S.I., Fortschr. Physik **1**, 367 (1954)
Pekar, S.I., Sov. Phys. JETP **6**, 785 (1958)
Pekar, S.I., J. Phys. Chem. Solids 5, **11** (1958)
Pekar, S.I., Sov. Phys. JETP **11**, 1286 (1960)
Pekar, S.I., Perlin, Yu.E., Phys. Stat. Sol. **6**, 615 (1964)
Pekar, S.I., Perlin, Yu.E., Sov. Phys. Solid State **6**, 2450 (1965)
Pekar, S.I., Sov. Phys. Solid State **8**, 890 (1966)
Pekar, S.I., Nuovo Cim. **60B**, 291 (1969)
Pekar, S.I., Sov. Phys. JETP **28**, 1054 (1969)
Pekar, S.I., Sheka, V.I., Dmitrenko, G.V., Sov. Phys. JETP **36**, 771 (1973)
Pekar, S.I., Khazan, L.S., Sheka, V.I., Sov. Phys. JETP **38**, 999 (1974)
Peletminskii, S.V., Theor. Math. Phys. **6**, 88 (1971)
Penchina, C.M., Phys. Rev. **138** A 924 (1965)
Perelman, N.F., Sov. Phys. Solid State **18**, 567 (1976)
Perlin, Yu.E., Sov. Phys. JETP **5**, 71 (1957)
Perlin, Yu.E., Opt. Spectrosc. **8**, 386 (1960)
Perlin, Yu.E., Palistrant, M.E., Opt. Spectrosc. **9**, 320 (1960)
Perlin, Yu.E., Marinchuk, A.E., Kon, L.Z., Sov. Phys. Solid State **3**, 1743 (1962)
Perlin, Yu.E., Cheban, A.G., Sov. Phys. Solid State **4**, 2358 (1963)
Perlin, Yu.E., Kovarskii, V.A., Tsukerblat, B.S., Opt. Spectrosc. **20**, 367 (1966)
Perlin, Yu.E., Gifeisman, Sh.N., Phys. Stat. Sol. **28**, 45 (1968)
Perlin, Yu.E., Sov. Phys. Solid State **10**, 1531 (1969)
Perlin, Yu.E., J. Luminescence **8**, 183 (1973)
Perlin, Yu.E., Tsukerblat, B.S., Perepelitsa, E.I., Sov. Phys. JETP **35**, 1185 (1973)
Perlin, Yu.E., Kharchenko, L.S., Man-Dyk, N., Opt. Spectrosc. **41**, 158 (1976)
Perrin, N., Budd, H., Phys. Rev. B **6**, 1359 (1972)
Perrot, F., Phys. Stat. Sol. (b) **52**, 163 (1972)
Perumareddi, J.R., Z. Naturforsch. **28a**, 1247 (1973)
Petersen, R.L., Patterson, J.D., Solid State Commun. **2**, 69 (1964)
Peterson, R.L., Phys. Rev. B **2**, 4135 (1970)
Peterson, R.L., Magnusson, B., Weissglas, P., Phys. Stat. Sol. (b) **46**, 729 (1971)
Peterson, R.L., Phys. Rev. B **5**, 3994 (1972)
Petrashen, M.I., Abarenkov, I.V., Berezin, A.A., Evarestov, R.A., Phys. Stat. Sol. **40**, 9 (1970)
Petrashen, M.I., Abarenkov, I.V., Berezin, A.A., Evarestov, R.A., Phys. Stat. Sol. **40**, 433 (1970)
Petritz, R.L., Scanlon, W.W., Phys. Rev. **97**, 1620 (1955)
Peuker, K., Trifonov, E.D., Phys. Stat. Sol. **30**, 479 (1968)
Peuker, K., Phys. Stat. Sol. **31**, 363 (1969)
Peuker, K., Enderlein, R., Phys. Stat. Sol. (b) **61**, 247 (1974)
Pfeffer, P., Zawadzki, W., Phys. Stat. Sol. (b) **88**, 247 (1978)
Pfister, G., Dreybrodt, W., Assmus, W., Phys. Stat. Sol. **36**, 351 (1969)
Pfleiderer, H., Z. Physik **193**, 134 (1966)
Phadke, U.P., Sharma, S., J. Phys. Chem. Solids **36**, 1 (1975)
Philipp, H.R., Ehrenreich, H., Phys. Rev. **129**, 1550 (1963)
Phillips, J.C., Phys. Rev. **104**, 1263 (1956)
Phillips, J.C., Kleinman, L., Phys. Rev. **116**, 287 (1959)
Phillips, J.C., Phys. Rev. **166**, 832 (1968)
Phillips, J.C., Phys. Rev. **168**, 905 (1968)
Phillips, J.C., Phys. Rev. Lett. **22**, 285 (1968)
Phillips, J.C., Phys. Rev. B **1**, 1540 (1970)
Phillips, J.C., Phys. Rev. B **1**, 1545 (1970)
Phillips, J.C., Covalent Bonding in Crystals and Molecules, University of Chicago Press, Chicago 1970
Phillips, J.C., Rev. Mod. Phys. **42**, 317 (1970)
Phillips, J.C., Science **169**, 1035 (1970)
Phillips, J.C., Festkörperprobleme **17**, 109 (1977) (Ed. J. Treusch, Braunschweig: Vieweg)
Pick, H., Struktur von Störstellen in Alkalihalogenidkristallen, in: Springer Tracts in Modern Physics 38, ed. G. Höhler, Berlin: Springer, 1965, p. 1
Pick, H., in: Optical Properties of Solids, ed. F. Abeles, Amsterdam, North-Holland, 1972, p. 653
Pickin, W., Solid State Commun. **26**, 765 (1978)
Pidgeon, C.R., Brown, R.N., Phys. Rev. **146** 575 (1966)
Pikus, G.E., Sov. Phys. Solid State **6**, 261 (1964)
Pikus, G.E., Bir, G.L., Sov. Phys. JETP **33**, 108 (1971)
Pikus, G.E., Bir, G.L., Sov. Phys. JETP **35**, 174 (1972)
Pikus, G.E., Bir, G.L., Sov. Phys. Semicond. 7, 81 (1973)
Pincherle, L., Proc. Roy. Soc. London A **64**, 648 (1951)
Pincherle, L., Proc. Phys. Soc. B **68**, 319 (1955)
Pincherle, L., Rep. Prog. Phys. **23**, 355 (1960)
Pinchuk, I.I., Korolyuk, S.L., Sov. Phys. Semicond. **11**, 136 (1977)
Pinchuk, I.I., Sov. Phys. Semicond. **11**, 840 (1977)
Pinchuk, I.I., Sov. Phys. Solid State **19**, 383 (1977)
Pinczuk, A., Abstreiter, G., Trommer, R., Cardona, M., Solid State Commun. **21**, 959 (1977)

Pines, D., Bohm, D., Phys. Rev. **85**, 338 (1952)
Pines, D., Phys. Rev. **92**, 626 (1953)
Pines, D., Solid State Physics, Vol. 1; ed. F. Seitz and D. Turnbull, New York: Academic Press 1955, p. 368
Piper, W.W., Williams, F.E., Phys. Rev. **98**, 1809 (1955)
Pirc, R., Krumhansl, J.A., Phys. Rev. B **11**, 4470 (1975)
Pirenne, J., Kartheuser, E., Physica **31**, 284 (1965)
Piric, M., Marinkovic, M.M., Tosic, B.S., Physica **90A**, 597 (1978)
Piskovoi, V.N., Sov. Phys. Solid State **4**, 1025 (1962)
Pitaevskii, L.P., Sov. Phys. JETP **43**, 382 (1976)
Placzek, G., in: Handbuch der Radiologie, Vol. VI/2, ed. G. Marx, Leipzig: Akademische Verlagsgesellschaft, 1934, p. 209
Platzmann, P.M., Phys. Rev. **125**, 1961 (1962)
Plavitu, C.N., Phys. Stat. Sol. **12**, 265 (1965)
Plavitu, C.N., Phys. Stat. Sol. **32**, 535 (1969)
Plavitu, C.N., Rev. Roumaine Phys. **19**, 1029 (1974)
Plavitu, C.N., St. Cerc. Fiz. (Bukarest) **28**, 441 (1976)
Plotnikov, V.G., Konoplev, G.G., Sov. Phys. Solid State **15**, 480 (1973)
Plotnikov, V.G., Konoplev, G.G., Opt. Spectrosc. **34**, 639 (1973)
Plotnikov, V.G., Konoplev, G.G., Chem. Phys. Lett. **23**, 541 (1973)
Pogrebnyak, V.A., Sov. Phys. Solid State **14**, 934 (1972)
Pohl, R.O., Phys. Rev. Letters **8**, 481 (1962)
Pokatilov, E.P., Tarakanova, L.V., Phys. Stat. Sol. (b) **75**, K143 (1976)
Pokatilov, E.P., Klyukanov, A.A., Muntyan, A.P., Sov. Phys. Semicond. **10**, 573 (1976)
Polder, D., Weiss, K., Phys. Rev. A **17**, 1478 (1978)
Polinger, V.Z., Rosenfeld, Yu.B., Vekhter, B.G., Tsukerblat, B.S., Phys. Stat. Sol. (b) **64**, 765 (1974)
Pollack, S.A., J. Chem. Phys. **38**, 2521 (1963)
Pollak, F.H., Higginbotham, C.W., Cardona, M., J Phys. Soc. Japan **21**, Suppl., 20 (1966)
Pollak, M., Geballe, T.H., Phys. Rev. **122**, 1742 (1961)
Pollak, M., Phil. Mag. **36**, 1157 (1977)
Pollmann, J., Büttner, H., Solid State Commun. **12**, 1105 (1973)
Pollmann, J., Phys. Stat. Sol. (b) **63**, 501 (1974)
Pollmann, J., Büttner, H., Phys. Rev. B **16**, 4480 (1977)
Polnikov, V.G., Sov. Phys. Solid State **16**, 360 (1974)
Polo, G.V., Mejia, C.R., J. Phys. Soc. Japan **45**, 191 (1978)
Polovinkin, V.G., Skok, E.M., Sov. Phys. Semicond. 8, 737 (1974)
Pomeranchuk, I., J. Phys. USSR **5**, 237 (1942)
Pomortsev, R.V., Sov. Phys. Solid State **8**, 2499 (1967)
Pompe, W., Voss, K., Ann. Physik (Germany) **7**, 261 (1967)
Ponath, H.E., Schubert, M.W., Ann. Physik (Leipzig) **33**, 413 (1976)
Poole, R.T., Jenkin, J.G., Liesegang, J., Leckey, R.C.G., Phys. Rev. B **11**, 5179 (1975)
Poole R.T., Liesegang, J., Leckey, R.C.G., Jenkin, J.G., Phys. Rev. B **11**, 5190 (1975)
Pooler, D.R., O'Brien, M.C.M., J. Phys. C: Solid State Phys. **10**, 3769 (1977)
Pooley, D., Proc. Phys. Soc. **87**, 245 (1966)
Pooley, D., Proc. Phys. Soc. **87**, 257 (1966)
Poplavnoi, A.S., Phys. Stat. Sol. **33**, 541 (1969)
Popovkin, I.V., Sov. Phys. Solid State **18**, 2016 (1976)
Porsch, M., Phys. Stat. Sol. **8**, 207 (1965)
Porsch, M., Phys. Stat. Sol. **41**, 151 (1970)
Posledovich, M., Winter, F.X., Borstel, G., Claus, R., Phys. Stat. Sol. (b) **55**, 711 (1973)
Poulet, H., Mathieu, J.P., Vibration Spectra and Symmetry of Crystals, New York: Gordon & Breach, 1976
Powell, B.M., Jandl, S., Brebner, J.L., Levy, F., J. Phys. C: Solid State Phys. **10**, 3039 (1977)
Praddaude, H.C., Phys. Rev. **140**, A 1292 (1965)
Prasad, P.N., Chem. Phys. Letters **20**, 507 (1973)
Prather, J.L., Atomic Energy Levels in Crystals, National Bureau of Standards Monograph 19, February 1961
Prevot, B., Carabatos, C., J. Physique **32**, 543 (1971)
Price, P.J., Phil. Mag. **46**, 1252 (1955)
Price, P.J., Phys. Rev. **102**, 1245 (1956)
Price, P.J., Phys. Rev. **104**, 1223 (1956)
Price, P.J., Solid State Electronics **21**, 9 (1978)
Prigogine, I., Ono, S., Physica **25**, 171 (1959)
Prigogine, I., Balescu, R., Physica **25**, 281 (1959)
Prigogine, I., Balescu, R., Physica **26**, 145 (1960)
Prigogine, I., Resibois, P., Physica **27**, 629 (1961)
Prodan, V.D., Rozneritsa, Y.A., Sov. Phys. Semicond. **10**, 1316 (1976)
Provotorov, B.N., Sov. Phys. Solid State **5**, 411 (1963)
Pryce, M.H.L., Proc. Phys. Soc. A **63**, 25 (1950)
Pryce, M.H.L., Phonons in Perfect Lattices and in Lattices with Point Imperfections, Edinburgh: Oliver and Boyd, 1966
Putley, E.H., Proc. Phys. Soc. **73**, 280 (1959)

Putley, E.H., The Hall Effect and Related Phenomena, London: Butterworths, 1960
Pytte, E., Phys. Rev. B **8**, 3954 (1973)

Quattropani, A., Forney, J.J., Bassani, F., Phys. Stat. Sol. (b) **70**, 497 (1975)
de Queiroz, S.L.A., Koiller, B., Maffeo, B., Brandi, H.S., Phys. Stat. Sol. (b) **87**, 351 (1978)
Queisser, H.J., Appl. Phys. **10**, 275 (1976)

Rabii, S., Phys. Rev. **167**, 801 (1968)
Rabii, S., Phys. Rev. **182**, 821 (1969)
Rabinovich, R.I., Sov. Phys. Solid State **12**, 440 (1970)
Rabinovitch, A., Zak, J., Phys. Rev. B **4**, 2358 (1971)
Rabinovitch, A., Phys. Letters **48A**, 149 (1974)
Radhakrishna, S., Nigam, N.N., Sivasankar, V.S., Phys. Rev. B **15**, 1187 (1977)
Radosevich, L.G., Walker, C.T., Phys. Rev. **171**, 1004 (1968)
Raghavacharyulu, I.V.V., J. Phys. C: Solid State Phys. **6**, L455 (1973)
Rahman, S.M.M., Rashid, A.M.H., Chowdhury, S.M.M.R., Nuovo Cim. **25B**, 803 (1975)
Rahman, S.M.M., Rashid, A.M.H., Chowdhury, S.M.M.R., Phys. Rev. B **14**, 2613 (1976)
Rajagopal, A.K., Grest, G.S., Ruvalds, J., Phys. Rev. B **14**, 67 (1976)
Ralph, H.I., J. Phys. C: Solid State Phys. **1**, 378 (1968)
Ralph, H.I., Philips Res. Rep. **32**, 160 (1977)
Ram, P.N., Agrawal, B.K., J. Phys. Chem. Solids **33**, 957 (1972)
Ram, P.N., Agrawal, B.K., Phys. Rev. B **5**, 2335 (1972)
Ram, P.N., Agrawal, B.K., Solid State Commun. **13**, 1671 (1973)
Ram, P.N., Solid State Commun. **21**, 313 (1977)
Ramani, G., Rao, K.J., Chem. Phys. Letters **31**, 498 (1975)
Ramani, G., Rao, K.J., J. Solid State Chem. **16**, 63 (1976)
Ramaseshan, S., Narayan, R., Curr. Sci. (India) **45**, 357 (1976)
Ramdas, S., Shukla, A.K., Rao, C.N.R., Chem Phys. Letters **16**, 14 (1972)
Ramdas, S., Rao, C.N.R., Cryst. Latt. Def. **6**, 199 (1976)
Rampacher, H., Z. Naturforsch. **17a**, 1057 (1962)
Rampacher, H., Z. Naturforsch. **18a**, 777 (1963)
Rampacher, H., Z. Naturforsch. **20a**, 350 (1965)
Rampacher, H., Z. Naturforsch. **23a**, 401 (1968)
Ranfagni, A., Phys. Rev. Letters **28**, 743 (1972)
Ranfagni, A., Viliani, G., J. Phys. Chem. Solids **35**, 25 (1974)
Ranfagni, A., Pazzi, G.P., Fabeni, P., Viliani, G., Fontana, M.P., Solid State Commun. **14**, 1169 (1974)
Ranfagni, A., Viliani, G., Cetica, M., Molesini, G., Phys. Rev. B **16**, 890 (1977)
Ranfagni, A., Viliani, G., Phys. Stat. Sol. (b) **84**, 393 (1977)
Ranninger, J., Ann. Phys. (N.Y.) **45**, 452 (1967)
Ranninger, J., Ann. Phys. (N.Y.) **49**, 297 (1968)
Rao, K.J., Rao, C.N.R., Solid State Commun. **6**, 45 (1968)
Rao, K.J., Rao, C.N.R., Phys. Stat. Sol. **28**, 157 (1968)
Rashba, E.I., Sov. Phys. JETP **36**, 1213 (1959)
Rashba, E.I., Gurgenishvili, G.E., Sov. Phys. Solid State **4**, 759 (1962)
Rashba, E.I., Sheka, V.I., Sov. Phys. Solid State **6**, 114 (1964)
Rashba, E.I., JETP Letters **15**, 411 (1972)
Rashba, E.I., Edel'shtein, V.M., Sov. Phys. JETP **34**, 1379 (1972)
Rashba, E.I., Zimin, A.B., Sov. Phys. JETP **39**, 726 (1974)
Rashba, E.I., Sov. Phys. JETP **44**, 166 (1976)
Rastogi, A., Hawranek, J.P., Lowndes, R.P., Phys. Rev. B **9**, 1938 (1974)
Rath, K., Borstel, G., Lamprecht, G., Merten, L., Z. Naturforsch. **30a**, 1385 (1975)
Ratner, A.M., Zilberman, G.E., Sov. Phys. Solid State **1**, 1551 (1959)
Ratner, A.M., Sov. Phys. Solid State **1**, 1560 (1959)
Ratner, A.M., Zil'berman, G.E., Sov. Phys. Solid State **3**, 499 (1961)
Raunio, G., Rolandson, S., Phys. Rev. B **2**, 2098 (1970)
Raveche, H.J., J. Phys. Chem. Solids **26**, 2088 (1965)
Ravich, Yu.I., Efimova, B.A., Tamarchenko, V.I., Phys. Stat. Sol. (b) **43**, 453 (1971)
Ravich, Yu.I., Tamarchenko, V.I., Sov. Phys. Solid State **14**, 1945 (1973)
Ray, T., Proc. Roy. Soc. A **277**, 76 (1964)
Ray, T., Phys. Rev. B **5**, 1758 (1972)
Ray, T., Regnard, J.R., Phys. Rev. B **9**, 2110 (1974)
Ray, T., Ray, D.K., Sangster, M.J.L., Solid State Commun. **17**, 93 (1975)
Razbirin, B.S., Ural'tsev, I.N., Bogdanov, A.A., Sov. Phys. Solid State **15**, 604 (1973)
Rebane, K.K., Opt. Spectrosc. **9**, 295 (1960)
Rebane, K.K., Khizhnyakov, V.V., Opt. Spectrosc. **14**, 193 (1963)
Rebane, K.S., Opt. Spectrosc. **12**, 137 (1962)
Rebane, K.S., Opt. Spectrosc. **12**, 217 (1962)
Rebane, L.A., Zavt, G.S., Haller, K.E., Phys. Stat. Sol. (b) **81**, 57 (1977)
Redfield, A., Phys. Rev. **94**, 537 (1954)

Ree, F.H., Holt, A.C., Phys. Rev. B **8**, 826 (1973)
Rees, G.J., J. Phys. C: Solid State Phys. **5**, 549 (1972)
Rees, H.D., J. Phys. Chem. Solids **29**, 143 (1968)
Reik, H.G., Solid State Commun. **8**, 1737 (1970)
Reineker, P., Kühne, R., Z. Physik B **22**, 193 (1975)
Reinisch, R., Biraud-Laval, S., Paraire, N., J. Physique **37**, 227 (1976)
Reiss, H.R., Phys. Rev. A **1**, 803 (1970)
Reiss, H.R., Phys. Rev. D **4**, 3533 (1971)
Reiss, H.R., Phys. Rev. A **6**, 817 (1972)
Reissland, J.A., The Physics of Phonons, New York: J. Wiley & Sons, 1973
Reitz, J.R., J. Chem. Phys. **22**, 595 (1954)
Reitz, J.R., Seitz, R.N., Genberg, R.W., J. Phys. Chem. Solids **19**, 73 (1961)
Renn, W., Zustandsberechnungen bei F- und F'-Zentren in Ionenkristallen, Thesis, University of Tübingen, 1973
Renn, W., Phys. Cond. Matter **17**, 233 (1974)
Renn, W., Z. Physik B **22**, 319 (1975)
Renn, W., Excitons Bound at Isoelectronic Impurities, Report for the Deutsche Forschungsgemeinschaft (1976), unpublished
Renn, W., Personal communication (1977)
Renn, W., Z. Naturforsch. **33a**, 1261 (1978)
Renner, R., Z. Physik **92**, 172 (1934)
Rennie, R., Adv. Phys. **26**, 285 (1977)
Resibois, P., Physica **27**, 541 (1961)
Resibois, P., Physica **29**, 721 (1963)
Reuter, H., Hübner, K., Phys. Rev. B **4**, 2575 (1971)
Reynolds, R.W., Boatner, L.A., Phys. Rev. B **12**, 4735 (1975)
Rhodes, R.G., Imperfections and Active Centers in Semiconductors, Oxford: Pergamon Press, 1964
Rhzanov, A.V., Sov. Phys. Solid State **3**, 2680 (1962)
Ribeiro, S.C., Nascimento, O.R., Brandi, H.S., Phys. Rev. B **11**, 3163 (1975)
Richter, G., Phys. Stat. Sol. **5**, 463 (1964)
Richter, W., Resonant Raman Scattering in Semiconductors, in: Solid State Physics, Springer Tracts in Modern Physics 78, Berlin: Springer, 1976, p. 121
Richter, W., Zeyher, R., Festkörperprobleme **16**, 15 (1976) (Ed. J. Treusch, Braunschweig: Vieweg)
Rickayzen, G., Proc. Roy. Soc. London A **241**, 480 (1957)
Ridley, B.K., Leach, M.F., J. Phys. C: Solid State Phys. **10**, 1267 (1977)
Ridley, B.K., J. Appl. Phys. **48**, 754 (1977)
Ridley, B.K., J. Phys. C: Solid State Phys. **10**, 1589 (1977)
Ridley, B.K., J. Phys. C: Solid State Phys. **11**, 2323 (1978)
Rieckers, A., Stumpf, H., Thermodynamik II, Braunschweig: Vieweg, 1977
Riehl, N., Kallmann, H., (Eds.), International Symposium on Luminescence – The Physics and Chemistry of Scintillators, München 1965, München: Thiemig, 1966
Rimbey, P.R., Mahan, G.D., Phys. Rev. **B10**, 3419 (1974)
Ripka, G., Advances of Nuclear Physics, Vol. I, ed. M. Baranger and E. Vogt, New York: Plenum Press 1968, p. 183
Riseberg, L.A., Moos, H.W., Phys. Rev. **174**, 429 (1968)
Ritter, J.T., Markham, J.J., Phys. Rev. **185**, 1201 (1969)
Robbins, D.J., Thomson, A.J., Phil. Mag. **36**, 999 (1977)
Robbins, D.J., Landsberg, P.T., Schöll, E., Phys. Stat. Sol. (a) **65**, 353 (1981)
Robbins, D., Page, J.B., Phys. Rev. Letters **38**, 365 (1977)
Robert, J.L., Pistoulet, B., Raymond, A., Aulombard, R.L., Bernard, C., Bousquet, C., Rev. Physique Appl. **13**, 246 (1978)
Roberts, P.J., Proc. Phys. Soc. **89**, 269 (1966)
Roberts, S., Williams, F.E., J. Opt. Soc. Am. **40**, 516 (1950)
Robertson, B., Phys. Rev. **144**, 151 (1966)
Robertson, B., Phys. Rev. **160**, 175 (1967)
Robertson, N., Friedman, L., Phil. Mag. **33**, 753 (1976)
Robertson, N., Friedman, L., Phil. Mag. **36**, 1013 (1977)
Robinson, G.W., Frosch, R.P., J. Chem. Phys. **38**, 1187 (1963)
Robinson, G.W., Molecular Electronic Radiationless Transitions, in: Excited States, Vol. 1, ed. E.C. Lim, New York: Academic Press 1974, p. 1
Robinson, J.E., Rodriguez, S., Phys. Rev. **135**, A779 (1964)
Rode, D.L., Phys. Rev. B **2**, 1012 (1970)
Rode, D.L., Phys. Rev. B **2**, 4036 (1970)
Rode, D.L., Phys. Rev. B **3**, 3287 (1971)
Rode, D.L., Knight, S., Phys. Rev. B **3**, 2534 (1971)
Rode, D.L., Phys. Stat. Sol. (b) **55**, 687 (1973)
Rodriguez, S., Schultz, T.D., Phys. Rev. **178**, 1252 (1969)
Rodriguez, S., Resonant Electron-Phonon Interactions in Solids, in: Polarons in Ionic Crystals and Polar Semiconductors, Proceedings Antwerp 1971, ed. J.T. Devreese, Amsterdam: North-Holland, 1972, p. 289
Röpke, G., Zehe, A., Phys. Stat. Sol. (a) **23**, K137 (1974)
Röseler, J., Henneberger, K., Fischbeck, H.J., Phys. Stat. Sol. (b) **55**, 595 (1973)

Rössler, U., Lietz, M., Phys. Stat. Sol. **17**, 597 (1966)
Rössler, U., Phys. Stat. Sol. **34**, 207 (1969)
Rössler, U., Phys. Rev. **184**, 733 (1969)
Rogalla, W., Schmalzried, H., Ber. Bunsenges. **72**, 12 (1968)
Rohner, P.G., Phys. Rev. B **3**, 433 (1971)
Roitsin, A.B., Sov. Phys. Uspekhi **14**, 766 (1972)
Roitsin, A.B., Sov. Phys. JETP **37**, 504 (1973)
Roitsin, A.B., Sov. Phys. Semicond. **8**, 1 (1974)
Rojas, I., Übergangswahrscheinlichkeiten für strahlende und strahlungslose Übergänge in F- und F'-Zentren, Thesis, University of Tübingen 1978
Roman, P., Advanced Quantum Theory, Reading: Addison-Wesley, 1965
Romestain, R., d'Aubigné, Y.M., Phys. Rev. B **4**, 4611 (1971)
Ron, A., Phys. Rev. **134**, A70 (1964)
Rona, M., Whitfield, G., Phys. Rev. B **7**, 2727 (1973)
von Roos, O., Solid-State Electron. **21**, 633 (1978)
van Roosbroeck, W., Phys. Rev. **101**, 1713 (1956)
van Roosbroeck, W., Phys. Rev. **119**, 636 (1960)
van Roosbroeck, W., Phys. Rev. **123**, 474 (1961)
Rose, A., Phys. Rev. **97**, 322 (1955)
Rosenstock, H.B., Phys. Rev. **121**, 416 (1961)
Rosenstock, H.B., Phys. Rev. **131**, 1111 (1963)
Rosenstock, H.B., Phys. Rev. B **9**, 1963 (1974)
Rosenthal, W., Solid State Commun. **13**, 1215 (1973)
Rosenthal, W., Z. Naturforsch. **28a**, 1233 (1973)
Rosier, L.L., Sah, C.T., Solid State Electron. **14**, 41 (1971)
Roth, L.M., Lax, B., Zwerdling, S., Phys. Rev. **114**, 90 (1959)
Roth, L.M., J. Phys. Chem. Solids **23**, 433 (1962)
Roth, L.M., Phys. Rev. **133**, A 542 (1964)
Roussel, K.M., O'Connell, R.F., J. Phys. Chem. Solids **35**, 1429 (1974)
Rovere, M., Tosatti, E., Nuovo Cim. **39B**, 538 (1977)
Roy, C.L., Czech, J. Phys. B **27**, 769 (1977)
Roy, D., Basu, A.N., Sengupta, S., Phys. Stat. Sol. **35**, 499 (1969)
Roy, D., Ghosh, A.K., Phys. Rev. B **3**, 3510 (1971)
Roy, D., Sen, S.K., Manna, A., Phys. Stat. Sol. (b) **61**, 723 (1974)
van Royen, J., de Sitter, J., Lemmens, L.F., Devreese, J.T., Physica **89B**, 101 (1977)
Rozenfel'd, Yu.B., Vekhter, B.G., Tsukerblat, B.S., Sov. Phys. JETP **28**, 1195 (1969)
Rozenfel'd, Yu.B., Polinger, V.Z. Sov. Phys. JETP **43**, 310 (1976)
Rozneritsa, Y.A., Cheban, A.G., Sov. Phys. Solid State **8**, 1097 (1966)
Rubinshtein, A.I., Fain, V.M., Sov. Phys. Solid State **15**, 332 (1973)
Rueff, M., Wagner, M., J. Chem. Phys. **67**, 169 (1977)
Rueff, M., Sigmund, E., Phys. Stat. Sol. (b) **80**, 215 (1977)
Rueff, M., Sigmund, E., Wagner, M., Phys. Stat. Sol. (b) **81**, 511 (1977)
Ruffa, A.R., Phys. Rev. **130**, 1412 (1963)
Rumyantsev, E.L., Salikhov, K.M., Opt. Spectrosc. **35**, 332 (1973)
Rupprecht, H., Weber, R., Weiss, H., Z. Naturforsch. **15a**, 783 (1960)
Rustagi, K.C., Solid State Commun. **12**, 607 (1973)
Rustagi, K.C., Weber, W., Solid State Commun. **18**, 673 (1976)
Ruvalds, J., McClure, J.W., J. Phys. Chem. Solids **28**, 509 (1967)
Ruvalds, J., Zawadowski, A., Phys. Rev. B **2**, 1172 (1970)
Ryvkin, S.M., J. Phys. Chem. Solids **22**, 5 (1961)
Ryzhii, V.I., Sov. Phys. Semicond. **8**, 204 (1974)

Safaryan, F.P., Sov. Phys. Solid State **19**, 1140 (1977)
Saglam, M., Friedman, L., J. Phys. C: Solid State Phys. **8**, L245 (1975)
Sahoo, D., Venkataraman, G., Pramana (Indien) **5**, 175 (1975)
Sahoo, D., Venkataraman, G., Pramana (Indien) **5**, 185 (1975)
Sahyun, M.R.V., J. Chem. Educ. **54**, 143 (1977)
Sak, J., Phys. Stat. Sol. **27**, 521 (1968)
Sak, J., Phys. Rev. Letters **25**, 1654 (1970)
Sak, J., Phys. Letters **38A**, 273 (1972)
Sakamoto, A., Ogawa, T., J. Phys. Chem. Solids **36**, 583 (1975)
Sakoda, S., J. Phys. Soc. Japan **34**, 1254 (1973)
Sakoda, S., Toyozawa, Y., J. Phys. Soc. Japan **35**, 172 (1973)
Sakoda, S., J. Phys. Soc. Japan **44**, 211 (1978)
Sakun, V.P., Sov. Phys. Solid State **14**, 1906 (1973)
Sakun, V.P., Sov. Phys. Solid State **15**, 1522 (1974)
Sakun, V.P., Sov. Phys. Solid State **18**, 1470 (1976)
Samartsev, V.V., Kaveeva, Z.M., Opt. Spectrosc. **35**, 560 (1973)
Sammel, B., Phys. Stat. Sol. **34**, 409 (1969)
Samoilovich, A.G., Buda, I.S., Sov. Phys. Semicond. **9**, 977 (1976)

Sander, L.M., Shore, H.B., Phys. Rev. B **3**, 1472 (1971)
Sandiford, D.J., Phys. Rev. **105**, 524 (1957)
Sandrock, R., Festkörperprobleme **10**, 283 (1970) (Ed. O. Madelung, Braunschweig: Vieweg)
Sangster, M.J.L., Peckham, G., Saunderson, D.H., J. Phys. C: Solid State Phys. **3**, 1026 (1970)
Sangster, M.J.L., McCombie, C.W., J. Phys. C: Solid State Phys. **3**, 1498 (1970)
Sangster, M.J.L. Phys. Rev. B **6**, 254 (1972)
Sangster, M.J.L., J. Phys. Chem. Solids **34**, 355 (1973)
Sangster, M.J.L., Solid State Commun. **18**, 67 (1976)
Sangster, M.J.L., Schröder, U., Atwood, R.M., J. Phys. C: Solid State Phys. **11**, 1523 (1978)
Sangster, M.J.L., Atwood, R.M., J. Phys. C: Solid State Phys. **11**, 1541 (1978)
Sarkar, A.K., Sengupta, S., Phys. Stat. Sol. **36**, 359 (1969)
Sarkar, A.K., Sengupta, S., Phys. Stat. Sol. (b) **58**, 775 (1973)
Sarkar, A.K., Sengupta, S., Indian J. Theor. Phys. **24**, 107 (1976)
Sarkar, C.K., Basu, P.K., Indian J. Pure Appl. Phys. **15**, 595 (1977)
Sarkar, C.K., Basu, P.K., Chattopadhyay, D., J. Phys. C: Solid State Phys. **11**, 937 (1978)
Sarkar, C.K., Chattopadhyay, D., Phys. Stat. Sol. (b) **88**, K109 (1978)
Sarkar, R.L., Chatterjee, S., J. Phys. C: Solid State Phys. **10**, 57 (1977)
Sarkar, S., Chakrabarti, S.K., Z. Physik B **22**, 309 (1975)
Sarkar, S.K., Sengupta, S., Indian J. Phys. **49**, 836 (1975)
Sarkar, S.K., Sengupta, S., Phys. Stat. Sol. (b) **87**, 517 (1978)
Sati, R., Wang, S., Inoue, M., Can. J. Phys. **50**, 1370 (1972)
Sati, R., Phys. Stat. Sol. (b) **73**, 353 (1976)
Sato, H., J. Phys. Soc. Japan **27**, 1501 (1969)
Sauer, U., Scherz, U., Maier, H., Phys. Stat. Sol. (b) **62**, K71 (1974)
Saunders, I.J., J. Phys. C: Solid State Phys. **2**, 2181 (1969)
Savvinykh, S.K., Sov. Phys. Solid State **17**, 2357 (1976)
Sawaki, N., Nishinaga, T., J. Phys. C: Solid State Phys. **10**, 5003 (1977)
Saxena, K.N., Saxena, N.N., Anikhindi, R.G., Chem. Phys. Letters **31**, 563 (1975)
Scop, P.M., Phys. Rev. **139**, A 934 (1965)
Scott, J.F., Damen, T.C., Ruvalds, J., Zawadowski, A., Phys. Rev. B **3**, 1295 (1971)
Seeger, K., Z. Physik **156**, 582 (1959)
Seeger, K., Z. Physik **244**, 439 (1971)
Seeger, K., Semiconductor Physics, Wien: Springer, 1973
Segall, B., Mahan, G.D., Phys. Rev. **171**, 935 (1968)
Seidel, H., Z. Physik **165**, 218 (1961)
Seidel, H., Wolf, H.C., in: Physics of Color Centers, ed. W.B. Fowler, New York: Academic Press, 1968, Chapter 8
Seitz, F., The Modern Theory of Solids, New York: McGraw-Hill, 1940
Sen, K.D., Narasimhan, P.T., Phys. Rev. B **15**, 95 (1977)
Sergeev, M.V., Pokrovsky, L.A., Physica **70**, 83 (1973)
Seth, U., Chaney, R., Phys. Rev. B **12**, 5923 (1975)
Setser, G.G., Barksdale, A.O., Estle, T.L., Phys. Rev. B **12**, 4720 (1975)
Setser, G.G., Estle, T.L., Phys. Rev. B **17**, 999 (1978)
Sevenich, R.A., Kliewer, K.L., J. Chem. Phys. :x **48**, 3045 (1968)
Sewell, G.L., Phil. Mag. **3**, 1361 (1958)
Sewell, G.L., Phys. Rev. **129**, 597 (1963)
Shah, J., Solid-State Electron. **21**, 43 (1978)
Shah, R.M., Schetzina, J.F., Phys. Rev. B **5**, 4014 (1972)
Sham, L.J., Phys. Rev. **156**, 494 (1967)
Sham, L.J., Maradudin, A.A., Solid State Commun. **5**, 337 (1967)
Shanker, J., Agarwal, L.D., Indian J. Pure Appl. Phys. **11**, 448 (1973)
Shanker, J., Agarwal, S.K., J. Phys. Chem. Solids **37**, 443 (1976)
Shanker, J., Verma, M.P., J. Phys. Chem. Solids **37**, 883 (1976)
Shanker, J., Sharma, J.C., Sharma, O.P., Indian J. Pure Appl. Phys. **15**, 809 (1977)
Shanker, J., Agarwal, S.C., J. Phys. Chem. Solids **38**, 91 (1977)
Shanker, J., Gupta, A.P., Sharma, O.P., Phil. Mag. **37**, 329 (1978)
Shannon, R.D., Acta Cryst. A **32**, 751 (1976)
Shapiro, S.M., Axe, J.D., Phys. Rev. B **6**, 2420 (1972)
Sharma, J.C., Shanker, J., Goyal, S.C., J. Phys. Chem. Solids **38**, 327 (1977)
Sharma, M.N., Kant, A., Misra, K.D., Phys. Stat. Sol. (b) **79**, 359 (1977)
Sharma, M.N., Kant, A., Phys. Stat. Sol. (b) **84**, 427 (1977)
Sharma, P.K., Bahadur, R., Phys. Rev. B **12**, 1522 (1975)
Sharma, R.R., Rodriguez, S., Phys. Rev. **153**, 823 (1967)
Sharma, R.R., Das, T.P., Orbach, R., Phys. Rev. **149**, 257 (1966)
Sharma, R.R., Das, T.P., Orbach, R., Phys. Rev. **155**, 338 (1967)
Sharma, R.R., Phys. Rev. **170**, 770 (1968)
Sharma, R.R., Phys. Rev. B **3**, 76 (1971)

Sharma, S.K., Dubey, P.K., Phys. Stat. Sol. (a) **4**, 357 (1971)
Sharma, S.K., Agarwal, D.B., Phys. Stat. Sol. (b) **69**, 169 (1975)
Sharma, T.P., Solid State Commun. **15**, 1171 (1974)
Sheboul, M.I., Ekardt, W., Phys. Stat. Sol. (b) **73**, 165 (1976)
Sheinkman, M.K., Shik, A.Ya., Sov. Phys. Semicond. **10**, 128 (1976)
Sheka, V.I., Sheka, D.I., Sov. Phys. Solid State **11**, 891 (1969)
Sheka, V.I., Khazan, L.S., Sov. Phys. Solid State **16**, 1093 (1974)
Sheka, V.I., Khazan, L.S., Mozdor, E.V., Phys. Stat. Sol. (b) **72**, 833 (1975)
Sheka, V.I., Khazan, L.S., Sov. Phys. Solid State **16**, 1389 (1975)
Sheka, V.I., Khazan, L.S., Mozdor, E.V., Sov. Phys. Solid State **18**, 1890 (1976)
Shekhtman, V.L., Trifonov, E.D., Phys. Stat. Sol. **41**, 855 (1970)
Shekhtman, V.L., Opt. Spectrosc. **41**, 591 (1976)
Shekhtman, V.L., Opt. Spectrosc. **43**, 270 (1977)
Shen, Y.R., Bloembergen, N., Phys. Rev. **137**, A 1787 (1965)
Sheng, P., Dow, J.D., Phys. Stat. Sol. (b) **44**, K131 (1971)
Sher, A., Thornber, K.K., Appl. Phys. Letters **11**, 3 (1967)
Shibata, F., Takahashi, Y., Hashitsume, N., J. Statist. Phys. **17**, 171 (1977)
Shimizu, T., Ishii, N., Phys. Letters **62A**, 122 (1977)
Shindo, K., Morita, A., Kamimura, H., J. Phys. Soc. Japan **20**, 2054 (1965)
Shindo, K., J. Phys. Soc. Japan **29**, 287 (1970)
Shindo, K., Nara, H., J. Phys. Soc. Japan **40**, 1640 (1976)
Shklovskii, B.I., Sov. Phys. JETP **34**, 1084 (1972)
Shklovskii, B.I., Sov. Phys. Semicond. **6**, 1964 (1973)
Shklovskii, B.I., Sov. Phys. Semicond. **6**, 1053 (1973)
Shmelev, G.M., Zil'berman, R.D., Sov. Phys. Solid State **13**, 2866 (1972)
Shmelev, G.M., Sibirskii, V.A., Phys. Stat. Sol. (b) **65**, 333 (1974)
Shmelev, G.M., Sov. Phys. Solid State **16**, 1585 (1975)
Shockley, W., Phys. Rev. **51**, 130 (1937)
Shockley, W., Read, W.T., Phys. Rev. **87**, 835 (1952)
Shogenji, K., Okazaki, T., J. Phys. Chem. Solids **36**, 445 (1975)
Shore, H.B., Sander, L.M., Phys. Rev. B **6**, 1551 (1972)
Shore, H.B., Sander, L.M., Phys. Rev. B **12**, 1546 (1975)
Shrivastava, K.N., Phys. Stat. Sol. **29**, 737 (1968)
Shrivastava, K.N., J. Phys. C: Solid State Phys. **2**, 777 (1969)
Shrivastava, K.N., Phys. Rev. **187**, 446 (1969)
Shrivastava, K.N., Phys. Letters **31A**, 454 (1970)
Shrivastava, K.N., Chem. Phys. Letters **7**, 477 (1970)
Shuey, R.T., Zeller, H.R., Helv. Phys. Acta **40**, 873 (1967)
Shugard, M., Tully, J.C., Nitzan, A., J. Chem. Phys. **69**, 336 (1978)
Shukla, A.K., Ramdas, S., Rao, C.N.R., J. Phys. Chem. Solids **34**, 761 (1972)
Shukla, A.K., Chattopadhyay, S., Rao, C.N.R., Crystal Lattice Defects **6**, 191 (1976)
Shukla, R.C., Wilk, L., Phys. Rev. B **10**, 3660 (1974)
Shul'man, L.A., Sov. Phys. JETP **36**, 1217 (1959)
Shultz, M.J., Silbey, R., J. Phys. C: Solid State Phys. **7**, L325 (1974)
Shur, M.S., Sov. Phys. Semicond. **10**, 88 (1976)
Sieber, A., Stark-Effekt am F-Zentrum, Thesis, University of Tübingen, 1973
Sievers, A.J., Maradudin, A.A., Jaswal, S.S., Phys. Rev. **138**, A 272 (1965)
Sigmund, E., Wagner, M., Phys. Stat. Sol. (b) **57**, 635 (1973)
Sigmund, E., Wagner, M., Phys. Stat. Sol. (b) **76**, 325 (1976)
Sigmund, E., Phys. Rev. B **14**, 4702 (1976)
Sigmund, E., Z. Naturforsch. **31a**, 904 (1976)
Sigmund, E., J. Physique **37**, suppl. au no 12, C7-117 (1976)
Sigmund, E., Wagner, M., Birkhold, M., Solid State Commun. **22**, 719 (1977)
Sigmund, E., Z. Naturforsch. **32a**, 113 (1977)
Sigmund, E., Z. Physik B **26**, 239 (1977)
Siklos, T., Acta Phys. Hung. **30**, 193 (1971)
Siklos, T., Aksienov, V.L., Acta Phys. Hung. **31**, 335 (1972)
Siklòs, T., Aksienov, V.L., Acta Phys. Hung. **31**, 345 (1972)
Siklos, T., Aksienov, V.L., Phys. Stat. Sol. (b) **50**, 171 (1972)
Sild, O.I., Opt. Spectrosc. **15**, 258 (1963)
Silver, R.N., J. Phys. C: Solid State Phys. **9**, 2505 (1976)
Simanek, E., Orbach, R., Phys. Rev. **145**, 191 (1966)
Simmons, J.G., Taylor, G.W., Phys. Rev. B **4**, 502 (1971)
Simmons, J.G., Taylor, G.W., Phys. Rev. B **5**, 1619 (1972)
Simmons, J.G., Taylor, G.W., J. Phys. C: Solid State Phys. **8**, 3353 (1975)

Simons, S., J. Phys. C: Solid State Phys. **8**, 1147 (1975)
Simpson, J.H., Proc. Roy. Soc. London A **197**, 269 (1949)
Singh, A.K., Sharma, M.N., J. Phys. Soc. Japan **44**, 191 (1978)
Singh, D.P., Verma, G.S., Phys. Rev. B **4**, 4647 (1971)
Singh, D.P., Verma, G.S., Phys. Stat. Sol. (b) **59**, 291 (1973)
Singh, D.P., Indian J. Pure Appl. Phys. **14**, 158 (1976)
Singh, M., Phys. Stat. Sol. (b) **89**, 299 (1978)
Singh, R.K., Verma, M.P., Phys. Stat. Sol. **36**, 335 (1969)
Singh, R.K., Verma, M.P., Phys. Rev. B **2**, 4288 (1970)
Singh, R.K., Verma, M.P., Phys. Stat. Sol. **38**, 851 (1970)
Singh, R.K., Upadhyaya, K.S., Phys. Rev. B **6**, 1589 (1972)
Singh, R.K., J. Phys. C: Solid State Phys. **7**, 3473 (1974)
Singh, R.K., Agarwal, M.K., Solid State Commun. **17**, 991 (1975)
Singh, R.K., Gupta, H.N., Solid State Commun. **16**, 197 (1975)
Singh, R.K., Gupta, H.N., Proc. Roy. Soc. (London) A **349**, 289 (1976)
Singh, R.K., Chandra, K., Phys. Rev. B **14**, 2625 (1976)
Singh, R.K., Singh, K.P., Phys. Stat. Sol. (b) **78**, 677 (1976)
Singh, R.K., Nirwal, V.V.S., Lett. Nuovo Cim. **22**, 559 (1978)
Singh, R.S., Mitra, S.S., Phys. Rev. B **2**, 1070 (1970)
Singh, R.S., Mitra, S.S., Phys. Rev. B **5**, 733 (1972)
Singh, T.J., Verma, G.S., Phys. Rev. B **15**, 1219 (1977)
Sinha, A.K., Thakur, L., Indian J. Pure Appl. Phys. **15**, 115 (1977)
Sinha, K., Helv. Phys. Acta **45**, 619 (1972)
Sinha, S.K., Phys. Rev. **177**, 1256 (1969)
Sinha, S.K., Gupta, R.P., Price, D.L., Phys. Rev. B **9**, 2564 (1974)
Sinha, S.K., Gupta, R.P., Price, D.L., Phys. Rev. B **9**, 2573 (1974)
Sinyavskii, E.P., Sov. Phys. Solid State **13**, 1750 (1972)
Sinyavskii, E.P., Sov. Phys. Solid State **18**, 1556 (1976)
Sinyukov, M., Trommer, R., Cardona, M., Phys. Stat. Sol. (b) **86**, 563 (1978)
Sirko, R., Subbaswamy, K.R., Mills, D.L., Phys. Rev. B **18**, 851 (1978)
Sjölander, A., Arkiv Fysik **14**, 315 (1958)
Skal, A.S., Shklovskii, B.I., Efros, A.L., Sov. Phys. Solid State **17**, 316 (1975)
Skettrup, T., Suffczynski, M., Gorzkowski, W., Phys. Rev. B **4**, 512 (1971)
Skinner, H.W.B., Phil. Trans. A. **239**, 95 (1940)
Skrinjar, M.J., Setrajcic, J.P., Kapor, D.V., Phys. Stat. Sol. (b) **84**, K25 (1977)
Slater, J.C., Rev. Mod. Phys. **6**, 209 (1934)
Slater, J.C., Krutter, H.M., Phys. Rev. **47**, 559 (1935)
Slater, J.C., Phys. Rev. **76**, 1592 (1949)
Slater, J.C., Koster, G.F., Phys. Rev. **95**, 1167 (1954)
Slater, J.C., Quantum Theory of Atomic Structure, Vol. I and II, New York: McGraw-Hill 1960
Slichter, C.P., Principles of Magnetic Resonance, New York: Harper & Row, 1963
Slichter, C.P., Principles of Magnetic Resonance, Berlin: Springer, 1978, 2nd revised edition
Slifkin, L., Semicond. Insulators **3**, 393 (1978)
Slonczewski, J.C., Phys. Rev. **131**, 1596 (1963)
Smith, A.C., Janak, J.F., Adler, R.B., Electronic Conduction in Solids, New York: McGraw-Hill, 1967
Smith, C.S., Cain, L.S., J. Phys. Chem. Solids **36**, 205 (1975)
Smith, D.Y., Phys. Rev. **137**, A574 (1965)
Smith, D.Y., Solid State Commun. **8**, 1677 (1970)
Smith, D.Y., Phys. Rev. B **6**, 565 (1972)
Smith, D.Y., Phys. Rev. B **8**, 3939 (1973)
Smith, P.V., J. Phys. Chem. Solids **37**, 581 (1976)
Smith, P.V., J. Phys. Chem. Solids **37**, 589 (1976)
Smith, R.A., (Ed.) Semiconductors, Proceedings of the Intern. School of Physics „Enrico Fermi", Course XXII, Varenna 1961, New York: Academic Press, 1963
Smith, S.R.P., Dravnieks, F., Wertz, J.E., Phys. Rev. **178**, 471 (1969)
Smith, V.H., Petelenz, P., Phys. Rev. B **17**, 3253 (1978)
Sneh, Dayal, B., Phys. Stat. Sol. (b) **67**, 125 (1975)
Sodha, M.S., Tripathi, V.K., Kamal, J., J. Phys. Chem. Solids **32**, 2631 (1971)
Sodha, M.S., Phadke, U.P., Chakravarti, A.K., Phys. Stat. Sol. (b) **48**, 823 (1971)
Sodha, M.S., Phadke, U.P., Chakravarti, A.K., Phys. Rev. B **6**, 2414 (1972)
Sodha, M.S., Phadke, U.P., Chakravarti, A.K., Phys. Rev. B **6**, 4879 (1972)
Soma, T., Physica **92B**, 1 (1977)
Soma, T., J. Phys. C: Solid State Phys. **11**, 2669 (1978)
Sondheimer, E.H., Proc. Roy. Soc. London A **203**, 75 (1950)
Song, K.S., J. Phys. Soc. Japan **26**, 1131 (1969)
Song, K.S., J. Phys. Chem. Solids **31**, 1389 (1970)

Song, K.S., Stoneham, A.M., Harker, A.H., J. Phys. C: Solid State Phys. **8**, 1125 (1975)
Song, K.S., Stoneham, A.M., Harker, A.H., J. Luminesc. **12/13**, 303 (1976)
Song, K.S., Stoneham, A.M., Solid State Commun. **18**, 367 (1976)
Song, M.K.-S., J. Physique **28**, 195 (1967)
Soules, T.F., Duke, C.B., Phys. Rev. B **3**, 262 (1971)
Soulie, E., Goodman, G., Theoret. Chim. Acta **41**, 17 (1976)
Spaeth, J.M., Z. Physik **192**, 106 (1966)
Sparks, M., Sham, L.J., Phys. Rev. B **8**, 3037 (1973)
Spector, H.N., Phys. Rev. **137**, A311 (1965)
Spector, H.N., Mitra, S.S., Schmeising, H.N., J. Chem. Phys. **46**, 2676 (1967)
Spector, H.N., Can. J. Phys. **46**, 2659 (1968)
Spinolo, G., Fowler, W.B., Phys. Rev. **138**, A 661 (1965)
Spitzer, L., Physics of Fully Ionized Gases, New York: J. Wiley & Sons, 1962
Srivastava, G.P., Verma, G.S., Phys. Stat. Sol. (b) **47**, 669 (1971)
Srivastava, G.P., Singh, D.P., Verma, G.S., Phys. Rev. B **6**, 3053 (1972)
Srivastava, G.P., Phys. Stat. Sol. (b) **77**, 131 (1976)
Srivastava, S.P., Sharma, M.N., Madan, M.P., J. Phys. Soc. Japan **25**, 212 (1968)
Srivastava, V.P., Verma, G.S., Phys. Rev. B **10**, 219 (1974)
Subashiev, V.K., Chalikyan, G.A., Sov. Phys. Solid State **11**, 2014 (1970)
Suchet, J.P., J. Electrochem. Soc. **124**, 30C (1977)
Suffczynski, M., J. Chem. Phys. **38**, 1558 (1963)
Suffczynski, M., Swierkowski, L., Opt. Commun. **17**, 184 (1976)
Sullivan, J.J., J. Phys. Chem. Solids **25**, 1039 (1964)
Sumi, A., J. Phys. Soc. Japan **43**, 1286 (1977)
Sumi, H., J. Phys. Soc. Japan **32**, 616 (1972)
Sumi, H., J. Phys. Soc. Japan **33**, 1240 (1972)
Sumi, H., Solid State Commun. **17**, 701 (1975)
Sumi, H., J. Luminesc. **12/13**, 207 (1976)
Sumi, H., J. Phys. Soc. Japan **41**, 526 (1976)
Surda, A., Czech. J. Phys. B **27**, 439 (1977)
Sussmann, J.A., Ann. Phys. (Paris) **6**, 135 (1971)
Sutherland, F.R., Spector, H.N., Phys. Rev. B **17**, 2728 (1978)
Suzuki, A., Frood, D.G., J. Phys. Chem. Solids **39**, 1099 (1978)
Svidzinskii, K.K., Sov. Phys. Solid State **6**, 1852 (1965)
Swank, R.K., Brown, F.C., Phys. Rev. **130**, 34 (1963)
Swenson, R.J., J. Math. Phys. **3**, 1017 (1962)
Swierkowski, L., Nuovo Cim. **29B**, 340 (1975)
Syme, R.W.G., Haas, W.J., Spedding, F.H., Good, R.H., J. Chem. Phys. **48**, 2772 (1968)
Szigeti, B., Trans. Faraday Soc. **45**, 155 (1949)
Szigeti, B., Proc. Roy. Soc. A (London) **204**, 51 (1950)
Szigeti, B., Proc. Roy. Soc. A (London) **258**, 377 (1960)
Szudy, J., Baylis, W.E., J. Quant. Spectrosc. Radiat. Transfer **15**, 641 (1975)
Szymanska, W., Dietl, T., J. Phys. Chem. Solids **39**, 1025 (1978)
Szymanski, J., Phys. Stat. Sol. (b) **77**, 347 (1976)

Schäfer, K., Ludwig, H., Ber. Bunsenges. **72**, 1114 (1968)
Schattke, W., Birkner, G.K., Z. Physik **252**, 12 (1972)
Schechter, D., Phys. Rev. **180**, 896 (1969)
Scher, H., Lax, M., Phys. Rev. B **7**, 4491 (1973)
Scher, H., Lax, M., Phys. Rev. B **7**, 4502 (1973)
Scher, H., Montroll, E.W., Phys. Rev. B **12**, 2455 (1975)
Scherz, U., J. Phys. Chem. Solids **30**, 2077 (1969)
Scherz, U., J. Phys. Chem. Solids **31**, 995 (1970)
Schieve, W.C., in: Lectures in Statistical Physics, ed. W.C. Schieve and J.S. Turner, Lecture Notes in Physics, Vol. 28., Berlin: Springer 1974, p. 1
Schirmer, O.F., Koidl, P., Reik, H.G., Phys. Stat. Sol. (b) **62**, 385 (1974)
Schirmer, O.F., Z. Physik B **24**, 235 (1976)
Schlesinger, M., Nara, H., J. Phys. Chem. Solids **34**, 1827 (1973)
Schlögl, F., Z. Physik **253**, 147 (1972)
Schlottmann, P., Passeggi, M.C.G., Phys. Stat. Sol. (b) **52**, K107 (1972)
Schlüter, M., Nuovo Cim. **13 B**, 313 (1973)
Schlüter, M., Martinez, G., Cohen, M.L., Phys. Rev. B **11**, 3808 (1975)
Schlüter, M., Martinez, G., Cohen, M.L., Phys. Rev. B **12**, 650 (1975)
Schmid, A., Kelly, P., Bräunlich, P., Phys. Rev. B **16**, 4569 (1977)
Schmid, J., Phys. kondens. Materie **15**, 119 (1972)
Schmidlin, W.F., Phys. Rev. B **16**, 2362 (1977)
Schmidlin, W.F., Solid State Commun. **22**, 451 (1977)
Schmidt, P.P., J. Phys. C: Solid State Phys. **2**, 785 (1969)
Schmidt, P.P., Theor. Chim. Acta 40, 263 (1975)
Schnakenberg, J., Z. Physik **171**, 199 (1963)
Schnakenberg, J., Z. Physik **208**, 165 (1968)
Schnakenberg, J., Phys. Stat. Sol. **28**, 623 (1968)

Schober, C., Phys. Stat. Sol. **32**, K37 (1969)
Schöll, E., Untersuchungen zur F-F'-Reaktionskinetik, Diplomarbeit, University of Tübingen, 1976
Schöll, E., A Study of Non-Equilibrium Phase Transitions in Semiconductors, Thesis, University of Southampton 1978
Schöll, E., Proc. Roy. Soc. London A **365**, 511 (1979)
Schöll, E., Landsberg, P.T., Proc. Roy. Soc. London A **365**, 495 (1979)
Schöll, E., Landsberg, P.T., Proc. 14th Int. Conf. Physics of Semiconductors, Edinburgh, Inst. Phys. Conf. Ser. No. 43 (1979), p. 461
Schöll, E., Verhandl. DPG **3**, 364 (1980)
Schöll, E., J. Physique **42**, Colloque C7, 57 (1981)
Schöll, E., Z. Phys. B **46**, 23 (1982)
Schöll, E., Heisel, W., Prettl, W., Z. Physik B **47**, 285 (1982)
Schöll, E., Z. Physik B **48**, 153 (1982)
Schön, M., Physikalische Zeitschrift **39**, 940 (1938)
Schön, M., Ann. Phys. (Leipzig) **3**, 333 (1948)
Schön, M., Z. Naturforsch. **6a**, 251 (1951)
Schön, M., Z. Naturforsch. **6a**, 287 (1951)
Schön, M., Halbleiterprobleme IV, 282 (1958), (Ed. W. Schottky, Braunschweig: Vieweg)
Schöne, D., Z. Naturforsch. **24a**, 1752 (1969)
Schönhofer, A., Z. Physik **163**, 277 (1961)
Scholz, A., Phys. Stat. Sol. **7**, 973 (1964)
Scholz, A.H., U.K.A.E.A. Research Group, Report AERE-R 5449 (1967)
Scholz, A.H., Phys. Stat. Sol. **25**, 285 (1968)
Schotte, K.-D., Z. Physik **196**, 393 (1966)
Schottky, W., Halbleiterprobleme I, 139 (1954) (Ed. W. Schottky, Braunschweig, Vieweg)
Schottky, W., Festkörperprobleme **1**, 316 (1962), (Ed. F. Sauter, Braunschweig: Vieweg)
Schröder, U., Solid State Commun. **4**, 347 (1966)
Schröder, U., Festkörperprobleme **13**, 171 (1973) (Ed. H.J. Queisser, Braunschweig: Vieweg)
Schütz, O., Treusch, J., Solid State Commun. **13**, 1155 (1973)
Schultz, T.D., Phys. Rev. **116**, 526 (1959)
Schulz, K., Z. Physik **163**, 293 (1961)
Schulze, P.D., Hardy, J.R., Phys. Rev. B **6**, 1580 (1972)
Schulze, P.D., Hardy, J.R., Phys. Rev. B **5**, 3270 (1972)
Schumacher, D.P., Hollingsworth, C.A., J. Phys. Chem. Solids **27**, 749 (1966)
Schupfner, J., Z. Naturforsch. **35a**, 902 (1980)
Schupfner, J., Übergangswahrscheinlichkeiten von Elektron-Photon-Prozessen in polaren Halbleitern, Thesis, University of Tübingen, 1982
Schwendimann, P., Z. Physik B **26**, 63 (1977)
Starostin, N.V., Zakharov, V.K., Opt. i Spectrosk. **33**, 87 (1972)
Starostin, N.V., Ganin, V.A., Sov. Phys. Solid State **15**, 2265 (1974)
Stasiw, O., Elektronen und Ionenprozesse in Ionenkristallen, Berlin: Springer, 1959
Stebe, B., Certier, M., Phys. Stat. Sol. (b) **53**, K89 (1972)
Stebe, B., Munschy, G., Phys. Stat. Sol. (b) **60**, 133 (1973)
Stebe, B., Compte, C., Phys. Rev. B **15**, 3967 (1977)
Stedman, G.E., J. Phys. C: Solid State Phys. **5**, 121 (1972)
Stedman, G.E., J. Phys. C: Solid State Phys. **11**, 1017 (1978)
Steggles, P., J. Phys. C: Solid State Phys. **10**, 2817 (1977)
Steiner, K.H., Interactions Between Electromagnetic Fields and Matter, Vieweg Tracts in Pure and Applied Physics, Vol. 1, Braunschweig: Vieweg, 1973
Stephens, P.J., J. Chem. Phys. **51**, 1995 (1969)
Stevens, K.W.H., Persico, F., Nuovo Cim. **41 A**, 37 (1966)
Stevens, K.W.H., Rep. Prog. Phys. **30**, 189 (1967)
Stevens, K.W.H., J. Phys. C: Solid State Phys. **2**, 1934 (1969)
Stevenson, R.W.H. (Ed.), Phonons in Perfect Lattices and in Lattices with Point Imperfections, Edinburgh: Oliver and Boyd, 1966
Stinchcombe, R.B., Proc. Phys. Soc. **78**, 275 (1961)
Stocker, D., Z. Naturforsch. **23a**, 1158 (1968)
Stocker, H.J., Kaplan, H., Phys. Rev. **150**, 619 (1966)
Stöckmann, F., Z. Physik **128**, 185 (1950)
Stöckmann, F., Z. Physik **143**, 348 (1955)
Stöckmann, F., Z. Physik **147**, 544 (1957)
Stöckmann, F., Z. Angew. Phys. **11**, 68 (1959)
Stöckmann, F., Phys. Stat. Sol. **2**, 517 (1962)
Stöckmann, F., Festkörperprobleme **9**, 138 (1969) (Ed. O. Madelung, Braunschweig: Vieweg)
Stöckmann, F., Phys. Stat. Sol. **34**, 751 (1969)
Stojanovic, S.D., Setrajcic, J.P., Skrinjar, M.J., Tosic, B.S., Phys. Stat. Sol. (b) **79**, 433 (1977)
Stokoe, T.Y., Cornwell, J.F., Phys. Stat. Sol. (b) **49**, 209 (1972)
Stokoe, T.Y., Cornwell, J.F., Phys. Stat. Sol. (b) **58**, 267 (1973)
Stolz, H., Ann. Phys. (Leipzig) **19**, 394 (1957)
Stolz, H., Phys. Stat. Sol. **32**, 631 (1969)
Stone, A.J., Proc. Roy. Soc. London A **271**, 424 (1963)
Stoneham, A.M., Proc. Phys. Soc. **86**, 1163 (1965)
Stoneham, A.M., Proc. Phys. Soc. **89**, 909 (1966)
Stoneham, A.M., Phys. Stat. Sol. **19**, 787 (1967)

Stoneham, A.M., J. Phys. C: Solid State Phys. **1**, 565 (1968)
Stoneham, A.M., Hayes, W., Smith, P.H.S., Stott, J.P., Proc. Roy. Soc. A (London) **306**, 369 (1968)
Stoneham, A.M., J. Phys. C: Solid State Phys. **3**, L131 (1970)
Stoneham, A.M., Bartram, R.H., Phys. Rev. B **2**, 3403 (1970)
Stoneham, A.M., Phys. Stat. Sol. (b) **52**, 9 (1972)
Stoneham, A.M., J. Phys. C: Solid State Phys. **7**, 2476 (1974)
Stoneham, A.M., Theory of Defects in Solids, Oxford: Clarendon Press, 1975
Stoneham, A.M., Harker, A.H., J. Phys. C: Solid State Phys. **8**, 1102 (1975)
Stoneham, A.M., Harker, A.H., J. Phys. C: Solid State Phys. **8**, 1109 (1975)
Stoneham, A.M., Phil. Mag. **36**, 983 (1977)
Stoneham, A.M., AERE Harwell Rept. No. OX11 ORA (1978)
Stramska, H., Acta Phys. Polon. A **53**, 511 (1978)
Strauch, D., Phys. Stat. Sol. **30**, 495 (1968)
Strauss, A.J., J. Appl. Phys. **30**, 559 (1959)
Streetman, B.G., J. Appl. Phys. **37**, 3137 (1966)
Streitwolf, H.W., Phys. Stat. Sol. **2**, 1595 (1962)
Streitwolf, H.W., Gruppentheorie in der Festkörperphysik, Leipzig: Akad. Verlagsges. Geest & Portig, 1967
Strekalov, V.N., Sov. Phys. Solid State **17**, 94 (1975)
Strekalov, V.N., Sov. Phys. Solid State **17**, 1687 (1976)
Strekalov, V.N., Sov. Phys. Solid State **19**, 975 (1977)
Striefler, M.E., Jaswal, S.S., J. Phys. Chem. Solids **30**, 827 (1969)
Strizhevskii, V.L., Obukhovskii, V.V., Phys. Stat. Sol. (b) **53**, 603 (1972)
Strizhevskii, V.L., Obukhovskii, V.V., Ponath, H.E., Sov. Phys. JETP **34**, 286 (1972)
Strizhevskii, V.L., Sov. Phys. JETP **35**, 760 (1972)
Strizhevskii, V.L., Yashkir, Yu.N., Phys. Stat. Sol. (b) **61**, 353 (1974)
Strozier, J.A., Dick, B.G., Phys. Stat. Sol. **31**, 203 (1969)
Struck, C., Herzfeld, F., J. Chem. Phys. **44**, 464 (1966)
Stukel, D.J., Euwema, R.N., Collins, T.C., Herman, F., Kortum, R.L., Phys. Rev. **179**, 740 (1969)
Stukel, D.J., Phys. Rev. B **2**, 1852 (1970)
Stumpf, H., Z. Naturforsch. **10a**, 971 (1955)
Stumpf, H., Z. Naturforsch. **12a**, 465 (1957)
Stumpf, H., Z. Naturforsch. **12a**, 153 (1957)
Stumpf, H., Z. Naturforsch. **13a**, 621 (1958)
Stumpf, H., Z. Naturforsch. **13a**, 171 (1958)
Stumpf, H., Z. Naturforsch. **14a**, 659 (1959)
Stumpf, H., Z. Naturforsch. **14a**, 403 (1959)
Stumpf, H., Wagner, M., Z. Naturforsch. **15a**, 30 (1960)
Stumpf, H., Quantentheorie der Ionenrealkristalle, Berlin: Springer, 1961
Stumpf, H., Z. Physik **229**, 488 (1969)
Stumpf, H., Phys. kondens. Materie **13**, 9 (1971)
Stumpf, H., Phys. kondens. Materie **13**, 101 (1971)
Stumpf, H., Schuler, W., Elektrodynamik, Braunschweig: Vieweg 1973
Stumpf, H., Phys. cond. Matter **18**, 217 (1974)
Stumpf, H., Rieckers, A., Thermodynamik I, Wiesbaden: Vieweg, 1976
Stumpf, H., Kleih, W., Z. Naturforsch. **35a**, 1121 (1980)
Sturge, M.D., Phys. Rev. **140**, A 880 (1965)
Sturge, M.D., The Jahn-Teller Effect in Solids, in: Solid State Physics, Vol. 20, ed. F. Seitz, D. Turnbull, H. Ehrenreich, New York: Academic Press 1967, p. 91

Tachiki, M., Sroubek, Z., J. Chem. Phys. **48**, 2383 (1968)
Tait, W.C., Weiher, R.L., Phys. Rev. **178**, 1404 (1969)
Takada, S., Prog. Theor. Phys. **36**, 224 (1966)
Takaoka, Y., Suzuki, N., Motizuki, K., J. Phys. Soc. Japan **44**, 1168 (1978)
Takegahara, K., Kasuya, T., J. Phys. Soc. Japan **39**, 1292 (1975)
Takeno, S., Prog. Theor. Phys. **25**, 102 (1961)
Takeno, S., Prog. Theor. Phys. **28**, 33 (1962)
Takeno, S., Kashiwamura, S., Teramoto, E., Prog. Theor. Phys., Suppl. **23**, 124 (1962)
Takeno, S., Prog. Theor. Phys. **33**, 363 (1965)
Takeno, S., Prog. Theor. Phys. **38**, 995 (1967)
Takeno, S., Prog. Theor. Phys. **42**, 1003 (1969)
Takeno, S., Mabuchi, M., Prog. Theor. Phys. **50**, 1848 (1973)
Takeshima, M., J. Appl. Phys. **43**, 4114 (1972)
Takeshima, M., Phys. Rev. B **12**, 575 (1975)
Takeuti, Y., Prog. Theor. Phys. **18**, 421 (1957)
Takeuti, Y., Prog. Theor. Phys., Suppl. **12**, 75 (1959)
Talwar, D.N., Agrawal, B.K., Phys. Rev. **9**, 2539 (1974)
Talwar, D.N., Agrawal, B.K., Phys. Rev. B **12**, 1432 (1975)
Talwar, D.N., Agrawal, B.K., J. Phys. Chem. Solids **39**, 207 (1978)
Tamaschke, O., Ann. Physik **8**, 76 (1961)
Tanaka, K., Shinada, M., J. Phys. Soc. Japan **34**, 108 (1973)
Tarasenko, A.A., Chumak, A.A., Sov. Phys. JETP **46**, 327 (1977)

Tasker, P.W., Stoneham, A.M., J. Phys. Chem. Solids **38**, 1185 (1977)
Tatsuyama, C., Ichimura, S., J. Phys. Soc. Japan **44**, 575 (1978)
Tauc, J., Photo and Thermoelectric Effects in Semiconductors, Oxford: Pergamon Press, 1962
Tausendfreund, W., Z. Physik **269**, 11 (1974)
Tavernier, J., J. Physique **28**, Suppl. au no 2, C 1–40 (1967)
Taylor, D.W., Phys. Rev. **156**, 1017 (1967)
Taylor, D.W., Dynamics of Impurities in Crystals, in: Dynamical Properties of Solids, v. 2, ed. G.K. Horton and A.A. Maradudin, Amsterdam: North-Holland, 1975, p. 285
Taylor, G.W., Simmons, J.G., J. Phys. C: Solid State Phys. **8**, 3360 (1975)
Tejeda, J., Shevchik, N.J., Phys. Rev. B **13**, 2548 (1976)
Tekhver, I.Y., Khizhnyakov, V.V., Sov. Phys. JETP **42**, 305 (1976)
Telezhkin, V.A., Tolpygo, K.B., Sov. Phys. Solid State **19**, 1773 (1977)
Teltow, J., Halbleiterprobleme III, 26 (1956) (Ed. W. Schottky, Braunschweig: Vieweg)
Templeton, T.L., Clayman, B.P., Can. J. Phys. **54**, 2010 (1976)
Tenan, M.A., Miranda, L.C.M., J. Phys. C: Solid State Phys. **10**, L389 (1977)
Tenerz, E., Ark. Fys. **11**, 247 (1956)
Tenerz, E., Ark. Fys. **12**, 277 (1957)
Terrile, M.C., Panepucci, H., Carvalho, R.A., Phys. Rev. B **15**, 1110 (1977)
Terwiel, R.H., Mazur, P., Physica **32**, 1813 (1966)
Terzi, N., Many-Phonon Processes Induced by Light Absorption in Localized Centres in Non-Conducting Crystals, in: Trieste Winter College on Interaction of Radiation with Condensed Matter, Trieste 1976, Wien: IAEA 1977, p. 217
Tessmann, J.R., Kahn, A.H., Shockley, W., Phys. Rev. **92**, 890 (1953)
Tewari, S., Solid State Commun. **15**, 387 (1974)
Tewary, V.K., Proc. Phys. Soc. **92**, 987 (1967)
Tewary, V.K., Adv. Phys. **22**, 757 (1973)
Tewordt, L., Z. Physik **137**, 604 (1954)
Tewordt, L., Phys. Rev. **109**, 61 (1958)
Thacher, P.D., Phys. Rev. **156**, 975 (1967)
Thakur, K.P., Acta Cryst. A **31**, 540 (1975)
Thakur, K.P., Austr. J. Phys. **29**, 39 (1976)
Thakur, K.P., J. Inorg. Nucl. Chem. **38**, 1433 (1976)
Thakur, K.P., Austr. J. Phys. **30**, 325 (1977)
Thalmeier, P., Fulde, P., Z. Physik B **29**, 299 (1978)
Tharmalingam, K., Phys. Rev. **130**, 2204 (1963)
Tharmalingam, K., J. Phys. C: Solid State Phys. **3**, 1856 (1970)
Tharmalingam, K., Phil. Mag. **23**, 199 (1971)
Theimer, O., Phys. Rev. **109**, 1095 (1958)
Theimer, O., Phys. Rev. **112**, 1857 (1958)
Theimer, O., Paul, R., J. Appl. Phys. **36**, 3678 (1965)
Thellung, A., Weiss, K., Phys. kondens. Materie **9**, 300 (1969)
Thoma, K., Z. Physik **211**, 218 (1968)
Thomas, D.G., Hopfield, J.J., Augustyniak, W.M., Phys. Rev. **140**, 202 (1965)
Thomas, D.G., Hopfield, J.J., Frosch, C.J., Phys. Rev. Letters **15**, 857 (1965)
Thomas, D.G. (Ed.), II–VI Semiconducting Compounds, New York: Benjamin, 1967
Thomas, D.G., Hopfield, J.J., Colbow, K., in: 7th Intern. Conference on the Physics of Semiconductors, Paris 1974, Paris: Dunod, 1975
Thomas, P., Wuertz, D., Phys. Stat. Sol. (b) **86**, 541 (1978)
Thomchick, J., Whitfield, G., Phys. Rev. B **10**, 3458 (1974)
Thornber, K.K., Solid-State Electron. **21**, 259 (1978)
Thornton, W.A., Phys. Rev. **102**, 38 (1956)
Thorson, W.R., J. Chem. Phys. **29**, 938 (1958)
Thuau, M., Margerie, J., Phys. Stat. Sol. (b) **47**, 271 (1971)
Thurmond, C.D., J. Electrochem. Soc. **122**, 1133 (1975)
Tibbs, S.R., Trans. Faraday Soc. **35**, 1471 (1939)
Tiemann, J.J., Fritzsche, H., Phys. Rev. **137**, A 1910 (1965)
Timashev, S.F., Opt. Spectrosc. **35**, 339 (1973)
Timashev, S.F., Sov. Phys. Solid State **15**, 753 (1973)
Timashev, S.F., Sov. Phys. Semicond. **9**, 898 (1976)
Timerkaev, B.A., Khabibullin, B.M., Sov. Phys. Solid State **20**, 295 (1978)
Timusk, T., Martienssen, W., Phys. Rev. **128**, 1656 (1962)
Timusk, T., Klein, M.V., Phys. Rev. **141**, 664 (1966)
Tindemans-van Eijndhoven, J.C.M., Kroese, C.J., J. Phys. C: Solid State Phys. **8**, 3963 (1975)
Tinkham, M., Group Theory and Quantum Mechanics, New York: McGraw–Hill, 1964
Tjablikov, S.V., Z. Exp. Theor. Phys. **23**, 381 (1952)
Tjablikov, S.V., Z. Exp. Theor. Phys. **25**, 688 (1953)
To, K.C., Henderson, B., J. Phys. C: Solid State Phys. **4**, L216 (1971)
Toda, M., Kotera, T., Kogure, K., J. Phys. Soc. Japan **17**, 426 (1962)
Tolpygo, E.I., Tolpygo, K.B., Sheinkman, M.K., Sov. Phys. Solid State **7**, 1442 (1965)

Tolpygo, E.I., Tolpygo, K.B., Sheinkman, M.K., Sov. Phys. Semicond. **8**, 326 (1974)
Tolpygo, H.I., Phys. Stat. Sol. **42**, 155 (1970)
Tolpygo, H.I., Phys. Stat. Sol. (b) **61**, 311 (1974)
Tolpygo, K.B., J. Exp. Theor. Phys. (USSR) **20**, 497 (1950)
Tolpygo, K.B., Ukrainsk. Fiz. Zh. **2**, 242 (1957)
Tolpygo, K.B., Sov. Phys. Solid State **2**, 2367 (1961)
Tolpygo, K.B., Sheka, D.I., Sov. Phys. Solid State **5**, 1905 (1964)
Tolpygo, K.B., Sov. Phys. Solid State **6**, 577 (1964)
Tolpygo, K.B., Sheka, D.I., Sov. Phys. Solid State **8**, 2069 (1967)
Tolpygo, K.B., Sov. Phys. Solid State **17**, 1149 (1975)
Tolstoi, N.A., Abramov, A.P., Opt. Spectrosc. **20**, 273 (1966)
Tomchuk, P.M., Sov. Phys. Solid State **3**, 913 (1961)
Tosatti, E., Harbeke, G., Nuovo Cim. **22 B**, 87 (1974)
Tosatti, E., Phys. Rev. Letters **33**, 1092 (1974)
Tosatti, E., Plasmons in Crystalline Media, Particularly Semiconductors, in: Trieste Winter College on Interaction of Radiation with Condensed Matter, Trieste 1976, Wien: IAEA 1977 p. 281
Tosi, M.P., Fumi, F.G., Nuovo Cim. **VII**, 95 (1958)
Tosi, M.P., Airoldi, G., Nuovo Cim. **VIII**, 584 (1958)
Tosi, M.P., Doyama, M., Phys. Rev. **151**, 642 (1966)
Tosic, B.S., Davidovic-Ristovski, G.S., Ristovksi, L.M., Nuovo Cim. **25 B**, 216 (1975)
Townsend, P.D., J. Phys. C: Solid State Phys. **9**, 1871 (1976)
Toyabe, T., Asai, S., Phys. Rev. B **8**, 1531 (1973)
Toyozawa, Y., Prog. Theor. Phys. **12**, 421 (1954)
Toyozawa, Y., Prog. Theor. Phys. **20**, 53 (1958)
Toyozawa, Y., Prog. Theor. Phys. **22**, 455 (1959)
Toyozawa, Y., Prog. Theor. Phys., Suppl. **12**, 111 (1959)
Toyozawa, Y., Prog. Theor. Phys. **26**, 29 (1961)
Toyozawa, Y., Prog. Theor. Phys. **27**, 89 (1962)
Toyozawa, Y., J. Phys. Chem. Solids **25**, 59 (1964)
Toyozawa, Y., Inoue, M., J. Phys. Soc. Japan **21**, 1663 (1966)
Toyozawa, Y., Hermanson, J., Phys. Rev. Letters **21**, 1637 (1968)
Toyozawa, Y., J. Phys. Soc. Japan **41**, 400 (1976)
Toyozawa, Y., J. Phys. Soc. Japan **44**, 482 (1978)
Tran-Thoai, D.B., Z. Physik B **26**, 115 (1977)
Trautenberg, E., Phys. kondens. Materie **16**, 163 (1973)
Trebin, H.-R., Rössler, U., Phys. Stat. Sol. (b) **70**, 717 (1975)
Trebin, H.-R., Phys. Stat. Sol. (b) **81**, 527 (1977)
Trifonov, E.D., Sov. Phys. Solid State **6**, 366 (1964)
Trifonov, E.D., Peuker, K., J. Physique **26**, 738 (1965)
Trifonov, E.D., Sov. Phys. Solid State **9**, 2680 (1968)
Trifonov, E.D., Shekhtman, V.L., Sov. Phys. Solid State **11**, 2415 (1970)
Tripathi, B.B., Ramesh, N., Jogi, S., Lett. Nuovo Cim. **18**, 116 (1977)
Tripathi, R.S., Pathak, K.N., Nuovo Cim. **21 B**, 289 (1974)
Trlifaj, M., Czech. J. Phys. **3**, 267 (1953)
Trlifaj, M., Czech. J. Phys. **5**, 133 (1955)
Trlifaj, M., Czech. J. Phys. **5**, 463 (1955)
Trlifaj, M., Czech. J. Phys. **6**, 533 (1956)
Trlifaj, M., Czech. J. Phys. **7**, 1 (1957)
Trlifaj, M., Czech. J. Phys. **7**, 379 (1957)
Trlifaj, M., Czech. J. Phys. **7**, 657 (1957)
Trlifaj, M., Czech. J. Phys. **8**, 510 (1958)
Trlifaj, M., Czech. J. Phys. **9**, 446 (1959)
Trlifaj, M., Czech. J. Phys. **9**, 671 (1959)
Trlifaj, M., Czech. J. Phys. **9**, 562 (1959)
Trlifaj, M., Czech. J. Phys. B **15**, 780 (1965)
Tröster, F., Zur Berechnung von Auger-Prozessen in Ionenkristallen mit F-F'-Zentren, Diplomarbeit, University of Tübingen, 1976
Tröster, F., Z. Naturforsch. **33a**, 1251 (1978)
Tröster, F., Reaktionskinetik von Auger-Prozessen in Ionischen Halbleitern, Thesis, University of Tübingen, 1978
Trommer, R., Cardona, M., Phys. Rev. B **17**, 1865 (1978)
Tsang, Y.W., Cohen, M.L., Phys. Rev. B **3**, 1254 (1971)
Tsang, Y.W., Cohen, M.L., Solid State Commun. **9**, 261 (1971)
Tsang, Y.W., Cohen, M.L., Phys. Rev. B **9**, 3541 (1974)
Tsay, Y.-F., Bendow, B., Phys. Rev. B **16**, 2663 (1977)
Tsertsvadze, A.A., Sov. Phys. Solid State **3**, 370 (1961)
Tsertsvadze, A.A., Sov. Phys. Solid State **3**, 241 (1961)
Tsertsvadze, A.A., Sov. Phys. Solid State **3**, 1428 (1962)

Tsu, R., Döhler, G., Phys. Rev. B **12**, 680 (1975)
Tsuboi, M., J. Chem. Phys. **40**, 1326 (1964)
Tsukanov, V.D., Theor. Math. Phys. **22**, 247 (1975)
Tsukerblat, B.S., Perlin, Yu.E., Sov. Phys. Solid State 7, 2647 (1966)
Tsukerblat, B.S., Sov. Phys. JETP **24**, 554 (1967)
Tsukerblat, B.S., Rosenfeld, Yu.B., Vekhter, B.G., Phys. Letters **38A**, 333 (1972)
Tsukerblat, B.S., Vekhter, B.G., Sov. Phys. Solid State **14**, 2204 (1973)
Tsukerblat, V.S., Rozenfel'd, Yu.B., Polinger, V.Z., Vekhter, V.G., Sov. Phys. JETP **41**, 553 (1976)
Tsuzuki, T., J. Low Temp. Phys. **22**, 631 (1976)
Tulub, A.V., Nachr. Leningrader Univ. Ser. Phys. Chem. Nr. **4**, 53 (1957)
Tulub, A.V., Sov. Phys. JETP **36**, 392 (1959)
Tulub, A.V., Sov. Phys. JETP **36**, 1325 (1959)
Tulub, A.V., Sov. Phys. JETP **14**, 1301 (1962)
Tung, Y.W., Cohen, M.L., Phys. Rev. **180**, 823 (1969)
Tyner, C.E., Drotning, W.D., Drickamer, H.G., J. Appl. Phys. **47**, 1044 (1976)

Uba, S.M., Baranowski, J.M., Phys. Rev. B **17**, 69 (1978)
Ueba, H., Ichimura, S., Phys. Stat. Sol. (b) **75**, 501 (1976)
Ueta, M., Nishina, Y. (Eds.), Physics of Highly Excited States in Solids, Lecture Notes in Physics 57, Berlin: Springer, 1976
Ulrici, W., Phys. Stat. Sol. (b) **62**, 431 (1974)
Unger, B., Schaack, G., Phys. Stat. Sol. (b) **48**, 285 (1971)
Upadhyaya, K.S., Mahesh, P.S., Phys. Stat. Sol. (b) **59**, 279 (1973)
Upadhyaya, K.S., Singh, R.K., J. Phys. Chem. Solids **36**, 293 (1975)
Uritsky, S.I., Novikov, V.S., Phys. Stat. Sol. **11**, 101 (1965)
Uritsky, S.I., Shuster, G.V., Phys. Stat. Sol. **11**, 105 (1965)

Vail, J.M., Phys. Stat. Sol. (b) **44**, 443 (1971)
Vail, J.M., Phys. Rev. B **7**, 5359 (1973)
Vaitkus, Yu.Yu., Vishchakas, Yu.K., Sov. Phys. Solid State **12**, 435 (1970)
Vallin, J.T., Phys. Rev. B **2**, 2390 (1970)
Valton, M., Acta Electron. **12**, 131 (1969)
Vandevyver, M., Plumelle, P., J. Phys. Chem. Solids **38**, 765 (1977)
Vandevyver, M., Plumelle, P., Phys. Rev. B **17**, 675 (1978)
Vanhaelst, M., Matthys, P., Boesman, E., Phys. Stat. Sol. (b) **89**, 151 (1978)
Varga, B.B., Phys. Rev. **137**, A 1896 (1965)
Varotsos, P.A., J. Phys. Chem. Solids **39**, 513 (1977)
Varotsos, P., Alexopoulos, K., Phys. Rev. B **15**, 2348 (1977)
Varshni, Y.P., Phys. Stat. Sol. **20**, 9 (1967)
Vasileff, H.D., Phys. Rev. **96**, 603 (1954)
Vasileff, H.D., Phys. Rev. **97**, 891 (1955)
Vasil'ev, A.V., Malkin, B.Z., Natadze, A.L., Ryskin, A.I., Sov. Phys. JETP **44**, 624 (1976)
Vas'ko, F.T., Sov. Phys. Solid State **19**, 533 (1977)
Vassell, M.O., J. Math. Phys. **11**, 408 (1970)
Vassell, M.O., Ganguly, A.K., Conwell, E.M., Phys. Rev. B **2**, 948 (1970)
Vasudevan, K.N., Markham, J.J., Solid State Commun. **25**, 587 (1978)
Vaughan, R.A., Magnetic Res. Rev. **4**, 25 (1975)
van Vechten, J.A., Phys. Rev. **187**, 1007 (1969)
van Vechten, J.A., Phys. Rev. B **3**, 562 (1971)
van Vechten, J.A., J. Electrochem. Soc. **122**, 419 (1975)
van Vechten, J.A., J. Electrochem. Soc. **122**, 423 (1975)
Veelken, R., Z. Physik **142**, 476 (1955)
Veelken, R., Z. Physik **142**, 544 (1955)
Vekhter, B.G., Tsukerblat, B.S., Sov. Phys. Solid State **10**, 1250 (1968)
Vekhter, B.G., Perlin, Yu.E., Polinger, V.Z., Rosenfeld, Yu.B., Tsukerblat, B.S., Cryst. Latt. Def. **3**, 61 (1972)
Vekhter, B.G., Phys. Letters **45 A**, 133 (1973)
Vekhter, B.G., Kaplan, M.D., Phys. Letters **43 A**, 389 (1973)
Vekhter, B.G., Polinger, V.Z., Rozenfel'd, Yu.B., Tsukerblat, B.S., JETP Letters **20**, 36 (1974)
Vekhter, B.G., Kaplan, M.D., Sov. Phys. Solid State **15**, 1344 (1974)
Velicky, B., Sak, J., Phys. Stat. Sol. **16**, 147 (1966)
Verma, M.P., Dayal, B., Phys. Stat. Sol. **6**, 545 (1964)
Verma, M.P., Dayal, B., Phys. Stat. Sol. **19**, 751 (1967)
Verma, M.P., Singh, R.K., Phys. Stat. Sol. **33**, 769 (1969)
Verma, M.P., Singh, R.K., J. Phys. C: Solid State Phys. **4**, 2749 (1971)
Verma, M.P., Agarwal, S.K., Phys. Rev. B **8**, 4880 (1973)
Vershinin, Yu.N., Zotov, Yu.A., Sov. Phys. Solid State **17**, 2279 (1976)
Vetelino, J.F., Mitra, S.S., Solid State Commun. **7**, 1181 (1969)
Vetelino, J.F., Mitra, S.S., Brafman, O., Damen, T.C., Solid State Commun. **7**, 1809 (1969)

Vetelino, J.F., Mitra, S.S., Namjoshi, K.V., Phys. Rev. B **2**, 2167 (1970)
Vetelino, J.F., Namjoshi, K.V., Mitra, S.S., Phys. Rev. B 7, 4001 (1973)
Vinetskii, V.L., Sov. Phys. JETP **13**, 1023 (1961)
Vinetskii, V.L., Kravchenko, V.Ya., Sov. Phys. Solid State **6**, 123 (1964)
Vinetskii, V.L., Sov. Phys. Solid State **6**, 1492 (1964)
Vinetskii, V.L., Kravchenko, V.Ya., Sov. Phys. JETP **20**, 604 (1965)
Vinetskii, V.L., Kholodar, G.A., Phys. Stat. Sol. **19**, 41 (1967)
Vinetskii, V.L., Sov. Phys. Solid State **10**, 680 (1968)
Vinetskii, V.L., Itskovskii, M.A., Kukushkin, L.S., Sov. Phys. Solid State **13**, 60 (1971)
Vinetskii, V.L., Yaskovets, I.I., Sov. Phys. Solid State **14**, 2605 (1973)
Vinetskii, V.L., Kudykina, T.A., Phys. Stat. Sol. (b) **87**, 507 (1978)
Vineyard, G.H., Dienes, G.J., Phys. Rev. **93**, 265 (1954)
Vink, H.J., Festkörperprobleme **1**, 1 (1962) (Ed. F. Sauter, Braunschweig: Vieweg)
Vinogradov, V.S., Sov. Phys. Solid State **15**, 285 (1973)
Visscher, W.M., Phys. Rev. **134**, A 965 (1964)
Vlasov, G.K., Sov. Phys. Solid State **17**, 2215 (1976)
Van Vleck, J.H., J. Chem. Phys. **7**, 72 (1939)
van Vleck, J.H., Phys. Rev. **57**, 426 (1940)
van Vliet, K.M., Marshak, A.H., Phys. Stat. Sol. (b) **78**, 501 (1976)
van Vliet, K.M., Can. J. Phys. **56**, 1204 (1978)
Vogl, P., J. Phys. C: Solid State Phys. **11**, 251 (1978)
Voigt, J., Phys. Stat. Sol. **2**, 1689 (1962)
Voigt, J., Phys. Stat. Sol. **7**, 643 (1964)
Volovik, G.E., Mel'nikov, V.I., Edel'shtein, V.M., JETP Letters **18**, 81 (1973)
de Vooght, J.G., Bajaj, K.K., Solid State Commun. **11**, 1303 (1972)
de Vooght, J.G., Bajaj, K.K., Phys. Rev. B 7, 1472 (1973)
Vorobev, G.A., Ekhanin, S.G., Nesmelov, N.S., Sov. Phys. Solid State **18**, 110 (1976)
Vredevoe, L.A., Phys. Rev. **140**, A 930 (1965)

Wada, K., Science of Light **26**, 77 (1977)
Waeber, W.B., Stoll, E.P., J. Phys. C: Solid State Phys. **4**, L97 (1971)
Wagner, E., Bludau, W., Solid State Commun. **17**, 709 (1975)
Wagner, M., Z. Naturforsch. **14a**, 81 (1959)
Wagner, M., Z. Naturforsch. **15a**, 889 (1960)
Wagner, M., Z. Naturforsch. **16a**, 302 (1961)
Wagner, M., Z. Naturforsch. **16a**, 410 (1961)
Wagner, M., Phys. Rev. **131**, 1443 (1963)
Wagner, M., Ann. Physik **11**, 59 (1963)
Wagner, M., Phys. Rev. **131**, 2520(1963)
Wagner, M., Phys. Rev. **133**, A 750 (1964)
Wagner, M., J. Chem. Phys. **41**, 3939 (1964)
Wagner, M., Phys. kondens. Materie **4**, 71 (1965)
Wagner, M., Bron, W.E., Phys. Rev. **139**, A 223 (1965)
Wagner, M., Z. Physik **206**, 131 (1967)
Wagner, M., Z. Physik **214**, 78 (1968)
Wagner, M., Z. Physik **230**, 460 (1970)
Wagner, M., Z. Physik **244**, 275 (1971)
Wagner, M., Kühner, K., Z. Physik **251**, 300 (1972)
Wagner, M., J. Physique **34**, Suppl. au no. 11–12, C9–133 (1973)
Wahl, F., Z. Naturforsch. **19a**, 620, (1964)
Wahl, F., Z. Naturforsch. **19a**, 632 (1964)
Wallace, D.C., Phys. Rev. **131**, 2046 (1963)
Wallace, D.C., Phys. Rev. **139**, A 877 (1965)
Wallis, R.F., Bowlden, H.J., J. Phys. Chem. Solids **8**, 318 (1959)
Wallis, R.F., Maradudin, A.A., Prog. Theor. Phys. **24**, 1055 (1960)
Wallis, R.F., Maradudin, A.A., Phys. Rev. **125**, 1277 (1962)
Wallis, R.F. (Ed.), Localized Excitations in Solids, New York: Plenum Press, 1968
Wallis, R.F., Phonons and Polaritons, in: Trieste Winter College on Interaction of Radiation with Condensed Matter, Trieste 1976, Wien: IAEA 1977, p. 163
Walls, D.F., Z. Physik **237**, 224 (1970)
Walter, J.P., Cohen, M.L., Phys. Rev. B **4**, 1877 (1971)
Wang, C.C., Ressler, N.W., Phys. Rev. **188**, 1291 (1969)
Wang, J.S.Y., Schlüter, M., Cohen, M.L., Phys. Stat. Sol. (b) **77**, 295 (1976)
Wang, S., Phys. Stat. Sol. **23**, 387 (1967)
Wang, S., Phys. Stat. Sol. **23**, 401 (1967)
Wang, S., Matsuura, M., Wong, C.C., Inoue, M., Phys. Rev. B 7, 1695 (1973)
Wang, S., Mahutte, C.K., Matsuura, M., Phys. Stat. Sol. **51**, 11 (1972)
Wang, S., Matsuura, M., Phys. Rev. B **10**, 3330 (1974)
Wang, S.F., Phys. Rev. **132**, 573 (1963)
Wang, S.F., Prog. Theor. Phys. **33**, 1001 (1965)
Wang, S.F., Phys. Rev. **147**, 521 (1966)
Wang, S.F., Phys. Rev. **153**, 939 (1967)
Wang, S.F., Chu, C., Phys. Rev. **147**, 527 (1966)
Wang, S.F., Chu, C., Phys. Rev. **154**, 838 (1967)
Wang, S.F., Phys. Rev. **170**, 799 (1968)
Wannier, G.H., Phys. Rev. **52**, 191 (1937)
Wannier, G., Elements of Solid State Theory, London: Cambridge University Press, 1959

Wannier, G., Phys. Rev. **117**, 432 (1960)
Wannier, G.H., Rev. Mod. Phys. **34**, 645 (1962)
Wannier, G.H., Fredkin, D.R., Phys. Rev. **125**, 1910 (1962)
Warburton, W.K., J. Phys. Chem. Solids **34**, 451 (1973)
Ward, R.W., Clayman, B.P., Can. J. Phys. **52**, 1492 (1974)
Warren, R.W., Phys. Rev. **155**, 948 (1967)
Wassef, W.A., Kao, K.C., Phys. Rev. B **6**, 1425 (1972)
Wassef, W.A., Int. J. Electronics **41**, 517 (1976)
Watanabe, H., J. Phys. Chem. Solids **28**, 961 (1967)
Watanabe, H., Kishishita, H., Prog. Theor. Phys., Suppl. **46**, 1 (1970)
Watson, R.E., Phys. Rev. **117**, 742 (1960)
Weaver, L., Shultis, J.K., Faw, R.E., J. Appl. Phys. **48**, 2762 (1977)
Weber, J., Lacroix, R., Helv. Phys. Acta **44**, 181 (1971)
Weber, W., Dick. B.G., Phys. Stat. Sol. **36**, 723 (1969)
Weber, W., Phys. Rev. Lett. **33**, 371 (1974)
van Weert, Ch. G., De Boer, W.P.H., J. Statist. Phys. **18**, 271 (1978)
Wehner, R., Phys. Stat. Sol. **15**, 725 (1966)
Wei, W.F., Phys. Rev. B **15**, 2250 (1977)
Weijland, A., Physica **32**, 337 (1966)
Weijland, A., Physica **32**, 625 (1966)
Weiler, M.H., Reine, M., Lax, B., Phys. Rev. **171**, 949 (1968)
Weinberger, P., Schwarz, K., Neckel, A., J. Phys. Chem. Solids **32**, 2063 (1971)
Weiner, J.H., Phys. Rev. **169**, 570 (1968)
Weisberg, L.R., J. Appl. Phys. **39**, 6096 (1968)
Weissman, J., Jortner, J., Phil. Mag. B **37**, 21 (1978)
Weissmann, A., Demco, D., Czech. J. Phys. B **16**, 234 (1966)
Welker, H., Ergebnisse der exakten Naturwiss. XXIX, 275 (1956) (Ed. S. Flügge and F. Trendelenburg, Berlin: Springer)
Welker, H., Weiss, H., Solid State Physics **3**, 1 (1956) (Ed. F. Seitz and D. Turnbull, New York: Academic Press)
Wemple, S.H., Phys. Rev. B **7**, 4007 (1973)
Wepfer, G.G., Collins, T.C., Euwema, R.N., Phys. Rev. B **4**, 1296 (1971)
Werthamer, N.R., Am. J. Phys. **37**, 763 (1969)
Werthamer, N.R., Phys. Rev. B **1**, 572 (1970)
Wertheim, G.K., Phys. Rev. **109**, 1086 (1958)
de Wette, F.W., Physica **25**, 1225 (1959)
Wheeler, R.G., Dimmock, J.O., Phys. Rev. **125**, 1805 (1962)
White, W.W. III, Greene, A.C., Crystal Latt. Defects **1**, 83 (1969)
Whitfield, G., Platzman, P.M., Phys. Rev. B **6**, 3987 (1972)
Whitfield, G., Engineer, M., Phys. Rev. B **12**, 5472 (1975)
Wielopolski, P., Acta Phys. Polon. A **44**, 177 (1973)
Wielopolski, P., Acta Phys. Polon. A **44**, 185 (1973)
Wielopolski, P., Stecki, J., Acta Phys. Polon. A **44**, 195 (1973)
Wielopolski, P., Stecki, J., Acta Phys. Polon. A **44**, 381 (1973)
van Wieringen, J.S., Prog. Semicond. **6**, 199 (1961)
Wiley, J.D., DiDomenico, M., Phys. Rev. B **2**, 427 (1970)
Wiley, J.D., Phys. Rev. B **4**, 2485 (1971)
Willardson, R.K., Beer, A.C., (Eds.), Transport Phenomena; Semiconductors and Semimetals, Vol. 10, New York: Academic Press, 1975
Wille, H., Wahl, F., Z. Naturforsch. **21a**, 304 (1966)
Wille, H., Z. Naturforsch. **24a**, 64 (1969)
Williams, D.N., Commun. Math. Phys. **21**, 314 (1971)
Williams, F.E., Eyring, H., J. Chem. Phys. **15**, 289 (1947)
Williams, F.E., J. Opt. Soc. Am. **39**, 648 (1949)
Williams, R.H., McCanny, J.V., Murray, R.B., Ley, L., Kemeny, P.C., J. Phys. C: Solid State Phys. **10**, 1223 (1977)
Williams, R.T., Semicond. Insulators **3**, 251 (1978)
Willis, C.R., Picard, R.H., Phys. Rev. A **9**, 1343 (1974)
Willis, C.R., Gen. Relat. Gravitation **7**, 69 (1976)
Wilson, A.H., Theory of Metals, London: Cambridge University Press, 1953
Wilson, R.S., King, W.T., Kim, S.K., Phys. Rev. **175**, 1164 (1968)
Wilson, R.S., Kim, S.K., Phys. Rev. B **7**, 4674 (1973)
Wilson, W.D., Hatcher, R.D., Dienes, G.J., Smoluchowski, R., Phys. Rev. **161**, 888 (1967)
Wilson, W.D., Hatcher, R.D., Smoluchowski, R., Dienes, G.J., Phys. Rev. **184**, 844 (1969)
Winnacker, A., Mollenauer, L.F., Phys. Rev. B **6**, 787 (1972)
Winnacker, A., Mauser, K.E., Niesert, B., Z. Physik B **26**, 97 (1977)
van Winsum, J.A., Lee, T., den Hartog, H.W., Phys. Rev. B **18**, 173 (1978)
Wise, M.E., Physica **17**, 1011 (1951)
Witschel, W., J. Phys. B: Atom. Molec. Phys. **6**, 527 (1973)
Wolfe, C.M., Stillman, G.E., Dimmrock, J.O., J. Appl. Phys. **41**, 504 (1970)
Wolfe, C.M., Stillman, G.E., Lindley, W.T., J. Appl. Phys. **41**, 3088 (1970)
Wolff, P.A., Phys. Rev. **95**, 1415 (1954)
Wolga, G.J., Strandberg, M.W., J. Phys. Chem. Solids **9**, 309 (1959)

Wolman, O., Ron, A., Phys. Rev. **148**, 548 (1966)
Woo, C.H., Wang, S., Phys. Rev. B **7**, 2810 (1973)
Woo, C.H., Wang, S., Phys. Rev. B **7**, 2827 (1973)
Wood, R.F., Korringa, J., Phys. Rev. **123**, 1138 (1961)
Wood, R.F., Joy, H.W., Phys. Rev. **136**, A 451 (1964)
Wood, R.F., Meyer, A., Solid State Commun. **2**, 225 (1964)
Wood, R.F., J. Phys. Chem. Solids **26**, 615 (1965)
Wood, R.F., Phys. Rev. **151**, 629 (1966)
Wood, R.F., Öpik, U., Phys. Rev. **162**, 736 (1967)
Wood, R.F., Gilbert, R.L., Phys. Rev. **162**, 746 (1967)
Wood, R.F., Öpik, U., Phys. Rev. **179**, 772 (1969)
Wood, R.F., Öpik, U., Phys. Rev. **179**, 783 (1969)
Wood, R.F., Phys. Stat. Sol. **42**, 849 (1970)
Wood, R.F., Ganguly, B.N., Phys. Rev. B **7**, 1591 (1973)
Wood, R.F., Wilson, T.M., Solid State Commun. **16**, 545 (1975)
Wood, R.F., Wilson, T.M., Phys. Rev. B **15**, 3700 (1977)
Wood, V.E., Reitz, J.R., J. Phys. Chem. Solids **23**, 229 (1962)
Woods, A.D.B., Cochran, W., Brockhouse, B.N., Phys. Rev. **119**, 980 (1960)
Woods, A.D.B., Brockhouse, B.N., Cowley, R.A., Cochran, W., Phys. Rev. **131**, 1025 (1963)
Woodward, A.S., Chatterjee, R., J. Phys. C: Solid State Phys. **4**, 1378 (1971)
Wruck, D., Phys. Stat. Sol. (b) **48**, 181 (1971)
Wu, C.-C., Spector, H.N., Phys. Rev. B **3**, 3979 (1971)
Wu, C.-C., Chen, A., Appl. Phys. Letters **30**, 434 (1977)
Wu, S.-T., Prog. Theor. Phys. **54**, 999 (1975)
Wünsche, H.-J., Henneberger, K., Khartsiev, V.E., Phys. Stat. Sol. (b) **86**, 505 (1978)

Xinh, N.X., Maradudin, A.A., Coldwell-Horsfall, R.A., J. Physique **26**, 717 (1965)
Xinh, N.X., Solid State Commun. **4**, 9 (1966)
Xinh, N.X., Phys. Rev. **163**, 896 (1967)

Yacoby, Y., Phys. Rev. **140**, A 263 (1965)
Yacoby, Y., Phys. Rev. **169**, 610 (1968)
Yacoby, Y., Phonon Assisted Optical Processes in Solids, in: Optical Properties of Solids, ed. B.O. Seraphin, Amsterdam: North-Holland 1975, p. 731
Yakhot, V., Chem. Phys. **14**, 441 (1976)
Yakhot, V., Phys. Stat. Sol. (b) **74**, 451 (1976)
Yamada, E., Kurosawa, T., J. Phys. Soc. Japan **34**, 603 (1973)
Yamada, Y., J. Phys. Soc. Japan **35**, 1600 (1973)
Yamada, Y., Phys. Stat. Sol. (b) **85**, 723 (1978)
Yamaguchi, T., Kamimura, H., J. Phys. Soc. Japan **33**, 953 (1972)
Yamashita, J., J. Phys. Soc. Japan **7**, 284 (1952)
Yamashita, J., Kurosawa, T., J. Phys. Soc. Japan **10**, 610 (1955)
Yamashita, J., Kurosawa, T., J. Phys. Soc. Japan **15**, 802 (1960)
Yamashita, J., Nakamura, K., Proc. Int. Conf. Physics of Semicond., Kyoto 1966, J. Phys. Soc. Japan **21**, Suppl., 455 (1966)
Yang, E., Phys. Rev. B **4**, 2046 (1971)
Yang, E., J. Phys. Chem. Solids **36**, 1255 (1975)
Yao, T., Inagaki, K., Maekawa, S., J. Phys. Soc. Japan **38**, 1394 (1975)
Yartsev, V.M., Sov. Phys. Solid State **19**, 751 (1977)
Yasevichyute, Ya., Sov. Phys. Semicond. **7**, 369 (1973)
Yasevichyute, Ya., Sov. Phys. Semicond. **8**, 966 (1975)
Yasojima, Y., Inuishi, Y., Phys. Letters **34A**, 129 (1971)
Yasojima, Y., Ohmori, Y., Okumura, N., Inuishi, Y., Jap. J. Appl. Phys. **14**, 815 (1975)
Yee, J.H., Phys. Rev. B **3**, 355 (1971)
Yee, J.H., J. Phys. Chem. Solids **33**, 643 (1972)
Yee, J.H., Phys. Rev. B **6**, 2279 (1972)
Yip, K.L., Fowler, W.B., Phys. Stat. Sol. (b) **53**, 137 (1972)
Yokota, I., Prog. Theor. Phys., Suppl. **57**, 97 (1975)
Yu, H.L., De Siqueira, M., Conolly, J.W.D., Phys. Rev. B **14**, 772 (1976)
Yu, P.Y., Evangelisti, F., Solid State Commun. **27**, 87 (1978)

Zaiko, Yu.N., Sov. Phys. Solid State **18**, 545 (1976)
Zaitsev, A.N., Sov. Phys. Solid State **15**, 513 (1973)
Zaitsev, V.M., Mel'nikova, T.N., Sov. Phys. Solid State **8**, 488 (1966)
Zak, J., Phys. Rev. **136**, A 776 (1964)
Zak, J., Phys. Rev. **134**, A 1602 (1964)
Zak, J., Phys. Rev. **134**, A 1607 (1964)
Zak, J., Commun. Phys. **1**, 73 (1976)
Zak, J., Phys. Rev. B **16**, 4154 (1977)

Zaretskii, G.A., Kucher, T.I., Tolpygo, K.B., Sov. Phys. Solid State **17**, 187 (1975)
Zaslavskaya, I.G., Sov. Phys. Solid State **4**, 705 (1962)
Zavada, P., Czech. J. Phys. B **27**, 211 (1977)
Zavoiski, E.K., J. Exp. Theor. Phys. (USSR) **16**, 603 (1946)
Zavt, G.S., Sov. Phys. Solid State **7**, 1698 (1966)
Zavt, G.S., Kristofel', N.N., Khizhnyakov, V.V., Sov. Phys. Solid State **7**, 1973 (1966)
Zavt, G.S., Kristofel', N.N., Sov. Phys. Solid State **8**, 1813 (1967)
Zavt, G.S., Phys. Stat. Sol. (b) **80**, 399 (1977)
Zawadzki, W., Phys. Stat. Sol. **2**, 385 (1962)
Zawadzki, W., Phys. Stat. Sol. **3**, 1006 (1963)
Zawadzki, W., Phys. Stat. Sol. **3**, 692 (1963)
Zawadzki, W., Phys. Stat. Sol. **3**, 1421 (1963)
Zawadzki, W., Kolodziejczak, J., Phys. Stat. Sol. **6**, 419 (1964)
Zawadzki, W., Kolodziejczak, J., Phys. Stat. Sol. **6**, 409 (1964)
Zawadzki, W., Kowalski, J., Phys. Rev. Letters **27**, 1713 (1971)
Zawadzki, W., Szymanska, W., Phys. Stat. Sol. (b) **45**, 415 (1971)
Zawadzki, W., in: New Developments in Semiconductors, ed. P.R. Wallace, Leyden: Noordhoff 1973, p. 441
Zawadzki, W., Adv. Phys. **23**, 435 (1974)
Zawadzki, W., Wlasak, J., J. Phys. C: Solid State Phys. **9**, L663 (1976)
Zehe, A., Röpke, G., Phys. Stat. Sol. (b) **67**, 169 (1975)
Zeller, G.R., Phys. Stat. Sol. (b) **65**, 521 (1974)
Zenchenko, V.P., Sinyavskii, E.P., Sov. Phys. Solid State **17**, 2243 (1976)
Zernik, W., Phys. Rev. **139**, A 1010 (1965)
Zernik, W., Phys. Rev. **158**. 562 (1967)
Zernik, W., Rev. Mod. Phys. **39**, 432 (1967)
Zevin, V.Ya., Konovalov, V.I., Sov. Phys. Solid State **14**, 738 (1972)
Zevin, V., Phys. Rev. B **11**, 2447 (1975)
Zeyher, R., Bilz, H., Phys. Stat. Sol. **31**, 157 (1969)
Zeyher, R., Phys. Stat. Sol. (b) **48**, 711 (1971)
Zeyher, R., Phys. Rev. Letters **35**, 174 (1975)
Zeyher, R., Solid State Commun. **16**, 49 (1975)
Zeyher, R., Bilz, H., Cardona, M., Solid State Commun. **19**, 57 (1976)
Zgierski, M.Z., Phys. Stat. Sol. (b) **59**, 589 (1973)
Zhdanov, V.A., Polyakov, V.V., Konusov, V.F., Sov. Phys. Solid State **15**, 2295 (1974)
Zhdanov, V.A., Polyakov, V.V., Sov. Phys. Chrystallogr. **22**, 488 (1977)
Zheru, I.I., Shmiglyuk, M.I., Sov. Phys. Semicond. **6**, 1326 (1973)
Zhilich, A.G., Cherepanov, V.I., Kargapolov, Yu.A., Sov. Phys. Solid State **3**, 1317 (1961)
Zhilich, A.G., Sov. Phys. Solid State **7**, 538 (1965)
Zhilich, A.G., Makarov, V.P., Sov. Phys. Solid State **6**, 1624 (1965)
Zhilich, A.G., Sov. Phys. Solid State **13**, 2425 (1972)
Zhilich, A.G., Maksimov, O.A., Sov. Phys. Semicond. **9**, 616 (1975)
Zhilich, A.G., Monozon, B.S., Sov. Phys. Solid State **17**, 866 (1975)
Zhukov, V.P., Kuzin, A.N., Fedorov, G.B., Sov. Phys. Solid State **18**, 1897 (1976)
Ziman, J.M., Electrons and Phonons, London: Oxford University Press, 1960
Zimmermann, R., Phys. Stat. Sol. (b) **46**, K111 (1971)
Zimmermann, R., Phys. Stat. Sol. (b) **48**, 603 (1971)
Zlobin, A.M., Sov. Phys. Solid State **18**, 2010 (1976)
Zlobin, A.M., Sov. Phys. Semicond. **11**, 393 (1977)
Zook, J., Phys. Rev. **136**, A 869 (1964)
Zubkova, S.M., Tolpygo, K.B., Sov. Phys. Solid State **19**, 318 (1977)
Zukotynski, S., Kolodziejczak, J., Phys. Stat. Sol. **3**, 990 (1963)
Zukotynski, S., Saleh, N., Phys. Stat. Sol. **36**, 593 (1969)
Zukotynski, S., Howlett, W., Solid-State Electron. **21**, 35 (1978)
Zunger, A., Freeman, A.J., Phys. Rev. B **16**, 2901 (1977)
Zunger, A., Freeman, A.H., Phys. Rev. B **17**, 4850 (1978)
Zvereva, G.A., Makarov, V.P., Sov. Phys. Solid State **19**, 1454 (1977)
Zvyagin, I.P., Sov. Phys. Solid State **7**, 779 (1965)
Zvyagin, I.P., Sov. Phys. Solid State **8**, 2269 (1967)
Zvyagin, I.P., Phys. Stat. Sol. (b) **85**, 105 (1978)
Zvyagin, I.P., Phys. Stat. Sol. (b) **88**, 149 (1978)
Zwanzger, J., Muschik, W., Haug, A., Z. Naturforsch. **27a**, 749 (1972)
Zwanzig, R.W., Boulder Lectures in Theoretical Physics **3**, 106 (1960)
Zwanzig, R., J. Chem. Phys. **33**, 1338 (1960)
Zwanzig, R., Physica **30**, 1109 (1964)
Zwicker, R.D., Phys. Stat. Sol. (b) **64**, K87 (1974)
Zwicker, R.D., Phys. Rev. B **12**, 5502 (1975)

Author Index

In the author index names with prefixes (de, di, van, von, etc.) are alphabetically arranged corresponding to the initials of the prefixes.

Subject Index